INTRODUCTION TO

SOLID STATE PHYSICS

Charles Kittel

INTRODUCTION TO

SOLID STATE PHYSICS

FIFTH EDITION

JOHN WILEY & SONS
New York • Chichester • Brisbane • Toronto

Library of Congress Cataloging in Publication Data:

Kittel, Charles.
Introduction to solid state physics.

Includes bibliographies and indexes.
1. Solids. I. Title.
QC176.K5 1976 530.4'1 75-25936
ISBN 0-471-49024-5

Printed in the United States of America

10 9 8

About the Author

CHARLES KITTEL has taught solid state physics at the University of California at Berkeley since 1951, having previously been at the Bell Laboratories. His undergraduate work in physics was done at M.I.T. and at the Cavendish Laboratory of Cambridge University. His Ph.D. research was in theoretical nuclear physics with Professor Gregory Breit at the University of Wisconsin. He has been awarded three Guggenheim fellowships, the Oliver Buckley Prize for Solid State Physics, and the Distinguished Teaching Award at Berkeley. He is a member of the National Academy of Science and of the American Academy of Arts and Sciences. The first edition of *ISSP* appeared in 1953 as the first textbook to integrate the elementary parts of solid state physics for use by seniors and beginning graduate students; now in the fifth edition, *ISSP* is one of the most widely translated science textbooks. His recent research is in the field of electron-hole condensation in semiconductors.

Preface

This is an elementary textbook on solid state physics and materials science for senior undergraduate and beginning graduate students of science and engineering. Solid state physics is largely concerned with the remarkable properties exhibited by atoms and molecules because of their regular arrangement in crystals. The properties include electron energy bands and the elementary excitations of solids: phonons, plasmons, polarons, excitons, magnons, and polaritons. This book tells how they can be understood in terms of simple models. Real solids may be more complicated, but the power and utility of the elementary models cannot be overestimated.

The first edition appeared in 1953, five years after the publication of the discovery of the transistor. Today the solid state revolution is in full development. Because of the profitable interplay of experiment and theory, there is intellectual excitement in solid state physics—as witnessed by the Nobel Prize awards during the past decade. Throughout the world, more physicists are usefully employed in solid state physics than in any other field.

In this edition a new effort has been made to help undergraduates: the first half of the book now provides students with an elementary one-quarter or one-semester introduction to solid state physics through the physics of semiconductors in Chapter 8. This part makes only modest demands on the reader's background in electricity and magnetism. A short course might be completed by parts or all of Chapters 12 through 15. The second half of the book applies the material in the beginning chapters to well-defined areas. It contains matter that makes a frank appeal to electromagnetism at the intermediate level: plasmons, optical properties, superconductivity, ferroelectrics, and magnetism.

The discussion of superconductors is improved: simple estimates are given for the upper and lower critical fields, the Josephson-Anderson derivation of the London equation is adopted, and the lifetime of persistent currents is calculated to exceed the age of the universe. The most common and most serious defect in solid state textbooks is an inadequate derivation of the properties of holes, as in semiconductors; in this edition their properties are clearly and extensively argued. Among the new topics in this edition are the Peierls instability, magnetic bubbles, Gunn oscillators, solar cells, Alfvén waves, electron-hole drops, and thermoelectric effects. Many of the 54 tables have been revised; there are some 50 new drawings and numerous new problems.

Some instructors prefer to start their lectures with the Drude model of the free electron gas. To do this, start with the chapter sequence 6, 7, 2, 8, 9.

The selection of subjects should not be seen as a measure of the importance of various areas, any more than the selection of references measures individual contributions. A single textbook cannot represent the range of current activity. Reviews in the Seitz-Turnbull-Ehrenreich and other series may be consulted for subjects not treated here and for detailed bibliographies. There are in the literature perhaps 20,000 articles of high quality that could usefully be cited. I have tried to give a helpful, but small, sample of those most accessible in English. The translations of this book into French, German, Spanish, Italian, Japanese, Russian, Polish, Roumanian, Hungarian, Chinese, and Arabic often give further references in these languages; there is a good student's workbook in Swedish.

Important equations are repeated in SI and CGS-Gaussian units, where these differ. Exceptions to this rule are the figure captions, the chapter summaries, and any long section of text where a single indicated substitution, as of 1 for c or $1/4\pi\epsilon_0$ for 1, will translate from CGS to SI.

Tables are in conventional units. The contents pages of several chapters discuss conventions that were adopted to make parallel usage simple and natural. The symbol e denotes the charge on the proton and is positive. The notation (18) refers to Equation 18 of the current chapter, but (3.18) refers to Equation 18 of Chapter 3; figures are referred to in the same way. A caret over a vector, as in $\hat{\mathbf{k}}$, denotes a unit vector. Problems of appreciable length or difficulty are preceded by an asterisk; I like to assign them for extra credit. For production reasons the format of the book has been changed from recent editions. The present page is longer and the line is longer. Appendices in the last edition have either been incorporated into the chapters or, where rarely taught, they have been dropped.

The preparation of this edition was made possible by the cooperation of many colleagues and friends. M. L. Cohen and L. M. Falicov kindly reviewed several new chapters. Among my new debts are those to: D. Gugan, D. Penn, C. D. Jeffries, A. M. Portis, M. J. Buerger, Leo Brewer, F. L. Richards, F. C. Brown, A. Maradudin, B. T. Matthias, R. Dalven, W. M. Star, C. P. Slichter, J. A. Wilson, J. C. Hensel, and R. T. Schumacher. I am again grateful to the more than 150 people who contributed to the earlier editions. I thank Felix Cooper, who did the artwork, and Carol Tung for the preparation of the manuscript. Richard Robinson and Frank Velek of our Physics Library have been generous with their interest and their time.

C. Kittel

BERKELEY, CALIFORNIA

An instructor's manual is available to instructors who have adopted the text for classroom use. Requests may be directed to John Wiley & Sons, Inc., 605 Third Avenue, New York, New York 10016.

Contents

Guide to Tables

Some General References

Statistical physics background

C. Kittel, *Thermal physics,* Wiley, 1969. Cited as *TP.*

Problem collection

H. J. Goldsmid, ed., *Problems in solid state physics,* Academic Press, 1968.

Advanced series

F. Seitz, D. Turnbull and H. Ehrenreich, *Solid state physics, advances in research and applications,* Academic Press. Cited as *Solid state physics.*

Advanced texts

R. E. Peierls, *Quantum theory of solids,* Oxford, 1955.

C. Kittel, *Quantum theory of solids,* Wiley, 1963. Cited as *QTS.*

J. M. Ziman, *Principles of the theory of solids,* 2nd ed., Cambridge, 1972.

W. Jones and N. H. March, *Theoretical solid state physics,* Wiley, 1973, 2 vols.

J. Callaway, *Quantum theory of the solid state,* Academic Press, 1974, 2 vols.

N. W. Ashcroft and N. D. Mermin, *Solid state physics,* Holt, Rinehart, Winston, 1976.

Experimental

R. V. Coleman, ed., *Solid state physics,* vol. 11 of *Methods of experimental physics,* L. Marton, ed., Academic Press, 1974.

Table of values

American Institute of Physics Handbook, 3rd ed., McGraw-Hill, 1971.

Guides to the literature

The most valuable modern bibliographic aid is the *Scientific citation index.* For references to data on specific materials, consult the formula indices to *Chemical abstracts* and the subject indices to *Physics abstracts* and *Solid state abstracts*. Good bibliographies often accompany the review articles in *Reports on progress in physics, Critical reviews in solid state sciences, Solid state physics, Springer tracts in modern physics,* Reviews of Modern Physics, Soviet Physics (Uspekhi), and Advances in Physics. The *Solid state physics literature guides* (Plenum) are valuable aids.

INTRODUCTION TO

SOLID STATE PHYSICS

1 CRYSTAL STRUCTURE

UNITS: 1 Å = 1 angstrom = 10^{-8} cm = 10^{-10} m.

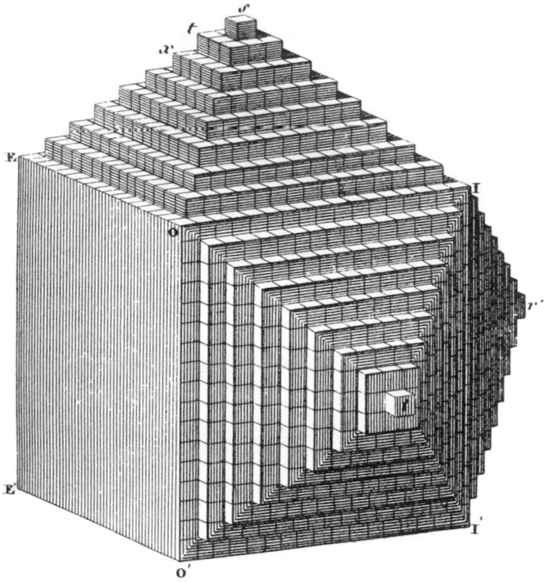

(a)

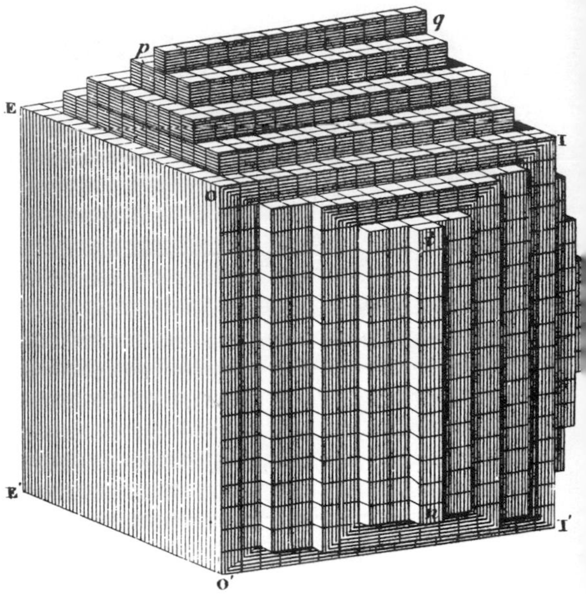

(b)

(c)

Figure 1 Relation of the external form of crystals to the form of the elementary building blocks. The building blocks are identical in (a) and (b), but different crystal faces are developed. (c) Cleaving a crystal of rocksalt.

1 CRYSTAL STRUCTURE

Solid state physics is devoted largely to the study of crystals and electrons in crystals. The study of solid state physics as an extension of atomic physics began in the early years of this century following the discovery of x-ray diffraction and the publication of a series of simple and successful calculations and predictions of crystalline properties.

The regularity of the external form of crystals disposed early observers to the belief that crystals are formed by the regular repetition of identical building blocks (Fig. 1). When a crystal grows in a constant environment, the shape remains unchanged during growth, as if identical building blocks are added continuously. The building blocks are atoms or groups of atoms: a crystal is a three-dimensional periodic array of atoms. This was known in the 18th century when mineralogists discovered that the index numbers of the directions of all faces of a crystal are exact integers. Haüy[1] showed that the arrangement of identical particles in a three-dimensional periodic array could account for a law of rational or integral indices.

On June 8, 1912, a paper entitled "Interference effects with Röntgen rays" was presented to the Bavarian Academy of Sciences in Munich. In the first part of the paper, Laue developed an elementary theory of diffraction of x-rays by a periodic array of atoms. In the second part, Friedrich and Knipping reported the first experimental observations of x-ray diffraction by crystals. The work proved decisively that crystals are composed of a periodic array of atoms[2] and marked the beginning of the field of solid state physics. With an established atomic model of a crystal, physicists could think much further.

[1]R. J. Haüy, *Essai d'une théorie sur la structure des cristaux*, Paris, 1784; *Traité de cristallographie*, Paris, 1801.

[2]For personal accounts of the early years of x-ray diffraction studies of crystals, see P. P. Ewald, ed., *Fifty years of x-ray diffraction*, A. Oosthoek's Uitgeversmij., Utrecht, 1962. Shrewd guesses about the structures of a number of crystals had been made much earlier by W. Barlow, Nature **29**, 186, 205, 404 (1883); he argued from considerations of symmetry and packing.

PERIODIC ARRAYS OF ATOMS

An ideal crystal is constructed by the infinite repetition in space of identical structural units. In the simplest crystals such as copper, silver, iron, aluminum, and the alkali metals, the structural unit is a single atom. Often the structural unit is several atoms or molecules, up to perhaps 100 in inorganic crystals[3] and 10,000 in protein crystals. The structure of all crystals is described in terms of a lattice with a group of atoms attached to each lattice point. The group is called the **basis**; it is repeated in space to form the crystal structure. We now make these definitions more precise.

Lattice Translation Vectors and Lattices

An ideal crystal is composed of atoms arranged on a lattice defined by three fundamental translation vectors **a, b, c** such that the atomic arrangement looks the same in every respect when viewed from any point **r** as when viewed from the point

$$\mathbf{r}' = \mathbf{r} + u\mathbf{a} + v\mathbf{b} + w\mathbf{c} \ , \qquad (1)$$

where u, v, w are arbitrary integers. The set of points $\mathbf{r}'$ specified by (1) for all values of the integers u, v, w defines a **lattice.** A lattice is a regular periodic arrangement of points in space. A lattice is a mathematical abstraction: the crystal structure is formed only when a basis of atoms is attached identically to each lattice point. The logical relation is

$$\textbf{lattice} + \textbf{basis} = \textbf{crystal structure} \ .$$

The lattice and the translation vectors **a, b, c** are said to be **primitive** if any two points **r, r′** from which the atomic arrangement looks the same always satisfy (1) with a suitable choice of the integers u, v, w. This definition of the primitive translation vectors guarantees that there is no cell of smaller volume that could serve as a building block for the structure. We often use primitive translation vectors to define the crystal axes, although nonprimitive crystal axes may be used when they are simpler. The crystal axes **a, b, c** form three adjacent edges of a parallelepiped. If there are lattice points only at the corners of the parallelepiped, then it is a primitive parallelepiped. A **lattice translation operation** is defined as the displacement of a crystal parallel to itself by a crystal translation vector

$$\mathbf{T} = u\mathbf{a} + v\mathbf{b} + w\mathbf{c} \ . \qquad (2)$$

Any two lattice points are connected by a vector of this form.

[3]The intermetallic compound $NaCd_2$ has a cubic cell of 1192 atoms as its smallest structural unit; S. Samson, Nature **195**, 259 (1962).

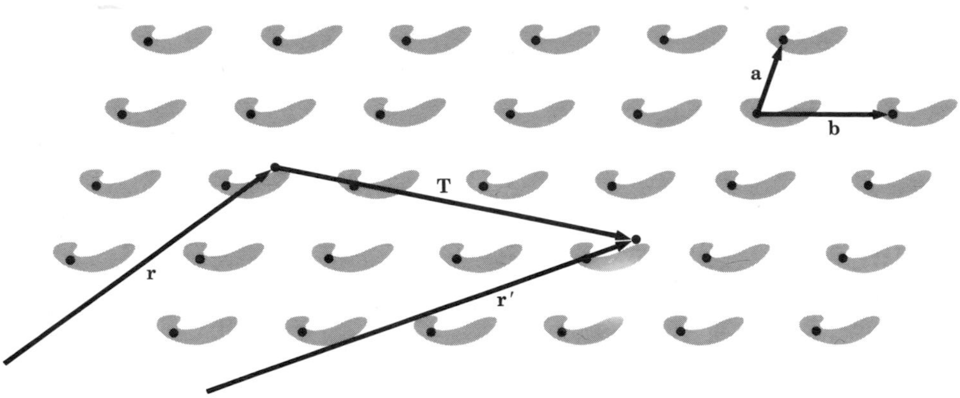

Figure 2 Portion of a crystal of an imaginary protein molecule, in a two-dimensional world. (We picked a protein molecule because it is not likely to have a special symmetry of its own.) The atomic arrangement in the crystal looks exactly the same to an observer at $\mathbf{r}'$ as to an observer at $\mathbf{r}$, provided that the vector $\mathbf{T}$ which connects $\mathbf{r}'$ and $\mathbf{r}$ may be expressed as an integral multiple of the vectors $\mathbf{a}$ and $\mathbf{b}$. In this illustration, $\mathbf{T} = -\mathbf{a} + 3\mathbf{b}$. The vectors $\mathbf{a}$ and $\mathbf{b}$ are primitive translation vectors of the two-dimensional lattice.

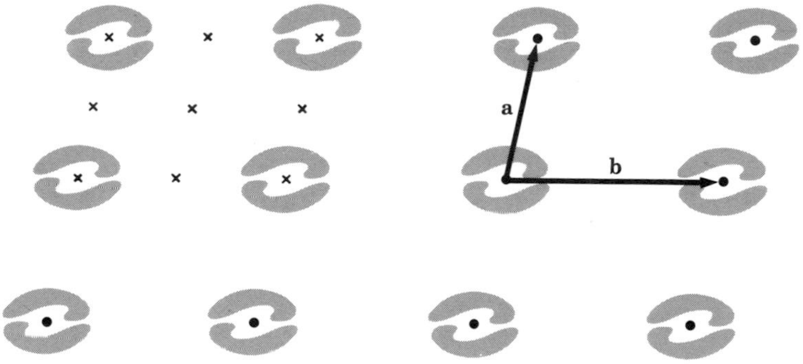

Figure 3 Similar to Fig. 2, but with protein molecules associated in pairs. The crystal translation vectors are $\mathbf{a}$ and $\mathbf{b}$. A rotation of π radians about any point marked $\times$ will carry the crystal into itself. This occurs also for equivalent points in other cells, but we have marked the points $\times$ only within one cell.

Symmetry Operations

In describing a crystal structure, there are three important questions to answer: What is the lattice? What crystal axes $\mathbf{a}, \mathbf{b}, \mathbf{c}$ do we wish to use to describe the lattice? What is the basis? More than one lattice is always possible for a given structure and more than one set of crystal axes is always possible for a given lattice. We cannot pick the basis until we have selected the lattice and the axes we wish to use. Everything (such as the x-ray diffraction pattern) works out the same in the end as long as the basis is chosen properly for whatever crystal axes we use.

The symmetry operations of a crystal carry the crystal structure into itself. These include the lattice translation operations (2). Further, there are rotation and reflection operations, called point operations. About lattice points or certain special points within an elementary parallelepiped it may be possible to apply rotations and reflections that carry the crystal into itself. Finally, there may exist compound operations made up of combined translation and point operations. Textbooks on crystallography are largely devoted to the description of symmetry operations. The crystal structure of Fig. 2 is drawn to have only translational symmetry operations. The crystal structure of Fig. 3 allows both translational and point symmetry operations.

The Basis and the Crystal Structure

Attached to every lattice point there is a basis of atoms, with every basis identical in composition, arrangement, and orientation. A crystal structure is formed by the addition of a basis to every lattice point, as in Fig. 4. In Figs. 2 and 3 the lattice is indicated by dots; in Fig. 4c the dots are omitted. The number of atoms in the basis may be as low as one, as for many metals and the inert gases, but there are structures for which the basis exceeds 1000 atoms. The position of the center of atom j of the basis is written as

$$\mathbf{r}_j = x_j\mathbf{a} + y_j\mathbf{b} + z_j\mathbf{c} \ , \tag{3}$$

relative to a lattice point. We may arrange matters so that $0 \le x_j, y_j, z_j \le 1$.

Primitive Lattice Cell

The parallelepiped (Fig. 5b) defined by primitive axes **a, b, c** is called a primitive cell. A primitive cell is a type of cell or unit cell. A cell will fill all space by the action of suitable crystal translation operations; **a primitive cell is a minimum-volume cell.** There are many ways of choosing the

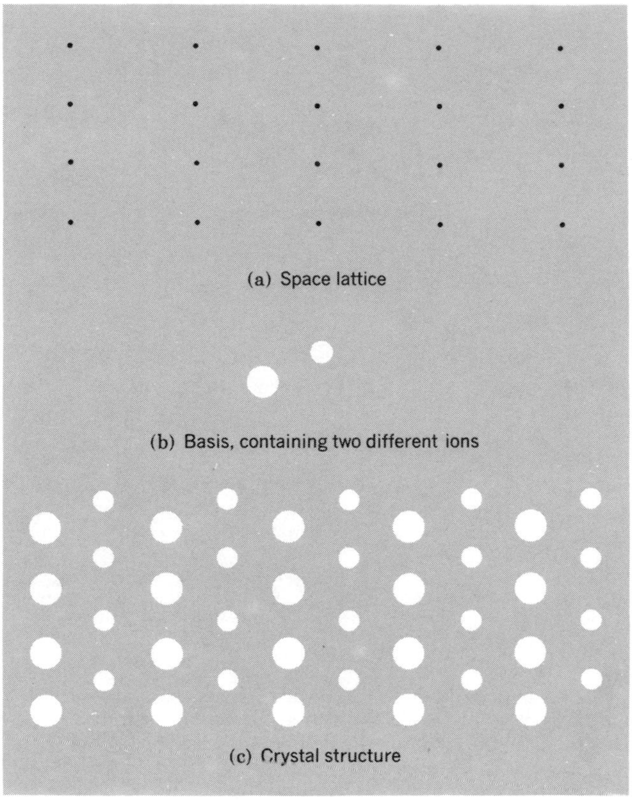

Figure 4 The crystal structure is formed by the addition of the basis (b) to every lattice point of the lattice (a). By looking at (c), you can recognize the basis and then you can abstract the space lattice. It does not matter where the basis is put in relation to a lattice point.

primitive axes and primitive cell for a given lattice. For all choices the number of atoms in the primitive basis will be the same. There is a density of one lattice point per primitive cell. There are lattice points at the eight corners of the parallelepiped, but each corner point is shared among the eight cells that touch there.

The volume of a cell defined by axes **a, b, c** is

$$V_c = |\mathbf{a} \times \mathbf{b} \cdot \mathbf{c}| \ , \qquad (4)$$

by elementary vector analysis. The basis associated with a lattice point of a primitive cell may be called a primitive basis. No basis contains fewer atoms than a primitive basis contains. Another way of choosing a cell of equal volume V_c is shown in Fig. 6. The cell formed in this way is known to physicists as a **Wigner-Seitz primitive cell.**

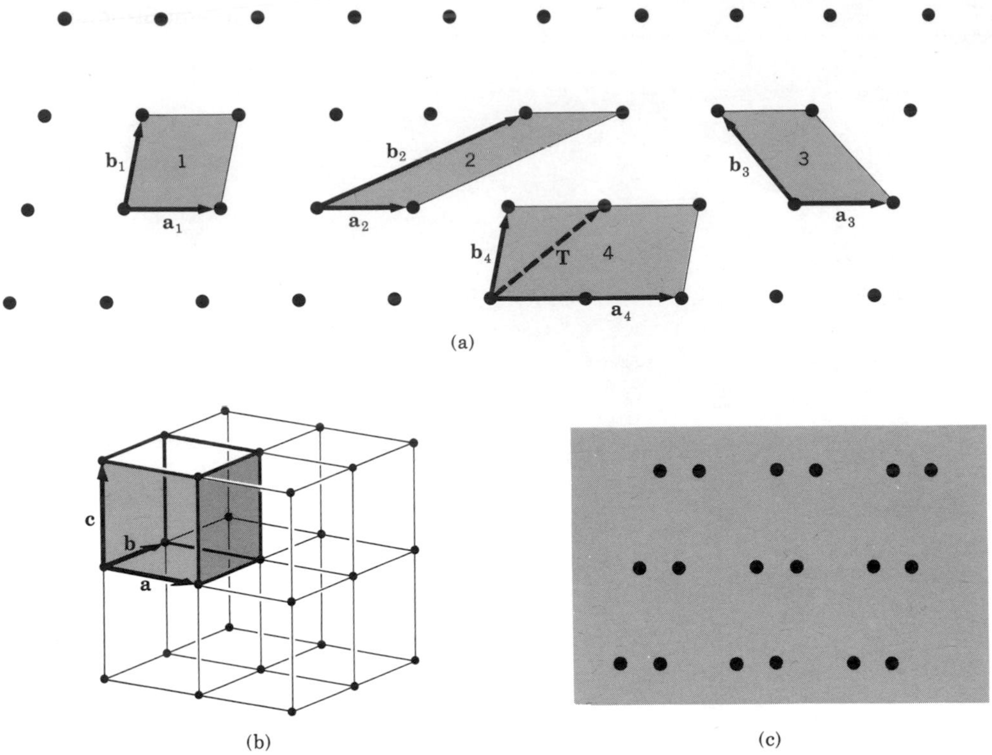

Figure 5a Lattice points of a space lattice in two dimensions. All pairs of vectors **a, b** are translation vectors of the lattice. But $\mathbf{a}_4$, $\mathbf{b}_4$ are not primitive translation vectors because we cannot form the lattice translation **T** from integral combinations of $\mathbf{a}_4$ and $\mathbf{b}_4$. All other pairs shown of **a** and **b** may be taken as the primitive translation vectors of the lattice. The parallelograms 1, 2, 3 are equal in area and any of them could be taken as the primitive cell. The parallelogram 4 has twice the area of a primitive cell.

Figure 5b Primitive cell of a space lattice in three dimensions.

Figure 5c Suppose these points are identical atoms: sketch in on the figure a set of lattice points, a choice of primitive axes, a primitive cell, and the basis of atoms associated with a lattice point.

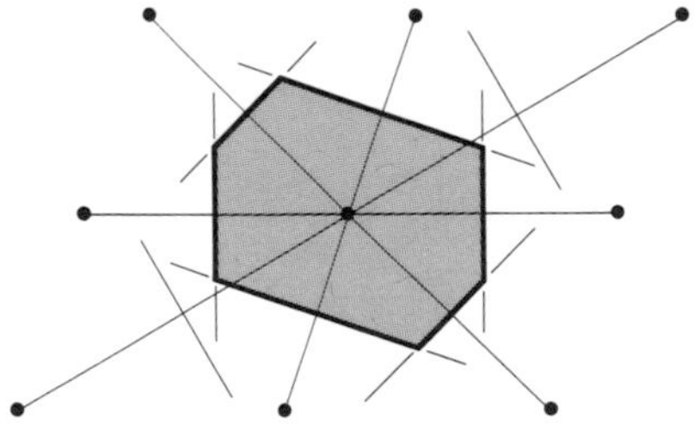

Figure 6 A primitive cell may also be chosen following this procedure: (1) draw lines to connect a given lattice point to all nearby lattice points; (2) at the midpoint and normal to these lines, draw new lines or planes. The smallest volume enclosed in this way is the Wigner-Seitz primitive cell. All space may be filled by these cells, just as by the cells of Fig. 5.

FUNDAMENTAL TYPES OF LATTICES

Crystal lattices can be carried into themselves by the lattice translations **T** and by various other symmetry operations. A typical symmetry operation is that of rotation about an axis that passes through a lattice point. Lattices can be found such that one-, two-, three-, four-, and sixfold rotation axes carry the lattice into itself, corresponding to rotations by 2π, $2\pi/2$, $2\pi/3$, $2\pi/4$, and $2\pi/6$ radians and by integral multiples of these rotations. The rotation axes are denoted by the symbols 1, 2, 3, 4, and 6. We cannot find a lattice that goes into itself under other rotations, such as by $2\pi/7$ radians or $2\pi/5$ radians. A single molecule properly designed can have any degree of rotational symmetry, but an infinite periodic lattice cannot. We can make a crystal from molecules that individually have a fivefold rotation axis, but we should not expect the lattice to have a fivefold rotation axis. In Fig. 7 we show what happens if we try to construct a periodic lattice having fivefold symmetry: the pentagons do not fit together neatly, showing that we cannot combine fivefold point symmetry with the required translational periodicity.

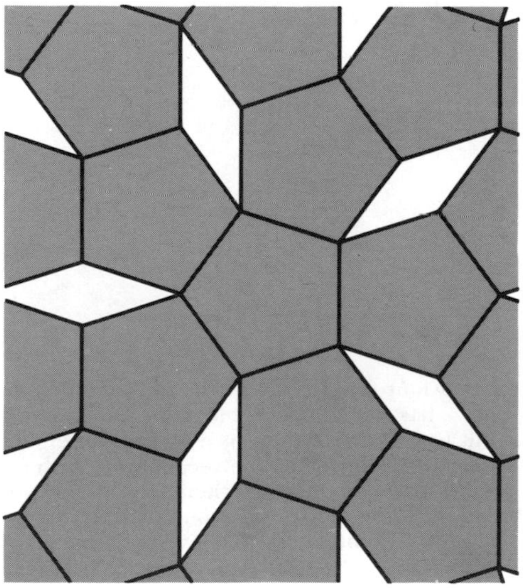

Figure 7 A fivefold axis of symmetry cannot exist in a lattice because it is not possible to fill all space with a connected array of pentagons.

A lattice point group is the collection of the symmetry operations which, when applied about a lattice point, leave the lattice invariant. The possible rotations have been given earlier. We can also have **mirror reflections** m about a plane through a lattice point. The **inversion operation** is made up of a rotation of π followed by reflection in a plane normal to the rotation axis; the total effect is to replace **r** by −**r**.

Locations of equivalent positions are shown for four point groups in Fig. 8. The symmetry operations of the group carry one point into all the equivalent positions. The equivalent points themselves must not be thought of as possessing symmetry elements; they may be scalene triangles or molecules with no symmetry elements. The symmetry axes and planes of a cube are shown in Fig. 9.

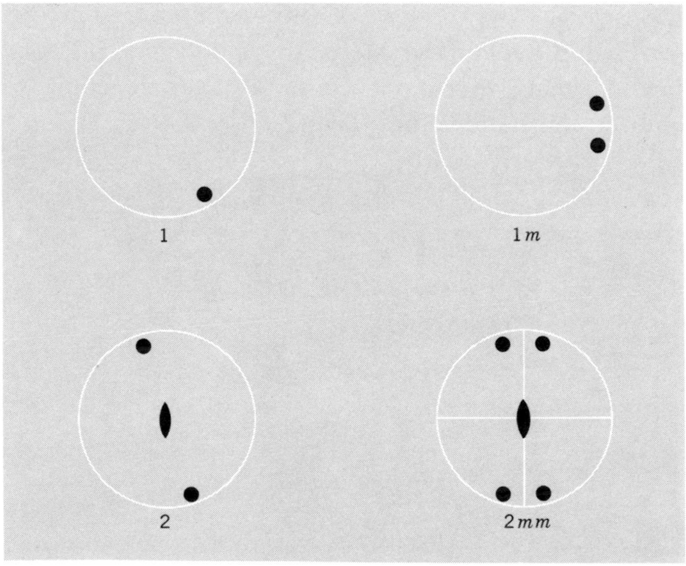

Figure 8 Diagrams illustrating four point groups. The dots denote equivalent points. The point group 1 has no symmetry elements, so here a point has no other point equivalent to it. For 1m there is a mirror plane: one dot on reflection across the mirror plane becomes the second dot. With a twofold axis 2, a rotation of π takes one dot into the other. With a twofold axis and one mirror plane there is automatically a second mirror plane normal to the first, and we have the point group 2mm with four equivalent points.

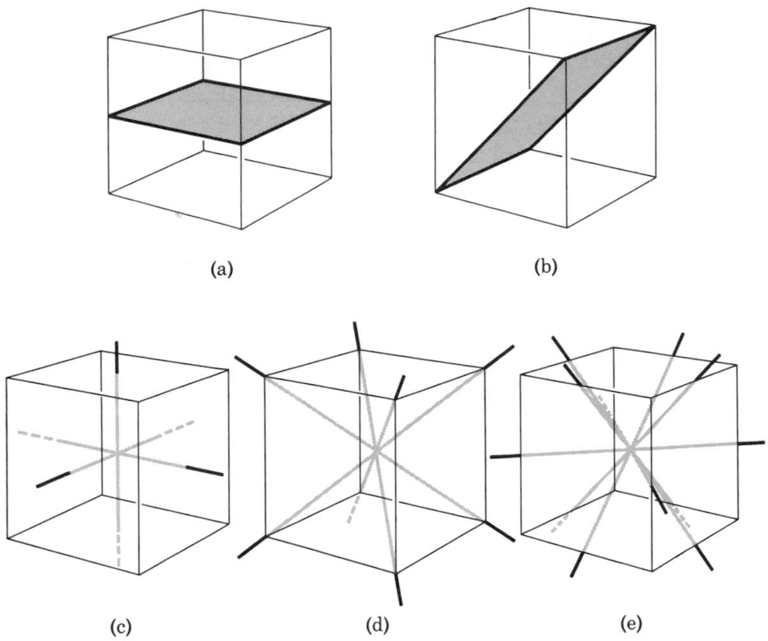

Figure 9 (a) A plane of symmetry parallel to the faces of a cube. (b) A diagonal plane of symmetry in a cube. (c) The three tetrad axes of a cube. (d) The four triad axes of a cube. (e) The six diad axes of a cube.

Two-Dimensional Lattice Types

There is an unlimited number of possible lattices because there is no natural restriction on the lengths a, b of the lattice translation vectors or on the angle φ between them. The lattice in Fig. 5a was drawn for arbitrary **a** and **b.** A general lattice such as this is known as an **oblique lattice** and is invariant only under rotation of π and 2π about any lattice point.

But special lattices of the oblique type can be invariant under rotation of $2\pi/3$, $2\pi/4$, or $2\pi/6$, or under mirror reflection. We must impose restrictive conditions on **a** and **b** if we want to construct a lattice that will be invariant under one or more of these new operations. There are four distinct types of restriction, and each leads to what we may call a **special lattice type.** Thus there are five distinct lattice types in two dimensions, the oblique lattice and the four special lattices. **Bravais lattice** is the common phrase for a distinct lattice type; we say that there are five Bravais lattices in two dimensions. We have not succeeded in finding or constructing a satisfactory definition that starts out "A Bravais lattice is . . ."; the sources say "That was a Bravais lattice."

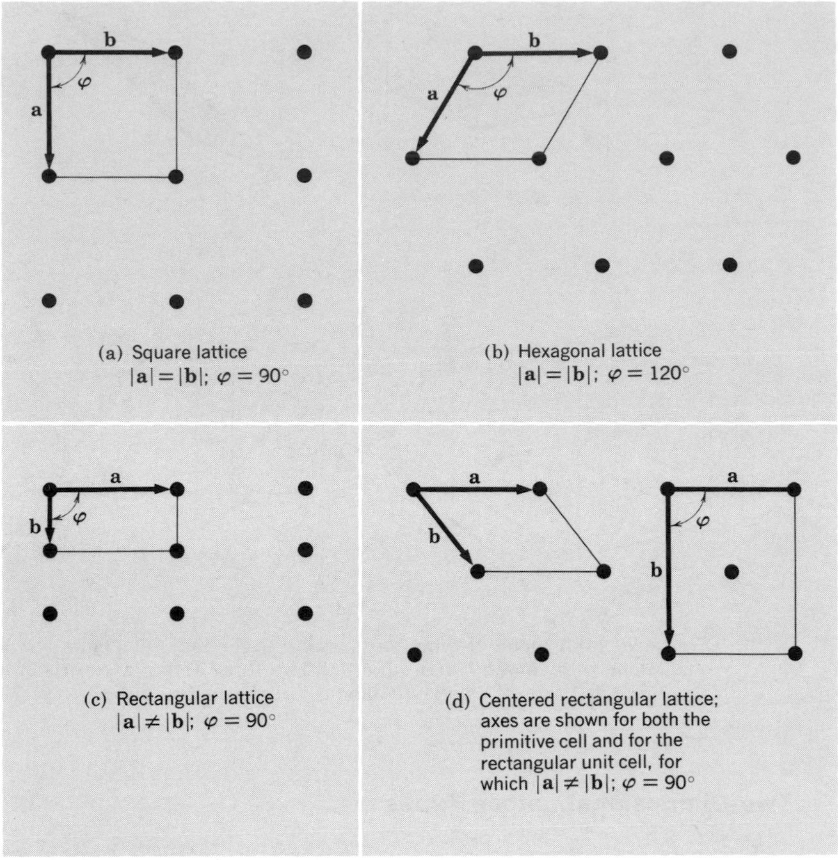

Figure 10

The point operation 4 requires a square lattice (Fig. 10a). The point operations 3 and 6 require a hexagonal lattice (Fig. 10b). This lattice is invariant under a rotation $2\pi/6$ about an axis through a lattice point and normal to the plane.

There are important consequences if the lattice allows the mirror reflection m. We write the primitive translation vectors **a**, **b** in terms of the unit vectors $\hat{\mathbf{x}}$, $\hat{\mathbf{y}}$ along the Cartesian x, y axes:

$$\mathbf{a} = a_x\hat{\mathbf{x}} + a_y\hat{\mathbf{y}} \; ; \qquad \mathbf{b} = b_x\hat{\mathbf{x}} + b_y\hat{\mathbf{y}} \; . \tag{5}$$

If the primitive vectors are mirrored in the x axis, then **a**, **b** are transformed by the reflection operation into new vectors **a′**, **b′** given by

$$\mathbf{a}' = a_x\hat{\mathbf{x}} - a_y\hat{\mathbf{y}} \; ; \qquad \mathbf{b}' = b_x\hat{\mathbf{x}} - b_y\hat{\mathbf{y}} \; . \tag{6}$$

If the lattice is invariant under the reflection, then **a′**, **b′** must be lattice vectors; that is, they must be of the form $n_1\mathbf{a} + n_2\mathbf{b}$, where n_1 and n_2 are integers. If we take

$$\mathbf{a} = a\hat{\mathbf{x}} \; ; \qquad \mathbf{b} = b\hat{\mathbf{y}} \; ; \tag{7}$$

then $\mathbf{a}' = \mathbf{a}$ and $\mathbf{b}' = -\mathbf{b}$, so that the lattice is carried into itself. The lattice defined by (7) is rectangular (Fig. 10c).

A second distinct possibility for **a** and **b** gives another type of lattice invariant under reflection. Note that $\mathbf{b}'$ will be a lattice vector if $\mathbf{b}' = \mathbf{a} - \mathbf{b}$; then using (6) we have

$$b_x' = a_x - b_x = b_x \ ; \qquad b_y' = a_y - b_y = -b_y \ . \tag{8}$$

These equations have a solution if $a_y = 0$; $b_x = \frac{1}{2}a_x$; thus a possible choice of primitive translation vectors for a lattice with mirror symmetry is

$$\mathbf{a} = a\hat{\mathbf{x}} \ ; \qquad \mathbf{b} = \tfrac{1}{2}a\hat{\mathbf{x}} + b_y\hat{\mathbf{y}} \ . \tag{9}$$

This choice gives a centered rectangular lattice (Fig. 10d). We have exhausted the two-dimensional Bravais lattices consistent with point-group operations applied to lattice points.

Three-Dimensional Lattice Types

In two dimensions the point groups are associated with five different types of lattices. In three dimensions the point symmetry groups require the 14 different (one general and 13 special) lattice types shown in Fig. 11 and listed in Table 1. The general lattice type is the triclinic lattice. The 14 lattice types are conveniently grouped into seven systems according to the seven types of conventional unit cells: triclinic, monoclinic, orthorhombic, tetragonal, cubic, trigonal, and hexagonal. The division into systems is summarized in terms of the special axial relations for the conventional unit cells. The axes **a, b, c** and angles α, β, γ are defined in Fig. 12. The magnitudes a, b, c are called the **lattice parameters.** The cells shown in Fig. 11 are the conventional cells, and they are not always primitive cells. Sometimes a nonprimitive cell has a more obvious connection with the point symmetry elements than has a primitive cell.

In the **cubic system** there are three lattices: the simple cubic (sc) lattice, the body-centered cubic (bcc) lattice, and the face-centered cubic (fcc) lattice. The characteristics of the three cubic lattices are summarized in Table 2. A primitive cell of the body-centered cubic lattice is shown in Fig. 13; the primitive translation vectors are shown in Fig. 14. The primitive translation vectors of the face-centered cubic lattice are shown in Fig. 15. The primitive cells contain only one lattice point, but the conventional cubic cells contain two lattice points (bcc) or four lattice points (fcc). The positions of a point in a cell is specified by (3) in terms of atomic coordinates x, y, z in which each coordinate is a fraction of the axial length a, b, or c in the direction of the coordinate, with the origin taken at a corner of a cell. Thus the coordinates of the body center of a cell are $\frac{1}{2}\frac{1}{2}\frac{1}{2}$. The face-centers include $\frac{1}{2}\frac{1}{2}0$; $0\frac{1}{2}\frac{1}{2}$; $\frac{1}{2}0\frac{1}{2}$. The coordinates of atoms in fcc and bcc lattices are usually given in terms of the conventional cubic cell.

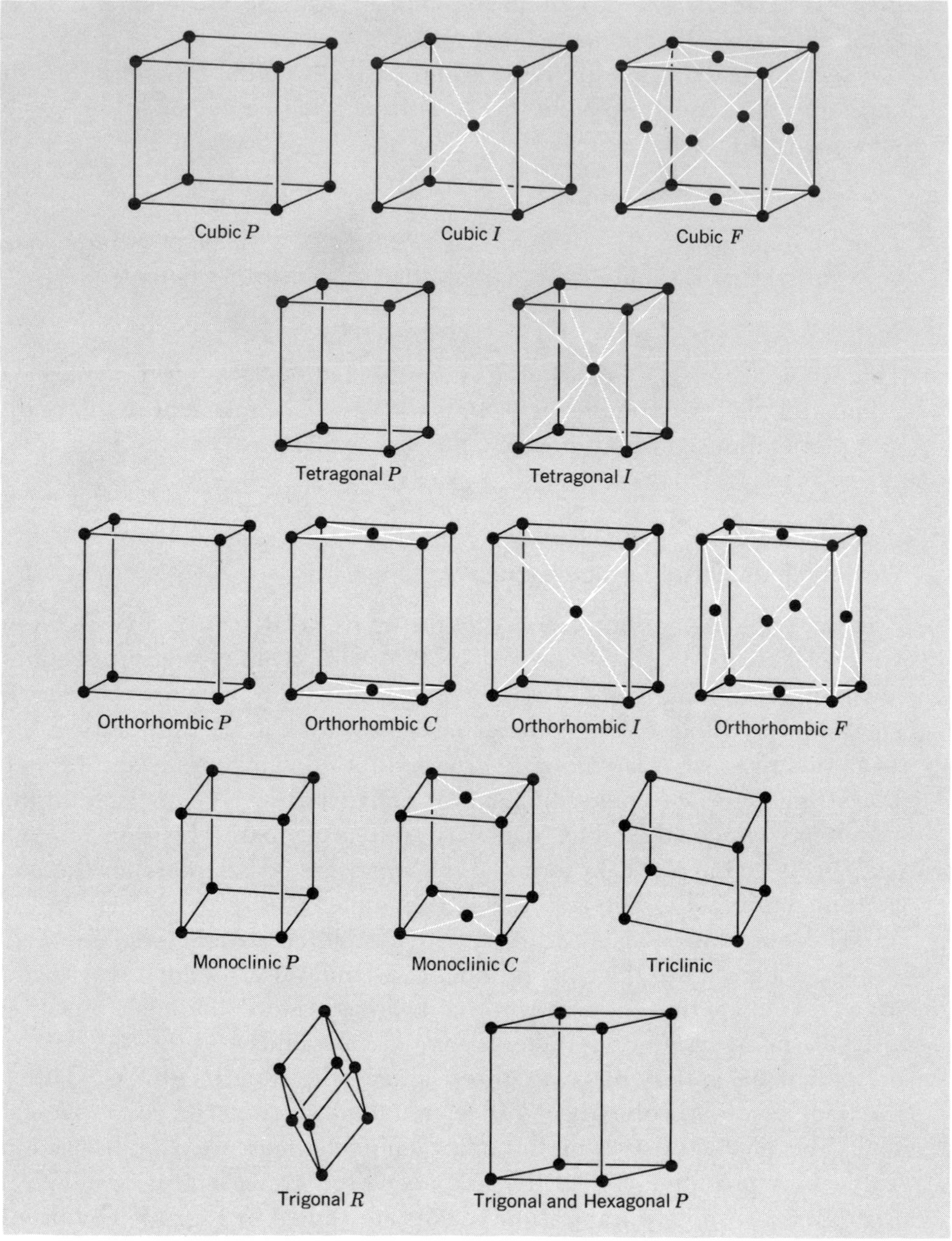

Figure 11 The 14 Bravais or space lattices. The cells shown are the conventional cells, which are not always the primitive cells.

TABLE 1 The 14 Lattice Types in Three Dimensions

System	Number of Lattices	Lattice Symbols	Restrictions on Conventional Cell Axes and Angles
Triclinic	1	P	$a \neq b \neq c$ $\alpha \neq \beta \neq \gamma$
Monoclinic	2	P, C	$a \neq b \neq c$ $\alpha = \gamma = 90° \neq \beta$
Orthorhombic	4	P, C, I, F	$a \neq b \neq c$ $\alpha = \beta = \gamma = 90°$
Tetragonal	2	P, I	$a = b \neq c$ $\alpha = \beta = \gamma = 90°$
Cubic	3	P or sc I or bcc F or fcc	$a = b = c$ $\alpha = \beta = \gamma = 90°$
Trigonal	1	R	$a = b = c$ $\alpha = \beta = \gamma < 120°, \neq 90°$
Hexagonal	1	P	$a = b \neq c$ $\alpha = \beta = 90°$ $\gamma = 120°$

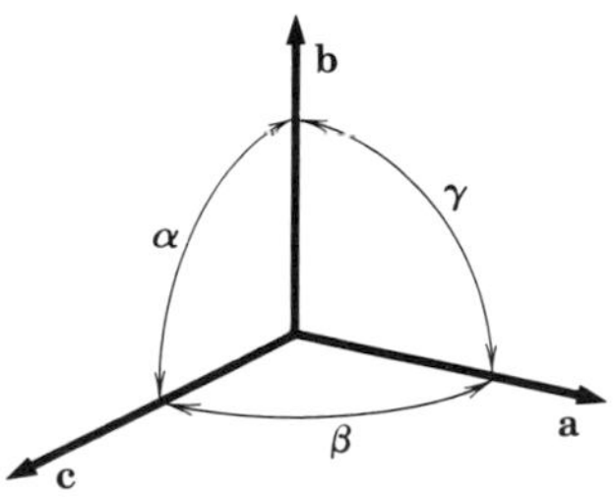

Figure 12 Crystal axes **a, b, c.** The angle α is included between **b** and **c.**

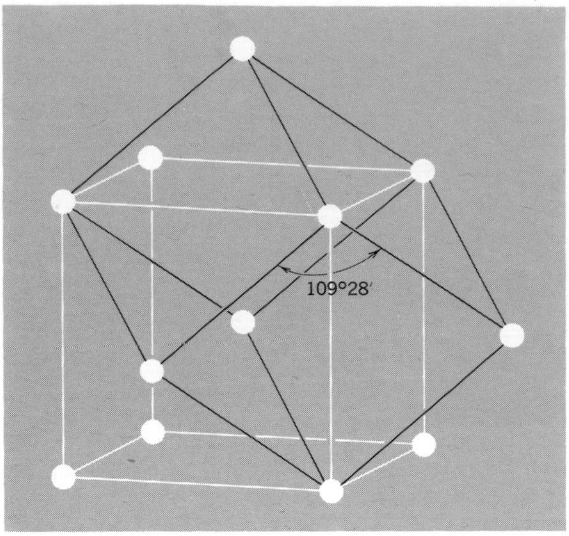

Figure 13 Body-centered cubic lattice, showing a primitive cell. The primitive cell shown is a rhombohedron of edge $\frac{1}{2}\sqrt{3}\,a$, and the angle between adjacent edges is 109°28′.

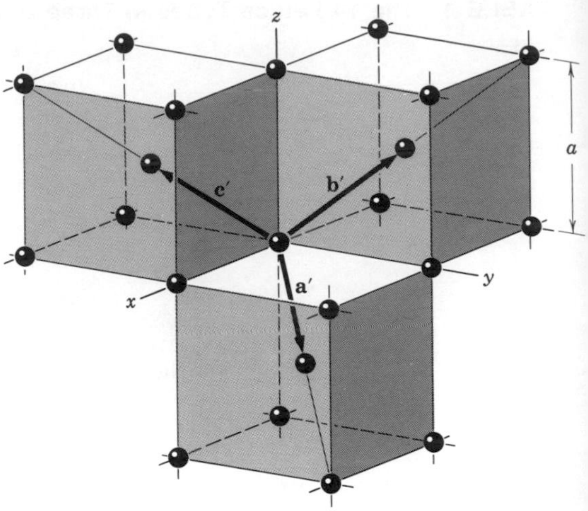

Figure 14 Primitive translation vectors of the body-centered cubic lattice; these vectors connect the lattice point at the origin to lattice points at the body centers. The primitive cell is obtained on completing the rhombohedron. In terms of the cube edge a the primitive translation vectors are

$$\mathbf{a}' = \frac{a}{2}(\hat{\mathbf{x}} + \hat{\mathbf{y}} - \hat{\mathbf{z}}) \; ; \qquad \mathbf{b}' = \frac{a}{2}(-\hat{\mathbf{x}} + \hat{\mathbf{y}} + \hat{\mathbf{z}}) \; ;$$

$$\mathbf{c}' = \frac{a}{2}(\hat{\mathbf{x}} - \hat{\mathbf{y}} + \hat{\mathbf{z}}) \; .$$

TABLE 2 Characteristics of Cubic Lattices[a]

	Simple	Body-Centered	Face-Centered
Volume, conventional cell	a^3	a^3	a^3
Lattice points per cell	1	2	4
Volume, primitive cell	a^3	$\frac{1}{2}a^3$	$\frac{1}{4}a^3$
Lattice points per unit volume	$1/a^3$	$2/a^3$	$4/a^3$
Number of nearest neighbors	6	8	12
Nearest-neighbor distance	a	$3^{1/2}a/2 = 0.866a$	$a/2^{1/2} = 0.707a$
Number of second neighbors	12	6	6
Second neighbor distance	$2^{1/2}a$	a	a
Packing fraction[b]	$\frac{1}{6}\pi$ $= 0.524$	$\frac{1}{8}\pi\sqrt{3}$ $= 0.680$	$\frac{1}{6}\pi\sqrt{2}$ 0.740

[a] The packing fraction is the maximum proportion of the available volume that can be filled hard spheres.

[b] Tables of numbers of neighbors and distances in sc, bcc, fcc, hcp, and diamond structures are given on pp. 1037–1039 of J. Hirschfelder, C. F. Curtis and R. B. Bird, *Molecular theory of gases and liquids*, Wiley 1964.

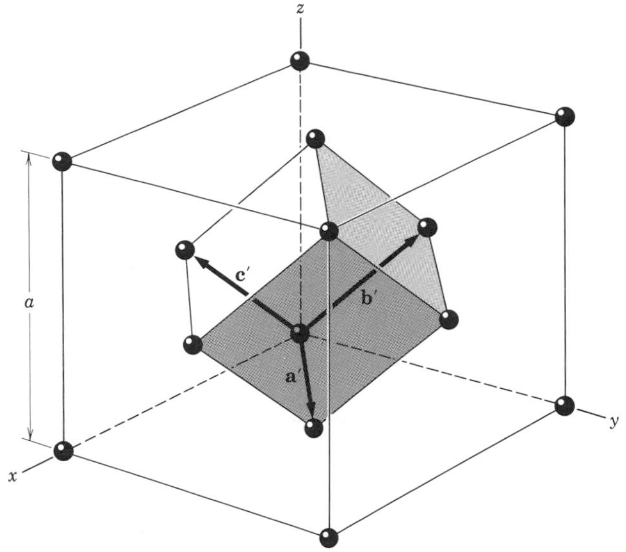

Figure 15 The rhombohedral primitive cell of the face-centered cubic crystal. The primitive translation vectors **a′**, **b′**, **c′** connect the lattice point at the origin with lattice points at the face centers. As drawn, the primitive vectors are:

$$\mathbf{a}' = \frac{a}{2}(\hat{\mathbf{x}} + \hat{\mathbf{y}}) \; ; \qquad \mathbf{b}' = \frac{a}{2}(\hat{\mathbf{y}} + \hat{\mathbf{z}}) \; ; \qquad \mathbf{c}' = \frac{a}{2}(\hat{\mathbf{z}} + \hat{\mathbf{x}}) \; .$$

The angles between the axes are 60°.

In the **hexagonal system** the conventional cell chosen is a right prism based on a rhombus with an angle of 60°. The lattice is primitive. The relation of the rhombic cell with a hexagonal prism is shown in Fig. 16.

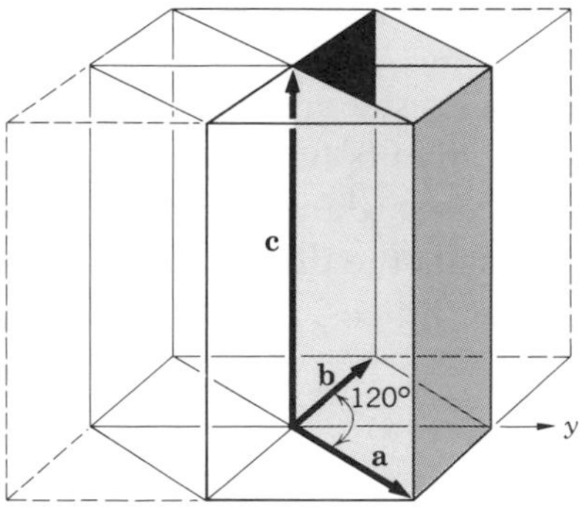

Figure 16 Relation of the primitive cell in the hexagonal system (heavy lines) to a prism of hexagonal symmetry. Here $a = b \neq c$. See Fig. 26.

INDEX SYSTEM FOR CRYSTAL PLANES

The position and orientation of a crystal plane are determined by any three points in the plane, provided the points are not collinear. If each point lies on a crystal axis, the plane could be specified by giving the positions of the points along the axes in terms of the lattice constants. If, for example, the points have coordinates (4, 0, 0), (0, 1, 0), (0, 0, 2) relative to axis vectors from some origin, the plane may be specified by the three numbers 4, 1, 2. But it is more useful for structure analysis to specify the orientation of a plane by indices determined by the following rules, exemplified in Fig. 17.

1. Find the intercepts on the axes **a, b, c** in terms of the lattice constants. The axes may be primitive or nonprimitive.
2. Take the reciprocals of these numbers and then reduce to three integers having the same ratio, usually the smallest three integers. The result is enclosed in parentheses: (hkl).

For the plane whose intercepts are 4, 1, 2, the reciprocals are $\frac{1}{4}$, 1, and $\frac{1}{2}$; the smallest three integers having the same ratio are (142). For an intercept at infinity, the corresponding index is zero. The indices of some important planes in a cubic crystal are illustrated by Fig. 18. The indices (hkl) may denote a single plane or a set of parallel planes. If a plane cuts an axis on the negative side of the origin, the corresponding index is negative, indicated by placing a minus sign above the index: $(h\bar{k}l)$. The cube faces of a cubic crystal are (100), (010), (001), $(\bar{1}00)$, $(0\bar{1}0)$, and $(00\bar{1})$. Planes equivalent by symmetry may be denoted by curly brackets (braces) around indices; the set of cube faces is {100}. When we speak of the (200) plane we mean a plane parallel to (100) but cutting the **a** axis at $\frac{1}{2}a$. The formation of a (111) plane of an fcc crystal structure, starting from (100) planes of atoms, is shown in Fig. 19.

The indices $[uvw]$ of a direction in a crystal are the set of the smallest integers that have the ratio of the components of a vector in the desired direction, referred to the axes. The **a** axis is the [100] direction; the $-$**b** axis is the $[0\bar{1}0]$ direction. In cubic crystals the direction $[hkl]$ is perpendicular to a plane (hkl) having the same indices, but this is not generally true in other crystal systems.

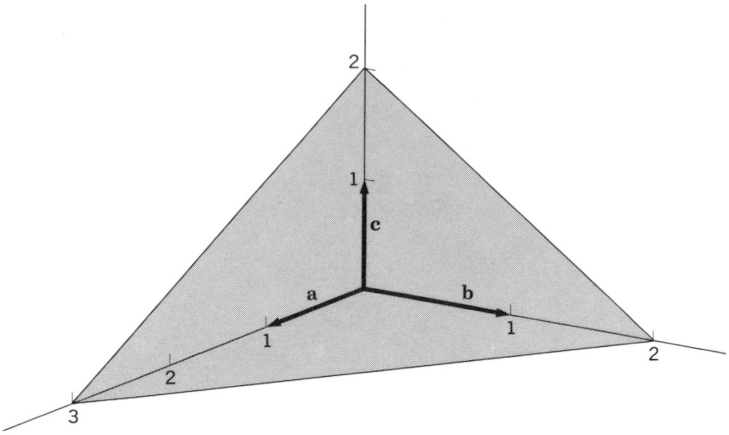

Figure 17 This plane intercepts the **a, b, c** axes at 3**a**, 2**b**, 2**c**. The reciprocals of these numbers are $\frac{1}{3}$, $\frac{1}{2}$, $\frac{1}{2}$. The smallest three integers having the same ratio are 2, 3, 3, and thus the indices of the plane are (233).

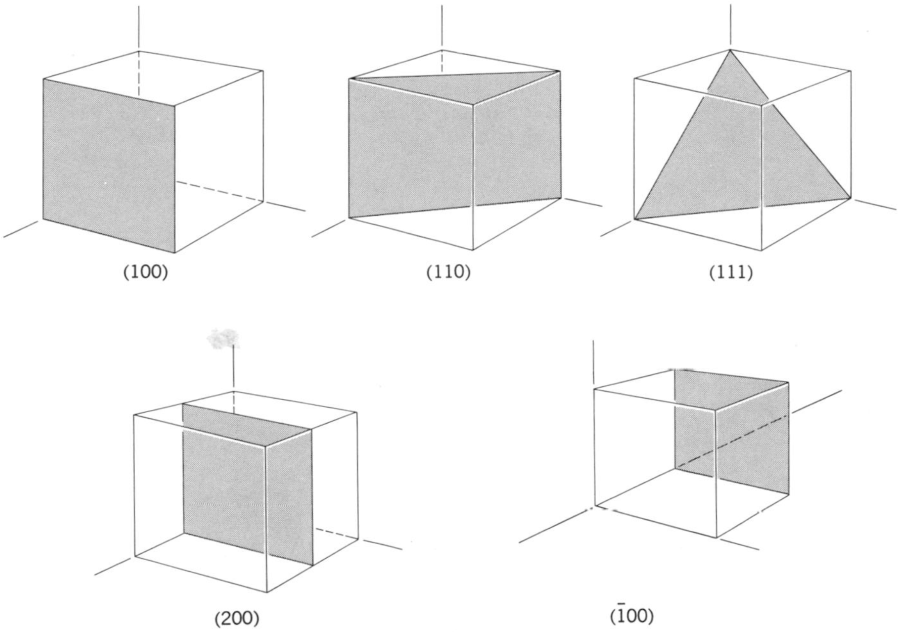

Figure 18 Indices of some important planes in a cubic crystal. The plane (200) is parallel to (100) and to $(\bar{1}00)$.

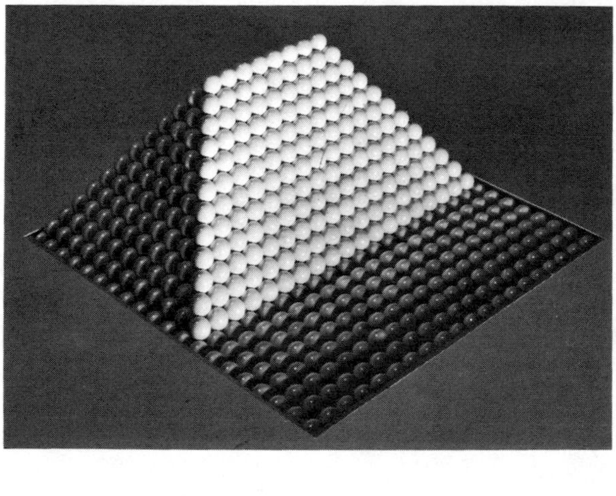

Figure 19 A (111) plane of an fcc crystal structure, based on (100) layers. (Photograph by J. F. Nicholas, *Atlas of models of crystal surfaces*, Gordon and Breach, 1965.)

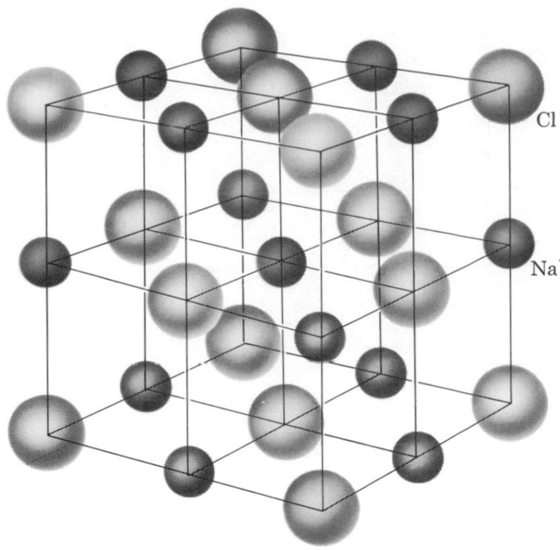

Figure 20 We may construct the sodium chloride crystal structure by arranging Na^+ and Cl^- ions alternately at the lattice points of a simple cubic lattice. In the crystal each ion is surrounded by six nearest neighbors of the opposite charge. The Na^+ ion carries a single positive charge, so that the electronic configuration is identical with neon, and the Cl^- ion carries a single negative charge and is in the argon configuration. The space lattice is fcc, and the basis has one Cl^- ion at 000 and one Na^+ ion at $\frac{1}{2}\frac{1}{2}\frac{1}{2}$. The figure shows one conventional cubic cell.

Figure 21 Model of sodium chloride. The sodium ions are smaller than the chlorine ions. (Courtesy of A. N. Holden and P. Singer, from *Crystals and crystal growing.*)

SIMPLE CRYSTAL STRUCTURES

We discuss briefly a small number of simple crystal structures of general interest, including the sodium chloride, cesium chloride, hexagonal close-packed, diamond, and cubic zinc sulfide structures.

Sodium Chloride Structure

The sodium chloride, NaCl, structure is shown in Figs. 20 and 21. The Bravais lattice is face-centered cubic; the basis consists of one Na atom and one Cl atom separated by one-half the body diagonal of a unit cube. There are four units of NaCl in each unit cube, with atoms in the positions

Cl:	000 ;	$\frac{1}{2}\frac{1}{2}0$;	$\frac{1}{2}0\frac{1}{2}$;	$0\frac{1}{2}\frac{1}{2}$.
Na:	$\frac{1}{2}\frac{1}{2}\frac{1}{2}$;	$00\frac{1}{2}$;	$0\frac{1}{2}0$;	$\frac{1}{2}00$.

Each atom has as nearest neighbors six atoms of the opposite kind. Representative crystals having the NaCl arrangement include those in the following table. The cube edge a is given in angstroms; $1\ \text{Å} \equiv 10^{-8}$ cm $\equiv 10^{-10}$ m.

Crystal	a	Crystal	a
LiH	4.08 Å	AgBr	5.77 Å
NaCl	5.63	MgO	4.20
KCl	6.29	MnO	4.43
PbS	5.92	KBr	6.59

Figure 22 is a photograph of crystals of galena (PbS) from Joplin, Missouri. The Joplin specimens form in beautiful cubes.

Figure 22 Natural crystals of lead sulfide, PbS, which has the NaCl crystal structure. (Photograph by B. Burleson.)

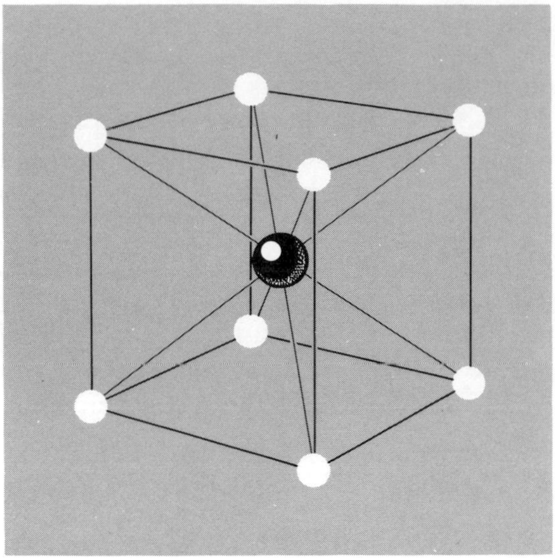

Figure 23 The cesium chloride crystal structure. The space lattice is simple cubic, and the basis has one Cs^+ ion at 000 and one Cl^- ion at $\frac{1}{2}\frac{1}{2}\frac{1}{2}$.

Cesium Chloride Structure

The cesium chloride, CsCl, structure is shown in Fig. 23. There is one molecule per primitive cell, with atoms at the corners 000 and body-centered positions $\frac{1}{2}\frac{1}{2}\frac{1}{2}$ of the simple cubic space lattice.[4] Each atom is at the center of a cube of atoms of the opposite kind, so that the coordination number is eight. Representative crystals having the CsCl arrangement include:

Crystal	a	Crystal	a
CsCl	4.11 Å	CuZn (β-brass)	2.94 Å
TlBr	3.97	AgMg	3.28
TlI	4.20	LiHg	3.29
NH_4Cl	3.87	AlNi	2.88
CuPd	2.99	BeCu	2.70

[4]Crystal structures with simple cubic Bravais lattices are not rare, but no common *element* crystallizes in a sc structure.

Hexagonal Close-packed Structure (hcp)

There are two ways (Fig. 24) of arranging equivalent spheres in a regular array to minimize the interstitial volume. One leads to the face-centered cubic (cubic close-packed) structure; the other has hexagonal symmetry and is called the hexagonal close-packed structure (Fig. 25). The fraction of the total volume filled by the spheres is 0.74 for both the fcc and hcp structures.

Spheres may be arranged in a single closest-packed layer by placing each sphere in contact with six others. Such a layer can be either the basal plane of a hcp structure or the (111) plane of the fcc structure. A second similar layer is added by placing each sphere in contact with three spheres of the bottom layer, as in Fig. 24. A third layer can be added in two ways: in the fcc structure the spheres in the third layer are placed over the holes in the first layer not occupied by the second layer; in the hexagonal structure the spheres in the third layer are placed directly over the spheres in the first layer. We say that the packing in the fcc structure is *ABCABC* . . . , whereas in the hcp structure the packing is *ABABAB*. . . . . The hcp structure has a hexagonal primitive cell; the

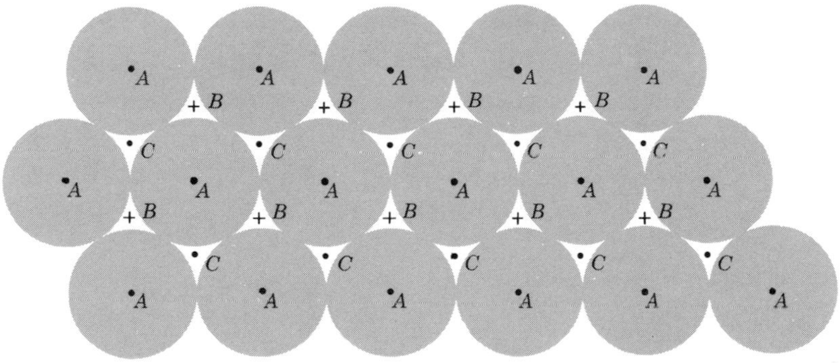

Figure 24 A close-packed layer of spheres is shown, with centers at points marked *A*. A second and identical layer of spheres can be placed over this, with centers over the points marked *B* (or, equivalently, over the points marked *C*). If the second layer goes in over *B*, there are two nonequivalent choices for a third layer. It can go in over *A* or over *C*. If it goes in over *A* the sequence is *ABABAB*. . . and the structure is hexagonal close-packed. If the third layer goes in over *C* the sequence is *ABCABCABC*. . . and the structure is face-centered cubic; the plane is a (111) plane, as in Fig. 19.

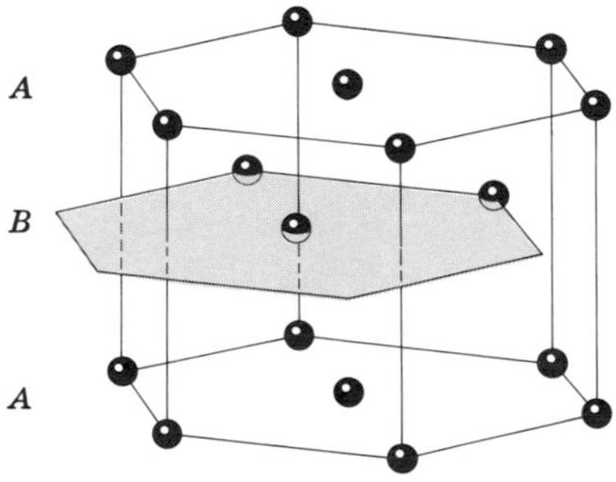

Figure 25 The hexagonal close-packed structure. The atom positions in this structure do not constitute a space lattice. The space lattice is simple hexagonal with a basis of two identical atoms associated with each lattice point.

basis contains two atoms, as shown in Fig. 26. The fcc primitive cell (Fig. 15) contains one atom.

The c/a ratio for hexagonal closest-packing of spheres is $(\frac{8}{3})^{\frac{1}{2}} = 1.633$. We refer to crystals as hcp even if the actual c/a ratio departs somewhat from the theoretical value. Thus zinc with $c/a = 1.86$ ($a = 2.66$ Å; $c = 4.94$ Å) is referred to commonly as hcp. Magnesium with $c/a = 1.623$ is close to ideal hcp. Many metals transform easily at appropriate temperatures between fcc and hcp structures. The **coordination number,** defined as the number of nearest-neighbor atoms, is 12 for the two structures. If the binding energy depended only on the number of nearest-neighbor bonds, then there would be no difference in energy between the fcc and hcp structures. Examples of hexagonal close-packed structures are

Crystal	c/a	Crystal	c/a	Crystal	c/a
He	1.633	Zn	1.861	Zr	1.594
Be	1.581	Cd	1.886	Gd	1.592
Mg	1.623	Co	1.622	Lu	1.586
Ti	1.586	Y	1.570		

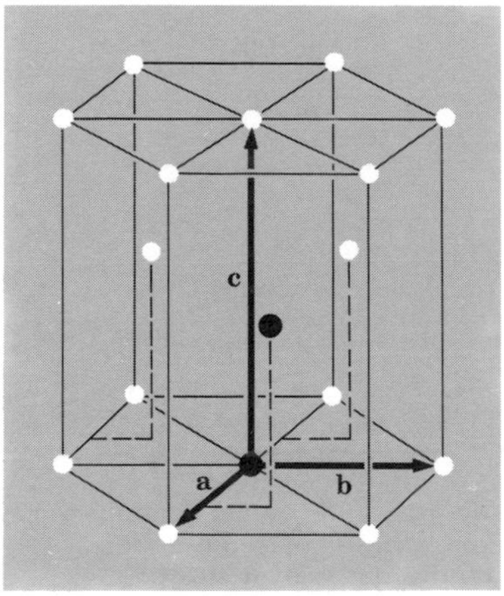

Figure 26 The primitive cell has $a = b$, with an included angle of 120°. The c axis is normal to the plane of a and b. In ideal hcp we have $c = 1.633\ a$. The two atoms of one basis are shown as solid in the figure. One atom of the basis is at the origin, 000; the other atom is at $\frac{2}{3}\frac{1}{3}\frac{1}{2}$, which means at $\mathbf{r} = \frac{2}{3}\mathbf{a} + \frac{1}{3}\mathbf{b} + \frac{1}{2}\mathbf{c}$.

Diamond Structure

The space lattice of diamond is face-centered cubic. A primitive basis of two identical atoms at 000; $\frac{1}{4}\frac{1}{4}\frac{1}{4}$ is associated with each lattice point,[5] as shown in Fig. 27. The tetrahedral bonding of the diamond structure is represented in Fig. 28. Each atom has four nearest neighbors and 12 next nearest neighbors. There are eight atoms in a unit cube. The diamond lattice is relatively empty; the maximum proportion of the available volume which may be filled by hard spheres is only 0.34, or about 46 percent of the filling factor for a closest-packed structure. Carbon, silicon, germanium, and gray tin crystallize in the diamond structure, with lattice constants $a = 3.56$, 5.43, 5.65, and 6.46 Å, respectively. The diamond structure is the result of directional covalent bonding.

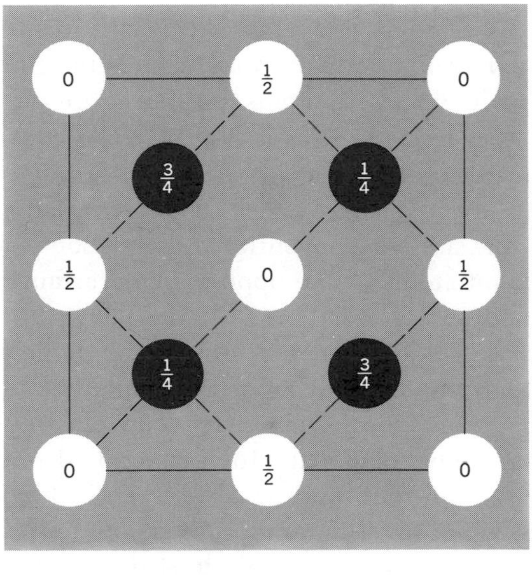

Figure 27 Atomic positions in the cubic cell of the diamond structure projected on a cube face; fractions denote height above the base in units of a cube edge. The points at 0 and $\frac{1}{2}$ are on the fcc lattice; those at $\frac{1}{4}$ and $\frac{3}{4}$ are on a similar lattice displaced along the body diagonal by one-fourth of its length. With a fcc space lattice, the basis consists of two identical atoms at 000; $\frac{1}{4}\frac{1}{4}\frac{1}{4}$.

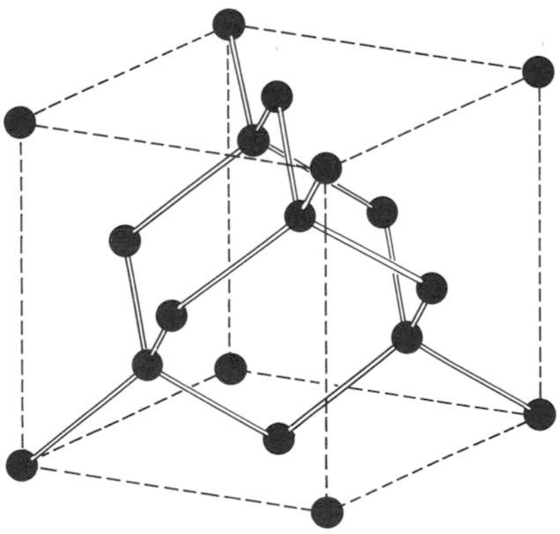

Figure 28 Crystal structure of diamond, showing the tetrahedral bond arrangement.

[5]The conventional unit cube contains eight atoms. There is no way of choosing a primitive cell such that the basis of diamond contains only one atom.

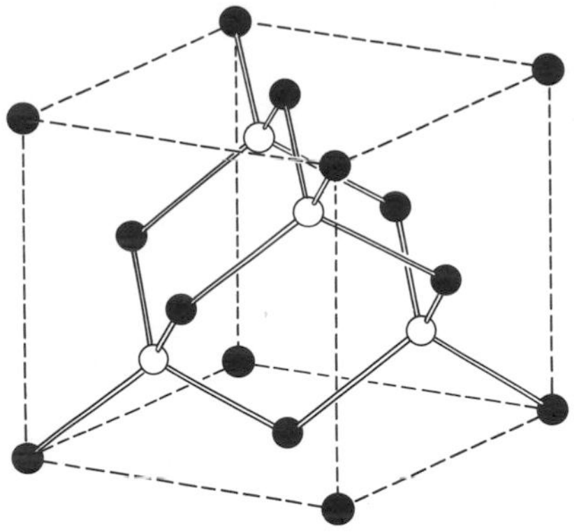

Figure 29 Crystal structure of cubic zinc sulfide.

Cubic Zinc Sulfide Structure

The diamond structure may be viewed as two fcc structures displaced from each other by one-quarter of a body diagonal. The cubic zinc sulfide (zinc blende) structure results when Zn atoms are placed on one fcc lattice and S atoms on the other fcc lattice, as in Fig. 29. The conventional cell is a cube. The coordinates of the Zn atoms are 000; $0\frac{1}{2}\frac{1}{2}$; $\frac{1}{2}0\frac{1}{2}$; $\frac{1}{2}\frac{1}{2}0$; the coordinates of the S atoms are $\frac{1}{4}\frac{1}{4}\frac{1}{4}$; $\frac{1}{4}\frac{3}{4}\frac{3}{4}$; $\frac{3}{4}\frac{1}{4}\frac{3}{4}$; $\frac{3}{4}\frac{3}{4}\frac{1}{4}$. The space lattice is fcc. There are four molecules of ZnS per conventional cell. About each atom there are four equally distant atoms of the opposite kind arranged at the corners of a regular tetrahedron.

The diamond structure possesses a center-of-inversion symmetry element at the midpoint of each line connecting nearest-neighbor atoms; the ZnS structure does not have inversion symmetry.[6] Consider the arrangement of atoms along a body diagonal. In diamond the order is CC··CC··CC, where dots represent vacancies. In ZnS the order is ZnS··ZnS··ZnS; this order is not invariant under inversion, for it becomes SZn··SZn··SZn. Examples of the cubic zinc sulfide structure are

Crystal	*a*	Crystal	*a*
CuF	4.26 Å	CdS	5.82 Å
CuCl	5.41	InAs	6.04
AgI	6.47	InSb	6.46
ZnS	5.41	SiC	4.35
ZnSe	5.65	AlP	5.42

[6]In the inversion operation we carry each point $\mathbf{r}$ into the point $-\mathbf{r}$. A tetrahedron does not have inversion symmetry about its center.

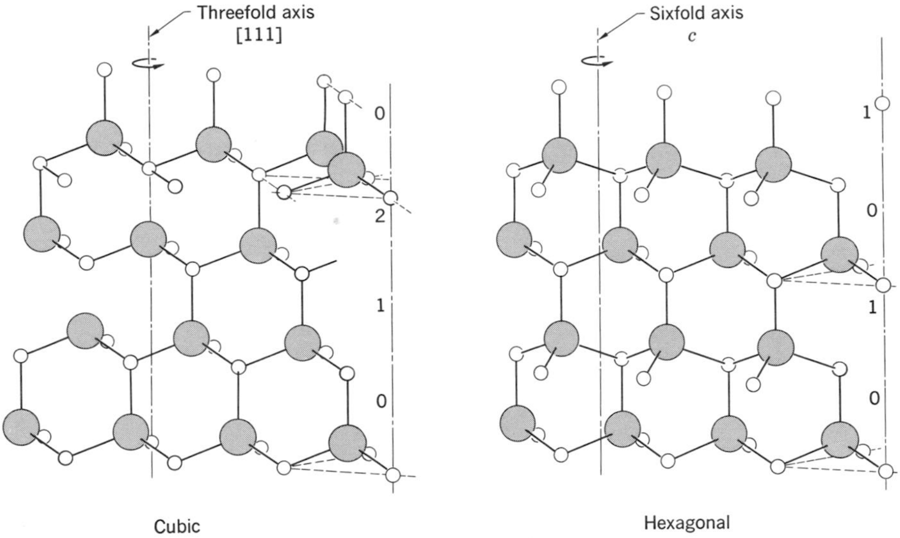

Figure 30 The stacking of tetrahedral layers in cubic and hexagonal ZnS. The large atoms are S; the small atoms are Zn. The vertical axis of hexagonal ZnS is a sixfold screw axis involving a translation of one-half *c* for each 60 degrees of rotation. The stacking sequences 012012. . . and 0101. . . are similar to the sequences of the fcc and hcp structures as shown in Fig. 24. (After Berry and Mason.)

Hexagonal Zinc Sulfide (Wurtzite) Structure

A hexagonal crystalline form of diamond was discovered in meteorites and has been synthesized in the laboratory.[7] The hexagonal diamond structure has tetrahedral covalent bonds just as does the cubic diamond structure, and the densities of the two forms are equal. How are the two structures related? The relation is the same as that of the cubic form of ZnS to the hexagonal[8] form of ZnS. The two forms of diamond are simulated by replacing all atoms by carbon. Figure 30 gives a view of cubic ZnS with a set of tetrahedral bonds in the upright position, and the figure also gives a view of hexagonal ZnS with a set of tetrahedral bonds in the upright position. Both cubic and hexagonal structures in this orientation are made up of pairs of layers, one plane Zn and the other plane S. The layer-pairs in both structures are stacked one above the other, and successive layer-pairs are displaced horizontally. A displacement is necessary to preserve the tetrahedral bond scheme: without a displacement, a sequence S-Zn-S would lie in a straight line.

In cubic ZnS the layer-pairs labeled 0, 3, 6, . . . lie exactly above each other; the stacking sequence is written as 012012012. . . . The stacking in hexagonal ZnS is such that layer-pairs 0, 2, 4, . . . lie exactly above each other, and the stacking sequence is written as 010101. . . . If we look

[7] R. E. Hanneman, H. M. Strong, and F. P. Bundy, Science **155,** 955 (1967).
[8] An excellent description of the two structures is given on pp. 308–313 of L. G. Berry and B. Mason, *Mineralogy*, Freeman, 1959.

only at the nearest neighbor atoms, we cannot tell whether we are in a cubic or hexagonal crystal, but when we look beyond to the second-nearest-neighbor atoms we can distinguish the cubic and hexagonal structures. The nearest-neighbor bond lengths Zn-S are closely equal in the two structures. Examples of the hexagonal zinc sulfide structures are

Crystal	a	c	Crystal	a	c
ZnO	3.25 Å	5.21 Å	SiC	3.08 Å	5.05 Å
ZnS	3.81	6.23	Hex. diamond	2.52	4.12
ZnSe	3.98	6.53	CdS	4.13	6.75
ZnTe	4.27	6.99	CdSe	4.30	7.02

OCCURRENCE OF NONIDEAL CRYSTAL STRUCTURES

The ideal crystal of classical crystallographers is formed by the periodic repetition in space of identical units. But no general proof exists that the ideal crystal is the state of minimum energy of the atoms at absolute zero. Many structures that occur in nature are regular without being entirely periodic. The crystallographers' ideal is not necessarily a law of nature. Some of the aperiodic structures may only be metastable, with very long lifetimes.

Random Stacking and Polytypism

The fcc and hcp structures consist of close-packed planes of atoms and differ only in the stacking sequence of the planes, fcc having the sequence *ABCABC* . . . and hcp having *ABABAB* . . . . Structures are also known in which the stacking sequence of close-packed layers is random[9] as in *ACBCABAC* . . . . This is called **random stacking.** It occurs for example in small particles of cobalt. Cobalt has two common crystal forms, hcp stable under 400°C and fcc stable above. The difference in energy of the two forms at room temperature is small. In some respects a random stacking structure may be thought of as crystalline in two dimensions and glasslike in the third. A glass has random elements in all three dimensions.

Polytypism[10] is characterized by a stacking sequence with a long repeat unit. The classic example is SiC, which not only occurs in the sequences 012012 . . . as for cubic ZnS and 010101 . . . as for hexagonal ZnS, but also in 45 other known stacking sequences of the hexagonal layers. The polytype of SiC known as 393R has a primitive cell with $a = 3.079$ Å and $c = 989.6$ Å. The longest primitive cell observed for SiC has a repeat distance of 594 layers. The mechanism that induces such long-range crystallographic order is not a long-range force, but the presence of spiral steps due to dislocations (Chapter 18) in the growth nucleus. The stacking sequences in polytypism are not random, since a given sequence is repeated many times within a single crystal.

GLASSES

A substance in the glassy state[11] is a material formed by cooling from the normal liquid state and shows no discontinuous change such as crystallization at any temperature, but has become more rigid through a progressive increase in its viscosity. Any liquid or supercooled liquid whose shear viscosity is greater than about 10^{13} poises (10^{12} N s m^{-2}) is called a glass. This value is taken to define the liquid-glass transformation temperature T_g and to define the boundary between the liquid state and the glassy state (Fig. 31).

Glasses, like ordinary liquids, possess atomic order only over a range of the order of one interatomic separation (Fig. 32). One or two diffuse rings

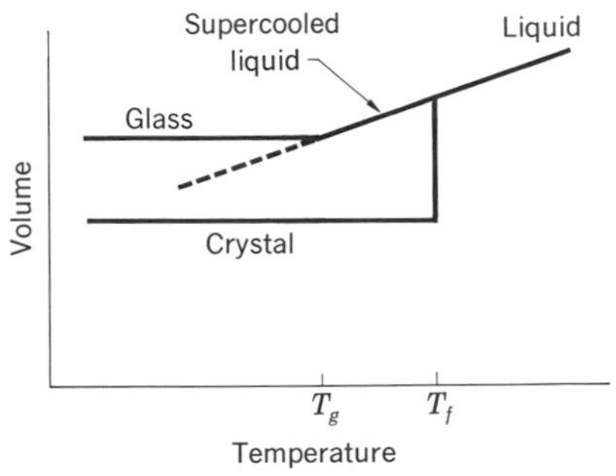

Figure 31 Volume-temperature relation between the glassy, liquid, and crystalline states. The usual freezing (melting) temperature is denoted by T_f; the glass transformation temperature is T_g. The liquid-glass cooling curve will differ somewhat at different cooling rates.

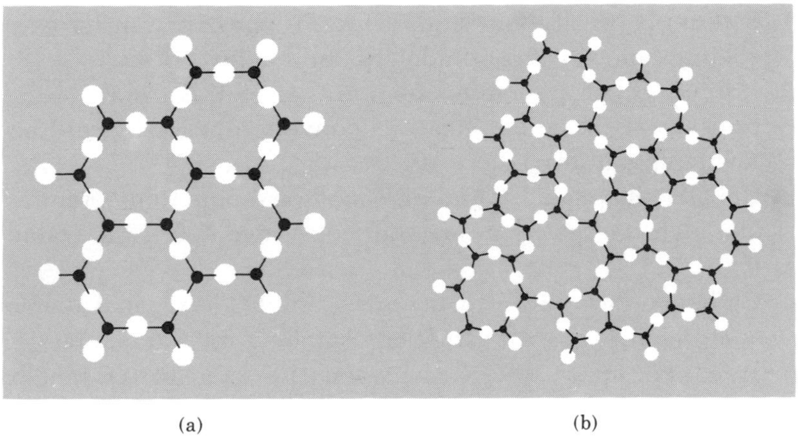

Figure 32 Schematic two-dimensional analogs, after Zachariasen, illustrating the difference between: (a) the regularly repeating structure of a crystal and (b) the random network of a glass.

---

[9]Notice that sequences such as *ACCB* . . . are excluded in practice: a close-packed plane cannot be put directly on top of another without a large increase in the energy.

[10]See A. R. Verma and P. Krishna, *Polymorphism and polytypism in crystals*, Wiley, 1966. A summary of polytypism in SiC is given by Wyckoff, Vol. I, pp. 113–120. The connection between polytypism and dislocations is treated by A. R. Verma in *Silicon carbide*, J. R. O'Connor and J. Smiltens, eds., Pergamon, 1960.

[11]A good general reference is G. O. Jones, *Glass*, 2nd ed., Chapman and Hall, 1971.

can be seen in the x-ray diffraction patterns of glasses and liquids, completely different from the sharp rings of polycrystalline aggregates. There is a gap between the finest grain size of polycrystalline material and the range of short-range order in a liquid. A supercooled liquid or a glass is not in absolute thermal equilibrium, although stable for hours or years with respect to transformation to the crystalline phase of lower free energy. A glass is in a local but not an absolute minimum of the free energy.

COLLECTIONS OF CRYSTAL STRUCTURE DATA

The reader who wishes to look up the crystal structure of a substance may profitably consult the excellent compilation by Wyckoff listed in the references. *Strukturbericht*, *Structure Reports*, and the journals Acta Crystallographica and Zeitschrift für Kristallographie are valuable aids.

In Table 3 we list for convenience the most common crystal structures and lattice constants of the elements. Values of the density and atomic concentration are given in Table 4.

Many elements occur in several crystal structures and transform from one to the other as the temperature is varied. Other transformations take place under pressure. Sometimes two structures will coexist at the same temperature, although one may be slightly more stable. The differences in the free energies of certain structures may be very small, often beyond our ability to calculate theoretically. Several examples of transformations follow:

1. Sodium is bcc at room temperature. It partially transforms to hcp on cooling below 36 K or on deformation below 51 K.
2. Lithium is bcc at room temperature. At 78 K both bcc and hcp structures coexist. The hcp phase is converted to fcc by cold working at low temperatures.
3. Cobalt: the stable form of cobalt at room temperature is hcp, although fine powders may be fcc. Above 400 K the stable form is fcc.
4. Carbon occurs in the diamond, graphite, hexagonal diamond, and amorphous forms, all essentially stable at room temperature.
5. Iron is bcc up to 910°C; fcc between 910°C and 1400°C; and bcc above 1400°C.

TABLE 3 Crystal Structures of the Elements

The data given are at room temperature for the most common form, or at the stated temperature in deg K. For further descriptions of the elements see Wyckoff, Vol. 1, Chap. 2. Structures labeled complex are described there. The notation *ABAC* refers to the sequence of close-packed planes.

Each cell gives: element (temperature) / crystal structure / *a* lattice parameter, in Å / *c* lattice parameter, in Å.

H^1 4K hcp 3.75 6.12																	He^4 2K hcp 3.57 5.83
Li 78K bcc 3.491	Be hcp 2.27 3.59											B rhomb.	C diamond 3.567	N 20K cubic 5.66 (N_2)	O complex (O_2)	F	Ne fcc 4.46
Na 5K bcc 4.225	Mg hcp 3.21 5.21	← Crystal structure →	← *a* lattice parameter, in Å →	← *c* lattice parameter, in Å →								Al fcc 4.05	Si diamond 5.430	P complex	S complex	Cl complex (Cl_2)	Ar 4K fcc 5.31
K 5K bcc 5.225	Ca fcc 5.58	Sc hcp 3.31 5.27	Ti hcp 2.95 4.68	V bcc 3.03	Cr bcc 2.88	Mn cubic complex	Fe bcc 2.87	Co hcp 2.51 4.07	Ni fcc 3.52	Cu fcc 3.61	Zn hcp 2.66 4.95	Ga complex	Ge diamond 5.658	As rhomb.	Se hex. chains	Br complex (Br_2)	Kr 4K fcc 5.64
Rb 5K bcc 5.585	Sr fcc 6.08	Y hcp 3.65 5.73	Zr hcp 3.23 5.15	Nb bcc 3.30	Mo bcc 3.15	Tc hcp 2.74 4.40	Ru hcp 2.71 4.28	Rh fcc 3.80	Pd fcc 3.89	Ag fcc 4.09	Cd hcp 2.98 5.62	In tetr. 3.25 4.95	Sn (α) diamond 6.49	Sb rhomb.	Te hex. chains	I complex (I_2)	Xe 4K fcc 6.13
Cs 5K bcc 6.045	Ba bcc 5.02	La hex. 3.77 *ABAC*	Hf hcp 3.19 5.05	Ta bcc 3.30	W bcc 3.16	Re hcp 2.76 4.46	Os hcp 2.74 4.32	Ir fcc 3.84	Pt fcc 3.92	Au fcc 4.08	Hg rhomb.	Tl hcp 3.46 5.52	Pb fcc 4.95	Bi rhomb.	Po sc 3.34	At —	Rn —
Fr —	Ra —	Ac fcc 5.31															

Ce fcc 5.16	Pr hex. 3.67 *ABAC*	Nd hex. 3.66	Pm —	Sm complex	Eu bcc 4.58	Gd hcp 3.63 5.78	Tb hcp 3.60 5.70	Dy hcp 3.59 5.65	Ho hcp 3.58 5.62	Er hcp 3.56 5.59	Tm hcp 3.54 5.56	Yb fcc 5.48	Lu hcp 3.50 5.55
Th fcc 5.08	Pa tetr. 3.92 3.24	U complex	Np complex	Pu complex	Am hex. 3.64 *ABAC*	Cm —	Bk —	Cf —	Es —	Fm —	Md —	No —	Lr —

TABLE 4 Density and Atomic Concentration of the Elements

The data are given at atmospheric pressure and room temperature, or at the stated temperature in deg K. (Crystal modifications as for Table 3.)

← Density in g cm^{-3} (10^3kg m^{-3}) →
← Concentration in 10^{22} cm^{-3} (10^{28} m^{-3}) →
← Nearest-neighbor distance, in Å (10^{-10}m) →

H 4K																	**He** 2K
0.088																	0.205 (at 37 atm)
Li 78K	**Be**											**B**	**C**	**N** 20K	**O**	**F**	**Ne** 4K
0.542	1.82											2.47	3.516	1.03			1.51
4.700	12.1											13.0	17.6				4.36
3.023	2.22												1.54			1.44	3.16
Na 5K	**Mg**											**Al**	**Si**	**P**	**S**	**Cl** 93K	**Ar** 4K
1.013	1.74											2.70	2.33			2.03	1.77
2.652	4.30											6.02	5.00				2.66
3.659	3.20											2.86	2.35			2.02	3.76
K 5K	**Ca**	**Sc**	**Ti**	**V**	**Cr**	**Mn**	**Fe**	**Co**	**Ni**	**Cu**	**Zn**	**Ga**	**Ge**	**As**	**Se**	**Br** 123K	**Kr** 4K
0.910	1.53	2.99	4.51	6.09	7.19	7.47	7.87	8.9	8.91	8.93	7.13	5.91	5.32	5.77	4.81	4.05	3.09
1.402	2.30	4.27	5.66	7.22	8.33	8.18	8.50	8.97	9.14	8.45	6.55	5.10	4.42	4.65	3.67	2.36	2.17
4.525	3.95	3.25	2.89	2.62	2.50	2.24	2.48	2.50	2.49	2.56	2.66	2.44	2.45	3.16	2.32		4.00
Rb 5K	**Sr**	**Y**	**Zr**	**Nb**	**Mo**	**Tc**	**Ru**	**Rh**	**Pd**	**Ag**	**Cd**	**In**	**Sn**	**Sb**	**Te**	**I**	**Xe** 4K
1.629	2.58	4.48	6.51	8.58	10.22	11.50	12.36	12.42	12.00	10.50	8.65	7.29	5.76	6.69	6.25	4.95	3.78
1.148	1.78	3.02	4.29	5.56	6.42	7.04	7.36	7.26	6.80	5.85	4.64	3.83	2.91	3.31	2.94	2.36	1.64
4.837	4.30	3.55	3.17	2.86	2.72	2.71	2.65	2.69	2.75	2.89	2.98	3.25	2.81	2.91	2.86	3.54	4.34
Cs 5K	**Ba**	**La**	**Hf**	**Ta**	**W**	**Re**	**Os**	**Ir**	**Pt**	**Au**	**Hg** 227	**Tl**	**Pb**	**Bi**	**Po**	**At**	**Rn**
1.997	3.59	6.17	13.20	16.66	19.25	21.03	22.58	22.55	21.47	19.28	14.26	11.87	11.34	9.80	9.31		
0.905	1.60	2.70	4.52	5.55	6.30	6.80	7.14	7.06	6.62	5.90	4.26	3.50	3.30	2.82	2.67	—	—
5.235	4.35	3.73	3.13	2.86	2.74	2.74	2.68	2.71	2.77	2.88	3.01	3.46	3.50	3.07	3.34		
Fr	**Ra**	**Ac**															
		10.07															
—	—	2.66															
		3.76															

Ce	**Pr**	**Nd**	**Pm**	**Sm**	**Eu**	**Gd**	**Tb**	**Dy**	**Ho**	**Er**	**Tm**	**Yb**	**Lu**	
6.77	6.78	7.00		7.54	5.25	7.89	8.27	8.53	8.80	9.04	9.32	6.97	9.84	
2.91	2.92	2.93	—	3.03	2.04	3.02	3.22	3.17	3.22	3.26	3.32	3.02	3.39	
3.65	3.63	3.66		3.59	3.96	3.58	3.52	3.51	3.49	3.47	3.54	3.88	3.43	
Th	**Pa**	**U**	**Np**	**Pu**	**Am**	**Cm**	**Bk**	**Cf**	**Es**	**Fm**	**Md**	**No**	**Lr**	
11.72	15.37	19.05	20.45	19.81	11.87									
3.04	4.01	4.80	5.20	4.26	2.96	—	—	—	—	—	—	—	—	
3.60	3.21	2.75	2.62	3.1	3.61									

SUMMARY

1. A lattice is an array of points related by the lattice translation operator $\mathbf{T} = u\mathbf{a} + v\mathbf{b} + w\mathbf{c}$, where u, v, w are integers and $\mathbf{a}$, $\mathbf{b}$, $\mathbf{c}$ are called the crystal axes.

2. To form a crystal we attach to every lattice point an identical basis of s atoms at the positions $\mathbf{r}_j = x_j\mathbf{a} + y_j\mathbf{b} + z_j\mathbf{c}$, with $j = 1,2, \ldots, s$. Here x, y, z may be selected to have values between 0 and 1.

3. The axes $\mathbf{a}$, $\mathbf{b}$, $\mathbf{c}$ are primitive for the minimum cell volume $|\mathbf{a} \cdot \mathbf{b} \times \mathbf{c}|$ for which the crystal structure can be constructed from a lattice translation operator $\mathbf{T}$ and a basis at every lattice point.

4. It may be convenient (particularly for cubic crystals) to describe a crystal in terms of a conventional cell of volume an integral multiple of the volume of a primitive cell.

5. A plane in the crystal is denoted by the indices (hkl) and a direction is denoted by $[uvw]$.

6. Important simple structures include the bcc, fcc, hcp, diamond, NaCl, CsCl, cubic ZnS, and hexagonal ZnS structures.

PROBLEMS

1. ***Tetrahedral angles.*** The angles between the tetrahedral bonds of diamond are the same as the angles between the body diagonals of a cube, as in Fig. 14. Use elementary vector analysis to find the value of the angle.
2. ***Indices of planes.*** Consider the planes with indices (100) and (001); the lattice is fcc, and the indices refer to the conventional cubic cell. What are the indices of these planes when referred to the primitive axes of Fig. 15?
3. ***Hcp structure.*** Show that the c/a ratio for an ideal hexagonal close-packed structure is $(\frac{8}{3})^{\frac{1}{2}} = 1.633$. If c/a is significantly larger than this value, the crystal structure may be thought of as composed of planes of closely packed atoms, the planes being loosely stacked.

References

Elementary

A. Holden, *Nature of solids*, Columbia University Press, 1965.
A. Holden and P. Singer, *Crystals and crystal growing* (Anchor S7) Doubleday, 1960.
B. Chalmers, J. G. Holland, K. A. Jackson, and R. B. Williamson, *Crystallography: a programmed course in three dimensions*, Books 1 through 6, Appleton, 1965.
W. B. Pearson, *Crystal chemistry and physics of metals and alloys*, Wiley, 1972.
E. A. Wood, *Crystal orientation manual*, Columbia University Press, 1963.
H. D. Megaw, *Crystal structures: a working approach*, Saunders, 1973.

Crystallography

M. J. Buerger, *Contemporary crystallography*, McGraw-Hill, 1970.
F. C. Phillips, *An introduction to crystallography*, 3rd ed., Wiley, 1963.
H. J. Juretscke, *Crystal physics: macroscopic physics of anisotropic solids*, Benjamin, 1974.

Geometry

D. Hilbert and S. Cohn-Vossen, *Geometry and the imagination*, Chelsea, New York, 1952. See especially Chap. 2.
H. S. M. Coxeter, *Regular polytopes*, Methuen, London, 1948. A treatment of the geometrical properties of polyhedra.

Crystal Growth

B. R. Pamplin, ed., *Crystal growth*, Pergamon, 1975.
J. J. Gilman, *The art and science of growing crystals*, Wiley, 1963.
W. G. Pfann, *Zone melting*, Wiley, 1966. The process of zone melting makes possible the production of materials of very high purity. See also the article by Pfann in Scientific American, December 1967.
Journal of Crystal Growth.

Classical Tables and Handbooks

P. H. Groth, *Chemische Krystallographie*, 5 volumes, W. Englemann, Leipzig, 1906.
International tables for x-ray crystallography, 3 volumes, Kynoch Press, Birmingham, 1952–1962.
C. Palache, H. Berman, and C. Frondel, *Dana's system of mineralogy*, 7th ed., Vols. I, II, and III, Wiley, 1944, 1951, 1962.
R. W. G. Wyckoff, *Crystal structures*, 2nd ed., Interscience, 1963.
W. B. Pearson, *Handbook of lattice spacings and structures of metals and alloys*, Pergamon, 1958, 1967 (2 vols.).
J. D. H. Donnay and G. Donnay, *Crystal data, determinative tables*, 2nd ed., Amer. Cryst. Assoc., 1963.
C. S. Barrett and T. B. Massalski, *Structure of metals: crystallographic methods, principles, data*, 3rd ed., McGraw-Hill, 1966.

2 CRYSTAL DIFFRACTION AND THE RECIPROCAL LATTICE

Figure 1 Field ion microscope image of atoms on tip of a fine iridium needle. (Courtesy of E. W. Müller.)

2 CRYSTAL DIFFRACTION AND THE RECIPROCAL LATTICE

X-ray diffraction is not a difficult branch of physics: on the contrary, it is easy to the point of tediousness. The widespread view that it is unintelligible has arisen because a certain intellectual effort is needed to grasp its mathematical foundations, and because it is supposed, incorrectly, that some special type of "three-dimensional imagination" is a prerequisite for understanding its methods and results.

F. H. C. Crick and J. C. Kendrew

We first ask a crystal: "Where are the atoms?" The methods developed in this chapter tell us how to map the atoms and the electron distribution around them. Rarely is it possible to form a direct microscope image of the structure of a crystal. An electron microscope with a resolving power of 2 Å can resolve prominent planes of laminar crystals such as graphite, but the resolution of the electron microscope does not yet permit the accurate direct determination of unknown crystal structures. A direct image of the atoms on the surface of a tip of iridium is shown in Fig. 1. To explore crystal structure we use diffraction patterns of waves that interact with atoms and that have a wavelength comparable with the interatomic spacing.

The Incident Beam

We study crystal structure through the diffraction of photons, neutrons, and electrons. The diffraction depends on the crystal structure and on the wavelength. At optical wavelengths such as 5000 Å the superposition of the waves scattered elastically by the individual atoms of a crystal results in ordinary optical refraction. But when the wavelength of the radiation is comparable with or smaller than the lattice constant, we may find diffracted beams in directions quite different from the incident direction.

X-Rays. The energy of an x-ray photon is related to its wavelength λ by $\epsilon = h\nu = hc/\lambda$, where $h = 6.62 \times 10^{-27}$ erg-s $= 6.62 \times 10^{-34}$ J-s is Planck's constant. In laboratory units,

$$\lambda(\text{Å}) = \frac{12.4}{\epsilon(\text{keV})} , \tag{1}$$

where λ is the wavelength in angstrom units ($1\ \text{Å} = 10^{-8}\ \text{cm} = 10^{-10}\ \text{m}$) and ϵ is the energy in kilo-electron-volts. Crystal studies require photon energies in the 10 to 50 keV range (Fig. 2). X-rays are generated both by the deceleration of electrons in metal targets and by the excitation of the core electrons in the atoms of the target. The first process gives a broad continuous spectrum; the second gives sharp lines. A copper target bom-

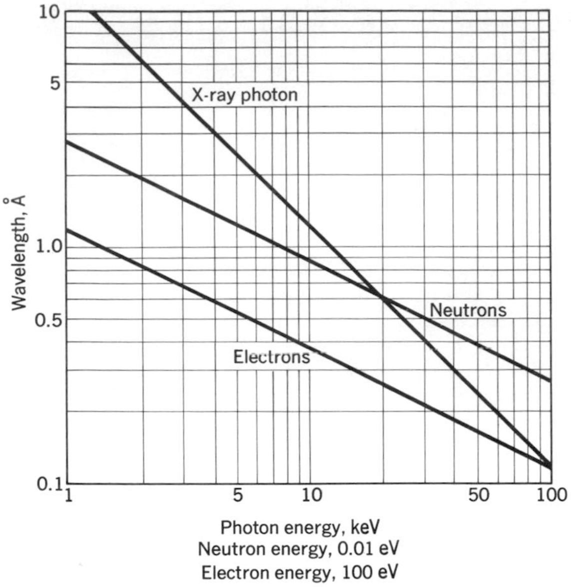

Figure 2 Wavelength versus particle energy, for photons, neutrons, and electrons.

barded by electrons has a strong line Cu$K\alpha$ at 1.5418 Å, in the middle of the important range. Copper makes a good target: it is an excellent heat conductor with a high melting point. Nuclei, because of their heavy mass, do not scatter x-rays effectively: x-rays see the electrons.

Neutrons. The energy of a neutron is related to its de Broglie wavelength λ by $\epsilon = h^2/2M_n\lambda^2$, where $M_n = 1.675 \times 10^{-24}$ g is the mass of the neutron. We recall that $\epsilon = p^2/2M_n$, and the wavelength λ is related to the momentum p by $\lambda = h/p$. In laboratory units,

$$\lambda(\text{Å}) \cong \frac{0.28}{[\epsilon(\text{eV})]^{1/2}} , \tag{2}$$

where ϵ is the neutron energy in eV. We have $\lambda = 1$ Å for $\epsilon \cong 0.08$ eV. Because of their magnetic moment, neutrons can interact with the magnetic electrons of a solid, and neutron methods are valuable in structural studies of magnetic crystals. In nonmagnetic materials the neutron interacts only with the nuclei of the constituent atoms.

Electrons. The energy of an electron is related to its de Broglie wavelength λ by $\epsilon = h^2/2m\lambda^2$, where $m = 0.911 \times 10^{-27}$ g is the mass of the electron. In laboratory units,

$$\lambda(\text{Å}) \cong \frac{12}{[\epsilon(\text{eV})]^{1/2}} . \tag{3}$$

Electrons interact strongly with matter because they are charged; they penetrate a relatively short distance into a crystal.

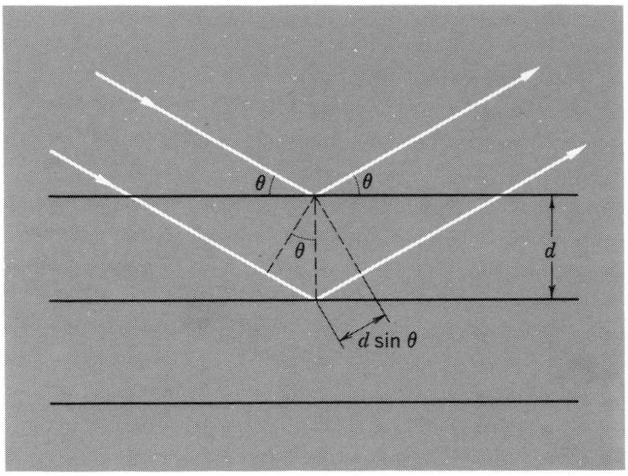

Figure 3 Derivation of the Bragg equation $2d \sin \theta = n\lambda$; here d is the spacing of parallel atomic planes and $2\pi n$ is the difference in phase between reflections from successive planes. The reflecting planes have nothing to do with the surface planes bounding the particular specimen.

Bragg Law

W. L. Bragg[1] presented a simple explanation of the diffracted beams from a crystal. Suppose that the incident waves are reflected specularly[2] from parallel planes of atoms in the crystal, with each plane reflecting only a very small fraction of the radiation, like a lightly silvered mirror. The diffracted beams are found when the reflections from parallel planes of atoms interfere constructively, as in Fig. 3. We treat elastic scattering in which the energy of the x-ray is not changed on reflection. Inelastic scattering, with the excitation of elastic waves, is discussed at the end of the chapter.

Consider parallel lattice planes spaced d apart, Fig. 4. The radiation is incident in the plane of the paper. The path difference for rays reflected from adjacent planes is $2d \sin \theta$, where θ is measured from the plane. Constructive interference of the radiation from successive planes occurs when the path difference is an integral number n of wavelengths λ, so that

$$2d \sin \theta = n\lambda \ . \qquad (4)$$

This is the Bragg law. Although the reflection from each plane is specular, for only certain values of θ will the reflections from all parallel planes add up in phase to give a strong reflected beam. Of course, if each plane were perfectly reflecting, only the first plane of a parallel set would see the radiation and any wavelength would be reflected. But each plane reflects 10^{-3} to 10^{-5} of the incident radiation.

[1]W. L. Bragg, Proc. Cambridge Phil. Soc. **17**, 43 (1913). The Bragg derivation is simple but is convincing only because it reproduces the correct result.

[2]In specular (mirrorlike) reflection the angle of incidence is equal to the angle of reflection.

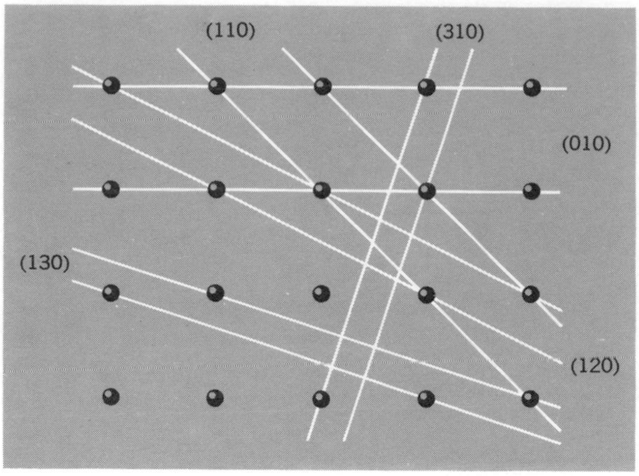

Figure 4 Several types of reflecting planes in a simple cubic crystal lattice. The planes shown are labeled by their indices. We have shown in each case a set of two parallel planes. The closest distance between parallel planes tends to decrease as the indices increase; thus high index reflections require shorter wavelengths.

The Bragg law is a consequence of the periodicity of the lattice. The law does not refer to the arrangement of atoms in the basis associated with each lattice point. The composition of the basis determines the relative intensity of the various orders n of diffraction from a given set of parallel planes. Bragg reflection can occur only for wavelength $\lambda \leq 2d$. This is why we cannot use visible light.

EXPERIMENTAL DIFFRACTION METHODS

The Bragg law (4) requires that θ and λ be matched: monochromatic x-rays of wavelength λ striking a three-dimensional crystal at an arbitrary angle of incidence will not in general be reflected. To satisfy the Bragg law requires an accident, and to create the accident it is necessary to scan in either wavelength or angle. The standard methods of diffraction used in crystal structure analysis are designed expressly to accomplish this. We describe three simple, older methods, still used by physicists; but for professional crystallography these techniques have been replaced by complicated precession camera methods.

Laue Method

In the Laue method (Fig. 5), a single crystal is stationary in a beam of x-ray or neutron radiation of continuous wavelength. The crystal selects and diffracts the discrete values of λ for which planes exist of spacing d and incidence angle θ satisfying the Bragg law. A source is used that produces a beam of x-rays over a wide range of wavelengths, perhaps from

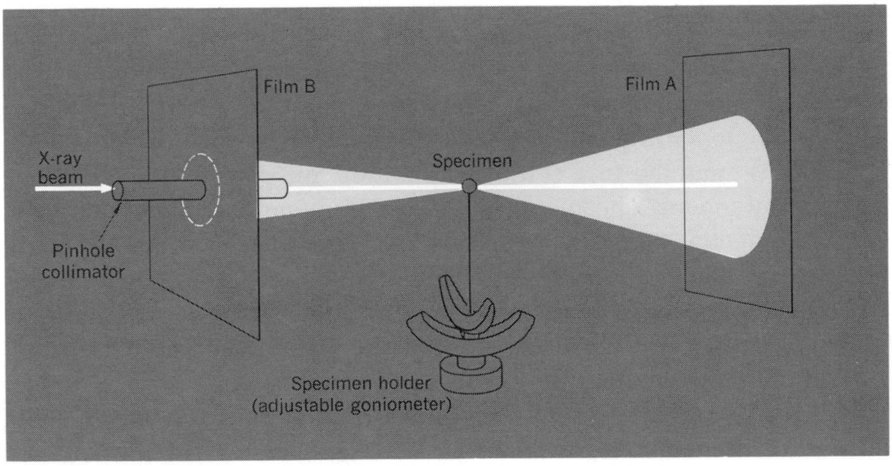

Figure 5 A flat plate camera. With a continuous spectrum x-ray beam and a single crystal specimen, the camera produces Laue patterns. The adjustable mount is convenient for the orientation of single crystals needed in other solid state experiments. The film B is used for back-reflection Laue patterns. (Courtesy of Philips Electronic Instruments.)

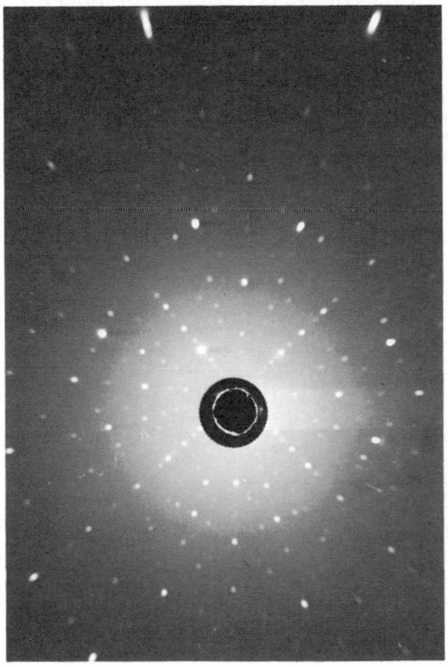

Figure 6 Laue pattern of a silicon crystal in approximately the [100] orientation. Note that the pattern is nearly invariant under a rotation of $2\pi/4$. The invariance follows from the fourfold symmetry of silicon about a [100] axis. The black center is a cut-out in the film. (Courtesy of J. Washburn.)

0.2 Å to 2 Å. A pinhole arrangement produces a well-collimated beam. The dimensions of the single-crystal specimen need not be greater than 1 mm. Flat film receives the diffracted beams. The diffraction pattern consists of a series of spots, Fig. 6. The pattern will show the symmetry of the crystal: if a crystal has a fourfold axis of symmetry parallel to the beam, the Laue pattern will show fourfold symmetry. The Laue method is widely used to orient crystals for solid state experiments.

Rotating-Crystal Method

In the rotating-crystal method (Fig. 7), a single crystal is rotated about a fixed axis in a beam of monoenergetic x-rays or neutrons. The variation in the angle θ brings different atomic planes into position for reflection. The film is mounted in a cylindrical holder concentric with a rotating spindle crystal mount. The incident x-ray beam is monochromatized by a filter or by reflection from an earlier crystal. The beam is diffracted from a given crystal plane when in the course of rotation the value of θ satisfies the Bragg equation. Beams from all planes parallel to the vertical rotation axis will lie in the horizontal plane. Planes with other orientations will reflect in layers above and below the horizontal plane.

The intensity distribution of the radiation from a 30 keV x-ray tube with a molybdenum target is shown by Fig. 8a. The distribution of neutrons emerging from a nuclear reactor is shown by Fig. 8b. If we reflect the beam from a monochromating crystal, as in Fig. 9, we get the cross-hatched distribution of Fig. 8b.

Several variations are in common use. In oscillating-crystal photographs the crystal is oscillated through a limited angular range, instead of being rotated through 360°. The limited range reduces the possibility of overlapping reflections. The precession camera developed by M. J. Buerger gives a picture of the various levels of the reciprocal lattice. Modern diffractometers use scintillation counters or proportional counter tubes to detect the diffracted radiation. These methods allow automatic collection of data, needed because complex structures may exhibit 10,000 diffracted rays.

Nearly all crystals with simple structures were solved by x-ray analysis long ago. One present center of interest in x-ray structure analysis is in the determination of the configuration of enzymes with a molecular weight between 10,000 and 100,000. The crystallization of an enzyme and the subsequent x-ray analysis of the structure of the crystal is the most effective method for the determination of the shape of the molecule. The coor-

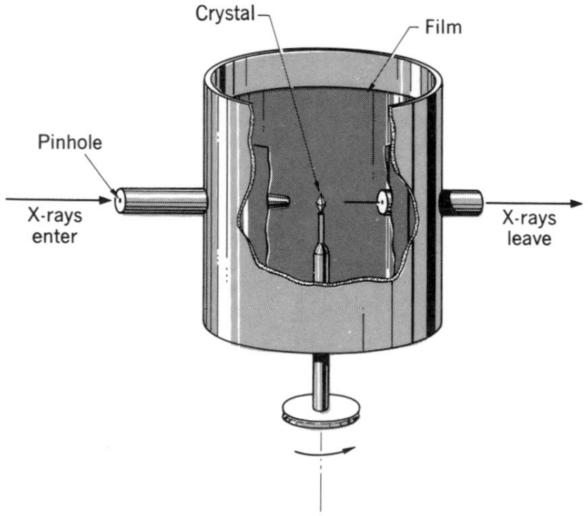

Figure 7 A rotating-crystal camera, with a crystal mounted on the rotating spindle. (By permission from *Structure of metals*, by C. S. Barrett.)

dinates of 500 to 5000 atoms in a cell are wanted, so at least this number of x-ray reflection lines is required. Computer programs have enormously simplified the problem of structure determination.

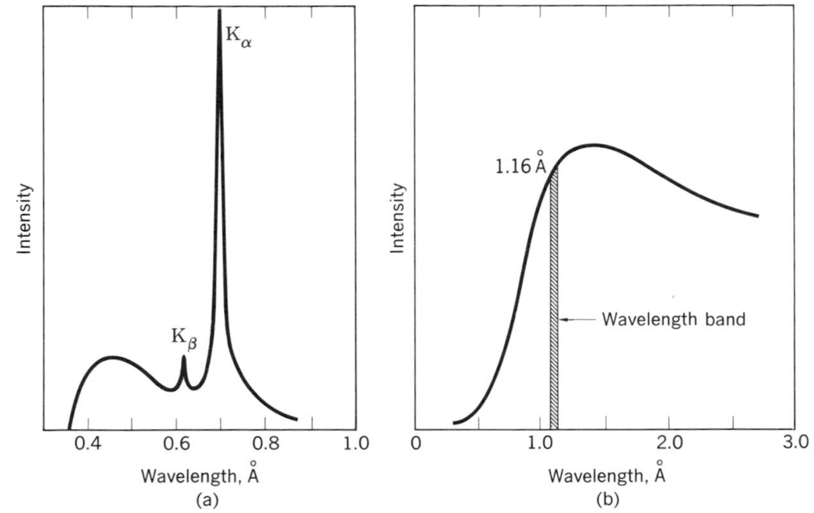

Figure 8 (a) Intensity versus wavelength distribution for x-rays from a Mo target bombarded by 30 keV electrons; (b) for the neutron beam emerging from a reactor with the wavelength band selected by a crystal monochromator. (After G. Bacon.)

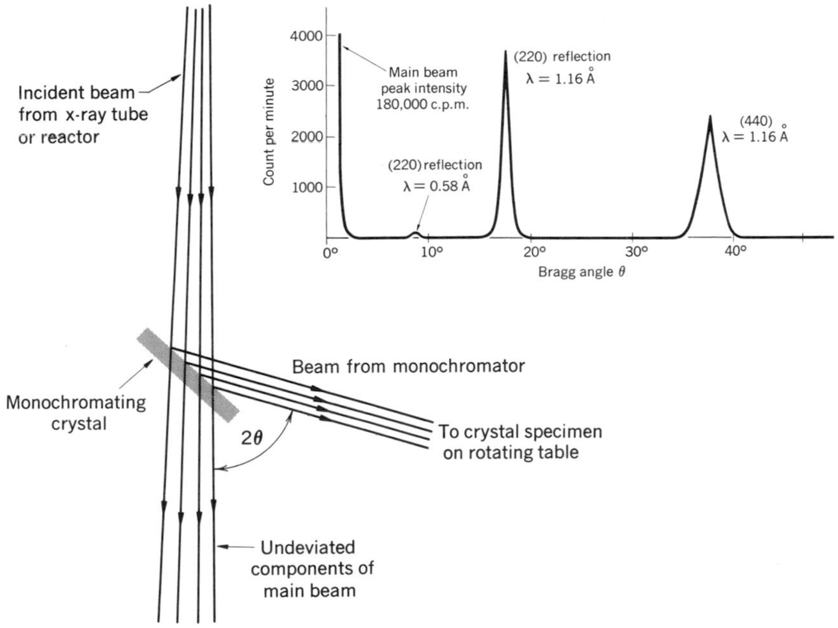

Figure 9 Sketch of a monochromator which by Bragg reflection selects a narrow spectrum of x-ray or neutron wavelengths from a broad spectrum incident beam. The upper part of the figure shows the analysis (obtained by reflection from a second crystal) of the purity of a 1.16 Å beam of neutrons from a calcium fluoride crystal monochromator. The main beam is that not reflected from the second crystal. (After G. Bacon.)

Powder Method

In the powder method (Fig. 10) the incident monochromatic radiation strikes a finely-powdered specimen or a fine-grained polycrystalline specimen contained in a thin-walled capillary tube. The distribution of crystallite orientations will be nearly continuous. The powder method is convenient precisely because single crystals are not required. Diffracted rays go out from individual crystallites that happen to be oriented with planes making an incident angle θ with the beam satisfying the Bragg equation. Diffracted rays leave the specimen along the generators of cones concentric with the original beam. The generators make an angle of 2θ with the direction of the original beam, where θ is the Bragg angle. The cones intercept the film in a series of concentric rings (Fig. 11).

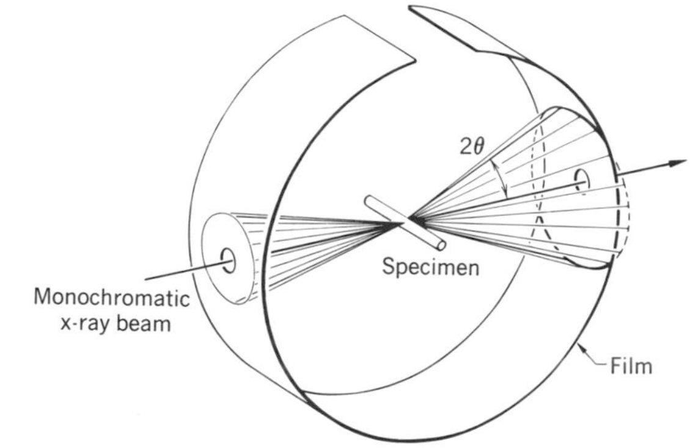

Figure 10 X-ray powder diffraction camera. The specimen is a polycrystalline powder. (Courtesy of Philips Electronic Instruments.)

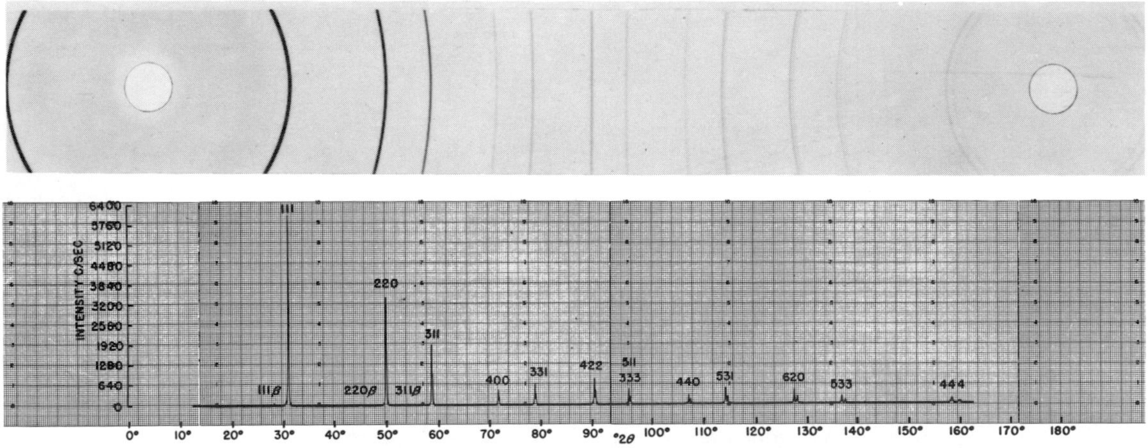

Figure 11 Powder camera and x-ray diffractometer recording of silicon. The upper figure is a film recording and the lower is a counter recording of the diffracted beams. (Courtesy of W. Parrish, Philips Laboratories and Philips Electronic Instruments.)

DERIVATION OF SCATTERED WAVE AMPLITUDE

The Bragg derivation of the diffraction condition gives a neat and clear statement of the condition for the constructive interference of waves scattered by point charges at the lattice points. To determine the intensity of scattering from a spatial distribution of electrons within each cell, we must carry out a deeper analysis.

We saw in Eq. (1.2) that a crystal is invariant under any translation of the form $\mathbf{T} = u\mathbf{a} + v\mathbf{b} + w\mathbf{c}$, where u, v, w are integers and $\mathbf{a}$, $\mathbf{b}$, $\mathbf{c}$ are the crystal axes. Any physical property of the crystal is invariant under $\mathbf{T}$. The charge concentration, the electron number density, the mass density, and the magnetic moment density are invariant under all translations $\mathbf{T}$. Thus the electron number density $n(\mathbf{r})$ is a periodic function of $\mathbf{r}$ with periods a, b, c in the directions of the three crystal axes, whence $n(\mathbf{r} + \mathbf{T}) = n(\mathbf{r})$.

Fourier Analysis

Most properties of a crystal can be related to the Fourier components of the electron density. The three-dimensional aspect of the periodicity causes no difficulties with the mathematics, but we consider first a function $n(x)$ with period a in one dimension. We expand $n(x)$ in a Fourier series of cosines and sines:

$$n(x) = n_0 + \sum_{p>0} [C_p \cos(2\pi px/a) + S_p \sin(2\pi px/a)] \ , \tag{5}$$

where the p's are positive integers and C_p, S_p are real constants, called the Fourier coefficients of the expansion. The factor $2\pi/a$ in the arguments ensures that $n(x)$ has the period a:

$$\begin{aligned} n(x + a) &= n_0 + \Sigma[C_p \cos(2\pi px/a + 2\pi p) + S_p \sin(2\pi px/a + 2\pi p)] \\ &= n_0 + \Sigma[C_p \cos(2\pi px/a) + S_p \sin(2\pi px/a)] = n(x) \ . \end{aligned} \tag{6}$$

We shall say that $2\pi p/a$ is a point in the reciprocal lattice or Fourier space of the crystal. In one dimension these points lie on a line. The reciprocal lattice points tell us the allowed terms in the Fourier series (5). A term is allowed if it is consistent with the periodicity of the crystal, as in Fig. 12; other points in the reciprocal space are not allowed in the Fourier expansion of a periodic function.

It is a great convenience to write the series (5) in the compact form

$$n(x) = \sum_p n_p e^{i2\pi px/a} \ , \tag{7}$$

where the sum is over all integers p: positive, negative, and zero. The

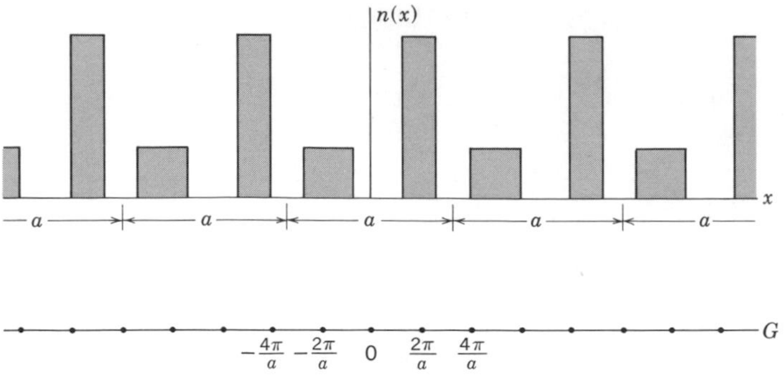

Figure 12 A periodic function $n(x)$ of period a, and the terms $2\pi p/a$ that may appear in the Fourier transform $n(x) = \Sigma\ n_p \exp(i2\pi px/a)$. The magnitudes of the individual terms n_p are not plotted.

coefficients n_p now are complex numbers. To ensure that $n(x)$ is a real function, we require

$$n^*_{-p} = n_p \ , \tag{8}$$

for then the sum of terms in p and $-p$ is real. With $\varphi \equiv 2\pi px/a$, this sum is

$$\begin{aligned} n_p(\cos\varphi + i\sin\varphi) &+ n_{-p}(\cos\varphi - i\sin\varphi) \\ &= (n_p + n_{-p})\cos\varphi + i(n_p - n_{-p})\sin\varphi \ , \end{aligned} \tag{9}$$

which in turn is equal to the real function

$$2\mathrm{Re}\{n_p\}\cos\varphi + 2\mathrm{Im}\{n_p\}\sin\varphi \ , \tag{10}$$

if (8) is satisfied. Here $\mathrm{Re}\{n_p\}$ and $\mathrm{Im}\{n_p\}$ denote the real and imaginary parts of n_p.

The extension of the Fourier analysis to periodic functions in three dimensions is not complicated. We find a set of vectors $\mathbf{G}$ such that

$$n(\mathbf{r}) = \sum_{\mathbf{G}} n_{\mathbf{G}}\, e^{i\mathbf{G}\cdot\mathbf{r}} \tag{11}$$

is invariant under all lattice translations $\mathbf{T}$ that leave the crystal invariant. We then prove that the amplitude of the elastic x-ray scattering from a crystal structure is determined by the magnitude of the Fourier coefficients $n_{\mathbf{G}}$ of the electron concentration. Herein lies the importance of the Fourier analysis.

Reciprocal Lattice Vectors

We construct the axis vectors of the reciprocal lattice:

$$\mathbf{A} = 2\pi \frac{\mathbf{b} \times \mathbf{c}}{\mathbf{a} \cdot \mathbf{b} \times \mathbf{c}} \; ; \qquad \mathbf{B} = 2\pi \frac{\mathbf{c} \times \mathbf{a}}{\mathbf{a} \cdot \mathbf{b} \times \mathbf{c}} \; ; \qquad \mathbf{C} = 2\pi \frac{\mathbf{a} \times \mathbf{b}}{\mathbf{a} \cdot \mathbf{b} \times \mathbf{c}} \; . \tag{12}$$

If **a, b, c** are primitive vectors of the crystal lattice, then **A, B, C** are primitive vectors of the reciprocal lattice. Each vector (12) is orthogonal to two of the axis vectors of the crystal lattice. Thus **A, B, C** have the property:

$$\begin{aligned}
&\mathbf{A} \cdot \mathbf{a} = 2\pi \; , \qquad && \mathbf{B} \cdot \mathbf{a} = 0 \; , \qquad && \mathbf{C} \cdot \mathbf{a} = 0 \; ; \\
&\mathbf{A} \cdot \mathbf{b} = 0 \; , && \mathbf{B} \cdot \mathbf{b} = 2\pi \; , && \mathbf{C} \cdot \mathbf{b} = 0 \; ; \\
&\mathbf{A} \cdot \mathbf{c} = 0 \; , && \mathbf{B} \cdot \mathbf{c} = 0 \; , && \mathbf{C} \cdot \mathbf{c} = 2\pi \; .
\end{aligned} \tag{13}$$

Any arbitrary set of primitive vectors **a, b, c** of a given crystal lattice leads to the same set of reciprocal lattice points

$$\mathbf{G} = h\mathbf{A} + k\mathbf{B} + l\mathbf{C}, \qquad (h,k,l = \text{integers}) \; . \tag{14}$$

Any vector **G** of this form is called a reciprocal lattice vector.

Every crystal structure has two lattices associated with it, the crystal lattice and the reciprocal lattice. A diffraction pattern of a crystal is a map of the reciprocal lattice of the crystal, in contrast to a microscope image, which is a map of the real crystal structure. The two lattices are related by the definitions (12). When we rotate a crystal, we rotate both the direct lattice and the reciprocal lattice. Vectors in the crystal lattice have the dimensions of [length]; vectors in the reciprocal lattice have the dimensions of $[\text{length}]^{-1}$. The crystal lattice is a lattice in real or ordinary space; the reciprocal lattice is a lattice in the associated Fourier space. The term Fourier space is motivated below. Wavevectors are always drawn in Fourier space. Every position in Fourier space may have a meaning, but there is a special importance to the points defined by the set of **G**'s.

The vectors **G** in the Fourier series (11) are the reciprocal lattice vectors, for then the Fourier series for the electron density has the desired invariance under any crystal translation $\mathbf{T} = u\mathbf{a} + v\mathbf{b} + w\mathbf{c}$:

$$n(\mathbf{r} + \mathbf{T}) = \sum_{\mathbf{G}} n_{\mathbf{G}} \, e^{i\mathbf{G}\cdot\mathbf{r}} e^{i\mathbf{G}\cdot\mathbf{T}} \; . \tag{15}$$

To prove that $\exp[i\mathbf{G} \cdot \mathbf{T}] = 1$, we form

$$\begin{aligned}
\exp[i\mathbf{G} \cdot \mathbf{T}] &= \exp[i(h\mathbf{A} + k\mathbf{B} + l\mathbf{C}) \cdot (u\mathbf{a} + v\mathbf{b} + w\mathbf{c})] \\
&= \exp[2\pi i(hu + kv + lw)] \; ,
\end{aligned} \tag{16}$$

using (13). The argument of the exponential has the form $2\pi i$ times an integer, because $hu + kv + lw$ is the sum of products of integers and is therefore an integer. Thus $\exp[i\mathbf{G} \cdot \mathbf{T}] = 1$, and from (15) it follows that $n(\mathbf{r} + \mathbf{T}) = \Sigma\, n_{\mathbf{G}} \exp[i\mathbf{G} \cdot \mathbf{r}] = n(\mathbf{r})$. This completes the proof that the Fourier analysis of a function periodic in the crystal lattice can contain components $n_{\mathbf{G}} \exp[i\mathbf{G} \cdot \mathbf{r}]$ only at the reciprocal lattice vectors **G** defined by $\mathbf{G} = h\mathbf{A} + k\mathbf{B} + l\mathbf{C}$.

Diffraction Conditions

Theorem The set of reciprocal lattice vectors **G** determines the possible x-ray reflections. You see in Fig. 13 that the difference in phase factors is $\exp[i(\mathbf{k} - \mathbf{k}') \cdot \mathbf{r}]$ between beams scattered from volume elements **r** apart. The wavevectors of the incoming and outgoing beams are **k** and **k′**. The amplitude of the wave scattered from a volume element is proportional to the local electron concentration. The total amplitude of the scattered wave in the direction of **k′** is proportional to the integral over the crystal of $n(\mathbf{r})\, dV$ times the phase factor $\exp[i(\mathbf{k}-\mathbf{k}') \cdot \mathbf{r}]$. In other words, the amplitude of the electric or magnetic field vectors in the scattered electromagnetic wave is proportional to the following integral which defines the quantity $\mathcal{a}$ that we will call the scattering amplitude:

$$\mathcal{a} = \int dV\, n(\mathbf{r}) \exp[i(\mathbf{k}-\mathbf{k}') \cdot \mathbf{r}] = \int dV\, n(\mathbf{r}) \exp(-i\Delta\mathbf{k} \cdot \mathbf{r}), \tag{17}$$

where

$$\mathbf{k} + \Delta\mathbf{k} = \mathbf{k}' \ . \tag{18}$$

Here $\Delta\mathbf{k}$ measures the change in wavevector on scattering and is called the **scattering vector** (Fig. 14). We add $\Delta\mathbf{k}$ to **k** to obtain the scattered wavevector **k′**.

We introduce the Fourier components of $n(\mathbf{r})$ into (17) to obtain

$$\mathcal{a} = \sum \int dV\, n_{\mathbf{G}} \exp[i(\mathbf{G} - \Delta\mathbf{k}) \cdot \mathbf{r}] \ . \tag{19}$$

When the scattering vector $\Delta\mathbf{k}$ is equal to a reciprocal lattice vector **G**, the argument of the exponential vanishes, and $\mathcal{a} = V n_{\mathbf{G}}$. It is a simple exercise (Problem 6) to show that $\mathcal{a}$ is negligible when $\Delta\mathbf{k}$ differs significantly from any reciprocal lattice vector.

In elastic scattering the photon energy $\hbar\omega$ is conserved, so that the frequency $\omega' = ck'$ of the emergent beam is equal to the frequency $\omega = ck$ of the incident beam. Thus the magnitudes of k and k' are equal, and $k'^2 = k^2$. From our result $\Delta\mathbf{k} = \mathbf{G}$ or $\mathbf{k} + \mathbf{G} = \mathbf{k}'$, the diffraction condition is $(\mathbf{k} + \mathbf{G})^2 = k^2$, or

$$2\mathbf{k} \cdot \mathbf{G} + G^2 = 0 \ . \tag{20}$$

This is the central result of the theory of elastic scattering in a periodic lattice. The identical result will occur in the theory of the electron energy band structure of crystals. Notice that if **G** is a reciprocal lattice vector, $-\mathbf{G}$ is also one; thus we can equally well write (20) as $2\mathbf{k} \cdot \mathbf{G} = G^2$.

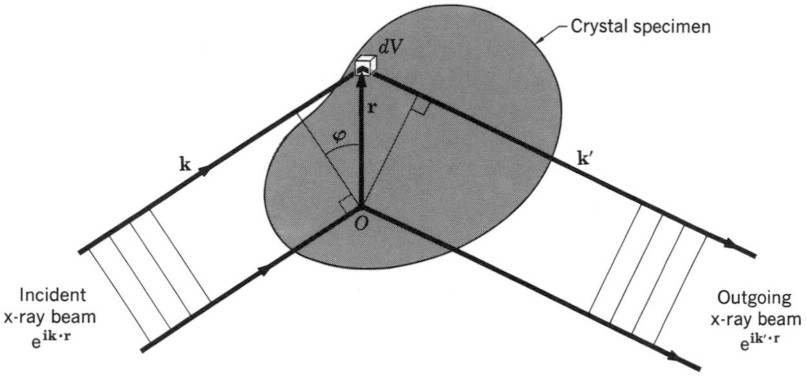

Figure 13 The difference in path length of the incident wave $\mathbf{k}$ at the points 0, r is $r \sin \varphi$, and the difference in phase angle is $(2\pi r \sin \varphi)/\lambda$, which is equal to $\mathbf{k} \cdot \mathbf{r}$. For the diffracted wave the difference in phase angle is $-\mathbf{k}' \cdot \mathbf{r}$. The total difference in phase angle is $(\mathbf{k} - \mathbf{k}') \cdot \mathbf{r}$, and the wave scattered from dV at $\mathbf{r}$ has the phase factor $\exp[i(\mathbf{k} - \mathbf{k}') \cdot \mathbf{r}]$ relative to the wave scattered from a volume element at the origin O.

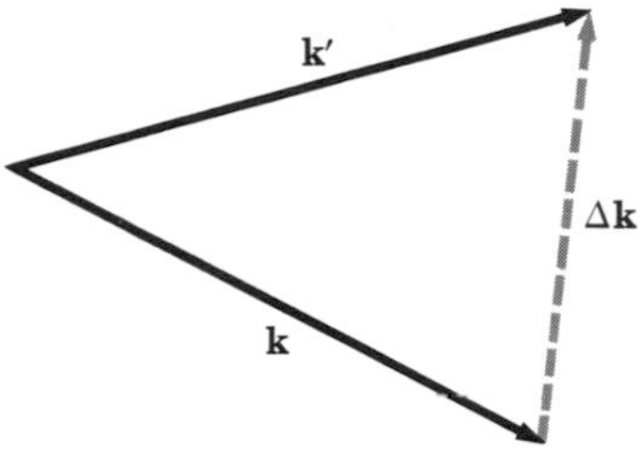

Figure 14 Definition of the scattering vector $\Delta\mathbf{k}$ such that $\mathbf{k} + \Delta\mathbf{k} = \mathbf{k}'$. In elastic scattering the magnitudes satisfy $\mathbf{k}' = \mathbf{k}$. Further, in Bragg scattering from a periodic lattice any allowed $\Delta\mathbf{k}$ must equal some reciprocal lattice vector $\mathbf{G}$.

Now Eq. (20) is another statement of the Bragg law: the result of Problem 2 is that the spacing $d(hkl)$ between parallel lattice planes normal to the direction $\mathbf{G} = h\mathbf{A} + k\mathbf{B} + l\mathbf{C}$ is $d(hkl) = 2\pi/|\mathbf{G}|$. Thus the result $2\mathbf{k} \cdot \mathbf{G} = G^2$ may be written as

$$2(2\pi/\lambda) \sin \theta = 2\pi/d\,(hkl) \tag{21}$$

or $2d(khl) \sin \theta = \lambda$. Here θ is the angle between the incident beam and the plane. The integers hkl that define $\mathbf{G}$ are not necessarily identical with the indices of an actual crystal plane. The integers that define $\mathbf{G}$ may contain a common factor n, while in the indices the common factor n has been eliminated. Thus we obtain the Bragg result:

$$2d \sin \theta = n\lambda \ , \tag{22}$$

where d is the spacing between adjacent planes with indices h/n, k/n, l/n.

The earlier result $\Delta\mathbf{k} = \mathbf{G}$ can be expressed in another way to give the Laue equations. Take the scalar product of both sides successively with **a, b, c** of the crystal lattice, and we obtain the three Laue conditions on the scattering vector:

$$\mathbf{a} \cdot \Delta\mathbf{k} = 2\pi h \ ; \qquad \mathbf{b} \cdot \Delta\mathbf{k} = 2\pi k \ ; \qquad \mathbf{c} \cdot \Delta\mathbf{k} = 2\pi l \ . \tag{23}$$

These equations have a clear geometrical interpretation. The first part $\mathbf{a} \cdot \Delta\mathbf{k} = 2\pi\mathbf{h}$ tells us that $\Delta\mathbf{k}$ lies on a certain cone about the direction of **a.** Further, $\Delta\mathbf{k}$ lies on a cone about **b,** and $\Delta\mathbf{k}$ on a cone about **c.** In three dimensions $\Delta\mathbf{k}$ at a reflection must satisfy all three equations, but only by accident can all three cones intercept to share a common ray. To create this accident, we have to sweep continuously in wavelength or in crystal orientation. A helpful geometrical construction, the Ewald construction, is exhibited in Fig. 15.

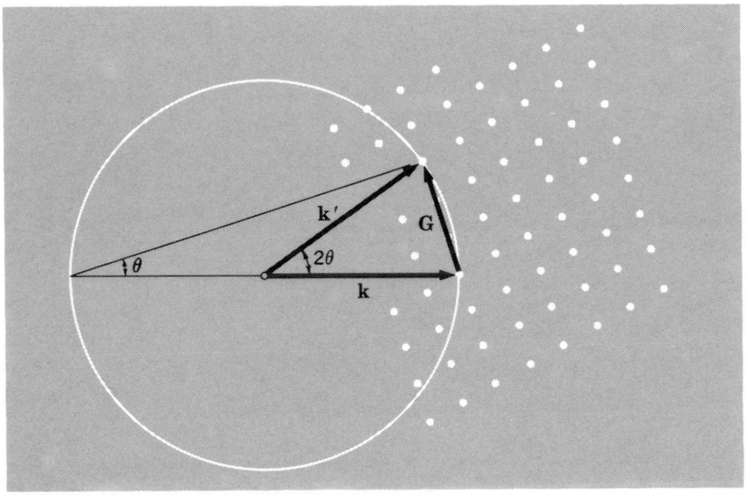

Figure 15 The points on the right-hand side are reciprocal lattice points of the crystal. The vector **k** is drawn in the direction of the incident x-ray beam and it terminates at any reciprocal lattice point. We draw a sphere of radius $k = 2\pi/\lambda$ about the origin of **k**. A diffracted beam will be formed if this sphere intersects any other point in the reciprocal lattice. The sphere as drawn intercepts a point connected with the end of **k** by a reciprocal lattice vector **G.** The diffracted x-ray beam is in the direction $\mathbf{k}' = \mathbf{k} + \mathbf{G}$. This construction is due to P. P. Ewald.

BRILLOUIN ZONES

For solid state physics the most important statement of the diffraction condition was given by Brillouin. This is the only construction used in electron energy band theory and in the expression of the elementary excitations of crystals. **A Brillouin zone is defined as a Wigner-Seitz cell in the reciprocal lattice.** The Wigner-Seitz cell of the direct lattice was described by Fig. 1.6. The Brillouin zone gives a vivid geometrical interpretation of the diffraction condition $2\mathbf{k} \cdot \mathbf{G} = G^2$ or

$$\mathbf{k} \cdot (\tfrac{1}{2}\mathbf{G}) = (\tfrac{1}{2}G)^2 \ . \tag{24}$$

We construct a plane normal to the vector $\mathbf{G}$ at the midpoint; any vector $\mathbf{k}$ from the origin to the plane will satisfy the diffraction condition. The plane thus described forms a part of the zone boundary, Fig. 16. An x-ray beam incident on the crystal will be diffracted if its wavevector has the magnitude and direction required by (24), and the diffracted beam will be in the direction of the vector $\mathbf{k} - \mathbf{G}$.

The set of planes that are the perpendicular bisectors of the reciprocal lattice vectors are of particular importance in the theory of wave propagation in crystals, because a wave whose wavevector drawn from the origin terminates on any of these planes will satisfy the conditions for diffraction. These planes divide the Fourier space of the crystal into odd bits and pieces, as in Fig. 17. The central square is a primitive cell of the reciprocal lattice.

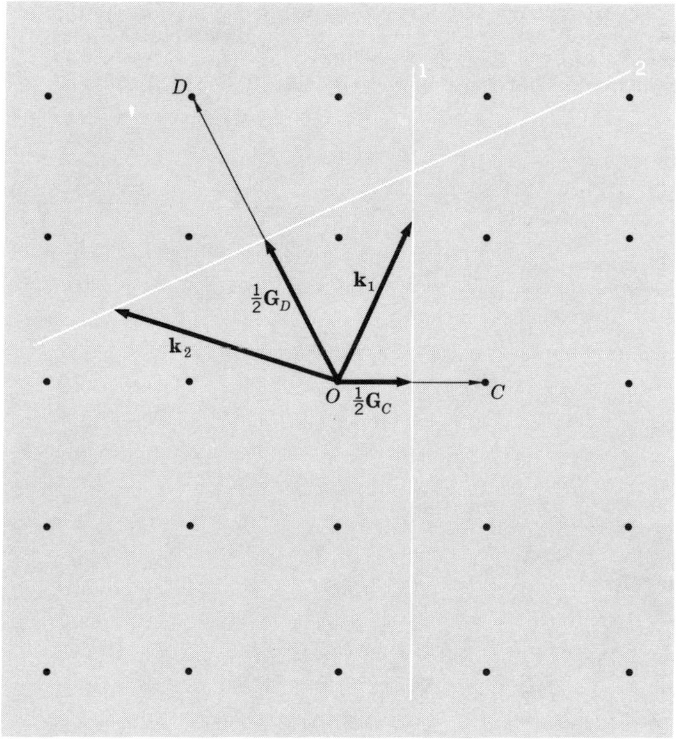

Figure 16 Reciprocal lattice points near the point O at the origin of the reciprocal lattice. The reciprocal lattice vector $\mathbf{G}_C$ connects points OC; and $\mathbf{G}_D$ connects OD. Two planes 1 and 2 are drawn which are the perpendicular bisectors of $\mathbf{G}_C$ and $\mathbf{G}_D$, respectively. Any vector from the origin to the plane 1, such as $\mathbf{k}_1$, will satisfy the diffraction condition $\mathbf{k}_1 \cdot (\frac{1}{2}\mathbf{G}_C) = (\frac{1}{2}\mathbf{G}_C)^2$. Any vector from the origin to the plane 2, such as $\mathbf{k}_2$, will satisfy the diffraction condition $\mathbf{k}_2 \cdot (\frac{1}{2}\mathbf{G}_D) = (\frac{1}{2}\mathbf{G}_D)^2$.

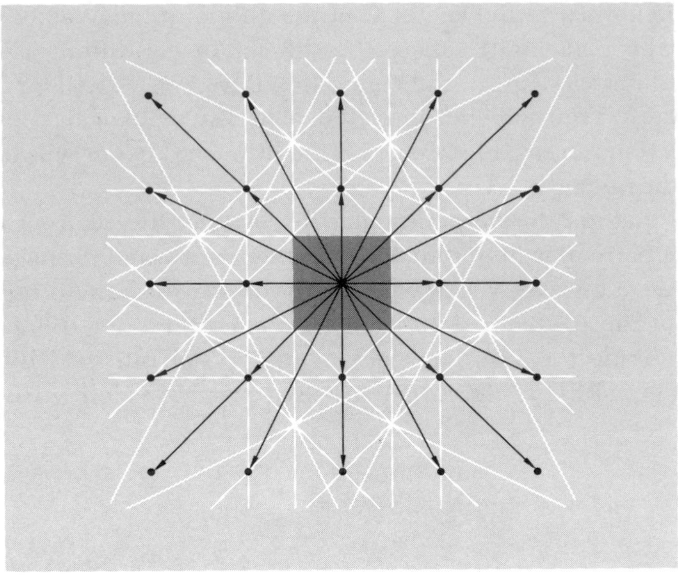

Figure 17 Square reciprocal lattice with reciprocal lattice vectors shown as fine black lines. The lines shown in white are perpendicular bisectors of the reciprocal lattice vectors. The central square is the smallest volume about the origin which is bounded entirely by white lines. The square is the Wigner-Seitz primitive cell of the reciprocal lattice. It is called the first Brillouin zone.

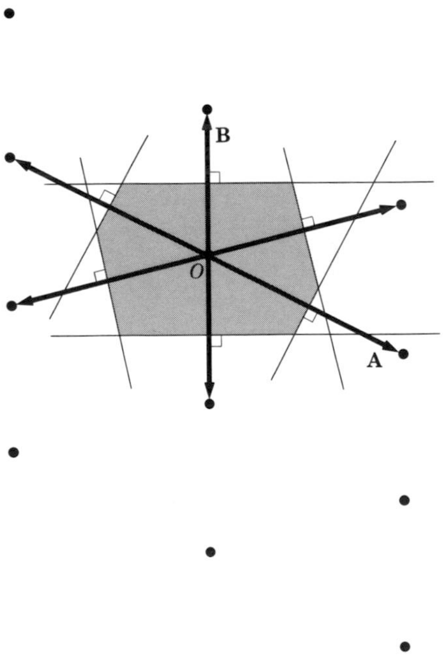

Figure 18 Construction of the first Brillouin zone for an oblique lattice in two dimensions. We first draw a number of vectors from O to nearby points in the reciprocal lattice. Next we construct lines perpendicular to these vectors at their midpoints. The smallest enclosed area is the first Brillouin zone.

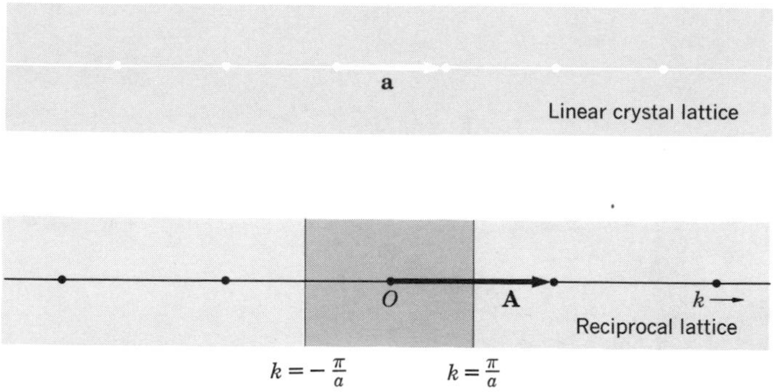

Figure 19 Crystal and reciprocal lattices in one dimension. The basis vector in the reciprocal lattice is **A**, of length equal to $2\pi/a$. The shortest reciprocal lattice vectors from the origin are **A** and −**A**. The perpendicular bisectors of these vectors form the boundaries of the first Brillouin zone. The boundaries are at $k=\pm\pi/a$.

The central cell in the reciprocal lattice is of special importance in the theory of solids, and we call it the first Brillouin zone. **The first Brillouin zone is the smallest volume entirely enclosed by planes that are the perpendicular bisectors of the reciprocal lattice vectors drawn from the origin.** The first Brillouin zone of an oblique lattice in two dimensions is constructed in Fig. 18 and of a linear lattice in one dimension in Fig. 19. The zone boundaries of the linear lattice are at $k=\pm\pi/a$, where a is the primitive axis of the crystal lattice. Historically, Brillouin zones are not part of the language of x-ray diffraction analysis of crystal structures, but the zones are an essential part of the analysis of the electronic energy-band structure of crystals. The special utility of the first zone is developed in Chapter 9. The Brillouin construction exhibits all the incident wavevectors **k** which can be Bragg-reflected by the crystal.

Reciprocal Lattice to sc Lattice. The primitive translation vectors of a simple cubic lattice may be taken as the set

$$\mathbf{a}=a\hat{\mathbf{x}}\ ;\qquad \mathbf{b}=a\hat{\mathbf{y}}\ ;\qquad \mathbf{c}=a\hat{\mathbf{z}}\ . \tag{25}$$

The volume of the cell is $\mathbf{a}\cdot\mathbf{b}\times\mathbf{c}=a^3$. The primitive translation vectors of the reciprocal lattice are found from the standard prescription (12):

$$\mathbf{A}=2\pi\frac{\mathbf{b}\times\mathbf{c}}{\mathbf{a}\cdot\mathbf{b}\times\mathbf{c}}=\frac{2\pi}{a}\hat{\mathbf{x}}\ ;\qquad \mathbf{B}=\frac{2\pi}{a}\hat{\mathbf{y}}\ ;\qquad \mathbf{C}=\frac{2\pi}{a}\hat{\mathbf{z}}\ . \tag{26}$$

Thus the reciprocal lattice is itself a simple cubic lattice, now of lattice constant $2\pi/a$.

The boundaries of the first Brillouin zones are the planes normal to the six reciprocal lattice vectors $\pm\mathbf{A}$, $\pm\mathbf{B}$, $\pm\mathbf{C}$ at their midpoints:

$$\pm\tfrac{1}{2}\mathbf{A}=\pm\frac{\pi}{a}\hat{\mathbf{x}}\ ;\qquad \pm\tfrac{1}{2}\mathbf{B}=\pm\frac{\pi}{a}\hat{\mathbf{y}}\ ;\qquad \pm\tfrac{1}{2}\mathbf{C}=\pm\frac{\pi}{a}\hat{\mathbf{z}}\ . \tag{27}$$

The six planes bound a cube of edge $2\pi/a$ and of volume $(2\pi/a)^3$; This cube is the first Brillouin zone of the sc crystal lattice.

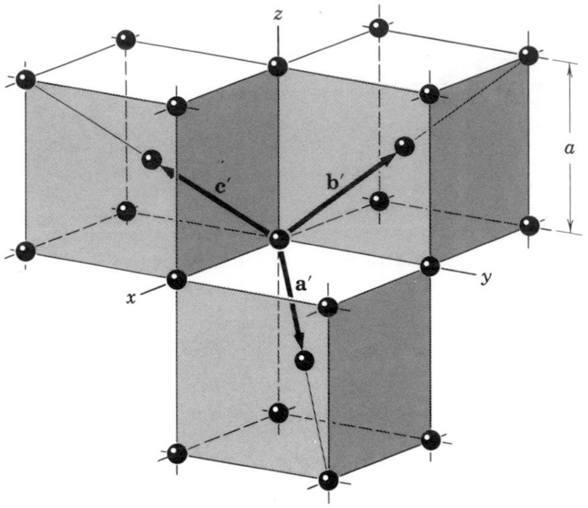

Figure 20 Primitive basis vectors of the body-centered cubic lattice.

Reciprocal Lattice to bcc Lattice. The primitive translation vectors of the bcc lattice shown in Fig. 20 are

$$\mathbf{a}' = \tfrac{1}{2}a(\hat{\mathbf{x}} + \hat{\mathbf{y}} - \hat{\mathbf{z}}) \; ; \quad \mathbf{b}' = \tfrac{1}{2}a(-\hat{\mathbf{x}} + \hat{\mathbf{y}} + \hat{\mathbf{z}}) \; ; \quad \mathbf{c}' = \tfrac{1}{2}a(\hat{\mathbf{x}} - \hat{\mathbf{y}} + \hat{\mathbf{z}}) \; , \tag{28}$$

where a is the side of the conventional cube and $\hat{\mathbf{x}}$, $\hat{\mathbf{y}}$, $\hat{\mathbf{z}}$ are orthogonal unit vectors parallel to the cube edges. The volume of the primitive cell is

$$V = |\mathbf{a}' \cdot \mathbf{b}' \times \mathbf{c}'| = \tfrac{1}{2}a^3 \; . \tag{29}$$

The primitive translations **A, B, C** of the reciprocal lattice are defined by (12). We have, using (28),

$$\mathbf{A} = \frac{2\pi}{a}(\hat{\mathbf{x}} + \hat{\mathbf{y}}) \; ; \qquad \mathbf{B} = \frac{2\pi}{a}(\hat{\mathbf{y}} + \hat{\mathbf{z}}) \; ; \qquad \mathbf{C} = \frac{2\pi}{a}(\hat{\mathbf{x}} + \hat{\mathbf{z}}) \; . \tag{30}$$

Note by comparison with Fig. 1.15 that these are just the primitive vectors of an fcc lattice; thus an fcc lattice is the reciprocal lattice of the bcc lattice.

If h, k, l are integers, the general reciprocal lattice vector is

$$\mathbf{G} = h\mathbf{A} + k\mathbf{B} + l\mathbf{C} = \frac{2\pi}{a}[(h + l)\hat{\mathbf{x}} + (h + k)\hat{\mathbf{y}} + (k + l)\hat{\mathbf{z}}] \; . \tag{31}$$

The shortest **G**'s are the following 12 vectors, where all choices of sign are independent:

$$\frac{2\pi}{a}(\pm\hat{\mathbf{x}} \pm \hat{\mathbf{y}}) \; ; \qquad \frac{2\pi}{a}(\pm\hat{\mathbf{y}} \pm \hat{\mathbf{z}}) \; ; \qquad \frac{2\pi}{a}(\pm\hat{\mathbf{x}} \pm \hat{\mathbf{z}}) \; . \tag{32}$$

The primitive cell of the reciprocal lattice is the parallelepiped described by the **A, B,** and **C** defined by (30). The volume of the primitive cell of the reciprocal lattice is $|\mathbf{A} \cdot \mathbf{B} \times \mathbf{C}| = 2(2\pi/a)^3$. The parallelepiped contains one reciprocal lattice point, because each of the eight corner points is shared among eight parallelepipeds: each parallelepiped contains one-eighth of each of eight corner points. It is customary in solid state physics to take the primitive cell of the reciprocal lattice as the first Brillouin zone; each such cell contains one lattice point at the center of the cell. This zone is bounded by the planes normal to the 12 vectors of Eq. (32) at their midpoints. The zone is a regular 12-faced solid, a rhombic dodecahedron, as shown in Fig. 21. The vectors from the origin to the center of each face are

$$\frac{\pi}{a}(\pm\mathbf{x} \pm \mathbf{y}) \; ; \qquad \frac{\pi}{a}(\pm\mathbf{y} \pm \mathbf{z}) \; ; \qquad \frac{\pi}{a}(\pm\mathbf{x} \pm \mathbf{z}) \; . \tag{33}$$

All choices of sign are independent, giving 12 vectors.

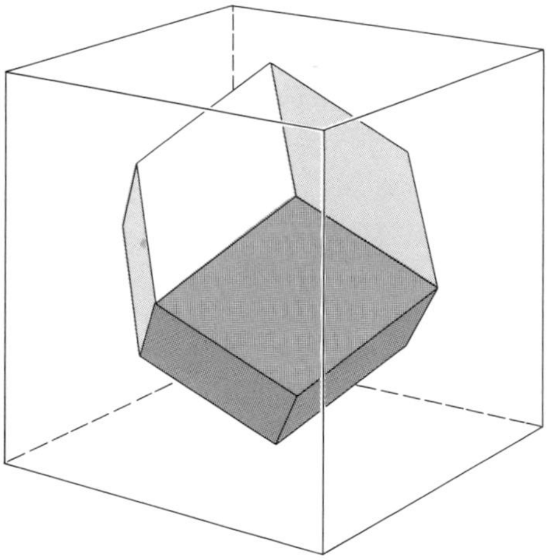

Figure 21 First Brillouin zone of the body-centered cubic lattice. The figure is a regular rhombic dodecahedron.

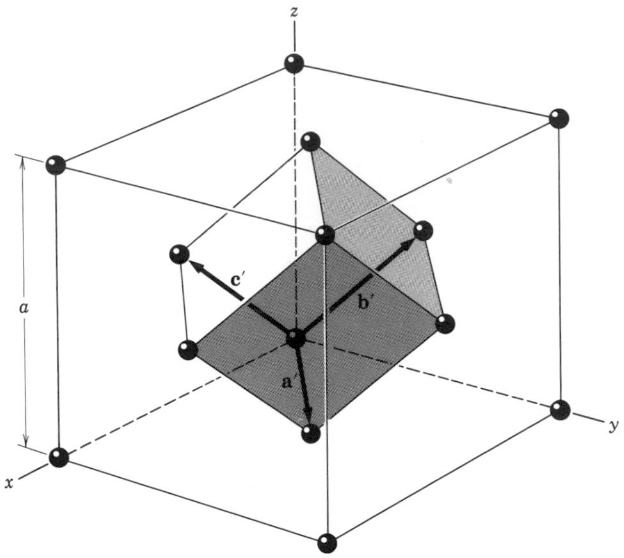

Figure 22 Primitive basis vectors of the face-centered cubic lattice.

Reciprocal Lattice to fcc Lattice. The primitive translation vectors of the fcc lattice shown in Fig. 22 are:

$$\mathbf{a}' = \tfrac{1}{2}a(\hat{\mathbf{x}} + \hat{\mathbf{y}}) \; ; \qquad \mathbf{b}' = \tfrac{1}{2}a(\hat{\mathbf{y}} + \hat{\mathbf{z}}) \; ; \qquad \mathbf{c}' = \tfrac{1}{2}a(\hat{\mathbf{z}} + \hat{\mathbf{x}}) \; . \tag{34}$$

The volume of the primitive cell is

$$V = |\mathbf{a}' \cdot \mathbf{b}' \times \mathbf{c}'| = \tfrac{1}{4}a^3 \; . \tag{35}$$

The primitive translation vectors **A, B, C** of the reciprocal lattice to the fcc lattice are

$$\mathbf{A} = \frac{2\pi}{a}(\hat{\mathbf{x}} + \hat{\mathbf{y}} - \hat{\mathbf{z}}) \; ; \quad \mathbf{B} = \frac{2\pi}{a}(-\hat{\mathbf{x}} + \hat{\mathbf{y}} + \hat{\mathbf{z}}) \; ; \quad \mathbf{C} = \frac{2\pi}{a}(\hat{\mathbf{x}} - \hat{\mathbf{y}} + \hat{\mathbf{z}}) \; . \tag{36}$$

These are primitive translation vectors of a bcc lattice, so that the bcc lattice is the reciprocal lattice of the fcc lattice. The volume of the primitive cell in the reciprocal lattice is $|\mathbf{A} \cdot \mathbf{B} \times \mathbf{C}| = 4(2\pi/a)^3$. The reciprocal lattice vectors are

$$\mathbf{G} = \frac{2\pi}{a}[(h - k + l)\hat{\mathbf{x}} + (h + k - l)\hat{\mathbf{y}} + (-h + k + l)\hat{\mathbf{z}}] \; , \tag{37}$$

where h, k, l are integers. The shortest **G**'s are the eight vectors:

$$\frac{2\pi}{a}(\pm\hat{\mathbf{x}} \pm \hat{\mathbf{y}} \pm \hat{\mathbf{z}}) \ . \tag{38}$$

The boundaries of a primitive cell in the reciprocal lattice are determined for the most part by the eight planes normal to these vectors at their midpoints. But the corners of the octahedron thus formed are cut by the planes that are the perpendicular bisectors of six other reciprocal lattice vectors:

$$\frac{2\pi}{a}(\pm 2\hat{\mathbf{x}}) \ ; \qquad \frac{2\pi}{a}(\pm 2\hat{\mathbf{y}}) \ ; \qquad \frac{2\pi}{a}(\pm 2\hat{\mathbf{z}}) \ . \tag{39}$$

Note that $(2\pi/a)(2\hat{\mathbf{x}})$ is a reciprocal lattice vector because it is equal to $\mathbf{A} + \mathbf{C}$. The first Brillouin zone is the smallest bounded volume about the origin, the truncated octahedron shown in Fig. 23. The six planes bound a cube of edge $4\pi/a$ and of volume $(4\pi/a)^3$.

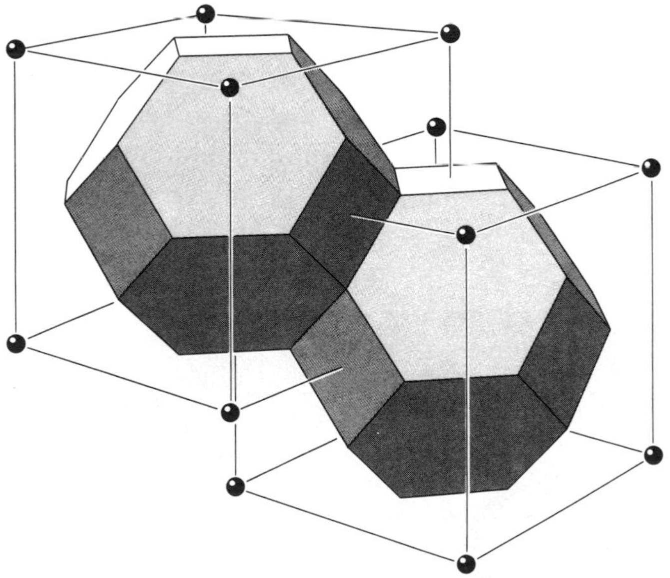

Figure 23 Brillouin zones of the face-centered cubic lattice. The cells are in reciprocal space, and the reciprocal lattice is body centered.

FOURIER ANALYSIS OF THE BASIS

When the diffraction condition $\Delta \mathbf{k} = \mathbf{G}$ is satisfied, the scattering amplitude is determined by (17), which for a crystal of N cells may be written as

$$F_{\mathbf{G}} = N \int_{cell} dV \, n(\mathbf{r}) \exp(-i\mathbf{G} \cdot \mathbf{r}) = N S_{\mathbf{G}} \tag{40}$$

The quantity $S_{\mathbf{G}}$ is called the **structure factor** and is defined as an integral over a single cell, with $\mathbf{r} = 0$ at one corner. Often it is useful to write the electron concentration as the superposition of electron concentration functions n_j associated with each atom j of the cell. If $\mathbf{r}_j$ is the vector to the center of atom j, then the function $n_j(\mathbf{r} - \mathbf{r}_j)$ defines the contribution of that atom to the electron concentration at $\mathbf{r}$. The total concentration at $\mathbf{r}$ due to all atoms in the cell is the sum

$$n(\mathbf{r}) = \sum_{j=1}^{s} n_j(\mathbf{r} - \mathbf{r}_j) \tag{41}$$

over the s atoms of the basis. The decomposition of $n(\mathbf{r})$ is not unique, for we cannot always say how much charge is associated with each atom. This is not an important difficulty.

The structure factor defined by (40) may now be written as integrals over the s atoms of a cell:

$$S_{\mathbf{G}} = \sum_j \int dV \, n_j(\mathbf{r} - \mathbf{r}_j) \exp(-i\mathbf{G} \cdot \mathbf{r}) =$$

$$\sum_j \exp(-i\mathbf{G} \cdot \mathbf{r}_j) \int dV \, n_j(\boldsymbol{\rho}) \exp(-i\mathbf{G} \cdot \boldsymbol{\rho}) \, , \tag{42}$$

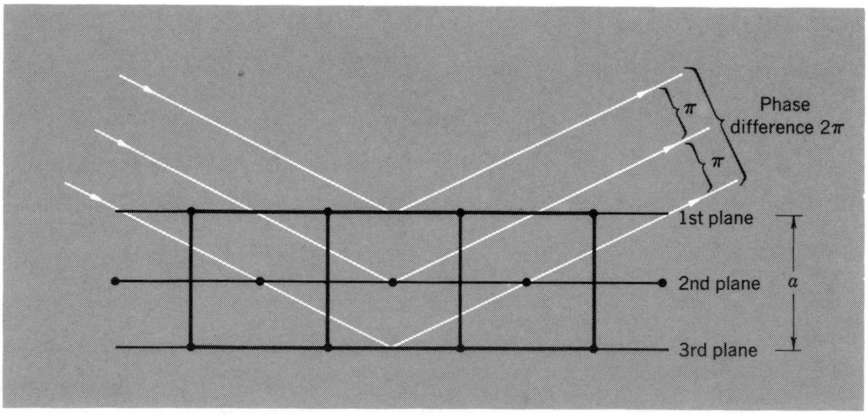

Figure 24 Explanation of the absence of a (100) reflection from a body-centered cubic lattice. The phase difference between successive planes is π, so that the reflected amplitude from two adjacent planes is $1 + e^{-i\pi} = 1 - 1 = 0$.

where $\boldsymbol{\rho} \equiv \mathbf{r} - \mathbf{r}_j$. We now define the atomic form factor as

$$f_j = \int dV \, n_j(\boldsymbol{\rho}) \exp(-i\mathbf{G} \cdot \boldsymbol{\rho}) \ , \tag{43}$$

integrated over all space. This is largely an atomic property. The structure factor of the basis becomes

$$\mathcal{S}_{\mathbf{G}} = \sum_j f_j \exp(-i\mathbf{G} \cdot \mathbf{r}_j) \ . \tag{44}$$

The usual form of this result follows on writing $\mathbf{r}_j = x_j\mathbf{a} + y_j\mathbf{b} + z_j\mathbf{c}$, as in (1.3). Then, for the reflection hkl,

$$\mathbf{r}_j \cdot \mathbf{G} = (x_j\mathbf{a} + y_j\mathbf{b} + z_j\mathbf{c}) \cdot (h\mathbf{A} + k\mathbf{B} + l\mathbf{C}) = 2\pi(x_jh + y_jk + z_jl) \ , \tag{45}$$

and

$$\mathcal{S}(hkl) = \sum_j f_j \exp[-i2\pi(x_jh + y_jk + z_jl)] \ . \tag{46}$$

The structure factor need not be real; the scattered power involves the real quantity $\mathcal{S}^*\mathcal{S}$, where $\mathcal{S}^*$ is the complex conjugate of $\mathcal{S}$. At a zero of $\mathcal{S}$ there is no intensity at a reflection $\mathbf{G}$ permitted by the space lattice. If we change our choice of cell, the basis is also changed, but the physical scattering must be unchanged. Thus for two choices, 1 and 2,

$$\mathcal{S}_1(\text{cell}) \times \mathcal{S}_1(\text{basis}) = \mathcal{S}_2(\text{cell}) \times \mathcal{S}_2(\text{basis}).$$

Structure Factor of the bcc Lattice. The basis of the bcc structure referred to the cubic cell has identical atoms at $x_1 = y_1 = z_1 = 0$ and at $x_2 = y_2 = z_2 = \frac{1}{2}$. Thus (46) becomes

$$\mathcal{S}(hkl) = f\{1 + \exp[-i\pi(h + k + l)]\} \ , \tag{47}$$

where f is the form factor of an atom. The value of $\mathcal{S}$ is zero whenever the exponential has the value -1, which is whenever its argument is $-i\pi$ times an odd integer. We have

$$\mathcal{S} = 0 \qquad \text{when } h + k + l = \text{odd integer} \ ;$$

$$\mathcal{S} = 2f \qquad \text{when } h + k + l = \text{even integer} \ .$$

Metallic sodium has a bcc structure. The diffraction spectrum does not contain lines such as (100), (300), (111), or (221), but lines such as (200), (110), and (222) will be present; here the indices (hkl) are referred to a cubic cell. What is the physical interpretation of the result that the (100) reflection vanishes? The (100) reflection normally occurs when reflections from the planes that bound the cubic cell differ in phase by 2π. In the bcc lattice there is an intervening plane (Fig. 24) of atoms, labeled the second plane in the figure, which is equal in scattering power to the other planes. Situated midway between them, it gives a reflection retarded in phase by π with respect to the first plane, thereby canceling the contribution from that plane. The cancellation of the (100) reflection occurs in the bcc lattice because the planes are identical in composition. A similar cancellation can easily be found in the hcp structure.

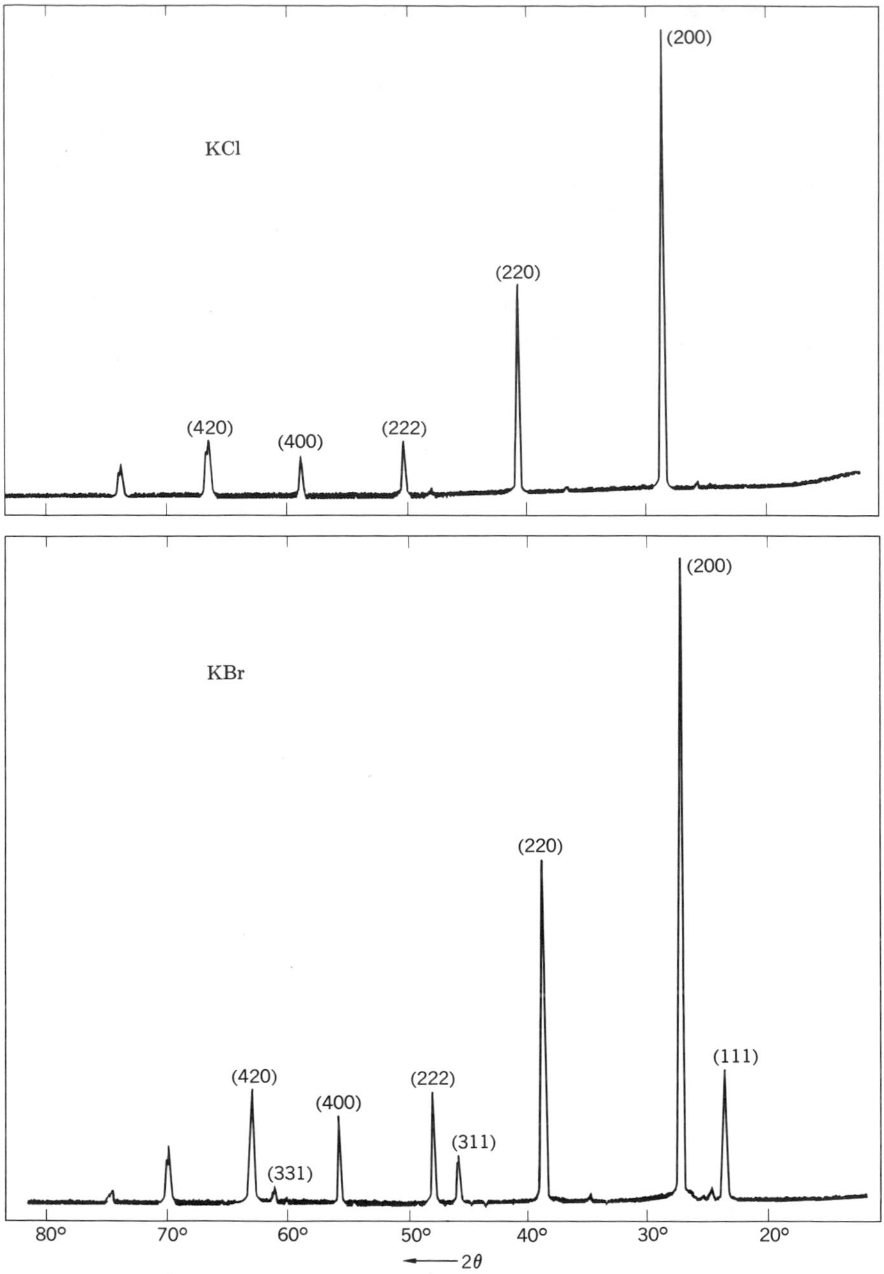

Figure 25 Comparison of x-ray reflections from KCl and KBr powders. In KCl the numbers of electrons of K^+ and Cl^- ions are equal. The scattering amplitudes $f(K^+)$ and $f(Cl^-)$ are almost exactly equal, so that the crystal looks to x-rays as if it were a monatomic simple cubic lattice of lattice constant $a/2$. Only even integers occur in the reflection indices when these are based on a cubic lattice of lattice constant a. In KBr the form factor of Br^- is quite different than that of K^+, and all reflections of the fcc lattice are present. (Courtesy of Robert van Nordstrand.)

Structure Factor of the fcc Lattice. The basis of the fcc structure referred to the cubic cell has identical atoms at 000; $0\frac{1}{2}\frac{1}{2}$; $\frac{1}{2}0\frac{1}{2}$; $\frac{1}{2}\frac{1}{2}0$. Thus (46) becomes

$$\mathcal{S}(hkl) = f\{1 + \exp[-i\pi(k+l)] + \exp[-i\pi(h+l)] + \exp[-i\pi(h+k)]\} \ . \quad (48)$$

If all indices are even integers, $\mathcal{S} = 4f$; similarly if all indices are odd integers. But if only one of the integers is even, two of the exponents will be odd multiples of $-i\pi$ and $\mathcal{S}$ will vanish. If only one of the integers is odd, the same argument applies and $\mathcal{S}$ will also vanish. Thus in the fcc lattice no reflections can occur for which the indices are partly even and partly odd. The point is beautifully illustrated by Fig. 25: both KCl and KBr have an fcc lattice, but KCl simulates an sc lattice because the K^+ and Cl^- ions have equal numbers of electrons.

Atomic Form Factor

In the expression (46) for the structure factor, there occurs the quantity f_j, which is a measure of the scattering power of the jth atom in the unit cell. The value of f involves the number and distribution of atomic electrons, and it also involves the wavelength and angle of scattering of the radiation. These factors enter because of interference effects arising from the finite extent of the atoms. We now give a classical calculation of the scattering factor.

The scattered radiation from a single atom takes account of interference effects within the atom. We defined in (43) the form factor

$$f_j = \int dV\, n_j(\mathbf{r}) \exp(-i\mathbf{G}\cdot\mathbf{r}) \ , \quad (49)$$

with the integral extended over the electron concentration associated with a single atom. Let $\mathbf{r}$ make an angle α with $\mathbf{G}$; then $\mathbf{G}\cdot\mathbf{r} = Gr\cos\alpha$. If the electron distribution is spherically symmetric about the origin, then

$$f_j = 2\pi \int dr\, r^2\, d(\cos\alpha)\, n_j(r)\, e^{-iGr\cos\alpha}$$
$$= 2\pi \int dr\, r^2 n_j(r) \cdot \frac{e^{iGr} - e^{-iGr}}{iGr} \ ,$$

after integration over $d(\cos\alpha)$ between -1 and 1. Thus the form factor is given by

$$f_j = 4\pi \int dr\, n_j(r) r^2 \frac{\sin Gr}{Gr} \ . \quad (50)$$

If the same total electron density were concentrated at $r = 0$, only $Gr = 0$ would contribute to the integrand. In this limit $(\sin Gr)/Gr = 1$, and

$$f_j = 4\pi \int dr\, n_j(r) r^2 = Z \ , \tag{51}$$

the number of atomic electrons. Therefore f is the ratio of the radiation amplitude scattered by the actual electron distribution in an atom to that scattered by one electron localized at a point. In the forward direction $G = 0$, and f reduces again to the value Z. Values of the atomic form factor f for atoms may be found in the *International tables for x-ray crystallography*, Vol. 3.

The overall electron distribution in a solid is fairly close to that of the appropriate free atoms. This statement does not mean that the outermost or valence electrons are not redistributed in forming the solid; it means only that the x-ray reflection intensities are represented well by the free atom values of the form factors and are not very sensitive to small redistributions of the electrons. As an example, Batterman and co-workers[3] find agreement within 1 percent in a comparison of the x-ray intensities of Bragg reflections of metallic iron, copper, and aluminum with the theoretical free atom values from wavefunction calculations. The results for aluminum are shown in Fig. 26.

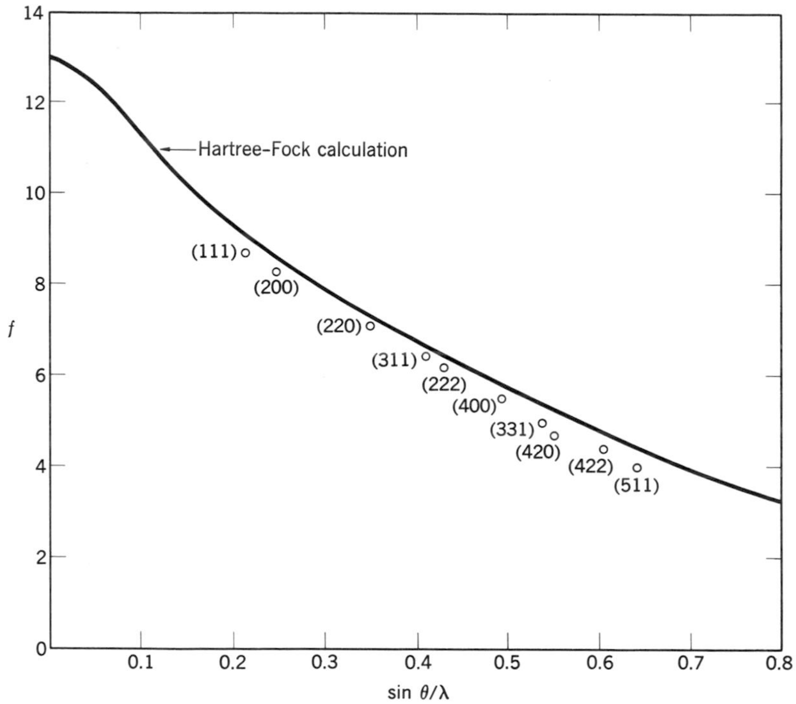

Figure 26 Absolute experimental atomic scattering factors for metallic aluminum, after Batterman, Chipman, and DeMarco. Each observed reflection is labeled. The incident radiation was Mo$K\alpha$ with $\lambda = 0.709$ Å. No reflections occur for indices partly even and partly odd, as predicted for an fcc crystal.

[3]B. W. Batterman, D. R. Chipman, and J. J. DeMarco, Phys. Rev. **122,** 68 (1961); see also L. D. Jennings et al., Phys. Rev. **135,** 1612 (1964).

There have been many attempts to obtain direct x-ray evidence about the actual electron distribution in a covalent chemical bond, particularly in crystals having the diamond structure.[4] The question lies near the limits of what can be explored by x-ray diffraction methods. In silicon midway between two nearest-neighbor atoms, there is an appreciable increase in electron concentration over what is expected from the overlap of the electron densities calculated for two free atoms.

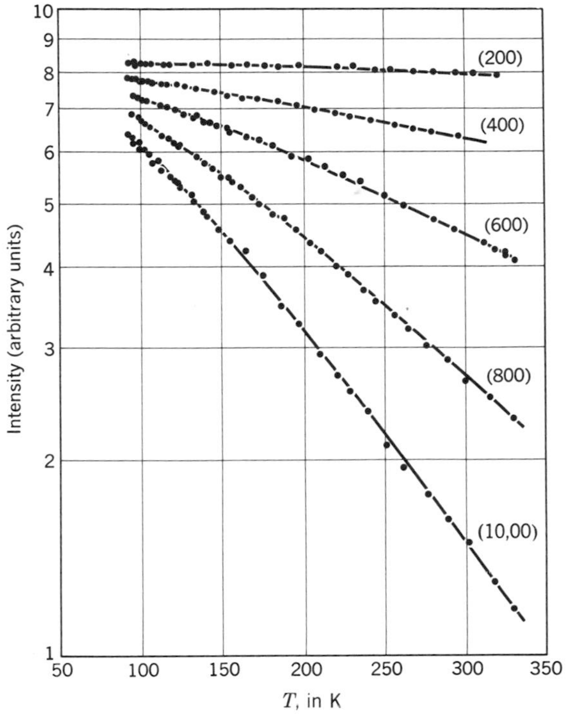

Figure 27 The dependence of intensity on temperature for the $(h00)$ x-ray reflections of aluminum. Reflections $(h00)$ with h odd are forbidden for an fcc structure. [After R. M. Nicklow and R. A. Young, Phys. Rev. **152**, 591 (1966).]

TEMPERATURE DEPENDENCE OF THE REFLECTION LINES

> . . . I came to the conclusion that the sharpness of the interference lines would not suffer but that their intensity should diminish with increasing angle of scattering, the more so the higher the temperature.
>
> P. Debye

As the temperature of the crystal is increased, the intensity of the Bragg-reflected beams decreases, but the angular width of the reflected line does not change. Experimental intensities for aluminum are shown in Fig. 27. It is surprising that we can get a sharp x-ray reflection from atoms undergoing large amplitude random thermal motion, with instantaneous nearest-neighbor spacings differing by 10 percent at room temperature.

[4]For diamond, see G. B. Carpenter, J. Chem. Phys. **32**, 525 (1960); for silicon, see S. Göttlicher et al., Z. f. physikalische Chemie **21**, 133 (1959).

Before the Laue experiment was done, but when the proposal was discussed[5] in a coffee house in Munich, the objection was made that the instantaneous positions of the atoms in a crystal at room temperature are far from a regular periodic array, because of the large thermal fluctuation. Therefore, the argument went, one should not expect a well-defined diffracted beam.

But a well-defined diffracted beam is found. The reason was given by Debye. Consider the expression (17) for the radiation amplitude scattered by a crystal: let the position of the atom nominally at $\mathbf{r}_j$ contain a term $\mathbf{u}(t)$ fluctuating in time: $\mathbf{r}(t) = \mathbf{r}_j + \mathbf{u}(t)$. We suppose each atom fluctuates independently about its own equilibrium position.[6] Then the thermal average of the structure factor (44) contains terms

$$f_j \exp(-i\mathbf{G}\cdot\mathbf{r}_j)\langle\exp(-i\mathbf{G}\cdot\mathbf{u})\rangle \ , \tag{52}$$

where $\langle\cdots\rangle$ denotes thermal average. The series expansion of the exponential is

$$\langle\exp(-i\mathbf{G}\cdot\mathbf{u})\rangle = 1 - i\langle\mathbf{G}\cdot\mathbf{u}\rangle - \tfrac{1}{2}\langle(\mathbf{G}\cdot\mathbf{u})^2\rangle + \cdots \ . \tag{53}$$

But $\langle\mathbf{G}\cdot\mathbf{u}\rangle = 0$, because $\mathbf{u}$ is a random thermal displacement uncorrelated with the direction of $\mathbf{G}$. Further,

$$\langle(\mathbf{G}\cdot\mathbf{u})^2\rangle = G^2\langle u^2\rangle\langle\cos^2\theta\rangle = \tfrac{1}{3}\langle u^2\rangle G^2 \ .$$

The factor $\frac{1}{3}$ arises as the geometrical average of $\cos^2\theta$ over a sphere.

The function

$$\exp(-\tfrac{1}{6}\langle u^2\rangle G^2) = 1 - \tfrac{1}{6}\langle u^2\rangle G^2 + \cdots \tag{54}$$

has the same series expansion as (53) for the first two terms shown here. For a harmonic oscillator all terms in the series (53) and (54) can be shown to be identical. Then the scattered intensity, which is the square of the amplitude, is

$$I = I_0 \exp(-\tfrac{1}{3}\langle u^2\rangle G^2) \ , \tag{55}$$

where I_0 is the scattered intensity from the rigid lattice. The exponential factor is the **Debye-Waller factor.**

Here $\langle u^2\rangle$ is the mean square displacement of an atom. The thermal average potential energy $\langle U\rangle$ of a classical harmonic oscillator in three dimensions is $\frac{3}{2}k_BT$, whence

$$\langle U\rangle = \tfrac{1}{2}C\langle u^2\rangle = \tfrac{1}{2}M\omega^2\langle u^2\rangle = \tfrac{3}{2}k_BT \ , \tag{56}$$

where C is the force constant, M is the mass of an atom, and ω is the frequency of the oscillator. We have used the result $\omega^2 = C/M$. Thus the scattered intensity is

$$I(hkl) = I_0 \exp(-k_BTG^2/M\omega^2) \ , \tag{57}$$

[5]P. P. Ewald, private communication.

[6]This is the Einstein model of a solid; it is not a very good model at low temperatures, but it works well at high temperatures. For a general treatment of scattering by thermal fluctuations, see *QTS*, Chap. 20.

where hkl are the indices of the reciprocal lattice vector **G**. This classical result is a good approximation at high temperatures.

For quantum oscillators $\langle u^2 \rangle$ does not vanish even at $T = 0$; there is zero-point motion. On the independent harmonic oscillator model the zero-point energy is $\frac{3}{2}\hbar\omega$; this is the energy of a three-dimensional quantum harmonic oscillator in its ground state referred to the classical energy of the same oscillator at rest. Half of the oscillator energy is potential energy, so that in the ground state

$$\langle U \rangle = \tfrac{1}{2}M\omega^2\langle u^2 \rangle = \tfrac{3}{4}\hbar\omega \ ; \qquad \langle u^2 \rangle = 3\hbar/2M\omega \ , \tag{58}$$

whence, by (55),

$$I(hkl) = I_0 \exp(-\hbar G^2/2M\omega) \tag{59}$$

at absolute zero. If $G = 10^9$ cm^{-1}, $\omega = 10^{14}$ s^{-1}, and $M = 10^{-22}$ g, the argument of the exponential is approximately 0.1, so that $I/I_0 \simeq 0.9$. At absolute zero 90 percent of the beam is elastically scattered and 10 percent is inelastically scattered.

We see from (57) and from Fig. 27 that the intensity of the diffracted line decreases, but not catastrophically, as the temperature is increased. Reflections of low G are affected less than reflections of high G. The intensity we have calculated is that of the coherent diffraction or the elastic scattering in the well-defined Bragg directions. The intensity lost from these directions appears as a diffuse background. In inelastic scattering the x-ray photon causes the excitation or de-excitation of a lattice vibration, and the photon changes direction and energy.

At a given temperature the Debye-Waller factor of a diffraction line decreases with an increase in the magnitude of the reciprocal lattice vector **G** associated with the reflection. The larger $|\mathbf{G}|$ is, the weaker the reflection at high temperatures. The temperature dependence of the reflected intensity for the (h00) reflections of aluminum is shown in Fig. 27. The theory we have worked out here for x-ray reflection applies equally well to neutron diffraction and to the **Mössbauer effect**, the recoilless emission of gamma rays by nuclei bound in crystals.

X-rays can also be absorbed in a crystal by the inelastic processes of photo-ionization of electrons and Compton scattering. In the photoeffect the x-ray photon is absorbed and an electron is ejected from an atom. In the Compton effect the photon is scattered inelastically by an electron: the photon loses energy and the electron is ejected from an atom. The depth of penetration[7] of the x-ray beam depends on the solid and on the photon energy, but 1 cm is typical. A diffracted beam in Bragg reflection may be formed in a much shorter distance, perhaps 10^{-3} cm in an ideal crystal.

[7]See W. Heitler, *Quantum theory of radiation*, 3rd ed., Oxford, 1954, p. 223.

SUMMARY

1. Various statements of the Bragg condition:

$$2d \sin \theta = n\lambda \ ; \qquad \Delta\mathbf{k} = \mathbf{G} \ ; \qquad 2\mathbf{k} \cdot \mathbf{G} = G^2 \ .$$

2. Laue conditions:

$$\mathbf{a} \cdot \Delta\mathbf{k} = 2\pi h \ ; \qquad \mathbf{b} \cdot \Delta\mathbf{k} = 2\pi k \ ; \qquad \mathbf{c} \cdot \Delta\mathbf{k} = 2\pi l \ .$$

3. The primitive translation vectors of the reciprocal lattice are

$$\mathbf{A} = 2\pi \frac{\mathbf{b} \times \mathbf{c}}{\mathbf{a} \cdot \mathbf{b} \times \mathbf{c}} \ ; \qquad \mathbf{B} = 2\pi \frac{\mathbf{c} \times \mathbf{a}}{\mathbf{a} \cdot \mathbf{b} \times \mathbf{c}} \ ; \qquad \mathbf{C} = 2\pi \frac{\mathbf{a} \times \mathbf{b}}{\mathbf{a} \cdot \mathbf{b} \times \mathbf{c}} \ .$$

Here **a, b, c** are the primitive translation vectors of the crystal lattice.

4. A reciprocal lattice vector has the form

$$\mathbf{G} = h\mathbf{A} + k\mathbf{B} + l\mathbf{C} \ ,$$

where h, k, l are integers or zero.

5. The scattered amplitude in the direction $\Delta\mathbf{k} = \mathbf{k}' - \mathbf{k} = \mathbf{G}$ is proportional to the geometrical structure factor:

$$\mathcal{S}_{\mathbf{G}} \equiv \Sigma f_j \exp(-i\mathbf{r}_j \cdot \mathbf{G}) = \Sigma f_j \exp[-i2\pi(x_j h + y_j k + z_j l)] \ ,$$

where j runs over the s atoms of the basis, and f_j is the atomic form factor (50) of the jth atom of the basis. The expression on the right-hand side is written for a reflection (hkl), for which $\mathbf{G} = h\mathbf{A} + k\mathbf{B} + l\mathbf{C}$.

6. Any function invariant under a lattice translation may be expanded in a Fourier series of the form

$$n(\mathbf{r}) = \sum_{\mathbf{G}} n_{\mathbf{G}} \exp(i\mathbf{G} \cdot \mathbf{r}) \ .$$

7. The first Brillouin zone is the Wigner-Seitz primitive cell of the reciprocal lattice. Any wave whose wavevector **k** drawn from the origin terminates on a surface of the Brillouin zone will be diffracted by the crystal.

8.

Crystal lattice	*First Brillouin zone*
Simple cubic	Cube
Body-centered cubic	Rhombic dodecahedron (Fig. 21)
Face-centered cubic	Truncated octahedron (Fig. 23)

9. Thermal motion does not broaden a diffraction line, but only reduces the intensity. The lost intensity reappears as long, low wings about the position of the diffraction line.

PROBLEMS

1. ***Inversion of Fourier series.*** (a) Show that the Fourier coefficient n_p in Eq. (7) is given by

$$n_p = a^{-1} \int_0^a dx\, n(x) \exp(-i2\pi px/a) \ . \tag{60}$$

(b) Show that the inversion of $n(\mathbf{r}) = \Sigma\, n_\mathbf{G} \exp(i\mathbf{G} \cdot \mathbf{r})$ gives

$$n_\mathbf{G} = V_c^{-1} \int_{\text{cell}} dV\, n(\mathbf{r}) \exp(-i\mathbf{G} \cdot \mathbf{r}) \ . \tag{61}$$

Here V_c is the volume of a cell of the crystal.

2. ***Interplanar separation.*** Consider a plane hkl in a crystal lattice. (a) Prove that the reciprocal lattice vector $\mathbf{G} = h\mathbf{A} + k\mathbf{B} + l\mathbf{C}$ is perpendicular to this plane. (b) Prove that the distance between two adjacent parallel planes of the lattice (planes through lattice points) is $d(hkl) = 2\pi/|\mathbf{G}|$, where $\mathbf{G} = h\mathbf{A} + k\mathbf{B} + l\mathbf{C}$. (c) Show for a simple cubic lattice that $d^2 = a^2/(h^2 + k^2 + l^2)$.

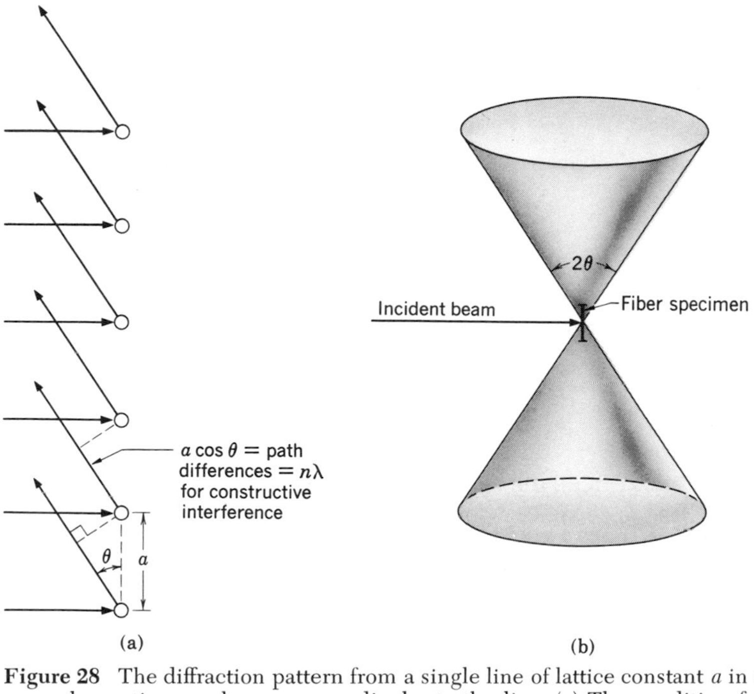

Figure 28 The diffraction pattern from a single line of lattice constant a in a monochromatic x-ray beam perpendicular to the line. (a) The condition for constructive interference is $a \cos \theta = n\lambda$, where n is an integer. (b) For given n the diffracted rays of constant λ lie on the surface of a cone.

3. ***Diffraction from a linear array and a square array.*** The diffraction pattern of a linear structure of lattice constant a is explained[8] in Fig. 28. Somewhat similar structures are important in molecular biology: DNA and many proteins are

[8]Another viewpoint is useful: for a linear lattice the diffraction pattern is described by the single Laue equation $\mathbf{a} \cdot \Delta\mathbf{k} = 2\pi q$, where q is an integer. The lattice sums which led to the other Laue equations do not occur for a linear lattice. Now $\mathbf{a} \cdot \Delta\mathbf{k} = \text{const.}$ is the equation of a plane; thus the reciprocal lattice becomes a set of parallel planes normal to the line of atoms. What happens to the reciprocal lattice for a single plane of atoms? Now there are two Laue equations. Care must be taken with the concept of reciprocal lattice points when discussing one- or two-dimensional structures, because the reciprocal lattice point becomes a reciprocal lattice line for a planar structure and a reciprocal lattice plane for a linear structure.

linear helices.[9] (a) A cylindrical film is exposed to the diffraction pattern of Fig. 28b; the axis of the cylinder is coincident with the axis of the linear structure or fiber. Describe the appearance of the diffraction pattern on the film. (b) A flat photographic plate is placed behind the fiber and normal to the incident beam. Sketch roughly the appearance of the diffraction pattern on the plate. (c) A single plane of atoms forms a square lattice of lattice constant a. The plane is normal to the incident x-ray beam. Sketch roughly the appearance of the

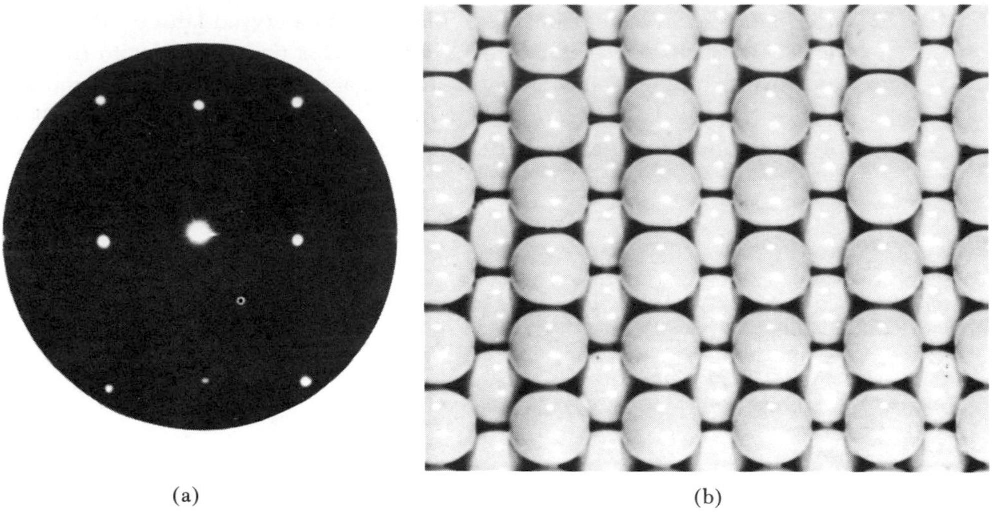

Figure 29 (a) Backward scattering pattern of 76 eV electrons incident normally on the (110) face of a nickel crystal; a model of the surface is shown in (b). (Courtesy of A. U. MacRae.)

diffraction pattern[10] on the photographic plate. Hint: The diffraction from a plane of atoms can be inferred from the patterns for two perpendicular lines of atoms. (d) Figure 29 shows the electron diffraction pattern in the backward direction from the nickel atoms on the (110) surface of a nickel crystal. Explain the orientation of the diffraction pattern in relation to the atomic positions of the surface atoms shown in the model. Assume that only the surface atoms are effective in the reflection of low-energy electrons.

4. ***Hexagonal space lattice.*** The primitive translation vectors of the hexagonal space lattice may be taken as

$$\mathbf{a} = (3^{\frac{1}{2}}a/2)\hat{\mathbf{x}} + (a/2)\hat{\mathbf{y}} \; ; \qquad \mathbf{b} = -(3^{\frac{1}{2}}a/2)\hat{\mathbf{x}} + (a/2)\hat{\mathbf{y}} \; ; \qquad \mathbf{c} = c\hat{\mathbf{z}} \; .$$

(a) Show that the volume of the primitive cell is $(3^{\frac{1}{2}}/2)a^2c$.
(b) Show that the primitive translations of the reciprocal lattice are

$$\mathbf{A} = (2\pi/3^{\frac{1}{2}}a)\hat{\mathbf{x}} + (2\pi/a)\hat{\mathbf{y}} \; ; \qquad \mathbf{B} = -(2\pi/3^{\frac{1}{2}}a)\hat{\mathbf{x}} + (2\pi/a)\hat{\mathbf{y}} \; ; \qquad \mathbf{C} = (2\pi/c)\hat{\mathbf{z}} \; ,$$

so that the lattice is its own reciprocal, but with a rotation of axes.
(c) Describe and sketch the first Brillouin zone of the hexagonal space lattice.

---

[9]The diffraction pattern of a helix is treated by W. Cochran, F. H. C. Crick, and V. Vand, Acta Cryst. **5,** 581 (1952); A. Klug, F. H. C. Crick, and H. W. Wyckoff, Acta Cryst. **11,** 199 (1958). An elementary account is given by C. Kittel, Amer. J. Phys. **36,** 610 (1968).
[10]See L. H. Germer, "Structure of crystal surfaces," Scientific American, March 1965, p. 32.

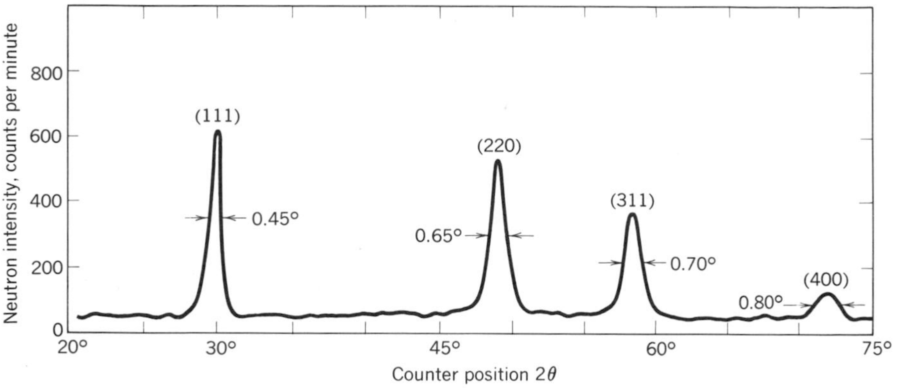

Figure 30 Neutron diffraction pattern for powdered diamond. (After G. Bacon.)

5. ***Volume of Brillouin zone.*** Show that the volume of the first Brillouin zone is $(2\pi)^3/V_c$, where V_c is the volume of a crystal primitive cell. Hint: The volume of a Brillouin zone is equal to the volume of the primitive parallelopiped in Fourier space. Recall the vector identity $(\mathbf{c} \times \mathbf{a}) \times (\mathbf{a} \times \mathbf{b}) = (\mathbf{c} \cdot \mathbf{a} \times \mathbf{b})\mathbf{a}$.

6. ***Width of diffraction maximum.*** We suppose that in a linear crystal there are identical point scattering centers at every lattice point $\boldsymbol{\rho}_m = m\mathbf{a}$, where m is an integer. By analogy with (17) the total scattered radiation amplitude will be proportional to $\mathcal{a} = \Sigma \exp[-im\mathbf{a} \cdot \Delta\mathbf{k}]$. The sum over M lattice points is

has the value
$$\mathcal{a} = \frac{1 - \exp[-iM(\mathbf{a} \cdot \Delta\mathbf{k}]}{1 - \exp[-i(\mathbf{a} \cdot \Delta\mathbf{k})]} ,$$

by the use of the series
$$\sum_{m-0}^{M-1} x^m = \frac{1 - x^M}{1 - x} .$$

(a) The scattered intensity is proportional to $|\mathcal{a}|^2$. Show that
$$|\mathcal{a}|^2 \equiv \mathcal{a}^* \mathcal{a} = \frac{\sin^2 \frac{1}{2} M(\mathbf{a} \cdot \Delta\mathbf{k})}{\sin^2 \frac{1}{2}(\mathbf{a} \cdot \Delta\mathbf{k})} .$$

(b) We know that a diffraction maximum appears when $\mathbf{a} \cdot \Delta\mathbf{k} = 2\pi h$, where h is an integer. We change $\Delta\mathbf{k}$ slightly and define ϵ in $\mathbf{a} \cdot \Delta\mathbf{k} = 2\pi h + \epsilon$ such that ϵ gives the position of the first zero in $\sin \frac{1}{2} M(\mathbf{a} \cdot \Delta\mathbf{k})$. Show that $\epsilon = 2\pi/M$, so that the width of the diffraction maximum is proportional to $1/M$ and can be extremely narrow for macroscopic values of M. The same result holds true for a three-dimensional crystal.

7. ***Neutron cutoff filter.*** For neutrons from a reactor, explain the operation of a cutoff filter that transmits thermal neutrons of long wavelengths. The filter is a pipe filled with a powdered crystal.

8. ***Structure factor of diamond.*** The crystal structure of diamond is described in Chapter 1. If the cell is taken as the conventional cube, the basis consists of eight atoms. (a) Find the structure factor of this basis. (b) Find the zeros of $\mathcal{S}$ and show that the allowed reflections of the diamond structure satisfy $h + k + l = 4n$, where all indices are even and n is any integer; or else all indices are odd (Fig. 30).

9. ***Form factor of atomic hydrogen.*** For the hydrogen atom in its ground state, the number density is $n(r) = (\pi a_0^3)^{-1} \exp(-2r/a_0)$, where a_0 is the Bohr radius. Show that the form factor is $f_G = 16/(4 + G^2a_0^2)^2$.

10. ***Diatomic line.*** Consider a line of atoms $ABAB \ldots AB$, with an A—B bond length of $\frac{1}{2}a$. The form factors are f_A, f_B for atoms A, B, respectively. The incident beam of x-rays is perpendicular to the line of atoms. (a) Show that the interference condition is $n\lambda = a \cos \theta$, where θ is the angle between the diffracted beam and the line of atoms. (b) Show that the intensity of the diffracted beam is proportional to $|f_A - f_B|^2$ for n odd, and to $|f_A + f_B|^2$ for n even. (c) Explain what happens if $f_A = f_B$.

References

X-ray Diffraction

C. S. Barrett and T. B. Massalski, *Structure of metals: crystallographic methods, principles, data,* 3rd ed., McGraw-Hill, 1966. Excellent guide to the practical solution of relatively simple structures.

M. J. Buerger, *Contemporary crystallography,* McGraw-Hill, 1970. A fine introduction to modern crystal structure determination.

N. N. Greenwood and T. C. Gibb, *Mössbauer spectroscopy,* Halsted, New York, 1971.

W. C. Hamilton, "The revolution in crystallography," Science **169,** 133 (1970).

R. W. James, *The optical principles of the diffraction of x-rays,* rev. ed., G. Bell and Sons, London, 1950.

R. W. James, "Dynamical theory of x-ray diffraction," *Solid state physics* **15,** 53–220 (1963).

H. Lipson and W. Cochran, *The determination of crystal structures,* 3rd ed., G. Bell and Sons, London, 1966.

M. M. Woolfson, "Direct methods in crystallography," *Repts. Prog. Phys.* **34,** 369 (1971).

W. H. Zachariasen, *Theory of x-ray diffraction in crystals,* Wiley, 1945. Advanced formal theory.

Neutron Diffraction

G. E. Bacon, *Neutron diffraction,* 2nd ed., Oxford, 1962.

A. Larose and J. Vanderwal, *Scattering of thermal neutrons, a bibliography (1932–1974),* Plenum, 1974.

W. Marshall and S. W. Lovesey, *Theory of thermal neutron scattering,* Oxford, 1971.

Historical

W. H. Bragg and W. L. Bragg, "Structure of the diamond," Proc. Roy. Soc. (London) **A89,** 277 (1913).

W. L. Bragg, "Structure of some crystals as indicated by their diffraction of x-rays," Proc. Roy. Soc. (London) **A89,** 248 (1913). A careful treatment and analysis; the first correct structure determinations are given here.

3 CRYSTAL BINDING

NOTATION: The choice of units is involved in this chapter only in the form of the coulomb interaction, $\pm q^2/r$ in CGS and $\pm q^2/4\pi\epsilon_0 r$ in SI, where ϵ_0 is the permittivity of free space. Where this choice makes a difference we write only the CGS form, so labeled, because the conversion to SI is trivial here.

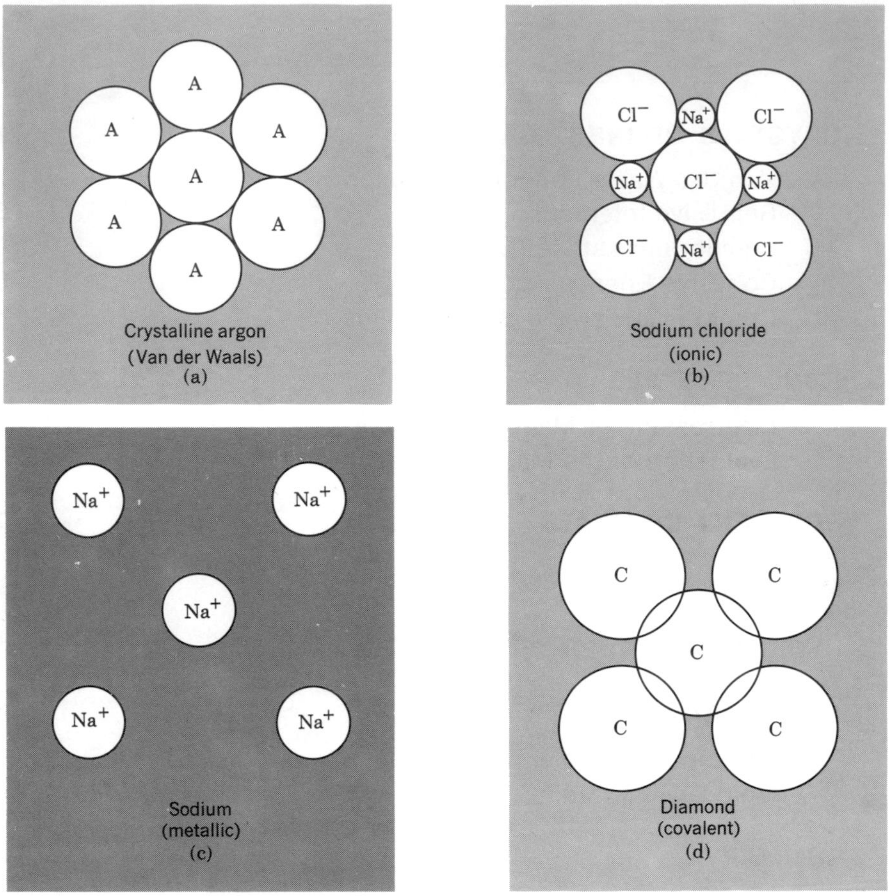

Figure 1 The principal types of crystalline binding. In (a) neutral atoms with closed electron shells are bound together weakly by the van der Waals forces associated with fluctuations in the charge distributions. In (b) electrons are transferred from the alkali atoms to the halogen atoms, and the resulting ions are held together by attractive electrostatic forces between the positive and negative ions. In (c) the valence electrons are taken away from each alkali atom to form a community electron sea in which the positive ions are dispersed. In (d) the neutral atoms appear to be bound together by the overlapping parts of their electron distributions.

3 CRYSTAL BINDING

In this chapter we are concerned with the question: What holds a crystal together? *The attractive electrostatic interaction between the negative charges of the electrons and the positive charges of the nuclei is entirely responsible for the cohesion of solids.* Magnetic forces have only a weak effect on cohesion, and gravitational forces are negligible. Specialized terms categorize distinctive situations: exchange energy, van der Waals forces, resonance stabilization energy, and covalent bonds. The observed differences between the forms of condensed matter are caused by differences in the distribution of the outermost electrons and the ion cores.

To define cohesion, we compare the total energy of the solid, kinetic plus potential, with the energy of the same number of free neutral atoms at infinite separation. A crystal can only be stable if its total energy is lower than the total energy of the atoms or molecules when free. The difference (free atom energy) − (crystal energy) is the **cohesive energy.** Values of the cohesive energy of the crystalline elements are given in Table 1. Notice the wide variation in cohesive energy between different columns of the table. The inert gas crystals are weakly bound, with cohesive energies less than a few percent of the elements in the C, Si, Ge . . . column. The alkali metal crystals have intermediate values of the cohesive energy. The transition element metals (in the middle columns) are quite strongly bound. In Fig. 1 we picture the principal types of crystalline binding.

TABLE 1 Cohesive Energies of the Elements

Energy required to form separated neutral atoms in their ground electronic state from the solid at 0 K at 1 atm. The data were supplied by Prof. Leo Brewer in units kcal per mole, revised to February 20, 1975, after LBL report 3720.

Li 158. 1.63 37.7	**Be** 320. 3.32 76.5											**B** 556. 5.77 133.	**C** 711. 7.37 170.	**N** 474. 4.92 113.4	**O** 251. 2.60 60.03	**F** 81.0 0.84 19.37	**Ne** 1.92 0.020 0.46
Na 107. 1.113 25.67	**Mg** 145. 1.51 34.7						← kJ/mol → ← eV/atom → ← kcal/mol →					**Al** 327. 3.39 78.1	**Si** 446. 4.63 106.7	**P** 331. 3.43 79.16	**S** 275. 2.85 65.75	**Cl** 135. 1.40 32.2	**Ar** 7.74 0.080 1.85
K 90.1 0.934 21.54	**Ca** 178. 1.84 42.5	**Sc** 37.6 3.90 89.9	**Ti** 468. 4.85 111.8	**V** 512. 5.31 122.4	**Cr** 395. 4.10 94.5	**Mn** 282. 2.92 67.4	**Fe** 413. 4.28 98.7	**Co** 424. 4.39 101.3	**Ni** 428. 4.44 102.4	**Cu** 336. 3.49 80.4	**Zn** 130. 1.35 31.07	**Ga** 271. 2.81 64.8	**Ge** 372. 3.85 88.8	**As** 285.3 2.96 68.2	**Se** 217. 2.25 51.8	**Br** 118. 1.22 28.18	**Kr** 11.2 0.116 2.68
Rb 82.2 0.852 19.64	**Sr** 166. 1.72 39.7	**Y** 422. 4.37 100.8	**Zr** 603. 6.25 144.2	**Nb** 730. 7.57 174.5	**Mo** 658. 6.82 173.3	**Tc** 661. 6.85 158.	**Ru** 650. 6.74 155.4	**Rh** 554. 5.75 132.5	**Pd** 376. 3.89 89.8	**Ag** 284. 2.95 68.0	**Cd** 112. 1.16 26.73	**In** 243. 2.52 58.1	**Sn** 303. 3.14 72.4	**Sb** 265. 2.75 63.4	**Te** 215. 2.23 51.4	**I** 107. 1.11 25.62	**Xe** 15.9 0.16 3.80
Cs 77.6 0.804 18.54	**Ba** 183. 1.90 43.7	**La** 431. 4.47 103.1	**Hf** 621. 6.44 148.4	**Ta** 782. 8.10 186.9	**W** 859. 8.90 205.2	**Re** 775. 8.03 185.2	**Os** 788. 8.17 188.4	**Ir** 670. 6.94 160.1	**Pt** 564. 5.84 134.7	**Au** 368. 3.81 87.96	**Hg** 65. 0.67 15.5	**Tl** 182. 1.88 43.4	**Pb** 196. 2.03 46.78	**Bi** 210. 2.18 50.2	**Po** 144. 1.50 34.5	**At**	**Rn** 19.5 0.202 4.66
Fr	**Ra** 160. 1.66 38.2	**Ac** 410. 4.25 98.															

Ce 417. 4.32 99.7	**Pr** 357. 3.70 85.3	**Nd** 328. 3.40 78.5	**Pm**	**Sm** 206. 2.14 49.3	**Eu** 179. 1.86 42.8	**Gd** 400. 4.14 95.5	**Tb** 391. 4.05 93.4	**Dy** 294. 3.04 70.2	**Ho** 302. 3.14 72.3	**Er** 317. 3.29 75.8	**Tm** 233. 2.42 55.8	**Yb** 154. 1.60 37.1	**Lu** 428. 4.43 102.2
Th 598. 6.20 142.9	**Pa**	**U** 536. 5.55 128.	**Np**	**Pu** 347. 3.60 83.0	**Am** 264. 2.73 63.	**Cm** 372. 3.86 89.	**Bk**	**Cf**	**Es**	**Fm**	**Md**	**No**	**Lr**

TABLE 2 Ionization Energies of the Elements

The total energy required to remove the first two electrons is the sum of the first and second ionization potentials. *Source:* National Bureau of Standards Circular 467.

← Energy to remove one electron, in eV →
← Energy to remove two electrons, in eV →

H 13.595																	**He** 24.58 78.98
Li 5.39 81.01	**Be** 9.32 27.53											**B** 8.30 33.45	**C** 11.26 35.64	**N** 14.54 44.14	**O** 13.61 48.76	**F** 17.42 52.40	**Ne** 21.56 62.63
Na 5.14 52.43	**Mg** 7.64 22.67											**Al** 5.98 24.80	**Si** 8.15 24.49	**P** 10.55 30.20	**S** 10.36 34.0	**Cl** 13.01 36.81	**Ar** 15.76 43.38
K 4.34 36.15	**Ca** 6.11 17.98	**Sc** 6.56 19.45	**Ti** 6.83 20.46	**V** 6.74 21.39	**Cr** 6.76 23.25	**Mn** 7.43 23.07	**Fe** 7.90 24.08	**Co** 7.86 24.91	**Ni** 7.63 25.78	**Cu** 7.72 27.93	**Zn** 9.39 27.35	**Ga** 6.00 26.51	**Ge** 7.88 23.81	**As** 9.81 30.0	**Se** 9.75 31.2	**Br** 11.84 33.4	**Kr** 14.00 38.56
Rb 4.18 31.7	**Sr** 5.69 16.72	**Y** 6.5 18.9	**Zr** 6.95 20.98	**Nb** 6.77 21.22	**Mo** 7.18 23.25	**Tc** 7.28 22.54	**Ru** 7.36 24.12	**Rh** 7.46 25.53	**Pd** 8.33 27.75	**Ag** 7.57 29.05	**Cd** 8.99 25.89	**In** 5.78 24.64	**Sn** 7.34 21.97	**Sb** 8.64 25.1	**Te** 9.01 27.6	**I** 10.45 29.54	**Xe** 12.13 33.3
Cs 3.89 29.0	**Ba** 5.21 15.21	**La** 5.61 17.04	**Hf** 7. 22.	**Ta** 7.88 24.1	**W** 7.98 25.7	**Re** 7.87 24.5	**Os** 8.7 26.	**Ir** 9.	**Pt** 8.96 27.52	**Au** 9.22 29.7	**Hg** 10.43 29.18	**Tl** 6.11 26.53	**Pb** 7.41 22.44	**Bi** 7.29 23.97	**Po** 8.43	**At**	**Rn** 10.74
Fr	**Ra** 5.28 15.42	**Ac** 6.9 19.0															

Ce 6.91	**Pr** 5.76	**Nd** 6.31	**Pm**	**Sm** 5.6	**Eu** 5.67	**Gd** 6.16	**Tb** 6.74	**Dy** 6.82	**Ho**	**Er**	**Tm**	**Yb** 6.2	**Lu** 5.0
Th	**Pa**	**U** 4.	**Np**	**Pu**	**Am**	**Cm**	**Bk**	**Cf**	**Es**	**Fm**	**Md**	**No**	**Lr**

CRYSTALS OF INERT GASES

The inert gases form the simplest crystals. The electron distribution is very close to that of the free atoms. Their properties at absolute zero are summarized in Table 2. The crystals are transparent insulators, weakly bound, with low melting temperatures. The atoms have very high ionization energies (see Table 3). The outermost electron shells of the atoms are completely filled, and the distribution of electron charge in the free atom is spherically symmetric. In the crystal the inert gas atoms pack together as closely as possible: the crystal structures (Fig. 2) are all cubic close-packed (fcc), except He^3 and He^4.

Zero-point motion of the atoms (kinetic energy at absolute zero) is a quantum effect that plays a dominant role in He^3 and He^4. They do not solidify at zero pressure even at absolute zero. The average fluctuation at 0 K of a He atom from its equilibrium position is of the order of 30 to 40 percent of the nearest-neighbor distance. The heavier the atom, the less important are the zero-point effects. If we omit zero-point motion, we calculate a molar volume of 9 $cm^3\ mol^{-1}$ for solid helium, as compared with the observed values of 27.5 and 36.8 $cm^3\ mol^{-1}$ for liquid He^4 and liquid He^3, respectively. In the ground state of helium we must take account of the zero-point motion of the atoms.

What holds an inert gas crystal together? The electron distribution in the crystal cannot be significantly distorted from the electron distribution around the free atoms, because the cohesive energy of an atom in the crystal is only 1 percent or less of the ionization energy of an atomic electron. Thus not much energy is available to distort the free atom charge distributions. Part of this distortion gives the van der Waals interaction.

TABLE 3 Properties of Inert Gas Crystals

(Extrapolated to 0 K and Zero Pressure)

	Nearest-Neighbor Distance, in Å	Experimental Cohesive Energy[a]		Melting Point, K	Ionization[b] Potential of Free Atom, eV	Parameters in Lennard-Jones Potential,[c] Eq. 10	
		kJ/mol	eV/atom			ϵ, in 10^{-16} erg	σ, in Å
He	(liquid at zero pressure)				24.58	14	2.56
Ne	3.13	1.88	0.02	24	21.56	50	2.74
Ar	3.76	7.74	0.080	84	15.76	167	3.40
Kr	4.01	11.2	0.116	117	14.00	225	3.65
Xe	4.35	16.0	0.17	161	12.13	320	3.98

[a] See E. R. Dobbs and G. O. Jones, Rpts. Prog. Phys. **20,** 516 (1957); the values for Ar and Kr were furnished by L. Brewer.

[b] See C. E. Moore, *Atomic energy levels,* Circular of the National Bureau of Standards 467 Vol. I, p. XL. These volumes are the canonical source for data about the electronic energy states of free atoms.

[c] N. Bernardes, Phys. Rev. **112,** 1534 (1958).

Figure 2 Cubic close-packed (fcc) crystal structure of the insert gases Ne, Ar, Kr, and Xe. The lattice parameters of the cubic cells are 4.46, 5.31, 5.64, and 6.13 Å, respectively, at 4 K.

Van der Waals-London Interaction

Consider two identical inert gas atoms at a separation R large in comparison with the radii of the atoms. What interactions exist between the two neutral atoms? If the charge distributions on the atoms were rigid, the interaction between atoms would be zero, because the electrostatic potential of a spherical distribution of electronic charge is canceled outside a neutral atom by the electrostatic potential of the charge on the nucleus. Then the inert gas atoms could show no cohesion and could not condense. But the atoms induce dipole moments in each other, and the induced moments cause an attractive interaction between the atoms.

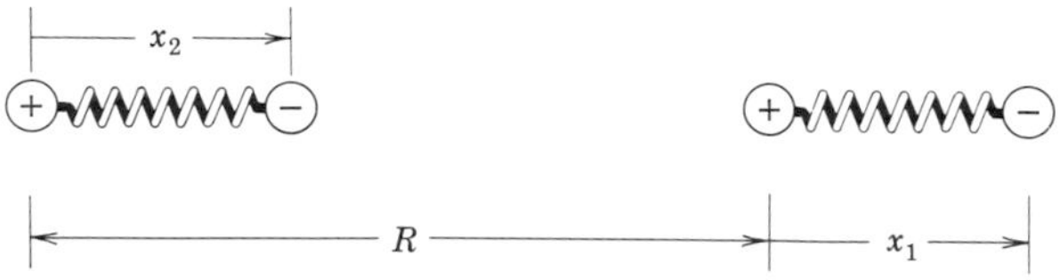

Figure 3 Coordinates of the two oscillators.

As a model we consider two identical linear harmonic oscillators 1 and 2 separated by R. Each oscillator bears charges $\pm e$ with separations x_1 and x_2, as in Fig. 3. The particles oscillate along the x axis. Let p_1 and p_2 denote the momenta. The force constant is C. Then the hamiltonian of the unperturbed system is

$$\mathcal{H}_0 = \frac{1}{2m} p_1^2 + \tfrac{1}{2}Cx_1^2 + \frac{1}{2m} p_2^2 + \tfrac{1}{2}Cx_2^2 \ . \tag{1}$$

The uncoupled oscillators each have the resonance frequency $\omega_0 = (C/m)^{1/2}$ appropriate to a simple harmonic oscillator.

Let $\mathcal{H}_1$ be the coulomb interaction energy of the two oscillators. The geometry is shown in the figure. The internuclear coordinate is R. Then

$$\mathcal{H}_1 = \frac{e^2}{R} + \frac{e^2}{R + x_1 - x_2} - \frac{e^2}{R + x_1} - \frac{e^2}{R - x_2} \ ; \tag{2}$$

in the approximation $|x_1|, |x_2| \ll R$ we expand (2) to obtain in lowest order:

$$\mathcal{H}_1 \cong -\frac{2e^2 x_1 x_2}{R^3} \ . \tag{3}$$

The total hamiltonian with the approximate form (3) for $\mathcal{H}_1$ can be diagonalized by the normal mode transformation

$$x_s \equiv \frac{1}{\sqrt{2}}(x_1 + x_2) \ ; \qquad x_a \equiv \frac{1}{\sqrt{2}}(x_1 - x_2) \ , \tag{4}$$

or, on solving for x_1 and x_2,

$$x_1 = \frac{1}{\sqrt{2}}(x_s + x_a) \ ; \qquad x_2 = \frac{1}{\sqrt{2}}(x_s - x_a) \ . \tag{5}$$

The subscript s and a denote symmetric and antisymmetric modes of motion. Further, we have the momenta p_s, p_a associated with the two modes:

$$p_1 \equiv \frac{1}{\sqrt{2}} (p_s + p_a) \; ; \qquad p_2 \equiv \frac{1}{\sqrt{2}} (p_s - p_a) \; . \tag{6}$$

The total hamiltonian $\mathcal{H}_0 + \mathcal{H}_1$ after the transformations (5) and (6) is

$$\mathcal{H} = \left[\frac{1}{2m} p_s^2 + \frac{1}{2}\left(C - \frac{2e^2}{R^3}\right) x_s^2\right] + \left[\frac{1}{2m} p_a^2 + \frac{1}{2}\left(C + \frac{2e^2}{R^3}\right) x_a^2\right] . \tag{7}$$

The two frequencies of the coupled oscillators are found by inspection of (7) to be

$$\omega = \left[\left(C \pm \frac{2e^2}{R^3}\right)\Big/ m\right]^{\frac{1}{2}} \cong \omega_0 \left[1 \pm \frac{1}{2}\left(\frac{2e^2}{CR^3}\right) - \frac{1}{8}\left(\frac{2e^2}{CR^3}\right)^2 + \cdots\right] , \tag{8}$$

with ω_0 given by $(C/m)^{\frac{1}{2}}$. In (8) we have expanded the square root.

The zero point energy of the system is $\frac{1}{2}\hbar(\omega_s + \omega_a)$; because of the interaction the sum is lowered from $2 \cdot \frac{1}{2}\hbar\omega_0$ by

$$\Delta U = \tfrac{1}{2}\hbar(\Delta\omega_s + \Delta\omega_a) = -\hbar\omega_0 \cdot \frac{1}{8}\left(\frac{2e^2}{CR^3}\right)^2 = -\frac{A}{R^6} \; . \tag{9}$$

This is an attractive interaction which varies as the minus sixth power of the separation of the two oscillators. This is the van der Waals interaction, known also as the London interaction or the induced dipole-dipole interaction. It is the principal attractive interaction in crystals of inert gases and also in crystals of many organic molecules. The interaction is a quantum effect, in the sense that $\Delta U \rightarrow 0$ as $\hbar \rightarrow 0$. Thus the zero point energy of the system is lowered by the dipole-dipole coupling of Eq. (3). The van der Waals interaction does not depend for its existence on any overlap of the change densities of the two atoms. Calculations of the van der Waals interaction between two H atoms are given in the references cited in footnotes 1 and 2.

[1]L. Pauling and E. B. Wilson, *Introduction to quantum mechanics,* McGraw-Hill, 1935, Sec. 47.
[2]H. Margenau, Revs. Mod. Phys. **11,** 1 (1939); see also *Advances in chemical physics* **12:** *Intermolecular forces,* J. O. Hirschfelder, ed., Interscience, 1967, and S. Doniach, Phil. Mag. **8,** 129 (1963).

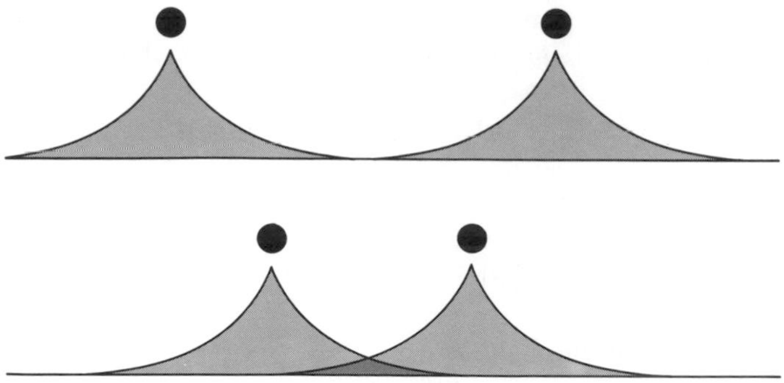

Figure 4 Electronic charge distributions overlap as atoms approach. The solid circles denote the nuclei.

Repulsive Interaction

As the two atoms are brought together their charge distributions gradually overlap (Fig. 4), thereby changing the electrostatic energy of the system. At sufficiently close separations the overlap energy is repulsive, in large part because of the **Pauli exclusion principle.** The elementary statement of the principle is that two electrons cannot have all their quantum numbers equal. When the charge distributions of two atoms overlap there is a tendency for electrons from atom B to occupy in part states of atom A already occupied by electrons of atom A, and vice versa. The Pauli principle prevents multiple occupancy, and electron distributions of atoms with closed shells can overlap only if accompanied by the partial promotion of electrons to unoccupied high energy states of the atoms. Thus the electron overlap increases the total energy of the system and gives a repulsive contribution to the interaction. An extreme example in which the overlap is complete is shown in Fig. 5.

We make no attempt here to evaluate the repulsive interaction[3] from first principles. Experimental data on the inert gases can be fitted well by an empirical repulsive potential of the form B/R^{12}, where B is a positive constant, when used together with a long-range attractive potential of the form of (9). The constants A and B are empirical parameters determined from independent measurements made in the gas phase; the data used include the virial coefficients and the viscosity. It is usual to write the

[3]The overlap energy naturally depends on the radial distribution of charge about each atom. The mathematical calculation is always complicated even if the charge distribution is known. For a clear discussion for two hydrogen atoms, see Chap. 12 of L. Pauling and E. B. Wilson, *Introduction to quantum mechanics*, McGraw-Hill, 1935.

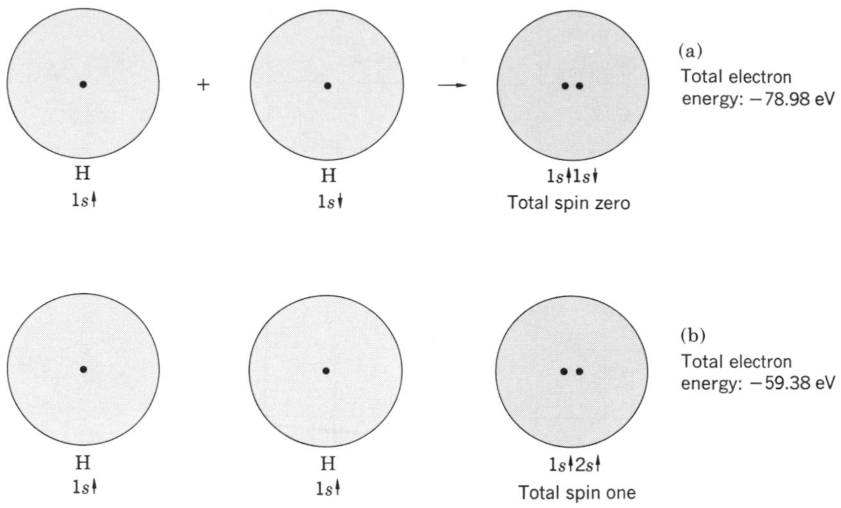

Figure 5 The effect of Pauli principle on the repulsive energy: in an extreme example, two hydrogen atoms are pushed together until the protons are almost in contact. The energy of the electron system alone can be taken from observations on atomic He, which has two electrons. In (a) the electrons have antiparallel spins and the Pauli principle has no effect: the electrons are bound by −78.98 eV. In (b) the spins are parallel: the Pauli principle forces the promotion of an electron from a 1s ↑ orbital of H to a 2s ↑ orbital of He. The electrons now are bound by −59.38 eV, less than (a) by 19.60 eV. This is the amount by which the Pauli principle has increased the repulsion. We have omitted the repulsive coulomb energy of the two protons, which is the same in both (a) and (b).

total potential energy of two atoms at separation R as

$$U(R) = 4\epsilon \left[\left(\frac{\sigma}{R}\right)^{12} - \left(\frac{\sigma}{R}\right)^{6} \right] , \qquad (10)$$

where ϵ and σ are the new parameters, with $4\epsilon\sigma^6 \equiv A$ and $4\epsilon\sigma^{12} \equiv B$. The potential (10) is known as the Lennard-Jones potential, Fig. 6. The force between the two atoms is given by $-dU/dR$. Values of ϵ and σ as recommended by Bernardes are given in Table 2; the values can be obtained from gas-phase data, so that calculations on properties of the solid do not involve any disposable parameters. Other empirical forms for the repulsive interaction are widely used,[4] in particular the exponential form $\lambda \exp(-R/\rho)$, where ρ is a measure of the range of the interaction. This is generally as easy to handle analytically as the inverse power law form.

[4]For discussions of repulsive terms, see M. P. Tosi, *Solid state physics* **16,** 1 (1964); Hirschfelder, Curtis, and Bird, *Molecular theory of gases and liquids,* Wiley, 1954; F. G. Fumi and M. P. Tosi, J. Phys. Chem. Solids **25,** 31, 45 (1964).

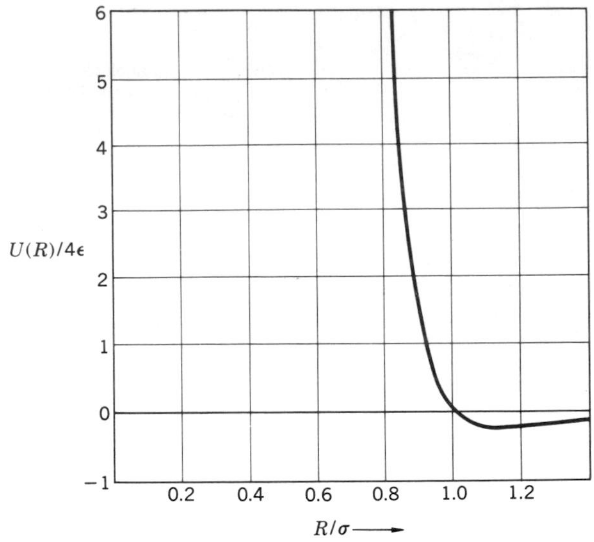

Figure 6 Form of the Lennard-Jones potential (10) which describes the interaction of two inert gas atoms. The minimum occurs at $R/\sigma = 2^{1/6} \cong 1.12$. Notice how steep the curve is inside the minimum, and how flat it is outside the minimum. The value of U at the minimum is $-\epsilon$; and $U = 0$ at $R = \sigma$.

Equilibrium Lattice Constants

If we neglect the kinetic energy of the inert gas atoms, the cohesive energy of an inert gas crystal is given by summing the Lennard-Jones potential (10) over all pairs of atoms in the crystal. If there are N atoms in the crystal, the total potential energy is

$$U_{\text{tot}} = \tfrac{1}{2}N(4\epsilon)\left[\sum_j{}'\left(\frac{\sigma}{p_{ij}R}\right)^{12} - \sum_j{}'\left(\frac{\sigma}{p_{ij}R}\right)^{6}\right] , \tag{11}$$

where $p_{ij}R$ is the distance between reference atom i and any other atom j, expressed in terms of the nearest neighbor distance R. The factor $\frac{1}{2}$ occurs with the N to avoid counting twice each pair of atoms. The summations in (11) have been evaluated,[5] and for the fcc structure

$$\sum_j{}' p_{ij}^{-12} = 12.13188 \ ; \qquad \sum_j{}' p_{ij}^{-6} = 14.45392 \ . \tag{12}$$

There are 12 nearest-neighbor sites in the fcc structure; we see that the series are rapidly converging and have values not far from 12. The nearest neighbors contribute most of the interaction energy of inert gas crystals. The corresponding sums for the hcp structure are 12.13229 and 14.45489.

If we take U_{tot} in (11) as the total energy of the crystal, the equilibrium value R_0 is given by requiring that U_{tot} be a minimum with respect to variations in the nearest neighbor distance R:

$$\frac{dU_{\text{tot}}}{dR} = 0 = -2N\epsilon\left[(12)(12.13)\,\frac{\sigma^{12}}{R^{13}} - (6)(14.45)\,\frac{\sigma^{6}}{R^{7}}\right] , \tag{13}$$

whence $$R_0/\sigma = 1.09 \ , \tag{14}$$

the same for all elements with an fcc structure. The observed values of R_0/σ, using the independently determined values of σ given in Table 3, are:

	Ne	Ar	Kr	Xe
R_0/σ	1.14	1.11	1.10	1.09

The agreement with (14) is remarkable. The slight departure of R_0/σ from the universal value 1.09 predicted for inert gases can be explained by quantum effects.[6] From measurements on the gas phase we have predicted the lattice constant of the crystal.

Cohesive Energy

The cohesive energy of inert gas crystals at absolute zero and at zero pressure is obtained by substituting (12) and (14) in (11):

$$U_{tot}(R) = 2N\epsilon\left[(12.13)\left(\frac{\sigma}{R}\right)^{12} - (14.45)\left(\frac{\sigma}{R}\right)^{6}\right] , \tag{15}$$

and, at $R = R_0$,

$$U_{tot}(R_0) = -(2.15)(4N\epsilon) \ , \tag{16}$$

the same for all inert gases. This is the calculated cohesive energy if the kinetic energy is zero. Bernardes has calculated quantum-mechanical corrections including kinetic energy contributions; the corrections act to reduce the binding by 28, 10, 6, and 4 percent of Eq. (16) for Ne, Ar, Kr, and Xe, respectively.

The heavier the atom, the smaller the quantum correction. We can understand the origin of the quantum correction by consideration of a simple model in which an atom is confined by fixed boundaries. If the particle has the quantum wavelength λ, where λ is determined by the boundaries, then the particle has kinetic energy $p^2/2M = (h/\lambda)^2/2M$ with the de Broglie relation $p = h/\lambda$ for the connection between the momentum and the wavelength of a particle. On this model the quantum zero-point correction to the energy is inversely proportional to the mass, in quite good agreement with the ratio of the corrections cited above for neon (atomic weight 20.2) and xenon (atomic weight 130). The final calculated cohesive energies agree with the experimental values of Table 3 within 1 to 7 percent. One consequence of the quantum kinetic energy is that a crystal of the isotope Ne^{20} is observed to have a larger lattice constant than a crystal of Ne^{22}. The higher quantum kinetic energy of the lighter isotope expands the lattice, because the kinetic energy is reduced by expansion. The observed[7] lattice constants (extrapolated to absolute zero from 2.5 K) are Ne^{20}, 4.4644 Å; Ne^{22}, 4.4559 Å.

[5]Summations for the exponents −4 to −30 are given on p. 1040 of ref. 4.
[6]N. Bernardes, Phys. Rev. **112,** 1534 (1958); see also L. H. Nosanow and G. L. Shaw, Phys. Rev. **128,** 546 (1962); T. R. Koehler, Phys. Rev. Letters **17,** 89 (1966).
[7]D. N. Batchelder, D. L. Losee, and R. O. Simmons, Phys. Rev. **173,** 873 (1968).

Compressibility and Bulk Modulus

An independent test of the theory is provided by the value of the **bulk modulus,** defined as

$$B = -V \frac{dp}{dV} , \tag{17}$$

where V is the volume and p the pressure. The compressibility is defined as the reciprocal of the bulk modulus. At absolute zero the entropy is constant, so that by the thermodynamic identity $dU = -p\, dV$ is the change in energy accompanying a change dV in volume. Thus $dp/dV = -d^2U/dV^2$, and

$$B = V \frac{d^2U}{dV^2} . \tag{18}$$

The bulk modulus is a measure of the stiffness of the crystal or of the energy required to produce a given deformation. The higher the bulk modulus, the stiffer is the crystal. Experimental values for the elements are given in Table 4.

The volume occupied by N atoms in an fcc lattice of lattice constant a is $V = \frac{1}{4}Na^3$, because $\frac{1}{4}a^3$ is the volume per atom. In terms of the nearest-neighbor distance $R = a/\sqrt{2}$, we have $V = NR^3/\sqrt{2}$. The potential energy (15) may be written as

$$U_{tot}(V) = \frac{b_{12}}{V^4} - \frac{b_6}{V^2} , \tag{19}$$

where the parameters b_{12} and b_6 follow from (11) and (12):

$$b_{12} \equiv \tfrac{1}{2}(12.13)N^5\epsilon\sigma^{12} \; ; \qquad b_6 \equiv (14.45)N^3\epsilon\sigma^6 \; .$$

At equilibrium under zero pressure

$$\frac{dU_{tot}}{dV} = 0 = -\frac{4b_{12}}{V^5} + \frac{2b_6}{V^3} , \tag{20}$$

whence the equilibrium volume is given by $V_0 = (2b_{12}/b_6)^{\frac{1}{2}}$. The bulk modulus is

$$B = \left(V \frac{d^2U}{dV^2}\right)_{V_0} = \frac{20b_{12}}{V_0^{\,5}} - \frac{6b_6}{V_0^{\,3}} = \sqrt{2}\,\frac{b_6^{\,5/2}}{b_{12}^{\,3/2}} , \tag{21}$$

of the order of magnitude ϵ/σ^3. With the empirical two-atom potential (10) inferred from gas-phase data we can give an excellent account of the observed properties of the inert gas crystals Ne, Ar, Kr, and Xe. Quantum corrections improve the agreement further.

TABLE 4 Isothermal Bulk Moduli and Compressibilities of the Elements at Room Temperature

After K. Gschneidner, Jr., *Solid state physics* **16**, 275–426 (1964); several data are from F. Birch, in *Handbook of physical constants*, Geological Society of America Memoir **97**, 107–173 (1966.) Original references should be consulted when values are needed for research purposes. Values in parentheses are estimates. Letters in parentheses refer to the crystal form. Letters in brackets refer to the temperature:

[a] = 77 K; [b] = 273 K; [c] = 1 K; [d] = 4 K; [e] = 81 K.

Bulk modulus in units 10^{12} dyn/cm² or 10^{11} N/m²
Compressibility in units 10^{-12} cm²/dyn or 10^{-11} m²/N

H [d] 0.002 500																	**He** [d] 0.00 1168
Li 0.116 8.62	**Be** 1.003 0.997											**B** 1.78 0.562	**C** (d) 5.45 0.183	**N** [e] 0.012 80	**O**	**F**	**Ne** [d] 0.010 100
Na 0.068 14.7	**Mg** 0.354 2.82											**Al** 0.722 1.385	**Si** 0.988 1.012	**P** (b) 0.304 3.29	**S** (r) 0.178 5.62	**Cl**	**Ar** [a] 0.016 93.8
K 0.032 31.	**Ca** 0.152 6.58	**Sc** 0.435 2.30	**Ti** 1.051 0.951	**V** 1.619 0.618	**Cr** 1.901 0.526	**Mn** 0.596 1.68	**Fe** 1.683 0.594	**Co** 1.914 0.522	**Ni** 1.86 0.538	**Cu** 1.37 0.73	**Zn** 0.598 1.67	**Ga** [b] 0.569 1.76	**Ge** 0.772 1.29	**As** 0.394 2.54	**Se** 0.091 11.0	**Br**	**Kr** [a] 0.018 56
Rb 0.031 32.	**Sr** 0.116 8.62	**Y** 0.366 2.73	**Zr** 0.833 1.20	**Nb** 1.702 0.587	**Mo** 2.725 0.366	**Tc** (2.97) (0.34)	**Ru** 3.208 0.311	**Rh** 2.704 0.369	**Pd** 1.808 0.553	**Ag** 1.007 0.993	**Cd** 0.467 2.14	**In** 0.411 2.43	**Sn** (g) 1.11 0.901	**Sb** 0.383 2.61	**Te** 0.230 4.35	**I**	**Xe**
Cs 0.020 50.	**Ba** 0.103 9.97	**La** 0.243 4.12	**Hf** 1.09 0.92	**Ta** 2.00 0.50	**W** 3.232 0.309	**Re** 3.72 0.269	**Os** (4.18) (0.24)	**Ir** 3.55 0.282	**Pt** 2.783 0.359	**Au** 1.732 0.577	**Hg** [c] 0.382 2.60	**Tl** 0.359 2.79	**Pb** 0.430 2.33	**Bi** 0.315 3.17	**Po** (0.26) (3.8)	**At**	**Rn**
Fr (0.020) (50.)	**Ra** (0.132) (7.6)	**Ac** (0.25) (4.)															

Ce (γ) 0.239 4.18	**Pr** 0.306 3.27	**Nd** 0.327 3.06	**Pm** (0.35) (2.85)	**Sm** 0.294 3.40	**Eu** 0.147 6.80	**Gd** 0.383 2.61	**Tb** 0.399 2.51	**Dy** 0.384 2.60	**Ho** 0.397 2.52	**Er** 0.411 2.43	**Tm** 0.397 2.52	**Yb** 0.133 7.52	**Lu** 0.411 2.43
Th 0.543 1.84	**Pa** (0.76) (1.3)	**U** 0.987 1.01	**Np** (0.68) (1.5)	**Pu** 0.54 1.9	**Am**	**Cm**	**Bk**	**Cf**	**Es**	**Fm**	**Md**	**No**	**Lr**

IONIC CRYSTALS

Ionic crystals are made up of positive and negative ions. The **ionic bond** results from the electrostatic interaction of oppositely charged ions. Two common crystal structures found for ionic crystals, the sodium chloride and the cesium chloride structures, were shown in Chapter 1.

The electronic configurations of all ions of a simple ionic crystal correspond to closed electronic shells, as in the inert gas atoms. In lithium fluoride the configuration of the neutral atoms are, according to the periodic table in the front endpapers of this book, Li: $1s^2 2s$, F: $1s^2 2s^2 2p^5$. The singly charged ions have the configurations Li^+: $1s^2$, F^-: $1s^2 2s^2 2p^6$, as for helium and neon, respectively. Inert gas atoms have closed shells, and the charge distributions are spherically symmetric. We expect that the charge distributions on each ion in an ionic crystal will have approximately spherical symmetry, with some distortion near the region of contact with neighboring atoms. This picture is confirmed by x-ray studies of electron distributions (Fig. 7).

A quick estimate suggests that we are not misguided in looking to electrostatic interactions for a large part of the binding energy of an ionic crystal. The distance between a positive ion and the nearest negative ion in crystalline sodium chloride is 2.81×10^{-8} cm, and the attractive coulomb part of the potential energy of the two ions by themselves is 5.1 eV. This value may be compared (Fig. 8) with the experimental value of 7.9 eV per molecule unit for the cohesive energy of crystalline NaCl with respect to separated Na^+ and Cl^- ions. We now calculate the lattice energy more closely.

Electrostatic or Madelung Energy

The long range interaction between ions with charge $\pm q$ is the electrostatic interaction $\pm q^2/r$ attractive between ions of opposite charge and repulsive between ions of the same charge. The ions arrange themselves in whatever crystal structure gives the strongest attractive interaction compatible with the repulsive interaction at short distances between ion cores. The repulsive interactions between ions with inert gas configurations are similar to those between inert gas atoms. The van der Waals part of the attractive interaction in ionic crystals makes a relatively small contribution to the cohesive energy in ionic crystals, of the order of 1 or 2 percent. The main contribution to the binding energy of ionic crystals is electrostatic and is called the **Madelung energy.**

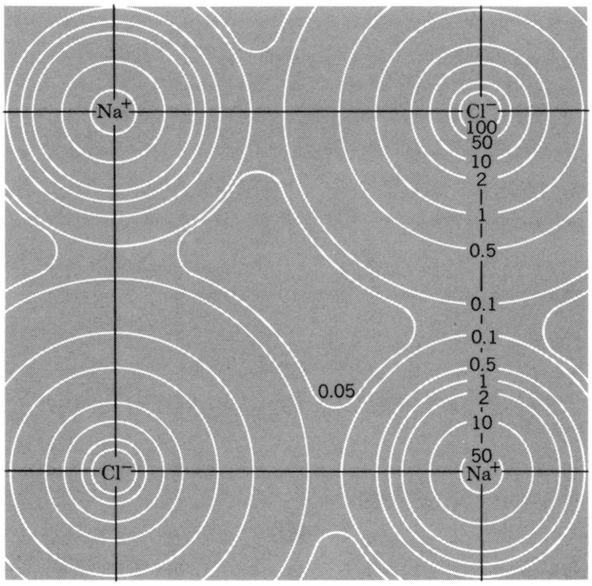

Figure 7 Electron density distribution in the base plane of NaCl, after x-ray studies by G. Schoknecht, Z. Naturforschg. **12a,** 983 (1957).

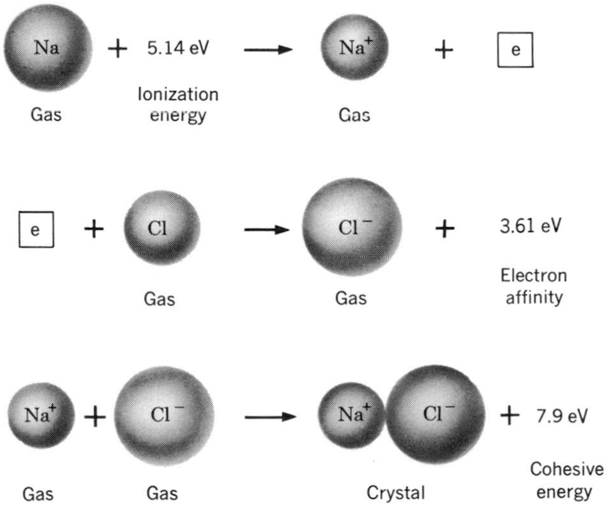

Figure 8 The energy per molecule unit of a crystal of sodium chloride is $(7.9 - 5.1 + 3.6) = 6.4$ eV lower than the energy of separated neutral atoms. The cohesive energy with respect to separated ions is 7.9 eV per molecule unit. All values on the figure are experimental. Values of the ionization energy are given in Table 3, and values of the electron affinity are given in Table 5.

If U_{ij} is the interaction energy between ions i and j, we define a sum U_i which includes all interactions involving the ion i:

$$U_i = \sum_j{}' U_{ij} \ , \tag{22}$$

where the summation includes all ions except $j = i$. We suppose that U_{ij} may be written as the sum of a central field repulsive potential of the form $\lambda \exp(-r/\rho)$, where λ and ρ are empirical parameters, and a coulomb potential $\pm q^2/r$. Thus

(CGS) $$U_{ij} = \lambda \exp(-r_{ij}/\rho) \pm q^2/r_{ij} \ , \tag{23}$$

where the + sign is taken for the like charges and the − sign for unlike charges. The repulsive term describes the fact that each ion resists overlap with the electron distributions of neighboring ions. We treat the strength λ and range ρ as constants to be determined from observed values of the lattice constant and compressibility; we have used the exponential form of the empirical repulsive potential rather than the R^{-12} form used for the inert gases. The change is made because it may give a better representation of the repulsive interaction. For the ions, we do not have gas-phase data available to permit the independent determination of λ and ρ. We note that ρ is a measure of the range of the repulsive interaction: when $r = \rho$, the repulsive interaction is reduced to e^{-1} of the value at $r = 0$. In SI units the coulomb interaction is $\pm q^2/4\pi\epsilon_0 r$; we write this section in CGS units in which the coulomb interaction is $\pm q^2/r$.

In the NaCl structure (Fig. 1.20) the value of U_i does not depend on whether the reference ion i is a positive or a negative ion. The sum (22) can be arranged to converge rapidly, so that its value will not depend on the site of the reference ion in the crystal, as long as it is not near the surface. Neglecting surface effects, we may write the total lattice energy U_{tot} of a crystal composed of N molecules or $2N$ ions as $U_{tot} = NU_i$. Here N, rather than $2N$, occurs because we must count each *pair* of interactions only once or each bond only once. The total lattice energy is the energy required to separate the crystal into individual ions at an infinite distance apart.

It is convenient again to introduce quantities p_{ij} such that $r_{ij} \equiv p_{ij}R$, where R is the nearest-neighbor separation in the crystal. If we include the repulsive interaction only among nearest neighbors, we have

(CGS) $$U_{ij} = \begin{cases} \lambda \exp(-R/\rho) - \dfrac{q^2}{R} & \text{(nearest neighbors)} \\ \pm \dfrac{1}{p_{ij}} \dfrac{q^2}{R} & \text{(otherwise).} \end{cases} \tag{24}$$

Thus

(CGS) $$U_{tot} = NU_i = N \left(z\lambda e^{-R/\rho} - \frac{\alpha q^2}{R} \right) \ . \tag{25}$$

TABLE 5 Electron Affinities of Negative Ions (Recommended Values)

The electron affinity is positive for a stable negative ion. Source: H. Hotop and W. C. Lineberger, J. Phys. Chem. Ref. Data **4**, 539 (1975).

Atom	Electron Affinity Energy, in eV
H	0.7542
Li	0.62
C	1.27
O	1.46
F	3.40
Na	0.55
Al	0.46
Si	1.39
P	0.74
S	2.08
Cl	3.61
Br	3.36
I	3.06
K	0.50

where z is the number of nearest neighbors of any ion and

$$\boxed{\alpha \equiv \sum_j \frac{(\pm)}{p_{ij}} \equiv \textbf{Madelung constant}\ .} \tag{26}$$

The sum should include the nearest-neighbor contribution, which is just z. The ($\pm$) sign is discussed just before (30). The value of the Madelung constant is of central importance in the theory of an ionic crystal. Methods for its calculation are discussed below.

At the equilibrium separation $dU_{tot}/dR = 0$, so that

(CGS) $$N\frac{dU_i}{dR} = -\frac{Nz\lambda}{\rho}\exp(-R/\rho) + \frac{N\alpha q^2}{R^2} = 0\ , \tag{27}$$

or

(CGS) $$R_0^2 \exp(-R_0/\rho) = \rho\alpha q^2/z\lambda\ . \tag{28}$$

This determines the equilibrium separation R_0 if the parameters ρ, λ of the repulsive interaction are known. For SI, replace q^2 by $q^2/4\pi\epsilon_0$.

The total lattice energy of the crystal of $2N$ ions at their equilibrium separation R_0 may be written, using (25) and (28), as

(CGS) $$U_{tot} = -\frac{N\alpha q^2}{R_0}\left(1 - \frac{\rho}{R_0}\right)\ . \tag{29}$$

The term $-N\alpha q^2/R_0$ is the Madelung energy. We shall find that ρ is of the order of $0.1R_0$, so that the repulsive interaction has a very short range.

Evaluation of the Madelung Constant[8]

The first calculation of the coulomb energy constant α was made by Madelung.[9] A powerful general method for lattice sum calculations was developed by Ewald,[10] and Evjen and Frank[11] have given simple methods that arrange the counting in rapidly convergent ways. Electronic computers are widely used.

The definition of the Madelung constant α is

$$\alpha = \sum_j{}' \frac{(\pm)}{p_{ij}} .$$

If we take the reference ion as a negative charge the plus sign will be used for positive ions and the minus sign for negative ions. An equivalent definition is

$$\frac{\alpha}{R} = \sum_j{}' \frac{(\pm)}{r_j} , \tag{30}$$

where r_j is the distance of the jth ion from the reference ion and R is the nearest-neighbor distance. It must be emphasized that the value given for α will depend on whether it is defined in terms of the nearest-neighbor distance R or in terms of the lattice parameter a or in terms of some other relevant length.

As an example, we compute the Madelung constant for the infinite line of ions of alternating sign in Fig. 9. Pick a negative ion as reference ion, and let R denote the distance between adjacent ions. Then

$$\frac{\alpha}{R} = 2\left[\frac{1}{R} - \frac{1}{2R} + \frac{1}{3R} - \frac{1}{4R} + \cdots\right] ,$$

or

$$\alpha = 2\left[1 - \frac{1}{2} + \frac{1}{3} - \frac{1}{4} + \cdots\right] ;$$

the factor 2 occurs because there are two ions, one to the right and one to the left, at equal distances r_j. We sum this series by the expansion

$$\ln(1 + x) = x - \frac{x^2}{2} + \frac{x^3}{3} - \frac{x^4}{4} + \cdots .$$

Thus the Madelung constant for the one-dimensional chain is $\alpha = 2 \ln 2$. In three dimensions the series presents greater difficulty. It is not possible to write down the successive terms by a casual inspection. More important, the series will not converge unless the successive terms in the series are arranged so that the contributions from the positive and negative terms nearly cancel.

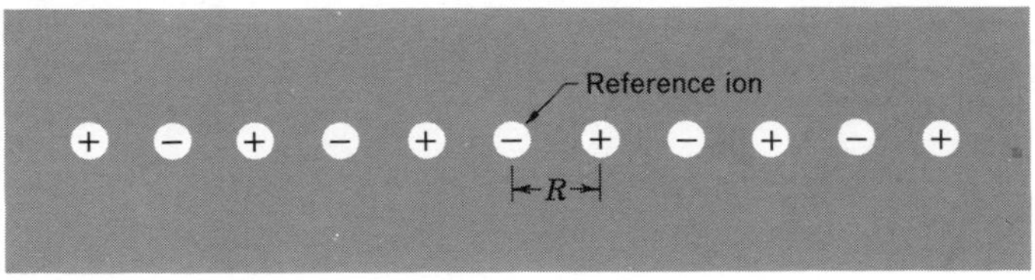

Figure 9 Line of ions of alternating signs, with distance R between ions.

Typical values of the Madelung constant are listed below, based on unit charges and *referred to the nearest-neighbor distance:*

Structure	α
Sodium chloride, NaCl	1.747565
Cesium chloride, CsCl	1.762675
Zinc blende, cubic ZnS	1.6381

The cesium chloride structure is shown in Fig. 1.23. Each ion is at the center of a cube formed by eight ions of the opposite charge. For the same nearest-neighbor distance the cesium chloride structure has a slightly (~ 1 percent) higher Madelung contribution to the cohesive energy than in the sodium chloride structure. But there are more nearest neighbors in CsCl, so that the repulsive energy is higher. Each ion has eight nearest neighbors contributing to the repulsive energy, whereas there are only six in sodium chloride. The repulsive energy is about 10 percent of the total energy of sodium chloride; roughly, we might expect the repulsive energy to be perhaps $\frac{8}{6} \times 10$ percent ≈ 13 percent of the total energy in cesium chloride. This increase outweighs the Madelung energy difference and may favor the sodium chloride structure by a small amount. Many more ionic crystals are known with the sodium chloride structure than with the cesium chloride structure, but the differences in binding energy are small. We can decide which structure will be stable for a particular salt only by a consideration of higher-order contributions to the energy. The Madelung and repulsive contributions to the binding of a KCl crystal are shown in Fig. 10. Properties of alkali halide crystals having the sodium chloride structure are given in Table 6. The calculated values of the cohesive energy are in exceedingly good agreement with the observed values.

[8]A detailed review and bibliography is given by M. P. Tosi, *Solid state physics* **16,** 1 (1964).
[9]E. Madelung, Physik. Z. **19,** 524 (1918).
[10]P. P. Ewald, Ann. Physik **64,** 253 (1921); see also Appendix A to the second edition of this book.
[11]H. M. Evjen, Phys. Rev. **39,** 675 (1932); F. C. Frank, Phil. Mag. **41,** 1287 (1950).

TABLE 6 Properties of Alkali Halide Crystals with the NaCl Structure

All values (except those in brackets) at room temperature and atmospheric pressure, with no correction for changes in R_0 and U from absolute zero. Values in brackets at absolute zero temperature and zero pressure, from private communication by L. Brewer.

	Nearest-Neighbor Separation R_0, in Å	Bulk Modulus B, in 10^{11} dyn/cm²	Repulsive Energy Parameter $z\lambda$, in 10^{-8} erg	Repulsive Range Parameter ρ, in Å	Cohesive Energy Compared to Free Ions, in kcal/mol	
					Experimental	Calculated
LiF	2.014	6.71	0.296	0.291	242.3[246.8]	242.2
LiCl	2.570	2.98	0.490	0.330	198.9[201.8]	192.9
LiBr	2.751	2.38	0.591	0.340	189.8	181.0
LiI	3.000	(1.71)	0.599	0.366	177.7	166.1
NaF	2.317	4.65	0.641	0.290	214.4[217.9]	215.2
NaCl	2.820	2.40	1.05	0.321	182.6[185.3]	178.6
NaBr	2.989	1.99	1.33	0.328	173.6[174.3]	169.2
NaI	3.237	1.51	1.58	0.345	163.2[162.3]	156.6
KF	2.674	3.05	1.31	0.298	189.8[194.5]	189.1
KCl	3.147	1.74	2.05	0.326	165.8[169.5]	161.6
KBr	3.298	1.48	2.30	0.336	158.5[159.3]	154.5
KI	3.533	1.17	2.85	0.348	149.9[151.1]	144.5
RbF	2.815	2.62	1.78	0.301	181.4	180.4
RbCl	3.291	1.56	3.19	0.323	159.3	155.4
RbBr	3.445	1.30	3.03	0.338	152.6	148.3
RbI	3.671	1.06	3.99	0.348	144.9	139.6

Data from various tables by M. P. Tosi, *Solid state physics* **16,** 1 (1964).

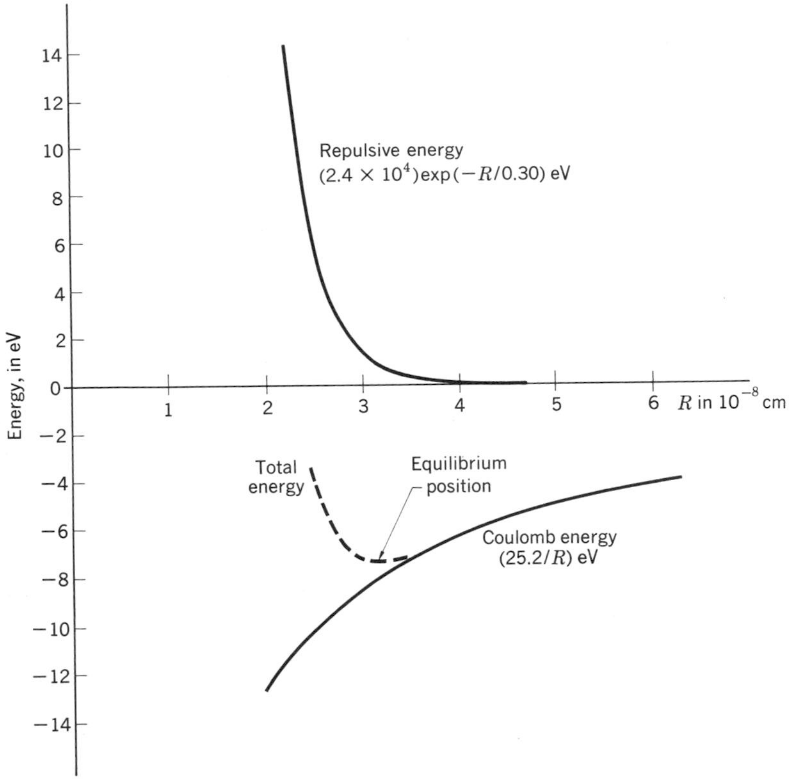

Figure 10 Energy per molecule of KCl crystal, showing Madelung and repulsive contributions.

COVALENT CRYSTALS

The covalent bond is the classical electron pair or homopolar bond of chemistry, particularly of organic chemistry. It is a strong bond: the bond between two carbon atoms in diamond has a cohesive energy of 7.3 eV with respect to separated neutral atoms. This is comparable with the bond strength in ionic crystals, although the covalent bond acts between neutral atoms. The covalent bond has strong directional properties. Thus carbon, silicon, and germanium have the diamond structure, with atoms joined to four nearest neighbors at tetrahedral angles, even though this arrangement gives a low filling of space, 0.34 of the available space, compared with 0.74 for a close-packed structure. The tetrahedral bond allows only four nearest neighbors, whereas a close-packed structure has 12. We should not overemphasize the similarity of the bonding of carbon and silicon. Carbon gives biology, but silicon gives geology.

The covalent bond is usually formed from two electrons, one from each atom participating in the bond. The electrons forming the bond tend to be partly localized in the region between the two atoms joined by the bond. The spins of the two electrons in the bond are antiparallel.

The Pauli principle gives a strong repulsive interaction between atoms with filled shells. If the shells are not filled, electron overlap can be accommodated without excitation of electrons to high energy states and the bond will be shorter. Compare the bond length (2 Å) of Cl_2 with the interatomic distance (3.76 Å) of Ar in solid Ar; also compare the cohesive energies given in Table 1. The difference between Cl_2 and Ar_2 is that the Cl atom has five electrons in the 3p shell and the Ar atom has six, filling the shell, so that the repulsive interaction is stronger in Ar than in Cl.

The elements C, Si, and Ge lack four electrons with respect to filled shells, and thus these elements (for example) can have an attractive interaction associated with charge overlap. The electron configuration of carbon is $1s^2 2s^2 2p^2$. Further details can be found in works on quantum chemistry, where it is shown that to form a tetrahedral system of covalent bonds the carbon atom must first be promoted to the electronic configuration $1s^2 2s 2p^3$. This promotion from the ground state requires 4 eV, an amount more than regained when the bonds are formed. The strength of the covalent bond is indicated in Table 7.

There is a continuous range of crystals between the ionic and the covalent limits. It is often important to estimate the extent a given bond is ionic or covalent. A semiempirical theory of the fractional ionic or covalent character of a bond in a dielectric crystal has been developed with considerable success by J. C. Phillips; some of his conclusions are given in Table 8. We think of NaCl as an ionic crystal and of SiC and GaAs as largely covalent. Atoms with nearly filled shells (Na, Cl) tend to be ionic, whereas atoms in columns III, IV, and V of the periodic table tend to be covalent (In, C, Ge, Si, As).

TABLE 7 Energy Values for Single Covalent Bonds

	Bond Energy			Bond Energy	
Bond	eV	kJ/mol	Bond	eV	kJ/mol
H—H	4.5	435	P—P	2.2	213
C—C	3.6	347	O—O	1.4	138
Si—Si	1.8	176	Te—Te	1.4	138
Ge—Ge	1.6	159	Cl—Cl	2.5	243

After L. Pauling.

TABLE 8 Fractional Ionic Character of Bonds in Binary Crystals

Crystal	Fractional Ionic Character	Crystal	Fractional Ionic Character
Si	0.00		
SiC	0.18	CuCl	0.75
Ge	0.00	CuBr	0.74
ZnO	0.62	AgCl	0.86
ZnS	0.62	AgBr	0.85
ZnSe	0.63	AgI	0.77
ZnTe	0.61		
		MgO	0.84
CdO	0.79	MgS	0.79
CdS	0.69	MgSe	0.79
CdSe	0.70		
CdTe	0.67	LiF	0.92
		NaCl	0.94
InP	0.42	RbF	0.96
InAs	0.36		
InSb	0.32		
GaAs	0.31		
GaSb	0.26		

After J. C. Phillips, *Bonds and bands in semiconductors*, Academic Press, 1973, Chap. 2.

The binding of molecular hydrogen is a simple example of a covalent bond.[12] The strongest binding (Fig. 11) occurs when the spins of the two electrons are antiparallel. The binding depends on the relative spin orientation not because there are strong magnetic dipole forces between the spins, but because the Pauli principle modifies the distribution of charge according to the spin orientation. This spin-dependent coulomb energy is called the **exchange interaction.**

What is the origin of the difference in chemical behavior between He, with a filled $1s^2$ shell and other atoms with filled s^2 outer shells, including Be($2s^2$), Mg($3s^2$), Ca($4s^2$), and Sr($5s^2$)? With respect to the possibility of ionic bonding in the divalent state, we see from Table 2 that the energy required to remove two electrons is much greater in He (79 eV) than in Be (28 eV), Mg (23 eV), Ca (18 eV), or Sr (17 eV). With respect to covalent bonding involving the promotion of one electron in an s orbital to the lowest available p orbital, the energy required is an order of magnitude higher in He (21 eV) than in Be (2.73 eV), Mg (2.71 eV), Ca (1.88 eV), and Sr (1.80 eV). These elements will not act like inert gases or like ionic crystals; for them the real competition is between the metallic state and the covalent state.

[12]L. Pauling and E. B. Wilson, *Introduction to quantum mechanics*, McGraw-Hill, 1935.

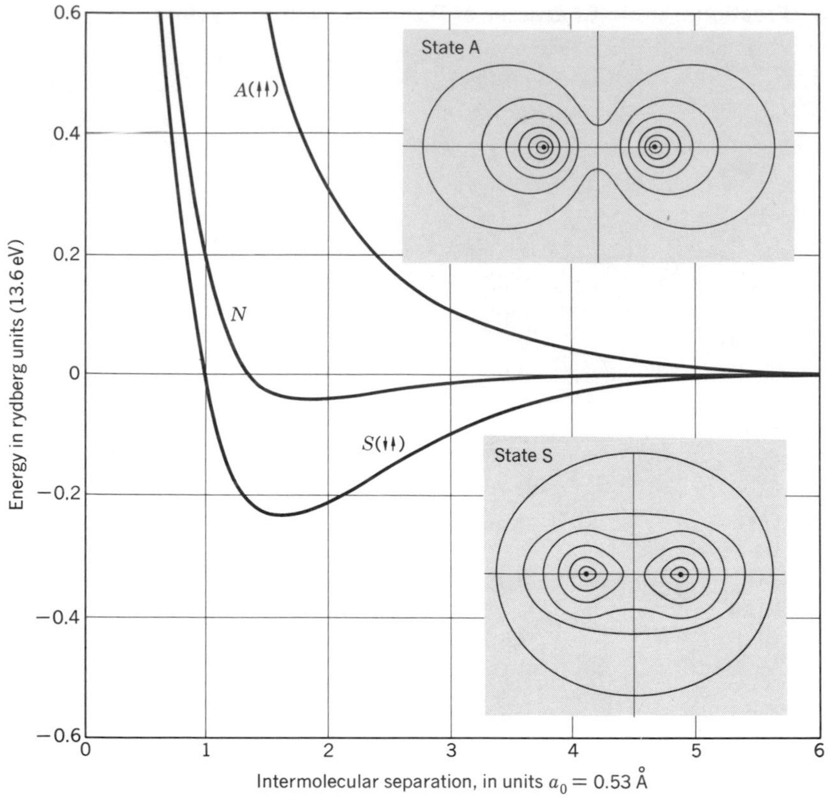

Figure 11 Energy of molecular hydrogen (H_2) referred to separated neutral atoms. A negative energy corresponds to binding. The curve *N* refers to a classical calculation with free atom charge densities; *A* is the result for parallel electron spins, taking the Pauli exclusion principle into account, and *S* (the stable state) for antiparallel spins. The density of charge is represented by contour lines for the states *A* and *S*.

METAL CRYSTALS

Metals are characterized by high electrical conductivity, and a large number of the electrons in a metal must be free to move about, usually one or two per atom. The electrons available to move about are called conduction electrons. In some metals the interaction of the ion cores with the conduction electrons always makes a large contribution to the binding energy, but, as compared with the free atom, the characteristic feature of metallic binding is the lowering of the energy of the valence electrons in the metal. The valence electrons of the atom become the conduction electrons of the metal.

We may think of an alkali metal crystal as an array of positive charges embedded in a nearly uniform sea of negative charge. In the transition metals there may be additional binding from inner electron shells. Transition metals and the metals immediately following them in the periodic table have large *d*-electron shells and are characterized by high binding energy (Table 1). This may be caused in part by covalent bonding and in part by the van der Waals interaction of the cores. In iron and tungsten, for example, the *d*-electrons make a substantial contribution to the binding energy.

The binding energy of an alkali metal crystal is considerably less than that of an alkali halide crystal: the bond formed by a conduction electron is not very strong. The interatomic distances are relatively large in the alkali metals because the kinetic energy of the conduction electrons is lower at large interatomic distances. This leads to weak binding. Metals tend to crystallize in relatively close-packed structures: hcp, fcc, bcc, and some other closely related structures, and not in loosely-packed structures such as diamond.

HYDROGEN-BONDED CRYSTALS

Because neutral hydrogen has only one electron, it should form a covalent bond with only one other atom. It is known, however, that under certain conditions an atom of hydrogen is attracted by rather strong forces to two atoms, thus forming what is called a **hydrogen bond**[13] between them, with a bond energy of the order of 0.1 eV. It is believed that the hydrogen bond is largely ionic in character, being formed only between the most electronegative atoms, particularly F, O, and N. In the extreme ionic form of the hydrogen bond, the hydrogen atom loses its electron to another atom in the molecule; the bare proton forms the hydrogen bond. The small size of the proton permits only two nearest-neighbor atoms, because the atoms adjacent to the proton are so close that more than two of them would get in each other's way; thus the hydrogen bond connects only two atoms (Fig. 12).

The hydrogen bond is an important part of the interaction between H_2O molecules and is responsible together with the electrostatic attraction of the electric dipole moments for the striking physical properties of water and ice. The hydrogen bond restrains protein molecules to their normal geometrical arrangements. It is also responsible for the polymerization of hydrogen fluoride and formic acid, for example. It is important in certain ferroelectric crystals, such as potassium dihydrogen phosphate.

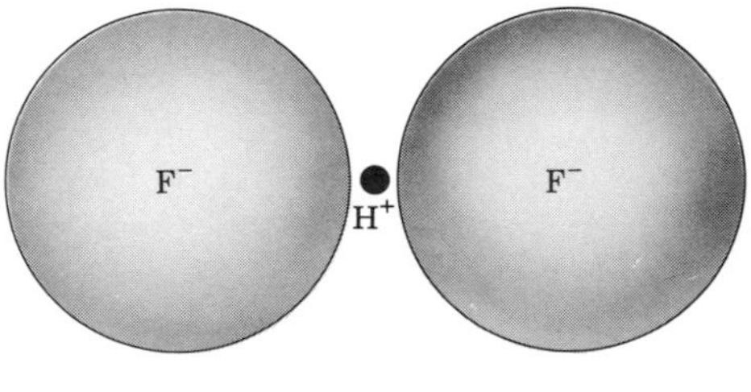

Figure 12 The hydrogen difluoride ion HF_2^- is stabilized by a hydrogen bond. The sketch is of an extreme model of the bond, extreme in the sense that the proton is shown bare of electrons.

[13] M. D. Joesten and L. Schaad, *Hydrogen bonding*, Dekker, 1974.

ATOMIC RADII

Distances between atoms in crystals can be measured very accurately by x-ray diffraction methods, often to 1 part in 10^5. Can we say that the observed distance between atoms or ions may be assigned, so much to atom A, and so much to atom B? Can a definite meaning be assigned to the radius of an atom or ion, irrespective of the nature and composition of the crystal?

Strictly, the answer is no. The charge distribution around an atom is not limited by a rigid spherical boundary. The size of a sodium atom depends on whether it is free, or in a metal, or in an ionic crystal. The radius of a sodium atom in metallic sodium might be taken as 1.86 Å, which is one-half the nearest-neighbor distance of 3.72 Å. Electron diffraction study of gaseous F_2 gives a F-F internuclear distance of 1.44 Å, one-half of which is 0.72 Å. Adding 1.86 Å and 0.72 Å gives 2.58 Å as an estimate of the Na-F bond length. The actual Na-F distance in crystalline sodium fluoride is rather smaller, 2.32 Å, so that taking the mean of the atomic radii is not very accurate here. The values of the ionic radii of Na^+ and F^- given in Table 9 are 1.24 Å and 1.16 Å, respectively. The sum of the ionic radii is 2.40 Å. The close agreement with the observed value of 2.32 Å for the crystal is not surprising: values of ionic radii in tables are usually selected so that their sums represent on the average the internuclear distances in crystals at room temperature.

But used with care and in the proper context, the concept of atomic radius can be useful and fruitful. The interatomic distance between C atoms in diamond is 1.54 Å; one-half of this is 0.77 Å. In silicon, which has the same crystal structure, one-half the interatomic distance is 1.17 Å. Now SiC crystallizes in two forms, in both of which each atom is surrounded by four atoms of the opposite kind. If we add the C and Si radii just given, we predict 1.94 Å for the length of the C-Si bond, in fair agreement with the 1.89 Å observed for the bond length. This is the kind of agreement (a few percent) that we shall find in using tables of atomic radii.[14]

TABLE 9 Average Crystal Radii of the Alkali and Halogen Ions in the NaCl-Type Alkali Halides (Å)

	Li^+	Na^+	K^+	Rb^+	Cs^+	F^-	Cl^-	Br^-	I^-
Huggins–Mayer form[a]	0.94	1.24	1.54	1.68	1.83	1.16	1.62	1.76	1.97

[a] M. P. Tosi and F. G. Fumi, J. Phys. Chem. Solids **25**, 31, 45 (1964).

Tetrahedral Covalent Radii

Pauling has proposed the set of empirical tetrahedral covalent atomic radii included in Table 10 for atoms in crystals with coordination number four, such as the diamond, cubic ZnS, and hexagonal ZnS structures. A large number of observed interatomic distances in appropriate compounds agree closely with the sums of the tetrahedral radii.

Ionic Crystal Radii

In Table 10 we include a set of ionic crystal radii in inert gas configurations, after Zachariasen. The ionic radii are to be used in conjunction with Table 11. Let us consider $BaTiO_3$, with a lattice constant of 4.004 Å at room temperature. Each Ba^{++} ion has 12 nearest O^{--} ions, so that the coordination number is 12 and the correction Δ_{12} of Table 11 applies. If we suppose that the structure is determined by the Ba-O contacts, we have $D_{12} = 1.29 + 1.46 + 0.19 = 2.94$ Å, or $a = 4.16$ Å; if the Ti-O contact determines the structure, we have $D_6 = 0.60 + 1.46 = 2.06$, or $a = 4.12$ Å. The actual lattice constant is somewhat smaller than the estimates and may perhaps suggest that the bonding is not purely ionic, but is partly covalent.

[14]For references on atomic and ionic radii, see L. Pauling, *The nature of the chemical bond*, 3rd ed., Cornell, 1960, Chap. 7, 11, 13; Landolt-Börnstein, *Tabellen* 1:4 (1950), p. 521 et seq; J. C. Slater, J. Chem. Phys. **41**, 3199 (1964); B. J. Austin and V. Heine, J. Chem. Phys. **45**, 928 (1966); R. G. Parsons and V. F. Weisskopf, Zeits. f. Physik **202**, 492 (1967); S. Geller, Z. Kristallographie **125**, 1 (1967); see Table 2 of H. G. F. Winkler, *Struktur und Eigenschaften der Kristalle,* 2nd ed., Springer, 1955. A detailed analysis of ionic radii in oxides and fluorides has been made by R. D. Shannon and C. T. Prewitt, Acta Cryst. **B25**, 925 (1969). A critical revision of the ionic radii in alkali halides has been proposed by M. P. Tosi and F. G. Fumi, Table 9.

TABLE 10 Atomic and Ionic Radii

Values approximate only: see Table 11 for use of the standard radii of ions; for alkali and halogen ions, see Table 9.
Units are 1 Å = 10^{-10} m.

⟵ Standard radii for ions in inert gas (filled shell) configuration ⟶
⟵ Radii of atoms when in tetrahedral covalent bonds ⟶
⟵ Radii of ions in valence state indicated in superscript ⟶

H																	**He**
Li 0.68	**Be** 0.30 1.06											**B** 0.16 0.88	**C** 0.77	**N** 0.70	**O** 1.46 0.66	**F** 1.33 0.64	**Ne** 1.58
Na 0.98	**Mg** 0.65 1.40											**Al** 0.45 1.26	**Si** 0.38 1.17	**P** 1.10	**S** 1.90 1.04	**Cl** 1.81 0.99	**Ar** 1.88
K 1.33	**Ca** 0.94	**Sc** 0.68	**Ti** 0.60 $^{2+}0.90$	**V** $^{2+}0.88$	**Cr** $^{2+}0.84$	**Mn** $^{2+}0.80$	**Fe** $^{2+}0.76$	**Co** $^{2+}0.74$	**Ni** $^{2+}0.72$	**Cu** 1.35 $^{+}0.96$	**Zn** 1.31 $^{2+}0.83$	**Ga** 1.26 $^{3+}0.62$	**Ge** 1.22 $^{4+}0.44$	**As** 1.18 $^{3+}0.69$	**Se** 1.14	**Br** 1.95 1.11	**Kr** 2.00
Rb 1.48	**Sr** 1.10	**Y** 0.88	**Zr** 0.77	**Nb** 0.67	**Mo**	**Tc**	**Ru**	**Rh**	**Pd** $^{2+}0.86$	**Ag** 1.52 $^{+}1.13$	**Cd** 1.48 $^{2+}1.03$	**In** 1.44 $^{3+}0.92$	**Sn** 1.40 $^{4+}0.74$	**Sb** 1.36 $^{3+}0.90$	**Te** 1.32	**I** 2.16 1.28	**Xe** 2.17
Cs 1.67	**Ba** 1.29	**La** 1.04	**Hf**	**Ta**	**W**	**Re**	**Os**	**Ir**	**Pt**	**Au** $^{+}1.37$	**Hg** 1.48 $^{2+}1.12$	**Tl** $^{3+}1.05$	**Pb** $^{4+}0.84$	**Bi**	**Po**	**At**	**Rn**
Fr 1.75	**Ra** 1.37	**Ac** 1.11															

Ce 0.92 $^{3+}1.11$	**Pr**	**Nd** $^{3+}1.08$	**Pm**	**Sm** $^{3+}1.04$	**Eu**	**Gd** $^{3+}1.02$	**Tb**	**Dy** $^{3+}0.99$	**Ho**	**Er** $^{3+}0.96$	**Tm**	**Yb** $^{3+}0.94$	**Lu**
Th 0.99	**Pa** 0.90	**U** 0.83 $^{4+}1.05$	**Np**	**Pu**	**Am**	**Cm**	**Bk**	**Cf**	**Es**	**Fm**	**Md**	**No**	**Lr**

TABLE 11 Use of the Standard Radii of Ions Given in Table 10

The interionic distance D is represented by $D_N = R_C + R_A + \Delta_N$, for ionic crystals, where N is the coordination number of the cation (positive ion), R_C and R_A are the standard radii of the cation and anion, and Δ_N is a correction for coordination number. Room temperature.

N	Δ_N(Å)	N	Δ_N(Å)	N	Δ_N(Å)
1	−0.50	5	−0.05	9	+0.11
2	−0.31	6	0	10	+0.14
3	−0.19	7	+0.04	11	+0.17
4	−0.11	8	+0.08	12	+0.19

SUMMARY

1. Crystals of inert gas atoms are bound by the van der Waals interaction (induced dipole-dipole interaction), and this varies with distance as $1/R^6$.

2. The repulsive interaction between atoms arises generally from the electrostatic repulsion of overlapping charge distributions, and the Pauli principle, which compels overlapping electrons of parallel spin to enter orbitals of higher energy.

3. Ionic crystals are bound by the electrostatic attraction of charged ions of opposite sign. The electrostatic energy of a structure of $2N$ ions of charge $\pm q$ is

$$\text{(CGS)} \qquad U = -N\alpha \frac{q^2}{R} = -N \sum \frac{(\pm) q^2}{r_{ij}} ,$$

where α is the Madelung constant and R is the distance between nearest neighbors.

4. Metals are bound in large measure by the reduction in the kinetic energy of the valence electrons in the metal as compared with the free atom.

5. A covalent bond is characterized by the overlap of charge distributions of antiparallel electron spin. The Pauli contribution to the repulsion is reduced for antiparallel spins, and this makes possible a greater degree of overlap. The overlapping electrons bind their associated ion cores by electrostatic attraction.

PROBLEMS

1. ***Quantum solid.*** In a quantum solid the dominant repulsive energy is the zero-point energy of the atoms. Consider a crude one-dimensional model of crystalline He^4 with each He atom confined to a line segment of length L. In the ground state the wave function within each segment is taken as a half wavelength of a free particle. (a) Find the zero-point kinetic energy per particle. (b) Find an expression for the force needed to keep the line from expanding. (c) In equilibrium the expansive tendency of the kinetic energy is balanced by the van der Waals (vdW) interaction. The vdW energy between nearest neighbors is given very roughly by $U(L) = -1.6\, L^{-6}\, 10^{-60}$ erg, where L is in cm. Find the equilibrium value of L.

2. ***Cohesive energy of bcc and fcc neon.*** Using the Lennard-Jones potential, calculate the ratio of the cohesive energies of neon in the bcc and fcc structures. The lattice sums for the bcc structures are:

$$\sum_j{}' p_{ij}^{-12} = 9.11418 \; ; \qquad \sum_j{}' p_{ij}^{-6} = 12.2533 \; .$$

3. ***Solid molecular hydrogen.*** For H_2 one finds from measurements on the gas that the Lennard-Jones parameters are $\epsilon = 50 \times 10^{-16}$ erg and $\sigma = 2.96$ Å. Find the cohesive energy in kJ per mole of H_2; do the calculation for an fcc structure. Treat each H_2 molecule as a sphere. The observed value of the cohesive energy is 0.751 kJ/mol, much less than we calculated, so that quantum corrections must be very important.

4. ***Possibility of ionic crystals R^+R^-.*** Imagine a crystal that exploits for binding the coulomb attraction of the positive and negative ions of the same atom or molecule R. This is believed to occur with certain organic molecules, but it is not found when R is a single atom. Use the data in Table 2 to evaluate the stability of such a form of Na in the NaCl structure relative to normal metallic sodium. Evaluate the energy at the observed interatomic distance in metallic sodium, and use 0.78 eV as the electron affinity of Na.

5. ***Linear ionic crystal.*** Consider a line of $2N$ ions of alternating charge $\pm q$ with a repulsive potential energy A/R^n between nearest neighbors. (a) Show that at the equilibrium separation

$$\text{(CGS)} \qquad U(R_0) = -\frac{2Nq^2 \ln 2}{R_0}\left(1 - \frac{1}{n}\right) \; .$$

(b) Let the crystal be compressed so that $R_0 \to R_0(1 - \delta)$. Show that the work done in compressing a unit length of the crystal has the leading term $\frac{1}{2}C\delta^2$, where

$$\text{(CGS)} \qquad C = \frac{(n-1)q^2 \ln 2}{R_0} \; .$$

To obtain the results in SI, replace q^2 by $q^2/4\pi\epsilon_0$. Note: We should not expect to obtain this result from the expression for $U(R_0)$, but we must use the complete expression for $U(R)$.

6. ***Cubic ZnS structure.*** Using λ and ρ from Table 6 and the Madelung constants given in the text, calculate the cohesive energy of KCl in the cubic ZnS structure described in Chapter 1. Compare with the value calculated for KCl in the NaCl structure.

7. ***Bulk modulus of LiF.*** From the experimental values of the cohesive energy and the nearest-neighbor distance, calculate the bulk modulus of LiF. Compare with the observed value.

8. ***Divalent ionic crystals.*** Barium oxide has the NaCl structure. Estimate the cohesive energies per molecule of the hypothetical crystals Ba^+O^- and $Ba^{++}O^{--}$ referred to separated neutral atoms. The observed nearest-neighbor internuclear distance is $R_0 = 2.76$ Å; the first and second ionization potentials of Ba are 5.19 and 9.96 eV; and the electron affinities of the first and second electrons added to the neutral oxygen atom are 1.5 and −9.0 eV. The first electron affinity of the neutral oxygen atom is the energy released in the reaction $O + e \rightarrow O^-$.

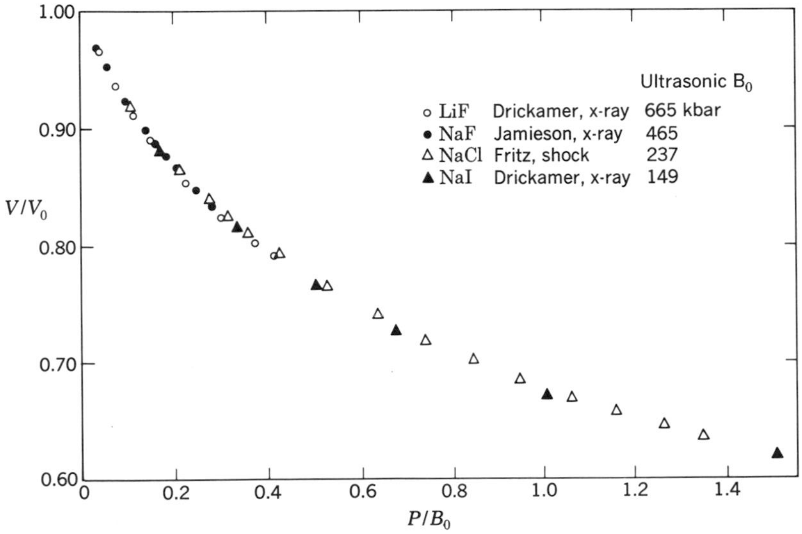

Figure 13 The composite reduced equation of state of LiF, NaF, NaCl, NaI at room temperature. [After C. S. Smith and K. O. McLean, J. Phys. Chem. Solids **34**, 1143 (1973).]

The second electron affinity is the energy released in the reaction $O^- + e \rightarrow O^{--}$. Which valence state do you predict will occur? Assume R_0 is the same for both forms, and neglect the repulsive energy.

9. ***High pressure experiments.*** Pressures up to 10^5 atmospheres (10^{11} dyn cm^{-2}) are commonly employed in static high pressure experiments, and pressures up to 10^7 atmospheres have been used in shock-wave experiments. The bulk modulus B of NaCl is 2.40×10^{11} dyn cm^{-2}. Estimate the fractional change in the nearest neighbor separation at 0.2 Mbar, assuming B to be constant. Experimental results are shown in Fig. 13.

References

M. Born and K. Huang, *Dynamical theory of crystal lattices,* Oxford, 1954. The classic work on ionic crystals.

G. Leibfried, in *Encyclo. of physics* **7**/1, 1955. Superb review of lattice properties.

M. P. Tosi, "Cohesion of ionic solids in the Born model," *Solid state physics* **16,** 1 (1964).

Advances in chemical physics **12:** *Intermolecular forces,* J. O. Hirschfelder, ed., Interscience, 1967.

L. Pauling, *Nature of the chemical bond,* 3rd ed., Cornell, 1960; see Chap. 13, "Sizes of ions and the structure of ionic crystals." Particularly valuable for the discussion of the principles determining the structure of complex ionic crystals, pp. 543–562.

J. Hirschfelder, C. F. Curtis, and R. B. Bird, *Molecular theory of gases and liquids,* Wiley, 1964.

G. K. Horton, "Ideal rare gas crystals," Amer. J. Phy. **36,** 93 (1968). A good review of the lattice properties of crystals of the inert gases.

T. K. Gaylord, "Tensor description of physical properties of crystals," Amer. J. Phys. **43,** 861 (1975).

M. D. Joesten and L. Schaad, *Hydrogen bonding,* Dekker, 1974.

4 PHONONS I. LATTICE VIBRATIONS

Phonons II (Chapter 5) treats the thermal properties of crystals. Chapter 10 treats the interaction of phonons and photons, and the interaction of phonons and electrons.

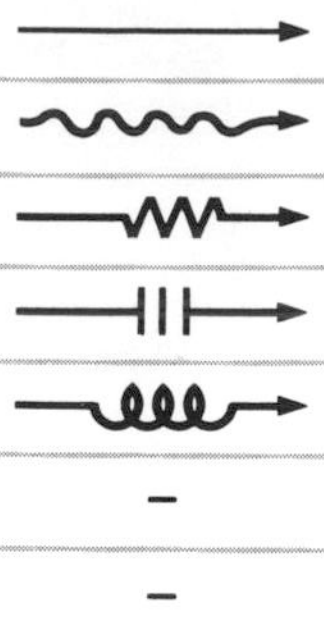

Name	Field
Electron	—
Photon	Electromagnetic wave
Phonon	Elastic wave
Plasmon	Collective electron wave
Magnon	Magnetization wave
Polaron	Electron + elastic deformation
Exciton	Polarization wave

Figure 1 Important elementary excitations in solids. The origins of the concepts and the names of the excitations are discussed by C. T. Walker and G. A. Slack, Am. J. Phys. **38,** 1380 (1970).

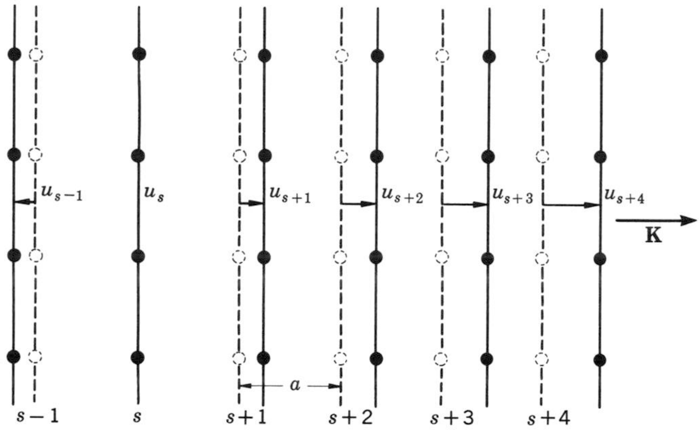

Figure 2 (Dashed lines) planes of atoms when in equilibrium. (Solid lines) planes of atoms when displaced as for a longitudinal wave. The coordinate u measures the displacement of the planes.

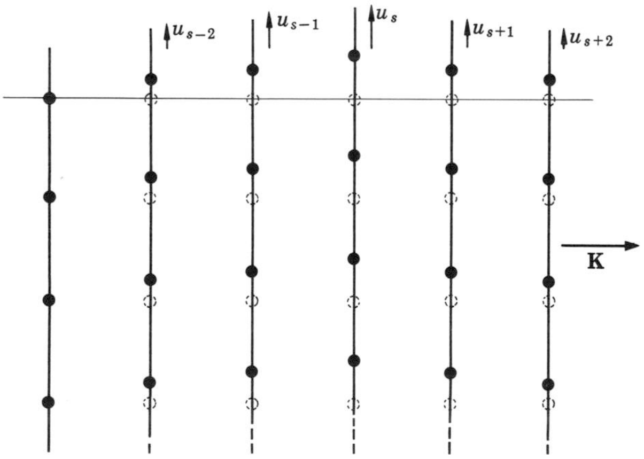

Figure 3 Planes of atoms as displaced during passage of a transverse wave.

4 PHONONS I. LATTICE VIBRATIONS

VIBRATIONS OF MONATOMIC LATTICES

Consider the elastic vibrations of a crystal with one atom in the primitive cell. We want to find the frequency of a lattice wave in terms of the wavevector that describes the wave. The mathematical solution is simplest in the [100], [110], and [111] propagation directions in cubic crystals. When a wave propagates along one of these directions, entire planes of atoms move in phase with displacements either parallel or perpendicular to the direction of the wavevector. We can describe with a single coordinate u_s the displacement of the plane s from its equilibrium position. The problem is then one dimensional. For each wavevector there are three modes, one of longitudinal polarization (Fig. 2) and two of transverse polarization (Fig. 3).

We assume that the force on the plane s caused by the displacement of the plane $s + p$ is proportional to the difference $u_{s+p} - u_s$ of their displacements. The total force on s is

$$F_s = \sum_p C_p(u_{s+p} - u_s) \ . \tag{1}$$

This expression is linear in the displacements and is of the form of Hooke's law. The constant C_p is the force constant between planes removed from each other by p and will differ for longitudinal and transverse waves. It is convenient hereafter to regard C_p as defined for one atom of the plane, so that F_s is the force on one atom in the plane s.

The equation of motion of the plane s is

$$M \frac{d^2u_s}{dt^2} = \sum_p C_p(u_{s+p} - u_s) \ , \tag{2a}$$

where M is the mass of an atom. In the summation p runs over positive and negative integers. We look for solutions with all displacements having the time dependence $e^{-i\omega t}$. Then $d^2u_s/dt^2 = -\omega^2 u_s$, and (2a) becomes

$$-M\omega^2 u_s = \sum_p C_p(u_{s+p} - u_s) \ . \tag{2b}$$

This is a difference equation in the displacements. It has traveling wave solutions:

$$u_{s+p} = ue^{i(s+p)Ka} \ , \tag{3}$$

where a is the spacing between planes and K is the wavevector. The value of the interplanar spacing for a given lattice will depend on the direction of K. Then (2b) reduces to

$$-\omega^2 Mu \ e^{isKa} = \sum_p C_p \ (e^{i(s+p)Ka} - e^{isKa})u \ . \tag{4}$$

We cancel ue^{isKa} from both sides, to leave

$$\omega^2 M = -\sum_p C_p(e^{ipKa} - 1) \ . \tag{5}$$

From the translational symmetry $C_p = C_{-p}$, and we regroup (5) as

$$\omega^2 M = -\sum_{p>0} C_p(e^{ipKa} + e^{-ipKa} - 2) \ . \tag{6}$$

With the identity $2 \cos pKa \equiv e^{ipKa} + e^{-ipKa}$ we have the relation that connects ω with K, called a dispersion relation:

$$\boxed{\omega^2 = \frac{2}{M} \sum_{p>0} C_p(1 - \cos pKa) \ .} \tag{7}$$

The slope of ω versus K is zero at the zone boundary $K = \pm\pi/a$. We have

$$\frac{d\omega^2}{dK} = \frac{2}{M} \sum_{p>0} paC_p \sin pKa = 0 \tag{8}$$

at $K = \pm\pi/a$, for here $\sin pKa = \sin(\pm p\pi) = 0$. There is a special significance to phonon wavevectors that lie on the Brillouin zone boundary; see (12) below.

If there are interactions only among nearest-neighbor planes, then (7) reduces to

$$\omega^2 = (2C_1/M)(1 - \cos Ka) \ . \tag{9a}$$

By a trigonometric identity this may be written as

$$\omega^2 = (4C_1/M) \sin^2 \tfrac{1}{2}Ka \ ; \qquad \omega = (4C_1/M)^{1/2} \left|\sin \tfrac{1}{2}Ka\right| \ . \tag{9b}$$

We plot ω versus K in Fig. 4.

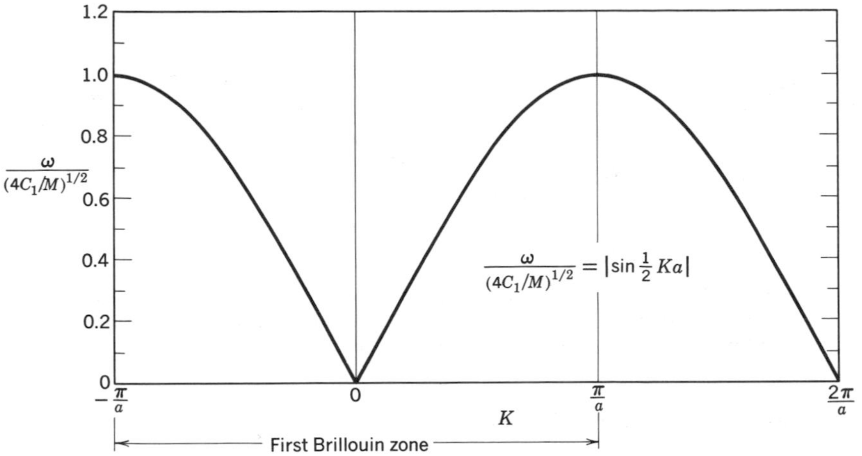

Figure 4 Plot of ω versus K. The region of $K \ll 1/a$ or $\lambda \gg a$ corresponds to the continuum approximation; here ω is directly proportional to K.

First Brillouin Zone

What range of K is physically significant for elastic waves? Only those in the first Brillouin zone. From (3) the ratio of the displacements of two successive planes is given by

$$\frac{u_{s+1}}{u_s} = \frac{u\ e^{i(s+1)Ka}}{u\ e^{isKa}} = e^{iKa} \ . \tag{10}$$

The range $-\pi$ to $+\pi$ for the phase Ka covers all independent values of the exponential. There is absolutely no point in saying that two adjacent atoms are out of phase by more than π: a relative phase of 1.2π is physically identical with a relative phase of -0.8π, and a relative phase of 4.2π is identical with 0.2π. We need both positive and negative values of K because waves can propagate to the right or to the left. The range of independent values is specified by

$$-\pi \leq Ka \leq \pi \ , \qquad \text{or} \qquad -\frac{\pi}{a} \leq K \leq \frac{\pi}{a} \ .$$

This range is the first Brillouin zone of the linear lattice, as defined in Chapter 2. The extreme values are $K_{max} = \pm\pi/a$. There is a real difference from an elastic continuum: in the continuum limit $a \rightarrow 0$ and $K_{max} \rightarrow \pm\infty$. Values of K outside of the first Brillouin zone (Fig. 5) merely reproduce lattice motions described by values within the limits $\pm\pi/a$.

We may treat a value of K outside these limits by subtracting the integral multiple of $2\pi/a$ that will give a wavevector inside these limits. Suppose K lies outside the first zone, but a related wavevector K' defined by $K' \equiv K - 2\pi n/a$ lies within the first zone, where n is an integer. Then the

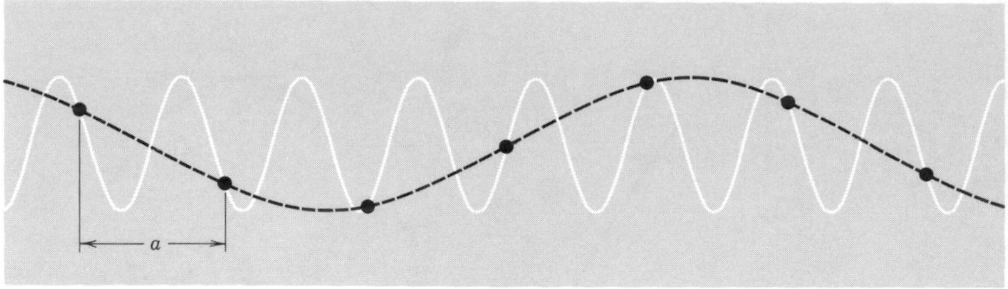

Figure 5 The wave represented by the solid curve conveys no information not given by the dashed curve. Only wavelengths longer than $2a$ are needed to represent the motion. (Courtesy of P. Hansma.)

displacement ratio (10) becomes

$$u_{s+1}/u_s = e^{iKa} \equiv e^{i2\pi n} \, e^{i(Ka-2\pi n)} \equiv e^{iK'a} \ , \tag{11}$$

because $e^{i2\pi n} = 1$. Thus the displacement can always be described by a wavevector within the first zone. We note that $2\pi n/a$ is a reciprocal lattice vector because $2\pi/a$ is a reciprocal lattice vector. By subtraction of an appropriate reciprocal lattice vector from K, we always obtain an equivalent wavevector in the first zone.

At the boundaries $K_{max} = \pm\pi/a$ of the Brillouin zone the solution $u_s = u \, e^{isKa}$ does not represent a traveling wave, but a standing wave. At the zone boundaries $sK_{max}a = \pm s\pi$, whence

$$u_s = u \, e^{\pm is\pi} = u \, (-1)^s \ . \tag{12}$$

This is a standing wave: alternate atoms oscillate in opposite phases, because $u_s = \pm 1$ according to whether s is an even or an odd integer. The wave moves neither to the right nor to the left. This situation is equivalent to Bragg reflection of x-rays: when the Bragg condition is satisfied a traveling wave cannot propagate in a lattice, but through successive reflections back and forth, a standing wave is set up. The critical value $K_{max} = \pm\pi/a$ found here satisfies the Bragg condition $2d \sin\theta = n\lambda$: we have $\theta = \frac{1}{2}\pi$, $d = a$, $K = 2\pi/\lambda$, $n = 1$, so that $\lambda = 2a$. With x-rays it is possible to have n equal to other integers besides unity because the amplitude of the electromagnetic wave has a meaning in the space between atoms, but the displacement amplitude of an elastic wave has a meaning only at the atoms themselves.

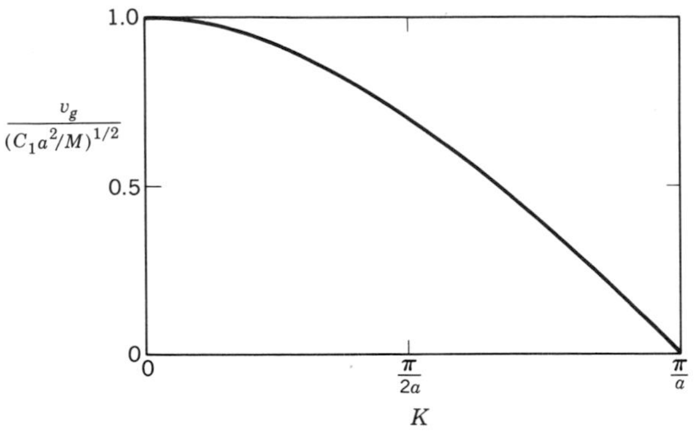

Figure 6 Group velocity v_g versus K, for model of Fig. 4. At the zone boundary the group velocity is zero. The range of K of most laboratory-produced ultrasonic waves is too limited at present to be visible at the left of the graph.

Group Velocity

The velocity of a wave packet is the group velocity, given as[1]

$$v_g = d\omega/dK, \qquad \text{or} \qquad \mathbf{v}_g = \text{grad}_{\mathbf{K}}\,\omega(\mathbf{K}) \ , \tag{13}$$

the gradient with respect to **K.** The group velocity is the velocity of energy transmission in the medium. For the dispersion relation (9) the group velocity (Fig. 6) is

$$v_g = (C_1a^2/M)^{1/2} \cos \tfrac{1}{2}Ka \ . \tag{14}$$

Our general result (8) shows that the group velocity is zero at the edge of the zone. This is what we expect from a standing wave.

Long Wavelength or Continuum Limit

For $pKa \ll 1$ we have $\cos pKa \cong 1 - \frac{1}{2}(pKa)^2$, and the dispersion relation (7) becomes

$$\omega^2 = K^2(a^2/M) \sum_{p>0} p^2C_p \ . \tag{15}$$

The result that the frequency is directly proportional to the wavevector is a statement that the velocity of sound is independent of frequency in the long wavelength limit. Here $\omega = vK$, exactly as in the continuum theory of elastic waves.

[1]A good explanation of group velocity is given by Frank S. Crawford, Jr., *Waves* (Vol. 3 of the Berkeley Physics Course), McGraw-Hill (1968), pp. 268–279.

Derivation of Force Constants from Experiment

In metals the effective forces may be of quite long range. Effects have been found that connect planes of atoms separated by as many as 20 planes.[2] We can make a statement about the range of the forces if we know the dispersion relation for ω. We solve for the C_p by multiplying both sides of (7) by cos rKa, where r is an integer, and integrating over the range of independent values of K:

$$M \int_{-\pi/a}^{\pi/a} dK\, \omega_K^2 \cos rKa = 2 \sum_{p>0} C_p \int_{-\pi/a}^{\pi/a} dK\,(1 - \cos pKa) \cos rKa$$
$$= -2\pi C_r/a \ . \tag{16}$$

The integral vanishes except for $p = r$. Thus

$$C_p = -\frac{Ma}{2\pi} \int_{-\pi/a}^{\pi/a} dK\, \omega_K^2 \cos pKa \ . \tag{17}$$

This result[3] expresses the force constant from the pth plane of atoms in terms of the Fourier cosine transform of ω^2 as a function of K. It holds only for monatomic lattices.

LATTICE WITH TWO ATOMS PER PRIMITIVE CELL

With crystals having more than one atom per primitive cell the vibrational spectrum shows new features. We consider two atoms per primitive cell, as in the NaCl structure or the diamond structure. For each polarization mode in a given propagation direction the dispersion relation ω versus K develops two branches, known as the acoustical and optical branches. We have longitudinal LA and transverse acoustical TA modes, and longitudinal LO and transverse optical TO modes, as in Fig. 7.

If there are p atoms in the primitive cell, there will be $3p$ branches to the dispersion relation: 3 acoustical branches and $3p - 3$ optical branches. Thus germanium (Fig. 8a) and KBr (Fig. 8b), each with two atoms in a primitive cell, have six branches: one LA, one LO, two TA, and two TO. The numerology of the branches follows from the number of degrees of freedom of the atoms. With p atoms in the primitive cell and N primitive cells, there are pN atoms. Each atom has three degrees of freedom, one for each of the x, y, z directions, making a total of $3pN$ degrees of freedom for the crystal. The number of allowed K values in a single branch is just N

[2] J. M. Rowe, B. N. Brockhouse, and E. C. Svensson, Phys. Rev. Letters **14,** 554 (1965).
[3] The result is due to A. J. E. Foreman and W. M. Lomer, Proc. Phys. Soc. (London) **B70,** 1143 (1957). The application to the observed dispersion relations of lead is considered by Brockhouse et al., Phys. Rev. **128,** 1099 (1962). The origin of the long-range forces is discussed by W. A. Harrison, Phys. Rev. **129,** 2512 (1963) and by S. H. Koenig, Phys. Rev. **135,** A1693 (1964).

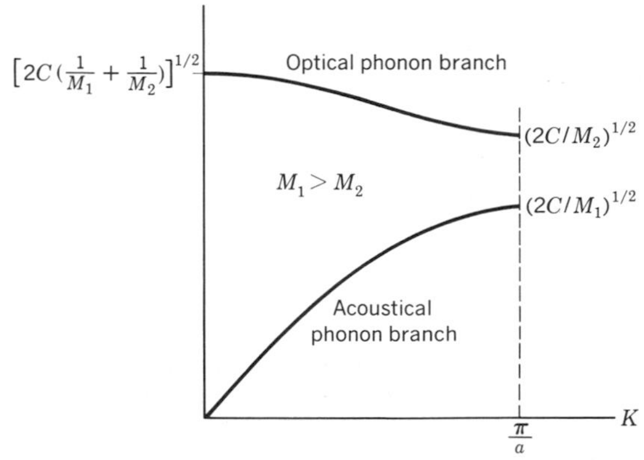

Figure 7 Optical and acoustical branches of the dispersion relation for a diatomic linear lattice, showing the limiting frequencies at $K = 0$ and $K = K_{max} = \pi/a$. The lattice constant is a.

for one Brillouin zone.[4] Thus the LA and the two TA branches have a total of $3N$ modes, thereby accounting for $3N$ of the total degrees of freedom. The remaining $(3p - 3)N$ degrees of freedom are accommodated by the optical branches.

We consider a cubic crystal where atoms of mass M_1 lie on one set of planes and atoms of mass M_2 lie on planes interleaved between those of the first set (Fig. 9). It is not essential that the masses be different, but either the force constants or the masses will be different if the two atoms of the basis are nonequivalent. Let a denote the repeat distance of the lattice in the direction normal to the lattice planes considered. We treat waves that propagate in a symmetry direction for which a single plane contains only a single type of ion; such directions are [111] in the NaCl structure and [100] in the CsCl structure.

[4]We show in Chapter 5 by application of periodic boundary conditions to the modes of the crystal that there is one **K** value in the volume $(2\pi)^3/V$ in Fourier space. The volume of a Brillouin zone is $(2\pi)^3/V_c$, where V_c is the volume of a crystal primitive cell (Problem 2.5). Thus the number of allowed **K** values in a Brillouin zone is V/V_c, which is just N, the number of primitive cells in the crystal.

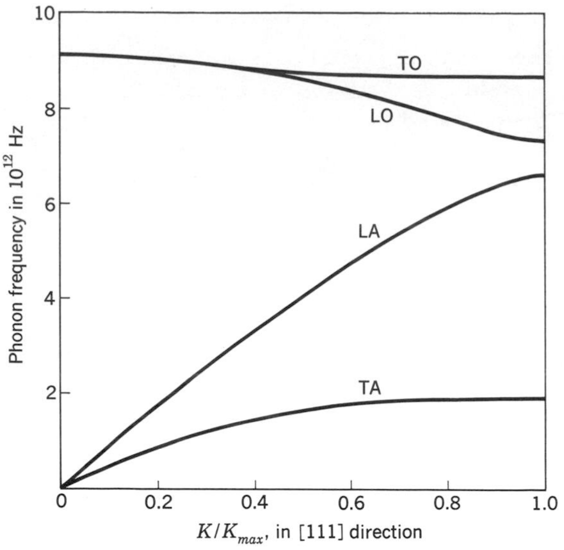

Figure 8a Phonon dispersion relations in the [111] direction in germanium at 80 K. The two TA phonon branches are horizontal at the zone boundary position, $K_{max} = (2\pi/a)(\frac{1}{2}\frac{1}{2}\frac{1}{2})$. The LO and TO branches coincide at $K = 0$; this also is a consequence of the crystal symmetry of Ge. The results were obtained with neutron inelastic scattering by G. Nilsson and G. Nelin, Phys. Rev. **B3**, 364 (1971).

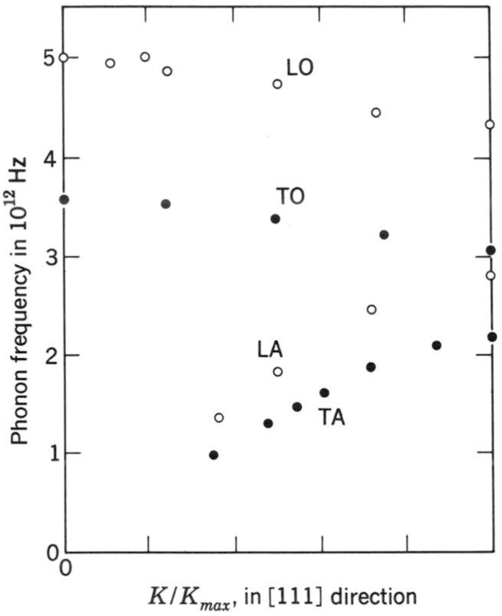

Figure 8b Dispersion curves in the [111] direction in KBr at 90 K, after A. D. B. Woods, B. N. Brockhouse, R. A. Cowley, and W. Cochran, Phys. Rev. **131**, 1025 (1963). The extrapolation to $K = 0$ of the TO, LO branches are called ω_T, ω_L; these are discussed in Chapter 10.

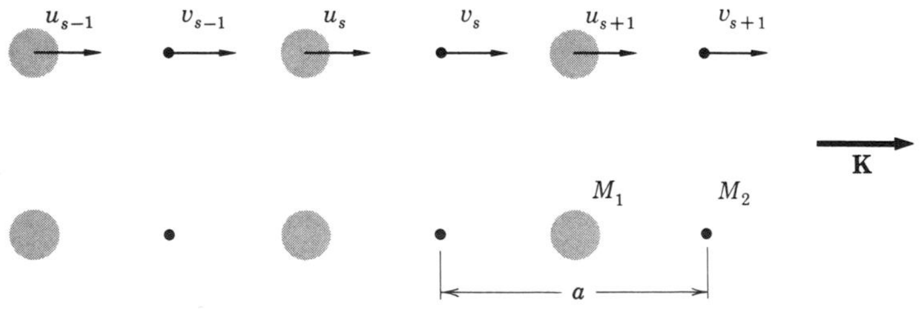

Figure 9 A diatomic crystal structure with masses M_1, M_2 connected by force constant C between adjacent planes. The displacements of atoms M_1 are denoted by u_{s-1}, u_s, u_{s+1}, . . . , and of atoms M_2 by v_{s-1}, v_s, v_{s+1}. The repeat distance is a in the direction of the wavevector K. The atoms are shown in their undisplaced positions.

We write the equations of motion under the assumption that each plane interacts only with its nearest-neighbor planes and that the force constants are identical between all pairs of nearest-neighbor planes. We refer to Fig. 9 to obtain

$$M_1 \frac{d^2u_s}{dt^2} = C(v_s + v_{s-1} - 2u_s) \ ;$$
$$M_2 \frac{d^2v_s}{dt^2} = C(u_{s+1} + u_s - 2v_s) \ . \tag{18}$$

We look for a solution in the form of a traveling wave, now with different amplitudes u, v on alternate planes:

$$u_s = u\, e^{isKa}\, e^{-i\omega t} \ ; \qquad v_s = v\, e^{isKa}\, e^{-i\omega t} \ . \tag{19}$$

On substitution in (18) we have

$$-\omega^2 M_1 u = Cv(1 + e^{-iKa}) - 2Cu \ ;$$
$$-\omega^2 M_2 v = Cu(e^{iKa} + 1) - 2Cv \ . \tag{20}$$

These homogeneous linear equations have a nontrivial solution only if the determinant of the coefficients of u and v vanishes:

$$\begin{vmatrix} 2C - M_1\omega^2 & -C(1 + e^{-iKa}) \\ -C(1 + e^{iKa}) & 2C - M_2\omega^2 \end{vmatrix} = 0 \ , \tag{21}$$

or

$$M_1M_2\omega^4 - 2C(M_1 + M_2)\omega^2 + 2C^2(1 - \cos Ka) = 0 \ . \tag{22}$$

We can solve this equation exactly for ω^2, but it is simpler to examine the limiting cases $Ka \ll 1$ and $Ka = \pm\pi$ at the zone boundary. For small Ka we have $\cos Ka \cong 1 - \frac{1}{2}K^2a^2 + \cdots$, and the two roots are

$$\omega^2 \cong 2C\left(\frac{1}{M_1} + \frac{1}{M_2}\right) \qquad \text{(optical branch)} \ ; \tag{23}$$

$$\omega^2 \cong \frac{2C}{M_1 + M_2} K^2a^2 \qquad \text{(acoustical branch)} \ . \tag{24}$$

The extent of the first Brillouin zone is $-\pi/a \le K \le \pi/a$, where a is the repeat distance of the lattice. At $K_{\rm max} = \pm\pi/a$ the roots are

$$\omega^2 = 2C/M_1 \ ; \qquad \omega^2 = 2C/M_2 \ . \tag{25}$$

The dependence of ω on K is shown in Fig. 7 for $M_1 > M_2$.

The particle displacements in the transverse acoustical (TA) and transverse optical (TO) branches are shown in Fig. 10. For the optical branch at $K = 0$ we find, on substitution of (23) in (20),

$$\frac{u}{v} = -\frac{M_2}{M_1} \ . \tag{26}$$

The atoms vibrate against each other, but their center of mass is fixed. If the two atoms carry opposite charges, we may excite a motion of this type

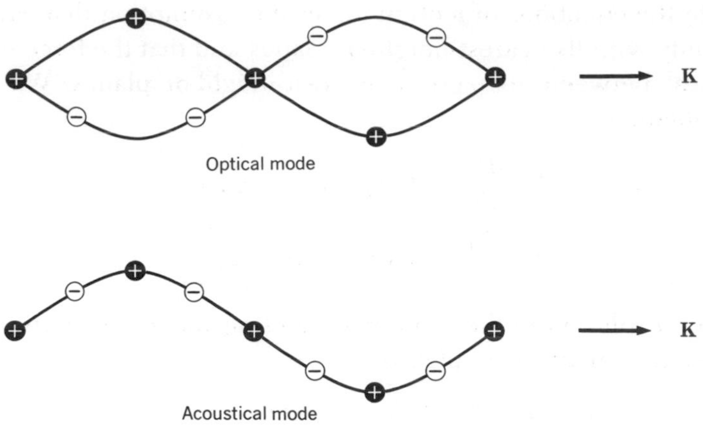

Figure 10 Transverse optical and transverse acoustical waves in a diatomic linear lattice, illustrated by the particle displacements for the two modes at the same wavelength.

with the electric field of a light wave, so that the branch is called the optical branch. At a general K the ratio u/v will be complex, as follows from either of the equations (20).

Figure 7 shows that wavelike solutions do not exist for certain frequencies, here between $(2C/M_1)^{1/2}$ and $(2C/M_2)^{1/2}$. This is a characteristic feature of elastic waves in polyatomic lattices. There is a frequency gap at the boundary $K_{max} = \pm\pi/a$ of the first Brillouin zone. If we look for solutions in the gap with ω real, then the wavevector K will be complex, so that the wave is damped in space. Another solution for the amplitude ratio at small K is $u = v$, obtained as the $K = 0$ limit of (24). The atoms (and their center of mass) move together, as in long wavelength acoustical vibrations, whence the term acoustical branch.

QUANTIZATION OF LATTICE VIBRATIONS

The energy of a lattice vibration is quantized. The quantum of energy is called a **phonon** in analogy with the photon of the electromagnetic wave. Elastic waves in crystals are made up of phonons. Thermal vibrations in crystals are thermally excited phonons, like the thermally excited photons of black-body electromagnetic radiation in a cavity. The energy of an elastic mode of angular frequency ω is

$$\epsilon = (n + \tfrac{1}{2})\hbar\omega \tag{27}$$

when the mode is excited to quantum number n; that is, when the mode is occupied by n phonons. Here $\hbar = h/2\pi = 1.05459 \times 10^{-27}$ erg s. The term $\frac{1}{2}\hbar\omega$ is the zero point energy of the mode. It occurs for both phonons and photons as a consequence of their equivalence to a quantum harmonic oscillator of frequency ω, for which the energy eigenvalues are also $(n + \frac{1}{2})\hbar\omega$.

We can easily find the mean square amplitude of the phonons. Consider the mode $u = u_0 \cos Kx \cos \omega t$, where u is the displacement of a volume element from its equilibrium position in the crystal. It is convenient to work with this real displacement, the superposition of the traveling wave complex modes of wavevectors $\pm K$. The energy of the oscillator is half kinetic and half potential, when averaged over time. The kinetic energy density is $\frac{1}{2}\rho(\partial u/\partial t)^2$, where ρ is the mass density. In a crystal of volume V, the volume integral of the kinetic energy is $\frac{1}{4}\rho V\omega^2 u_0^2 \cos^2 \omega t$. The time average kinetic energy is

$$\tfrac{1}{8}\rho V\omega^2 u_0^2 = \tfrac{1}{2}(n + \tfrac{1}{2})\hbar\omega \ , \tag{28}$$

and the square of the amplitude is

$$u_0^2 = 4(n + \tfrac{1}{2})\hbar/\rho V\omega \ . \tag{29}$$

This relates the displacement in a given mode to the phonon occupancy n of the mode.

What is the sign of ω? The equations of motion such as (2) are equations for ω^2, and if this is positive then ω can have either sign, + or −. But the energy of a phonon must be positive, so it is conventional and suitable to take ω as positive.[5] If the lattice is unstable, or becomes unstable through an unusual temperature dependence of the force constants, then ω^2 will be negative and ω will be imaginary. A mode with ω imaginary is unstable, and the crystal will deform spontaneously to a more stable structure. A mode with ω close to zero, unless it is simply an acoustical mode of small K, is called a soft mode. Phase transitions in crystals often involve soft modes. We discuss these modes in the chapter on ferroelectric crystals.

[5]This convention is for linearly polarized waves.

PHONON MOMENTUM

A phonon of wavevector K will interact with particles such as photons, neutrons, and electrons as if it had a momentum $\hbar K$. We show first, however, that a phonon does not carry physical momentum. The reason that phonons on a lattice do not carry momentum is that a phonon coordinate (except for $K=0$) involves relative coordinates of the atoms. Thus in an H_2 molecule the internuclear vibrational coordinate $\mathbf{r}_1 - \mathbf{r}_2$ is a relative coordinate and does not carry linear momentum; the center of mass coordinate $\frac{1}{2}(\mathbf{r}_1 + \mathbf{r}_2)$ corresponds to the uniform mode and can carry linear momentum. The physical momentum of a crystal is

$$p = M(d/dt)\ \Sigma\ u_s \ ; \tag{30}$$

when the crystal carries a phonon K,

$$p = M(du/dt) \sum_s e^{isKa} = M(du/dt)(1 - e^{iNKa})/(1 - e^{iKa}) \ , \tag{31}$$

for N atoms. We have used the series

$$\sum_{s=0}^{N-1} x^s = (1 - x^N)/(1 - x) \ . \tag{32}$$

In the next chapter we enumerate the discrete values of K compatible with the boundary conditions, and we shall see that $K = \pm 2\pi p/Na$, where p is an integer. Thus $e^{iNKa} = e^{\pm i2\pi p} = 1$, and from (31) it follows that the crystal momentum is zero:

$$p = M(du/dt)\ \Sigma\ e^{isKa} = 0 \ . \tag{33}$$

The only exception is for the uniform mode $K=0$, for which all u_s equal u, so that $p = NM(du/dt)$. This mode represents a uniform translation of the crystal as a whole, and such a translation does carry momentum.

All the same, for most practical purposes a phonon acts as if its momentum were $\hbar\mathbf{K}$, sometimes called the **crystal momentum.** In crystals there exist wavevector selection rules for allowed transitions between quantum states. We saw that the elastic scattering of an x-ray photon by a crystal is governed by the wavevector selection rule

$$\mathbf{k}' = \mathbf{k} + \mathbf{G} \ , \tag{34}$$

where $\mathbf{G}$ is a vector in the reciprocal lattice; $\mathbf{k}$ is the wavevector of the incident photon, and $\mathbf{k}'$ is the wavevector of the scattered photon. In the reflection process the crystal as a whole will recoil with momentum $-\hbar\mathbf{G}$, but this uniform mode momentum is rarely considered explicitly. Equation (34) is an example of the rule that the total wavevector of interacting waves is conserved in a periodic lattice, with the possible addition of a reciprocal lattice vector. The true momentum of the whole system always is rigorously conserved.

If the scattering of the photon is inelastic, with the creation of a phonon of wavevector $\mathbf{K}$, then the wavevector selection rule becomes[6]

$$\mathbf{k}' + \mathbf{K} = \mathbf{k} + \mathbf{G} \ . \tag{35}$$

If a phonon $\mathbf{K}$ is absorbed in the process, we have instead the relation

$$\mathbf{k}' = \mathbf{k} + \mathbf{K} + \mathbf{G} \ . \tag{36}$$

Relations (35) and (36) are the natural extensions of (34).

---

[6]We exhibit the mathematics involved in the wavevector selection rule. Suppose two phonons $\mathbf{K}_1$, $\mathbf{K}_2$ interact through cubic terms in the elastic energy to create a third phonon $\mathbf{K}_3$. The probability of the collision will involve the product of the three phonon wave amplitudes, summed over all lattice sites:

(phonon $\mathbf{K}_1$ in)(phonon $\mathbf{K}_2$ in)(phonon $\mathbf{K}_3$ out)

$$\propto \sum_n \exp(-i\mathbf{K}_1 \cdot \mathbf{r}_n)\exp(-i\mathbf{K}_2 \cdot \mathbf{r}_n)\exp(i\mathbf{K}_3 \cdot \mathbf{r}_n) = \sum_n \exp\,[i(\mathbf{K}_3 - \mathbf{K}_1 - \mathbf{K}_2)\cdot \mathbf{r}_n] \ .$$

This sum in the limit of a large number of lattice sites approaches zero unless $\mathbf{K}_3 = \mathbf{K}_1 + \mathbf{K}_2$ or unless $\mathbf{K}_3 = \mathbf{K}_1 + \mathbf{K}_2 + \mathbf{G}$. If either of these conditions is satisfied, of which the first is merely a special case of the second, the sum is equal to the number of lattice sites N. A similar sum was considered in Problem 2.6.

INELASTIC SCATTERING OF NEUTRONS BY PHONONS

Neutron inelastic scattering is the most common method for the experimental determination of phonon dispersion relations. The method is not applicable when the absorption of neutrons by the nuclei of the crystal is high. It is also possible to obtain phonon lifetimes from the angular width of the scattered neutron beam. A neutron sees the crystal lattice chiefly by interaction with the nuclei of the atoms. The kinematics of the scattering of a neutron beam by a crystal lattice are described by the general wavevector selection rule:

$$\mathbf{k} + \mathbf{G} = \mathbf{k}' \pm \mathbf{K} \ , \tag{37}$$

and by the requirement of conservation of energy. Here **K** is the wavevector of the phonon created (+) or absorbed (−) in the process, and **G** is any reciprocal lattice vector. We choose **G** such that **K** lies in the first Brillouin zone, as it must for a phonon.

The kinetic energy of the incident neutron is $p^2/2M_n$, where M_n is the mass of the neutron. The momentum **p** is given by $\hbar\mathbf{k}$, where **k** is the wavevector of the neutron. Thus $\hbar^2k^2/2M_n$ is the kinetic energy of the incident neutron. If **k**′ is the wavevector of the scattered neutron, the energy of the scattered neutron is $\hbar^2k'^2/2M_n$. The statement of conservation of energy is

$$\frac{\hbar^2k^2}{2M_n} = \frac{\hbar^2k'^2}{2M_n} \pm \hbar\omega \ , \tag{38}$$

where $\hbar\omega$ is the energy of the phonon created (+) or absorbed (−) in the process. To determine the dispersion relation using (37) and (38) it is necessary in the experiment to find the energy gain or loss of the scattered neutrons as a function of the scattering direction $\mathbf{k} - \mathbf{k}'$. Results for germanium and KBr are given in Fig. 8; results for sodium are given in Fig. 11. A spectrometer used for phonon studies is shown in Fig. 12.

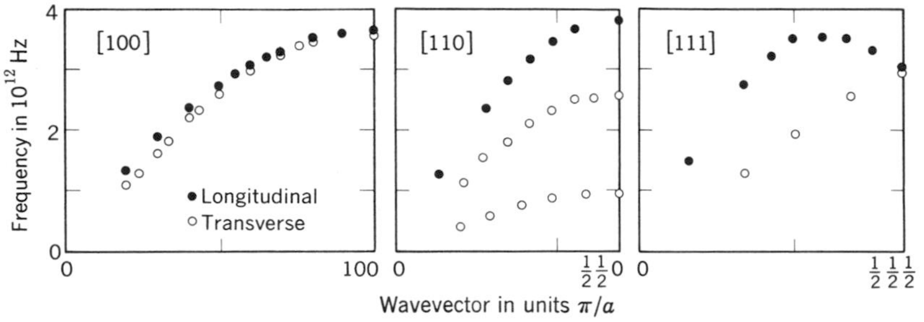

Figure 11 The dispersion curves of sodium for phonons propagating in the [001], [110], and [111] directions at 90 K, as determined by inelastic scattering of neutrons. [Woods, Brockhouse, March and Bowers, Proc. Phys. Soc. London 79, pt. 2, 440 (1962).]

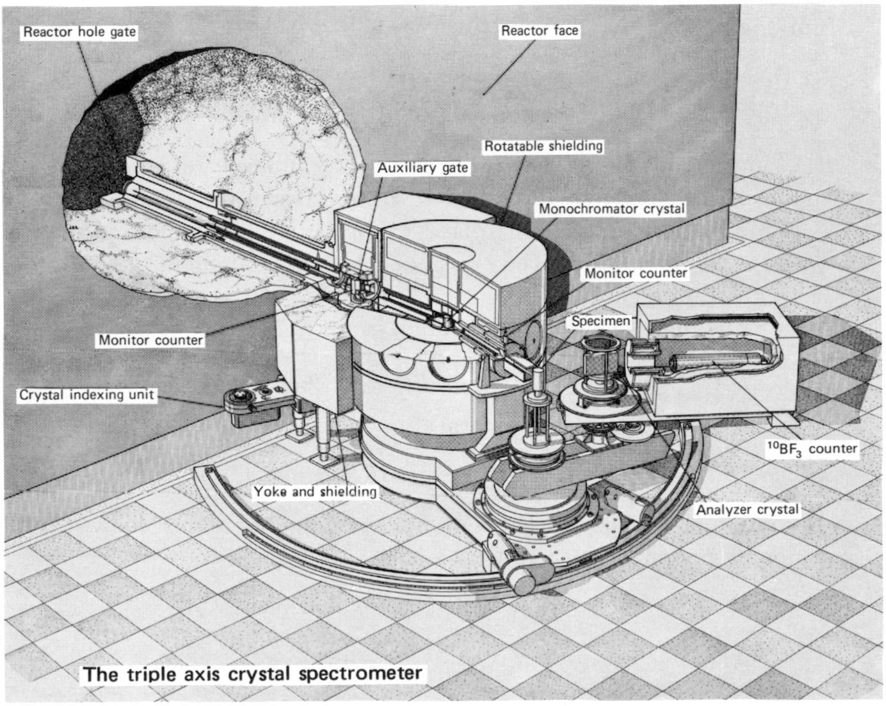

Figure 12 Cutaway drawing of the Brockhouse triple-axis crystal spectrometer at Chalk River.

SUMMARY

1. The quantum unit of a lattice vibration is a phonon. If the angular frequency is ω, the energy of the phonon is $\hbar\omega$.

2. When a phonon of wavevector $\mathbf{K}$ is created by the inelastic scattering of a photon or neutron from wavevector $\mathbf{k}$ to $\mathbf{k}'$, the wavevector selection rule that governs the process is

$$\mathbf{k} = \mathbf{k}' + \mathbf{K} + \mathbf{G} \ ,$$

where $\mathbf{G}$ is a reciprocal lattice vector.

3. All lattice waves can be described by wavevectors that lie within the first Brillouin zone in reciprocal space.

4. If there are p atoms in the primitive cell, the phonon dispersion relation will have 3 acoustical phonon branches and $3p - 3$ optical phonon branches.

PROBLEMS

1. ***Vibrations of square lattice.*** We consider transverse vibrations of a planar square lattice of rows and columns of identical atoms, and let $u_{l,m}$ denote the displacement normal to the plane of the lattice of the atom in the lth column and mth row (Fig. 13). The mass of each atom is M, and C is the force constant for nearest neighbor atoms.
(a) Show that the equation of motion is

$$M(d^2u_{lm}/dt^2) = C\,[(u_{l+1,m} + u_{l-1,m} - 2u_{lm}) + (u_{l,m+1} + u_{l,m-1} - 2u_{lm})]$$

(b) Assume solutions of the form

$$u_{lm} = u(0)\exp[i(lK_xa + mK_ya - \omega t)] \ ,$$

where a is the spacing between nearest-neighbor atoms. Show that the equation of motion is satisfied if

$$\omega^2 M = 2C(2 - \cos K_xa - \cos K_ya) \ .$$

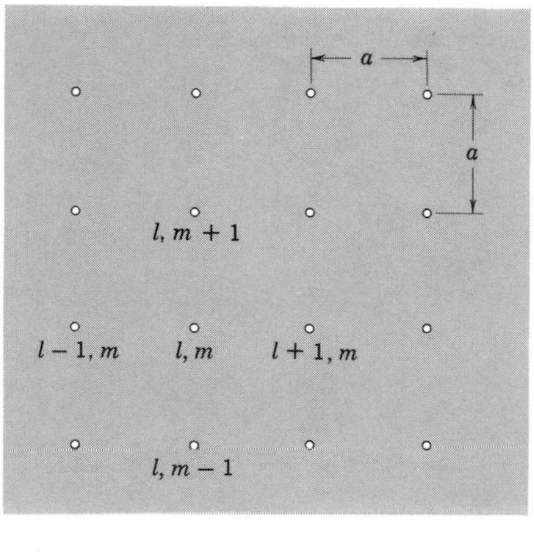

Figure 13 Square array of lattice constant a. The displacements considered are normal to the plane of the lattice.

This is the dispersion relation for the problem. (c) Show that the region of **K** space for which independent solutions exist may be taken as a square of side $2\pi/a$. This is the first Brillouin zone of the square lattice. Sketch ω versus K for $K = K_x$ with $K_y = 0$, and for $K_x = K_y$. (d) For $Ka \ll 1$, show that

$$\omega = (Ca^2/M)^{\frac{1}{2}}(K_x^2 + K_y^2)^{\frac{1}{2}} = (Ca^2/M)^{\frac{1}{2}}K \ ,$$

so that in this limit the velocity is constant.

2. *Monatomic linear lattice.* Consider a longitudinal wave

$$u_s = u \cos(\omega t - sKa)$$

which propagates in a monatomic linear lattice of atoms of mass M, spacing a, and nearest-neighbor interaction C.
(a) Show that the total energy of the wave is

$$E = \tfrac{1}{2}M \sum_s (du_s/dt)^2 + \tfrac{1}{2}C \sum_s (u_s - u_{s+1})^2 \ ,$$

where s runs over all atoms.
(b) By substitution of u_s in this expression, show that the time-average total energy per atom is

$$\tfrac{1}{4}M\omega^2u^2 + \tfrac{1}{2}C(1 - \cos Ka)u^2 = \tfrac{1}{2}M\omega^2u^2 \ ,$$

where in the last step we have used the dispersion relation (9) for this problem.

3. *Continuum wave equation.* Show that for long wavelengths the equation of motion (2) reduces to the continuum elastic wave equation

$$\frac{\partial^2 u}{\partial t^2} = v^2 \frac{\partial^2 u}{\partial x^2} \ ,$$

where v is the velocity of sound.

4. *Basis of two unlike atoms.* For the problem treated by (18) to (26), find the amplitude ratios u/v for the two branches at $K_{\max} = \pi/a$. Show that at this value of K the two lattices act as if decoupled: one lattice remains at rest while the other lattice moves.

5. *Kohn anomaly.* We suppose that the interplanar force constant C_p in Eq. (1) is of the form

$$C_p = A \frac{\sin pk_0a}{pa} \ ,$$

where A and k_0 are constants and p runs over all integers. Such a form is expected in metals. Use this and Eq. (7) to find an expression for ω^2 and also for $\partial\omega^2/\partial K$. Prove that $\partial\omega^2/\partial K$ is infinite when $K = k_0$. Thus a plot of ω^2 versus K or of ω versus K has a vertical tangent at k_0: there is a kink at k_0 in the phonon dispersion relation $\omega(K)$. A related effect was predicted by W. Kohn, Phys. Rev. Letters **2**, 393 (1959).

6. *Diatomic chain.* Consider the normal modes of a linear chain in which the force constants between nearest-neighbor atoms are alternately C and $10C$. Let the masses be equal, and let the nearest-neighbor separation be $a/2$. Find $\omega(K)$ at $K = 0$ and $K = \pi/a$. Sketch in the dispersion relation by eye. This problem simulates a crystal of diatomic molecules such as H_2.

7. ***Atomic vibrations in a metal.*** We can make a crude model of the Debye frequency in a simple metal: Consider point ions of mass M and charge e immersed in a uniform sea of conduction electrons. The ions are imagined to be in stable equilibrium when at regular lattice points. If one ion is displaced a small distance r from its equilibrium position, the restoring force is largely due to the electric charge within the sphere of radius r centered at the equilibrium position. Take the number density of ions (or of conduction electrons) as $3/4\pi R^3$, which defines R. (a) Show that the frequency of a single ion set into oscillation is $\omega = (e^2/MR^3)^{1/2}$. (b) Estimate the value of this frequency for sodium, roughly. (c) From (a), (b), and some common sense, estimate the order of magnitude of the velocity of sound in the metal.

8. ***Soft phonon modes.*** Consider a line of ions of equal mass but alternating in charge, with $e_p = e(-1)^p$ as the charge on the pth ion. The interatomic potential is the sum of two contributions: (1) a short-range interaction of force constant $C_{1R} = \gamma$ that acts between nearest neighbors only, and (2) a coulomb interaction between all ions. (a) Show that the contribution of the coulomb interaction to the atomic force constants is $C_{pC} = 2(-1)^p\, e^2/p^3a^3$, where a is the equilibrium nearest-neighbor distance. (b) From (7) show that the dispersion relation may be written as

$$\omega^2/\omega_0^2 = \sin^2 \tfrac{1}{2}Ka + \sigma \sum_{p=1}^{\infty} (-1)^p\, (1 - \cos pKa)p^{-3} \ ,$$

where $\omega_0^2 \equiv 4\gamma/M$ and $\sigma = e^2/\gamma a^3$. (c) Show that ω^2 is negative (unstable mode) at the zone boundary $Ka = \pi$ if $\sigma > 0.475$ or $4/7\zeta(3)$, where ζ is a Riemann zeta function. Show further that the speed of sound at small Ka is imaginary if $\sigma > (2 \ln 2)^{-1} = 0.721$. Thus ω^2 goes to zero and the lattice is unstable for some value of Ka in the interval $(0,\pi)$ if $0.475 < \sigma < 0.721$. Notice that the phonon spectrum is not that of a diatomic lattice because the interaction of any ion with its neighbors is the same as that of any other ion.

References

A. A. Maradudin, E. W. Montroll, G. H. Weiss, and I. P. Ipatova, "Theory of lattice dynamics in the harmonic approximation," *Solid state physics*, 2nd ed., Supp. **3** (1971).

S. S. Mitra, "Vibrational spectra of solids," *Solid state physics* **13**, 1–80 (1962).

W. Cochran, *Dynamics of atoms in crystals*, Crane, Russak, N. Y., 1973.

M. Born and K. Huang, *Dynamical theory of crystal lattices*, Oxford, 1954.

G. Leibfried, "Gittertheorie der mechanischen und thermischen Eigenschaften der Kristalle," *Encyclo. of physics*, **7/1**, 104–324 (1955).

S. K. Joshi and A. K. Rajagopal, "Lattice dynamics of metals," *Solid state physics* **22**, 160–312 (1968).

J. Ziman, *Electrons and phonons*, Oxford, 1960.

J. A. Reissland, *Physics of phonons*, Wiley, 1973.

A. A. Maradudin and S. H. Vosko, "Symmetry properties of the normal vibrations of a crystal," *Rev. Mod. Phys.* **40**, 1 (1968).

G. H. Horton and A. A. Maradudin, eds., *Dynamics of crystalline solids*, North-Holland, 1974.

B. T. M. Willis and A. W. Pryor, *Thermal vibrations in crystallography*, Cambridge, 1975.

Bibliography

A. Larose and J. Vanderwal, eds., *Bibliography of papers relevant to the scattering of thermal neutrons (1963–1972)*, Plenum, 1975.

5 PHONONS II. THERMAL PROPERTIES

TABLE 1 Debye Temperature and Thermal Conductivity[a]

Low temperature limit of θ, in Kelvin
Thermal conductivity at 300 K, in W $cm^{-1}K^{-1}$

Li	**Be**											**B**	**C**	**N**	**O**	**F**	**Ne**
344	1440												2230				75
0.85	2.00											0.27	1.29				
Na	**Mg**											**Al**	**Si**	**P**	**S**	**Cl**	**Ar**
158	400											428	645				92
1.41	1.56											2.37	1.48				
K	**Ca**	**Sc**	**Ti**	**V**	**Cr**	**Mn**	**Fe**	**Co**	**Ni**	**Cu**	**Zn**	**Ga**	**Ge**	**As**	**Se**	**Br**	**Kr**
91	230	360.	420	380	630	410	470	445	450	343	327	320	374	282	90		72
1.02		0.16	0.22	0.31	0.94	0.08	0.80	1.00	0.91	4.01	1.16	0.41	0.60	0.50	0.02		
Rb	**Sr**	**Y**	**Zr**	**Nb**	**Mo**	**Tc**	**Ru**	**Rh**	**Pd**	**Ag**	**Cd**	**In**	**Sn** *w*	**Sb**	**Te**	**I**	**Xe**
56	147	280	291	275	450		600	480	274	225	209	108	200	211	153		64
0.58		0.17	0.23	0.54	1.38	0.51	1.17	1.50	0.72	4.29	0.97	0.82	0.67	0.24	0.02		
Cs	**Ba**	**La** β	**Hf**	**Ta**	**W**	**Re**	**Os**	**Ir**	**Pt**	**Au**	**Hg**	**Tl**	**Pb**	**Bi**	**Po**	**At**	**Rn**
38	110	142	252	240	400	430	500	420	240	165	71.9	78.5	105	119			
0.36		0.14	0.23	0.58	1.74	0.48	0.88	1.47	0.72	3.17		0.46	0.35	0.08			
Fr	**Ra**	**Ac**															

Ce	**Pr**	**Nd**	**Pm**	**Sm**	**Eu**	**Gd**	**Tb**	**Dy**	**Ho**	**Er**	**Tm**	**Yb**	**Lu**
						200		210				120	210
0.11	0.12	0.16		0.13		0.11	0.11	0.11	0.16	0.14	0.17	0.35	0.16
Th	**Pa**	**U**	**Np**	**Pu**	**Am**	**Cm**	**Bk**	**Cf**	**Es**	**Fm**	**Md**	**No**	**Lr**
163		207											
0.54		0.28	0.06	0.07									

[a] Most of the θ values were supplied by N. Pearlman; references are given the *A.I.P. Handbook*, 3rd ed, the thermal conductivity values are from R. W. Powell and Y. S. Touloukian, Science **181**, 999 (1973).

5 PHONONS II. THERMAL PROPERTIES

In this chapter we discuss the Einstein and Debye approximations to the phonon heat capacity; the features of more exact calculations are indicated. We then consider effects of anharmonic lattice interactions, including thermal expansion, the Grüneisen relation, and the thermal conductivity of insulators. The thermal properties of metals are considered in Chapter 6, of superconductors in Chapter 12, and the thermal properties particular to magnetic materials are considered in Chapters 14 and 15.

LATTICE HEAT CAPACITY

By heat capacity we shall usually mean the heat capacity at constant volume, which is more fundamental than the heat capacity at constant pressure, which is what the experiments determine.[1] The heat capacity at constant volume is defined as

$$C_V \equiv T \sum \left(\frac{\partial S}{\partial T}\right)_V = \left(\frac{\partial U}{\partial T}\right)_V , \tag{1}$$

where S is the entropy, U the energy, and T the temperature.

The experimental facts about the heat capacity of representative inorganic solids are these:

1. In the room-temperature range the value of the heat capacity of nearly all monatomic solids is close to $3Nk_B$, or 25 J mol^{-1} deg^{-1}.
2. At lower temperatures the heat capacity drops markedly (Fig. 1) and approaches zero as T^3 in insulators and as T in metals. If the metal becomes a superconductor, the drop is faster than T.
3. In magnetic solids there is a large contribution to the heat capacity over the temperature range in which the magnetic moments become ordered. A change in the degree of order means a change in the entropy and thus a contribution to the heat capacity. Below 0.1 K the ordering of nuclear moments may give very large heat capacities.

[1]A simple thermodynamic relation gives $C_p - C_V = 9\alpha^2 BVT$, where α is the temperature coefficient of linear expansion, V the volume, and B the bulk modulus. The fractional difference between C_p and C_V is usually small and often may be neglected, particularly below room temperature.

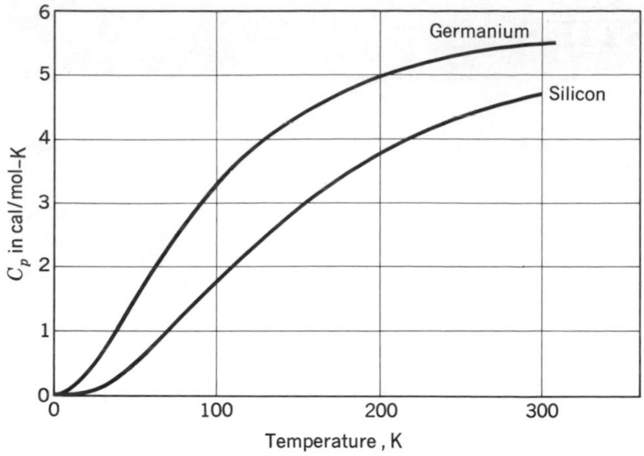

Figure 1 Heat capacity of silicon and germanium. Note the decrease at low temperatures. To convert cal/mol-K to J/mol-K, multiply by 4.186.

The normal modes of vibration of a lattice are independent if Hooke's law is applicable. Thus the energy of a lattice mode depends only on its frequency ω and phonon occupancy n, and is independent of the occupancy of the other lattice modes. In thermal equilibrium at temperature T the occupancy of a phonon mode is given by the Planck distribution

$$\langle n \rangle = \frac{1}{e^{\hbar\omega/k_BT} - 1} \ , \qquad \text{or} \qquad \langle n \rangle + \frac{1}{2} = \frac{1}{2} \operatorname{ctnh} \frac{\hbar\omega}{2k_BT} \ , \tag{2}$$

where the $\langle \cdot \ \cdot \ \cdot \rangle$ denotes the average in thermal equilibrium and k_B is the Boltzmann constant. A graph of $\langle n \rangle$ is given in Fig. 2.

Planck Distribution.[2] Consider a set of identical harmonic oscillators in thermal equilibrium. The ratio of the number of oscillators in their $(n + 1)$th quantum state of excitation to the number in the nth quantum state is

$$N_{n+1}/N_n = e^{-\hbar\omega/\tau} \ , \qquad \tau \equiv k_BT \ , \tag{3}$$

according to the Boltzmann factor. Thus the fraction of the total number of oscillators in the nth quantum state is

$$\frac{N_n}{\sum\limits_{s=0}^{\infty} N_s} = \frac{e^{-n\hbar\omega/\tau}}{\sum\limits_{s=0}^{\infty} e^{-s\hbar\omega/\tau}} \ . \tag{4}$$

[2]See *TP*, Chapters 9, 15, 16.

We see that the average excitation quantum number of an oscillator is

$$\langle n \rangle = \frac{\sum_s s e^{-s\hbar\omega/\tau}}{\sum_s e^{-s\hbar\omega/\tau}} \quad . \tag{5}$$

The summations in (5) are

$$\sum_s x^s = \frac{1}{1-x} \; ; \qquad \sum_s s x^s = x \frac{d}{dx} \sum_s x^s = \frac{x}{(1-x)^2} \; , \tag{6}$$

with $x = \exp(-\hbar\omega/\tau)$. Thus we may rewrite (5) as the Planck distribution:

$$\langle n \rangle = \frac{x}{1-x} = \frac{1}{e^{\hbar\omega/\tau} - 1} \quad . \tag{7}$$

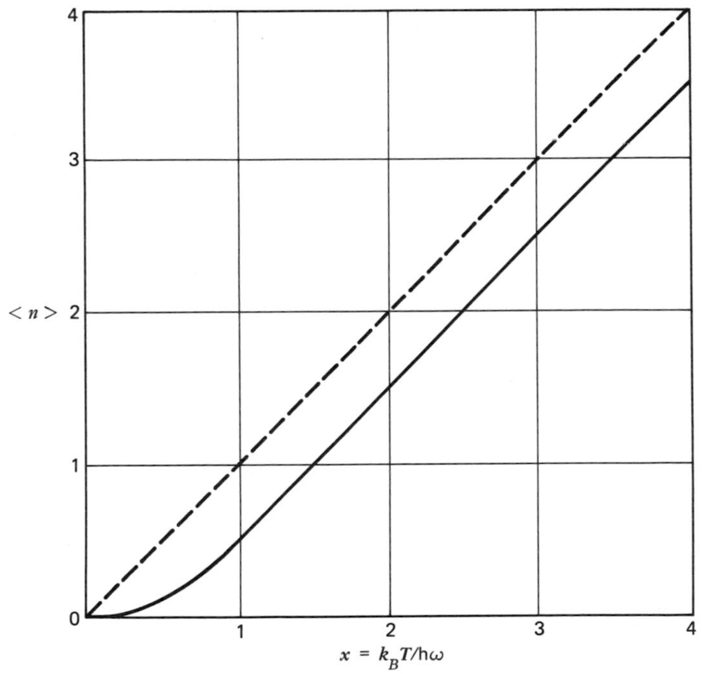

Figure 2 Plot of Planck distribution function. Note that at high temperatures the occupancy of a state is approximately linear in the temperature. The function $\langle n \rangle + \frac{1}{2}$, which is not plotted, approaches the dashed line as asymptote at high temperatures. The dashed line is the classical limit.

Einstein Model

The average thermal energy of an oscillator of frequency ω is $\langle n \rangle \hbar\omega$. For N oscillators in one dimension, all having the same frequency, the thermal energy is

$$U = N\langle n \rangle \hbar\omega = \frac{N\hbar\omega}{e^{\hbar\omega/\tau} - 1} \ . \tag{8}$$

The heat capacity of the oscillators is

$$C_V = \left(\frac{\partial U}{\partial T}\right)_V = Nk_B \left(\frac{\hbar\omega}{\tau}\right)^2 \frac{e^{\hbar\omega/\tau}}{(e^{\hbar\omega/\tau} - 1)^2} \ , \tag{9}$$

as plotted in Fig. 3. This is the result of the Einstein model for the contribution of N oscillators of the same frequency to the heat capacity of a solid. In three dimensions N is replaced by $3N$ because each of the N atoms has three degrees of freedom; then the high-temperature limit of C_V becomes $3Nk_B$, the Dulong and Petit value. At low temperatures (9) decreases as in Fig. 3 but gives $C_V \propto e^{-\hbar\omega/\tau}$, whereas the experimental phonon contribution is known to be $C_V \propto T^3$, which can be accounted for by the Debye model discussed later. The limitation of the Einstein model is that all the elastic waves in a solid do not have the same frequency. At the time Einstein wanted to show that mechanical oscillators should be quantized just as Planck had quantized radiation oscillators. Einstein's example was a quick demonstration of why the heat capacity of solids dropped to zero as $T \rightarrow 0$. The Einstein model is often used to approximate part of the phonon spectrum, particularly the contribution of optical phonons.

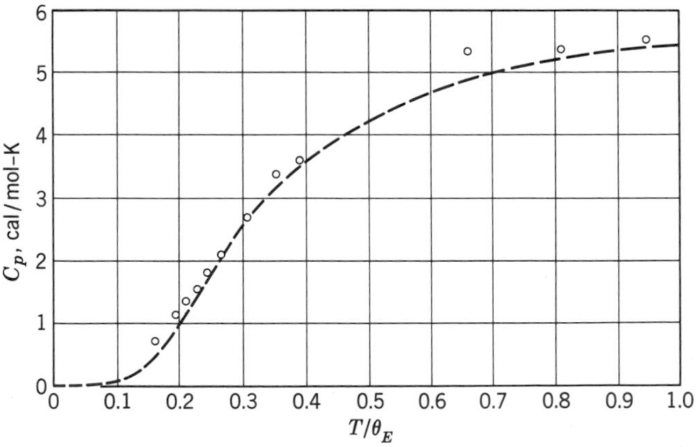

Figure 3 Comparison of experimental values of the heat capacity of diamond with values calculated on the Einstein model, using the characteristic temperature $\theta_E = \hbar\omega/k_B = 1320$ K. To convert to J/mol-deg, multiply by 4.186. [After A. Einstein, Ann. Physik **22**, 180 (1907).]

Enumeration of Normal Modes

The energy in thermal equilibrium of a collection of oscillators of different frequencies $\omega_{\mathbf{K}}$ is

$$U = \sum_{\mathbf{K}} n_{\mathbf{K}} \hbar \omega_{\mathbf{K}} \ , \tag{10}$$

where each n is related to $\omega_{\mathbf{K}}$ by the Planck distribution. It is often convenient to replace the summation by an integral. Suppose that the crystal has $\mathfrak{D}(\omega)\,d\omega$ modes of vibration in the frequency range ω to $\omega + d\omega$. Then the energy is

$$U = \int d\omega \ \mathfrak{D}(\omega)\ n(\omega,T)\ \hbar\omega \ ; \tag{11}$$

the heat capacity is found from this by differentiation of $n(\omega,T)$ with respect to temperature. The central problem is to find $\mathfrak{D}(\omega)$, the number of modes per unit frequency range. This function is also called the density of modes. The best practical way to obtain the density of modes is to measure the dispersion relation in selected crystal directions by inelastic neutron scattering and then to make a theoretical analytic fit to give the dispersion relation in a general direction, from which $\mathfrak{D}(\omega)$ may be calculated by an electronic computer.

Density of Modes in One Dimension

Consider the elastic problem for a one-dimensional line (Fig. 4) of length L carrying $N+1$ particles at separation a. We suppose that the particles $s = 0$ and $s = N$ at the ends are held fixed. Each normal mode is a standing wave:

$$u_s = u(0)\ e^{-i\omega_K t} \sin sKa \ , \tag{12}$$

where ω_K is related to K by the appropriate dispersion relation. As in Fig. 5, K is restricted by the fixed-end boundary conditions to the values

$$K = \frac{\pi}{L}\ , \quad \frac{2\pi}{L}\ , \quad \frac{3\pi}{L}\ , \ \ldots\ , \ \frac{(N-1)\pi}{L}\ . \tag{13}$$

The solution for $K = \pi/L$ has

$$u_s \propto \sin\,(s\pi a/L) \tag{14}$$

and vanishes for $s = 0$ and $s = N$ as required. The solution for $K = N\pi/L = \pi/a = K_{\max}$ has $u_s \propto \sin s\pi$; this permits no motion of any atom, because $\sin s\pi$ vanishes at each atom. Thus there are $N-1$ allowed independent values of K in (13). This number is equal to the number of particles allowed to move. Each allowed value of K is associated with a standing wave. For the one-dimensional line there is one mode for each interval $\Delta K = \pi/L$, so that the number of modes per unit range of K is L/π for $K \le \pi/a$, and 0 for $K > \pi/a$.

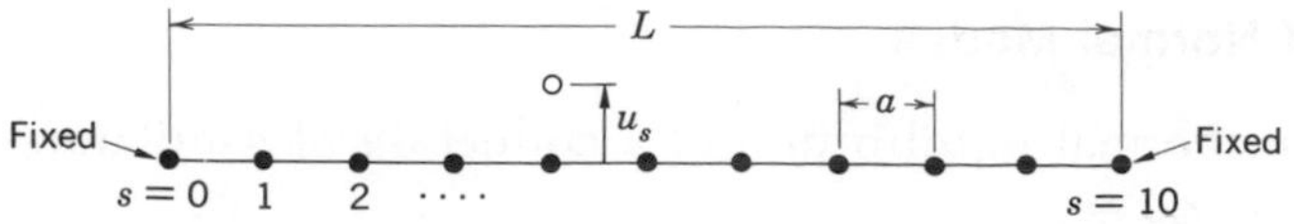

Figure 4 Elastic line of $N + 1$ atoms, with $N = 10$, for boundary conditions that the end atoms $s = 0$ and $s = 10$ are fixed. The particle displacements in the normal modes for either longitudinal or transverse displacements are of the form $u_s \propto \sin sKa$. This form is automatically zero at the atom at the end $s = 0$, and we choose K to make the displacement zero at the end $s = 10$.

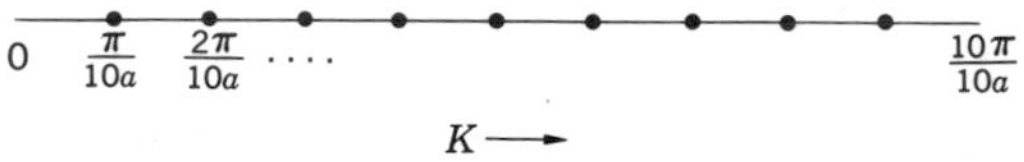

Figure 5 The boundary condition $\sin sKa = 0$ for $s = 10$ can be satisfied by choosing $K = \pi/10a, 2\pi/10a, \ldots, 9\pi/10a$, where $10a$ is the length L of the line. The present figure is in K space. The dots are not atoms but are the allowed values of K. Of the $N + 1$ particles on the line, only $N - 1$ are allowed to move, and their most general motion can be expressed in terms of the $N - 1$ allowed values of K. This quantization of K has nothing to do with quantum mechanics but follows classically from the boundary conditions that the end atoms be fixed.

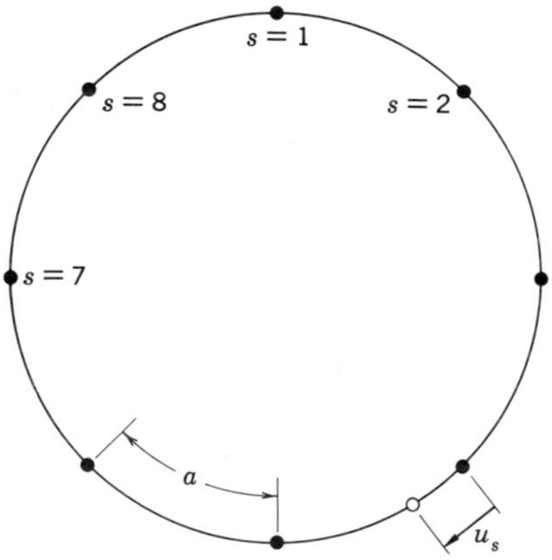

Figure 6 Consider N particles constrained to slide on a circular ring. The particles can oscillate if connected by elastic springs. In a normal mode the displacement u_s of atom s will be of the form $\sin sKa$ or $\cos sKa$: these are independent modes. By the geometrical periodicity of the ring the boundary condition is that $u_{N+s} = u_s$ for all s, so that NKa must be an integral multiple of 2π. For $N = 8$ the allowed independent values of K are 0, $2\pi/8a$, $4\pi/8a$, $6\pi/8a$, and $8\pi/8a$. The value $K = 0$ is meaningless for the sine form, because $\sin s0a = 0$. The value $8\pi/8a$ also has a meaning only for the cosine form, because $\sin(s8\pi a/8a) = \sin s\pi = 0$. The three other values of K are allowed for both the sine and cosine modes, giving a total of eight allowed modes for the eight particles. Thus the periodic boundary condition leads to one allowed mode per particle, exactly as for the fixed-end boundary condition of Fig. 5. If we had taken the modes in the complex form $\exp(isKa)$, the periodic boundary condition would lead to the eight modes with $K = 0, \pm 2\pi/Na$, $\pm 4\pi/Na$, $\pm 6\pi/Na$, and $8\pi/Na$, as in Eq. (15).

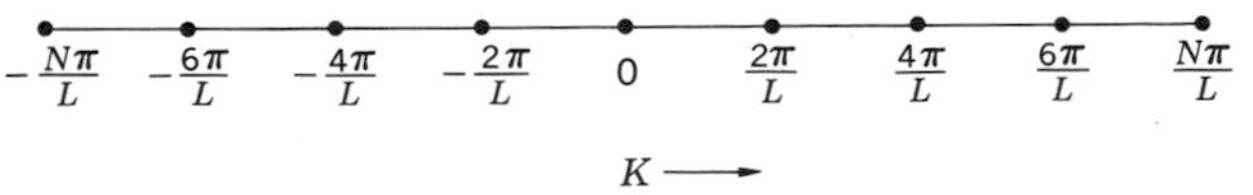

Figure 7 Allowed values of wavevector K for periodic boundary conditions applied to a linear lattice of periodicity $N = 8$ atoms on a line of length L. The $K = 0$ solution is the uniform mode. The special points $\pm N\pi/L$ represent only a single solution because $e^{i\pi s}$ is identical to $e^{-i\pi s}$; thus there are eight allowed modes, with displacements of the sth atom proportional to 1, $e^{\pm i\pi s/4}$, $e^{\pm i\pi s/2}$, $e^{\pm i3\pi s/4}$, $e^{i\pi s}$.

There is another device for enumerating modes that is often used and that is equally valid. We consider the medium as unbounded, but require that the solutions be periodic over a large distance L, so that $u(sa) = u(sa + L)$. The method of **periodic boundary conditions** (Figs. 6 and 7) does not change the physics of the problem in any essential respect for a large system. Then in the running wave solution $u_s = u(0)\exp[i(sKa - \omega_K t)]$ the allowed values of K are

$$K = 0 \ , \quad \pm\frac{2\pi}{L} \ , \quad \pm\frac{4\pi}{L} \ , \quad \pm\frac{6\pi}{L} \ , \quad \ldots \ , \quad \frac{N\pi}{L} \ . \tag{15}$$

This method of enumeration gives the same number of modes (one per mobile atom) as given by (13), but we have now both plus and minus values of K, with the interval $\Delta K = 2\pi/L$ between successive values of K. For periodic boundary conditions the number of modes per unit range of K is $L/2\pi$ for $-\pi/a \le K \le \pi/a$, and 0 otherwise. The situation in a two-dimensional lattice is portrayed in Fig. 8.

We need to know $\mathcal{D}(\omega)$, the number of modes per unit frequency range. The number of modes $\mathcal{D}(\omega)\, d\omega$ in $d\omega$ at ω is given in one dimension by

$$\mathcal{D}(\omega)\, d\omega = \frac{L}{\pi}\frac{dK}{d\omega}\, d\omega = \frac{L}{\pi} \cdot \frac{d\omega}{d\omega/dK} \ . \tag{16}$$

We can obtain the group velocity $d\omega/dK$ from the dispersion relation ω versus K. There is a singularity in $\mathcal{D}(\omega)$ whenever the dispersion relation $\omega(K)$ is horizontal; that is, whenever the group velocity is zero.

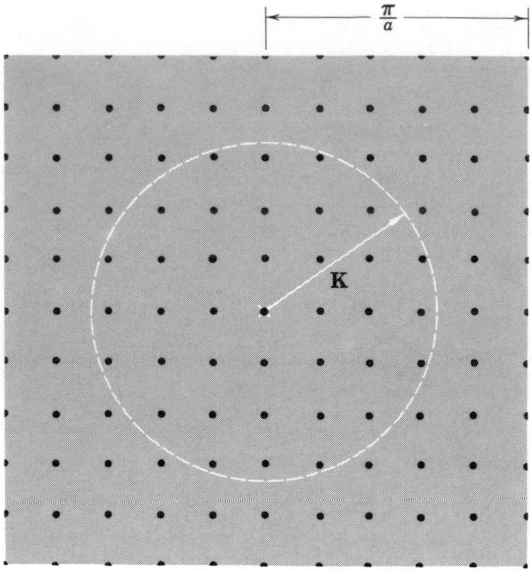

Figure 8 Allowed values in Fourier space of the phonon wavevector K for a square lattice of lattice constant a, with periodic boundary conditions applied over a square of side $L = 10a$. The uniform mode is marked with a cross. There is one allowed value of K per area $(2\pi/10a)^2 = (2\pi/L)^2$, so that within the circle of area πK^2 the smoothed number of allowed points is $\pi K^2(L/2\pi)^2$.

Density of Modes in Three Dimensions

We apply periodic boundary conditions over N^3 primitive cells within a cube of side L, so that $\mathbf{K}$ is determined by the condition

$$\exp[i(K_x x + K_y y + K_z z)] \equiv \exp\{i[K_x(x+L) + K_y(y+L) + K_z(z+L)]\} \ , \tag{17}$$

whence

$$K_x, K_y, K_z = 0 \ ; \quad \pm\frac{2\pi}{L} \ ; \quad \pm\frac{4\pi}{L} \ ; \quad \ldots \ ; \quad \frac{N\pi}{L} \ . \tag{18}$$

Therefore there is one allowed value of $\mathbf{K}$ per volume $(2\pi/L)^3$ in $\mathbf{K}$ space, or

$$\left(\frac{L}{2\pi}\right)^3 = \frac{V}{8\pi^3} \tag{19}$$

allowed values of $\mathbf{K}$ per unit volume of $\mathbf{K}$ space, for each polarization and for each branch. The volume of the specimen is $V = L^3$.

In the continuum or Debye approximation the velocity of sound is constant: $\omega = vK$. The total number of modes with wavevector less than K is, by (19), given by $(L/2\pi)^3$ times the volume of a sphere of radius K, so that

$$N = \left(\frac{L}{2\pi}\right)^3 \frac{4\pi}{3} K^3 = \left(\frac{L}{2\pi}\right)^3 \frac{4\pi\omega^3}{3v^3} = \frac{V\omega^3}{6\pi^2 v^3} \tag{20}$$

for each polarization type. The density of modes for each polarization type is

$$D(\omega) = \frac{dN}{d\omega} = \frac{V\omega^2}{2\pi^2 v^3} \ . \tag{21}$$

If there are N primitive cells in the specimen, the total number of acoustic phonon modes is N, and a cutoff frequency ω_D is determined by (20) as

$$\omega_D{}^3 = 6\pi^2 v^3 N/V \ . \tag{22}$$

To this frequency there corresponds a cutoff wavevector in **K** space:

$$K_D = \omega_D/v = (6\pi^2 N/V)^{1/3} \ . \tag{23}$$

On the Debye model we do not allow modes of wavevector larger than K_D; the number of modes of $K \leq K_D$ exhausts the number of degrees of freedom of a monatomic lattice.

EXAMPLE:[3] ***General result for $\mathfrak{D}(\omega)$.*** We want to find a general expression for $\mathfrak{D}(\omega)$, the number of modes per unit frequency range, given the phonon dispersion relation $\omega(\mathbf{K})$. The number of allowed values of **K** for which the phonon frequency is between ω and $\omega + d\omega$ is

$$\mathfrak{D}(\omega)\, d\omega = \left(\frac{L}{2\pi}\right)^3 \int_{\text{shell}} d^3K \ , \tag{24}$$

by (19); here the integral is extended over the volume of the shell in **K** space bounded by the two surfaces on which the phonon frequency is constant, one surface on which the frequency is ω and the other on which the frequency is $\omega + d\omega$.

The real problem is to evaluate the volume of this shell. We let dS_ω denote an element of area on the surface in **K** space of the selected constant frequency ω. The element of volume between the constant frequency surfaces ω and $\omega + d\omega$ is a right cylinder of base dS_ω and altitude $dK_\perp$, so that

$$\int_{\text{shell}} d^3K = \int dS_\omega dK_\perp \ . \tag{25}$$

Here $dK_\perp$ is the perpendicular distance between the surface ω constant and the surface $\omega + d\omega$ constant. The value of $dK_\perp$ will vary from one point to another on the surface.

The gradient of ω, which is $\nabla_{\mathbf{K}}\omega$, is also normal to the surface ω constant, and the quantity

$$|\nabla_{\mathbf{K}}\,\omega|\ dK_\perp = d\omega \ ,$$

is the difference in frequency between the two surfaces connected by $dK_\perp$. Thus the element of the volume is

$$dS_\omega\, dK_\perp = dS_\omega \frac{d\omega}{|\nabla_{\mathbf{K}}\,\omega|} = dS_\omega \frac{d\omega}{v_g} \ ,$$

where $v_g = |\nabla_{\mathbf{K}}\,\omega|$ is the magnitude of the group velocity of a phonon. For (24) we have

$$\mathfrak{D}(\omega)\, d\omega = \left(\frac{L}{2\pi}\right)^3 \int \frac{dS_\omega}{v_g}\, d\omega \ .$$

[3]This example may be skipped in a first reading.

We divide both sides by $d\omega$ and write $V = L^3$ for the volume of the crystal: the result for the density of modes is

$$\boxed{\mathfrak{D}(\omega) = \frac{V}{(2\pi)^3} \int \frac{dS_\omega}{v_g}} \tag{26}$$

The integral is taken over the area of the surface ω constant, in **K** space. The result refers to a single branch of the dispersion relation. We can use this result also in electron band theory.

There is a special interest in the contribution to $\mathfrak{D}(\omega)$ from points at which the group velocity is zero. Such critical points produce singularities (known as Van Hove[4] singularities) in the distribution function. Saddle points in $\omega(\mathbf{K})$ are of particular importance (Fig. 9).

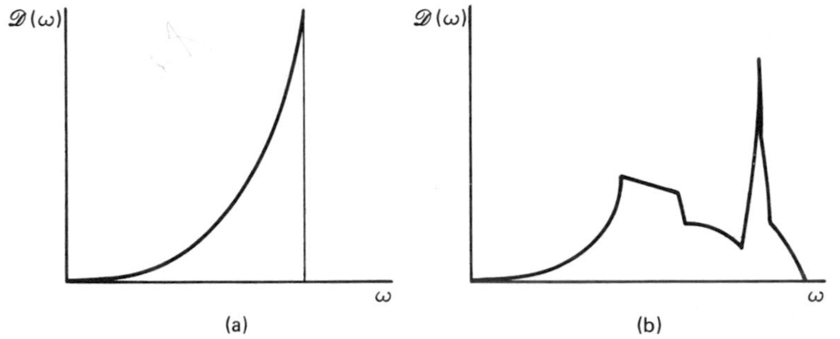

Figure 9 Density of modes as a function of frequency for (a) the Debye solid and (b) an actual crystal structure. The spectrum for the crystal starts as ω^2 for small ω, but discontinuities develop at singular points.

Debye Model of the Lattice Heat Capacity

The thermal energy (11) is given by

$$U = \int d\omega\, \mathfrak{D}(\omega)\, n(\omega)\, \hbar\omega = \int_0^{\omega_D} d\omega \left(\frac{V\omega^2}{2\pi^2 v^3}\right)\left(\frac{\hbar\omega}{e^{\hbar\omega/\tau} - 1}\right) , \tag{27}$$

for each polarization type. For brevity we assume that the phonon velocity is independent of the polarization, so that we multiply by the factor 3 to obtain

$$U = \frac{3V\hbar}{2\pi^2 v^3} \int_0^{\omega_D} d\omega \frac{\omega^3}{e^{\hbar\omega/\tau} - 1} = \frac{3Vk_B{}^4T^4}{2\pi^2 v^3 \hbar^3} \int_0^{x_D} dx \frac{x^3}{e^x - 1} , \tag{28}$$

where $x \equiv \hbar\omega/\tau \equiv \hbar\omega/k_BT$ and

$$x_D \equiv \hbar\omega_D/k_BT \equiv \theta/T . \tag{29}$$

[4]L. Van Hove, Phys. Rev. **89,** 1189 (1953); see also H. P. Rosenstock, Phys. Rev. **97,** 290 (1955); J. C. Phillips, Phys. Rev. **104,** 1263 (1956); and C. Kittel, *Introduction to solid state physics*, 4th ed., Wiley, 1971, Advanced Topic C.

This defines the **Debye temperature** θ in terms of ω_D defined by (23). We may express θ as

$$\theta = \frac{\hbar v}{k_B} \cdot \left(\frac{6\pi^2 N}{V}\right)^{\frac{1}{3}} \ , \tag{30}$$

so that

$$U = 9Nk_BT\left(\frac{T}{\theta}\right)^3 \int_0^{x_D} dx\, \frac{x^3}{e^x - 1} \ , \tag{31}$$

where N is the number of atoms in the specimen and $x_D = \theta/T$.

The heat capacity is found most easily by differentiating the middle expression of (28) with respect to temperature. Then

$$C_V = \frac{3V\hbar^2}{2\pi^2 v^3 k_B T^2} \int_0^{\omega_D} d\omega\, \frac{\omega^4\, e^{\hbar\omega/\tau}}{(e^{\hbar\omega/\tau} - 1)^2} = 9Nk_B\left(\frac{T}{\theta}\right)^3 \int_0^{x_D} dx\, \frac{x^4\, e^x}{(e^x - 1)^2} \ . \tag{32}$$

Tables calculated on the Debye theory for U, C_V, and other quantities are given in the invaluable U.S. National Bureau of Standards *Handbook of mathematical functions* AMS 55 and also in the Jahnke-Emde-Lösch tables. The heat capacity is plotted in Fig. 9a. At $T \gg \theta$ the heat capacity approaches the classical value of $3Nk_B$.

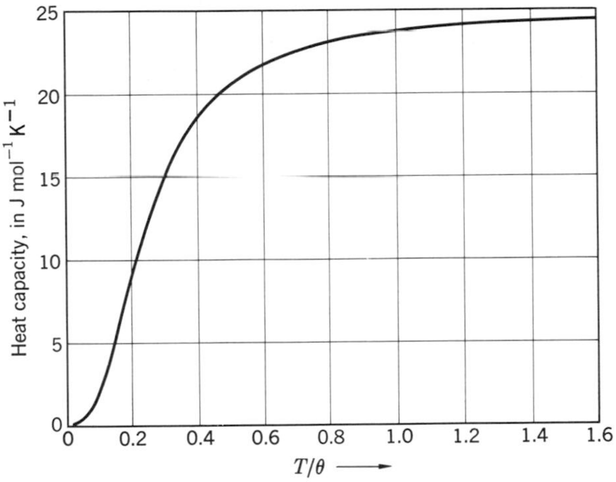

Figure 9a. Heat capacity C_V of a solid, according to the Debye approximation. The vertical scale is in J mol^{-1} K^{-1}. The horizontal scale is the temperature normalized to the Debye temperature θ. The region of the T^3 law is below 0.1θ. The asymptotic value at high values of T/θ is 24.943 J mol^{-1} deg^{-1}.

Debye T^3 Law. At very low temperatures we may approximate (31) by letting the upper limit go to infinity. We have

$$\int_0^\infty dx \frac{x^3}{e^x - 1} = \int_0^\infty dx\, x^3 \sum_{s=1}^{\infty} e^{-sx} = 6 \sum_1^\infty \frac{1}{s^4} = \frac{\pi^4}{15} , \tag{33}$$

where the sum over s^{-4} is to be found in standard tables. Thus $U \cong 3\pi^4 N k_B T^4 / 5\theta^3$ for $T \ll \theta$, and

$$C_V \cong \frac{12\pi^4}{5} N k_B \left(\frac{T}{\theta}\right)^3 = 234\, N k_B \left(\frac{T}{\theta}\right)^3 . \tag{34}$$

This exhibits the Debye T^3 approximation. For sufficiently low temperatures the Debye approximation should be quite good, as here only long wavelength acoustic modes are excited. These are just the modes that may be treated as in an elastic continuum with macroscopic elastic constants. The energy of short wavelength modes is too high to be populated at low temperatures.

We can understand the T^3 region by a simple argument (Fig. 10). Only those lattice modes having $\hbar\omega < k_B T$ will be excited to any appreciable extent at a low temperature T. The excitation of these modes will be approximately classical with an energy close to $k_B T$, according to Fig. 2. Of the allowed volume in $\mathbf{K}$ space, the fraction occupied by the excited modes is of the order of $(\omega_T/\omega_D)^3$ or $(K_T/K_D)^3$, where K_T is a wavevector defined such that $\hbar v K_T = k_B T$ and K_D is the Debye cutoff wavevector, or $(T/\theta)^3$ of the total volume in $\mathbf{K}$ space. Thus there are of the order of $N(T/\theta)^3$ excited modes, each having energy $k_B T$. The energy is $\sim N k_B T (T/\theta)^3$, and the heat capacity is $\sim 4 N k_B (T/\theta)^3$. For actual lattices the

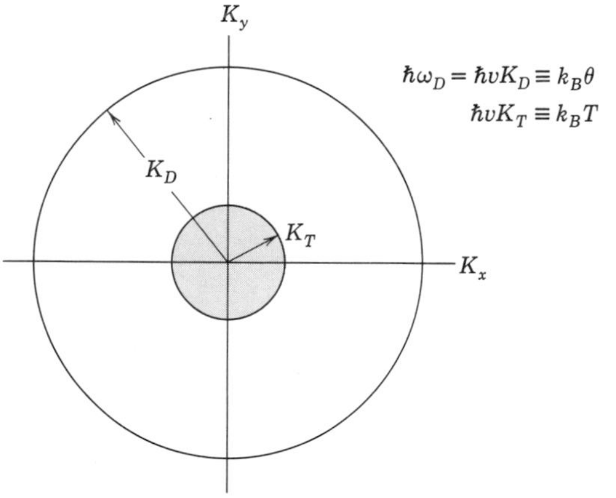

Figure 10 To obtain a qualitative explanation of the Debye T^3 law, we suppose that all phonon modes of wavevector less than K_T have the classical thermal energy $k_B T$ and that modes between K_T and the Debye cutoff K_D are not excited at all. Of the $3N$ possible modes, the fraction excited is $(K_T/K_D)^3 = (T/\theta)^3$, because this is the ratio of the volume of the inner sphere to the outer sphere. The energy is $U \approx k_B T \cdot 3N(T/\theta)^3$, and the heat capacity is $C_V = \partial U/\partial T \approx 4 N k_B (T/\theta)^3$.

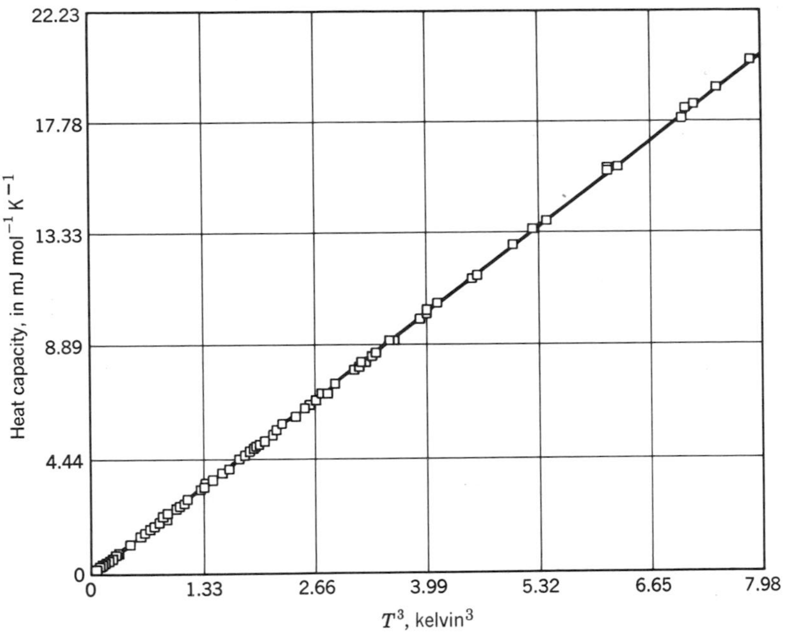

Figure 11 Low temperature heat capacity of solid argon, plotted against T^3. In this temperature region the experimental results are in excellent agreement with the Debye T^3 law with $\theta_0 = 92.0$ K. (Courtesy of L. Finegold and N. E. Phillips.)

temperatures at which the T^3 approximation holds true are quite low. It may be necessary to be below $T = \theta/50$ to get reasonably pure T^3 behavior.

Selected values of θ are given in Table 1. Note, for example, in the alkali metals that the heavier atoms have the lowest θ's, because the velocity of sound decreases as the density increases. Experimental results for argon, a dielectric crystal with one atom per primitive cell, are given in Fig. 11.

Heat Capacity of Glasses and Amorphous Solids. The Debye theory predicts that the leading term in the heat capacity of all dielectric solids at sufficiently low temperatures will be a term in T^3. This is in fact observed, but with the surprising exception of glasses and amorphous solids, where the leading term is linear in T. Observations on SiO_2 glass were reported by the Giauque group[5] and on amorphous GeO_2, Se, and Pyrex by Zeller and Pohl.[6] A plausible explanation of the linear term has been proposed[7] as a generalization of the model developed by Marshall[8] to account for the anomalously large linear term in the heat capacity of dilute magnetic alloys.

[5] E. W. Hornung, R. A. Fisher, G. E. Brodale, and W. F. Giauque, J. Chem. Phys. **50,** 4878 (1969); see also O. L. Anderson, J. Phys. Chem. Solids **12,** 41 (1959).
[6] R. C. Zeller and R. O. Pohl, Phys. Rev. **B4,** 2029 (1971).
[7] P. W. Anderson, B. I. Halperin, and C. M. Varma, Phil. Mag. **25,** 1 (1972).
[8] W. Marshall, Phys. Rev. **118,** 1519 (1960).

ANHARMONIC CRYSTAL INTERACTIONS

The theory of lattice vibrations discussed in this and in the preceding chapter has been limited in the potential energy to terms quadratic in the interatomic displacements. This is the harmonic theory; among its consequences are:

1. There is no thermal expansion.
2. Adiabatic and isothermal elastic constants are equal.
3. The elastic constants are independent of pressure and temperature.
4. The heat capacity becomes constant at high temperatures $T > \theta$.
5. Two lattice waves do not interact; a single wave does not decay or change form with time.

In real crystals none of these consequences is satisfied accurately. The deviations may be attributed to the neglect of anharmonic (higher than quadratic) terms in the interatomic displacements. We shall discuss some of the simpler aspects of anharmonic effects.

Beautiful demonstrations of anharmonic effects are the experiments on the interaction of two phonons to produce a third phonon at a frequency $\omega_3 = \omega_1 + \omega_2$. Shiren[9] described an experiment in which a beam of longitudinal phonons of frequency 9.20 GHz interacts in an MgO crystal with a parallel beam of longitudinal phonons at 9.18 GHz. The interaction of the two beams produced a third beam of longitudinal phonons at $9.20 + 9.18 = 18.38$ GHz. Rollins[10] et al. carried out the elegant experiment pictured in Fig. 12. Two narrow ultrasonic beams are allowed to interact near the center of a large circular disk. They verified that a third beam is generated at the volume of interaction when energy and wavevector are conserved between the three beams.

The three-phonon processes are caused by third-order terms in the lattice potential energy. A typical term might be $U_3 = Ae_{xx}e_{yy}e_{zz}$, where the e's are strain components and A is a constant. The A's have the same dimensions as elastic stiffness constants but may have values perhaps an order of magnitude larger. The physics of the phonon interaction can be stated simply: the presence of one phonon causes a periodic elastic strain which (through the anharmonic interaction) modulates in space and time the elastic constant of the crystal. A second phonon perceives the modulation of the elastic constant and thereupon is scattered to produce a third phonon, just as from a moving three-dimensional grating. In the absence of anharmonic terms the acoustic modulation does not occur.

[9]N. S. Shiren, Phys. Rev. Letters **11**, 3 (1963).
[10]F. R. Rollins, Jr., L. H. Taylor, and P. H. Todd, Jr., Phys. Rev. **136**, A597 (1964).

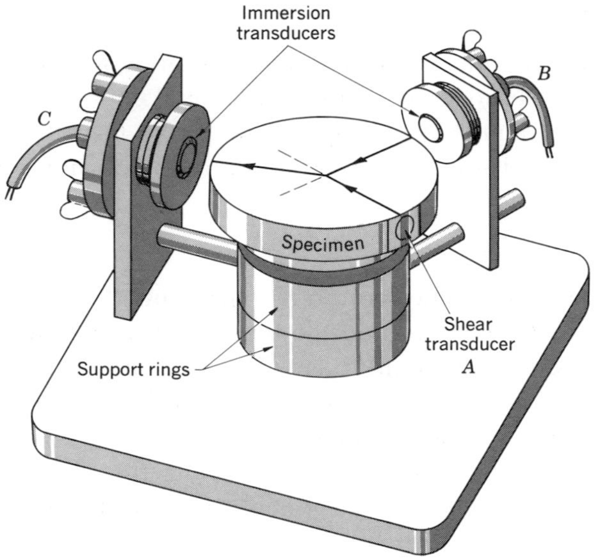

Figure 12 Ultrasonic experiment on three-phonon interactions. In a typical experiment the transducer A generates a 10 MHz shear wave, which interacts near the center of the disc-shaped specimen with a 15 MHz longitudinal wave generated by B, to produce by their interaction a 25 MHz longitudinal wave detected by the transducer C: $L(15) + T(10) \rightarrow L(25)$. The wavevectors satisfy $\mathbf{K}_{15} + \mathbf{K}_{10} = \mathbf{K}_{25}$. The apparatus is immersed in a suitable fluid to provide coupling between the immersion transducers and the specimen. (After Rollins, Taylor, and Todd.)

Thermal Expansion

We may understand thermal expansion by considering for a classical oscillator the effect of anharmonic terms in the potential energy on the mean separation of a pair of atoms at a temperature T. We take the potential energy of the atoms at a displacement x from their equilibrium separation at absolute zero as

$$U(x) = cx^2 - gx^3 - fx^4 \ , \tag{35}$$

with c, g, and f all positive. The term in x^3 represents the asymmetry of the mutual repulsion of the atoms and the term in x^4 represents the softening of the vibration at large amplitudes. The minimum at $x = 0$ is not an absolute minimum, but for small oscillations the form is an adequate representation of an interatomic potential.

We calculate the average displacement by using the Boltzmann distribution function, which weights the possible values of x according to their thermodynamic probability:

$$\langle x \rangle = \frac{\int_{-\infty}^{\infty} dx \ x \ e^{-\beta U(x)}}{\int_{-\infty}^{\infty} dx \ e^{-\beta U(x)}} \ ,$$

with $\beta \equiv 1/k_B T$. For displacements such that the anharmonic terms in the energy are small in comparison with $k_B T$, we may expand the integrands as

$$\int dx \ x \ e^{-\beta U} \cong \int dx \ e^{-\beta c x^2} (x + \beta g x^4 + \beta f x^5) = (3\pi^{1/2}/4)(g/c^{5/2})\beta^{-3/2} \ ; \tag{35a}$$
$$\int dx \ e^{-\beta U} \cong \int dx \ e^{-\beta c x^2} = (\pi/\beta c)^{1/2} \ ,$$

whence

$$\langle x \rangle = \frac{3g}{4c^2} k_B T$$

in the classical region. Note that in (35a) we have left cx^2 in the exponential, but we have expanded $\exp(\beta g x^3 + \beta f x^4) \cong 1 + \beta g x^3 + \beta f x^4 + \cdots$. Measurements of the lattice constant of solid argon are shown in Fig. 13. The slope of the curve is proportional to the thermal expansion coefficient. The expansion coefficient vanishes as $T \rightarrow 0$, as we expect from Problem 5. In lowest order the thermal expansion does not involve the symmetric term fx^4 in $U(x)$, but only the antisymmetric term gx^3.

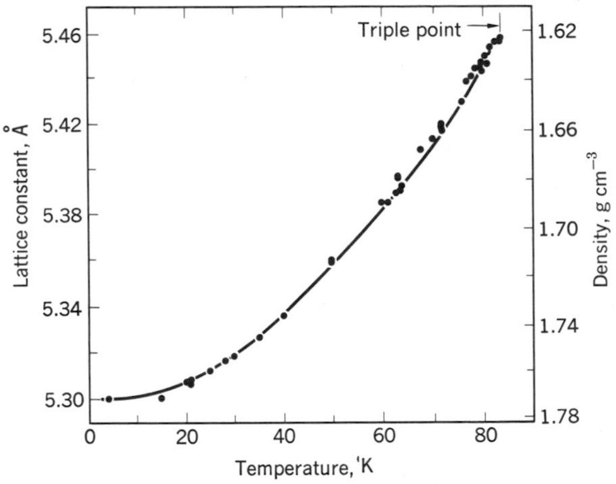

Figure 13 Lattice constant of solid argon as a function of temperature, as measured by various workers [O. G. Peterson, D. N. Batchelder, and R. O. Simmons, Phys. Rev. **150**, 703 (1966)].

THERMAL CONDUCTIVITY

The thermal conductivity coefficient K of a solid is most easily defined with respect to the steady-state flow of heat down a long rod with a temperature gradient dT/dx:

$$j_U = -K\frac{dT}{dx} , \tag{36}$$

where j_U is the flux of thermal energy, or the energy transmitted across unit area per unit time. The form of the equation which defines the conductivity implies that the process of thermal energy transfer is a random process. The energy does not simply enter one end of the specimen and proceed directly in a straight path to the other end, but rather the energy diffuses through the specimen, suffering frequent collisions. If the energy were propagated directly through the specimen without deflection, then the expression for the thermal flux would not depend on the temperature gradient, but only on the difference in temperature ΔT between the ends of the specimen, regardless of the length of the specimen. The random nature of the conductivity process brings the temperature gradient and, as we shall see, a mean free path into the expression for the thermal flux.

From the kinetic theory of gases we find below in a certain approximation the following expression for the thermal conductivity:

$$K = \tfrac{1}{3}Cv\ell , \tag{37}$$

where C is the heat capacity per unit volume, v is the average particle velocity, and ℓ is the mean free path of a particle between collisions. This result was applied first by Debye to describe thermal conductivity in

TABLE 2 Phonon Mean Free Paths

[Calculated from (39), taking $v = 5 \times 10^5$ cm/sec as a representative sound velocity. The ℓ's obtained in this way refer to umklapp processes.]

Crystal	T, °C	C, in J $\mathrm{cm^{-3}deg^{-1}}$	K, W $\mathrm{cm^{-1}deg^{-1}}$	ℓ, in Å
Quartz[a]	0	2.00	0.13	40
	−190	0.55	0.50	540
NaCl	0	1.88	0.07	23
	−190	1.00	0.27	100

[a] Parallel to optic axis.

dielectric solids, with C as the heat capacity of the phonons, v the phonon velocity, and ℓ the phonon mean free path. Several representative values of the mean free path are given in Table 2.

We give first the elementary kinetic theory which leads to (37). The flux of particles in the x direction is $\frac{1}{2}n\langle|v_x|\rangle$, where n is the concentration of molecules; in equilibrium there is a flux of equal magnitude in the opposite direction. The $\langle\cdot\ \cdot\ \cdot\rangle$ denote average value. If c is the heat capacity of a particle, then in moving from a region at local temperature $T + \Delta T$ to a region at local temperature T a particle will give up energy $c\,\Delta T$. Now ΔT between the ends of a free path of the particle is given by

$$\Delta T = \frac{dT}{dx}\ell = \frac{dT}{dx}\, v_x\tau \ ,$$

where τ is the average time between collisions.

The net flux of energy (from both senses of the particle flux) is therefore

$$j_U = -n\langle v_x{}^2\rangle c\tau\,\frac{dT}{dx} = -\tfrac{1}{3}n\langle v^2\rangle\ c\tau\,\frac{dT}{dx} \ . \tag{38}$$

If, as for phonons, v is constant, we may write (38) as

$$j_U = -\tfrac{1}{3}Cv\ell\,\frac{dT}{dx} \ , \tag{39}$$

with $\ell \equiv v\tau$ and $C \equiv nc$. Thus $K = \frac{1}{3}Cv\ell$.

Lattice Thermal Resistivity

The phonon mean free path ℓ is determined principally by two processes, geometrical scattering and scattering by other phonons. If the forces between atoms were purely harmonic, there would be no mechanism for collisions between different phonons, and the mean free path would be limited solely by collisions of a phonon with the crystal boundary, and by lattice imperfections. There are situations where these effects are dominant. With anharmonic lattice interactions, there is a coupling between different phonons which limits the value of the mean free path.

The exact normal modes of the anharmonic system are no longer like pure phonons. We consider first the thermal resistivity from phonon interactions.

The theory of the effect of anharmonic coupling on thermal resistivity is a complicated problem.[11] One finds ℓ is proportional to $1/T$ at high temperatures, in agreement with many experiments. We can understand this dependence in terms of the number of phonons with which a given phonon can interact: at high temperature the total number of excited phonons is proportional to T. The collision frequency of a given phonon should be proportional to the number of phonons with which it can collide, whence $\ell \propto 1/T$.

To define a thermal conductivity there must exist mechanisms in the crystal whereby the distribution of phonons may be brought locally into thermal equilibrium. Without such mechanisms we may not speak of the phonons at one end of the crystal as being in thermal equilibrium at a temperature T_2 and those at the other end in equilibrium at T_1. It is not sufficient for thermal conductivity to have only a way of limiting the mean free path, but there must also be a way of establishing a true equilibrium distribution of phonons. Phonon collisions with a static imperfection or a crystal boundary will not by themselves establish thermal equilibrium, because such collisions do not change the energy of individual phonons: the frequency ω_2 of the scattered phonon is equal to the frequency ω_1 of the incident phonon.

It is rather remarkable also that a three-phonon collision process

$$\mathbf{K}_1 + \mathbf{K}_2 = \mathbf{K}_3 \tag{40}$$

will not establish equilibrium, but for a subtle reason: the total momentum of the phonon gas is not changed by such a collision. An equilibrium distribution of phonons at a temperature T can move down the crystal with a drift velocity which is not disturbed by three-phonon collisions of the form (40). For such collisions the phonon momentum

$$\mathbf{J} = \sum_{\mathbf{K}} n_{\mathbf{K}} \hbar \mathbf{K} \tag{41}$$

is conserved, because on collision the change in $\mathbf{J}$ is $\mathbf{K}_3 - \mathbf{K}_2 - \mathbf{K}_1 = 0$. Here $n_{\mathbf{K}}$ is the number of phonons having wavevector $\mathbf{K}$. For a distribution with $\mathbf{J} \neq 0$, collisions such as (40) are incapable of establishing complete thermal equilibrium because they leave $\mathbf{J}$ unchanged. If we start a distribution of hot phonons down a rod with $\mathbf{J} \neq 0$, the distribution will propagate down the rod with $\mathbf{J}$ unchanged. Therefore there is no thermal resistance. The problem as illustrated in Fig. 14 is like that of the collisions between molecules of a gas in a straight tube with frictionless walls.

[11]See J. M. Ziman, *Electrons and phonons*, Oxford, 1960; R. Berman, "Heat conductivity of non-metallic crystals," Contemp. Phys. **14**, 101 (1973).

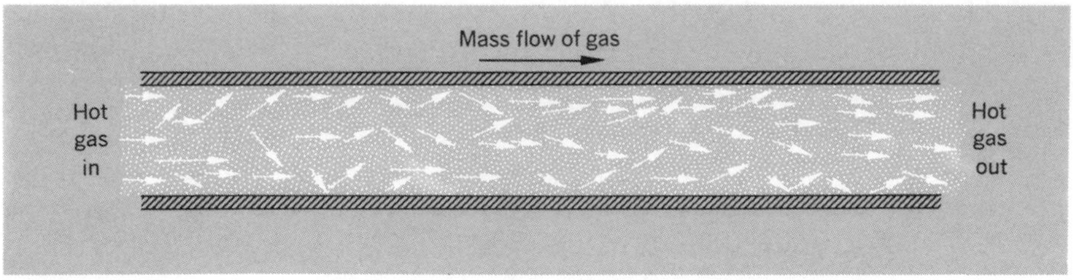

Figure 14a Flow of gas molecules in a state of drifting equilibrium down a long open tube with frictionless walls. Elastic collision processes among the gas molecules do not change the momentum or energy flux of the gas because in each collision the velocity of the center of mass of the colliding particles and their energy remain unchanged. Thus energy is transported from left to right without being driven by a temperature gradient. Therefore the thermal resistivity is zero and the thermal conductivity is infinite.

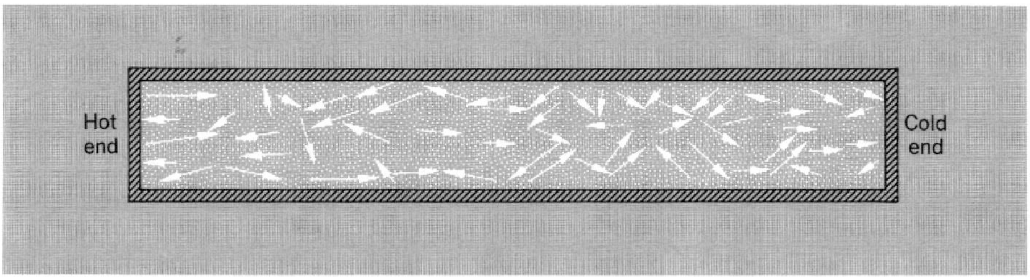

Figure 14b The usual definition of thermal conductivity in a gas refers to a situation where no mass flow is permitted. Here the tube is closed at both ends, preventing the escape or entrance of molecules. With a temperature gradient the colliding pairs with above-average center of mass velocities will tend to be directed to the right, those with below-average velocities will tend to be directed to the left. A slight concentration gradient, high on the right, will be set up to enable the net mass transport to be zero while allowing a net energy transport from the hot to the cold end.

Umklapp Processes

Peierls pointed out that the important three-phonon processes for thermal resistivity are not of the form $\mathbf{K}_1 + \mathbf{K}_2 = \mathbf{K}_3$, as in (40), but are of the form

$$\mathbf{K}_1 + \mathbf{K}_2 = \mathbf{K}_3 + \mathbf{G} \ , \tag{42}$$

where $\mathbf{G}$ is a reciprocal lattice vector (Fig. 15). We recall that $\mathbf{G}$ may occur in all momentum conservation laws in crystal lattices. We have already encountered examples of wave interaction processes in crystals for which the total wavevector change need not be zero, but may be a reciprocal lattice vector. Such processes are always possible in periodic lattices. The argument is particularly strong for phonons: the only meaningful phonon $\mathbf{K}$'s lie in the first Brillouin zone, so that any longer $\mathbf{K}$ produced in a collision must be brought back into the first zone by addition of a $\mathbf{G}$.

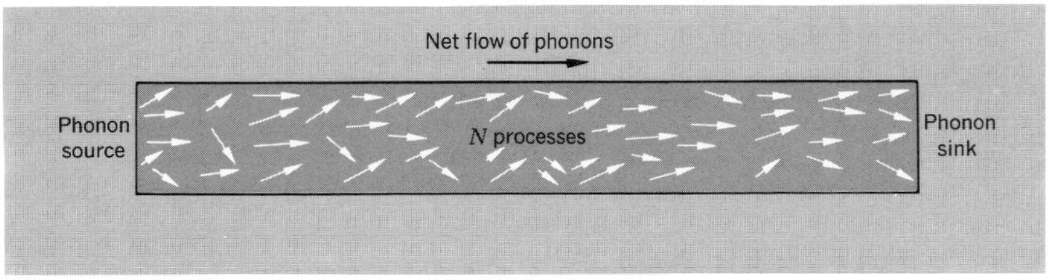

Figure 14c In a crystal we may arrange to create phonons chiefly at one end, as by illuminating the left end with a lamp. From that end there will be a net flux of phonons toward the right end of the crystal. If only *N* processes ($\mathbf{K}_1 + \mathbf{K}_2 = \mathbf{K}_3$) occur, the phonon flux is unchanged in momentum on collision and some phonon flux will persist down the length of the crystal. On arrival of phonons at the right end we can arrange in principle to convert most of their energy to radiation, thereby creating a sink for the phonons. Just as in (a) the thermal resistivity is zero.

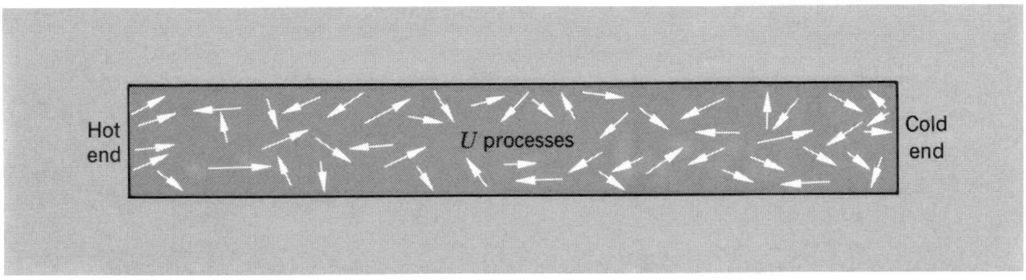

Figure 14d In *U* processes there is a large net change in phonon momentum in each collision event. An initial net phonon flux will rapidly decay as we move to the right. The ends may act as sources and sinks. Net energy transport under a temperature gradient occurs as in (b).

Processes or collisions in which $\mathbf{G} \neq 0$ are called **umklapp processes,** after the German for "flipping over." This term refers to the circumstance (as in Fig. 15b) that a collision of two phonons both having a positive K_x can by umklapp give after collision a phonon with a negative K_x. Umklapp processes are also called *U* processes. Collisions in which $\mathbf{G} = 0$ are called **normal processes** or *N* processes. At high temperatures ($T > \theta$) all phonons are excited because $k_B T > \hbar\omega_{\max}$. A substantial proportion of all phonon collisions will then be *U* processes, with the attendant high momentum change in the collision. We can estimate the thermal resistivity without particular distinction between *N* and *U* processes; by the earlier argument about nonlinear effects we expect to find a lattice thermal resistivity $\propto T$ at high temperatures.

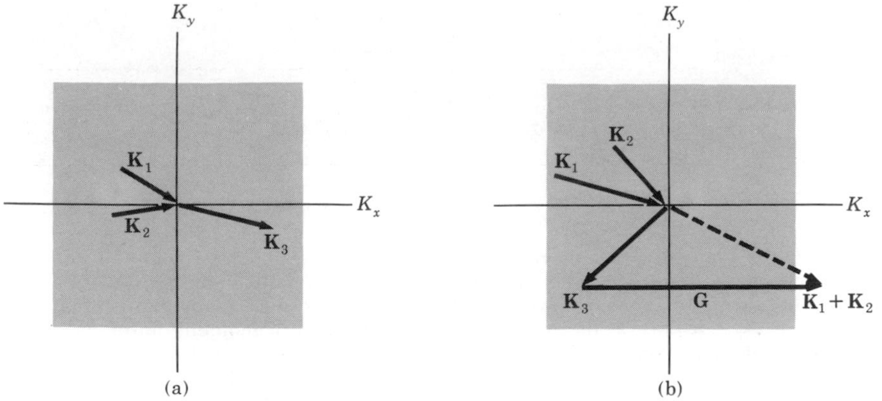

Figure 15 (a) Normal $\mathbf{K}_1 + \mathbf{K}_2 = \mathbf{K}_3$ and (b) umklapp $\mathbf{K}_1 + \mathbf{K}_2 = \mathbf{K}_3 + \mathbf{G}$ phonon collision processes in a two-dimensional square lattice. The square in each figure represents the first Brillouin zone in the phonon **K** space; this zone contains all the possible independent values of the phonon wavevector. Vectors **K** with arrowheads at the center of the zone represent phonons absorbed in the collision process; those with arrowheads away from the center of the zone represent phonons emitted in the collision. We see in (b) that in the umklapp process the direction of the x-component of the phonon flux has been reversed. The reciprocal lattice vector **G** as shown is of length $2\pi/a$, where a is the lattice constant of the crystal lattice, and is parallel to the K_x axis. For all processes, N or U, energy must be conserved, so that $\omega_1 + \omega_2 = \omega_3$.

The energy of phonons $\mathbf{K}_1$, $\mathbf{K}_2$ suitable for umklapp to occur is of the order of $\frac{1}{2}k_B\theta$, because each of the phonons 1 and 2 must have wavevectors of the order of $\frac{1}{2}G$ in order for the collision (42) to be possible. (If both phonons have low K, and therefore low energy, there is no way to get from their collision a phonon of wavevector outside the first zone. The umklapp process must conserve energy, just as for the normal process.) At low temperatures the number of suitable phonons of the high energy $\frac{1}{2}k_B\theta$ required may be expected to vary roughly as $e^{-\theta/2T}$, according to the Boltzmann factor. The exponential form is in good agreement with experiment. In summary, the phonon mean free path which enters (37) is the mean free path for umklapp collisions between phonons and not for all collisions between phonons.

Imperfections

Geometrical effects may also be important in limiting the mean free path. We must consider scattering by crystal boundaries, the distribution of isotopic masses in natural chemical elements, chemical impurities, lattice imperfections, and amorphous structures.

When at low temperatures the mean free path ℓ becomes comparable with the width of the test specimen, the value of ℓ is limited by the width, and the thermal conductivity becomes a function of the dimensions of the

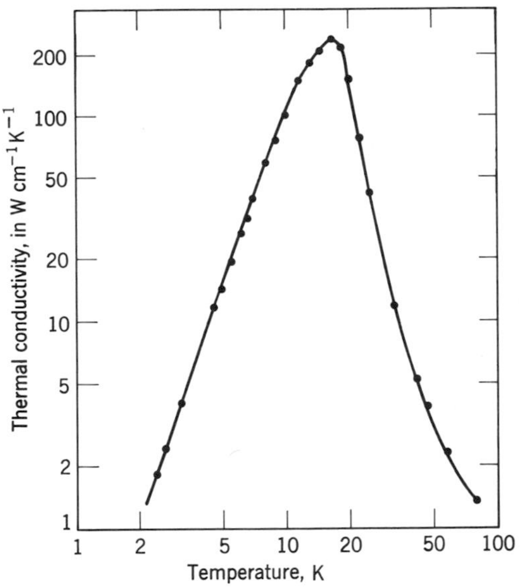

Figure 16 Thermal conductivity of a highly purified crystal of sodium fluoride, after H. E. Jackson, C. T. Walker, and T. F. McNelly, Phys. Rev. Letters **25**, 26 (1970).

specimen. This effect was discovered by de Haas and Biermasz.[12] The abrupt decrease in thermal conductivity of pure crystals at low temperatures is caused by the size effect. At low temperatures the umklapp process becomes ineffective in limiting the thermal conductivity, and the size effect becomes dominant, as shown in Fig. 16. One would expect then that the phonon mean free path would be constant and of the order of the diameter D of the specimen, so that

$$K \approx CvD \ . \tag{43}$$

The only temperature-dependent term on the right is C, the heat capacity, which varies as T^3 at low temperatures. We may therefore expect the thermal conductivity to vary as T^3 at low temperatures. The size effect enters whenever the phonon mean free path becomes comparable with the diameter of the specimen.

[12]W. J. de Haas and T. Biermasz, Physica **2,** 673 (1935); **4,** 752 (1937); **5,** 47, 320, 619 (1938); see also R. Berman, Proc. Roy. Soc. (London) **A208,** 90 (1951); P. D. Thacher, Phys. Rev. **156,** 975 (1967).

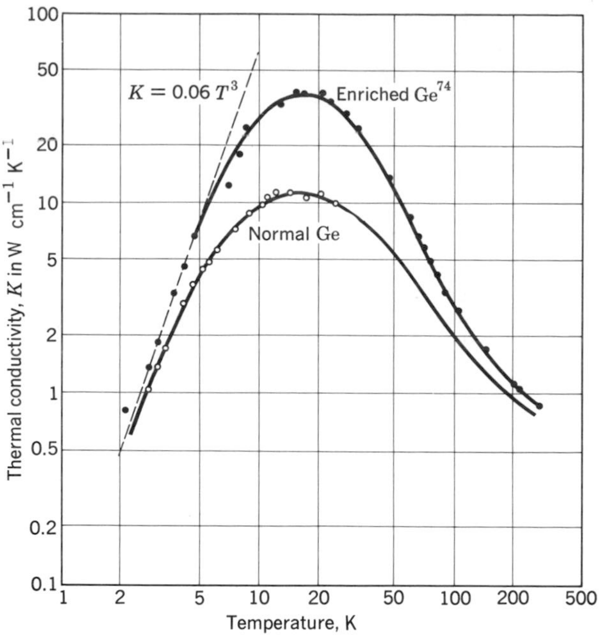

Figure 17 Isotope effect on thermal conduction in germanium, amounting to a factor of three at the conductivity maximum. The enriched specimen is 96 percent Ge^{74}; natural germanium is 20 percent Ge^{70}, 27 percent Ge^{72}, 8 percent Ge^{73}, 37 percent Ge^{74}, and 8 percent Ge^{76}. Below 5 K the enriched specimen has $K = 0.060\ T^3$, which agrees well with Casimir's theory for thermal resistance caused by boundary scattering. [After T. H. Geballe and G. W. Hull, Phys. Rev. **110,** 773 (1958).]

In an otherwise perfect crystal the distribution of isotopes of the chemical elements often provides an important mechanism for phonon scattering. (The random distribution of isotopes disturbs the periodicity of the density as seen by an elastic wave.) In some substances the importance of scattering of phonons by isotopes is comparable to scattering by other phonons. Results for germanium are shown in Fig. 17.

In glasses the thermal conductivity (Fig. 18) decreases as the temperature is lowered. The values of the thermal conductivity at room temperature run about an order of magnitude lower for glasses than for crystals. The mean free path in quartz glass at room temperature is 8 Å, which is of the order of magnitude of the dimensions of a silicon dioxide tetrahedron (7 Å).

Dielectric crystals may have thermal conductivities as high as metals. Synthetic sapphire (Al_2O_3) has one of the highest values of the conductivity: nearly 200 W cm^{-1} K^{-1} at 30 K. The maximum of the thermal conductivity in sapphire is greater than the maximum of 100 W cm^{-1} K^{-1} in copper.[13] Metallic gallium, however, has a conductivity[14] of 845 W cm^{-1} K^{-1} at 1.8 K. The electronic contribution to the thermal conductivity of metals is treated in Chapter 6.

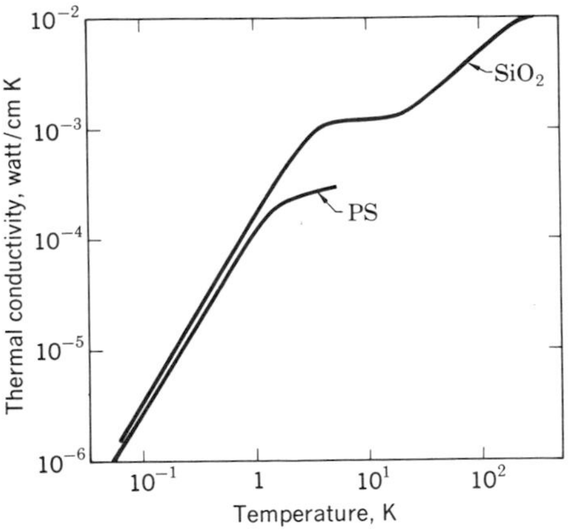

Figure 18 Thermal conductivity of fused silica (SiO_2) and amorphous polystyrene (PS), after R. B. Stephens, G. S. Cieloszyk, and G. L. Salinger, Phys. Letters **38A**, 215 (1972). The measurements on fused silica were by R. C. Zeller and R. O. Pohl, Phys. Rev. **B4**, 2029 (1971).

PROBLEMS

1. ***Singularity in density of modes.*** (a) From the dispersion relation derived in Chapter 4 for a monatomic linear lattice of N atoms with nearest neighbor interactions, show that the density of modes is

$$\mathfrak{D}(\omega) = \frac{2N}{\pi} \cdot \frac{1}{(\omega_m{}^2 - \omega^2)^{1/2}} ,$$

where ω_m is the maximum frequency. (b) Suppose that an optical phonon branch has the form $\omega(K) = \omega_0 - AK^2$, near $K = 0$ in three dimensions. Show that $\mathfrak{D}(\omega) = (L/2\pi)^3(2\pi/A^{3/2})(\omega_0 - \omega)^{1/2}$ for $\omega < \omega_0$ and $\mathfrak{D}(\omega) = 0$ for $\omega > \omega_0$. Here the density of modes is discontinuous.

2. ***Rms thermal dilation of crystal cell.*** (a) Estimate for 300 K the root mean square thermal dilation $\Delta V/V$ for a primitive cell of sodium. Take the bulk modulus as 7×10^{10} erg cm^{-3}. Note that the Debye temperature 158 K is less than 300 K, so that the thermal energy is of the order of k_BT. (b) Use this result to estimate the root mean square thermal fluctuation $\Delta a/a$ of the lattice parameter.

3. ***Zero point lattice displacement and strain.*** (a) In the Debye approximation, show that the mean square displacement of an atom at absolute zero is $\langle R^2 \rangle = 3\hbar\omega_D{}^2/8\pi^2\rho v^3$, where v is the velocity of sound. Start from the result (4.29) summed over the independent lattice modes: $\langle R^2 \rangle = (\hbar/2\rho V)\Sigma\omega^{-1}$. We have included a factor of $\frac{1}{2}$ to go from square amplitude to square displacement. (b) Show that $\Sigma\omega^{-1}$ and $\langle R^2 \rangle$ diverge for a one-dimensional lattice, but that the mean square strain is finite. Consider $\langle (\partial R/\partial x)^2 \rangle = \frac{1}{2}\Sigma K^2 u_0{}^2$ as the mean square strain, and show that it is equal to $\hbar\omega_D{}^2L/4\pi MNv^3$ for a line of N atoms each of mass M, counting longitudinal modes only. The divergence of R^2 is not significant for any physical measurement.

[13]R. Berman and D. K. C. MacDonald, Proc. Roy. Soc. (London) **A211,** 122 (1952); the peak value of the conductivity of NaF in the experiment of Fig. 16 is 240 W cm^{-1} K^{-1}.
[14]R. I. Boughton and M. Yaqub, Phys. Rev. Letters **20,** 108 (1968).

4. ***Heat capacity of one-dimensional lattice and layer lattice.*** (a) Show that the heat capacity of a monatomic lattice in one dimension in the Debye approximation is proportional to T/θ for low temperatures $T \ll \theta$, where θ is the effective Debye temperature in one dimension defined as $\theta = \hbar\omega_m/k_B = \hbar\pi v_0/k_B a$; here k_B is the Boltzmann constant and a the interatomic separation. (b) Consider a dielectric crystal made up of layers of atoms, with adjacent layers very weakly bound to each other. What form would you expect the heat capacity to approach at very low temperatures?

*5. ***Gruneisen constant.*** (a) Show that the free energy of a phonon mode of frequency ω is $k_B T \ln [2 \sinh (\hbar\omega/2k_B T)]$. It is necessary to retain the zero-point energy $\frac{1}{2}\hbar\omega$ to obtain this result. (b) If Δ is the fractional volume change, then the free energy of the crystal may be written as

$$F(\Delta,T) = \tfrac{1}{2}B\Delta^2 + k_B T \sum \ln [2 \sinh (\hbar\omega_{\mathbf{K}}/2k_B T)] \ ,$$

where B is the bulk modulus. Assume that the volume dependence of $\omega_{\mathbf{K}}$ is $\delta\omega/\omega = -\gamma\Delta$, where γ is known as the Grüneisen constant. If γ is taken as independent of the mode **K**, show that F is a minimum with respect to Δ when $B\Delta = \gamma\Sigma\frac{1}{2}\hbar\omega \coth (\hbar\omega/2k_B T)$, and show that this may be written in terms of the thermal energy density as $\Delta = \gamma U(T)/B$. (c) Show that on the Debye model $\gamma = -\partial \ln \theta/\partial \ln V$. Note: Many approximations are involved in this theory: the result (a) is valid only if ω is independent of temperature; γ may be quite different for different modes.

†6. ***Density of modes of square lattice.*** The dispersion relation of a square lattice with nearest-neighbor interactions was found in Problem 4.1 to be $\omega^2 M = 2C(2 - \cos K_x a - \cos K_y a)$. Note that the Brillouin zone can be divided into eight equivalent sectors. If you have access to a small computer, divide a sector into around 100 squares, evaluate ω at the center of each square (or at common corners), and plot a histogram of the number of squares per unit frequency range. Sketch $\mathfrak{D}(\omega)$ as a function of ω. In the calculation take $2C = 1$ and $M = 1$.

References

R. A. Cowley, "Anharmonic crystals," Repts. Prog. Phys. **31**, pt. 1, 123–166 (1968).

G. Leibfried and W. Ludwig, "Theory of anharmonic effects in crystals," *Solid state physics* **12**, 276–444 (1961).

T. Riste, ed., *Anharmonic lattices, structural transitions and melting*, Noordhoff, 1964.

B. Yates, *Thermal expansion*, Plenum, 1972.

Thermal Conductivity

J. E. Parrott and A. D. Stuckes, *Thermal conductivity of Solids*, Academic Press, 1975.

P. G. Klemens, "Thermal conductivity and lattice vibration modes," *Solid state physics* **7**, 1–98 (1958); see also *Encyclo. of physics* **14**, 198 (1956).

C. Y. Ho, R. W. Powell and P. E. Liley, *Thermal conductivity of the elements: A comprehensive review*, (J. of Phys. and Chem. Ref. Data, Vol. 3, Supplement 1), AIP and ACS for the National Bureau of Standards, 1975.

H. M. Rosenberg, *Low temperature solid state physics*, Oxford, 1963, Chap. 3.

R. P. Tye, ed., *Thermal conductivity*, Academic Press, 1969.

J. M. Ziman, *Electrons and phonons*, Oxford, 1960, Chap. 8.

*This problem is somewhat difficult.

†This problem is tedious without a computer.

6 FREE ELECTRON FERMI GAS

TABLE 1 Calculated Free Electron Fermi Surface Parameters for Metals at Room Temperature
(Except for Na, K, Rb, Cs at 5 K and Li at 78 K)

Valency	Metal	Electron Concentration, in cm^{-3}	Radius[a] Parameter r_s	Fermi Wavevector, in cm^{-1}	Fermi Velocity, in $cm\ s^{-1}$	Fermi Energy, in eV	Fermi Temperature, $T_F \equiv \epsilon_F/k_B$, in deg K
1	Li	4.70×10^{22}	3.25	1.11×10^{8}	1.29×10^{8}	4.72	5.48×10^{4}
	Na	2.65	3.93	0.92	1.07	3.23	3.75
	K	1.40	4.86	0.75	0.86	2.12	2.46
	Rb	1.15	5.20	0.70	0.81	1.85	2.15
	Cs	0.91	5.63	0.64	0.75	1.58	1.83
	Cu	8.45	2.67	1.36	1.57	7.00	8.12
	Ag	5.85	3.02	1.20	1.39	5.48	6.36
	Au	5.90	3.01	1.20	1.39	5.51	6.39
2	Be	24.2	1.88	1.93	2.23	14.14	16.41
	Mg	8.60	2.65	1.37	1.58	7.13	8.27
	Ca	4.60	3.27	1.11	1.28	4.68	5.43
	Sr	3.56	3.56	1.02	1.18	3.95	4.58
	Ba	3.20	3.69	0.98	1.13	3.65	4.24
	Zn	13.10	2.31	1.57	1.82	9.39	10.90
	Cd	9.28	2.59	1.40	1.62	7.46	8.66
3	Al	18.06	2.07	1.75	2.02	11.63	13.49
	Ga	15.30	2.19	1.65	1.91	10.35	12.01
	In	11.49	2.41	1.50	1.74	8.60	9.98
4	Pb	13.20	2.30	1.57	1.82	9.37	10.87
	Sn(*w*)	14.48	2.23	1.62	1.88	10.03	11.64

[a] The dimensionless radius parameter is defined as $r_s = r_0/a_H$, where a_H is the first Bohr radius and r_0 is the radius of a sphere that contains one electron.

6 FREE ELECTRON FERMI GAS

In a theory which has given results like these, there must certainly be a great deal of truth.

H. A. Lorentz

We can understand a number of important physical properties of metals, particularly the simple metals, in terms of the free electron model. According to this model the most weakly bound electrons of the constituent atoms move about freely through the volume of the metal. The valence electrons of the atoms become the conduction electrons. Forces between conduction electrons and ion cores are neglected in the free electron approximation: all calculations proceed as if the conduction electrons were free to move everywhere within the specimen. The total energy is all kinetic energy; the potential energy is neglected.

Even in metals for which the free electron model works well, the actual charge distribution of the conduction electrons is known to reflect the strong electrostatic potential of the ion cores. The usefulness of the free electron model is greatest for experiments that depend essentially upon the kinetic properties of the conduction electrons. Later chapters are concerned with the conduction electron interaction with the lattice.

The simplest metals are the alkali metals (lithium, sodium, potassium, cesium, and rubidium). Conduction electrons in all metals are observed to act very much like free electrons, except those metals where the d or f shell electrons overlap or fall close in energy to the conduction band. Electrons in d shells tend to be more localized and less mobile than electrons in s and p shells. Other simple metals are Be, Mg, Ca, Sr, Ba, Al, Ga, In, Tl, and Pb. Metals that are not simple are the noble metals (copper, silver, gold), the transition metals, and the lanthanide and actinide metals.

Conduction electrons in a simple metal arise from the valence electrons of the constituent atoms. In a sodium atom the valence electron is in a $3s$ state; in the metal this electron becomes a conduction electron, roving throughout the crystal. A monovalent crystal which contains N atoms will have N conduction electrons and N positive ion cores. The 10 electrons of the Na^+ ion core fill the $1s$, the $2s$, and the $2p$ states in the free ion; the distribution of core electrons is essentially the same in the metal as in the free ion. The ion cores fill only about 15 percent of the volume of a

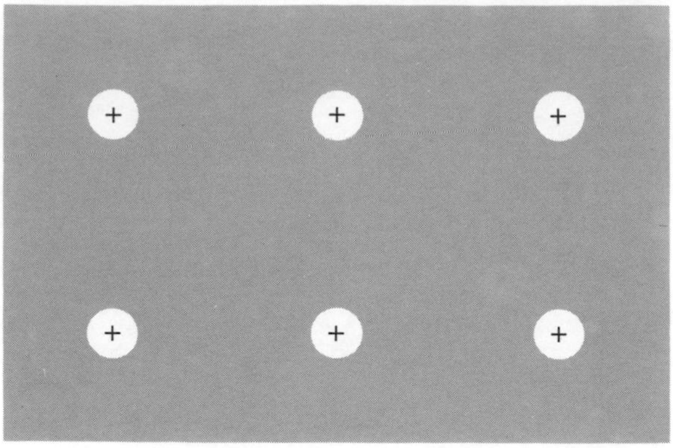

Figure 1 Schematic model of a crystal of sodium metal. The atomic cores are Na^+ ions; they are immersed in a sea of conduction electrons. The conduction electrons are derived from the 3*s* valence electrons of the free atoms. The atomic cores contain 10 electrons in the configuration $1s^2\ 2s^2\ 2p^6$. In an alkali metal the atomic cores occupy a relatively small part (~15 percent) of the total volume of the crystal, but in a noble metal (Cu, Ag, Au) the atomic cores are relatively larger and may be in contact with each other. The common crystal structure at room temperature is bcc for the alkali metals and fcc for the noble metals.

sodium crystal, as in Fig. 1. The radius of the free Na^+ ion is 0.98 Å, whereas one-half of the nearest-neighbor distance of the metal is 1.83 Å.

The interpretation of metallic properties in terms of the motion of free electrons was developed long before the invention of quantum mechanics. The classical theory had several conspicuous successes, notably the derivation of the form of Ohm's law and the relation between the electrical and thermal conductivity. The classical theory fails to explain the heat capacity and the magnetic susceptibility of the conduction electrons.

There is a further difficulty. From many types of experiments it is clear that a conduction electron in a metal can move freely in a straight path over many atomic distances, undeflected by collisions with other conduction electrons or by collisions with the atom cores. In a very pure specimen at low temperatures the mean free path may be as long as 10^8 interatomic spacings (more than 1 cm). Why is condensed matter so transparent to conduction electrons? The answer to our question contains two parts: (a) A conduction electron is not deflected by ion cores arranged on a *periodic* lattice because matter waves propagate freely in a periodic structure. We showed free propagation of x-rays in periodic lattices in Chapter 2. (b) A conduction electron is scattered only infrequently by other conduction electrons. This property we shall show is a consequence of the Pauli exclusion principle. By a **free electron Fermi gas,** we mean a gas of free and noninteracting electrons subject to the Pauli principle.

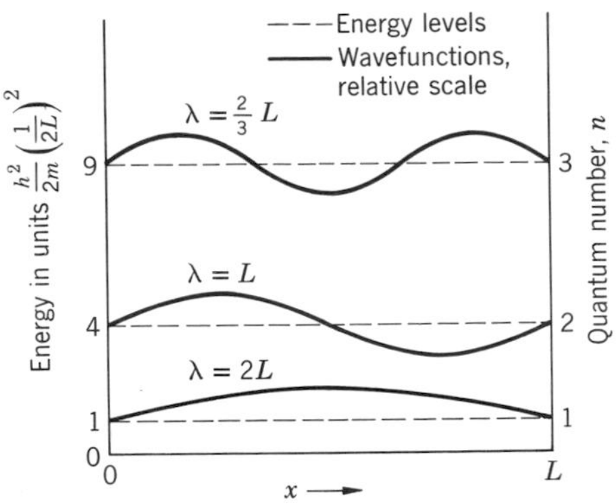

Figure 2 First three energy levels and wavefunctions of a free electron of mass m confined to a line of length L. The energy levels are labeled according to the quantum number n which gives the number of half-wavelengths in the wavefunction. The wavelengths are indicated on the wavefunctions. The energy ϵ_n of the level of quantum number n is equal to $(h^2/2m)(n/2L)^2$.

ENERGY LEVELS AND DENSITY OF ORBITALS IN ONE DIMENSION

Consider a free electron gas in one dimension, taking account of quantum theory and of the Pauli principle. An electron of mass m is confined to a length L by infinite barriers (Fig. 2). The wavefunction $\psi_n(x)$ of the electron is a solution of the Schrödinger equation $\mathcal{H}\psi = \epsilon\psi$; with the neglect of potential energy we have $\mathcal{H} = p^2/2m$, where p is the momentum. In quantum theory p may be represented by $-i\hbar\, d/dx$, so that

$$\psi_n = -\frac{\hbar^2}{2m}\frac{d^2\psi_n}{dx^2} = \epsilon_n\psi_n \ , \tag{1}$$

where ϵ_n is the energy of the electron in the orbital. We use the term **orbital** to denote a solution of the wave equation for a system of only one electron. The term allows us to distinguish between an exact quantum state of the wave equation of a system of N electrons and an approximate quantum state which we construct by assigning the N electrons to N different orbitals, where each orbital is a solution of a wave equation for one electron. The orbital model is exact only if there are no interactions between electrons.

The boundary conditions are $\psi_n(0) = 0$; $\psi_n(L) = 0$, as imposed by the infinite potential energy barrier; they are satisfied if the wavefunction is sinelike with an integral number n of half-wavelengths between 0 and L:

$$\psi_n = A\sin\left(\frac{2\pi}{\lambda_n}x\right) \ ; \qquad \tfrac{1}{2}n\lambda_n = L \ , \tag{2}$$

where A is a constant. We see that (2) is a solution of (1), because

$$\frac{d\psi_n}{dx} = A\left(\frac{n\pi}{L}\right)\cos\left(\frac{n\pi}{L}x\right) \ ; \qquad \frac{d^2\psi_n}{dx^2} = -A\left(\frac{n\pi}{L}\right)^2\sin\left(\frac{n\pi}{L}x\right) \ ,$$

whence the energy ϵ_n is given by

$$\epsilon_n = \frac{\hbar^2}{2m}\left(\frac{n\pi}{L}\right)^2 . \tag{3}$$

We want to accommodate N electrons on the line. According to the **Pauli exclusion principle** no two electrons can have all their quantum numbers identical. That is, each orbital can be occupied by at most one electron. This applies to electrons in atoms, molecules, or solids. In a linear solid the quantum numbers of a conduction electron orbital are n and m_s, where n is any positive integer and $m_s = \pm\frac{1}{2}$, according to the spin orientation. A pair of orbitals labeled by the quantum number n can accommodate two electrons, one with spin up and one with spin down. If there are six electrons, then in the ground state of the system the filled orbitals are those given in the table:

n	m_s	*Electron Occupancy*	n	m_s	*Electron Occupancy*
1	↑	1	3	↑	1
1	↓	1	3	↓	1
2	↑	1	4	↑	0
2	↓	1	4	↓	0

More than one orbital may have the same energy. The number of orbitals with the same energy is called the **degeneracy.**

Let n_F denote the topmost filled energy level, where we start filling the levels from the bottom ($n = 1$) and continue filling the higher levels with electrons until all N electrons are accommodated. It is convenient to suppose that N is an even number. The condition $2n_F = N$ determines n_F, the value of n for the uppermost filled level. The **Fermi energy** ϵ_F is defined as the energy of the topmost filled level in the ground state. By (3) with $n = n_F$ we have in one dimension:

$$\epsilon_F = \frac{\hbar^2}{2m}\left(\frac{n_F\pi}{L}\right)^2 = \frac{\hbar^2}{2m}\left(\frac{N\pi}{2L}\right)^2 . \tag{4}$$

EFFECT OF TEMPERATURE ON THE FERMI-DIRAC DISTRIBUTION

The ground state is the state of the system at absolute zero. What happens as the temperature is increased? This is a standard problem in elementary statistical mechanics and the solution is given by the Fermi-Dirac distribution function. The kinetic energy of the electron gas increases as the temperature is increased: some energy levels are occupied which were vacant at absolute zero, and some levels are vacant which were occupied at absolute zero (Fig. 3). The **Fermi-Dirac distribution**

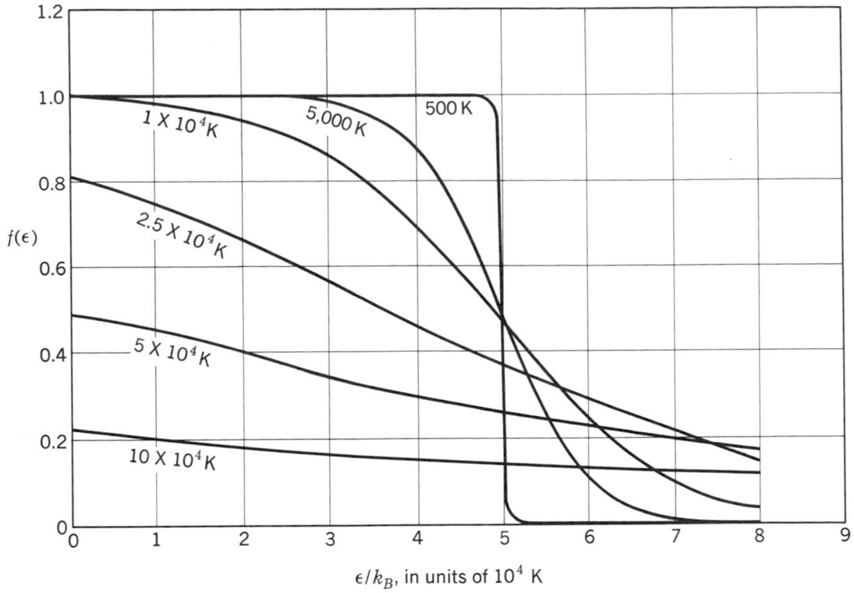

Figure 3 Fermi-Dirac distribution function at various temperatures, for $T_F \equiv \epsilon_F/k_B = 50{,}000$ K. The results apply to a gas in three dimensions. The total number of particles is constant, independent of temperature. (Courtesy of B. Feldman.)

gives the probability that an orbital at energy ϵ will be occupied in an ideal electron gas in thermal equilibrium:

$$f(\epsilon) = \frac{1}{e^{(\epsilon-\mu)/k_BT} + 1} \quad . \tag{5}$$

The quantity μ is a function of the temperature; μ is to be chosen for the particular problem in such a way that the total number of particles in the system comes out correctly—that is, equal to N. At absolute zero $\mu = \epsilon_F$, because in the limit $T \rightarrow 0$ the function $f(\epsilon)$ changes discontinuously from the value 1 (filled) to the value 0 (empty) at $\epsilon = \epsilon_F = \mu$. *At all temperatures $f(\epsilon)$ is equal to $\frac{1}{2}$ when $\epsilon = \mu$*, for then the denominator of (5) has the value 2. The quantity μ is called the **chemical potential**,[1] and we see that at absolute zero the chemical potential is equal to the Fermi energy. The Fermi energy was defined above as the energy of the topmost filled orbital at absolute zero.

The high energy tail of the distribution is that part for which $\epsilon - \mu \gg k_BT$; here the exponential term is dominant in the denominator of (5), so that $f(\epsilon) \cong e^{(\mu-\epsilon)/k_BT}$. This is the Boltzmann distribution.

[1]In the presence of an external field ϵ is changed but so also is μ. The new value of μ is usually called the electrochemical potential. The term chemical potential is often reserved for the electrochemical potential minus the potential energy of the particle due to the external field. Such a usage is common in semiconductor physics in connection with junction devices. In *TP*, chemical potential means the same thing as the electrochemical potential. Thermodynamic functions of a Fermi gas are tabulated in *TP*, Appendix C.

FREE ELECTRON GAS IN THREE DIMENSIONS

The free-particle Schrödinger equation in three dimensions is

$$-\frac{\hbar^2}{2m}\left(\frac{\partial^2}{\partial x^2}+\frac{\partial^2}{\partial y^2}+\frac{\partial^2}{\partial z^2}\right)\psi_{\mathbf{k}}(\mathbf{r})=\epsilon_{\mathbf{k}}\,\psi_{\mathbf{k}}(\mathbf{r}) \ . \tag{6}$$

If the electrons are confined to a cube of edge L, the wavefunction is the standing wave

$$\psi_n(\mathbf{r})=A\sin(\pi n_x x/L)\sin(\pi n_y y/L)\sin(\pi n_z z/L) \ , \tag{7}$$

where n_x, n_y, n_z are positive integers.

It is convenient to introduce wavefunctions that satisfy periodic boundary conditions, as we did for phonons in Chapter 5. We now require the wavefunctions to be periodic in x, y, z with period L. Thus

$$\psi(x+L,y,z)=\psi(x,y,z) \ , \tag{8}$$

and similarly for the y and z coordinates. Wavefunctions satisfying the free-particle Schrödinger equation and the periodicity condition are of the form of a traveling plane wave:

$$\boxed{\psi_{\mathbf{k}}(\mathbf{r})=\exp(i\mathbf{k}\cdot\mathbf{r}) \ ,} \tag{9}$$

provided that the components of the wavevector $\mathbf{k}$ satisfy

$$k_x=0 \ ; \quad \pm\frac{2\pi}{L} \ ; \quad \pm\frac{4\pi}{L} \ ; \quad \ldots \ , \tag{10}$$

and similarly for k_y and k_z. That is, any component of $\mathbf{k}$ is of the form $2n\pi/L$, where n is a positive or negative integer. The components of $\mathbf{k}$ are the quantum numbers of the problem, along with the quantum number m_s for the spin direction. We confirm that these values of k_x satisfy (8), for

$$\begin{aligned}\exp[ik_x(x+L)]&=\exp[i2n\pi(x+L)/L]\\&=\exp(i2n\pi x/L)\exp(i2n\pi)=\exp(i2n\pi x/L)=\exp(ik_x x) \ .\end{aligned} \tag{11}$$

On substituting (9) in (6) we have the energy $\epsilon_{\mathbf{k}}$ of the orbital with wavevector $\mathbf{k}$:

$$\epsilon_{\mathbf{k}}=\frac{\hbar^2}{2m}k^2=\frac{\hbar^2}{2m}(k_x{}^2+k_y{}^2+k_x{}^2) \ . \tag{12}$$

The magnitude of the wavevector is related to the wavelength by $k=2\pi/\lambda$.

The linear momentum $\mathbf{p}$ may be represented in quantum mechanics by the operator $\mathbf{p}=-i\hbar\nabla$, whence for the orbital (9)

$$\mathbf{p}\psi_{\mathbf{k}}(\mathbf{r})=-i\hbar\nabla\psi_{\mathbf{k}}(\mathbf{r})=\hbar\mathbf{k}\psi_{\mathbf{k}}(\mathbf{r}) \ , \tag{13}$$

so that the plane wave $\psi_{\mathbf{k}}$ is an eigenfunction of the linear momentum with the eigenvalue $\hbar\mathbf{k}$. The particle velocity in the orbital $\mathbf{k}$ is given by $\mathbf{v} = \hbar\mathbf{k}/m$.

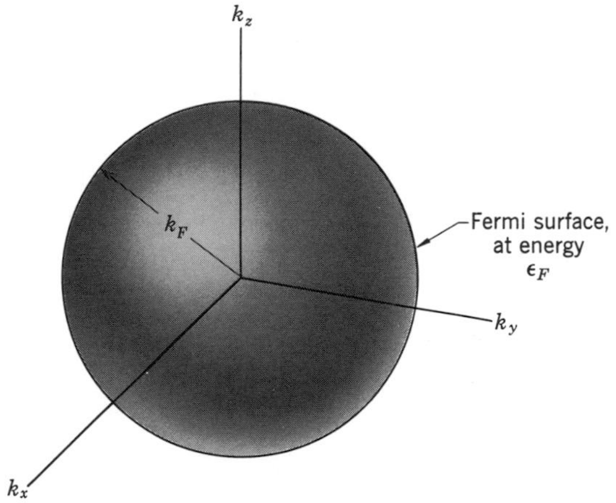

Figure 4 In the ground state of a system of N free electrons the occupied orbitals of the system fill a sphere of radius k_F, where $\epsilon_F = \hbar^2 k_F^2/2m$ is the energy of an electron having a wavevector k_F.

In the ground state of a system of N free electrons the occupied orbitals may be represented as points inside a sphere in $\mathbf{k}$ space. The energy at the surface of the sphere is the Fermi energy; the wavevectors at the Fermi surface have a magnitude k_F such that (Fig. 4):

$$\epsilon_F = \frac{\hbar^2}{2m} k_F^2 \ . \tag{14}$$

From (10) we see that there is one allowed wavevector — that is, one distinct triplet of quantum numbers k_x, k_y, k_z — for the volume element $(2\pi/L)^3$ of $\mathbf{k}$ space. Thus in the sphere of volume $4\pi k_F^3/3$ the total number of orbitals is

$$2 \cdot \frac{4\pi k_F^3/3}{(2\pi/L)^3} = \frac{V}{3\pi^2} k_F^3 = N \ , \tag{15}$$

where the factor 2 on the left comes from the two allowed values of m_s, the spin quantum number, for each allowed value of $\mathbf{k}$. Then

$$k_F = \left(\frac{3\pi^2 N}{V}\right)^{1/3} \ , \tag{16}$$

which depends only on the particle concentration and not on the mass.

Using (14),
$$\boxed{\epsilon_F = \frac{\hbar^2}{2m}\left(\frac{3\pi^2 N}{V}\right)^{2/3}} \tag{17}$$

This relates the Fermi energy to the electron concentration N/V. The electron velocity v_F at the Fermi surface is

$$v_F = \left(\frac{\hbar k_F}{m}\right) = \left(\frac{\hbar}{m}\right)\left(\frac{3\pi^2 N}{V}\right)^{1/3} \tag{18}$$

Calculated values of k_F, v_F, and ϵ_F are given in Table 1 for selected metals; also given are values of the quantity T_F which is defined as ϵ_F/k_B. (The quantity T_F has nothing to do with the temperature of the electron gas!)

We now find an expression for the number of orbitals per unit energy range, $\mathfrak{D}(\epsilon)$, often called the density of states. We use (17) for the total number of orbitals of energy $\leq \epsilon$:

$$N = \frac{V}{3\pi^2}\left(\frac{2m\epsilon}{\hbar^2}\right)^{3/2} , \tag{19}$$

so that the density of orbitals (Fig. 5) is

$$\boxed{\mathfrak{D}(\epsilon) \equiv \frac{dN}{d\epsilon} = \frac{V}{2\pi^2} \cdot \left(\frac{2m}{\hbar^2}\right)^{3/2} \cdot \epsilon^{1/2} .} \tag{20}$$

This result may be obtained and expressed most simply by writing (19) as

$$\ln N = \frac{3}{2}\ln \epsilon + \text{constant} \; ; \qquad \frac{dN}{N} = \frac{3}{2} \cdot \frac{d\epsilon}{\epsilon} ,$$

whence

$$\mathfrak{D}(\epsilon) \equiv \frac{dN}{d\epsilon} = \frac{3N}{2\epsilon} . \tag{21}$$

Within a factor of the order of unity, the number of orbitals per unit energy range at the Fermi energy is just the total number of conduction electrons divided by the Fermi energy.

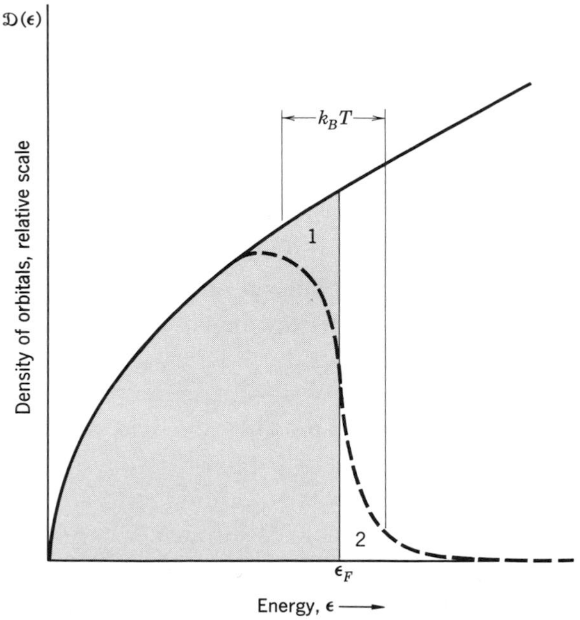

Figure 5 Density of single-particle states as a function of energy, for a free electron gas in three dimensions. The dashed curve represents the density $f(\epsilon,T)\mathfrak{D}(\epsilon)$ of filled orbitals at a finite temperature, but such that k_BT is small in comparison with ϵ_F. The shaded area represents the filled orbitals at absolute zero. The average energy is increased when the temperature is increased from 0 to T, for electrons are thermally excited from region 1 to region 2.

HEAT CAPACITY OF THE ELECTRON GAS

The question that caused the greatest difficulty in the early development of the electron theory of metals concerns the heat capacity of the conduction electrons. Classical statistical mechanics predicts that a free particle should have a heat capacity of $\frac{3}{2}k_B$, where k_B is the Boltzmann constant. If N atoms each give one valence electron to the electron gas, and the electrons are freely mobile, then the electronic contribution to the heat capacity should be $\frac{3}{2}Nk_B$. But the observed electronic contribution at room temperature is usually less than 0.01 of this value. This discrepancy distracted the early workers, such as Lorentz: how can the electrons participate in electrical conduction processes as if they were mobile, while not contributing to the heat capacity? The question was answered only upon the discovery of the Pauli exclusion principle and the Fermi distribution function. Fermi found the correct equation, and he wrote, "One recognizes that the specific heat vanishes at absolute zero and that at low temperatures it is proportional to the absolute temperature."

When we heat the specimen from absolute zero not every electron gains an energy $\sim k_BT$ as expected classically, but only those electrons in orbitals within an energy range k_BT of the Fermi level are excited thermally; these electrons gain an energy which is itself of the order of k_BT, as in Fig. 5. This gives an immediate qualitative solution to the problem of the heat capacity of the conduction electron gas. If N is the total number of electrons, only a fraction of the order of T/T_F can be excited thermally at temperature T, because only these lie within an energy range of the order of k_BT of the top of the energy distribution. Each of these NT/T_F electrons has a thermal energy of the order of k_BT, and so the total electronic thermal energy U is of the order of

$$U \approx (NT/T_F)k_BT \ .$$

The electronic heat capacity is given by

$$C_{el} = \partial U/\partial T \approx Nk_B(T/T_F) \tag{22}$$

and is directly proportional to T, in agreement with the experimental results discussed in the following section. At room temperature C_{el} is smaller than the classical value $\frac{3}{2}Nk_B$ by a factor of the order of 0.01 or less, for $T_F \sim 5 \times 10^4$ K.

We now derive a quantitative expression for the electronic heat capacity valid at low temperatures $k_BT \ll \epsilon_F$. The argument is simple, but devious. The increase U in the total energy (Fig. 5) of a system of N electrons when heated from 0 to T is

$$U = \int_0^\infty d\epsilon \ \epsilon \ \mathfrak{D}(\epsilon) f(\epsilon) - \int_0^{\epsilon_F} d\epsilon \ \epsilon \ \mathfrak{D}(\epsilon) \ . \tag{23}$$

Here $f(\epsilon)$ is the Fermi-Dirac function and $\mathcal{D}(\epsilon)$ is the number of orbitals per unit energy range. We multiply the number of particles

$$N = \int_0^\infty d\epsilon \; \mathcal{D}(\epsilon)\, f(\epsilon)$$

by ϵ_F to obtain

$$\epsilon_F N = \epsilon_F \int_0^\infty d\epsilon \; \mathcal{D}(\epsilon)\, f(\epsilon) \; , \qquad (24)$$

independent of temperature. We now differentiate (23) and (24):

$$C_{el} = \frac{\partial U}{\partial T} = \int_0^\infty d\epsilon \; \epsilon \; \mathcal{D}(\epsilon) \frac{\partial f}{\partial T} \; ;$$

$$0 = \epsilon_F \frac{\partial N}{\partial T} = \int_0^\infty d\epsilon \; \epsilon_F \; \mathcal{D}(\epsilon) \frac{\partial f}{\partial T} \; ,$$

and subtract the second line from the first to find the electronic heat capacity in the form

$$C_{el} = \int_0^\infty d\epsilon \; (\epsilon - \epsilon_F) \; \mathcal{D}(\epsilon) \frac{\partial f}{\partial T} \; .$$

At the low temperatures ($k_BT/\epsilon_F < 0.01$) of interest the derivative $\partial f/\partial T$ is large only at energies near ϵ_F, so that we may evaluate $\mathcal{D}(\epsilon)$ at ϵ_F and take it outside of the integrand:

$$C_{el} \cong \mathcal{D}(\epsilon_F) \int_0^\infty d\epsilon \; (\epsilon - \epsilon_F) \frac{\partial f}{\partial T} \; .$$

Examination[2] of Fig. 3 suggests that to the first order in T we may replace the chemical potential μ by the constant Fermi energy ϵ_F as defined by $\epsilon_F = \mu(0)$. Then

$$\frac{\partial f}{\partial T} = \frac{\epsilon - \epsilon_F}{k_B T^2} \cdot \frac{e^{(\epsilon - \epsilon_F)/k_BT}}{[e^{(\epsilon - \epsilon_F)/k_BT} + 1]^2} \; ,$$

$$\boxed{C_{el} = \tfrac{1}{3}\pi^2 \mathcal{D}(\epsilon_F) k_B^2 T \; .} \qquad (25)$$

From (21) we have for a free electron gas $\mathcal{D}(\epsilon_F) = 3N/2\epsilon_F$. Thus

$$C_{el} = \frac{1}{2}\, \pi^2 N k_B \cdot \frac{k_B T}{\epsilon_F} = \frac{1}{2}\, \pi^2 N k_B \cdot \frac{T}{T_F} \; .$$

The electronic heat capacity of one- and two-dimensional electron gases are also linearly proportional to T at temperatures $T \ll T_F$. The dimensionality of the gas does not enter the argument except by a numerical factor.

[2]The present derivation was suggested by J. Twidell, private communication.

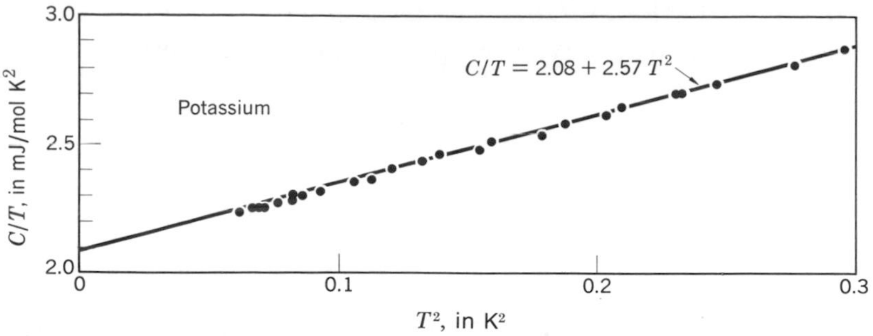

Figure 6 Experimental heat capacity values for potassium, plotted as C/T versus T^2. The solid points were determined with an adiabatic demagnetization cryostat. [After W. H. Lien and N. E. Phillips, Phys. Rev. **133**, A1370 (1964).]

Experimental Heat Capacity of Metals

At temperatures much below both the Debye temperature and the Fermi temperature, the heat capacity of metals may be written as the sum of electronic and lattice contributions: $C = \gamma T + AT^3$, where γ and A are constants characteristic of the material, given by (25) and (5.34). The electronic term is linear in T and is dominant at sufficiently low temperatures. It is convenient to exhibit the experimental values of C as a plot of C/T versus T^2:

$$C/T = \gamma + AT^2 \ ,$$

for then the points should lie on a straight line with slope A and intercept γ. Such a plot for potassium is shown in Fig. 6. Observed values of γ are given in Table 2.

The observed values of the coefficient γ are of the expected magnitude, but often do not agree very closely with the value calculated for free electrons of mass m by use of the relation (25). It is common practice to express the ratio of the observed to the free electron values of the electronic heat capacity as a ratio of a **thermal effective mass** m_{th} to the electron mass m, where m_{th} is defined by the relation

$$\frac{m_{th}}{m} \equiv \frac{\gamma(\text{observed})}{\gamma(\text{free})} \ . \qquad (26)$$

TABLE 2 Experimental and Free Electron Values of Electronic Heat Constant γ of Metals

(From compilations kindly furnished by N. Phillips and N. Pearlman. The thermal effective mass is defined by Eq. (26).)

Observed γ in mJ mol^{-1} K^{-2}.
Calculated free electron γ in mJ mol^{-1} K^{-2}
m_{th}/m = (observed γ)/(free electron γ).

	Be											B	C	N
53	0.17													
749	0.500													
18	0.34													
a	Mg											Al	Si	P
38	1.3											1.35		
094	0.992											0.912		
26	1.3											1.48		
	Ca	Sc	Ti	V	Cr	Mn(γ)	Fe	Co	Ni	Cu	Zn	Ga	Ge	As
08	2.9	10.7	3.35	9.26	1.40	9.20	4.98	4.73	7.02	0.695	0.64	0.596		0.19
568	1.511									0.505	0.753	1.025		
25	1.9									1.38	0.85	0.58		
	Sr	Y	Zr	Nb	Mo	Tc	Ru	Rh	Pd	Ag	Cd	In	Sn (w)	Sb
41	3.6	10.2	2.80	7.79	2.0	—	3.3	4.9	9.42	0.646	0.688	1.69	1.78	0.11
911	1.790									0.645	0.948	1.233	1.410	
26	2.0									1.00	0.73	1.37	1.26	
	Ba	La	Hf	Ta	W	Re	Os	Ir	Pt	Au	Hg(α)	Tl	Pb	Bi
20	2.7	10.	2.16	5.9	1.3	2.3	2.4	3.1	6.8	0.729	1.79	1.47	2.98	0.008
238	1.937									0.642	0.952	1.29	1.509	
43	1.4									1.14	1.88	1.14	1.97	

This form arises in a natural way because ϵ_F in the expression (25) is inversely proportional to the mass of the electron, whence $\gamma \propto m$. Values of the ratio are given in Table 2. The departure from unity involves three separate effects:

1. The interaction of the conduction electrons with the periodic potential of the rigid crystal lattice. The effective mass of an electron in this potential is called the band effective mass and is treated later.
2. The interaction of the conduction electrons with phonons. An electron tends to polarize or distort the lattice in its neighborhood, so that the moving electron tries to drag nearby ions along, thereby increasing the effective mass of the electron.
3. The interaction of the conduction electrons with themselves. A moving electron causes an inertial reaction in the surrounding electron gas, thereby increasing the effective mass of the electron.

ELECTRICAL CONDUCTIVITY AND OHM'S LAW

The momentum of a free electron is related to the wavevector by $m\mathbf{v} = \hbar\mathbf{k}$. In an electric field $\mathbf{E}$ and magnetic field $\mathbf{B}$ the force $\mathbf{F}$ on an electron of charge $-e$ is $-e[\mathbf{E} + (1/c)\mathbf{v} \times \mathbf{B}]$, so that Newton's second law of motion becomes

$$\text{(CGS)} \qquad \boxed{\mathbf{F} = m\,\frac{d\mathbf{v}}{dt} = \hbar\,\frac{d\mathbf{k}}{dt} = -e\left(\mathbf{E} + \frac{1}{c}\,\mathbf{v} \times \mathbf{B}\right)\ .} \tag{27}$$

In the absence of collisions the Fermi sphere (Fig. 7) in $\mathbf{k}$ space is displaced at a uniform rate by a constant applied electric field. We integrate with $\mathbf{B} = 0$ to obtain

$$\mathbf{k}(t) - \mathbf{k}(0) = -e\mathbf{E}t/\hbar\ . \tag{28}$$

If the field is applied at time $t = 0$ to an electron gas that fills the Fermi sphere centered at the origin of $\mathbf{k}$ space, then at a later time t the sphere will be displaced to a new center at

$$\delta\mathbf{k} = -e\mathbf{E}t/\hbar\ . \tag{29}$$

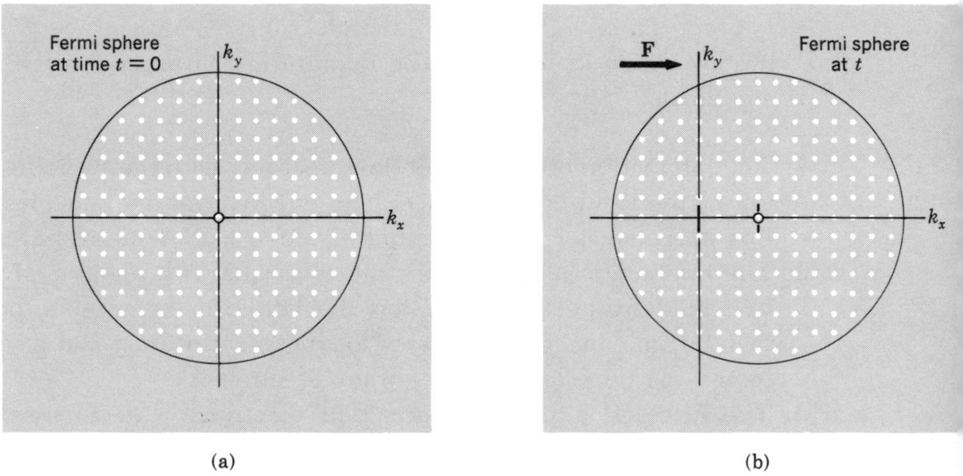

Figure 7 (a) The Fermi sphere encloses the occupied electron orbitals in $\mathbf{k}$ space in the ground state of the electron gas. The net momentum is zero, because for every orbital $\mathbf{k}$ there is an occupied orbital at $-\mathbf{k}$. (b) Under the influence of a constant force $\mathbf{F}$ acting for a time interval t every orbital has its $\mathbf{k}$ vector increased by $\delta\mathbf{k} = \mathbf{F}t/\hbar$. This is equivalent to a displacement of the whole Fermi sphere by $\delta\mathbf{k}$. The total momentum is $N\hbar\delta\mathbf{k}$, if there are N electrons present. The application of the force increases the energy of the system by $N(\hbar\delta\mathbf{k})^2/2m$.

Because of collisions of electrons with impurities, lattice imperfections, and phonons, the displaced sphere may be maintained in a steady state in an electric field. If collision time is τ, the displacement of the Fermi sphere in the steady state is given by (29). The incremental velocity[3] is $\mathbf{v} = -e\mathbf{E}\tau/m$. If in a constant electric field $\mathbf{E}$ there are n electrons of charge $q = -e$ per unit volume, the electric current density is

$$\mathbf{j} = nq\mathbf{v} = ne^2\tau\mathbf{E}/m \ . \tag{30}$$

This is in the form of Ohm's law. The electrical conductivity σ is defined by $\mathbf{j} = \sigma\mathbf{E}$, so that

$$\boxed{\sigma = \frac{ne^2\tau}{m}} \ . \tag{31}$$

The electrical resistivity ρ is defined as the reciprocal of the conductivity, so that

$$\rho = m/ne^2\tau \ . \tag{32}$$

Values of the electrical conductivity and resistivity of the elements are given in Table 3.

It is easy to understand the result (31) for the conductivity. We expect the charge transported to be proportional to the charge density $-ne$; the factor $-e/m$ enters because the acceleration in a given electric field is proportional to e and inversely proportional to the mass m; and the time τ describes the free time during which the field acts on the carrier.

It is possible to obtain crystals of copper so pure that their conductivity at liquid helium temperatures (4 K) is nearly 10^5 times that at room temperature; for these conditions $\tau \approx 2 \times 10^{-9}$ s at 4 K. The mean free path ℓ of a conduction electron is defined as

$$\ell = v_F\tau \ , \tag{33}$$

where v_F is the velocity at the Fermi surface, because all collisions involve only electrons near the Fermi surface. From Table 1 we have $v_F = 1.57 \times 10^8$ cm s^{-1} for Cu; the mean free path is

$$\ell(300\ \mathrm{K}) \approx 3 \times 10^{-6}\ \mathrm{cm} \ ; \qquad \ell(4\ \mathrm{K}) \approx 0.3\ \mathrm{cm} \ . \tag{34}$$

Mean free paths as long as 10 cm have been observed in very pure metals in the liquid helium temperature range.

[3]Here $\mathbf{v}$ is the average electron velocity, also called the drift velocity. Such a treatment of a transport process in a solid is known as a drift velocity theory. This model does not apply when the current flow proceeds by hopping of carriers from site to site, a process treated in Chapter 17 for the diffusion of ions in a crystal.

TABLE 3 Electrical Conductivity and Resistivity of Metals at 295 K

(Resistivity values as given by G. T. Meaden, *Electrical resistance of metals*, Plenum, 1965; residual resistivities have been subtracted.)

Conductivity in units of 10^5 (ohm-cm)$^{-1}$.
Resistivity in units of 10^{-6} ohm-cm.

Li 1.07 9.32	Be 3.08 3.25											B	C	N	O	F	Ne
Na 2.11 4.75	Mg 2.33 4.30											Al 3.65 2.74	Si	P	S	Cl	Ar
K 1.39 7.19	Ca 2.78 3.6	Sc 0.21 46.8	Ti 0.23 43.1	V 0.50 19.9	Cr 0.78 12.9	Mn 0.072 139.	Fe 1.02 9.8	Co 1.72 5.8	Ni 1.43 7.0	Cu 5.88 1.70	Zn 1.69 5.92	Ga 0.67 14.85	Ge	As	Se	Br	Kr
Rb 0.80 12.5	Sr 0.47 21.5	Y 0.17 58.5	Zr 0.24 42.4	Nb 0.69 14.5	Mo 1.89 5.3	Tc $\sim$0.7 $\sim$14.	Ru 1.35 7.4	Rh 2.08 4.8	Pd 0.95 10.5	Ag 6.21 1.61	Cd 1.38 7.27	In 1.14 8.75	Sn (w) 0.91 11.0	Sb 0.24 41.3	Te	I	Xe
Cs 0.50 20.0	Ba 0.26 39.	La 0.13 79.	Hf 0.33 30.6	Ta 0.76 13.1	W 1.89 5.3	Re 0.54 18.6	Os 1.10 9.1	Ir 1.96 5.1	Pt 0.96 10.4	Au 4.55 2.20	Hg liq. 0.10 95.9	Tl 0.61 16.4	Pb 0.48 21.0	Bi 0.086 116.	Po 0.22 46.	At	Rn
Fr	Ra	Ac															

Ce 0.12 81.	Pr 0.15 67.	Nd 0.17 59.	Pm	Sm 0.10 99.	Eu 0.11 89.	Gd 0.070 134.	Tb 0.090 111.	Dy 0.11 90.0	Ho 0.13 77.7	Er 0.12 81.	Tm 0.16 62.	Yb 0.38 26.4	Lu 0.19 53.
Th 0.66 15.2	Pa	U 0.39 25.7	Np 0.085 118.	Pu 0.070 143.	Am	Cm	Bk	Cf	Es	Fm	Md	No	Lr

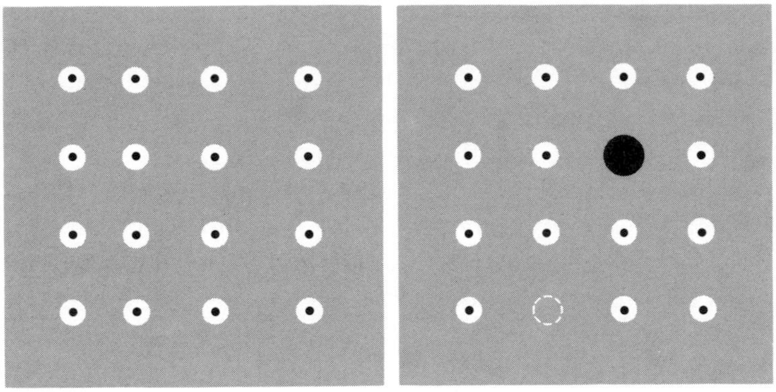

(a) (b)

Figure 8 Electrical resistivity in most metals arises from collisions of electrons with irregularities in the lattice, as in (a) by phonons and in (b) by impurities and vacant lattice sites.

Experimental Electrical Resistivity of Metals

The electrical resistivity of most metals is dominated at room temperature (300 K) by collisions of the conduction electrons with lattice phonons and at liquid helium temperature (4 K) by collisions with impurity atoms and mechanical imperfections in the lattice (Fig. 8). The rates of these collisions are often independent to a good approximation, so that if the electric field were switched off the momentum distribution would relax back to its ground state with the net relaxation time

$$\frac{1}{\tau} = \frac{1}{\tau_L} + \frac{1}{\tau_i} , \tag{35}$$

where τ_L and τ_i are the collision times for scattering by phonons and by imperfections, respectively. The net resistivity is given by

$$\rho = \rho_L + \rho_i , \tag{36}$$

where ρ_L is the resistivity caused by the thermal phonons, and ρ_i is the resistivity caused by scattering of the electron waves by static defects that disturb the periodicity of the lattice. Often ρ_L is independent of the number of defects when their concentration is small, and often ρ_i is independent of temperature. This empirical observation expresses **Matthiessen's rule,** which is convenient in analyzing experimental data, as in Fig. 9.

The residual resistivity, $\rho_i(0)$, is the extrapolated resistivity at 0 K because ρ_L vanishes as $T \rightarrow 0$. The lattice resistivity, $\rho_L(T) = \rho - \rho_i(0)$, is the same for different specimens of a metal, even though $\rho_i(0)$ may itself vary widely. The resistivity ratio of a specimen is usually defined as the ratio of its resistivity at room temperature to its residual resistivity. It is a convenient approximate indicator of sample purity: for many materials an impurity in solid solution causes a residual resistivity of about 10^{-6} ohm cm per atomic percent. A specimen of copper with a resistivity ratio of 1000 will have a residual resistivity of 1.7×10^{-9} ohm cm and an impurity level of about 20 ppm. In exceptional specimens the resistivity ratio may be as high as 10^6, while in some alloys (e.g., manganin) it is as low as 1.1.

Although the most characteristic feature of a metal, the electrical resistivity is not at all easy to treat theoretically. The collision time depends on the details of the electron-phonon interaction. One feature that emerges is that τ_L and τ_i are different functions of the electron wavevector **k**. Thus the contributions to the net relaxation time $\tau(\mathbf{k})$ are no longer simply additive as in (35). Matthiessen's rule is not always valid. Calculations of the phonon resistivity, ρ_L, have been quite successful for some metals. One finds empirically that ρ_L can be rather well represented by a universal function, $\rho_L \propto (T/M\theta_R^2)f(T/\theta_R)$, where the function f goes to unity at high temperatures and is proportional to $(T/\theta_R)^4$ at very low temperatures. The proportionality of ρ_L to $T/M\theta_R^2$ at high temperatures follows because the scattering is proportional to the mean square thermal strain. The resistive characteristic temperature θ_R is close to the Debye temperature for the metal; equality of θ_R and θ is not expected because longitudinal and transverse phonons contribute to the resistivity and to the heat capacity in different ways.

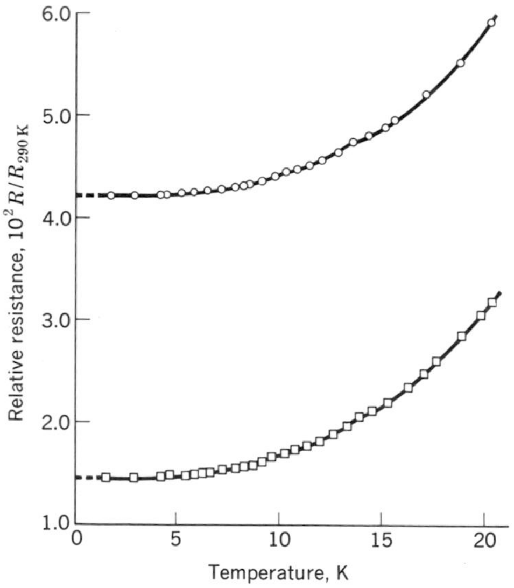

Figure 9 Resistance of potassium below 20 K, as measured on two specimens by D. K. C. MacDonald and K. Mendelssohn, Proc. Roy. Soc. (London) **A202**, 103 (1950). The different intercepts at 0 K are attributed to different concentrations of impurities and static imperfections in the two specimens. For measurements below 4.2 K, see D. Gugan, Proc. Roy. Soc. (London) **A325**, 223 (1971).

MOTION IN MAGNETIC FIELDS

By the argument of (27) and (29) we are led to the equation of motion for the displacement $\delta\mathbf{k}$ of a Fermi sphere of particles acted on by a force $\mathbf{F}$ and by friction as represented by collisions:

$$\hbar\left(\frac{d}{dt}+\frac{1}{\tau}\right)\delta\mathbf{k}=\mathbf{F}\ . \tag{37}$$

The free particle acceleration term is $(\hbar d/dt)\,\delta\mathbf{k}$ and the effect of collisions (the friction) is represented by $\hbar\delta\mathbf{k}/\tau$, where τ is the collision time. Consider now the motion of the system in a uniform magnetic field $\mathbf{B}$. The Lorentz force on an electron is

(CGS) $$\mathbf{F}=-e\left(\mathbf{E}+\frac{1}{c}\mathbf{v}\times\mathbf{B}\right)\ ; \tag{38}$$

(SI) $$\mathbf{F}=-e(\mathbf{E}+\mathbf{v}\times\mathbf{B})\ .$$

If $m\mathbf{v}=\hbar\delta\mathbf{k}$, then the equation of motion is

(CGS) $$m\left(\frac{d}{dt}+\frac{1}{\tau}\right)\mathbf{v}=-e\left(\mathbf{E}+\frac{1}{c}\mathbf{v}\times\mathbf{B}\right)\ . \tag{39}$$

An important situation is the following: let a static magnetic field $\mathbf{B}$ lie along the z axis. Then the equations of motion are

(CGS) $$m\left(\frac{d}{dt}+\frac{1}{\tau}\right)v_x=-e\left(E_x+\frac{B}{c}v_y\right)\ ;$$

$$m\left(\frac{d}{dt}+\frac{1}{\tau}\right)v_y=-e\left(E_y-\frac{B}{c}v_x\right)\ ;$$

$$m\left(\frac{d}{dt}+\frac{1}{\tau}\right)v_z=-eE_z\ . \tag{40}$$

The results in SI are obtained by replacing c by 1. In the steady state in a static electric field the time derivatives are zero, so that

$$v_x=-\frac{e\tau}{m}E_x-\omega_c\tau v_y\ ;\qquad v_y=-\frac{e\tau}{m}E_y+\omega_c\tau v_x\ ;\qquad v_z=-\frac{e\tau}{m}E_z\ , \tag{41}$$

where $\omega_c\equiv eB/mc$, the **cyclotron frequency.**

Hall Effect

The Hall field is the electric field developed across two faces of a conductor, in the direction $\mathbf{j} \times \mathbf{B}$, when a current $\mathbf{j}$ flows across a magnetic field $\mathbf{B}$. Consider a rod-shaped specimen in a longitudinal electric field E_x and a transverse magnetic field, as in Fig. 10. If current cannot flow out of the rod in the y direction we must have $\delta v_y = 0$. From (41) this is possible only if there is a transverse electric field

(CGS) $$E_y = -\omega_c \tau E_x = -\frac{eB\tau}{mc} E_x \ ; \tag{42}$$

(SI) $$E_y = -\omega_c \tau E_x = -\frac{eB\tau}{m} E_x \ .$$

The quantity

$$\boxed{R_H = \frac{E_y}{j_x B}} \tag{43}$$

is called the **Hall coefficient.** To evaluate it on our simple model we use $j_x = ne^2\tau E_x/m$ and obtain

(CGS) $$R_H = -\frac{eB\tau E_x/mc}{ne^2\tau E_x B/m} = -\frac{1}{nec} \ ; \tag{44}$$

(SI) $$R_H = -\frac{1}{ne} \ .$$

This is negative for free electrons, for e is positive by definition. The lower the carrier concentration, the greater the magnitude of the Hall coefficient. Measuring R_H is an important way of measuring the carrier concentration.

The simple result (44) follows from the assumption that all relaxation times are equal, independent of the velocity of the electron. A numerical factor of order unity enters if the relaxation time is a function of the velocity. The expression becomes somewhat more complicated if both electrons and holes contribute to the conductivity. The theory of the Hall effect again becomes simple in high magnetic fields[4] such that $\omega_c \tau \gg 1$, where ω_c is the cyclotron frequency and τ the relaxation time.

[4]A review of galvanomagnetic effects in high fields is given by E. Fawcett, Advances in Physics **13,** 139 (1964); see *QTS*, pp. 241–244.

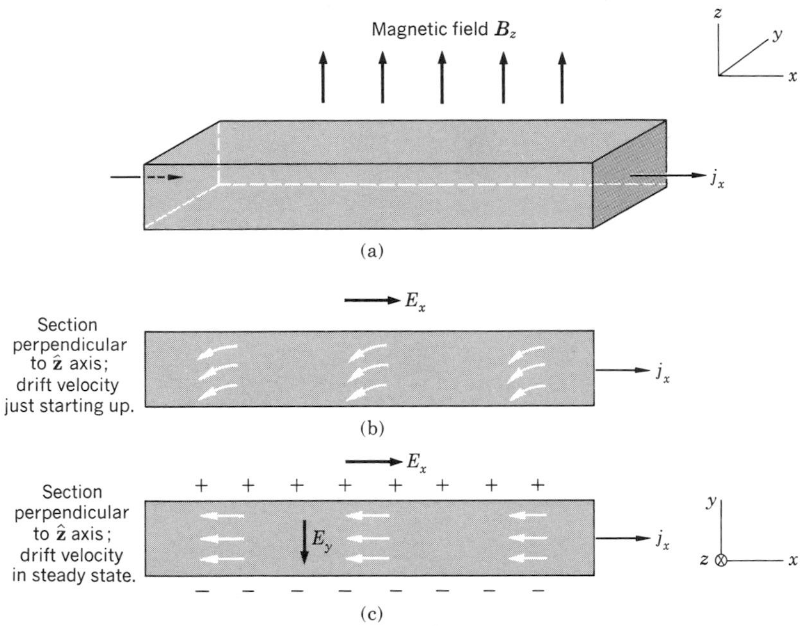

Figure 10 The standard geometry for the Hall effect: a rod-shaped specimen of rectangular cross-section is placed in a magnetic field B_z, as in (a). An electric field E_x applied across the end electrodes causes an electric current density j_x to flow down the rod. The drift velocity of the electrons immediately after the electric field is applied as shown in (b). The deflection in the y direction is caused by the magnetic field. Electrons accumulate on one face of the rod and a positive ion excess is established on the opposite face until, as in (c), the transverse electric field (Hall field) just cancels the force due to the magnetic field.

Observed values of the Hall coefficients for several metals are compared in Table 4 with values calculated directly from the concentration of charge carriers. The most accurate measurements are made on pure specimens at low temperatures and strong magnetic fields by the method of helicon resonance, discussed later. The accurate values for the monovalent metals sodium and potassium are in excellent agreement with the values calculated for one valence electron per atom, using (44). Notice the values for the trivalent elements aluminum and indium: the experimental values are in close concord with the values calculated for one positive charge carrier (or hole) per atom and would disagree in sign and magnitude with values calculated for three electrons per atom. The problem of the positive signs arises also for Be and As in the table, for a positive Hall coefficient must be associated with the motion of positive charge carriers. The problem was recognized in the early work on the Hall effect; as Lorentz wrote, "It seems to prove that we must imagine two kinds of free electrons, the motion of the positive ones predominating in one body and that of the negative ones in the other." We need to go beyond the free electron Fermi gas to explain positive Hall coefficients and also to explain the large magnitude of the Hall coefficients in As, Sb, and Bi. We need energy band theory, Chapters 7–9.

TABLE 4 Comparison of Observed Hall Coefficients with Those Calculated on Free Electron Theory

[The experimental values of R_H as obtained by conventional methods are summarized from data at room temperature presented in the Landolt-Börnstein tables, Vol. II.6, p. 161 (1959); the values obtained by the helicon wave method at 4 K are by J. M. Goodman, Phys. Rev. **171,** 641 (1968). The values of the carrier concentration n are from Table 1.4 except for Na, K, Al, In, where Goodman's values are used. To convert the value of R_H in CGS units to the value in volt-cm/amp-gauss, multiply by 9×10^{11}; to convert R_H in CGS to m³/coulomb, multiply by 9×10^{13}.]

Metal	Method	Experimental R_H, in 10^{-24} CGS Units	Assumed Carriers per Atom	Calculated $-1/nec$, in 10^{-24} CGS Units
Li	conv.	−1.89	1 electron	−1.48
Na	helicon	−2.619	1 electron	−2.603
	conv.	−2.3		
K	helicon	−4.946	1 electron	−4.944
	conv.	−4.7		
Rb	conv.	−5.6	1 electron	−6.04
Cu	conv.	−0.6	1 electron	−0.82
Ag	conv.	−1.0	1 electron	−1.19
Au	conv.	−0.8	1 electron	−1.18
Be	conv.	+2.7	—	—
Mg	conv.	−0.92	—	—
Al	helicon	+1.136	1 hole	+1.135
In	helicon	+1.774	1 hole	+1.780
As	conv.	+50.	—	—
Sb	conv.	−22.	—	—
Bi	conv.	−6000.	—	—

THERMAL CONDUCTIVITY OF METALS

In Chapter 5 we found an expression $K = \frac{1}{3}Cv\ell$ for the thermal conductivity of particles of velocity v, heat capacity C per unit volume, and mean free path ℓ. The thermal conductivity of a Fermi gas follows from (25) for the heat capacity, and with $\epsilon_F = \frac{1}{2}mv_F^2$:

$$K_{el} = \frac{\pi^2}{3} \cdot \frac{nk_B^2T}{mv_F^2} \cdot v_F \cdot \ell = \frac{\pi^2 nk_B^2 T\tau}{3m} \; . \qquad (45)$$

Here $\ell = v_F\tau$; the electron concentration is n, and τ is the collision time.

Do the electrons or the phonons carry the greater part of the heat current in a metal? At room temperature normal pure metals tend to have values of the thermal conductivity one or two orders of magnitude higher than for dielectric solids, so that under these conditions the electrons must carry almost all the heat current. In pure metals the electronic contribution is dominant at all temperatures. In impure metals or in disordered alloys the phonon contribution may be comparable with the electronic contribution.

Measurements on copper are shown in Fig. 11. Experimental curves for many metals at low temperatures are given by H. M. Rosenberg, Phil. Trans. Roy. Soc. (London) **A247**, 441–497 (1955), and R. L. Powell and W. A. Blanpied, "Thermal conductivities of metals and alloys at low temperatures," National Bureau of Standards Circular 556.

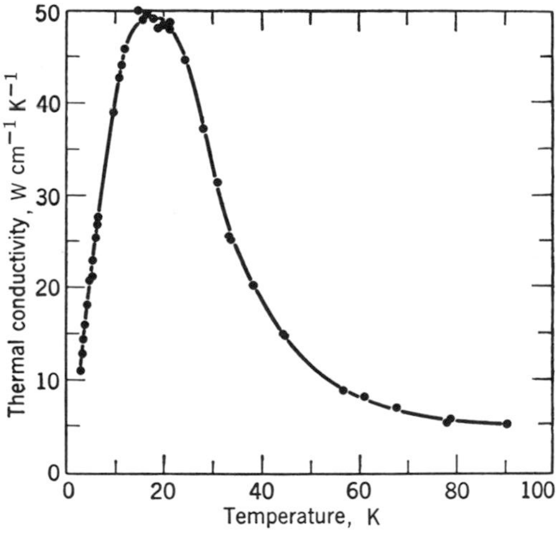

Figure 11 The thermal conductivity of copper, after Berman and MacDonald.

Ratio of Thermal to Electrical Conductivity

The **Wiedemann-Franz law** states that for metals at not too low temperatures the ratio of the thermal conductivity to the electrical conductivity is directly proportional to the temperature, with the value of the constant of proportionality independent of the particular metal. This result was most important in the history of the theory of metals, for it supported the picture of an electron gas. It can be explained by using (31) for σ and (45) for K:

$$\frac{K}{\sigma} = \frac{\pi^2 k_B^2 T n\tau/3m}{ne^2\tau/m} = \frac{\pi^2}{3}\left(\frac{k_B}{e}\right)^2 T \ . \qquad (46)$$

The **Lorenz number** L is defined as

$$L \equiv K/\sigma T \ , \qquad (47)$$

and according to (46) should have the value

$$L = \frac{\pi^2}{3}\left(\frac{k_B}{e}\right)^2 = 2.72 \times 10^{-13}\ \text{esu/deg}^2 = 2.45 \times 10^{-8}\ \text{watt-ohm/deg}^2 \ . \qquad (48)$$

This remarkable result involves neither n nor m. It does not involve τ if the relaxation times are identical for electrical and thermal processes. Experimental values of L at 0°C and at 100°C as given in Table 5 are in good agreement with (48). On purely classical theory with a Maxwellian distribution of velocities, the result is $L = 3(k_B/e)^2$, very close to (48), and also in fair agreement with experiment. The quotation from Lorentz at the beginning of this chapter refers to this happy agreement.

At low temperatures ($T \ll \theta$) the Lorenz number tends to decrease; for pure copper near 15 K the value is an order of magnitude smaller than (48). The reason is attributed to a difference in the collision averages involved in the thermal and electrical conductivities; the thermal and electrical relaxation times are not identical.

TABLE 5 Experimental Lorenz Numbers

	$L \times 10^8$ watt-ohms/deg^2			$L \times 10^8$ watt-ohms/deg^2	
Metal	0°C	100°C	Metal	0°C	100°C
Ag	2.31	2.37	Pb	2.47	2.56
Au	2.35	2.40	Pt	2.51	2.60
Cd	2.42	2.43	Sn	2.52	2.49
Cu	2.23	2.33	W	3.04	3.20
Ir	2.49	2.49	Zn	2.31	2.33
Mo	2.61	2.79			

PROBLEMS

1. ***Kinetic energy of electron gas.*** Show that the kinetic energy of a three-dimensional gas of N free electrons at 0 K is

$$U_0 = \tfrac{3}{5}N\epsilon_F \ . \tag{49}$$

2. ***Pressure and bulk modulus of an electron gas.*** (a) Derive a relation connecting the pressure and volume of an electron gas at 0 K. Hint: Use the result of Problem 1 and the relation between ϵ_F and electron concentration. The result may be written as $p = \tfrac{2}{3}(U_0/V)$. (b) Show that the bulk modulus $B = -V(\partial p/\partial V)$ of an electron gas at 0 K is $B = 5p/3 = 10U_0/9V$. (c) Estimate for potassium, using Table 1, the value of the electron gas contribution to B.

3. ***Chemical potential in two dimensions.*** Show that the chemical potential of a Fermi gas in two dimensions is given by:

$$\mu(T) = k_BT\ \ln[\exp(\pi n\hbar^2/mk_BT) - 1] \ , \tag{50}$$

for n electrons per unit area. Note: The density of orbitals of a free electron gas in two dimensions is independent of energy: $\mathfrak{D}(\epsilon) = m/\pi\hbar^2$, per unit area of specimen.

4. ***Fermi gases in astrophysics.*** (a) Given $M_\odot = 2 \times 10^{33}$ g for the mass of the Sun, estimate the number of electrons in the Sun. In a white dwarf star this number of electrons may be ionized and contained in a sphere of radius 2×10^9 cm; find the Fermi energy of the electrons in electron volts. (b) The energy of an electron in the relativistic limit $\epsilon \gg mc^2$ is related to the wavevector as $\epsilon \cong pc = \hbar kc$. Show that the Fermi energy in this limit is $\epsilon_F \approx \hbar c(N/V)^{1/3}$, roughly. (c) If the above number of electrons were contained within a pulsar of radius 10 km, show that the Fermi energy would be $\approx 10^8$ eV. This value explains why pulsars are believed to be composed largely of neutrons rather than of protons and electrons, for the energy release in the reaction $n \rightarrow p + e^-$ is only 0.8×10^6 eV, which is not large enough to enable many electrons to form a Fermi sea. The neutron decay proceeds only until the electron concentration builds up enough to create a Fermi level of 0.8×10^6 eV, at which point the neutron, proton, and electron concentrations are in equilibrium.

5. ***Liquid He^3.*** The atom He^3 has spin $\tfrac{1}{2}$ and is a fermion. The density of liquid He^3 is 0.081 g cm^{-3} near absolute zero. Calculate the Fermi energy ϵ_F and the Fermi temperature T_F. (For a survey of the properties of liquid He^3, see J. Wilks, *Properties of liquid and solid helium*, Oxford, 1967.)

6. ***Frequency dependence of the electrical conductivity.*** Use the equation $m(dv/dt + v/\tau) = -eE$ for the electron drift velocity v to show that the conductivity at frequency ω is

$$\sigma(\omega) = \sigma(0)\left(\frac{1 + i\omega\tau}{1 + (\omega\tau)^2}\right) \ , \tag{51}$$

where $\sigma(0) = ne^2\tau/m$.

*7. ***Dynamic magnetoconductivity tensor for free electrons.*** A metal with a concentration n of free electrons of charge $-e$ is in a static magnetic field $B\hat{\mathbf{z}}$. The electric current density in the xy plane is related to the electric field by

$$j_x = \sigma_{xx}E_x + \sigma_{xy}E_y \; ; \qquad j_y = \sigma_{yx}E_x + \sigma_{yy}E_y \; .$$

Assume that the frequency $\omega \gg \omega_c$ and $\omega \gg 1/\tau$, where $\omega_c \equiv eB/mc$ and τ is the collision time. (a) Solve the drift velocity equation (40) to find the components of the magnetoconductivity tensor:

$$\sigma_{xx} = \sigma_{yy} = i\omega_p{}^2/4\pi\omega \; ; \qquad \sigma_{xy} = -\sigma_{yx} = \omega_c\omega_p{}^2/4\pi\omega^2 \; ,$$

where $\omega_p{}^2 \equiv 4\pi ne^2/m$. (b) Note from a Maxwell equation that the dielectric function tensor of the medium is related to the conductivity tensor as $\boldsymbol{\epsilon} = \mathbf{1} + i(4\pi/\omega)\boldsymbol{\sigma}$. Consider an electromagnetic wave with wavevector $\mathbf{k} = k\hat{\mathbf{z}}$. Show that the dispersion relation for this wave in the medium is

$$c^2k^2 = \omega^2 - \omega_p{}^2 \pm \omega_c\omega_p{}^2/\omega \; . \tag{52}$$

At a given frequency there are two modes of propagation with different wavevectors and different velocities. The two modes correspond to circularly polarized waves. Because a linearly polarized wave can be decomposed into two circularly polarized waves, it follows that the plane of polarization of a linearly polarized wave will be rotated by the magnetic field.

8. ***Cohesive energy of free electron Fermi gas.*** We define the dimensionless length r_s as r_0/a_H, where r_0 is the radius of a sphere that contains one electron, and a_H is the Bohr radius $\hbar^2/e^2m$. (a) Show that the average kinetic energy per electron in a free electron Fermi gas at 0 K is $2.21/r_s{}^2$, where the energy is expressed in rydbergs, with 1 Ry $= me^4/2\hbar^2$. (b) Show that the coulomb energy of a point positive charge e interacting with the uniform electron distribution of one electron in the volume of radius r_0 is $-3e^2/2r_0$, or $-3/r_s$ in rydbergs. (c) Show that the coulomb self-energy of the electron distribution in the sphere is $3e^2/5r_0$, or $6/5r_s$ in rydbergs. (d) The sum of (b) and (c) gives $-1.80/r_s$ for the total coulomb energy per electron. Show that the equilibrium value of r_s is 2.45. Will such a metal be stable with respect to separated H atoms?

9. ***Static magnetoconductivity tensor.*** For the drift velocity theory of (40), show that the static current density can be written in matrix form as

$$\begin{pmatrix} j_x \\ j_y \\ j_z \end{pmatrix} = \frac{\sigma_0}{1 + (\omega_c\tau)^2} \begin{pmatrix} 1 & -\omega_c\tau & 0 \\ \omega_c\tau & 1 & 0 \\ 0 & 0 & 1 + (\omega_c\tau)^2 \end{pmatrix} \begin{pmatrix} E_x \\ E_y \\ E_z \end{pmatrix} . \tag{53}$$

The transverse magnetoresistivity of a solid is defined as E_x/j_x for the standard geometry of Fig. 10. Show that (53) leads to $j_x = \sigma_0 E_x$, because $j_y = 0$ for this geometry. Thus the resistivity on this model is independent of the magnetic field, whereas experiments generally show a resistivity which increases as the magnetic field is increased.

*This problem is somewhat difficult.

References

H. M. Rosenberg, *Low temperature solid state physics*, Oxford, 1963, Chaps. 4 and 5.

J. M. Ziman, *Electrons and Phonons*, Oxford, 1960, Chap. 9.

D. N. Langenberg, "Resource letter OEPM-1 on the ordinary electronic properties of metals," Amer. J. Phys. **36**, 777 (1968). An excellent bibliography on transport effects, anomalous skin effect, Azbel-Kaner cyclotron resonance, magnetoplasma waves, size effects, conduction electron spin resonance, optical spectra and photoemission, quantum oscillations, magnetic breakdown, ultrasonic effects, and the Kohn effect.

G. T. Meaden, *Electrical resistance of metals*, Plenum, New York, 1965; "Conduction electron scattering and the resistance of the magnetic elements," Contemp. Phys. **12**, 313 (1971).

C. M. Hurd, *The Hall effect in metals and alloys*, Plenum, 1972.

Historical Development

H. A. Lorentz, *Collected papers*, vol. 8, Martinus Nijhoff, The Hague, 1935. A good view of the historical development of the prequantum electron theory of metals.

P. Drude, "Zur Elektronentheorie der Metalle," Annalen der Physik **1**, 566 (1900); **3**, 369 (1900). Two pioneer papers in the electron theory of the transport properties of metals; the earlier paper contains the first derivation of the Wiedemann-Franz ratio.

L. L. Campbell, *Galvanomagnetic and thermomagnetic effects*, Longmans, Green, 1923. Includes an account of the early history of the Hall effect.

7 ENERGY BANDS

Energy bands of semiconductors are discussed in Chapter 8; energy bands of metals are discussed in Chapter 9.

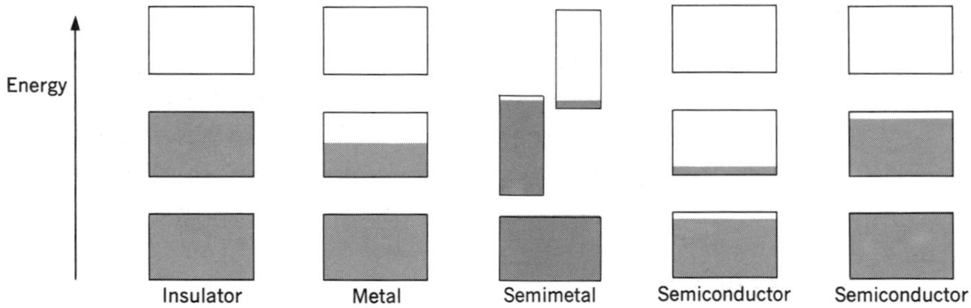

Figure 1 Schematic electron occupancy of allowed energy bands for an insulator, metal, semimetal, and semiconductor. The vertical extent of the boxes indicates the allowed energy regions; the shaded areas indicate the regions filled with electrons. In a **semimetal** (such as bismuth) one band is almost filled and another band is nearly empty at absolute zero, but a pure **semiconductor** (such as silicon) becomes an insulator at absolute zero. The left of the two semiconductors shown is at a finite temperature, with carriers excited thermally. The other semiconductor is electron-deficient because of impurities.

7 ENERGY BANDS

When I started to think about it, I felt that the main problem was to explain how the electrons could sneak by all the ions in a metal.... By straight Fourier analysis I found to my delight that the wave differed from the plane wave of free electrons only by a periodic modulation.

F. Bloch

The free electron model of metals gives us good insight into the heat capacity, thermal conductivity, electrical conductivity, magnetic susceptibility, and electrodynamics of metals. But the model fails to help us with other large questions: the distinction between metals, semimetals, semiconductors, and insulators; the occurrence of positive values of the Hall coefficient; the relation of conduction electrons in the metal to the valence electrons of free atoms; and many detailed transport properties, particularly magnetotransport. We need a less naïve theory, and fortunately it turns out that almost any simple attempt to improve upon the free electron model is enormously profitable.

The difference between a good conductor and a good insulator is striking. The electrical resistivity of a pure metal may be as low as 10^{-10} ohm-cm at a temperature of 1 K, apart from the possibility of superconductivity. The resistivity of a good insulator may be as high as 10^{22} ohm-cm. This range of 10^{32} may be the widest of any common physical property of solids.

Every solid contains electrons. The important question for electrical conductivity is how the electrons respond to an applied electric field. We shall see that electrons in crystals are arranged in energy bands (Fig. 1) separated by regions in energy for which no wavelike electron orbitals exist. Such forbidden regions are called **energy gaps** or **band gaps,** and result from the interaction of the conduction electron waves with the ion cores of the crystal. The crystal behaves as an insulator if the allowed energy bands are either filled or empty, for then no electrons can move in an electric field. The crystal behaves as a metal if one or more bands are partly filled, say 10 to 90 percent filled. The crystal is a semiconductor or a semimetal if all bands are entirely filled, except for one or two bands slightly filled or slightly empty.

To understand the difference between insulators and conductors, we must extend the free electron model to take account of the periodic lattice of the solid. The possibility of a band gap is the most important new property that emerges. We shall encounter other quite remarkable properties of electrons in crystals. For example, they respond to applied electric or magnetic fields as if the electrons were endowed with an effective mass m^*, which may be larger or smaller than the free electron mass, or may even be negative. Electrons in crystals respond to applied fields as if endowed with negative or positive charges, $-e$ or $+e$, and herein lies the explanation of the negative and positive values of the Hall coefficient.

NEARLY FREE ELECTRON MODEL

On the free electron model the allowed energy values are distributed continuously from zero to infinity. We saw in Chapter 6 that

$$\epsilon_{\mathbf{k}} = \frac{\hbar^2}{2m} (k_x^2 + k_y^2 + k_z^2) \ , \tag{1}$$

where, for periodic boundary conditions over a cube of side L,

$$k_x, \ k_y, \ k_z = 0 \ ; \quad \pm\frac{2\pi}{L} \ ; \quad \pm\frac{4\pi}{L} \ ; \quad \ldots . \tag{2}$$

The free electron wavefunctions are of the form

$$\psi_{\mathbf{k}}(\mathbf{r}) = \exp(i\mathbf{k}\cdot\mathbf{r}) \ ; \tag{3}$$

they represent running waves and carry momentum $\mathbf{p} = \hbar\mathbf{k}$.

The band structure of a crystal can often be described by the nearly free electron model for which the band electrons are treated as perturbed only weakly by the periodic potential of the ion cores. Often the band structure can be explained on this model; we shall also find some situations where it is not applicable. But this model answers almost all the qualitative questions about the behavior of electrons in metals.

We know that Bragg reflection is a characteristic feature of wave propagation in crystals. Bragg reflection of electron waves in crystals is the cause of energy gaps. There arise substantial regions of energy in which wavelike solutions of the Schrödinger equation do not exist, as in Fig. 2. These energy gaps are of decisive significance in determining whether a solid is an insulator or a conductor. We explain physically the origin of energy gaps in the simple problem of a linear solid of lattice constant a. The low energy portions of the band structure are shown qualitatively in Fig. 2, in (a) for entirely free electrons and in (b) for electrons that are

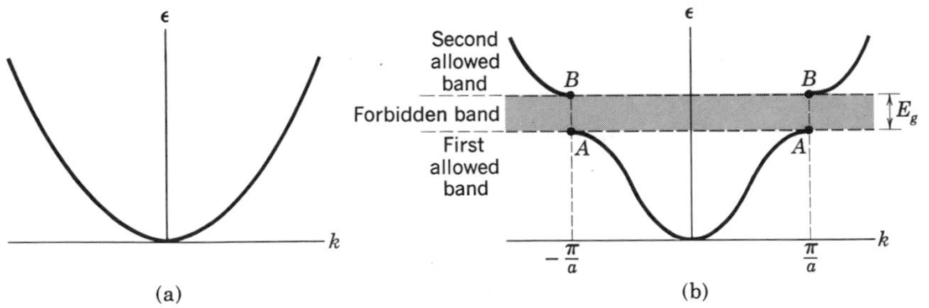

Figure 2 (a) Plot of energy ϵ versus wavevector k for a free electron. (b) Plot of energy versus wavevector for an electron in a monatomic linear lattice of lattice constant a. The energy gap E_g shown is associated with the first Bragg reflection at $k = \pm\pi/a$; other gaps are found at $\pm n\pi/a$, for integral values of n.

nearly free, but with an energy gap at $k = \pm\pi/a$. The Bragg condition $(\mathbf{k} + \mathbf{G})^2 = k^2$ for diffraction of a wave of wavevector $\mathbf{k}$ becomes in one dimension

$$k = \pm\tfrac{1}{2}G = \pm n\pi/a \ , \tag{4}$$

where $G = 2\pi n/a$ is a reciprocal lattice vector and n is an integer. The first reflections and the first energy gap occur at $k = \pm\pi/a$. Other energy gaps occur for other values of the integer n. The reflection at $k = \pm\pi/a$ arises because the wave reflected from one atom in the linear lattice interferes constructively with phase difference 2π with the wave reflected from a nearest-neighbor atom. The region in $\mathbf{k}$ space between $-\pi/a$ and π/a is the **first Brillouin zone** of this lattice.

At $k = \pm\pi/a$ the wavefunctions are not the traveling waves $e^{i\pi x/a}$ and $e^{-i\pi x/a}$ of the free electron model. The solutions at these particular k values are made up *equally* of waves traveling to the right and to the left: the solutions are standing waves. We first give a plausibility argument. When the Bragg condition is satisfied a wave traveling in one direction is Bragg-reflected and then travels in the opposite direction. Each subsequent Bragg reflection reverses the direction of travel again. The only time-independent situation is formed by standing waves. We can form two different standing waves from the traveling waves $e^{i\pi x/a}$ and $e^{-i\pi x/a}$:

$$\psi(+) = e^{i\pi x/a} + e^{-i\pi x/a} = 2\cos(\pi x/a) \ ;$$
$$\psi(-) = e^{i\pi x/a} - e^{-i\pi x/a} = 2i\sin(\pi x/a) \ . \tag{5}$$

The standing waves are made up each of equal portions of right- and left-directed traveling waves. The standing waves are labeled $(-)$ or $(+)$ according to whether they do or do not change sign when $-x$ is substituted for x.

Origin of the Energy Gap

The two standing waves $\psi(+)$ and $\psi(-)$ pile up electrons at different regions, and therefore the two waves have different values of the potential energy. This is the origin of the energy gap. The probability density ρ of a particle is $|\psi|^2$. For a pure traveling wave e^{ikx}, we have $\rho = e^{-ikx}\, e^{ikx} = 1$, so that the charge density is constant. The charge density is not constant for linear combinations of plane waves. Consider the standing wave $\psi(+)$ in (5); for this we have

$$\rho(+) = |\psi(+)|^2 \propto \cos^2 \pi x/a \ ,$$

which piles up negative charge on the positive ions centered at $x = 0, a, 2a, \ldots$, where the potential energy is lowest. Figure 3a indicates the variation of the electrostatic potential energy of a conduction electron in the field of the positive ion cores of the monatomic linear lattice. The ion cores bear a positive charge, because in the metal each atom has given up one or more valence electrons to form the conduction band. The potential energy of an electron in the field of a positive ion is negative, that is, attractive. In (b) we sketch the distribution of electron density corresponding to the standing waves $\psi(+)$, $\psi(-)$, and to a traveling wave.

For the standing wave $\psi(-)$ we have the probability density

$$\rho(-) = |\psi(-)|^2 \propto \sin^2 \pi x/a \ ,$$

which distributes electrons preferentially midway between ion cores, that is, away from the ion cores. On calculating average values of the potential energy over the three charge distributions, we expect to find the potential energy of $\rho(+)$ lower than that of a traveling wave, whereas the potential energy of $\rho(-)$ is higher than that of a traveling wave. If the energies of $\rho(-)$ and $\rho(+)$ differ by E_g, we have an energy gap of width E_g. The wavefunction just below the energy gap at points A in Fig. 2 will be $\psi(+)$, and the wavefunction just above the energy gap at points B will be $\psi(-)$.

Magnitude of the Energy Gap. The wavefunctions at the zone boundary are $\sqrt{2}\cos \pi x/a$ and $\sqrt{2}\sin \pi x/a$, normalized over unit length. We write the crystal potential as $U(x) = U \cos 2\pi x/a$. The energy difference in first order between the two states is

$$E_g = 2U \int_0^1 dx\,(\cos 2\pi x/a)\,(\cos^2 \pi x/a - \sin^2 \pi x/a) = U \ . \tag{6}$$

The gap is equal to the amplitude of the Fourier component of the crystal potential.

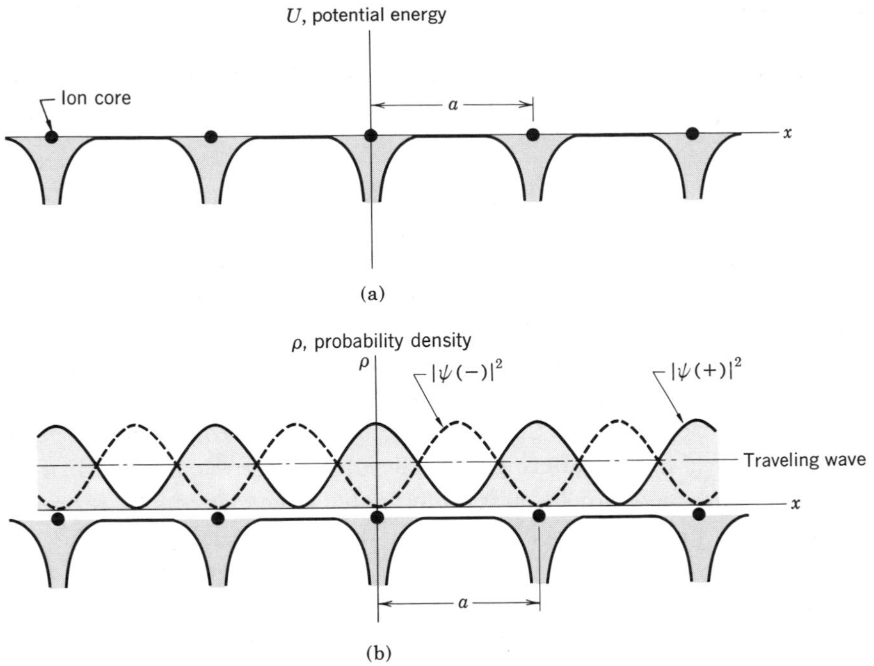

Figure 3 (a) Variation of potential energy of a conduction electron in the field of the ion cores of a linear lattice. (b) Distribution of probability density ρ in the lattice for $|\psi(-)|^2 \propto \sin^2 \pi x/a$; $|\psi(+)|^2 \propto \cos^2 \pi x/a$; and for a traveling wave. The wavefunction $\psi(+)$ piles up electronic charge on the cores of the positive ions, thereby lowering the potential energy in comparison with the average potential energy seen by a traveling wave. The wavefunction $\psi(-)$ piles up charge in the region between the ions and removes it from the ion cores; thereby raising the potential energy in comparison with that seen by a traveling wave. This figure is the key to understanding the origin of the energy gap.

BLOCH FUNCTIONS

F. Bloch proved the important theorem that the solutions of the Schrödinger equation for a periodic potential must be of a special form:

$$\boxed{\psi_{\mathbf{k}}(\mathbf{r}) = u_{\mathbf{k}}(\mathbf{r}) \exp(i\mathbf{k} \cdot \mathbf{r})} \tag{7}$$

where $u_{\mathbf{k}}(\mathbf{r})$ has the period of the crystal lattice, with $u_{\mathbf{k}}(\mathbf{r}) = u_{\mathbf{k}}(\mathbf{r} + \mathbf{T})$. The result (7) is a statement of the Bloch theorem which states that the eigenfunctions of the wave equation for a periodic potential are of the form of the product of a plane wave $\exp(i\mathbf{k} \cdot \mathbf{r})$ times a function $u_{\mathbf{k}}(\mathbf{r})$ with the periodicity of the crystal lattice. The subscript **k** indicates that the function $u_{\mathbf{k}}(\mathbf{r})$ depends on the wavevector **k**. An orbital of the form (7) is known as a Bloch function. These solutions are composed of traveling waves, and they can be assembled into wave packets to represent electrons that propagate freely through the potential field of the ion cores.

We give here a proof valid when $\psi_{\mathbf{k}}$ is nondegenerate, when there is no other wavefunction with the same energy and wavevector as $\psi_{\mathbf{k}}$. The general case will be handled by another method below in this chapter. We consider N identical lattice points on a ring of length Na. The potential is periodic, with $U(x) = U(x + sa)$, where s is an integer. Because of the symmetry of the ring we look for solutions such that

$$\psi(x + a) = C\psi(x) \ , \tag{8}$$

where C is a constant. Then, on going once around the ring,

$$\psi(x + Na) = \psi(x) = C^N \psi(x) \ ,$$

because $\psi(x)$ must be single-valued. It follows that C is one of the N roots of unity, or

$$C = \exp[i2\pi s/N] \ ; \qquad s = 0, 1, 2, \ldots, N-1 \ . \tag{9}$$

We see that

$$\psi(x) = u_k(x) \exp[i2\pi sx/Na] \tag{10}$$

satisfies (8), provided always that $u_k(x) = u_k(x + a)$. With $k = 2\pi s/Na$, we have the Bloch result (7). For another derivation, see Eq. (29).

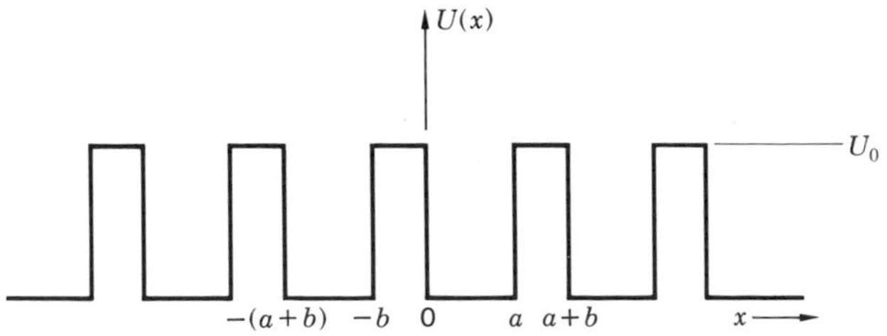

Figure 4 Square-well periodic potential introduced by R. de L. Kronig and W. G. Penney, Proc. Roy. Soc. (London) **A130,** 499 (1931).

KRONIG-PENNEY MODEL*

A periodic potential problem for which the wave equation can be solved in terms of simple analytic functions is the square-well array of Fig. 4. The wave equation is

$$-\frac{\hbar^2}{2m}\frac{d^2\psi}{dx^2}+U(x)\psi=\epsilon\psi \ , \qquad (11)$$

where $U(x)$ is the potential energy and ϵ is the energy eigenvalue. In the region $0 < x < a$ in which $U=0$, the eigenfunction is a linear combination,

$$\psi = A\,e^{iKx} + B\,e^{-iKx} \ , \qquad (12)$$

of plane waves traveling to the right and to the left, with energy

$$\epsilon = \hbar^2K^2/2m \ . \qquad (13)$$

In the region $a < x < a + b$ within the barrier the solution is of the form

$$\psi = C\,e^{Qx} + D\,e^{-Qx} \ , \qquad (14)$$

with

$$U_0 - \epsilon = \hbar^2Q^2/2m \ . \qquad (15)$$

The constants A, B, C, D are chosen so that ψ and $d\psi/dx$ are continuous at $x=0$:

$$A + B = C + D \ ; \qquad (16)$$

$$iKA - iKB = QC - QD \ . \qquad (17)$$

There must also be continuity at $x = a$:

$$A\,e^{iKa} + B\,e^{-iKa} = C\,e^{Qa} + D\,e^{-Qa} \ ; \qquad (18)$$

$$iKAe^{iKa} - iKB\,e^{-iKa} = QC\,e^{Qa} - QD\,e^{-Qa} \ . \qquad (19)$$

A final condition is that the values of ψ and $d\psi/dx$ at $x=a$ be equal to those at $x=-b$, but advanced by the phase factor $\exp[ik(a+b)]$. This

* This section is optional.

condition is a consequence of the form (7) of Bloch functions. The left-hand side of (18) is replaced by

$$A \exp[iKa - ik(a+b)] + B \exp[-iKa - ik(a+b)] \ ,$$

and the corresponding change is made in the exponential factors on the left-hand side of (19). The four equations (16) to (19) have a solution only if the determinant of the coefficients of A, B, C, D vanishes, or if

$$[(Q^2 - K^2)/2QK] \sinh Qb \sin Ka + \cosh Qb \cos Ka = \cos k(a+b) \ . \quad (20)$$

It is rather tedious to obtain this equation.

The result is simplified if we represent the potential by the periodic delta function obtained when we pass to the limit $b = 0$ and $U_0 = \infty$ in such a way that $Q^2ba/2 = P$, a finite quantity. Then (20) reduces to

$$(P/Ka)\sin Ka + \cos Ka = \cos ka \ . \quad (21)$$

The ranges of K for which this equation has solutions are plotted in Fig. 5, for the case $P = 3\pi/2$. The corresponding values of the energy are plotted in Fig. 6. Note the energy gaps at the zone boundaries. The wavevector k of the Bloch function is the important index, not the wavevector K in (12). Equation (21) may also be obtained by passing to the delta-function limit before writing down the boundary conditions, as was done in the second edition of this book.

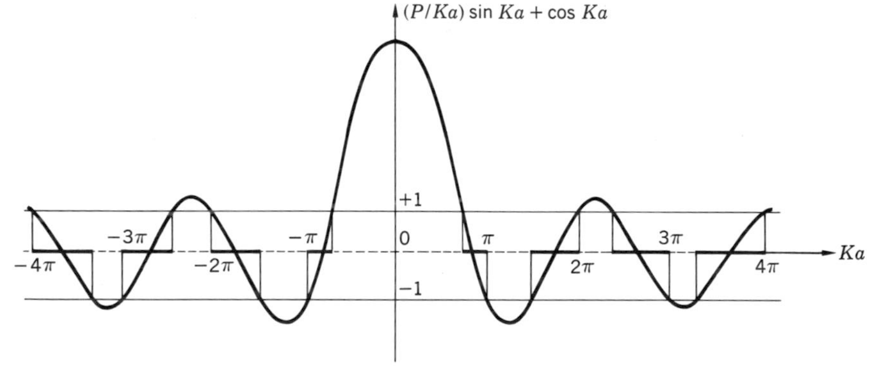

Figure 5 Plot of the function $(P/Ka) \sin Ka + \cos Ka$, for $P = 3\pi/2$. The allowed values of the energy ϵ are given by those ranges of $K = (2m\epsilon/\hbar^2)^{1/2}$ for which the function lies between +1 and −1. (After Kronig and Penney.)

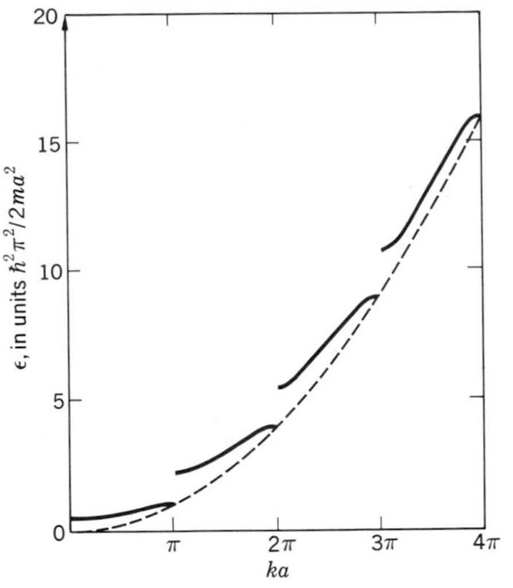

Figure 6 Plot of energy vs. wavenumber for the Kronig-Penney potential, with $P = 3\pi/2$. (After Sommerfeld and Bethe.)

WAVE EQUATION OF ELECTRON IN A PERIODIC POTENTIAL

We considered above the approximate form we expect the solution of the Schrödinger equation to have if the wavevector is at a zone boundary, as at $k = \pi/a$. We now treat in detail the wave equation for a general potential at general values of k. Let $U(x)$ denote the potential energy of an electron in a linear lattice of lattice constant a. We know that the potential energy is invariant under a crystal lattice translation: $U(x) = U(x + a)$. A function invariant under a crystal lattice translation may be expanded as a Fourier series in the reciprocal lattice vectors G. We write the Fourier series for the potential energy as

$$U(x) = \sum_G U_G \, e^{iGx} \ . \tag{22}$$

The values of the coefficients U_G for actual crystal potentials tend to decrease rapidly with increasing magnitude of G. For a bare coulomb potential U_G decreases as $1/G^2$.

We want the potential energy $U(x)$ to be a real function:

$$U(x) = \sum_{G>0} U_G (e^{iGx} + e^{-iGx}) = 2 \sum_{G>0} U_G \cos Gx \ . \tag{23}$$

We set $U_0 = 0$ for convenience.

The wave equation of an electron in the crystal is $\mathcal{H}\psi = \epsilon\psi$, where $\mathcal{H}$ is the hamiltonian and ϵ is the energy eigenvalue. The solutions ψ are called eigenfunctions or orbitals. Explicitly, the wave equation is

$$\left(\frac{1}{2m}\,p^2 + U(x)\right)\psi(x) = \left(\frac{1}{2m}\,p^2 + \sum_G U_G\,e^{iGx}\right)\psi(x) = \epsilon\psi(x) \ . \tag{24}$$

Equation (24) is written in the one-electron approximation in which the orbital $\psi(x)$ describes the motion of one electron in the potential of the ion cores and in the average potential of the other conduction electrons.

The wavefunction $\psi(x)$ may be expressed as a Fourier series summed over all values of the wavevector permitted by the boundary conditions, so that

$$\psi = \sum_K C(K)\,e^{iKx} \ , \tag{25}$$

where K is real. The set of values of K is the form $2\pi n/L$, for these values satisfy the periodic boundary conditions over the length L; here n is any integer, positive or negative. We do not assume that $\psi(x)$ is periodic in the fundamental lattice translation a; the translational properties of $\psi(x)$ are determined by the Bloch theorem (7). Not all wavevectors of the set $2\pi n/L$ enter the Fourier expansion of any particular Bloch function. If one wavevector k is allowed—that is, if k is known to be contained in a particular ψ—then the other wavevectors in the expansion of this ψ will be shown to have the form $k + G$, where G is any reciprocal lattice vector. We can label a wavefunction ψ that contains a Fourier component k as $\psi(k)$ or equally well as $\psi(k+G)$, for if k enters the Fourier series then $k + G$ also enters. The wavevectors $k + G$, running over G, are a very restricted subset of the wavevectors $2\pi n/L$, as in Fig. 7. We shall generally choose k to lie in the first Brillouin zone. When other conventions are used, we shall say so. This is not like the phonon problem, where there are no components of the ion motion outside the first zone, but it is like the x-ray diffraction problem.

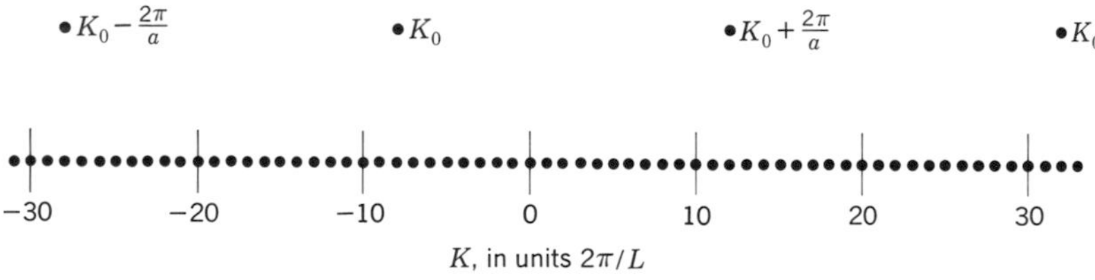

Figure 7 The lower points represent values of the wavevector $K = 2\pi n/L$ allowed by the periodic boundary condition on the wavefunction over a ring of circumference L composed of 20 primitive cells. The allowed values continue to $\pm\infty$. The upper points represent the first few wavevectors which may enter into the Fourier expansion of a wavefunction $\psi(x)$, starting from a particular wavevector $k = K_0 = -8(2\pi/L)$. The shortest reciprocal lattice vector is $2\pi/a = 20(2\pi/L)$.

To solve the wave equation, substitute (25) in (24) to obtain a set of linear algebraic equations for the Fourier coefficients. The kinetic energy term is

$$\frac{1}{2m}\,p^2\psi(x) = \frac{1}{2m}\left(-i\hbar\frac{d}{dx}\right)^2 \psi(x) = -\frac{\hbar^2}{2m}\frac{d^2\psi}{dx^2} = \frac{\hbar^2}{2m}\sum_K K^2 C(K) e^{iKx} \ ;$$

and the potential energy term is

$$\left(\sum_G U_G\, e^{iGx}\right)\psi(x) = \sum_G\sum_K U_G\, e^{iGx} C(K)\, e^{iKx} \ .$$

The wave equation is obtained as the sum:

$$\sum_K \frac{\hbar^2}{2m} K^2 C(K)\, e^{iKx} + \sum_G\sum_K U_G C(K)\, e^{i(K+G)x} = \epsilon \sum_K C(K)\, e^{iKx} \ . \tag{26}$$

Each Fourier component must have the same coefficient on both sides of the equation. Thus

$$\boxed{(\lambda_K - \epsilon) C(K) + \sum_G U_G C(K-G) = 0 \ ,} \tag{27}$$

with the notation

$$\lambda_K = \hbar^2 K^2/2m \ . \tag{28}$$

Equation (27) is a useful form of the wave equation in a periodic lattice, although unfamiliar because a set of algebraic equations has taken the place of the usual differential equation. The set appears formidable because there are, in principle, an infinite number of $C(K-G)$ to be determined. In practice a small number will often suffice, perhaps two or four.

Restatement of the Bloch Theorem. Once we determine the C's from (27), the wavefunction is given as

$$\psi_k(x) = \sum_G C(k-G)\, e^{i(k-G)x} \ , \tag{29}$$

which may be rearranged as

$$\psi_k(x) = \left(\sum_G C(k-G)\, e^{-iGx}\right) e^{ikx} = e^{ikx} u_k(x) \ ,$$

with the definition

$$u_k(x) \equiv \sum_G C(k-G)\, e^{iGx} \ .$$

Because $u_k(x)$ is a Fourier series over the reciprocal lattice vectors, it is invariant under a crystal lattice translation T, so that $u_k(x) = u_k(x+T)$.

We verify this directly by evaluating $u_k(x+T)$:

$$u_k(x+T) = \Sigma\, C(k-G)\, e^{-iG(x+T)} = e^{-iGT}[\Sigma\, C(k-G)\, e^{-iGx}] \;.$$

Because $\exp(-iGT) = 1$, it follows that $u_k(x+T) = u_k(x)$, thereby establishing the periodicity of u_k. This is an alternate and exact proof of the Bloch theorem and is valid even when the ψ_k are degenerate.

Crystal Momentum of an Electron

What is the significance of the wavevector **k** used to label the Bloch function? It has several properties:

(a) Under a crystal lattice translation which carries **r** to **r** + **T** we have

$$\psi_{\mathbf{k}}(\mathbf{r}+\mathbf{T}) = e^{i\mathbf{k}\cdot\mathbf{T}}\, e^{i\mathbf{k}\cdot\mathbf{r}}\, u_{\mathbf{k}}(\mathbf{r}+\mathbf{T}) = e^{i\mathbf{k}\cdot\mathbf{T}}\psi_{\mathbf{k}}(\mathbf{r}) \;, \tag{30}$$

because $u_{\mathbf{k}}(\mathbf{r}+\mathbf{T}) = u_{\mathbf{k}}(\mathbf{r})$. Thus $e^{i\mathbf{k}\cdot\mathbf{T}}$ is the phase factor[1] by which a Bloch function is multiplied when we make a crystal lattice translation **T.**

(b) If the lattice potential vanishes, the central equation (27) reduces to $(\lambda_{\mathbf{k}} - \epsilon)C(\mathbf{k}) = 0$, so that all $C(\mathbf{k}-\mathbf{G})$ are zero except $C(\mathbf{k})$, and thus $u_{\mathbf{k}}(\mathbf{r})$ is constant. We have $\psi_{\mathbf{k}}(\mathbf{r}) = e^{i\mathbf{k}\cdot\mathbf{r}}$, just as for a free electron. (This assumes we have had the foresight to pick the "right" **k** as the label; for many purposes other choices of **k**, as we shall see, are more convenient.)

(c) The value of **k** enters into the conservation laws for collision processes of electrons in crystals. For this reason $\hbar\mathbf{k}$ is called the **crystal momentum** of the electron. When an electron **k** collides with a phonon of wavevector **q**, the selection rule is $\mathbf{k}+\mathbf{q} = \mathbf{k}' + \mathbf{G}$, if the phonon is absorbed in the collision. The electron has been scattered from a state **k** to a state **k**′, and **G** is a reciprocal lattice vector.

Solution of the Central Equation

The central equation

$$(\lambda_K - \epsilon)C(K) + \sum_G U_G C(K-G) = 0 \tag{31}$$

represents a set of simultaneous linear equations connecting the $C(K-G)$ for all reciprocal lattice vectors G. These equations are consistent if the determinant of the coefficients vanishes. We write out the equations for an explicit problem. Let g denote the shortest G, and suppose that the potential energy contains only one Fourier component $U_g = U_{-g}$,

[1]We may also say that $\exp(i\mathbf{k}\cdot\mathbf{T})$ is the eigenvalue of the crystal translation operation **T**, and $\psi_{\mathbf{k}}$ is the eigenvector. That is, $\mathbf{T}\psi_{\mathbf{k}}(x) = \psi_{\mathbf{k}}(x+\mathbf{T}) = \exp(i\mathbf{k}\cdot\mathbf{T})\psi_{\mathbf{k}}(x)$, so that **k** is a suitable label for the eigenvalue. Here we have used the Bloch theorem.

denoted by U. Then a block of the determinant is given by:

$$\begin{array}{ccccc} \lambda_{K-2g}-\epsilon & U & 0 & 0 & 0 \\ U & \lambda_{K-g}-\epsilon & U & 0 & 0 \\ 0 & U & \lambda_K-\epsilon & U & 0 \\ 0 & 0 & U & \lambda_{K+g}-\epsilon & U \\ 0 & 0 & 0 & U & \lambda_{K+2g}-\epsilon \end{array} \qquad (32)$$

To see this, write out five successive equations of the set (31). The determinant in principle is infinite in extent, but it will often be sufficient to set equal to zero the portion we have shown. If the potential contains two important Fourier components, $U_1 = U_g = U_{-g}$ and $U_2 = U_{2g} = U_{-2g}$, the determinant will have other entries. The third line above will become:

$$U_2 \qquad U_1 \qquad \lambda_K - \epsilon \qquad U_1 \qquad U_2 \;.$$

At a given K, each root ϵ or ϵ_K lies on a different energy band, except in case of coincidence. The solution of the determinant (32) gives a set of energy eigenvalues ϵ_{nK}, where n is an index for ordering the energies and K is the wavevector that labels C_K. Most often K will be taken in the first zone, to reduce possible confusion in the labeling. If we chose a K different from the original by some reciprocal lattice vector, we would have obtained the same set of equations in a different order—but the same energy spectrum.

Empty Lattice Approximation. Actual band structures are usually shown in the first Brillouin zone. Because the band energies are sometimes approximated rather well by free electron energies, it is helpful to show the free electron energies also in the first zone. The free electron energy is $\epsilon_{\mathbf{K}} = \hbar^2K^2/2m$. Here there is no restriction on the wavevector **K**, but we can always write this in terms of a wavevector **k** restricted to the first zone. For any **K** we can find some reciprocal lattice vector **G** such that $\mathbf{K} = \mathbf{k} + \mathbf{G}$, with **k** in the first zone. The energy is written

$$\begin{aligned} \epsilon(k_x,k_y,k_z) &= (\hbar^2/2m)(\mathbf{k}+\mathbf{G})^2 \\ &= (\hbar^2/2m)\,[(k_x+G_x)^2 + (k_y+G_y)^2 + (k_z+G_z)^2] \;. \end{aligned} \qquad (33)$$

We consider the low-lying free electron bands of a simple cubic lattice. Suppose we want to exhibit the energy as a function of **k** in the [100] direction. For convenience, choose units such that $\hbar^2/2m = 1$. We show several low-lying bands in this empty lattice approximation with their energies at $\mathbf{k} = 0$ and along the k_x axis in the first zone:

Band	$Ga/2\pi$	$\epsilon(000)$	$\epsilon(k_x00)$
1	000	0	k_x^2
2,3	$100,\bar{1}00$	$(2\pi/a)^2$	$(k_x \pm 2\pi/a)^2$
4,5,6,7	$010,0\bar{1}0,001,00\bar{1}$	$(2\pi/a)^2$	$k_x^2 + (2\pi/a)^2$
8,9,10,11	$110,101,1\bar{1}0,10\bar{1}$	$2(2\pi/a)^2$	$(k_x + 2\pi/a)^2 + (2\pi/a)^2$
12,13,14,15	$\bar{1}10,\bar{1}01,\bar{1}\bar{1}0,\bar{1}0\bar{1}$	$2(2\pi/a)^2$	$(k_x - 2\pi/a)^2 + (2\pi/a)^2$
16,17,18,19	$011,0\bar{1}1,01\bar{1},0\bar{1}\bar{1}$	$2(2\pi/a)^2$	$k_x^2 + 2(2\pi/a)^2$

These bands are plotted in Fig. 8. It is a good exercise to plot the same bands for **k** parallel to the [111] direction of wavevector space.

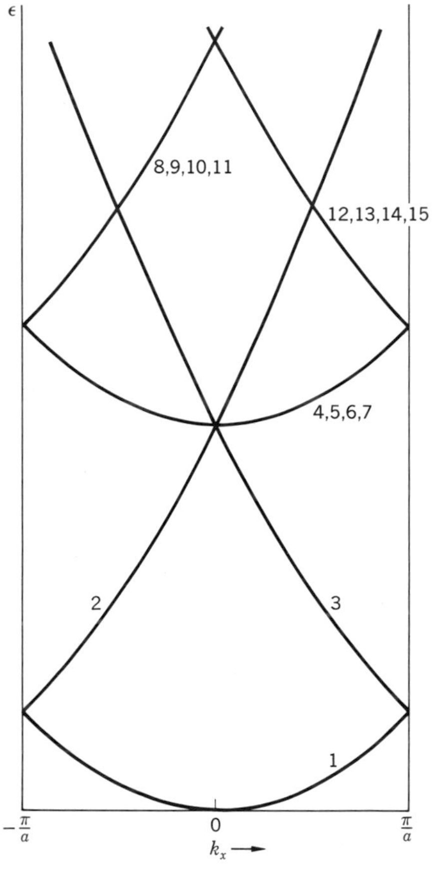

Figure 8 Empty lattice electron energy. The bold curves are in the first Brillouin zone, with $-\pi/a \leq k_x \leq \pi/a$. Energy bands drawn in this zone are said to be in the reduced zone scheme. Low-lying free electron energy bands of simple cubic lattice as transformed to first Brillouin zone and plotted versus (k_x00).

Approximate Solution Near a Zone Boundary

We suppose that the Fourier components U_G of the potential energy are small in comparison with the kinetic energy of a free electron at the zone boundary. We first consider a wavevector exactly at the zone boundary at $\frac{1}{2}G$, that is, at π/a. Here

$$k^2 = (\tfrac{1}{2}G)^2 \; ; \qquad (k - G)^2 = (\tfrac{1}{2}G - G)^2 = (\tfrac{1}{2}G)^2 \; ,$$

so that at the zone boundary the kinetic energy of the two component waves $k = \pm\frac{1}{2}G$ are equal. If $C(\frac{1}{2}G)$ is an important coefficient in the orbital (29) at the zone boundary, then $C(-\frac{1}{2}G)$ is also an important coefficient. This result also follows from the discussion of (5).

We retain only those equations in the central equation that contain both coefficients $C(\frac{1}{2}G)$ and $C(-\frac{1}{2}G)$, and neglect all other coefficients. Then one equation of (31) becomes, with $K = \frac{1}{2}G$ and $\lambda \equiv \hbar^2(\frac{1}{2}G)^2/2m$,

$$(\lambda - \epsilon)C(\tfrac{1}{2}G) + UC(-\tfrac{1}{2}G) = 0 \; . \tag{34}$$

Another equation of (31) becomes

$$(\lambda - \epsilon)C(-\tfrac{1}{2}G) + UC(\tfrac{1}{2}G) = 0 \; . \tag{35}$$

These two equations have nontrivial solutions for the two coefficients if the energy ϵ satisfies

$$\begin{vmatrix} \lambda - \epsilon & U \\ U & \lambda - \epsilon \end{vmatrix} = 0 \; , \tag{36}$$

whence

$$(\lambda - \epsilon)^2 = U^2 \; ; \qquad \epsilon = \lambda \pm U = \frac{\hbar^2}{2m}(\tfrac{1}{2}G)^2 \pm U \; . \tag{37}$$

The energy has two roots, one lower than the free electron kinetic energy by U, and one higher by U. Thus the potential energy $2U \cos Gx$ has created an energy gap of extent $2U$ at the zone boundary.

The ratio of the C's may be found from either (34) or (35):

$$\frac{C(-\frac{1}{2}G)}{C(\frac{1}{2}G)} = \frac{\epsilon - \lambda}{U} = \pm 1 \; , \tag{38}$$

where the last step uses (37). Thus the Fourier expansion of $\psi(x)$ has the two solutions

$$\psi(x) = \exp(iGx/2) \pm \exp(-iGx/2) \; .$$

These orbitals are identical to (5). One solution gives the wavefunction at the bottom of the energy gap; the other gives the wavefunction at the top of the gap. Which solution has the lower energy depends on the sign of U.

We now solve for the orbitals with a wavevector k near the zone boundary $\frac{1}{2}G$. We use the same two-component approximation, now with a wavefunction of the form

$$\psi(x) = C(k)\,e^{ikx} + C(k - G)\,e^{i(k-G)x} \; . \tag{39}$$

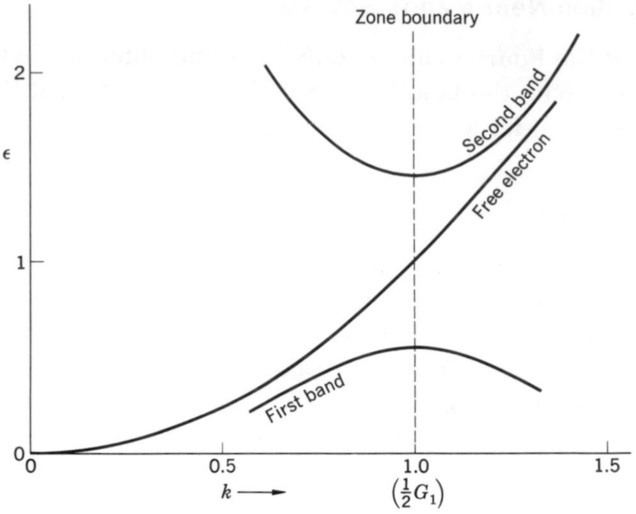

Figure 9 Solutions of (40) in the periodic zone scheme, in the region near a boundary of the first Brillouin zone. The units are such that $U = -0.45$; $G = 2$; and $\hbar^2/m = 1$. The free electron curve is drawn for comparison. The energy gap at the zone boundary is 0.90. The value of U has deliberately been chosen large for this illustration, too large for the two-term approximation to be accurate.

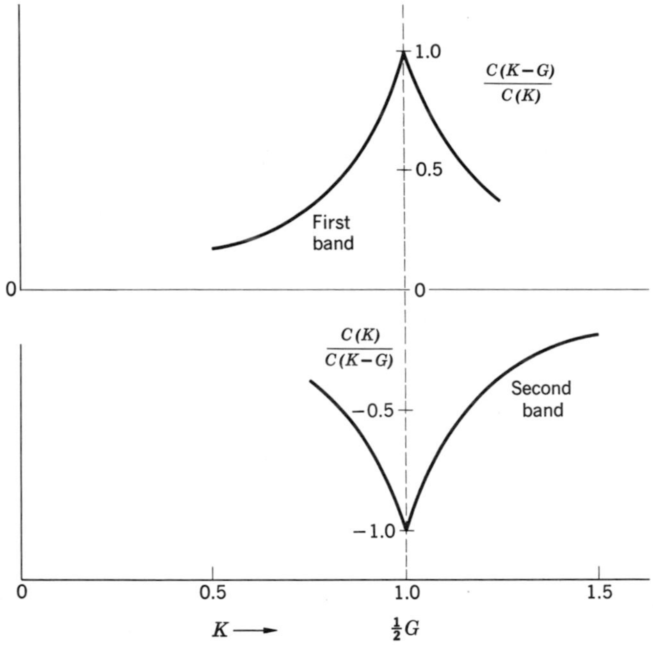

Figure 10 Ratio of the coefficients in $\psi(x) = C(K)e^{iKx} + C(K-G)e^{i(K-G)x}$ as calculated near the boundary of the first Brillouin zone. One component dominates as we leave the boundary.

As directed by the central equation, we solve the pair of equations

$$(\lambda_k - \epsilon)C(k) + UC(k-G) = 0 \ ;$$
$$(\lambda_{k-G} - \epsilon)C(k-G) + UC(k) = 0 \ ,$$

with λ_k defined as $\hbar^2k^2/2m$. These equations have a solution if the energy ϵ satisfies

$$\begin{vmatrix} \lambda_k - \epsilon & U \\ U & \lambda_{k-G} - \epsilon \end{vmatrix} = 0 \ ,$$

whence $\epsilon^2 - \epsilon(\lambda_{k-G} + \lambda_k) + \lambda_{k-G}\lambda_k - U^2 = 0$.

The energy has two roots:

$$\epsilon = \tfrac{1}{2}(\lambda_{k-G} + \lambda_k) \pm [\tfrac{1}{4}(\lambda_{k-G} - \lambda_k)^2 + U^2]^{\frac{1}{2}} \ , \qquad (40)$$

and each root describes an energy band. The two roots are plotted in Fig. 9. It is convenient to expand the energy in terms of a quantity $\tilde{K}$ (the mark over the $\tilde{K}$ is called a tilde), which measures the difference $\tilde{K} \equiv k - \frac{1}{2}G$ in wavevector between k and the zone boundary:

$$\epsilon_{\tilde{K}} = (\hbar^2/2m)(\tfrac{1}{4}G^2 + \tilde{K}^2) \pm [4\lambda(\hbar^2\tilde{K}^2/2m) + U^2]^{\frac{1}{2}}$$
$$\simeq (\hbar^2/2m)(\tfrac{1}{4}G^2 + \tilde{K}^2) \pm U[1 + 2(\lambda/U^2)(\hbar^2\tilde{K}^2/2m)] \ , \qquad (41)$$

in the region $\hbar^2G\tilde{K}/2m \ll |U|$. Here $\lambda = (\hbar^2/2m)\ (\frac{1}{2}G)^2$ as before. Writing the two roots of (37) as $\epsilon_1(\pm)$, we may write (41) as

$$\epsilon_{\tilde{K}}(\pm) = \epsilon_1(\pm) + \frac{\hbar^2\tilde{K}^2}{2m}\left(1 \pm \frac{2\lambda}{U}\right) \ . \qquad (42)$$

These are the roots for the energy when the wavevector is very close to the zone boundary at $\frac{1}{2}G$. Note the quadratic dependence of the energy on the wavevector $\tilde{K}$. For U negative, the solution $\epsilon(-)$ corresponds to the upper of the two bands, and $\epsilon(+)$ to the lower of the two bands. The two C's are plotted in Fig. 10.

NUMBER OF ORBITALS IN A BAND

Consider a linear crystal constructed of primitive cells of lattice constant a. In order to count states we apply periodic boundary conditions to the wavefunctions over the length of the crystal. The allowed values of the electron wavevector k in the first Brillouin zone are given by (2):

$$k = 0 \ ; \quad \pm\frac{2\pi}{L} \ ; \quad \pm\frac{4\pi}{L} \ ; \quad \ldots \ ; \quad \frac{N\pi}{L} \ . \qquad (43)$$

We cut off the series at $N\pi/L = \pi/a$, for this is the zone boundary. The point $-N\pi/L = -\pi/a$ is not to be counted as an independent point because it is connected by a reciprocal lattice vector with π/a. The total number of points is exactly N, the number of primitive cells. **Each primitive cell contributes exactly one independent value of k to each energy band.** This result carries over into three dimensions. With account taken of the two

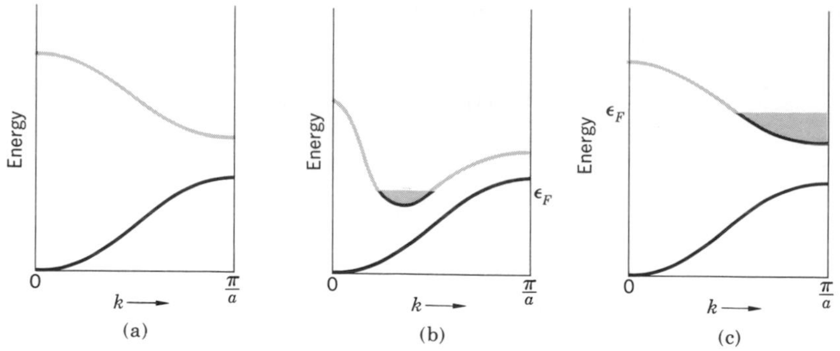

Figure 11 Occupied states and band structures giving (a) an insulator, (b) a metal or a semimetal because of band overlap, and (c) a metal because of electron concentration. In (b) the overlap need not occur along the same directions in the Brillouin zone. If the overlap is small, with relatively few states involved, we speak of a semimetal.

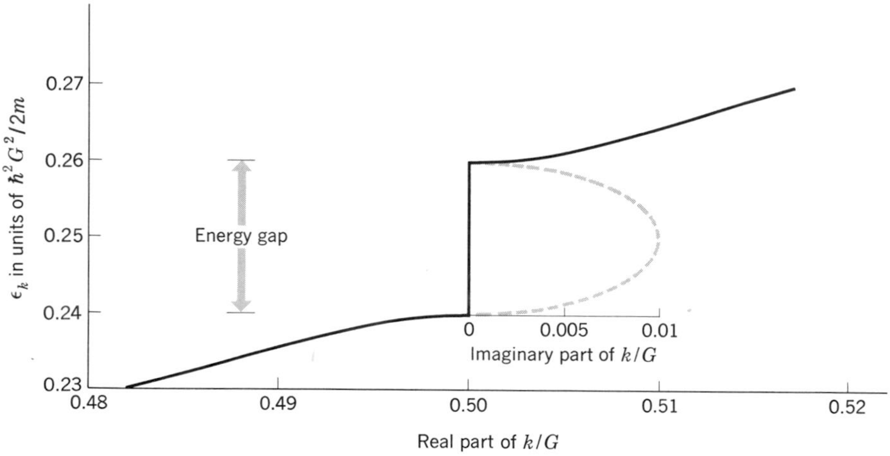

Figure 12 In the energy gap there exist solutions of the wave equation for complex values of the wavevector. At the boundary of the first zone the real part of the wavevector is $\frac{1}{2}G$. The imaginary part is plotted in the approximation of two plane waves, for $U = 0.01\ \hbar^2G^2/2m$. In an infinite unbounded crystal the wavevector must be real, or else the amplitude will increase without limit. But on a surface or at a junction there can exist solutions with complex wavevector.

independent orientations of the electron spin, **there are $2N$ independent orbitals in each energy band.** If there is a single atom of valence one in each primitive cell, the band can be half filled with electrons. If each atom contributes two valence electrons to the band, the band can be exactly filled. If there are two atoms of valence one in each primitive cell, the band can also be exactly filled.

Metals and Insulators

If the valence electrons exactly fill one or more bands, leaving others empty, the crystal will be an insulator. An external electric field will not cause the flow of current. We suppose that the electric field is not strong enough to disrupt the electronic structure. Because a filled band is separated by an energy gap from the next higher band, there is no continuous way to change the total momentum of the electrons if every accessible state is filled. Nothing changes when the field is applied. This is quite unlike the situation for free electrons.

A crystal can be an insulator only if the number of valence electrons in a primitive cell of the crystal is an even integer. (An exception must be made sometimes for electrons in tightly bound inner shells which cannot be treated by band theory.) If a crystal has an even number of valence electrons per primitive cell, it is necessary to consider whether or not the bands overlap in energy. If the bands overlap, then instead of one filled band giving an insulator, we can have two partly filled bands giving a metal (Fig. 11).

The alkali metals and the noble metals have one valence electron per primitive cell, so that they have to be metals. The alkaline earth metals have two valence electrons per primitive cell; they could be insulators, but the bands overlap in energy to give metals, but not very good metals. Diamond, silicon, and germanium each have two atoms of valence four (or eight valence electrons) per primitive cell; the bands do not overlap, and the pure crystals are insulators at absolute zero.

SUMMARY

1. The solutions of the wave equation in a periodic lattice are of the Bloch form $\psi_{\mathbf{k}}(\mathbf{r}) = e^{i\mathbf{k}\cdot\mathbf{r}} u_{\mathbf{k}}(\mathbf{r})$, where $u_{\mathbf{k}}(\mathbf{r})$ is invariant under a crystal lattice translation. The **K**'s which appear in a Fourier series expansion

$$\psi_{\mathbf{k}}(\mathbf{r}) = \sum_{\mathbf{K}} C(\mathbf{K}) \exp(i\mathbf{K}\cdot\mathbf{r})$$

are all of the form $\mathbf{k} + \mathbf{G}$, where **G** runs over all reciprocal lattice vectors.

2. There are regions of energy for which no Bloch function solutions of the wave equation exist. These energies form forbidden regions in which the wavefunctions are damped in space and the values of the **K**'s are complex, as pictured in Fig. 12. The existence of insulators is due to the existence of forbidden regions of energy.

3. Energy bands may often be approximated by one or two plane waves: for example, $\psi_k(x) \cong C(k)e^{ikx} + C(k-G)e^{i(k-G)x}$ near the zone boundary at $\frac{1}{2}G$.

4. The number of orbitals in a band is $2N$, where N is the number of primitive cells in the specimen.

PROBLEMS

1. ***Square lattice, free electron energies.*** (a) Show for a simple square lattice (two dimensions) that the kinetic energy of a free electron at a corner of the first zone is higher than that of an electron at midpoint of a side face of the zone by a factor of 2. (b) What is the corresponding factor for a simple cubic lattice

(three dimensions)? (c) What bearing might the result of (b) have on the conductivity of divalent metals?

2. ***Free electron energies in reduced zone.*** Consider the free electron energy bands of an fcc crystal lattice in the approximation of an empty lattice, but in the reduced zone scheme in which all **k**'s are transformed to lie in the first Brillouin zone. Plot roughly in the [111] direction the energies of all bands up to six times the lowest band energy at the zone boundary at $\mathbf{k} = (2\pi/a)(\frac{1}{2}, \frac{1}{2}, \frac{1}{2})$. Let this be the unit of energy. This problem shows why band edges need not necessarily be at the zone center. Several of the degeneracies (band crossings) will be removed when account is taken of the crystal potential.

3. ***Kronig-Penney model.*** (a) For the delta-function potential and with $P \ll 1$, find at $k = 0$ the energy of the lowest energy band. (b) For the same problem find the band gap at $k = \pi/a$.

4. ***Potential energy in the diamond structure.*** (a) Show that for the diamond structure the Fourier component $U_{\mathbf{G}}$ of the crystal potential seen by an electron is equal to zero for $\mathbf{G} = 2\mathbf{A}$, where $\mathbf{A}$ is a basis vector in the reciprocal lattice referred to the conventional cubic cell. (b) Show that in the usual first-order approximation to the solutions of the wave equation in a periodic lattice the energy gap vanishes at the zone boundary plane normal to the end of the vector **A**.

*5. ***Complex wavevectors in the energy gap.*** Find an expression for the imaginary part of the wavevector in the energy gap at the boundary of the first Brillouin zone, in the approximation that led to Eq. (36). Give the result for the Im(k) at the center of the energy gap. The result for small Im(k) is

$$(\hbar^2/2m)[\mathrm{Im}(k)]^2 \approx 2mU^2/\hbar^2G^2 \ .$$

The form as plotted in Fig. 12 is of importance in the theory of Zener tunneling from one band to another in the presence of a strong electric field. [Experimental confirmation of the predicted form of Im(k) is reported by G. H. Parker and C. A. Mead, "Energy-momentum relationship in InAs," Phys. Rev. Letters **21,** 605 (1968); see also S. Kurtin, T. C. McGill, and C. A. Mead, Phys. Rev. Letters **25,** 756 (1970).]

6. ***Square lattice.*** Consider a square lattice in two dimensions with the crystal potential

$$U(x,y) = -4U \cos(2\pi x/a) \cos(2\pi y/a) \ .$$

Apply the central equation to find approximately the energy gap at the corner point $(\pi/a, \pi/a)$ of the Brillouin zone. It will suffice to solve a 2×2 determinantal equation.

References

J. M. Ziman, *Principles of the theory of solids,* Cambridge, 2nd ed., 1972.

N. F. Mott and H. Jones, *Theory of the properties of metals and alloys,* Oxford, 1936. (Dover paperback reprint.)

(Further references on band theory are given at the end of Chapter 9.)

*This problem is somewhat difficult.

8 SEMICONDUCTOR CRYSTALS

NOTE: Material on the motion of electrons and holes in applied electric and magnetic fields has been collected together in this chapter. The discussion of orbits is continued in Chapter 9.

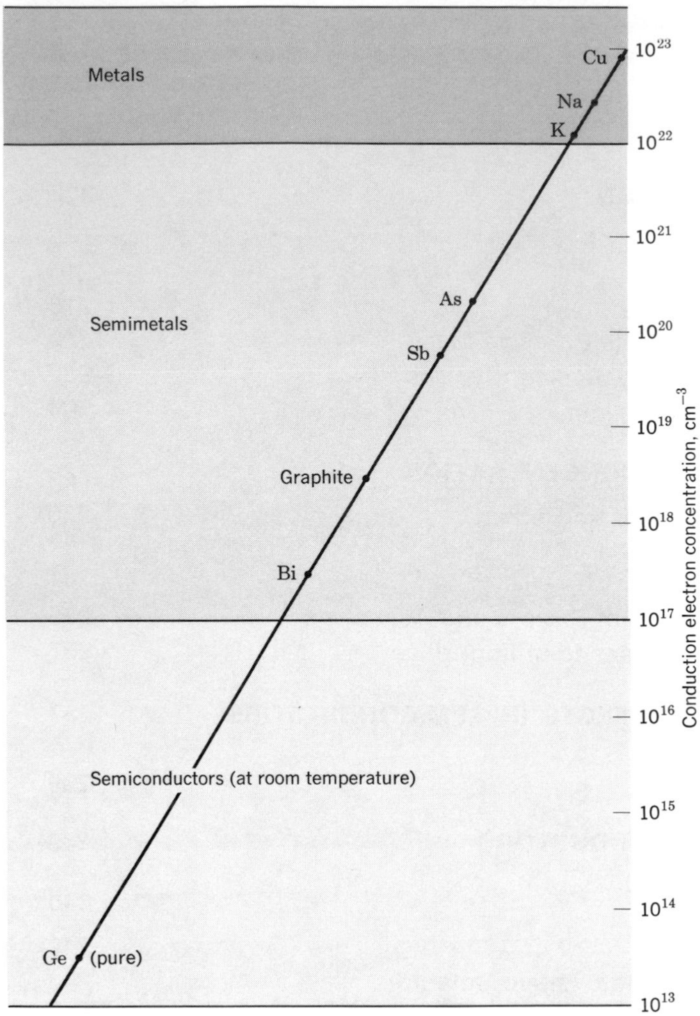

Figure 1 Carrier concentrations for metals, semimetals, and semiconductors. The semiconductor range may be extended upward by increasing the concentration of impurity atoms. (The horizontal axis of this plot is for display purposes and has no significance.)

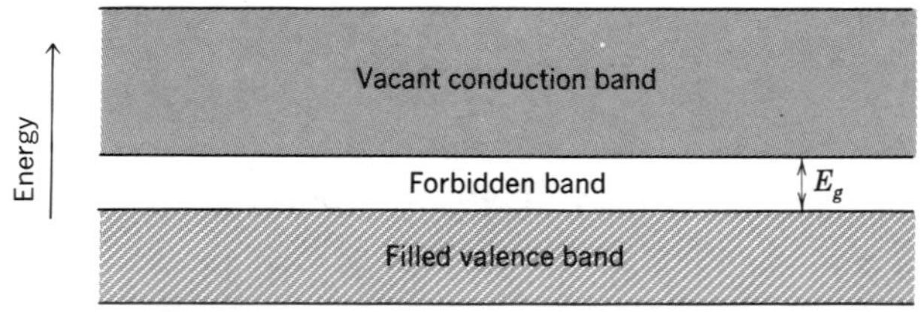

Figure 2 Band scheme for intrinsic conductivity in a semiconductor. At 0 K the conductivity is zero because all states in the valence band are filled and all states in the conduction band are vacant. As the temperature is increased, electrons are thermally excited from the valence band to the conduction band, where they become mobile.

8 SEMICONDUCTOR CRYSTALS

There is an essential difference between a semiconductor, such as germanium, and a good conductor, such as silver. . . . The resistance of a good conductor falls rapidly with falling temperature, while that of a poor conductor rises and becomes very large as the temperature approaches the absolute zero.

A. H. Wilson

At absolute zero a pure, perfect crystal of most semiconductors will be an insulator. The characteristic semiconducting properties are brought about by thermal excitation, impurities, lattice defects, or departure from nominal chemical composition. Semiconductors are electronic conductors with electrical resistivity values generally in the range of 10^{-2} to 10^{9} ohm-cm at room temperature, intermediate between good conductors (10^{-6} ohm-cm) and insulators (10^{14} to 10^{22} ohm-cm). Carrier concentrations representative of metals, semimetals, and semiconductors are shown in Fig. 1.

The electrical resistivity of a semiconductor may be strongly dependent on temperature. Devices based on the properties of semiconductors include transistors, rectifiers, modulators, detectors, thermistors, and photocells. We discuss in this chapter the central physical features of semiconductor crystals, particularly silicon and germanium; other important crystals include cuprous oxide (Cu_2O), selenium, PbTe, PbS, SiC, InSb, GaAs, and graphite.

Some useful nomenclature: the semiconductor compounds of chemical formula AB, where A is a trivalent element and B is a pentavalent element, are called III-V (three-five) compounds. Examples are indium antimonide and gallium arsenide. Where A is divalent and B is hexavalent, the compound is called a II-VI compound; examples are zinc sulfide and cadmium sulfide. Silicon and germanium are sometimes called diamond-type semiconductors, because they have the crystal structure of diamond. Diamond itself is more an insulator rather than a semiconductor. Silicon carbide SiC is a IV-IV compound.

A highly purified semiconductor exhibits intrinsic conductivity, as distinguished from the impurity conductivity of less pure specimens. In the **intrinsic temperature range** the electrical properties of a semiconductor are not essentially modified by impurities in the crystal. An electronic band scheme leading to intrinsic conductivity is indicated in Fig. 2. The conduction band is vacant at absolute zero and is separated by an energy gap E_g from the filled valence band. The **band gap** is the difference in energy between the lowest point of the conduction band and the highest point of the valence band. The lowest point in the conduction band is called the **conduction band edge;** the highest point in the valence band is called the **valence band edge.**

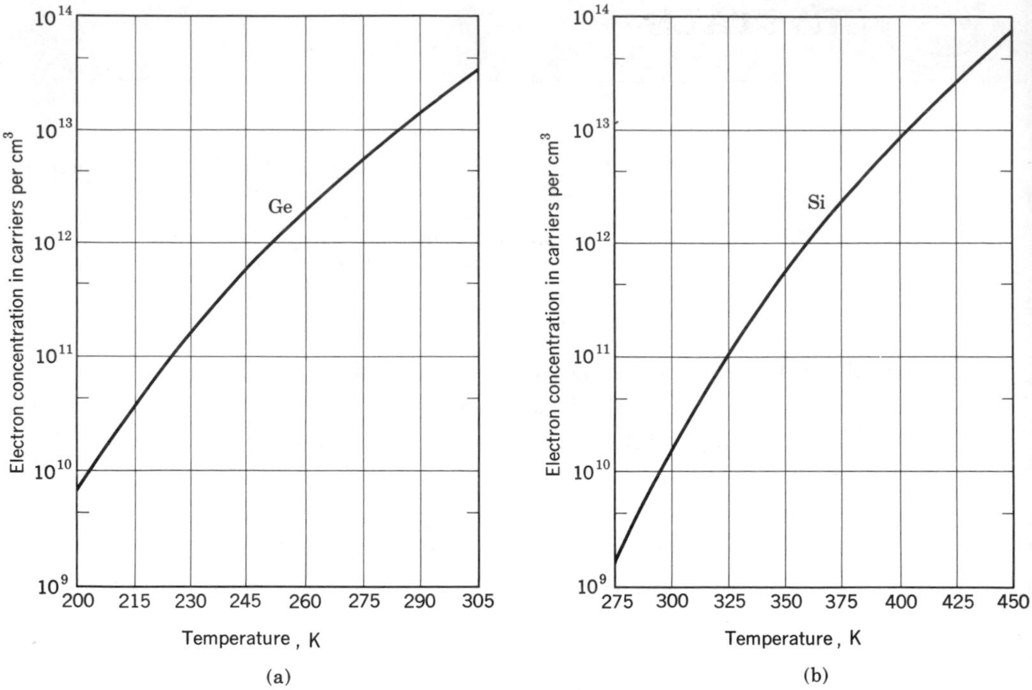

Figure 3 Intrinsic electron concentration as a function of temperature for (a) germanium and (b) silicon. Under intrinsic conditions the hole concentration is equal to the electron concentration. The intrinsic concentration at a given temperature is higher in Ge than in Si because the energy gap is narrower in Ge (0.67 eV) than in Si (1.14 eV). (After W. C. Dunlap.)

As the temperature is increased, electrons are thermally excited from the valence band to the conduction band (Fig. 3). Both the electrons in the conduction band and the vacant orbitals or holes left behind in the valence band contribute to the electrical conductivity.

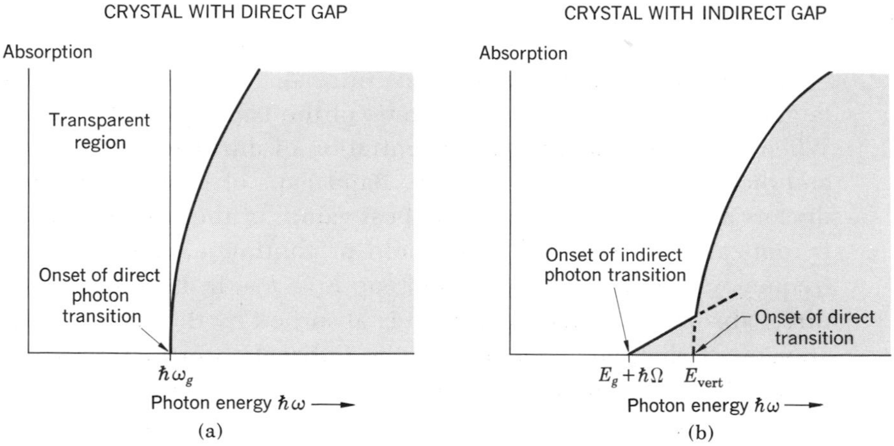

Figure 4 Optical absorption in pure insulators at absolute zero. In (a) the threshold determines the energy gap as $E_g = \hbar\omega_g$. In (b) the optical absorption is weaker near the threshold: at $\hbar\omega = E_g + \hbar\Omega$ a photon is absorbed with the creation of three particles: a free electron, a free hole, and a phonon of energy $\hbar\Omega$. In (b) the energy E_{vert} marks the threshold for the creation of a free electron and a free hole, with no phonon involved. Such a transition is called vertical; it is similar to the direct transition in (a). These plots do not show absorption lines that sometimes are seen lying just to the low energy side of the threshold. Such lines are due to the creation of a bound electron-hole pair, called an exciton.

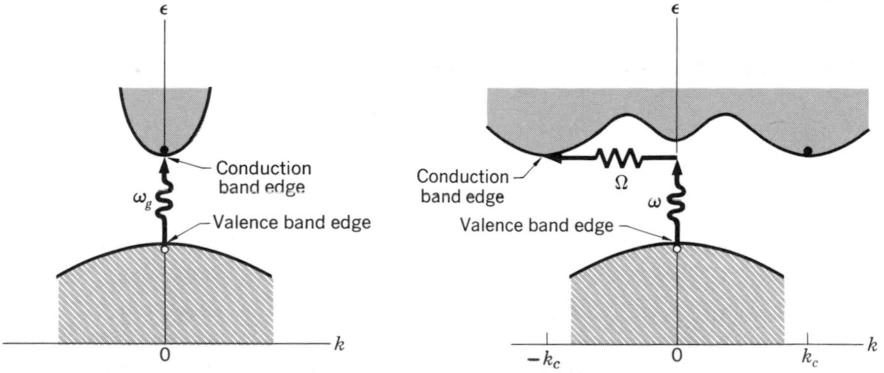

Figure 5 In (a) the lowest point of the conduction band occurs at the same value of **k** as the highest point of the valence band. A direct optical transition is drawn vertically with no significant change of **k**, because the absorbed photon has a very small wavevector. The threshold frequency ω_g for absorption by the direct transition determines the energy gap $E_g = \hbar\omega_g$. The indirect transition in (b) involves both a photon and a phonon because the band edges of the conduction and valence bands are widely separated in **k** space. The threshold energy for the indirect process in (b) is greater than the true band gap. The absorption threshold for the indirect transition between the band edges is at $\hbar\omega = E_g + \hbar\Omega$, where Ω is the frequency of an emitted *phonon* of wavevector $\mathbf{K} \cong -\mathbf{k}_c$. At higher temperatures phonons are already present; if a phonon is absorbed along with a photon, the threshold energy is $\hbar\omega = E_g - \hbar\Omega$. *Note:* The figure shows only the threshold transitions. Transitions occur generally between almost all points of the two bands for which the wavevectors and energy can be conserved.

BAND GAP

The intrinsic conductivity and intrinsic carrier concentrations are largely controlled by E_g/k_BT, the ratio of the band gap to the temperature. When this ratio is large, the concentration of intrinsic carriers will be low and the conductivity will be low. Band gaps of representative semiconductors are given in Table 1. The best values of the band gap are obtained by optical absorption. The threshold of continuous optical absorption at frequency ω_g determines the band gap $E_g = \hbar\omega_g$ in Figs. 4a and 5a. In the **direct absorption process** a photon is absorbed by the crystal with the creation of an electron and a hole. In the **indirect absorption process** in Figs. 4b and 5b the minimum energy gap of the band structure involves electrons and holes separated by a substantial wavevector $\mathbf{k}_c$. Here a direct photon transition at the energy of the minimum gap cannot satisfy the requirement of conservation of wavevector, because photon wavevectors are negligible at the energy range of interest. But if a phonon of wavevector $\mathbf{K}$ and frequency Ω is created in the process, then we can have

$$\mathbf{k}(\text{photon}) = \mathbf{k}_c + \mathbf{K} \cong 0 \; ; \qquad \hbar\omega = E_g + \hbar\Omega \; ,$$

as required by the conservation laws. The phonon energy $\hbar\Omega$ will gener-

TABLE 1 Energy Gap Between the Valence and Conduction Bands

(i = indirect gap; d = direct gap)

Crystal	Gap	E_g, eV 0 K	E_g, eV 300 K	Crystal	Gap	E_g, eV 0 K	E_g, eV 300 K
Diamond	i	5.4		HgTe[a]	d	−0.30	
Si	i	1.17	1.14	PbS	d	0.286	0.34–0.37
Ge	i	0.744	0.67	PbSe	d	0.165	0.27
αSn	d	0.00	0.00	PbTe	d	0.190	0.30
InSb	d	0.24	0.18	CdS	d	2.582	2.42
InAs	d	0.43	0.35	CdSe	d	1.840	1.74
InP	d	1.42	1.35	CdTe	d	1.607	1.45
GaP	i	2.32	2.26	ZnO		3.436	3.2
GaAs	d	1.52	1.43	ZnS		3.91	3.6
GaSb	d	0.81	0.78	SnTe	d	0.3	0.18
AlSb	i	1.65	1.52	AgCl		—	3.2
SiC(hex)		3.0	—	AgI		—	2.8
Te	d	0.33	—	Cu_2O		2.172	—
ZnSb		0.56	0.56	TiO_2		3.03	—

[a] HgTe is a semimetal; the bands overlap.
General references: D. Long, *Energy bands in semiconductors*, Interscience, 1968; also the *A.I.P. Handbook*, 3rd ed., Sec. 9.

ally be much less than E_g: a phonon even of high wavevector is an inexpensive source of crystal momentum because the phonon energies are characteristically small (~0.01 to 0.03 eV) in comparison with the energy gap. If the temperature is so high that the necessary phonon is already thermally excited in the crystal, it is possible also to have a photon absorption process in which the phonon is absorbed.

The band gap may also be deduced from the temperature dependence of the conductivity or of the carrier concentration in the intrinsic range. The carrier concentration is obtained from measurements of the Hall voltage (Chapter 6), sometimes supplemented by conductivity measurements. Optical measurements determine whether the gap is direct or indirect. The band edges in Ge and in Si are connected by indirect transitions; the band edges in InSb are connected by a direct transition (Fig. 6). The gap in αSn is direct and is exactly zero;[1] HgTe and HgSe are semimetals and have negative gaps—the bands overlap.

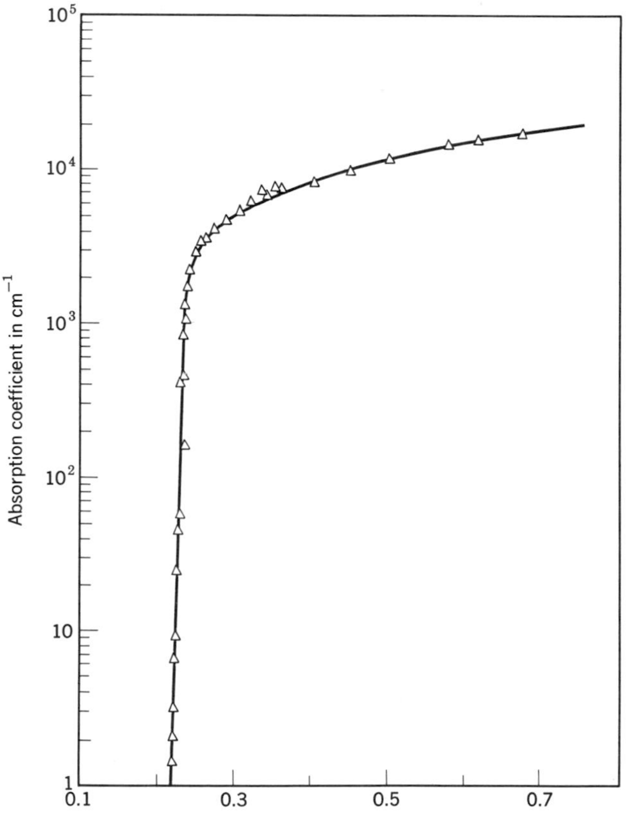

Figure 6 Optical absorption in pure indium antimonide, InSb. The transition is direct because both conduction and valence band edges are at the center of the Brillouin zone, $k = 0$. Notice the sharp threshold. (After G. W. Gobeli and H. Y. Fan.)

[1] S. Groves and W. Paul, Phys. Rev. Letters **11**, 194 (1963).

EQUATIONS OF MOTION

We derive the equation of motion of an electron in an energy band. We look at the motion of a wave packet in an applied electric field. Suppose that the wave packet is made up of wavefunctions near a particular wavevector k. The group velocity is $v_g = d\omega/dk$. The frequency associated with a wavefunction of energy ϵ is $\omega = \epsilon/\hbar$, and so

$$v_g = \hbar^{-1}\, d\epsilon/dk \ . \tag{1}$$

The effects of the crystal on the electron motion are contained in the dispersion relation $\epsilon(k)$. The work $\delta\epsilon$ done on the electron by the electric field E in the interval δt is

$$\delta\epsilon = -eEv_g\, \delta t \ . \tag{2}$$

We observe that

$$\delta\epsilon = (d\epsilon/dk)\delta k = \hbar v_g\, \delta k \ , \tag{3}$$

using (1). On comparing (2) with (3) we have

$$\delta k = -(eE/\hbar)\delta t \ , \tag{4}$$

whence $\hbar dk/dt = -eE$, the same relation as for free electrons.

We may write (4) in terms of the external force **F** as

$$\boxed{\hbar \frac{d\mathbf{k}}{dt} = \mathbf{F} \ .} \tag{5}$$

This is an important relation: In a crystal $\hbar d\mathbf{k}/dt$ is equal to the external force on the electron. In free space $d(m\mathbf{v})/dt$ is equal to the force. We have not overthrown Newton's second law of motion: the electron in the crystal is subject to forces from the crystal lattice as well as from external sources. The result also applies to the Lorentz force on an electron in a magnetic field, under ordinary conditions where the magnetic field is not so strong that it breaks down the band structure. Thus the equation of motion of an electron of group velocity **v** in a constant magnetic field **B** is

$$\text{(CGS)} \quad \hbar \frac{d\mathbf{k}}{dt} = -\frac{e}{c}\, \mathbf{v} \times \mathbf{B} \ ; \qquad \text{(SI)} \quad \hbar \frac{d\mathbf{k}}{dt} = -e\mathbf{v} \times \mathbf{B} \ , \tag{6}$$

where the right-hand side is the Lorentz force on the electron. With the group velocity $\hbar\mathbf{v} = \mathrm{grad}_{\mathbf{k}}\epsilon$, the rate of change of the wavevector is

$$\text{(CGS)} \quad \frac{d\mathbf{k}}{dt} = -\frac{e}{\hbar^2 c}\, \nabla_{\mathbf{k}}\epsilon \times \mathbf{B} \ ; \qquad \text{(SI)} \quad \frac{d\mathbf{k}}{dt} = -\frac{e}{\hbar^2}\, \nabla_{\mathbf{k}}\epsilon \times \mathbf{B} \ , \tag{7}$$

where now both sides of the equation refer to the coordinates in **k** space.

We see from the vector cross-product in (7) that in a magnetic field an electron moves in $\mathbf{k}$ space in a direction normal to the direction of the gradient of the energy ϵ, so that the **electron moves on a surface of constant energy.** The value of the projection k_B of $\mathbf{k}$ on $\mathbf{B}$ is arbitrary, but constant during the motion. The motion in $\mathbf{k}$ space is on a plane normal to the direction of $\mathbf{B}$, and the orbit is defined by the intersection of this plane with a surface of constant energy.

Fuller Derivation of $\hbar\dot{k} = F$

We consider the Bloch eigenfunction $\psi_{\mathbf{k}}$ belonging to the energy eigenvalue $\epsilon_{\mathbf{k}}$ and wavevector $\mathbf{k}$:

$$\psi_{\mathbf{k}} = \sum_{\mathbf{G}} C(\mathbf{k}+\mathbf{G}) \exp[i(\mathbf{k}+\mathbf{G})\cdot\mathbf{r}] \ . \tag{7a}$$

The expectation value of the momentum of an electron in the state $\mathbf{k}$ is

$$\begin{aligned}\mathbf{p}_{\mathrm{el}} &= (\mathbf{k}|-i\hbar\nabla|\mathbf{k}) = \sum_{\mathbf{G}} \hbar(\mathbf{k}+\mathbf{G})|C(\mathbf{k}+\mathbf{G})|^2 \\ &= \hbar(\mathbf{k} + \sum_{\mathbf{G}} \mathbf{G}|C(\mathbf{k}+\mathbf{G})|^2) \ ,\end{aligned} \tag{7b}$$

using $\Sigma|C(\mathbf{k}+\mathbf{G})|^2 = 1$. We examine the transfer of momentum between the electron and the lattice when the state $\mathbf{k}$ of the electron is changed to $\mathbf{k}+\Delta\mathbf{k}$ by the application of an external force.

We imagine an insulating crystal electrostatically neutral except for a single electron in the state $\mathbf{k}$ of an otherwise empty band. We suppose that a weak external force is applied for a time interval such that the total impulse given to the entire crystal system is $\mathbf{J} = \int \mathbf{F}dt$. If the conduction electron were free ($m^* = m$), the total momentum imparted to the crystal system by the impulse would appear in the change of momentum of the conduction electron:

$$\mathbf{J} = \Delta\mathbf{p}_{\mathrm{tot}} = \Delta\mathbf{p}_{\mathrm{el}} = \hbar\Delta\mathbf{k} \ . \tag{7c}$$

The neutral crystal suffers no net interaction with the electric field, either directly or indirectly through the free electron.

If the conduction electron interacts with the periodic potential of the crystal lattice, we must have

$$\mathbf{J} = \Delta\mathbf{p}_{\mathrm{tot}} = \Delta\mathbf{p}_{\mathrm{lat}} + \Delta\mathbf{p}_{\mathrm{el}} \ . \tag{7d}$$

From the result for $\mathbf{p}_{\mathrm{el}}$ we have

$$\Delta\mathbf{p}_{\mathrm{el}} = \hbar\Delta\mathbf{k} + \sum_{\mathbf{G}} \hbar\mathbf{G}[(\nabla_{\mathbf{k}}|C(\mathbf{k}+\mathbf{G})|^2)\cdot\Delta\mathbf{k}] \ . \tag{7e}$$

The change $\Delta \mathbf{p}_{lat}$ in the lattice momentum resulting from the change of state of the electron may be derived by an elementary physical consideration. An electron reflected by the lattice transfers momentum to the lattice. If an incident electron with plane wave component of momentum $\hbar\mathbf{k}$ is reflected with momentum $\hbar(\mathbf{k}+\mathbf{G})$, the lattice acquires the momentum $-\hbar\mathbf{G}$, as required by momentum conservation. The momentum transfer to the lattice when the state ψ goes over to $\psi_{\mathbf{k}+\Delta\mathbf{k}}$ is

$$\Delta \mathbf{p}_{lat} = -\hbar \sum_{\mathbf{G}} \mathbf{G}[(\nabla_{\mathbf{k}}|C(\mathbf{k}+\mathbf{G})|^2 \cdot \Delta\mathbf{k}] \ , \tag{7f}$$

as the portion

$$\nabla_{\mathbf{k}}|C(\mathbf{k}+\mathbf{G})|^2 \cdot \Delta\mathbf{k} \tag{7g}$$

of each individual component of the initial state is reflected during the state change $\Delta\mathbf{k}$.

The total momentum change is therefore

$$\Delta \mathbf{p}_{el} + \Delta \mathbf{p}_{lat} = \mathbf{J} = \hbar \Delta\mathbf{k} \ , \tag{7h}$$

exactly as for free electrons, Eq. (7c). Thus from the definition of $\mathbf{J}$, we have

$$\hbar d\mathbf{k}/dt = \mathbf{F} \ , \tag{7i}$$

derived in (5) by a different method.

Holes

The properties of vacant orbitals in an otherwise filled band are important in semiconductor physics and in solid state electronics. Vacant orbitals in a band are commonly called holes. A hole acts in applied electric and magnetic fields as if it has a positive charge $+e$. The reason is given in five steps in the boxes that follow. (One test of a solid state course is that it should treat holes correctly.)

1. $$\mathbf{k}_h = -\mathbf{k}_e \ . \tag{8}$$

The total wavevector of the electrons in a filled band is zero: $\Sigma\mathbf{k} = 0$. This result follows from the geometrical symmetry of the Brillouin zone: every fundamental lattice type has symmetry under the inversion operation $\mathbf{r} \rightarrow -\mathbf{r}$ about any lattice point; it follows that the Brillouin zone of the lattice also has inversion symmetry. If the band is filled all pairs of orbitals $\mathbf{k}$ and $-\mathbf{k}$ are filled, and the total wavevector is zero. If an electron is missing from an orbital of wavevector $\mathbf{k}_e$, the total wavevector of the system is $-\mathbf{k}_e$ and is attributed to the hole. This result is surprising: the electron is missing from $\mathbf{k}_e$ and the position of the hole is usually indicated graphically as situated at $\mathbf{k}_e$, as in Fig. 7. But the true wavevector $\mathbf{k}_h$ of the hole is $-\mathbf{k}_e$, which is the wavevector of the point G if the hole is at E. The wavevector $-\mathbf{k}_e$ enters into selection rules for photon absorption. The hole is an alternate description of a band with one missing electron, and we either say that the hole has wavevector $-\mathbf{k}_e$ or that the band with one missing electron has total wavevector $-\mathbf{k}_e$.

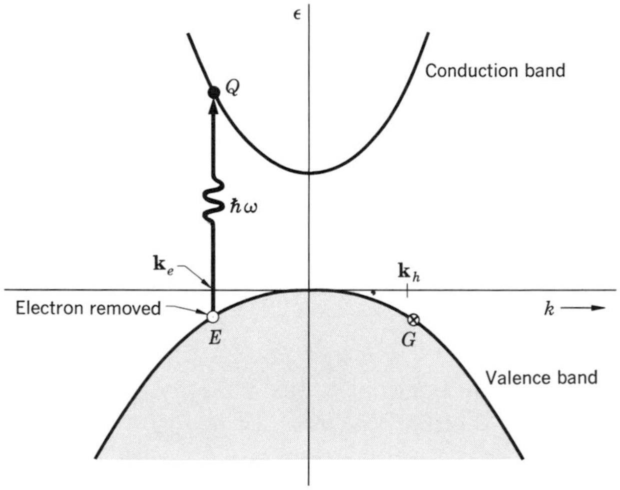

Figure 7 Absorption of a photon of energy $\hbar\omega$ and negligible wavevector takes an electron from E in the filled valence band to Q in the conduction band. If $\mathbf{k}_e$ was the wavevector of the electron at E, it becomes the wavevector of the electron at Q. The total wavevector of the valence band after the absorption is $-\mathbf{k}_e$, and this is the wavevector we must ascribe to the hole if we describe the valence band as occupied by one hole. Thus $\mathbf{k}_h = -\mathbf{k}_e$; the wavevector of the hole is the same as the wavevector of the electron which remains at G. For the entire system the total wavevector after the absorption of the photon is $\mathbf{k}_e + \mathbf{k}_h = 0$, so that the total wavevector is unchanged by the absorption of the photon and the creation of a free electron and free hole.

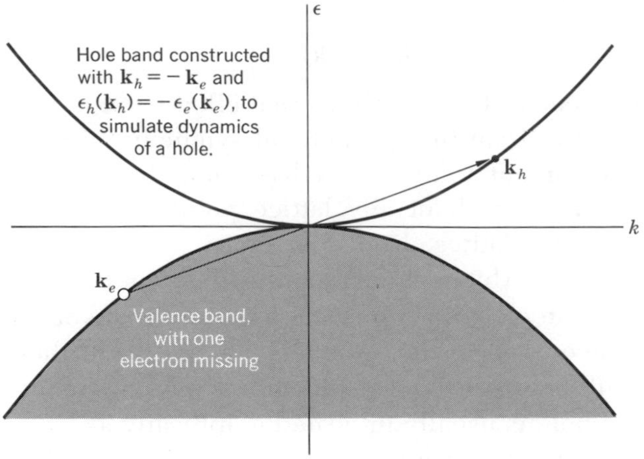

Figure 8 The upper half of the figure shows the hole band that simulates the dynamics of a hole, constructed by inversion of the valence band in the origin. The wavevector and energy of the hole are equal, but opposite in sign, to the wavevector and energy of the empty electron orbital in the valence band. We do not show the disposition of the electron removed from the valence band at $\mathbf{k}_e$.

2.
$$\epsilon_h(\mathbf{k}_h) = -\epsilon_e(\mathbf{k}_e) \ . \tag{9}$$

Let the zero of energy of the valence band be at the top of the band. The lower in the band the missing electron lies, the higher the energy of the system. The energy of the hole is opposite in sign to the energy of the missing electron, because it takes more work to remove an electron from a low orbital than from a high orbital. Thus if the band is symmetric,[2] $\epsilon_e(\mathbf{k}_e) = \epsilon_e(-\mathbf{k}_e) = -\epsilon_h(-\mathbf{k}_e) = -\epsilon_h(\mathbf{k}_h)$. We construct in Fig. 8 a band scheme to represent the properties of a hole. This hole band is a helpful representation because it appears right side up.

3.
$$\mathbf{v}_h = \mathbf{v}_e \ . \tag{10}$$

The velocity of the hole is equal to the velocity of the missing electron. From Fig. 8 we see that $\nabla\epsilon_h(\boldsymbol{k}_h) = \nabla\epsilon_e(\boldsymbol{k}_e)$, so that $\boldsymbol{v}_h(\boldsymbol{k}_h) = \boldsymbol{v}_e(\boldsymbol{k}_e)$.

4.
$$m_h = -m_e \ . \tag{11}$$

We show in (19) below that the effective mass is inversely proportional to the curvature $d^2\epsilon/dk^2$, and for the hole band this has the opposite sign to that for an electron in the valence band. Near the top of the valence band m_e is negative, so that m_h is positive.

[2] Bands are always symmetric under the inversion $\mathbf{k} \rightarrow -\mathbf{k}$ if the spin-orbit interaction is neglected. Even with spin-orbit interaction, bands are always symmetric if the crystal structure permits the inversion operation. Without a center of symmetry, but with spin-orbit interaction, the bands are symmetric if we compare subbands for which the spin direction is reversed: $\epsilon(\boldsymbol{k}, \uparrow) = \epsilon(-\boldsymbol{k}, \downarrow)$. See *QTS*, Chap. 9.

5.

$$\hbar \frac{d\mathbf{k}_h}{dt} = e(\mathbf{E} + \frac{1}{c}\mathbf{v}_h \times \mathbf{B}) \ . \qquad (12)$$

This comes from the equation of motion

(CGS)
$$\hbar \frac{d\mathbf{k}_e}{dt} = -e(\mathbf{E} + \frac{1}{c}\mathbf{v}_e \times \mathbf{B}) \qquad (13)$$

that applies to the missing electron when we substitute $-\mathbf{k}_h$ for $\mathbf{k}_e$ and $\mathbf{v}_h$ for $\mathbf{v}_e$. **The equation of motion for a hole is that of a particle of positive charge *e*.** The positive charge is consistent with the electric current carried by the valence band of Fig. 9: the current is carried by the unpaired electron in the orbital G:

$$\mathbf{j} = (-e)\mathbf{v}(G) = (-e)[-\mathbf{v}(E)] = e\mathbf{v}(E) \ , \qquad (14)$$

which is just the current of a positive charge moving with the velocity ascribed to the missing electron at E. The current is shown in Fig. 10.

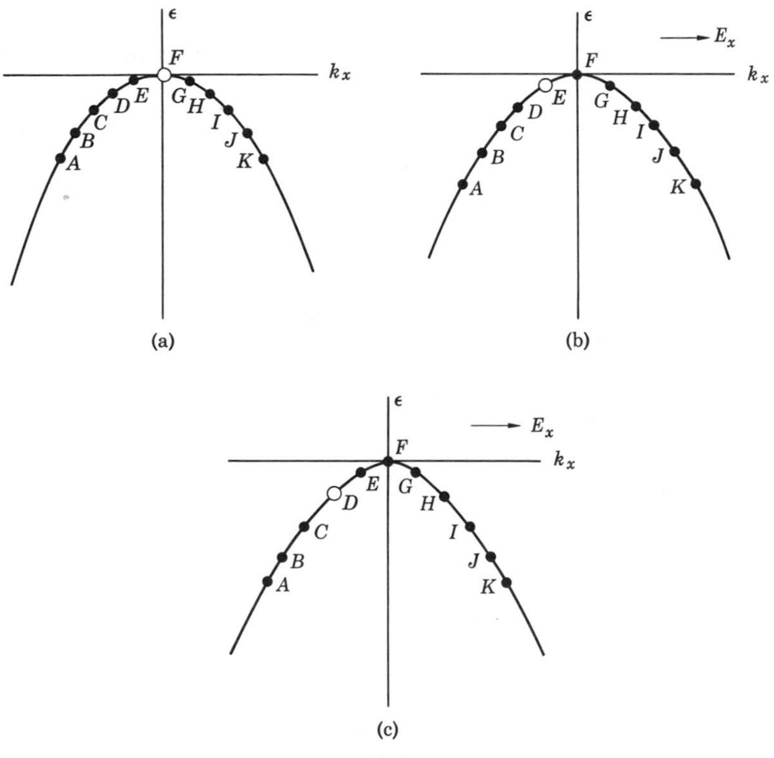

Figure 9 (a) At $t = 0$ all states are filled except F at the top of the band; the velocity v_x is zero at F because $d\epsilon/dk_x = 0$. (b) An electric field E_x is applied in the $+x$ direction. The force on the electrons is in the $-k_x$ direction and all electrons make transitions together in the $-k_x$ direction, moving the hole to the state E. (c) After a further interval the electrons move farther along in k space and the hole is now at D.

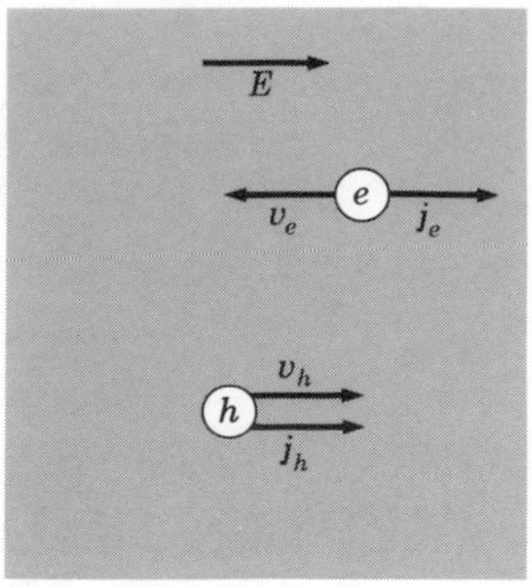

Figure 10 Motion of electrons in the conduction band and holes in the valence band in the electric field E. The hole and electron drift velocities are in opposite directions, but their electric currents are in the same direction, the direction of the electric field.

Effective Mass

When we look at the energy-wavevector relation $\epsilon = (\hbar^2/2m)k^2$ for free electrons, we see that the coefficient of k^2 determines the curvature of ϵ versus k. Turned about, we can say that $1/m$, the reciprocal mass, determines the curvature. For electrons in a band there can be regions of unusually high curvature near the band gap at the zone boundary, as we see from the solutions of the wave equation near the zone boundary. If the energy gap is small in comparison with the free electron energy λ at the boundary, the curvature is enhanced by the factor λ/E_g and the reciprocal mass is enhanced by the same factor. In semiconductors the band width, which is like the free electron energy, is of the order of 20 eV, while the band gap is of the order of 0.2 to 2 eV. Thus the reciprocal mass is enhanced by a factor 10 to 100, and the effective mass is reduced to 0.1–0.01 of the free electron mass. These values apply near the band gap; as we go away from the gap the curvatures are likely to approach those of free electrons. To summarize the earlier solutions, an electron near the lower edge of the second band has an energy that may be written as

$$\epsilon(K) = \epsilon(-) + (\hbar^2/2m_e)K^2 \ ; \qquad m_e/m = 1/[1 - (2\lambda/U)] \ . \tag{15}$$

Here K is the wavevector measured from the zone boundary, and m_e denotes the effective mass of the electron near the edge of the second band. An electron near the top of the first band has the energy

$$\epsilon(K) = \epsilon(+) - (\hbar^2/2m_h)K^2 \ ; \qquad m_h/m = 1/[1 + (2\lambda/U)] \ . \tag{16}$$

The curvature and hence the mass will be negative near the top of the first band, but we have introduced a minus sign into (16) in order that the symbol m_h for the hole mass will have a positive value—see (11) above.

The crystal does not weigh any less if the effective mass of a carrier is less than the free electron mass, nor is Newton's second law violated for the crystal *taken as a whole*. The important point is that an electron in a periodic potential is accelerated relative to the lattice in an applied electric or magnetic field as if the mass of the electron were equal to an effective mass which we now define. We differentiate the result (1) for the group velocity to obtain

$$\frac{dv_g}{dt} = \hbar^{-1}\frac{d^2\epsilon}{dk\,dt} = \hbar^{-1}\left(\frac{d^2\epsilon}{dk^2}\frac{dk}{dt}\right) . \tag{17}$$

We know from (5) that $dk/dt = F/\hbar$, whence

$$\frac{dv_g}{dt} = \left(\frac{1}{\hbar^2}\frac{d^2\epsilon}{dk^2}\right)F \; ; \qquad \text{or} \qquad F = \frac{\hbar^2}{d^2\epsilon/dk^2}\frac{dv_g}{dt} . \tag{18}$$

If we identify $\hbar^2/(d^2\epsilon/dk^2)$ as a mass, then (18) assumes the form of Newton's second law. We define the **effective mass** m^* by

$$\boxed{\frac{1}{m^*} = \frac{1}{\hbar^2}\frac{d^2\epsilon}{dk^2} .} \tag{19}$$

It is easy to generalize this to take account of an anisotropic energy surface. We introduce the components of the reciprocal effective mass tensor

$$\left(\frac{1}{m^*}\right)_{\mu\nu} = \frac{1}{\hbar^2}\frac{d^2\,\epsilon_k}{dk_\mu\,dk_\nu} \; ; \qquad \frac{dv_\mu}{dt} = \left(\frac{1}{m^*}\right)_{\mu\nu} F_\nu \; , \tag{20}$$

where μ, ν are Cartesian coordinates.

Physical Basis of the Effective Mass

How can an electron of mass m when put into a crystal respond to applied fields as if the mass were m^*? It is helpful to think of the process of Bragg reflection of electron waves in a lattice. Consider the weak interaction approximation. Near the bottom of the lower band the orbital is represented quite adequately by a plane wave $\exp(ikx)$ with momentum $\hbar k$; the wave component $\exp[i(k-G)x]$ with momentum $\hbar(k-G)$ is small and increases only slowly as k is increased, and in this region $m^* \simeq m$. An increase in the reflected component $\exp[i(k-G)x]$ as k is increased represents momentum transfer to the electron from the lattice. Near the boundary the reflected component is quite large; at the boundary it becomes equal in amplitude to the forward component, at which point the eigenfunctions are standing waves, rather than running waves. Here the momentum component $\hbar(-\frac{1}{2}G)$ cancels the momentum component $\hbar(\frac{1}{2}G)$.

It is not surprising to find negative values for m^* just below a zone boundary. A single electron in an energy band may have positive or negative effective mass: the states of positive effective mass occur near the bottom of a band because positive effective mass means that the band has upward curvature ($d^2\epsilon/dk^2$ is positive). States of negative effective mass occur near the top of the band. A negative effective mass means that on going from state k to state $k + \Delta k$, the momentum transfer to the lattice from the electron is larger than the momentum transfer from the applied force to the electron. Although k is increased by Δk by the applied electric field, the approach to Bragg reflection can give an overall decrease in the forward momentum of the electron; if this happens the effective mass is negative (Fig. 11).

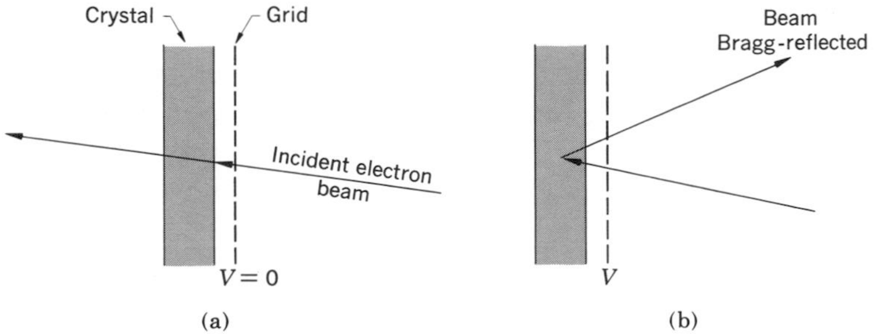

Figure 11 Explanation of negative effective masses which occur near a Brillouin zone boundary. In (a) the energy of the electron beam incident on a thin crystal is slightly too low to satisfy the condition for Bragg reflection and the beam is transmitted through the crystal. The application of a small voltage across the grid may, as in (b), cause the Bragg condition to be satisfied, and the electron beam will then be reflected from the appropriate set of crystal planes.

As we proceed in the second band away from the boundary, the amplitude of $\exp[i(k-G)x]$ decreases rapidly and m^* assumes a small positive value. Here the increase in electron velocity resulting from a given external impulse is larger than that which a free electron would experience. The lattice makes up the difference through the recoil it experiences when the amplitude of $\exp[i(k-G)x]$ is diminished.

If the energy in a band depends only slightly on k, then the effective mass will be very large. That is, $m^*/m \gg 1$ because $d^2\epsilon/dk^2$ is very small. The tight-binding approximation discussed in Chapter 9 gives quick insight into the formation of narrow bands. If the wavefunctions centered on neighboring atoms overlap very little, then the overlap integral will be small; the width of the band narrow; and the effective mass large. The overlap of wavefunctions centered on neighboring atoms is small for the inner or core electrons. The $4f$ electrons of the rare earth metals, for example, overlap very little. The overlap integral determines the rate of quantum tunneling of an electron from one ion to another. When the effective mass is heavy, the electron tunnels slowly from one ion to an adjacent ion in the lattice.

Effective Masses in Semiconductors

In many semiconductors it has been possible to determine by cyclotron resonance the form of the energy surfaces of the conduction and valence bands near the band edges. The determination of the energy surface is equivalent to a determination of the effective mass tensor (20). Cyclotron resonance in a semiconductor is carried out with centimeter wave or millimeter wave radiation at low carrier concentration. The current carriers are accelerated in helical orbits about the axis of a static magnetic field. The angular rotation frequency ω_c is

$$\text{(CGS)} \quad \omega_c = \frac{eB}{m^*c} , \qquad \text{(SI)} \quad \omega_c = \frac{eB}{m^*} ,$$

where m^* is the appropriate cyclotron effective mass. Resonant absorption of energy from an rf electric field perpendicular to the static magnetic field (Fig. 12) occurs when the rf frequency is equal to the cyclotron frequency. Holes and electrons rotate in opposite senses in a magnetic field.

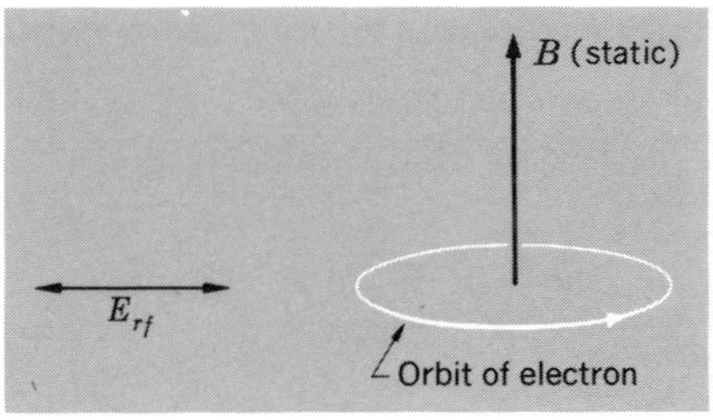

Figure 12 Arrangement of fields in a cyclotron resonance experiment in a semiconductor. The sense of the circulation is opposite for electrons and holes.

We consider the experiment for $m^*/m \simeq 0.1$. At $f_c = 24$ GHz, or $\omega_c = 1.5 \times 10^{11}$ s^{-1}, we have $B \simeq 860$ G at resonance. The line width is determined by the collision relaxation time τ, and to obtain a distinctive resonance it is necessary that $\omega_c \tau \geq 1$. The mean free path must be long enough to permit the average carrier to get one radian around a circle between collisions. At room temperature the relaxation times of carriers in crystals are too short, commonly in the range 10^{-13} to 10^{-15} s. It is usually necessary to work with high-purity crystals in liquid helium to obtain long enough relaxation times. The requirements are relaxed with the use of higher frequency radiation and higher magnetic fields.

In direct-gap semiconductors with band edges at the center of the Brillouin zone, the bands have the structure shown in Fig. 13. The conduction band edge is spherical with the effective mass m_e:

$$\epsilon_c = E_g + \hbar^2 k^2 / 2m_e , \tag{21}$$

referred to the valence band edge. The valence bands are characteristically threefold near the edge, with the heavy hole *hh* and light hole *lh*

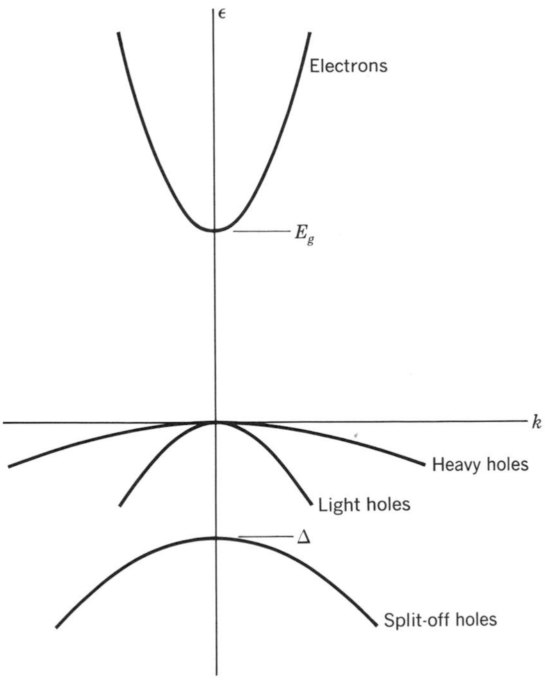

Figure 13 Simplified view of the band edge structure of a direct gap semiconductor.

bands degenerate at the center, and a band *soh* split off by the spin-orbit splitting Δ:

$$\epsilon_v(hh) \cong -\hbar^2k^2/2m_{hh} \ ; \qquad \epsilon_v(lh) \cong -\hbar^2k^2/2m_{lh} \ ; \tag{22}$$
$$\epsilon_v(soh) \cong -\Delta - \hbar^2k^2/2m_{soh} \ .$$

Values of the mass parameters are given in Table 2. The forms (22) are only approximate, because even close to $k = 0$ the heavy and light hole bands are not spherical—see the discussion below for Ge and Si.

TABLE 2 Effective Masses of Electrons and Holes in Direct Gap Semiconductors

Crystal	Electron m_e/m	Heavy Hole m_{hh}/m	Light Hole m_{lh}/m	Split-Off Hole m_{soh}/m	Spin-Orbit Δ, eV
InSb	0.015	0.39	0.021	(0.11)	0.82
InAs	0.026	0.41	0.025	0.08	0.43
InP	0.073	0.4	(0.078)	(0.15)	0.11
GaSb	0.047	0.3	0.06	(0.14)	0.80
GaAs	0.07	0.68	0.12	0.20	0.34

For spin-orbit splittings, see the table on p. 70 of M. Cardona, *Modulation spectroscopy*, Academic Press, 1969.

The perturbation theory of band edges (Problem 9.8) suggests that the electron effective mass should be inversely proportional to the band gap, approximately, for a direct gap crystal. We use Tables 1 and 2 to find $m_e/mE_g = 0.063$, 0.060, and 0.051 in $(\mathrm{eV})^{-1}$ for the series InSb, InAs, and InP, in agreement with this suggestion.

Special Crystals. In gray tin (αSn, cubic) the mass of what we might call the light hole changes sign and the energy surface becomes concave upward at the zone center. There it touches the heavy hole band which is concave downward. The valence band is filled just up to the contact point, so that the energy gap $E_g = 0$. The "light hole" band has become the conduction band, with an effective mass 0.02 m.

The lead chalcogenides PbS, PbSe, PbTe are direct gap semiconductors and have the NaCl crystal structure. The band edges are all at the zone boundary in the [111] direction, where the bands are ellipsoidal, as discussed below for Ge and Si. All gaps are less than 0.30 eV.

The alloy $Hg_{1-x}Cd_xTe$ is an insulator when $x = 1$, for CdTe has $E_g = 1.607$ eV, and a semimetal when $x = 0$, for the conduction and valence bands of HgTe overlap by 0.30 eV, or $E_g = -0.30$ V. Near $x = 0.2$ the gap goes through zero. At $x = 0.212$ the gap is 0.093 eV and $m_e = 0.005\ m$.

Silicon and Germanium

The conduction and valence bands of germanium below in Fig. 15, based on a combination of theoretical and experimental results. The valence band edge in both crystals is at $\mathbf{k} = 0$ and is derived from $p_{3/2}$ and $p_{1/2}$ states of the free atoms. This is clear from the tight-binding approximation (Chap. 9) to the wavefunctions. The $p_{3/2}$ level is fourfold degenerate as in the atom; the four states correspond to m_J values $\pm\frac{3}{2}$ and $\pm\frac{1}{2}$. The $p_{1/2}$ level is doubly degenerate, with $m_J = \pm\frac{1}{2}$. The $p_{3/2}$ states are higher in energy than the $p_{1/2}$ states; the energy difference Δ is a measure of the spin-orbit interaction.

The valence band edges in germanium and silicon are complicated. The holes near the band edge are characterized by two effective masses, light and heavy. These arise from the two bands formed from the $p_{3/2}$ level of the atom. There is also a spin-orbit split-off band formed from the $p_{1/2}$ level. The energy surfaces are warped (*QTS*, p. 271):

$$\epsilon(\mathbf{k}) = Ak^2 \pm [B^2k^4 + C^2(k_x^2k_y^2 + k_y^2k_z^2 + k_z^2k_x^2)]^{\frac{1}{2}} \ . \tag{23}$$

The choice of sign distinguishes the two masses. The split-off band has $\epsilon(k) = -\Delta + Ak^2$. The experiments give, in units $\hbar^2/2m$,

Si: $A = -4.29$; $|B| = 0.68$; $|C| = 4.87$; $\Delta = 0.044$ eV

Ge: $A = -13.38$; $|B| = 8.48$; $|C| = 13.15$; $\Delta = 0.29$ eV

Roughly, the light and heavy holes in germanium have masses 0.043 m and 0.34 m; in silicon 0.16 m and 0.52 m; in diamond 0.7 m and 2.12 m.

The conduction band edges in Ge are at the equivalent points L of the Brillouin zone, Fig. 14a. Each band edge has a spheroidal energy surface oriented along a $\langle 111 \rangle$ crystal axis, with a longitudinal mass $m_l = 1.59\ m$ and a transverse mass $m_t = 0.082\ m$. For a static magnetic field at an angle θ with the longitudinal axis of a spheroid, the effective cyclotron mass m_c is

$$\frac{1}{m_c{}^2} = \frac{\cos^2\theta}{m_t{}^2} + \frac{\sin^2\theta}{m_t m_l} \ . \tag{24}$$

Results for Ge are shown in Fig. 16.

In silicon the conduction band edges are spheroids oriented along the equivalent $\langle 100 \rangle$ directions in the Brillouin zone, with mass parameters $m_l = 0.92\ m$ and $m_t = 0.19\ m$, as in Fig. 17. The band edges lie along the lines Δ in the zone of Fig. 14, a little way in from the points X. All mass parameters quoted in this section are due to J. C. Hensel and coworkers.

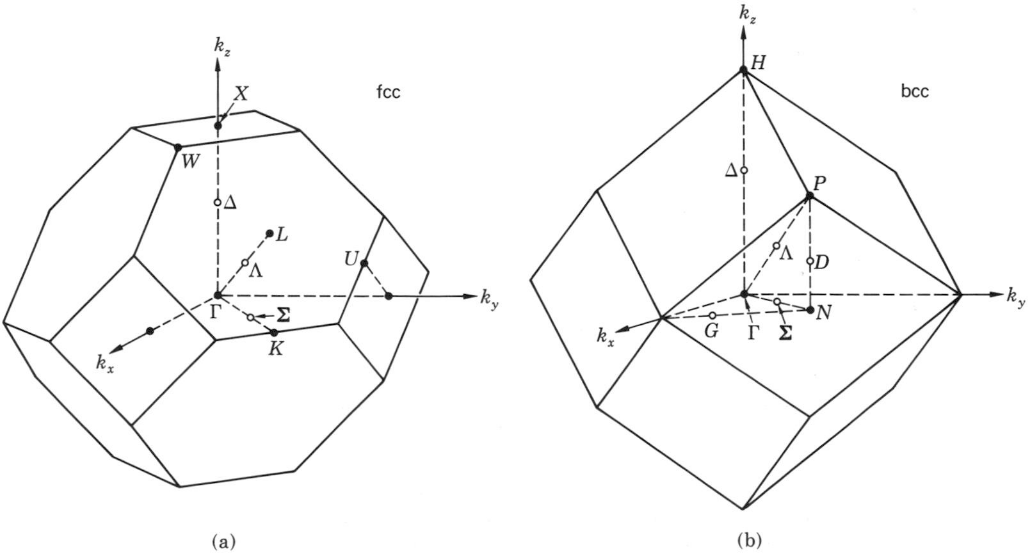

Figure 14 Standard labels of the symmetry points and axes of the Brillouin zones of the fcc and bcc lattices. The zone centers are Γ. In (a) the boundary point at $(2\pi/a)(100)$ is X; the boundary point at $(2\pi/a)(\frac{1}{2}\frac{1}{2}\frac{1}{2})$ is L; the line Δ runs between Γ and X. In (b) the corresponding symbols are H, P and Δ.

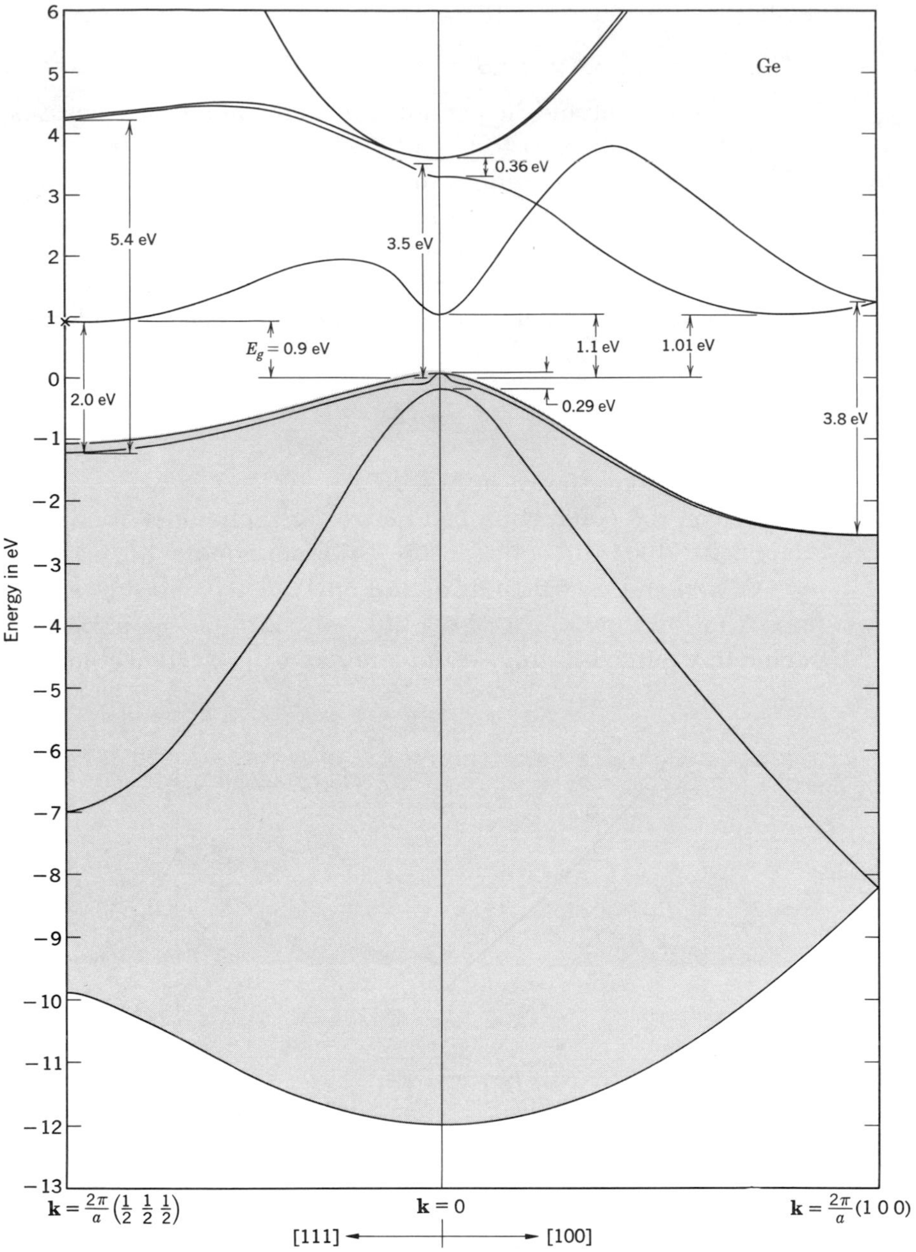

Figure 15 Calculated band structure of germanium, after C. Y. Fong. The general features are in good agreement with experiment. The four valence bands are shown in gray. The fine structure of the valence band edge is caused by spin-orbit splitting. The energy gap is indirect; the conduction band edge is at the point $(2\pi/a)(\frac{1}{2}\frac{1}{2}\frac{1}{2})$. The constant energy surfaces around this point are ellipsoidal.

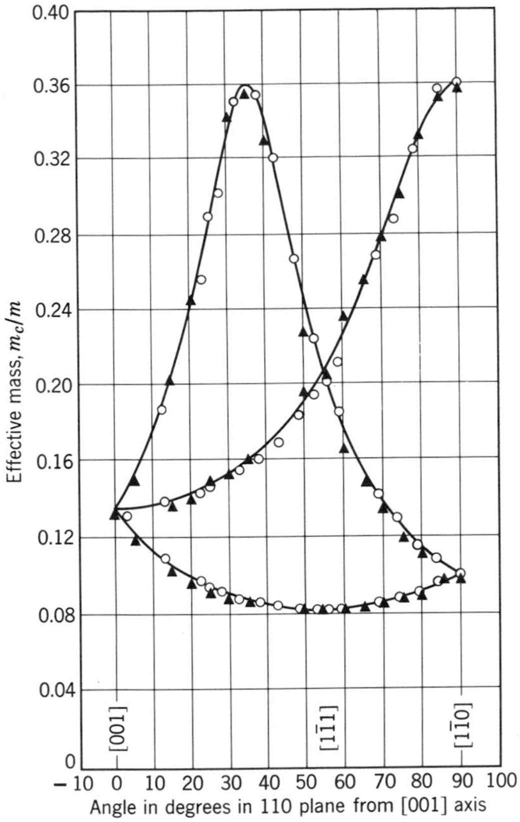

Figure 16 Effective cyclotron mass of electrons in germanium at 4 K for magnetic field directions in a 110 plane. There are four independent mass spheroids in Ge, one along each 111 axis, but viewed in the 110 plane two spheroids always appear equivalent. (After Dresselhaus, Kip, and Kittel.)

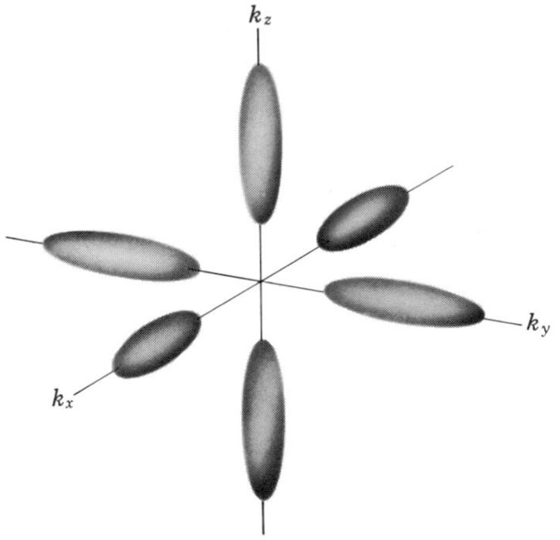

Figure 17 Constant energy ellipsoids for electrons in silicon, drawn for $m_l/m_t = 5$.

INTRINSIC CARRIER CONCENTRATION

We want the concentration of intrinsic carriers in terms of the band gap. We first calculate in terms of the chemical potential μ the number of electrons excited to the conduction band at temperature T. In semiconductor physics μ is called the Fermi level. We measure the energy ϵ from the top of the valence band, as in Fig. 18. At the temperatures of interest we may suppose for the conduction band of a semiconductor that $\epsilon - \mu \gg k_B T$, and the Fermi-Dirac distribution function reduces to

$$f \simeq e^{(\mu-\epsilon)/k_B T} . \tag{25}$$

This is the probability that a conduction electron orbital is occupied, in an approximation valid when $f \ll 1$. The energy of an electron in the conduction band is

$$\epsilon_k = E_g + \hbar^2 k^2/2m_e , \tag{26}$$

where m_e is the effective mass of an electron. Thus from (6.20) the density of orbitals at ϵ is

$$\mathfrak{D}_e(\epsilon) = \frac{1}{2\pi^2}\left(\frac{2m_e}{\hbar^2}\right)^{3/2}(\epsilon - E_g)^{\frac{1}{2}} . \tag{27}$$

The concentration of electrons in the conduction band is

$$n = \int_{E_g}^{\infty} \mathfrak{D}_e(\epsilon) f_e(\epsilon) d\epsilon = \frac{1}{2\pi^2}\left(\frac{2m_e}{\hbar^2}\right)^{3/2} e^{\mu/k_B T}\int_{E_g}^{\infty}(\epsilon - E_g)^{\frac{1}{2}}\, e^{-\epsilon/k_B T} d\epsilon , \tag{28}$$

which integrates to give[3]

$$n = 2\left(\frac{m_e k_B T}{2\pi\hbar^2}\right)^{3/2} e^{(\mu - E_g)/k_B T} . \tag{29}$$

The problem is solved when μ is known. It is useful to calculate the equilibrium concentration of holes p. The distribution function f_h for holes is related to the electron distribution function f_e by $f_h = 1 - f_e$, because a hole is the absence of an electron. We have

$$f_h = 1 - \frac{1}{e^{(\epsilon-\mu)/k_B T} + 1} = \frac{1}{e^{(\mu-\epsilon)/k_B T} + 1} \cong e^{(\epsilon-\mu)/k_B T} , \tag{30}$$

provided $(\mu - \epsilon) \gg k_B T$. If the holes near the top of the valence band behave as particles with effective mass m_h, the density of hole orbitals is given by

$$\mathfrak{D}_h(\epsilon) = \frac{1}{2\pi^2}\left(\frac{2m_h}{\hbar^2}\right)^{3/2}(-\epsilon)^{\frac{1}{2}} , \tag{31}$$

with the energy measured positive upwards from the top of the valence band. Proceeding as in (28) we obtain

[3]In the notation of *TP*, Chap. 11, the result may be written as $n = 2\lambda e^{-E_g/k_B T}/V_Q$, where $V_Q \equiv (2\pi\hbar^2/m_e k_B T)^{3/2}$ is the quantum volume of a conduction electron and λ is the absolute activity $e^{\mu/k_B T}$.

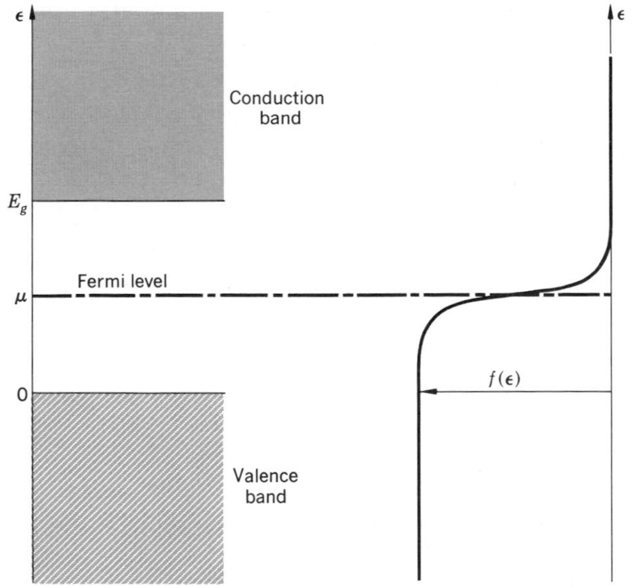

Figure 18 Energy scale for statistical calculations. The Fermi distribution function is shown on the same scale, for a temperature $k_BT \ll E_g$. The Fermi level μ is taken to lie well within the band gap, as for an intrinsic semiconductor. If $\epsilon = \mu$, then $f = \frac{1}{2}$.

$$p = \int_{-\infty}^{0} \mathfrak{D}_h(\epsilon) f_h(\epsilon) d\epsilon = 2 \left(\frac{m_h k_B T}{2\pi\hbar^2} \right)^{3/2} e^{-\mu/k_BT} \tag{32}$$

for the concentration p of holes in the valence band.

We multiply together the expressions for n and p to obtain the equilibrium relation

$$np = 4 \left(\frac{k_B T}{2\pi\hbar^2} \right)^3 (m_e m_h)^{3/2} e^{-E_g/k_BT} , \tag{33}$$

of the form of the **law of mass action.** This useful result does not involve the Fermi level μ. It is an expression of the law of mass action. We have nowhere assumed in the derivation that the material is intrinsic: the result holds in the presence of impurities as well. The only assumption made is that the distance of the Fermi level from the edge of both bands should be large in comparison with k_BT. At 300 K the value of np is 3.6×10^{27} cm^{-6} for Ge and 4.6×10^{19} cm^{-6} for Si, all calculated with $m_e = m_h = m$.

A simple kinetic argument shows why the product np is constant at a given temperature. Suppose that the equilibrium population of electrons and holes is maintained by black-body photon radiation. The photons generate electron-hole pairs at a rate $A(T)$, while $B(T)np$ is the rate of the recombination reaction $e + h =$ photon. Then

$$dn/dt = A(T) - B(T)np = dp/dt \ . \tag{34}$$

In equilibrium $dn/dt = 0$; $dp/dt = 0$, whence $np = A(T)/B(T)$.

Because the product of the electron and hole concentrations is a constant independent of impurity concentration at a given temperature, the introduction of a small proportion of a suitable impurity to increase n, say, must decrease p. This result is important in practice—we can reduce the total carrier concentration $n + p$ in an impure crystal, sometimes enormously, by the controlled introduction of suitable impurities. Such a reduction is called **compensation** of one impurity type by adding another.

In an intrinsic semiconductor the number of electrons is equal to the number of holes, because the thermal excitation of an electron leaves behind a hole in the valence band. Thus from (33) we have, letting the subscript i denote intrinsic,

$$n_i = p_i = 2\left(\frac{k_B T}{2\pi\hbar^2}\right)^{3/2} (m_e m_h)^{3/4} e^{-E_g/2k_B T} \ . \tag{35}$$

The intrinsic carrier excitation depends exponentially on $E_g/2k_BT$, where E_g is the energy gap. We set (29) equal to (32) to obtain

$$e^{2\mu/k_BT} = (m_h/m_e)^{3/2} e^{E_g/k_BT} \tag{36}$$

or, for the Fermi level,

$$\mu = \tfrac{1}{2}E_g + \tfrac{3}{4}k_BT \ln (m_h/m_e) \ . \tag{37}$$

If $m_h = m_e$, then $\mu = \frac{1}{2}E_g$ and the Fermi level is in the middle of the forbidden gap.

Mobility in the Intrinsic Region

The mobility is the magnitude of the drift velocity per unit electric field:

$$\mu = |v|/E \ . \tag{38}$$

The mobility is defined to be positive for both electrons and holes, although their drift velocities are opposite. By writing μ_e or μ_h for the electron or hole mobility we can avoid any confusion between μ as the chemical potential and as the mobility. In an ideal intrinsic semiconductor the mobility is determined by collisions between electrons and phonons.

The electrical conductivity is the sum of the electron and hole contributions:

$$\sigma = (ne\mu_e + pe\mu_h) \ , \tag{39}$$

where n and p are the concentrations of electrons and holes. In Chapter 6 the drift velocity of a charge q was found to be $v = q\tau E/m$, whence

$$\mu_e = e\tau_e/m_e \ ; \qquad \mu_h = e\tau_h/m_h \ . \tag{40}$$

The mobilities depend on temperature as a modest power law. The tem-

TABLE 3 Carrier Mobilities at Room Temperature, in cm^2/V-s

Crystal	Electrons	Holes	Crystal	Electrons	Holes
Diamond	1800	1200	GaSb	4000	1400
Si	1300	500	PbS	550	600
Ge	4500	3500	PbSe	1020	930
InSb	77,000	750	PbTe	1620	750
InAs	33,000	460	AgCl	50	—
InP	4600	150	KBr(100 K)	100	—
GaAs	8800	400	SiC	100	50

perature dependence of the conductivity in the intrinsic region will be dominated by the exponential dependence $\exp(-E_g/2k_BT)$ of the carrier concentration, Eq. (35).

Experimental values of the mobility at room temperature are given in Table 3. By comparison, the mobility in metallic copper is only 35 cm^2/V-s at room temperature. In most substances the values quoted are representative of lattice scattering of carriers by phonons. There is a tendency for crystals with small energy gaps to have high values of the electron mobility. Small gaps lead to small effective masses, which by (40) favor high mobilities. The highest mobility observed in a semiconductor is 5×10^6 cm^2/V-s for electrons in PbTe at 4 K. The mobility in CGS is expressed in cm^2/statV-s and is 300 times higher than the mobility in practical units expressed in cm^2/V-s.

IMPURITY CONDUCTIVITY

Certain impurities and imperfections drastically affect the electrical properties of a semiconductor. The addition of boron to silicon in the proportion of 1 boron atom to 10^5 silicon atoms increases the conductivity of pure silicon by a factor of 10^3 at room temperature. In a compound semiconductor a stoichiometric deficiency of one constituent will act as an impurity; such semiconductors are known as **deficit semiconductors.** The deliberate addition of impurities to a semiconductor is called **doping.**

We consider the effect of impurities in silicon and germanium. These elements crystallize in the diamond structure. Each atom forms four covalent bonds, one with each of its nearest neighbors, corresponding to the chemical valence four. If an impurity atom of valence five, such as phosphorus, arsenic, or antimony, is substituted in the lattice in place of a normal atom, there will be one valence electron from the impurity atom left over after the four covalent bonds are established with the nearest neighbors, that is, after the impurity atom has been accommodated in the structure with as little disturbance as possible.

Donor States. The structure in Fig. 19 has a positive charge on the impurity atom (which has lost one electron). Lattice constant studies have verified that the pentavalent impurities enter the lattice by substitution for normal atoms, and not in interstitial positions. Impurity atoms that can give up an electron are called **donors.** The crystal as a whole remains neutral because the electron remains in the crystal.

The electron moves in the coulomb potential $e/\epsilon r$ of the impurity ion, where ϵ in a covalent crystal is the static dielectric constant of the medium. The factor $1/\epsilon$ takes account of the reduction in the coulomb force between charges caused by the electronic polarization of the medium. This treatment is valid for orbits large in comparison with the distance between atoms, and for slow motions of the electron such that the orbital frequency is low in comparison with the frequency ω_g corresponding to the energy gap. These conditions are satisfied quite well in Ge and Si by the donor electron of P, As, or Sb.

We estimate the binding energy of the donor impurity. The Bohr theory of the hydrogen atom may readily be modified to take into account both the dielectric constant of the medium and the effective mass of an electron in the periodic potential of the crystal. The binding energy of atomic hydrogen is $-e^4m/2\hbar^2$ in CGS and $-e^4m/2(4\pi\epsilon_0\hbar)^2$ in SI. In the semiconductor we replace e^2 by e^2/ϵ and m by the effective mass m_e to obtain

$$\text{(CGS)}\quad E_d = \frac{e^4 m_e}{2\epsilon^2\hbar^2}\ ; \qquad \text{(SI)}\quad E_d = \frac{e^4 m_e}{2(4\pi\epsilon\epsilon_0\hbar)^2}\ , \tag{41}$$

as the donor ionization energy of the semiconductor. The Bohr radius of the ground state of hydrogen is $\hbar^2/me^2$ in CGS or $4\pi\epsilon_0\hbar^2/me^2$ in SI. Thus the Bohr radius of the donor is

$$\text{(CGS)}\quad a_d = \frac{\epsilon\hbar^2}{m_e e^2}\ ; \qquad \text{(SI)}\quad a_d = \frac{4\pi\epsilon\epsilon_0\hbar^2}{m_e e^2}\ . \tag{42}$$

The application to germanium and silicon is complicated by the anisotropic effective mass of the conduction electrons. But the dielectric constant has the most important effect on the donor energy because it enters as the square, whereas the effective mass enters only as the first power. To obtain a general impression of the impurity levels we use $m_e \approx 0.1\ m$ for electrons in germanium and $m_e \approx 0.2\ m$ in silicon. The static dielectric constant is given in Table 4. The ionization energy of the free hydrogen atom is 13.6 eV. For germanium the donor ionization energy E_d on our model is 6 meV, reduced with respect to hydrogen by the factor $m_e/m\epsilon^2 = 4 \times 10^{-4}$. The corresponding result for silicon is 20 meV. Calculations using the correct anisotropic mass tensor predict 9.05 meV for germanium and 29.8 meV for silicon. Observed values of donor ionization energies are given in Table 5. Recall that 1 meV $\equiv 10^{-3}$ eV.

TABLE 4 Static Dielectric Constant of Semiconductors

Crystal	ϵ_0	Crystal	ϵ_0
Diamond	5.5	GaSb	15.69
Si	11.7	GaAs	13.13
Ge	15.8	GaP	10.18
InSb	17.88	PbS	205.
InAs	14.55	PbSe	280.
InP	12.37	PbTe	400–1300.

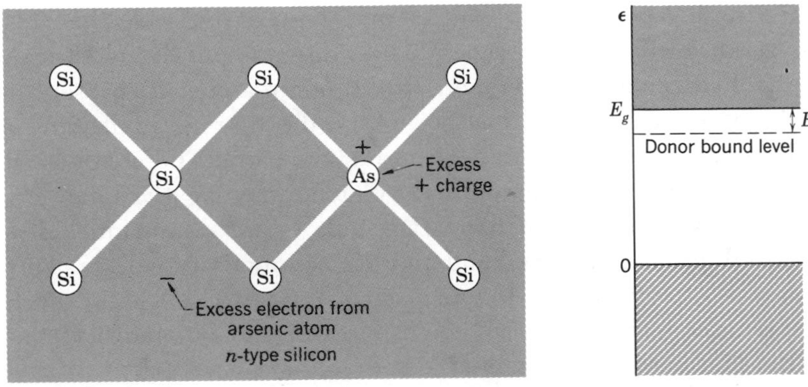

Figure 19 Charges associated with an arsenic impurity atom in silicon. Arsenic has five valence electrons, but silicon has only four valence electrons. Thus four electrons on the arsenic form tetrahedral covalent bonds similar to silicon, and the fifth electron is available for conduction. The arsenic atom is called a **donor** because when ionized it donates an electron to the conduction band.

TABLE 5 Donor Ionization Energies E_d of Pentavalent Impurities in Germanium and Silicon, in meV

	P	As	Sb
Si	45.	49.	39.
Ge	12.0	12.7	9.6

The radius of the first Bohr orbit is increased by $\epsilon m/m_e$ over the value 0.53 Å for the free hydrogen atom. The corresponding radius is (160)(0.53) $\simeq$ 80 Å in germanium and (60)(0.53) $\simeq$ 30 Å in silicon. These are large radii, so that impurity orbits overlap at relatively low impurity concentrations.

Acceptor States. Just as an electron may be bound to a pentavalent impurity, a hole may be bound to a trivalent impurity in germanium or silicon (Fig. 20). Trivalent impurities such as B, Al, Ga, and In are called acceptors because they can accept electrons from the valence band, leaving holes in the band. When we ionize an acceptor, we free a hole, which requires an input of energy. On the usual energy band diagram, an electron rises when it gains energy, whereas a hole sinks in gaining energy. Experimental ionization energies of acceptors in germanium and silicon are given in Table 6. The Bohr model applies qualitatively for holes just as for electrons, but the degeneracy at the top of the valence band complicates the effective mass problem.

The tables show that donor and acceptor ionization energies in Si are comparable with $k_B T$ at room temperature (26 meV), so that the thermal ionization of donors and acceptors is important in the electrical conductivity of silicon at room temperature. If donor atoms are present in considerably greater numbers than acceptors, the thermal ionization of donors will release electrons into the conduction band. The conductivity of the specimen then will be controlled by electrons (negative charges), and the material is said to be n type. If acceptors are dominant, holes will be released into the valence band and the conductivity will be controlled by holes (positive charges): the material is p type. The sign of the Hall voltage is a rough test for n or p type. Another handy laboratory test is the sign of the thermoelectric potential, discussed below.

The numbers of holes and electrons are equal in the intrinsic regime. The intrinsic electron concentration n_i at 300 K is 6×10^{13} cm^{-3} in germanium and 7×10^{9} cm^{-3} in silicon; the electrical resistivity of intrinsic material is 43 ohm-cm for germanium and 2.6×10^{5} ohm-cm for silicon. The lowest impurity concentrations attained at present are of the order of 10^{10} cm^{-3}, as in Fig. 21. Impurities and lattice defects that do not ionize have no effect on the carrier concentration and may be present in higher proportions without being detected in electrical measurements.

TABLE 6 Acceptor Ionization Energies E_a of Trivalent Impurities in Germanium and Silicon, in meV

	B	Al	Ga	In
Si	45.	57.	65.	16.
Ge	10.4	10.2	10.8	11.2

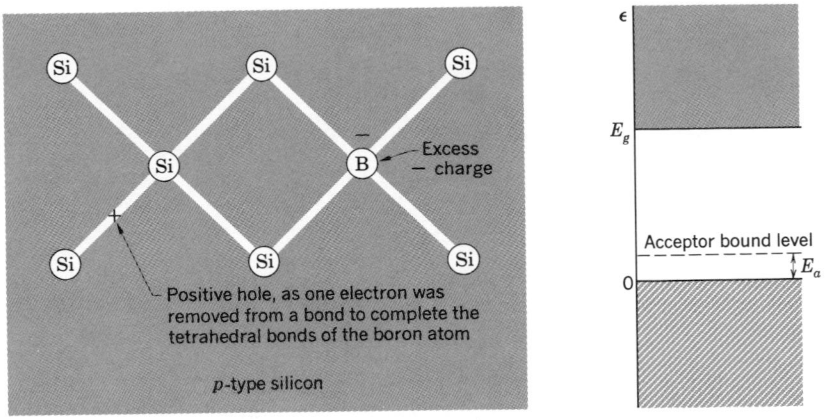

Figure 20 Boron has only three valence electrons; it can complete its tetrahedral bonds only by taking an electron from a Si-Si bond, leaving behind a hole in the silicon valence band. The positive hole is then available for conduction. The boron atom is called an **acceptor** because when ionized it accepts an electron from the valence band. At 0 K the hole is bound: remember that holes float.

Thermal Ionization of Donors and Acceptors

The calculation of the equilibrium concentration of conduction electrons from ionized donors is identical with the standard calculation in statistical mechanics of the thermal ionization of hydrogen atoms. If there are no acceptors present, the result in the low temperature limit $k_BT \ll E_d$ is

$$\boxed{n \cong (n_0 N_d)^{\frac{1}{2}}\, e^{-E_d/2k_BT}\ ,} \tag{43}$$

with $n_0 \equiv 2(m_e k_B T/2\pi\hbar^2)^{3/2}$; here N_d is the concentration of donors. To obtain (43) we apply the laws of chemical equilibria to the concentrations $[e][N_d^+]/[N_d]$ = function of the temperature, and then set $[N_d^+] = [e] = n$. Identical results hold for acceptors, under the assumption of no donor atoms. If the donor and acceptor concentrations are comparable, affairs are quite complicated and the equations are solved by numerical methods. But the law of mass action requires the np product to be constant at a given temperature. An excess of donors will increase the electron concentration and decrease the hole concentration; the sum $n + p$ will increase. The conductivity will increase as $n + p$ if the mobilities are equal, Fig. 22.

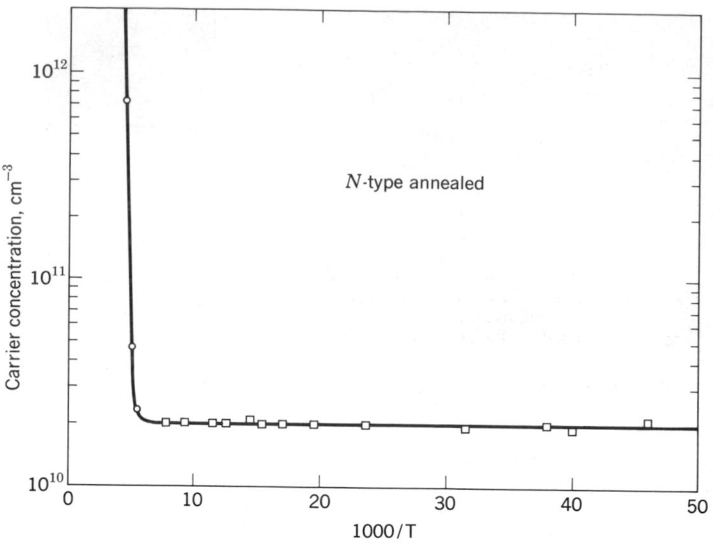

Figure 21 Temperature dependence of the free carrier concentration in ultrapure Ge, after R. N. Hall. The net concentration of electrically active impurities is 2×10^{10} cm^{-3}, as determined by Hall coefficient measurements. The rapid onset of intrinsic excitation is evident at low values of $1/T$. The carrier concentration is closely constant between 20 K and 200 K.

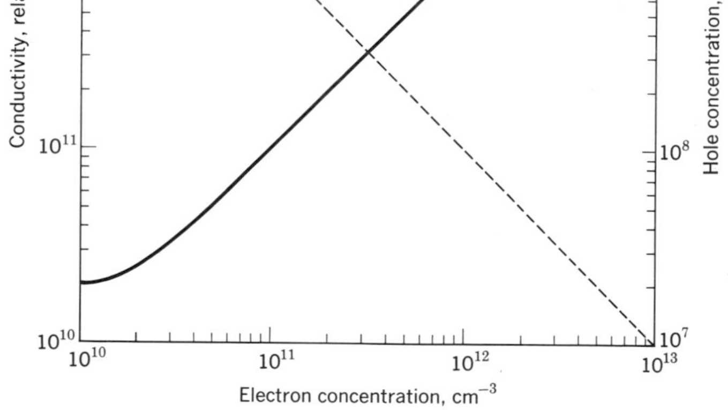

Figure 22 Electrical conductivity and hole concentration p calculated as a function of electron concentration n for a semiconductor at a temperature such that $np = 10^{20}$ cm^{-6}. The conductivity is symmetrical about $n = 10^{10}$ cm^{-3}. For $n > 10^{10}$ the specimen is n type; for $n < 10^{10}$, it is p type. We have taken $\mu_n = \mu_p$.

Mobility in the Presence of Impurities

Scattering by phonons determines the mobility of carriers when relatively few impurity atoms are present, or at high temperatures. Scattering by impurity atoms may be important at higher impurity concentrations. The scattering will depend on whether the impurity is neutral or ionized. The neutral atom problem is equivalent to the scattering of an electron by a hydrogen atom. We note that the area of the first Bohr orbit is increased by $(\epsilon m/m^*)^2$. An exact solution for the neutral scattering cross section is quite difficult in the energy range of interest in semiconductors. The scattering of carriers by ionized donors or acceptors has been treated by Conwell and Weisskopf, who utilized the Rutherford scattering formula.

THERMOELECTRIC EFFECTS IN SEMICONDUCTORS

Consider a semiconductor maintained at a constant temperature while an electric field drives through it an electric current density j_q. If the current is carried only by electrons, the charge flux is

$$j_q = n(-e)(-\mu_n)E = ne\mu_e E \ . \tag{44}$$

The average energy transported by an electron is, referred to the Fermi level μ,

$$(E_c - \mu) + \tfrac{3}{2}k_B T \ ,$$

where E_c is the energy at the conduction band edge. We refer the energy to the Fermi level because different conductors in contact have the same Fermi level; we can use our results to treat contact problems. The energy flux that accompanies the charge flux is

$$j_U = n(E_c - \mu + \tfrac{3}{2}k_B T)(-\mu_e)E \ . \tag{45}$$

The **Peltier coefficient** Π is defined by the relation $j_U = \Pi j_q$; it is the energy carried per unit charge. For electrons,

$$\Pi_e = -(E_c - \mu + \tfrac{3}{2}k_B T)/e \tag{46}$$

and is negative because the energy flux is opposite to the charge flux. For holes

$$j_q = pe\mu_h E \ ; \qquad j_U = p(\mu - E_v + \tfrac{3}{2}k_B T)/e \tag{47}$$

where E_v is the energy at the valence band edge. Thus

$$\Pi_h = (\mu - E_v + \tfrac{3}{2}k_B T)/e \tag{48}$$

and is positive. Equations (46) and (48) are the result of our simple drift velocity theory; a treatment by the Boltzmann transport equation gives minor differences.[4]

[4] R. A. Smith, *Semiconductors*, Cambridge, 1959, pp. 180–184; H. Fritzsche, Solid State Comm. **9**, 1813 (1971). A simple discussion of Boltzmann transport theory is given in Kittel, *TP*.

The **absolute thermoelectric power** Q is defined from the open circuit electric field created by a temperature gradient:

$$E = Q \text{ grad } T \ . \tag{49}$$

The Peltier coefficient is related to the thermoelectric power by

$$\Pi = QT \ . \tag{50}$$

This is the famous Kelvin relation of irreversible thermodynamics.[5] A measurement of the sign of the voltage across a semiconductor specimen, one end of which is heated, is a rough and ready way to tell if the specimen is n type or p type (Fig. 23).

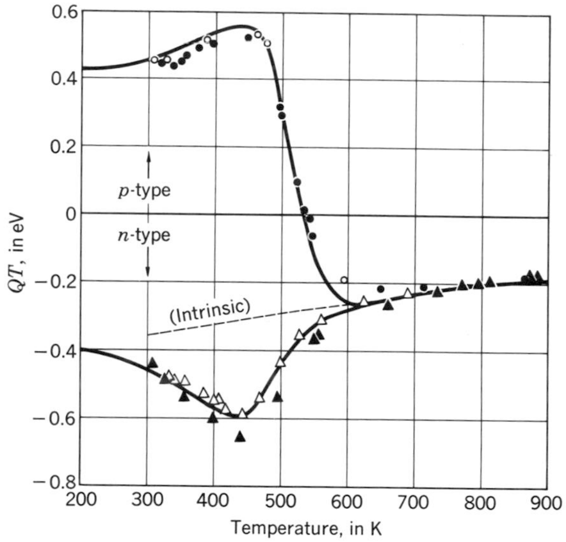

Figure 23 Thermoelectric power of n and p silicon as a function of temperature. Above 600 K the specimens become intrinsic. The curves are calculated and the points are measured, after T. H. Geballe and G. W. Hull, Phys. Rev. **98**, 940 (1955).

TABLE 7 Electron and Hole Concentrations in Semimetals

Semimetal	n_e, in cm^{-3}	n_h, in cm^{-3}
Arsenic	$(2.12 \pm 0.01) \times 10^{20}$	$(2.12 \pm 0.01) \times 10^{20}$
Antimony	$(5.54 \pm 0.05) \times 10^{19}$	$(5.49 \pm 0.03) \times 10^{19}$
Bismuth	2.88×10^{17}	3.00×10^{17}
Graphite	2.72×10^{18}	2.04×10^{18}

Courtesy of L. Falicov.

SEMIMETALS

In semimetals the conduction band edge is very slightly lower in energy than the valence band edge. A small overlap in energy of the conduction and valence bands leads to small concentration of holes in the valence band and of electrons in the conduction band (Table 7). Three of the semimetals, arsenic, antimony, and bismuth, are in group V of the periodic table. Their atoms associate in pairs in the crystal lattice, with two ions and ten valence electrons per primitive cell. Small overlaps of the fifth and sixth energy bands create equal numbers of holes and electrons in small nearly ellipsoidal pockets in the Brillouin zone. Like semiconductors, the semimetal elements may be doped with suitable impurities to vary the relative concentrations of holes and electrons. The absolute concentrations may also be varied by application of pressure, for the overlap of the band edges varies with pressure.

AMORPHOUS SEMICONDUCTORS

Amorphous semiconductors[6] can be prepared in thin film form by evaporation or sputtering, or in some cases as bulk glasses by supercooling the melt. What happens to the band model in a solid without regular crystalline order? Although the electron states cannot be described by well-defined k values, allowed bands and energy gaps still occur, the form of the density of states often being determined almost exclusively by local bonding configurations. In an amorphous semiconductor, both electrons and holes can carry current but the carriers are subjected to strong scattering by the disordered structure, the mean free path being of the order of the average interatomic spacing. Near the band extremities the states are localized and conduction occurs by a thermally-assisted hopping process.

Although these materials appear to behave almost like intrinsic semiconductors, with the Fermi level lying near to the center of the gap, this property is now known to arise from the presence of defects (e.g. dangling bonds and other misfits in the structure) which produce localized states in the gap. The presence of these states, along with the tendency in covalent alloys for local valence requirements to be satisfied, necessitates the use of special techniques for the preparation of doped amorphous semiconductors.

[5]An excellent discussion is given by H. B. Callen, *Thermodynamics*, Wiley, 1960, Chap. 17.

[6]N. F. Mott and E. A. Davis, *Electronic processes in non-crystalline materials*, Oxford, 2nd ed., 1976.

p-n JUNCTIONS

What happens if we place in contact two pieces of a semiconductor, one p type and the other n type, as in Fig. 24? On the p side of the junction there are free holes and an equal concentration of (−) ionized acceptor impurity atoms that maintain charge neutrality. On the n side of the junction there are free electrons and an equal number of (+) ionized donor impurity atoms. Thus the majority carriers are holes on the p side and electrons on the n side, Fig. 25a. In thermal equilibrium with the majority carriers there are small concentrations of the minority carrier types, exaggerated in the figure. The holes concentrated on the p side would like to diffuse to fill the crystal uniformly. Electrons would like to diffuse from the n side. But diffusion upsets electrical neutrality. As soon as a small charge transfer by diffusion has taken place, there is left behind on the p side an excess of (−) ionized acceptor atoms and on the n side an excess of (+) ionized donor atoms (Fig. 25b). This double layer of charge creates an electric field directed from n to p that inhibits diffusion and maintains the separation of the two carrier types. Because of the double layer the electrostatic potential in the crystal takes a jump at the junction region.

The electrochemical potential[7] is constant across the crystal and the junction even though there is a jump in the electrostatic potential at the junction. The net particle flow of holes or electrons is zero in thermal equilibrium, for the current is proportional to the gradient of the electrochemical potential and not to the gradient of the electrostatic potential alone. The concentration gradient exactly cancels the electrostatic potential gradient. A voltmeter measures the difference of electrochemical potential, so that a voltmeter connected across the crystal reads zero.

Even in thermal equilibrium there will be a small flow of electrons J_{nr} from the n region into the p region where the electrons end their lives by recombination with holes. This recombination current is balanced by a current J_{ng} of electrons generated thermally in the p region and which diffuse to the n region:

$$J_{nr}(0) + J_{ng}(0) = 0 \; ; \tag{51}$$

otherwise electrons would pile up on one side of the barrier.

[7]The electrochemical potential of each carrier type is everywhere constant in thermal equilibrium; see *TP* Eq. (11.85). For holes $k_BT \ln p(\mathbf{r}) + e\varphi(\mathbf{r}) =$ constant, where p is the hole concentration and φ is the electrostatic potential. We see p will be low where φ is high. Similarly, for electrons $k_BT \ln n(\mathbf{r}) - e\varphi(\mathbf{r}) =$ constant, and n will be low where φ is low.

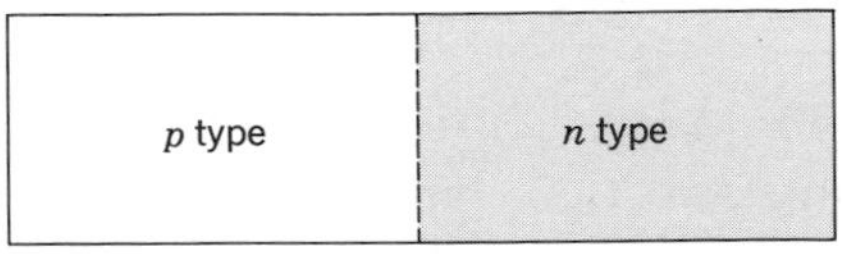

Figure 24 A $p-n$ junction is made from a single crystal modified in two separate regions. Acceptor impurity atoms are incorporated into one part to produce the p region in which the majority carriers are holes. Donor impurity atoms in the other part produce the n region in which the majority carriers are electrons. The interface region may be less than 10^{-4} cm thick.

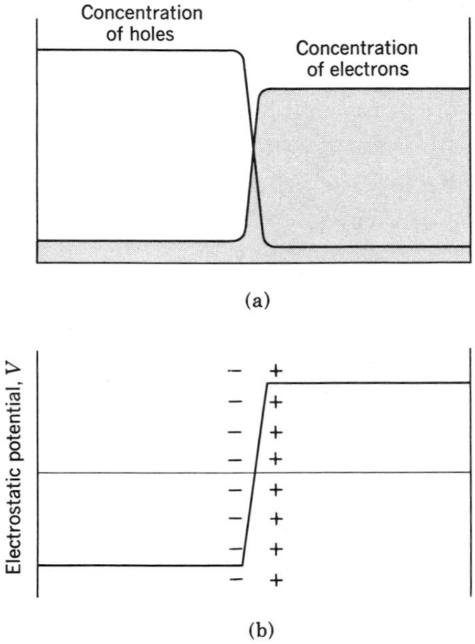

Figure 25 (a) Variation of the hole and electron concentrations across an unbiased (zero applied voltage) junction. The carriers are in thermal equilibrium with the acceptor and donor inpurity atoms, so that the product pn of the hole and electron concentrations is constant throughout the crystal in conformity with the law of mass action. (b) Electrostatic potential from acceptor (−) and donor (+) ions near the junction. The potential gradient inhibits diffusion of holes from the p side to the n side, and it inhibits diffusion of electrons from the n side to the p side. The electric field in the junction region is called the built-in electric field.

Rectification

A p-n junction can act as a rectifier. A large current will flow if we apply a voltage across the junction in one direction, but if the voltage is in the opposite direction only a very small current will flow. If an alternating voltage is applied across the junction the current will flow chiefly in one direction—the junction has rectified the current (Fig. 26). For back voltage bias a negative voltage is applied to the p region and a positive voltage to the n region, thereby increasing the potential difference between the two regions. Now practically no electrons can climb the potential energy hill from the low side of the barrier to the high side. The recombination current is reduced by the Boltzmann factor:

$$J_{nr}(V \text{ back}) = J_{nr}(0) \exp{(-e|V|/k_B T)} \ . \tag{52}$$

The Boltzmann factor controls the number of electrons with enough energy to get over the barrier. The thermal generation current of electrons is not particularly affected by the back voltage because the generation electrons flow downhill (from p to n) anyway:

$$J_{ng}(V \text{ back}) = J_{ng}(0) \ . \tag{53}$$

We saw in (51) that $J_{nr}(0) = -J_{ng}(0)$; thus by (52) the generation current dominates the recombination current for a back bias.

When a forward voltage is applied, the recombination current increases because the potential energy barrier is lowered, thereby enabling more electrons to flow from the n side to the p side:

$$J_{nr}(V \text{ forward}) = J_{nr}(0) \exp{(e|V|/k_B T)} \ . \tag{54}$$

Again the generation current is unchanged:

$$J_{ng}(V \text{ forward}) = J_{ng}(0) \ . \tag{55}$$

The hole current flowing across the junction behaves similarly to the electron current. The applied voltage which lowers the height of the barrier for electrons also lowers it for holes, so that large numbers of electrons flow from the n region under the same voltage conditions that produce large hole currents in the opposite direction. The electrical currents due to holes and electrons add. The total forward electric current, including the effects of both holes and electrons, is given by

$$I = I_s(e^{eV/k_B T} - 1) \ , \tag{56}$$

where I_s is the sum of the two generation currents. As shown in Fig. 26, this equation is well satisfied for p-n junctions in germanium.

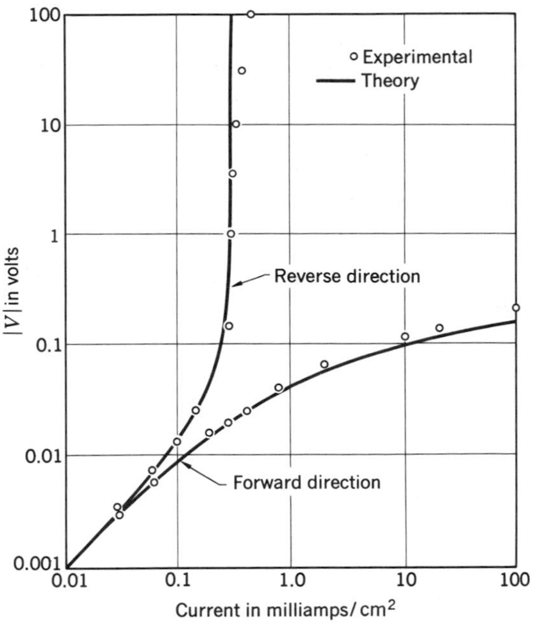

Figure 26 Rectification characteristic of a $p - n$ junction in germanium, after Shockley. The voltage is plotted horizontally and the current vertically.

Solar Cells and Photovoltaic Detectors

Let us shine light on a *p-n* junction without an external bias voltage. Each absorbed photon creates an electron and a hole. When these carriers diffuse to the junction, the built-in electric field of the junction sweeps them down the energy barrier. The separation of the carriers produces a forward voltage across the barrier: forward, because the electric field of the photo-excited carriers is opposite to the built-in field of the junction. The appearance of a forward voltage across an illuminated junction is called the **photovoltaic effect.** An illuminated junction can deliver power to an external circuit. Large area *p-n* junctions of silicon can be used to convert solar photons to electrical energy. Considerations governing the choice of the optimum semiconductor for photovoltaic solar energy conversion are reviewed by Loferski.[8]

[8] J. J. Loferski, J. Appl. Phys. **27**, 777 (1956).

Schottky Barrier

When a semiconductor is brought into contact with a metal, there is formed in the semiconductor a barrier layer from which charge carriers are severely depleted. The barrier layer is also called a depletion layer or exhaustion layer. In Fig. 27 an *n*-type semiconductor is brought into contact with a metal. The Fermi levels are coincident after the transfer of electrons from the conduction band of the metal. Positively charged donor ions are left behind in this region that is practically stripped of electrons. Here the Poisson equation is

(CGS) $$\mathrm{div}\,\mathbf{D} = 4\pi Ne \ , \tag{57}$$

where N is the donor concentration. The electrostatic potential is determined by

(CGS) $$\frac{d^2\phi}{dx^2} = -\frac{4\pi Ne}{\epsilon} \ , \tag{58}$$

which has a solution of the form

$$\phi = -2\pi Nex^2/\epsilon \ . \tag{59}$$

The origin of x has been taken for convenience at the right-hand edge of the barrier. The contact is at $-x_b$ and here the potential energy relative to the right-hand side is $-e\phi_0$, whence the thickness of the barrier is

(CGS) $$x_b = (\epsilon|\phi_0|/2\pi Ne)^{1/2} \tag{60}$$

With $\epsilon = 16$, $e\phi_0 = 0.5$ eV, $N = 10^{16}$ cm^{-3}, we find $x_b = 0.3\ \mu$m. This is a simplified view of actual metal-semiconductor contacts; further details are given in the book by H. P. Wolf cited in the references.

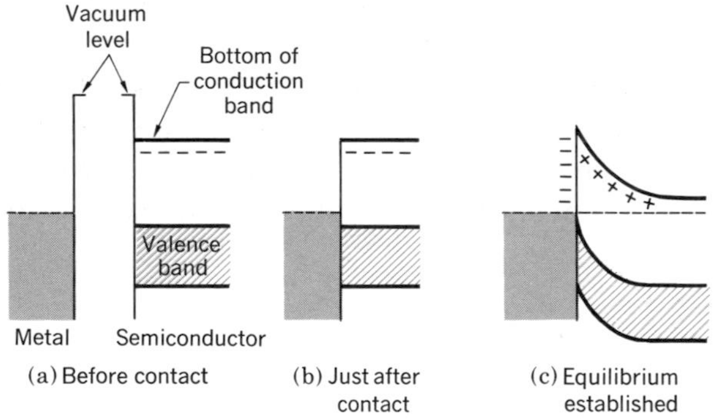

Figure 27 Rectifying barrier between a metal and an *n*-type semiconductor. The Fermi level is shown as a broken line.

Gunn Effect Oscillators

In 1963, J. B. Gunn[9] discovered that microwave oscillations occur in a simple slab of gallium arsenide under an applied dc bias voltage of several thousand volts. Frequencies as high as 150 GHz have been achieved with Gunn oscillators, which now are an important class of solid state devices. The oscillations are a consequence of the negative differential mobility of the conduction electrons in GaAs. As shown in Fig. 28, the electron drift velocity increases with the applied electric field until they reach a velocity of 2×10^7 cm s^{-1} at an electric field of 3300 V cm^{-1}. At higher electric fields the velocity decreases; in this region the differential mobility is negative. A similar effect is observed also in InP and CdTe.

The differential electrical conductance can also be negative. The current density is the sum of the contributions from two conduction bands, the lower of light mass and the upper of heavier mass:

$$j = \sigma(E)E = (n_l e\mu_l + n_h e\mu_h)E \ , \tag{61}$$

and the differential conductivity is

$$\frac{dj}{dE} = (n_l e\mu_l + n_h e\mu_h) + \left(\frac{dn_l}{dE}\,\mu_l + \frac{dn_h}{dE}\,\mu_h\right) eE \ . \tag{62}$$

If the number of carriers is constant, $dn_l/dE = -dn_h/dE$, whence

$$\frac{dj}{dE} = (n_l e\mu_l + n_h e\mu_h) - (\mu_l - \mu_h)\frac{dn_h}{dE}\,E \ . \tag{63}$$

This can be negative if the second term on the right is larger than the first

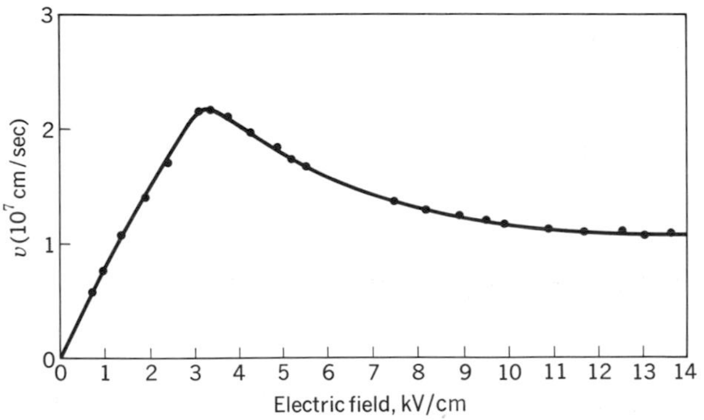

Figure 28 Electron drift velocity in GaAs as function of applied electric field intensity, at room temperature. Electrons slow down at fields above 3300 V cm^{-1}. In this region the differential mobility dv/dE is negative, leading to the possibility of electrical oscillations. In GaAs the heavy electron band ($m_h \approx m$) is 0.36 eV above the light mass electron band ($m_l = 0.07\ m$). The mobility at low fields is 7500 cm^2/V-s. [After J. G. Ruch and G. S. Kino, Phys. Rev. **174,** 921 (1968).]

[9] J. B. Gunn, Solid State Comm. **1,** 88 (1963); see also H. Kroemer, "Gunn effect—bulk instabilities," in *Topics in solid state and quantum electronics,* W. D. Hershberger, ed., Wiley, 1972, pp. 20–98.

term. Such a condition is favored by high E; by $\mu_l \gg \mu_h$, likely if $m_l \ll m_h$; and by strong interband scattering processes to increase the population in the heavy electron band when a strong field is applied.

It is observed that a uniform field distribution is not stable in a negative conductivity regime. Space-charge accumulation layers form in domains that propagate across the crystal with the drift velocity of the carriers. The frequency of the oscillations associated with this motion is simply $f = v/d$, where the drift velocity $v \sim 10^7$ cm s^{-1} and the devices may have a thickness $d \sim 10^{-3}$ cm; thus $f \sim 10^{10}$ Hz.

SUMMARY

1. The motion of a wave packet centered at wavevector $\mathbf{k}$ is described by $\mathbf{F} = \hbar\, d\mathbf{k}/dt$, where $\mathbf{F}$ is the applied force. The motion in real space is obtained from the group velocity $\mathbf{v}_g = \hbar^{-1}\nabla_{\mathbf{k}}\epsilon(\mathbf{k})$.

2. The effective mass m^* of an electron at $\mathbf{k}$ is given by

$$\left(\frac{1}{m^*}\right)_{\mu\nu} = \frac{1}{\hbar^2}\frac{\partial^2\epsilon}{\partial k_\mu \partial k_\nu} \ .$$

The smaller the energy gap, the smaller is $|m^*|$ near the gap.

3. A crystal with one hole has one empty electron state in an otherwise filled band. The properties of the hole are those of the $N-1$ electrons: (a) If the electron is missing from the state of wavevector $\mathbf{k}_e$, then the wavevector of the hole is

$$\mathbf{k}_h = -\mathbf{k}_e \ .$$

(b) The rate of change of $\mathbf{k}_h$ in an applied field requires the assignment of a positive charge to the hole: $e_h = e = -e_e$, whence

(CGS)
$$\hbar\frac{d\mathbf{k}_h}{dt} = e\left(\mathbf{E} + \frac{1}{c}\mathbf{v}_h \times \mathbf{B}\right) \ .$$

(c) If $\mathbf{v}_e$ is the velocity an electron would have in the state $\mathbf{k}_e$, then the velocity to be ascribed to the hole of wavevector $\mathbf{k}_h = -\mathbf{k}_e$ is

$$\mathbf{v}_h = \mathbf{v}_e \ .$$

(d) The energy of the hole referred to zero for a filled band is positive and is

$$\epsilon_h(\mathbf{k}_h) = -\epsilon(\mathbf{k}_e) \ .$$

(e) The effective mass of a hole is opposite to the effective mass of an electron at the same point on an energy band:

$$m_h = -m_e \ .$$

4. The product np of the electron and hole concentrations in a semiconductor is constant at a given temperature, independent of the purity of the crystal.

5. Acceptor impurities lead to an excess of holes over electrons. Donor impurities lead to an excess of electrons. In an intrinsic semiconductor the number of electrons is equal to the number of holes.

PROBLEMS

1. ***Impurity orbits.*** Indium antimonide has $E_g = 0.23$ eV; dielectric constant $\epsilon = 18$; electron effective mass $m_e = 0.015\ m$. Calculate (a) the donor ionization energy; (b) the radius of the ground state orbit. (c) At what minimum donor concentration will appreciable overlap effects between the orbits of adjacent impurity atoms occur? This overlap tends to produce an impurity band—a band of energy levels which permit conductivity presumably by a hopping mechanism in which electrons move from one impurity site to a neighboring ionized impurity site.

2. ***Ionization of donors.*** In a particular semiconductor there are 10^{13} donors/cm^3 with an ionization energy E_d of 1 meV and an effective mass $0.01\ m$. (a) Estimate the concentration of conduction electrons at 4 K. (b) What is the value of the Hall coefficient? Assume no acceptor atoms are present and that $E_g \gg k_BT$.

3. ***Hall effect with two carrier types.*** Assuming concentrations n,p; relaxation times τ_e, τ_h; and masses m_e, m_h, show that the Hall coefficient in the drift velocity approximation is

   (CGS) $$R_H = \frac{1}{ec} \cdot \frac{p - nb^2}{(p + nb)^2} ,$$

   where $b = \mu_e/\mu_h$ is the mobility ratio. In the derivation neglect terms of order B^2. In SI we drop the c. Hint: In the presence of a longitudinal electric field, find the transverse electric field such that the transverse current vanishes. The algebra may seem tedious, but the result is worth the trouble. Use (6.53), but for two carrier types; neglect $(\omega_c\tau)^2$ in comparison with $\omega_c\tau$.

4. ***Cyclotron resonance for a spheroidal energy surfacc.*** Consider the energy surface

   $$\epsilon(\mathbf{k}) = \hbar^2 \left(\frac{k_x^2 + k_y^2}{2m_t} + \frac{k_z^2}{2m_l} \right) ,$$

   where m_t is the transverse mass parameter and m_l is the longitudinal mass parameter. A surface on which $\epsilon(\mathbf{k})$ is constant will be a spheroid. Use the equation of motion (6), with $\mathbf{v} = \hbar^{-1}\nabla_{\mathbf{k}}\epsilon$, to show that $\omega_c = eB/(m_lm_t)^{1/2}c$ when the static magnetic field B lies in the x,y plane. This result agrees with (24) when $\theta = \pi/2$. The result is in CGS; to obtain SI, omit the c.

5. ***Donor ionization.*** Consider a semiconductor with a donor concentration N_D. Let $k_BT \ll E_D$, the donor ionization energy. (a) Derive (43). (b) Find the position of the Fermi level.

6. ***Magnetoresistance with two carrier types.*** Problem 6.9 shows that the drift velocity approximation to the motion of charge carriers in electric and magnetic fields does not lead to transverse magnetoresistance. The result is different with two carrier types. Consider a conductor with a concentration n of electrons of effective mass m_e and relaxation time τ_e; and a concentration p of holes of effective mass m_h and relaxation time τ_h. Treat the limit of very strong magnetic fields, $\omega_c\tau \gg 1$. (a) Show in this limit that $\sigma_{yx} = (n - p)ec/B$. (b) Show that the Hall field is given by, with $Q \equiv \omega_c\tau$,

$$E_y = -(n - p)\left(\frac{n}{Q_e} + \frac{p}{Q_h}\right)^{-1} E_x \ ,$$

which vanishes if $n = p$. (c) Show that the effective conductivity in the x direction is

$$\sigma_{\text{eff}} = \frac{ec}{B}\left[\left(\frac{n}{Q_e} + \frac{p}{Q_h}\right) + (n - p)^2\left(\frac{n}{Q_e} + \frac{p}{Q_h}\right)^{-1}\right] .$$

If $n = p$, $\sigma \propto B^{-2}$. If $n \neq p$, σ saturates in strong fields; that is, it approaches a limit independent of B as $B \to \infty$.

References

D. L. Carter and R. T. Bate, eds., *Physics of metals and narrow-gap semiconductors*, Pergamon, 1971.

M. L. Cohen, "Electronic charge densities in semiconductors," Science **179,** 1189 (1973).

R. Dalven, "Electronic structure of PbS, PbSe, and PbTe," *Solid state physics* **28,** 179 (1973).

S. Deme, *Semiconductor detectors for nuclear radiation measurement*, Wiley, 1972.

H. J. Goldsmid, *Applications of thermoelectricity*, Wiley, 1960.

F. Gutmann and L. E. Lyons, *Organic semiconductors*, Wiley, 1967.

D. Long, *Energy bands in semiconductors*, Wiley, 1968.

J. P. McKelvey, *Solid state and semiconductor physics*, Harper & Row, 1966.

A. E. Owen, "Semiconducting glasses," Contemp. Physics **11,** 227, 257 (1970).

J. C. Phillips, *Bonds and bands in semiconductors*, Academic Press, 1973.

R. A. Smith, *Semiconductors*, Cambridge, 1959.

B. G. Streetman, *Solid state electronic devices*, Prentice-Hall, 1972. Good introduction to devices.

S. M. Sze, *Physics of semiconductor devices*, Wiley, 1969.

H. F. Wulf, *Semiconductors*, Wiley, 1971. Concise summary of physical and engineering properties of semiconductors of practical interest.

Series

A. F. Gibson, P. Aigrain, and R. E. Burgess, eds., *Progress in semiconductors*, Heywood, London.

H. J. Queisser, ed., *Festkoerper Probleme*, Vieweg, Pergamon.

R. K. Willardson and A. C. Beer, eds., *Semiconductors and semimetals*, Academic Press.

Proceedings of the Biennial International Conference on the Physics of Semiconductors. Moscow, 1968; Boston, 1970; Warsaw, 1972; Stuttgart, 1974; Rome, 1976.

9 FERMI SURFACES AND METALS

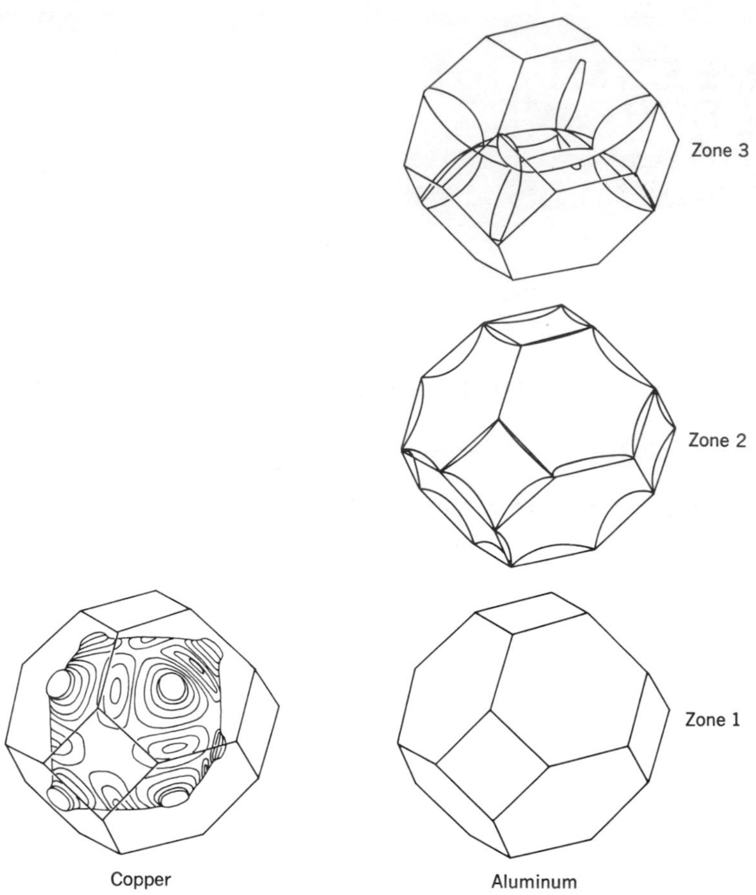

Figure 1 Free electron Fermi surfaces for fcc metals with one (Cu) and three (Al) valence electrons per primitive cell. The Fermi surface shown for copper has been deformed from a sphere to agree with the experimental results. The second zone of aluminum is nearly filled with electrons. The complex structure in the third zone of aluminum is shown only in part: There are two similar structures in the same zone oriented so that the set of three structures has cubic symmetry. (Redrawn from A. R. Mackintosh, "Fermi surface of metals," Scientific American, July 1963, pp. 110–120.)

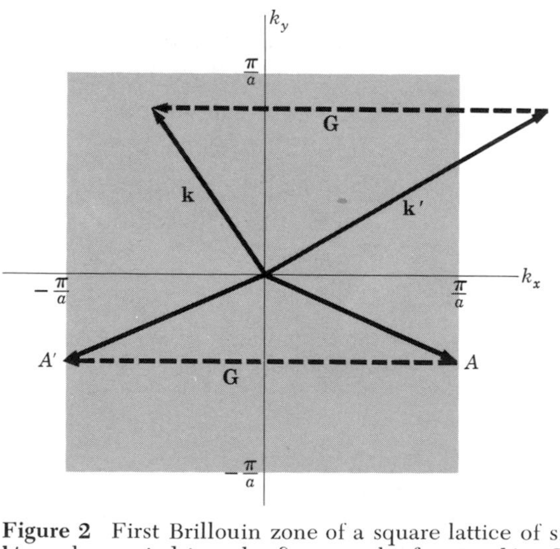

Figure 2 First Brillouin zone of a square lattice of side a. The wavevector $\mathbf{k}'$ can be carried into the first zone by forming $\mathbf{k}' + \mathbf{G}$. The wavevector at a point A on the zone boundary is carried by $\mathbf{G}$ to the point A' on the opposite boundary of the same zone. Shall we count both A and A' as lying in the first zone? We count them as *one* identical point in the zone.

9 FERMI SURFACES AND METALS

Few people would define a metal as "a solid with a Fermi surface." This may nevertheless be the most meaningful definition of a metal one can give today; it represents a profound advance in the understanding of why metals behave as they do. The concept of the Fermi surface, as developed by quantum physics, provides a precise explanation of the main physical properties of metals.

A. R. Mackintosh

The Fermi surface is the surface of constant energy ϵ_F in **k** space. The Fermi surface separates the unfilled orbitals from the filled orbitals, at absolute zero. The electrical properties of the metal are determined by the shape of the Fermi surface, because the current is due to changes in the occupancy of states near the Fermi surface. The shape may be very intricate in a reduced zone scheme and yet have a simple interpretation in terms of a sphere. We exhibit in Fig. 1 the free electron Fermi surfaces constructed for two metals that have the face-centered cubic crystal structure: copper, with one valence electron, and aluminum, with three. The free electron Fermi surfaces were developed from spheres of radius k_F determined by the valence electron concentration. How do we start from a sphere and attain these surfaces? These constructions require the reduced and periodic zone schemes.

Reduced Zone Scheme

It is always possible to select the wavevector index **k** of the Bloch function to lie within the first Brillouin zone. This procedure is known as mapping the band in the reduced zone scheme. If we encounter a Bloch function written as $\psi_{\mathbf{k}'}(\mathbf{r}) = e^{i\mathbf{k}'\cdot\mathbf{r}} u_{\mathbf{k}'}(\mathbf{r})$, with $\mathbf{k}'$ outside the first zone, as in Fig. 2, we may always find a suitable reciprocal lattice vector **G** such that $\mathbf{k} = \mathbf{k}' + \mathbf{G}$ lies within the first Brillouin zone. Then

$$\psi_{\mathbf{k}'}(\mathbf{r}) = e^{i\mathbf{k}'\cdot\mathbf{r}} u_{\mathbf{k}'}(\mathbf{r}) = e^{i\mathbf{k}\cdot\mathbf{r}} (e^{-i\mathbf{G}\cdot\mathbf{r}} u_{\mathbf{k}'}(\mathbf{r})) \equiv e^{i\mathbf{k}\cdot\mathbf{r}} u_{\mathbf{k}}(\mathbf{r}) = \psi_{\mathbf{k}}(\mathbf{r}) \ , \tag{1}$$

where $u_{\mathbf{k}}(\mathbf{r}) \equiv e^{-i\mathbf{G}\cdot\mathbf{r}} u_{\mathbf{k}'}(\mathbf{r})$. Both $e^{-i\mathbf{G}\cdot\mathbf{r}}$ and $u_{\mathbf{k}'}(\mathbf{r})$ are periodic in the crystal lattice, so $u_{\mathbf{k}}(\mathbf{r})$ is also, whence $\psi_k(\mathbf{r})$ is of the Bloch form. Even with free electrons it is useful to work in the reduced zone scheme, as in Fig. 3. Any energy $\epsilon_{\mathbf{k}'}$ for $\mathbf{k}'$ outside the first zone is equal to an $\epsilon_{\mathbf{k}}$ in the first zone, where $\mathbf{k} = \mathbf{k}' + \mathbf{G}$. Thus we need solve for the energy only in the first Brillouin zone, for each band. An energy band is a single branch of the $\epsilon_{\mathbf{k}}$ versus **k** surface.

In the reduced zone scheme we should not be surprised to find different energies at the same value of the wavevector. Each different energy characterizes a different band. Two wavefunctions at the same **k** but of different energies will be independent of each other: the wavefunctions will be made up of different combinations of the plane wave components $e^{i(\mathbf{k}+\mathbf{G})\cdot\mathbf{r}}$ in the expansion of Chapter 7. Because the values of the coefficients $C(\mathbf{k}+\mathbf{G})$ defined there are different for the two bands we should add a symbol, say n, to the C's to serve as a band index: $C_n(\mathbf{k}+\mathbf{G})$. Thus the Bloch function for a state of wavevector **k** in the band n can be written as

$$\psi_{n,\mathbf{k}} = e^{i\mathbf{k}\cdot\mathbf{r}} u_{n,\mathbf{k}}(\mathbf{r}) = \sum_{\mathbf{G}} C_n(\mathbf{k}+\mathbf{G})\; e^{i(\mathbf{k}+\mathbf{G})\cdot\mathbf{r}}$$

Periodic Zone Scheme

We can repeat a given Brillouin zone periodically through all of wavevector space. To repeat a zone, we translate the zone by a reciprocal lattice vector. If we can translate a band from other zones into the first zone, we can translate a band in the first zone into every other zone. In this scheme the energy $\epsilon_{\mathbf{k}}$ of a band is a periodic function in the reciprocal lattice:

$$\epsilon_{\mathbf{k}} = \epsilon_{\mathbf{k}+\mathbf{G}} \quad . \tag{2}$$

Here $\epsilon_{\mathbf{k}+\mathbf{G}}$ is understood to refer to the same energy band as $\epsilon_{\mathbf{k}}$. The result of this construction is known as the **periodic zone scheme;** it is helpful in exhibiting the connectivity of electron orbits in a magnetic field. The periodic property of the energy also can be seen easily from the central equation (7.27).

Consider as an example an energy band of a simple cubic lattice as calculated in the tight-binding approximation:

$$\epsilon_k = E_0 - \alpha - 2\gamma\,(\cos k_x a + \cos k_y a + \cos k_z a) \ , \tag{3}$$

where E_0, α, and γ are constants. A reciprocal lattice vector of the sc lattice is $\mathbf{G} = (2\pi/a)\hat{\mathbf{x}}$; if we add this vector to **k** the only change in (3) is

$$\cos k_x a \rightarrow \cos\,(k_x + 2\pi/a)a = \cos\,(k_x a + 2\pi) \ ,$$

but this is identically equal to $\cos k_x a$. The energy is unchanged when the wavevector is increased by a reciprocal lattice vector, so that the energy is a periodic function of the wavevector.

Three different zone schemes are useful (Fig. 4):

1. the **extended zone scheme** in which different bands are drawn in different zones in wavevector space;
2. the **reduced zone scheme** in which all bands are drawn in the first Brillouin zone;
3. the **periodic zone scheme** in which every band is drawn in every zone.

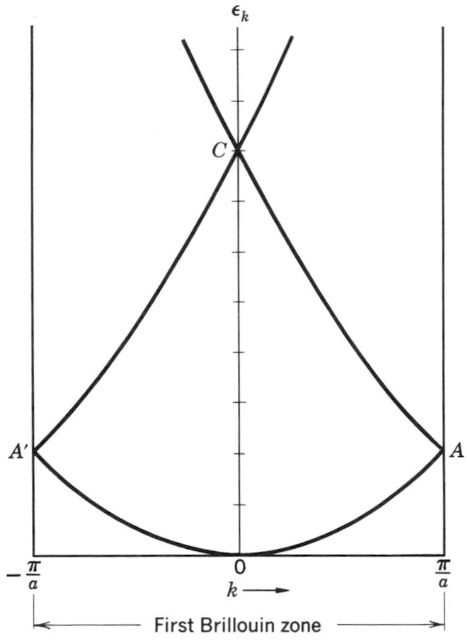

Figure 3 Energy-wavevector relation $\epsilon_k = \hbar^2 k^2/2m$ for free electrons as drawn in the reduced zone scheme. This construction often gives a useful idea of the overall appearance of the band structure of a crystal. The branch AC if displaced by $-2\pi/a$ gives the usual free electron curve for negative k. The branch $A'C$ if displaced by $2\pi/a$ gives the usual curve for positive k. A crystal potential $U(x)$ will introduce band gaps at the edges of the zone (as at A and A') and at the center of the zone (as at C). The point C when viewed in the extended zone scheme falls at the edges of the second zone. The overall width and gross features of the band structure are often indicated properly by such free electron bands in the reduced zone scheme.

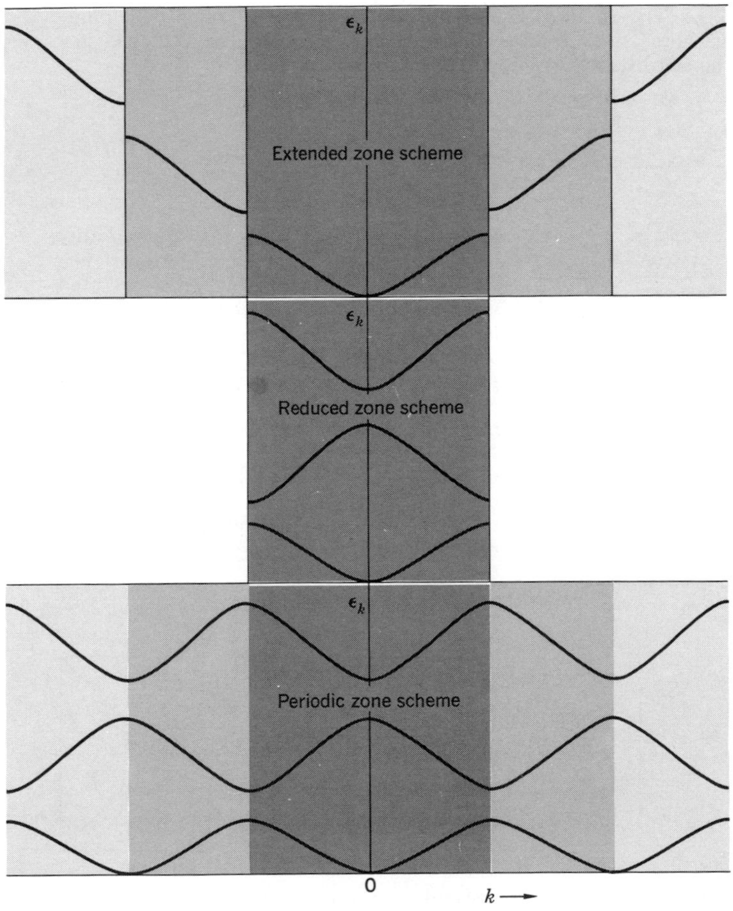

Figure 4 Three energy bands of a linear lattice plotted in the extended (Brillouin), reduced, and periodic zone schemes.

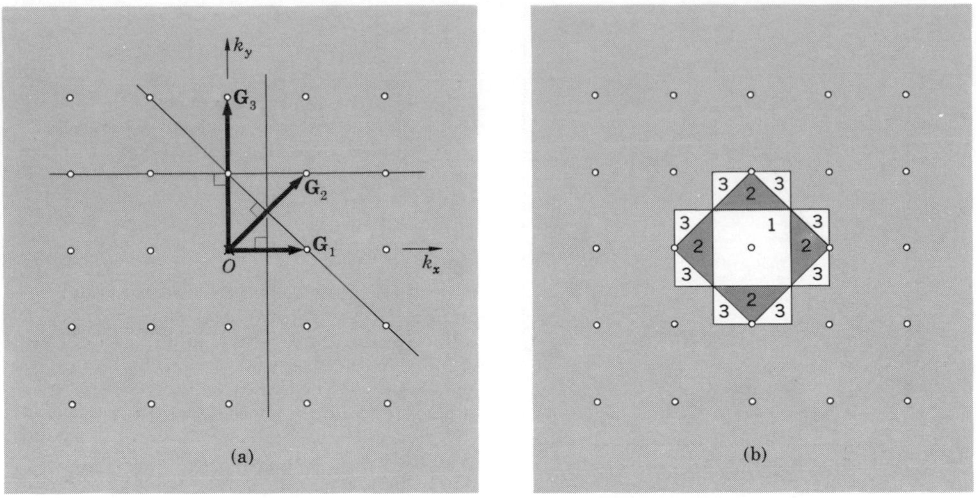

Figure 5 (a) Construction in **k** space of the first three Brillouin zones of a square lattice. The three shortest forms of the reciprocal lattice vectors are indicated as $\mathbf{G}_1$, $\mathbf{G}_2$, and $\mathbf{G}_3$. The lines drawn are the perpendicular bisectors of these **G**'s. (b) On constructing all lines equivalent by symmetry to the three lines in (a) we obtain the regions in **k** space which form the first three Brillouin zones. The numbers denote the zone to which the regions belong; the numbers here are ordered according to the length of the vector **G** involved in the construction of the outer boundary of the region.

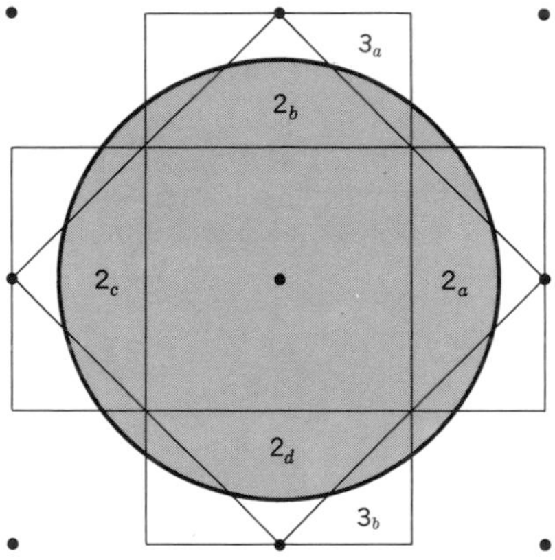

Figure 6 Brillouin zones of a square lattice in two dimensions. The circle shown is a surface of constant energy for free electrons; it will be the Fermi surface for some particular value of the electron concentration. The total area of the filled region in **k** space depends only on the electron concentration and is independent of the interaction of the electrons with the lattice. The shape of the Fermi surface depends on the lattice interaction, and the shape will not be an exact circle in an actual lattice. The labels within the sections of the second and third zones refer to Fig. 7.

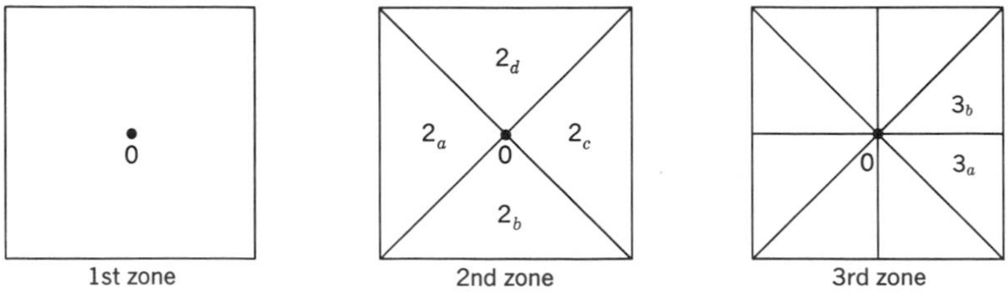

Figure 7 Mapping of the first, second, and third Brillouin zones in the reduced zone scheme. The sections of the second zone in Fig. 6 are put together into a square by translation through an appropriate reciprocal lattice vector. A different **G** is needed for each piece of a zone.

CONSTRUCTION OF FERMI SURFACES

We consider the analysis of a square lattice in Fig. 5. The equation for the zone boundaries is $2\mathbf{k} \cdot \mathbf{G} + G^2 = 0$, satisfied if $\mathbf{k}$ terminates on the plane normal to $\mathbf{G}$ at the midpoint of $\mathbf{G}$. The first Brillouin zone of the square lattice is the area enclosed by the perpendicular bisectors of $\mathbf{G}_1$ and of the three reciprocal lattice vectors equivalent by symmetry to $\mathbf{G}_1$ in Fig. 5a. These four reciprocal lattice vectors are $\pm(2\pi/a)\hat{\mathbf{k}}_x$ and $\pm(2\pi/a)\hat{\mathbf{k}}_y$.

The second zone is constructed from $\mathbf{G}_2$ and the three vectors equivalent to it by symmetry, and similarly for the third zone. The pieces of the second and third zones are drawn in Fig. 5b. To determine the boundaries of some zones we have to consider sets of several nonequivalent reciprocal lattice vectors. Thus the boundaries of section 3_a of the third zone are formed from the perpendicular bisectors of three $\mathbf{G}$'s, namely $(2\pi/a)\hat{\mathbf{k}}_x$; $(4\pi/a)\hat{\mathbf{k}}_y$; and $(2\pi/a)(\hat{\mathbf{k}}_x + \hat{\mathbf{k}}_y)$.

The free electron Fermi surface for an arbitrary electron concentration is shown in Fig. 6. It is inconvenient to have sections of the Fermi surface that belong to the same zone appear detached from one another. The detachment can be repaired by a transformation to the reduced zone scheme. We take the triangle labeled 2_a and move it by a reciprocal lattice vector $\mathbf{G} = -(2\pi/a)\hat{\mathbf{k}}_x$ such that the triangle reappears in the area of the first Brillouin zone (Fig. 7). Other reciprocal lattice vectors will shift the triangles 2_b, 2_c, 2_d to other parts of the first zone, completing the mapping of the second zone into the reduced zone scheme. The parts of the Fermi surface falling in the second zone are now connected, as shown in Fig. 8. A third zone is assembled into a square in Fig. 8, but the parts of the Fermi surface still appear disconnected. When we look at it in the periodic zone scheme (Fig. 9), the Fermi surface forms a lattice of rosettes.

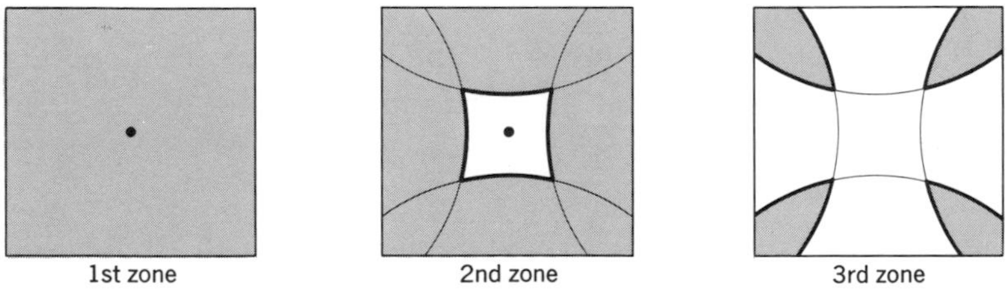

Figure 8 The free electron Fermi surface of Fig. 6, as viewed in the reduced zone scheme. The shaded areas represent occupied electron states. Parts of the Fermi surface fall in the second, third, and fourth zones. The fourth zone is not shown. The first zone is entirely occupied.

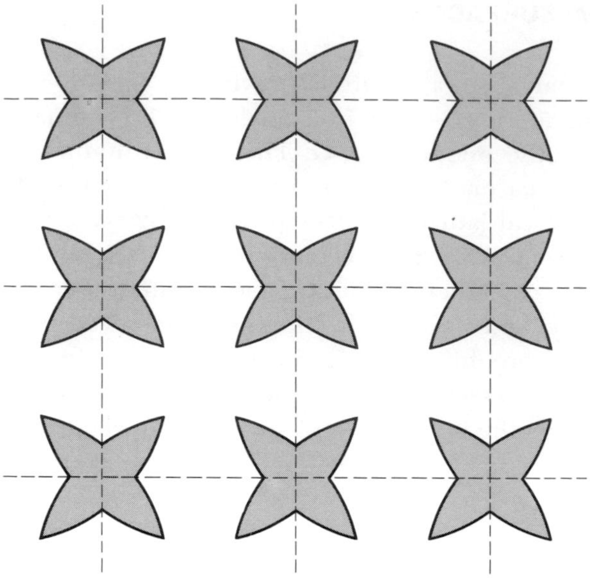

Figure 9 The Fermi surface in the third zone as drawn in the periodic zone scheme. The figure was constructed by repeating the third zone as shown in Fig. 8 in the reduced zone scheme.

How do we go from Fermi surfaces for free electrons to Fermi surfaces for nearly free electrons? We can make approximate constructions freehand by the use of four facts:

1. The interaction of the electron with the periodic potential of the crystal causes energy gaps at the zone boundaries.
2. Almost always the Fermi surface will intersect zone boundaries perpendicularly (see below).
3. The crystal potential will round out sharp corners in the Fermi surfaces.
4. The total volume enclosed by the Fermi surface depends only on the electron concentration and is independent of the details of the lattice interaction.

We cannot make quantitative statements without calculation, but qualitatively we expect the Fermi surfaces in the second and third zones of Fig. 8 to be changed as shown in Fig. 10.

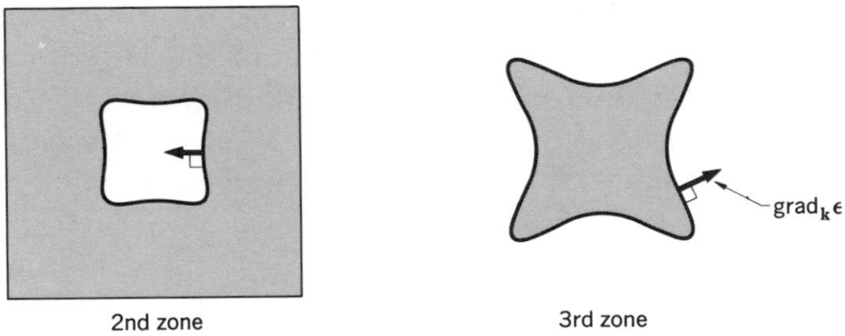

Figure 10 Qualitative impression of the effect of a weak periodic crystal potential on the Fermi surface of Fig. 8. At one point on each Fermi surface we have shown the vector $\text{grad}_{\mathbf{k}}\ \epsilon$. In the second zone the energy increases toward the interior of the figure, and in the third zone the energy increases toward the exterior. The shaded regions are filled with electrons and are lower in energy than the unshaded regions. We shall see that a Fermi surface like that of the third zone is electronlike, whereas one like that of the second zone is holelike.

Freehand impressions of the Fermi surfaces derived from free electron surfaces are useful. Fermi surfaces for free electrons are constructed by a procedure credited to Harrison, Fig. 11. The reciprocal lattice points are determined, and a free-electron sphere of radius appropriate to the electron concentration is drawn around each point. Any point in **k** space within at least one sphere corresponds to an occupied state in the first zone. Points within at least two spheres correspond to occupied states in the second zone, and similarly for points in three or more spheres.

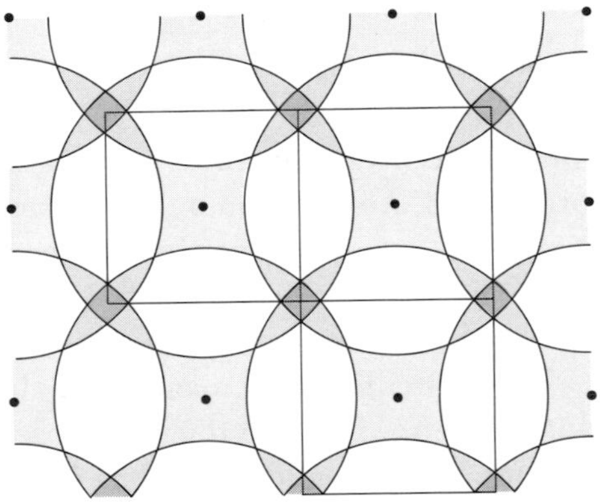

Figure 11 Harrison construction of free electron Fermi surfaces in the second, third, and fourth zones for a square lattice. The Fermi surface does not intercept the first zone, which therefore is filled with electrons.

Early in Chapter 6 we said that the alkali metals are the simplest metals. They have one valence electron and weak lattice interactions, so the Fermi surface of each is a slightly deformed sphere: Na is closely spherical; Cs is deformed $\approx$ 10 percent. The divalent metals Be and Mg also have weak lattice interactions and nearly spherical Fermi surfaces, but with twice the volume, so the Fermi surface extends out of the first zone. In Al, which is trivalent, the free electron monster of the third zone of Fig. 1 becomes dismembered by lattice interactions.[1]

[1]The Fermi surfaces of the simple metals are reviewed in Chapter 4 of A. P. Cracknell and K. C. Wong, *Fermi surface*, Oxford, 1973. For nonsimple metals, see their Chap. 5.

Zero Slope of Bands at Zone Boundaries. We have drawn energy bands ϵ versus k that come in horizontally at the zone boundary, with $\partial\epsilon/\partial k = 0$ at the boundary. Why is this so? We consider those directions of **k** such that the crystal structure has a reflection plane of symmetry normal to **k.** Then $\epsilon(\mathbf{k})$ must be symmetrical along this direction, so that

$$\epsilon(\mathbf{k}) = \epsilon(-\mathbf{k}) \ ; \qquad \left.\frac{\partial \epsilon}{\partial k}\right|_{\mathbf{k}} = -\left.\frac{\partial \epsilon}{\partial k}\right|_{-\mathbf{k}} \ . \tag{4a}$$

But $\epsilon(\mathbf{k})$ is a periodic function of **k** with period **G,** so that

$$\epsilon(\mathbf{k}) = \epsilon(\mathbf{k} + \mathbf{G}) \ ; \qquad \left.\frac{\partial \epsilon}{\partial k}\right|_{\mathbf{k}} = \left.\frac{\partial \epsilon}{\partial k}\right|_{\mathbf{k}+\mathbf{G}} \ . \tag{4b}$$

At a zone boundary $\mathbf{k} = \pm\frac{1}{2}\mathbf{G}$, and from (4a)

$$\left.\frac{\partial \epsilon}{\partial k}\right|_{\frac{1}{2}\mathbf{G}} = -\left.\frac{\partial \epsilon}{\partial k}\right|_{-\frac{1}{2}\mathbf{G}} \ ,$$

whereas from (4b)

$$\left.\frac{\partial \epsilon}{\partial k}\right|_{\frac{1}{2}\mathbf{G}} = \left.\frac{\partial \epsilon}{\partial k}\right|_{-\frac{1}{2}\mathbf{G}} \ .$$

These two results are compatible only if $\partial\epsilon/\partial k = 0$ at the zone boundary. The argument can fail if bands are degenerate at the boundary. If $\partial\epsilon/\partial k = 0$ all along a boundary face, any Fermi surface that makes contact with this face must come in at normal incidence to the face.

ELECTRON ORBITS, HOLE ORBITS, AND OPEN ORBITS

We saw in Eq. (8.7) that electrons in a static magnetic field move on a curve of constant energy on a plane normal to **B.** An electron on the Fermi surface will move in a curve on the Fermi surface, because this is a surface of constant energy. Three types of orbits in a magnetic field are shown in Fig. 12. The closed orbits of (a) and (b) are traversed in opposite senses. Because particles of opposite charge circulate in a magnetic field in opposite senses, we say that one orbit is electronlike and the other orbit is holelike. Electrons in holelike orbits move in a magnetic field as if endowed with a positive charge. This is consistent with the treatment of holes in Chapter 8. In (c) the orbit is not closed: the particle on reaching the zone boundary at A is instantly umklapped back to B, where B is equivalent to B' because they are connected by a reciprocal lattice vector. Such an orbit is called an open orbit. Open orbits have an important effect on the magnetoresistance. Vacant orbitals near the top of an otherwise filled band give rise to holelike orbits, as in Figs. 13 and 14. A view of a possible energy surface in three dimensions is given in Fig. 15. Orbits that enclose filled states are electron orbits; orbits that enclose empty states are hole orbits; and orbits that move from zone to zone without closing are open orbits.

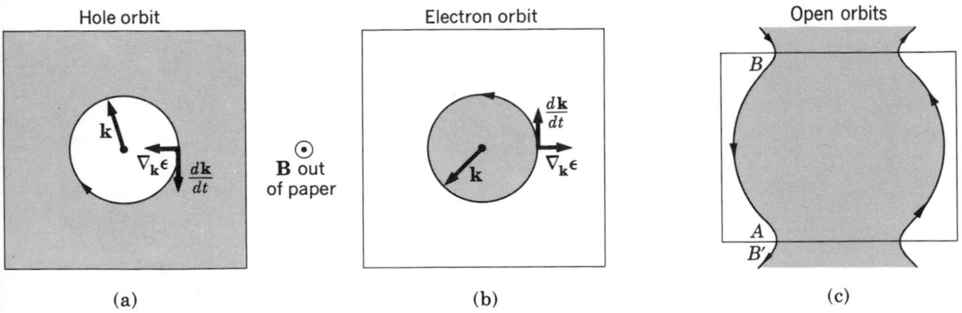

Figure 12 Motion in a magnetic field of the wavevector of an electron on the Fermi surface, in (a) and (b) for Fermi surfaces topologically equivalent to those of Fig. 10. In (a) the wavevector moves around the orbit in a clockwise direction; in (b) the wavevector moves around the orbit in a counterclockwise direction. The direction in (b) is what we expect for a free electron of charge $-e$: the smaller **k** values have the lower energy, so that the filled electron states lie inside the Fermi surface. We call the orbit in (b) **electronlike.** The sense of the motion in a magnetic field is opposite in (a) to that in (b), so that we refer to the orbit in (a) as **holelike.** A hole moves as a particle of positive charge e. In (c) for a rectangular zone we show the motion on an **open orbit** in the periodic zone scheme. This is topologically intermediate between a hole orbit and an electron orbit.

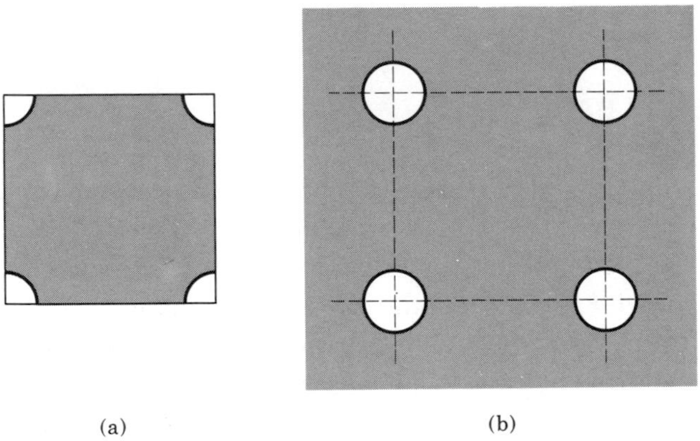

Figure 13 (a) Vacant states at the corners of an almost-filled band, drawn in the reduced zone scheme. (b) In the periodic zone scheme the various parts of the Fermi surface are connected. Each circle forms a holelike orbit. The different circles are entirely equivalent to each other, and the density of states is that of a single circle. (The orbits need not be true circles: for the lattice shown it is only required that the orbits have fourfold symmetry.)

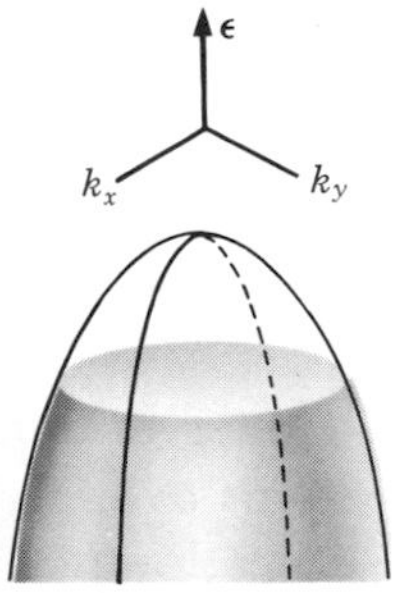

Figure 14 Vacant states near the top of an almost filled band in a two-dimensional crystal. This figure is equivalent to Fig. 12a.

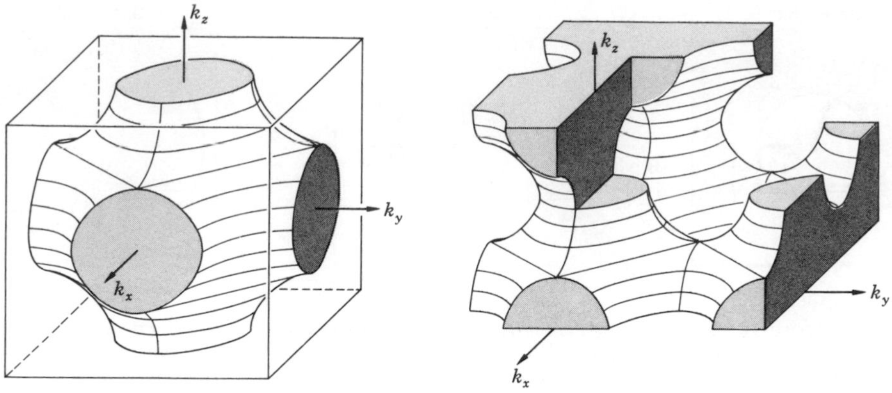

(a) (b)

Figure 15 Constant energy surface in the Brillouin zone of a simple cubic lattice, for the assumed energy band $\epsilon_k = -\alpha - 2\gamma(\cos k_xa + \cos k_ya + \cos k_za)$. (a) Constant energy surface $\epsilon = -\alpha$. The filled volume contains one electron per primitive cell. (b) The same surface exhibited in the periodic zone scheme. The connectivity of the orbits is clearly shown. Can you find electron, hole, and open orbits for motion in a magnetic field $B\hat{z}$? (A. Sommerfeld and H. A. Bethe.)

CALCULATION OF ENERGY BANDS

Few of the journeymen and masters of energy band calculation learned their methods entirely from books. Band calculation is a craft learned by experience, often developed in groups, and needing access to computers. Wigner and Seitz, who performed the first serious band calculations in 1933, refer to afternoons spent on the manual desk calculators of those days, using one afternoon for a trial wavefunction. Modern computers have eased the pain. However, the formulation of the problem requires great care, and the computer programs are not trivial. Here we limit ourselves to three methods useful to beginners: the tight-binding method, useful for interpolation; the Wigner-Seitz method, useful for the visualization and understanding of the alkali metals; and the pseudopotential method, utilizing the general theory of Chapter 7, which shows the simplicity of many problems and provides empirical parameters helpful in the analysis of charge distributions and optical spectra. Cited at the end of this chapter are reviews of these and other methods, as well as the bibliography compiled by Slater.

Tight Binding Method for Energy Bands

Let us start with neutral separated atoms and watch the changes in the atomic energy levels as the charge distributions of adjacent atoms overlap when the atoms are brought together to form a crystal. Consider two hydrogen atoms, each with an electron in the $1s$ ground state. The wavefunctions ψ_A, ψ_B on the separated atoms are shown in Fig. 16a. As the atoms are brought together and their wavefunctions overlap, we consider the two combinations $\psi_A \pm \psi_B$. Each combination shares the electrons with the two protons, but an electron in the state $\psi_A + \psi_B$ will have a somewhat lower energy than in the state $\psi_A - \psi_B$. In $\psi_A + \psi_B$ in (b) the electron spends part of the time in the region midway between the two protons, and in this region it is in the attractive potential of both protons at once, thereby increasing the binding energy. In $\psi_A - \psi_B$ in (c) the probability density vanishes midway between the nuclei; an extra binding does not appear. As two atoms are brought together, two separated energy levels are formed for each level of the isolated atom. For N atoms, N orbitals are formed for each orbital of the isolated atom (Fig. 17).

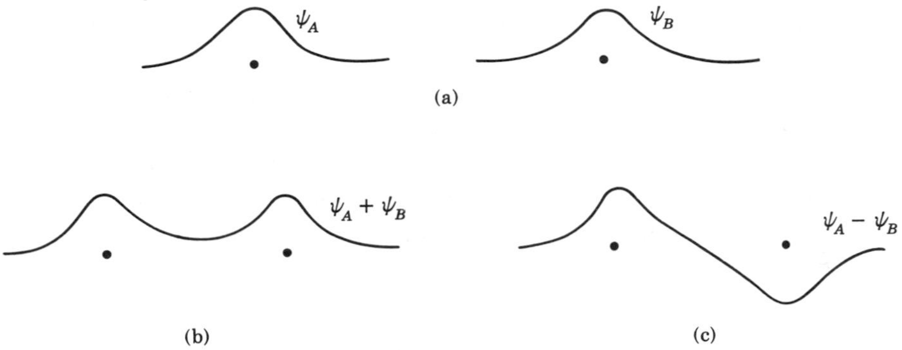

Figure 16 (a) Schematic drawing of wavefunctions of electrons on two hydrogen atoms at large separation. (b) Ground state wavefunction at closer separation. (c) Excited state wavefunction.

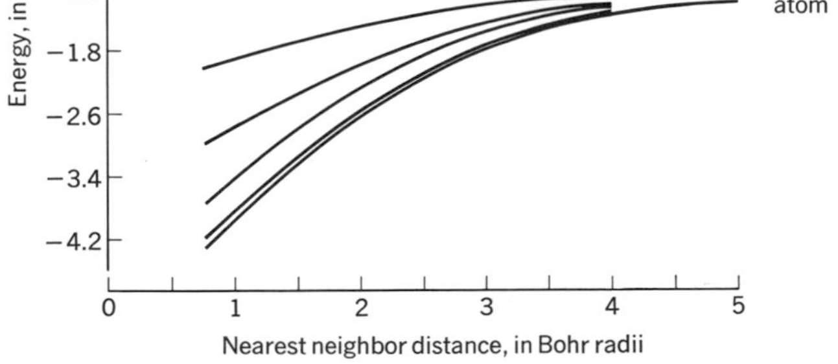

Figure 17 The $1s$ band of a ring of 20 hydrogen atoms; the one-electron energy calculated in the tight-binding approximation with the nearest-neighbor overlap integral of Eq. (9).

As free atoms are brought together, the coulomb interaction between the atom cores and the electron splits the energy levels, spreading them into bands. Each state of given quantum number of the free atom is spread in the crystal into a band of energies. The width of the band is proportional to the strength of the overlap interaction between neighboring atoms. There will be bands formed from p, d, . . . states ($l = 1,2, \ldots$) of the free atoms. States degenerate in the free atom will form different bands. Each will not have the same energy as any other band over any substantial range of the wavevector. Bands may coincide in energy at certain values of $\mathbf{k}$ in the Brillouin zone.

The approximation that starts out from the wavefunctions of the free atoms is known as the tight binding approximation or the LCAO (linear combination of atomic orbitals) approximation. The approximation is quite good for the inner electrons of atoms, but it is not often a good description of the conduction electrons themselves. It is used to describe approximately the d bands of the transition metals and the valence bands of diamondlike and inert gas crystals.

Suppose that the ground state of an electron moving in the potential $U(\mathbf{r})$ of an isolated atom is $\phi(\mathbf{r})$ and that the energy is E_0; suppose ϕ is an s state. The treatment of bands arising from degenerate ($p,d, \ldots$) atomic levels is more complicated. If the influence of one atom on another is small, we obtain an approximate wavefunction for one electron in the whole crystal by taking

$$\psi_{\mathbf{k}}(\mathbf{r}) = \sum_j C_{\mathbf{k}j}\phi(\mathbf{r} - \mathbf{r}_j) \ ,$$

where the sum is over all lattice points. We assume the primitive basis contains one atom. This function is of the Bloch form if $C_{\mathbf{k}j} = N^{-\frac{1}{2}} e^{i\mathbf{k}\cdot\mathbf{r}_j}$, which gives, for a crystal of N atoms,

$$\psi_{\mathbf{k}}(\mathbf{r}) = N^{-\frac{1}{2}} \sum_j e^{i\mathbf{k}\cdot\mathbf{r}_j} \phi(\mathbf{r} - \mathbf{r}_j) \ . \tag{5}$$

We prove (5) is of the Bloch form. Consider a translation $\mathbf{T}$ connecting two lattice points:

$$\begin{aligned}\psi_{\mathbf{k}}(\mathbf{r} + \mathbf{T}) &= N^{-\frac{1}{2}} \sum_j e^{i\mathbf{k}\cdot\mathbf{r}_j} \phi(\mathbf{r} + \mathbf{T} - \mathbf{r}_j) \\ &= e^{i\mathbf{k}\cdot\mathbf{T}} N^{-\frac{1}{2}} \sum_j e^{i\mathbf{k}\cdot(\mathbf{r}_j - \mathbf{T})} \phi[\mathbf{r} - (\mathbf{r}_j - \mathbf{T})] = e^{i\mathbf{k}\cdot\mathbf{T}} \psi_{\mathbf{k}}(\mathbf{r}) \ ,\end{aligned} \tag{6}$$

exactly the Bloch requirement.

We find the first-order energy by calculating the diagonal matrix elements of the hamiltonian of the crystal:

$$\langle \mathbf{k}|H|\mathbf{k}\rangle = N^{-1} \sum_j \sum_m e^{i\mathbf{k}\cdot(\mathbf{r}_j - \mathbf{r}_m)} \langle \varphi_m|H|\varphi_j\rangle \ , \tag{7}$$

where $\varphi_m \equiv \varphi(\mathbf{r}-\mathbf{r}_m)$. Writing $\boldsymbol{\rho}_m = \mathbf{r}_m - \mathbf{r}_j$,

$$\langle \mathbf{k}|H|\mathbf{k}\rangle = \sum_m e^{-i\mathbf{k}\cdot\boldsymbol{\rho}_m} \int dV\, \varphi^*(\mathbf{r}-\boldsymbol{\rho}_m)H\varphi(\mathbf{r}) \ . \tag{8}$$

We now neglect all integrals in (8) except those on the same atom and those between nearest neighbors connected by $\boldsymbol{\rho}$. We write

$$\int dV\, \varphi^*(\mathbf{r})H\varphi(\mathbf{r}) = -\alpha \ ; \qquad \int dV\, \varphi^*(\mathbf{r}-\boldsymbol{\rho})H\varphi(\mathbf{r}) = -\gamma \ ; \tag{9}$$

and we have the first-order energy, provided $\langle \mathbf{k}|\mathbf{k}\rangle = 1$:

$$\langle \mathbf{k}|H|\mathbf{k}\rangle = -\alpha - \gamma \sum_m e^{-i\mathbf{k}\cdot\boldsymbol{\rho}_m} = \epsilon_{\mathbf{k}} \ . \tag{10}$$

The dependence of the overlap energy γ on the interatomic separation ρ can be evaluated explicitly for two hydrogen atoms in $1s$ states. In rydberg energy units, $\mathrm{Ry} = me^4/2\hbar^2$, we have

$$\gamma(\mathrm{Ry}) = 2(1+\rho/a_0)\exp(-\rho/a_0) \ , \tag{11}$$

where $a_0 = \hbar^2/me^2$. The overlap energy decreases exponentially with the separation.

For a simple cubic structure the nearest-neighbor atoms are at

$$\boldsymbol{\rho}_m = (\pm a,0,0) \ ; \qquad (0,\pm a,0) \ ; \qquad (0,0,\pm a) \ , \tag{12}$$

so that (10) becomes

$$\epsilon_{\mathbf{k}} = -\alpha - 2\gamma(\cos k_x a + \cos k_y a + \cos k_z a) \ . \tag{13}$$

Thus the energies are confined to a band of width 12γ. The weaker the overlap, the narrower is the energy band. A constant energy surface is shown in Fig. 15. For $ka \ll 1$, $\epsilon_k \simeq -\alpha - 6\gamma + \gamma k^2a^2$. The effective mass is $m^* = \hbar^2/2\gamma a^2$. When the overlap integral γ is small, the band is narrow and the effective mass is high.

We considered one orbital of each free atom and obtained one band $\epsilon_{\mathbf{k}}$. The number of orbitals in the band that corresponds to a nondegenerate atomic level is $2N$, for N atoms. We see this directly: values of $\mathbf{k}$ within the first Brillouin zone define independent wavefunctions. The simple cubic zone has $-\pi/a < k_x < \pi/a$, etc. The zone volume is $8\pi^3/a^3$. The number of orbitals (counting both spin orientations) per unit volume of $\mathbf{k}$ space is $V/4\pi^3$, so that the number of orbitals is $2V/a^3$. Here V is the volume of the crystal, and $1/a^3$ is the number of atoms per unit volume. Thus there are $2N$ orbitals.

For the bcc structure with eight nearest neighbors,

$$\epsilon_{\mathbf{k}} = -\alpha - 8\gamma \cos \tfrac{1}{2}k_x a \cos \tfrac{1}{2}k_y a \cos \tfrac{1}{2}k_z a \ . \tag{14}$$

For the fcc structure with 12 nearest neighbors,

$$\epsilon_{\mathbf{k}} = -\alpha - 4\gamma(\cos \tfrac{1}{2}k_y a \cos \tfrac{1}{2}k_z a + \cos \tfrac{1}{2}k_z a \cos \tfrac{1}{2}k_x a + \cos \tfrac{1}{2}k_x a \cos \tfrac{1}{2}k_y a) \ . \tag{15}$$

A constant energy surface is shown in Fig. 18.

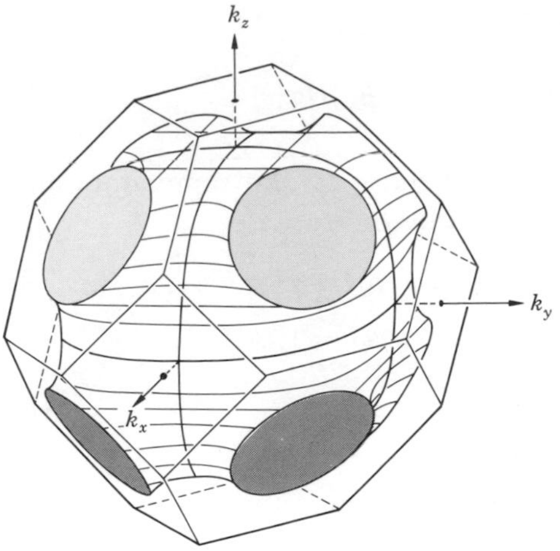

Figure 18 A constant energy surface of an fcc crystal structure, in the nearest-neighbor tight-binding approximation. The surface shown has $\epsilon = -\alpha + 2|\gamma|$.

Wigner-Seitz Method

Wigner and Seitz showed that at least for the alkali metals there is no inconsistency between the electron wavefunctions in free atoms and the nearly free electron model of the band structure of a crystal. Over most of a band the energy may depend on the wavevector nearly as for a free electron, but the wavefunction, unlike a plane wave, will pile up charge on the positive ion cores as in the atom.

A Bloch function satisfies the wave equation

$$\left(\frac{1}{2m}\,\mathbf{p}^2 + U(\mathbf{r})\right) e^{i\mathbf{k}\cdot\mathbf{r}}u_{\mathbf{k}}(\mathbf{r}) = \epsilon_{\mathbf{k}}\, e^{i\mathbf{k}\cdot\mathbf{r}}u_{\mathbf{k}}(\mathbf{r}) \ . \tag{16}$$

With $\mathbf{p} \equiv -i\hbar$ grad, we have

$$\mathbf{p}\ e^{i\mathbf{k}\cdot\mathbf{r}}u_{\mathbf{k}}(\mathbf{r}) = \hbar\mathbf{k}\ e^{i\mathbf{k}\cdot\mathbf{r}}u_{\mathbf{k}}(\mathbf{r}) + e^{i\mathbf{k}\cdot\mathbf{r}}\mathbf{p}u_{\mathbf{k}}(\mathbf{r}) \ ;$$

$$\mathbf{p}^2\ e^{i\mathbf{k}\cdot\mathbf{r}}u_{\mathbf{k}}(\mathbf{r}) = (\hbar k)^2\ e^{i\mathbf{k}\cdot\mathbf{r}}u_{\mathbf{k}}(\mathbf{r}) + e^{i\mathbf{k}\cdot\mathbf{r}}\,(2\hbar\mathbf{k}\cdot\mathbf{p})\,u_{\mathbf{k}}(\mathbf{r}) + e^{i\mathbf{k}\cdot\mathbf{r}}\mathbf{p}^2u_{\mathbf{k}}(\mathbf{r}) \ ;$$

thus (16) may be written as an equation for $u_{\mathbf{k}}$:

$$\left(\frac{1}{2m}\,(\mathbf{p} + \hbar\mathbf{k})^2 + U(\mathbf{r})\right) u_{\mathbf{k}}(\mathbf{r}) = \epsilon_{\mathbf{k}}u_{\mathbf{k}}(\mathbf{r}) \ . \tag{17}$$

At $\mathbf{k} = 0$ we have $\psi_0 = u_0(\mathbf{r})$, where $u_0(\mathbf{r})$ has the periodicity of the lattice, sees the ion cores, and near them will look like the wavefunction of the free atom.

It is much easier to find a solution at $\mathbf{k} = 0$ than for a general $\mathbf{k}$, because at $\mathbf{k} = 0$ a nondegenerate solution will have the full symmetry of the crystal. We can then use $u_0(\mathbf{r})$ to construct the function

$$\psi_{\mathbf{k}} = e^{i\mathbf{k}\cdot\mathbf{r}}u_0(r) \ . \tag{18}$$

This is of the Bloch form, but u_0 is not an exact solution of (17); it is a solution only if we drop the term[2] in $\mathbf{k}\cdot\mathbf{p}$. Because it takes account of the ion

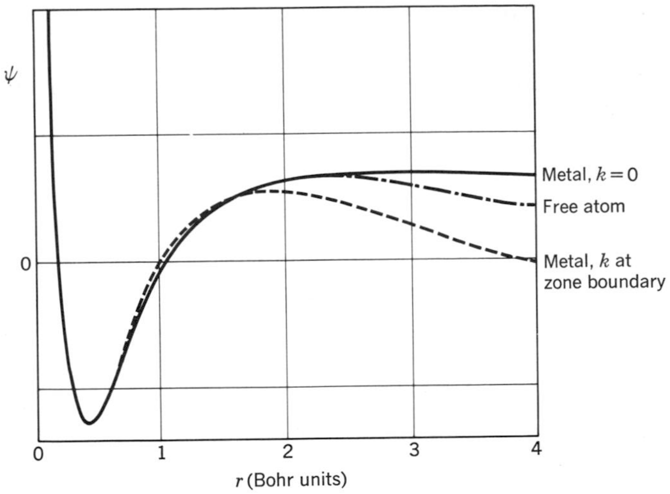

Figure 19 Radial wavefunctions for the 3s orbital of free sodium atom and for 3s conduction band in sodium metal. The wavefunctions are found by integrating the Schrödinger equation for an electron in the potential well of an Na^+ ion core. For the free atom the wavefunction is integrated subject to the usual Schrödinger boundary condition $\psi(r) \rightarrow 0$ as $r \rightarrow \infty$; the energy eigenvalue is −5.15 eV. The wavefunction for wavevector $k = 0$ in the metal is subject to the Wigner-Seitz boundary condition that $d\psi/dr = 0$ when r is midway between neighboring atoms; the energy of this orbital is −8.2 eV, considerably lower than for the free atom. The orbitals at the zone boundary are not filled in sodium; their energy is +2.7 eV. [After E. Wigner and F. Seitz, Phys. Rev. **34,** 804 (1933).]

core potential the function is a much better approximation than a plane wave to the correct wavefunction. The energy of the approximate solution depends on **k** as $(\hbar k)^2/2m$, exactly as for the plane wave, even though the modulation represented by $u_0(\mathbf{r})$ may be very strong. Because u_0 is a solution of

$$\left(\frac{1}{2m}\,\mathbf{p}^2 + U(\mathbf{r})\right) u_0(\mathbf{r}) = \epsilon_0 u_0(\mathbf{r}) \ , \qquad (19)$$

the function (18) has the energy expectation value $\epsilon_0 + (\hbar^2 k^2/2m)$. The function $u_0(\mathbf{r})$ often will give us a good picture of the charge distribution within a cell.

Wigner and Seitz developed a simple and fairly accurate method of calculating $u_0(\mathbf{r})$. Figure 19 shows the Wigner-Seitz wavefunction for $\mathbf{k} = 0$ in the 3s conduction band of metallic sodium. The function is practically constant over 0.9 of the atomic volume. To the extent that the solutions for higher **k** may be approximated by $e^{i\mathbf{k}\cdot\mathbf{r}}u_0(\mathbf{r})$, the wavefunctions in the conduction band will be similar to plane waves over most of the atomic volume, but increase markedly and oscillate within the ion core.

[2]Often this term is taken into account by treating it as a perturbation, giving $\mathbf{k}\cdot\mathbf{p}$ perturbation theory, especially useful in finding the effective mass at a band edge.

Cohesive Energy. The stability of the simple metals with respect to free atoms is caused by the lowering of the energy of the Bloch orbital with $\mathbf{k} = 0$ in the crystal compared to the ground valence orbital of the free atom. The effect is illustrated in Fig. 19 for sodium and in Fig. 20 for a linear periodic potential of attractive square wells. The ground orbital energy is much lower at the actual spacing in the metal than for isolated atoms.

A decrease in ground orbital energy tends to increase the binding. Some of the increase is offset by the Fermi energy contribution to the kinetic energy of the valence electrons, Fig. 21. The decrease in ground orbital energy when atoms are assembled in a periodic array is a consequence of the change in the boundary condition on the wavefunction: The Schrödinger boundary condition for the free atom is $\psi(\mathbf{r}) \rightarrow 0$ as $r \rightarrow \infty$. In the periodic crystal the $\mathbf{k} = 0$ wavefunction $u_0(\mathbf{r})$ has the symmetry of the lattice and is symmetric about $\mathbf{r} = 0$. To have this, the normal derivative of ψ must vanish across every plane midway between adjacent atoms. In a spherical approximation to the smallest Wigner-Seitz cell we have the Wigner-Seitz boundary condition

$$(d\psi/dr)_{r_0} = 0 \ , \tag{20}$$

where r_0 is the radius of a sphere equal in volume to a primitive cell of the lattice. In sodium, $r_0 = 3.95$ Bohr units, or 2.08 Å; the half-distance to a nearest neighbor is 1.86 Å. The spherical approximation is not bad for fcc and bcc structures. The boundary condition allows the ground orbital wavefunction to have much less curvature than the free atom boundary condition. Much less curvature means much less kinetic energy.

In sodium the other filled orbitals in the conduction band can be represented in a rough approximation by wavefunctions of the form (18), with

$$\psi_{\mathbf{k}} = e^{i\mathbf{k}\cdot\mathbf{r}} u_0(\mathbf{r}) \ ; \qquad \epsilon_{\mathbf{k}} = \epsilon_0 + \frac{\hbar^2 k^2}{2m} \ .$$

The Fermi energy is 3.1 eV, from Table 6.1. The average kinetic energy per electron is 0.6 of the Fermi energy, or 1.9 eV. Because $\epsilon_0 = -8.2$ eV at $\mathbf{k} = 0$, the average electron energy is $\langle \epsilon_{\mathbf{k}} \rangle = -8.2 + 1.9 = -6.3$ eV, compared with -5.15 eV for the valence electron of the free atom, Fig. 21. We therefore estimate that sodium metal is stable by about 1.1 eV with respect to the free atom. This result agrees well with the experimental value 1.13 eV. We have neglected several corrections whose overall effect in sodium is small.

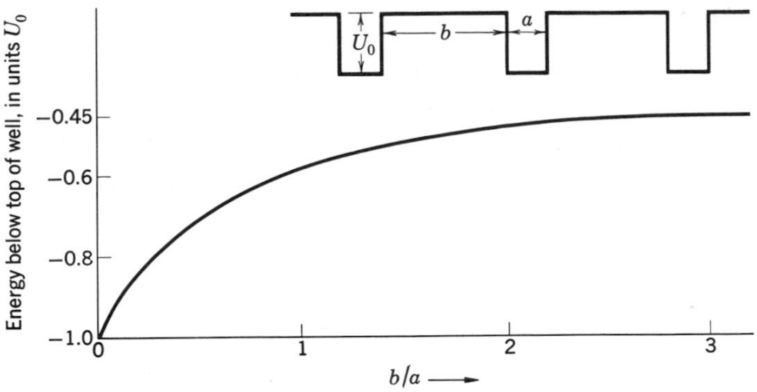

Figure 20 Ground orbital ($k = 0$) energy for an electron in a periodic square well potential of depth $|U_0| = 2\hbar^2/ma^2$. The energy is lowered as the wells come closer together. Here a is held constant and b is varied. Large b/a corresponds to separated atoms. (Courtesy of C. Y. Fong.)

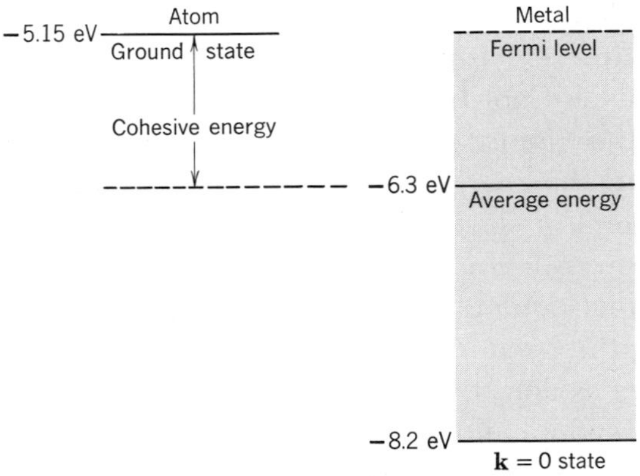

Figure 21 Cohesive energy of sodium metal is the difference between the average energy of an electron in the metal (−6.3 eV) and the ground state energy (−5.15 eV) of the valence 3s electron in the free atom, referred to an Na^+ ion plus free electron at infinite separation.

Pseudopotentials

The orbitals of conduction electrons in metals are simple and smooth in the region between the ion cores, but are complicated by a nodal structure in the region of the cores, as shown for the ground orbital of sodium in Fig. 19. In the outer region the potential energy of a conduction electron is relatively weak: it is the coulomb potential energy of the positive charge of the ions, but reduced by the electrostatic screening by the other conduction electrons (Chap. 10). In the outer region the wavefunctions are somewhat like plane waves, complicated neither by the strong and rapidly varying potential near the nuclei nor by the requirement of orthogonality to the wavefunctions of the core electrons. The nodes in the wavefunction in the core region are caused by orthogonality: the $3s$ conduction band orbital of sodium has two nodes that enable it to be orthogonal to the $1s$ and $2s$ electrons of the ion core.

If the conduction orbitals in the outer region are approximately plane waves, the energy must depend on the wavevector approximately as the relation $\epsilon_{\mathbf{k}} = \hbar^2 k^2/2m$ for free electrons. But how do we handle the core region, where the orbitals are not like plane waves and the potential is strong? What goes on in the core is largely irrelevant to the dependence of ϵ on $\mathbf{k}$, for we can calculate the energy by applying the hamiltonian operator to an orbital at any point in space, and in the outer region this operation will give nearly the free electron kinetic energy.

We can replace the actual potential energy in the core region by an effective potential energy, the pseudopotential, that gives the same wavefunctions outside the core as does the actual potential.[3] It is startling that in the core region the pseudopotential is nearly zero. This conclusion is derived from a large amount of practical experience with pseudopotentials as well as from theoretical arguments referred to as the cancellation theorem. The pseudopotential is not unique: in the empty core model we can even take the unscreened pseudopotential to be zero inside of some radius R_e:

$$U(r) = \begin{cases} 0 \ , & \text{for } r < R_e \ ; \\ -\dfrac{e^2}{r} \ , & \text{for } r > R_e \ . \end{cases} \tag{21}$$

This potential is to be screened; that is, each Fourier component $U(\mathbf{K})$ of $U(r)$ must be divided by the dielectric function $\epsilon(\mathbf{K})$ of the electron gas, derived in Chapter 10. If we use (just as an example) the Thomas-Fermi dielectric function, we obtain the pseudopotential plotted in Fig. 22. The pseudopotential is much weaker than the true potential, but the wavefunctions in the outer regions are nearly identical for both potentials. It can be viewed as a matter of adjusting phase shifts in the pseudowavefunction to match the true wavefunction.

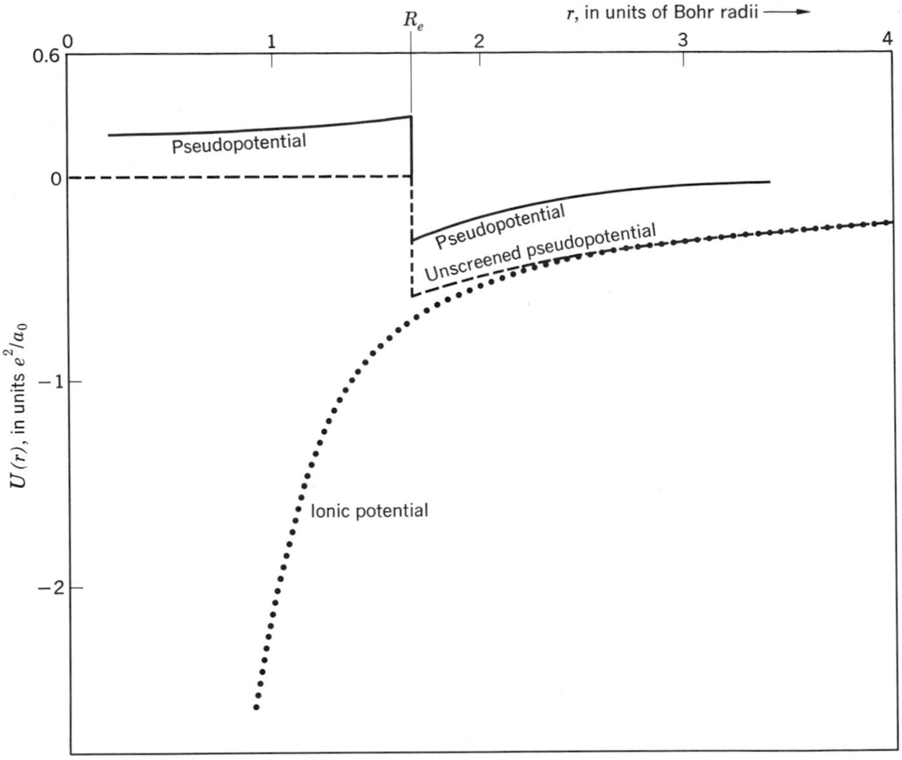

Figure 22 Pseudopotential for metallic sodium, based on the empty core model and screened by the Thomas-Fermi dielectric function. The calculations were made for an empty core radius $R_e = 1.66a_0$, where a_0 is the Bohr radius, and for a screening parameter $k_s a_0 = 0.79$. The dashed curve shows the assumed unscreened potential, as from (21). The dotted curve is the actual potential of the ion core; other values of $U(r)$ are -50.4, -11.6, and -4.6, for $r = 0.15$, 0.4, and 0.7, respectively. Thus the actual potential of the ion (chosen to fit the energy levels of the free atom) is very much larger than the pseudopotential, over 200 times larger at $r = 0.15$.

The calculation of the band structure of a crystal involves only the values $U(\mathbf{G})$ of the Fourier coefficients of the potential at the reciprocal lattice vectors. Often only a few coefficients are required for the determination of the band structure to good accuracy. These coefficients are sometimes calculated from model potentials, and sometimes they are deduced from optical experiments. In the very successful empirical pseudopotential method, the band structure is calculated using a few $U(\mathbf{G})$ obtained from theoretical fits to observations on the optical reflectance and absorption of crystals, as discussed in Chapter 11. Values of $U(\mathbf{G})$ for the elements are tabulated and plotted in the review by Cohen and Heine.

[3]J. C. Phillips and L. Kleinman, Phys. Rev. **116,** 287 (1959); E. Antoncik, J. Phys. Chem. Solids **10,** 314 (1959). The general theory of pseudopotentials is discussed by B. J. Austin, V. Heine, and L. J. Sham, Phys. Rev. **127,** 276 (1962); see also Vol. 24 of *Solid state physics*. The utility of the empty core model has been known for many years; it goes back to H. Hellmann, Acta Physiochimica URSS **1,** 913 (1935) and H. Hellmann and W. Kassatotschkin, J. Chem. Phys. **4,** 324 (1936), who wrote "Since the field of the ion determined in this way runs a rather flat course, it is sufficient in the first approximation to set the valence electron in the lattice equal to a plane wave."

EXPERIMENTAL METHODS IN FERMI SURFACE STUDIES

Powerful experimental methods have been developed for the determination of Fermi surfaces. The methods include magnetoresistance, anomalous skin effect, cyclotron resonance, magneto-acoustic geometric effects, the Schubnikow-de Haas effect, and the de Haas-van Alphen effect. Further information on the momentum distribution is given by positron annihilation, Compton scattering, and the Kohn effect. We propose to study one method rather thoroughly. All the methods are useful, but all need detailed theoretical analysis and are not for beginners. We select the de Haas-van Alphen effect because it exhibits very well the characteristic periodicity in $1/B$ of the properties of a metal in a uniform magnetic field.

Quantization of Orbits in a Magnetic Field

The momentum $\mathbf{p}$ of a particle in a magnetic field is the sum of two parts, the kinetic momentum $\mathbf{p}_{\text{kin}} = m\mathbf{v} = \hbar\mathbf{k}$ and the potential momentum or field momentum $\mathbf{p}_{\text{field}} = q\mathbf{A}/c$, where q is the charge and the vector potential is related to the magnetic field by $\mathbf{B} = \text{curl}\ \mathbf{A}$. The total momentum is

(CGS) $$\mathbf{p} = \mathbf{p}_{\text{kin}} + \mathbf{p}_{\text{field}} = \hbar\mathbf{k} + q\mathbf{A}/c \ . \tag{22}$$

In SI the term c^{-1} is omitted. Following the semiclassical approach of L. Onsager and I. M. Lifshitz, we assume that the orbits in a magnetic field are quantized by the Bohr-Sommerfeld relation

$$\oint \mathbf{p} \cdot d\mathbf{r} = (n + \gamma)2\pi\hbar \ , \tag{23}$$

where n is an integer and γ is a phase correction that for free electrons has the value $\frac{1}{2}$. Then

$$\oint \mathbf{p} \cdot d\mathbf{r} = \oint \hbar\mathbf{k} \cdot d\mathbf{r} + \frac{q}{c}\oint \mathbf{A} \cdot d\mathbf{r} \ . \tag{24}$$

The equation of motion of a particle of charge q in a magnetic field is

$$\hbar\frac{d\mathbf{k}}{dt} = \frac{q}{c}\frac{d\mathbf{r}}{dt} \times \mathbf{B} \ ; \tag{25}$$

this may be integrated with respect to time to give

$$\hbar\mathbf{k} = \frac{q}{c}\mathbf{r} \times \mathbf{B} \ ,$$

apart from an additive constant which does not contribute to the final result. Thus one of the path integrals in (24) is

$$\oint \hbar \mathbf{k} \cdot d\mathbf{r} = \frac{q}{c} \oint \mathbf{r} \times \mathbf{B} \cdot d\mathbf{r} = -\frac{q}{c} \mathbf{B} \cdot \oint \mathbf{r} \times d\mathbf{r} = -\frac{2q}{c} \Phi ,$$

where Φ is the magnetic flux contained within the orbit in real space. We have used the geometrical result that

$$\oint \mathbf{r} \times d\mathbf{r} = 2 \times (\text{area enclosed by the orbit}) .$$

The other path integral in (24) is

$$\frac{q}{c} \oint \mathbf{A} \cdot d\mathbf{r} = \frac{q}{c} \int \operatorname{curl} \mathbf{A} \cdot d\boldsymbol{\sigma} = \frac{q}{c} \int \mathbf{B} \cdot d\boldsymbol{\sigma} = \frac{q}{c} \Phi ,$$

by the Stokes theorem; here $d\boldsymbol{\sigma}$ is the area element in real space. The momentum path integral

$$\oint \mathbf{p} \cdot d\mathbf{r} = -\frac{q}{c} \Phi = (n + \gamma) 2\pi\hbar , \tag{26}$$

so that the orbit of an electron is quantized in such a way that the flux through it is

$$\Phi_n = (n + \gamma)(2\pi\hbar c/e) . \tag{27}$$

The flux unit $2\pi\hbar c/e = 4.14 \times 10^{-7}$ gauss cm^2 or T m^2.

In the de Haas-van Alphen effect we need the area of the orbit in wavevector space. We obtained in (27) the flux through the orbit in real space. By (25) we know that a line element Δr in the plane normal to $\mathbf{B}$ is related to Δk by $\Delta r = (\hbar c/eB)\, \Delta k$, so that the area S_n in $\mathbf{k}$ space is related to the area A_n of the orbit in $\mathbf{r}$ space by

$$A_n = (\hbar c/eB)^2 S_n . \tag{28}$$

It follows that

$$\Phi_n = \left(\frac{\hbar c}{e}\right)^2 \frac{1}{B} S_n = (n + \gamma) \frac{2\pi\hbar c}{e} , \tag{29}$$

from (27), whence the area in $\mathbf{k}$ space of an orbit will satisfy

$$S_n = (n + \gamma) \frac{2\pi e}{\hbar c} B . \tag{30}$$

In Fermi surface experiments we are often interested in the increment ΔB, for which two successive orbits, n and $n + 1$, have the same area in $\mathbf{k}$ space on the Fermi surface. The areas are equal when

$$S \left(\frac{1}{B_{n+1}} - \frac{1}{B_n}\right) = \frac{2\pi e}{\hbar c} , \tag{31}$$

from (30). We have the important result that equal increments of $1/B$ reproduce identical orbits—this periodicity in $1/B$ is a striking feature of magneto-oscillatory effects in metals at low temperatures: resistivity, susceptibility, heat capacity. The population of orbits on or near the Fermi surface oscillates as B is varied, causing a wide variety of effects. From the period of the oscillation, we reconstruct the Fermi surface. It should be said that the result (30) is independent of the gauge of the vector potential used in the expression (22) for momentum; that is, $\mathbf{p}$ is not gauge invariant, but S_n is. Gauge invariance is discussed more in Chapter 12.

De Haas-van Alphen Effect

The de Haas-van Alphen effect is the oscillation of the magnetic moment of a metal as a function of the static magnetic field intensity. The effect can be observed in pure specimens at low temperatures in strong magnetic fields: we do not want the quantization of the electron orbits to be blurred by collisions, and we do not want the population oscillations to be averaged out by thermal population of adjacent orbits. Our analysis is for absolute zero. We neglect the electron spin, for brevity. We start with a two-dimensional model. An elementary explanation of the dHvA effect is offered in Fig. 23.

The area in $\mathbf{k}$ space of an orbit is quantized according to (30). The area between successive orbits, at constant magnetic field, is

$$\Delta S = S_{n+1} - S_n = 2\pi eB/\hbar c \ . \tag{32}$$

The area of $\mathbf{k}$ space occupied by a single orbital is $(2\pi/L)^2$, neglecting spin, for a square specimen of side L. Thus the degeneracy of the state labeled n is

$$D = (2\pi eB/\hbar c)\ (L/2\pi)^2 = \rho B \ , \tag{33}$$

where ρ is a constant of proportionality equal to $eL^2/2\pi\hbar c$.

Where does the Fermi level lie? For a system of N electrons at absolute zero the energy levels are filled up to some quantum number λ, and orbitals at the level $(\lambda + 1)$ will be partly filled to the extent needed to accommodate the electrons. The degeneracy increases with B: the higher the field, the lower the value of λ. The distribution of electrons is shown in Fig. 24. As B is increased there occur values at which λ decreases abruptly by unity. At the critical values B_λ no level is partly occupied, so that

$$\rho B_\lambda \lambda = N \ . \tag{34}$$

This states that the degeneracy times the number of filled levels λ is equal to the number of particles N.

To go further, it is convenient to use the free electron result that the energy of the orbital of quantum number n is $E_n = (n + \frac{1}{2})\hbar\omega_c$, where ω_c is the cyclotron frequency eB/m^*c. The result for E_n follows from the analogy between the cyclotron problem and the simple harmonic oscillator.

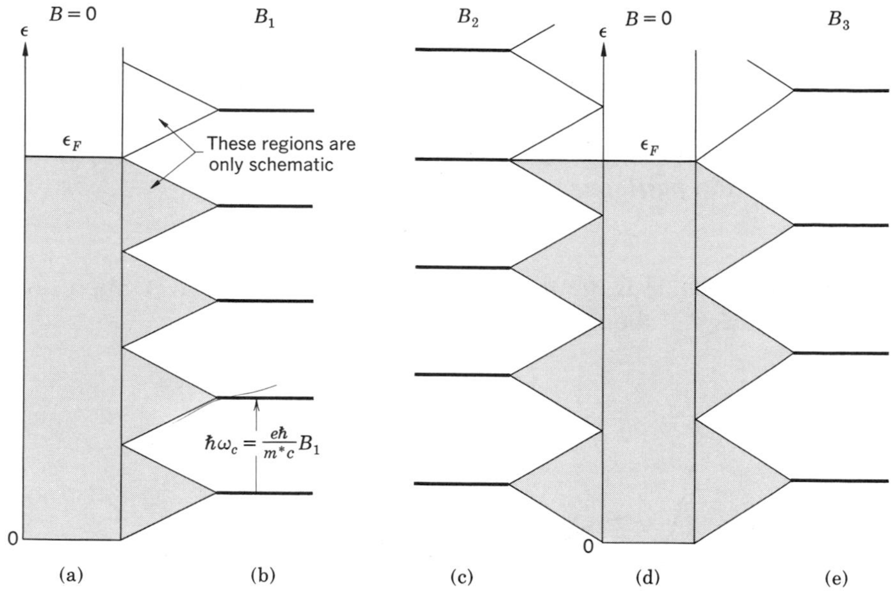

Figure 23 Explanation of the de Haas-van Alphen effect for a free electron gas in two dimensions in a magnetic field. The filled orbitals of the Fermi sea in the absence of a magnetic field are shaded in *a* and *d*. The energy levels in a magnetic field are shown in *b*, *c*, and *e*. In *b* the field has a value B_1 such that the total energy of the electrons is the same as in the absence of a magnetic field: as many electrons have their energy raised as lowered by the orbital quantization in the magnetic field B_1. When we increase the field to B_2 the total electron energy is increased, because the uppermost electrons have their energy raised. In *e* for field B_3 the energy is again equal to that for the field $B = 0$. The total energy is a minimum at points such as B_1, B_3, B_5, . . . , and a maximum near points such as B_2, B_4, ...

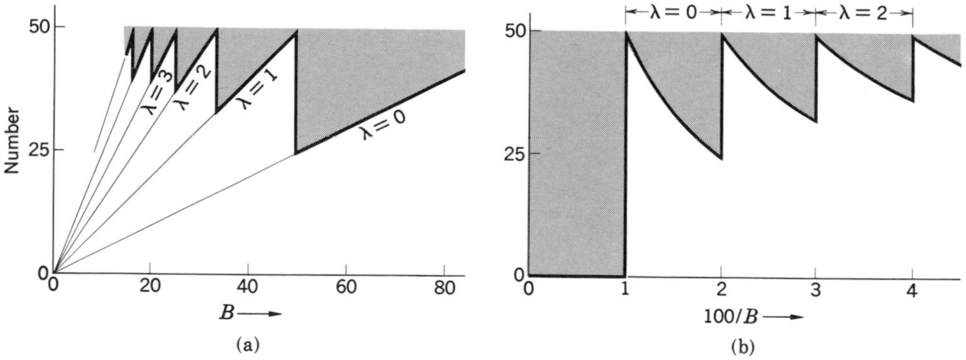

Figure 24 (a) The heavy line gives the number of particles in levels which are completely occupied in a magnetic field B, for a two-dimensional system with $N = 50$ and $\rho = 0.50$. The shaded area gives the number of particles in levels partially occupied. The value of λ denotes the quantum number of the highest level which is completely filled. Thus at $B = 40$ we have $\lambda = 1$: the levels $n = 0$ and $n = 1$ are filled and there are 10 particles in the level $n = 2$. At $B = 50$ the level $n = 2$ is empty. (b) The periodicity in $1/B$ is evident when the same points are plotted against $1/B$.

The energy of the electrons in *fully occupied* levels is

$$\sum_{l=0}^{\lambda} D\hbar\omega_c(l+\tfrac{1}{2}) = \tfrac{1}{2}D\hbar\omega_c\,(\lambda+1)^2 \ ,$$

where D is the number of electrons in each level. The energy of the electrons in the *partly occupied* level $\lambda + 1$ is

$$\hbar\omega_c(\lambda+\tfrac{3}{2})\,[N - D(\lambda+1)] \ ,$$

where $D(\lambda + 1)$ is the number of electrons in the lower filled levels. The total energy of the system is the sum, Fig. 25.

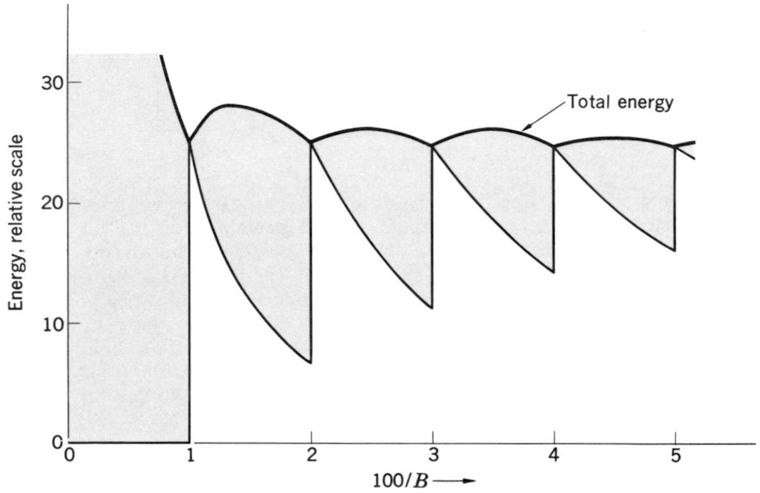

Figure 25 The upper curve is the total electronic energy versus $1/B$. The oscillations in the energy U may be detected by measurement of the magnetic moment, given by $-\partial U/\partial B$: The thermal and transport properties of the metal also oscillate as successive orbital levels cut through the Fermi level when the field is increased. The shaded region in the figure gives the contribution to the energy from levels that are only partly filled. The parameters for the figure are the same as for Fig. 24, and we have taken the units of B such that $B = \hbar\omega_c$.

The magnetic moment μ of a system at absolute zero is given by $\mu = -\partial U/\partial B$. The moment here is an oscillatory function of $1/B$, Fig. 26. This oscillatory magnetic moment of the Fermi gas at low temperatures is the de Haas-van Alphen effect. From (31) we see that the oscillations occur at equal intervals of $1/B$ such that

$$\Delta\left(\frac{1}{B}\right) = \frac{2\pi e}{\hbar c S} , \tag{34a}$$

where S is the extremal area (see below) of the Fermi surface normal to the direction of **B**. From measurements of $\Delta(1/B)$, we deduce the corresponding extremal areas S; thereby much can be inferred about the shape and size of the Fermi surface.[4]

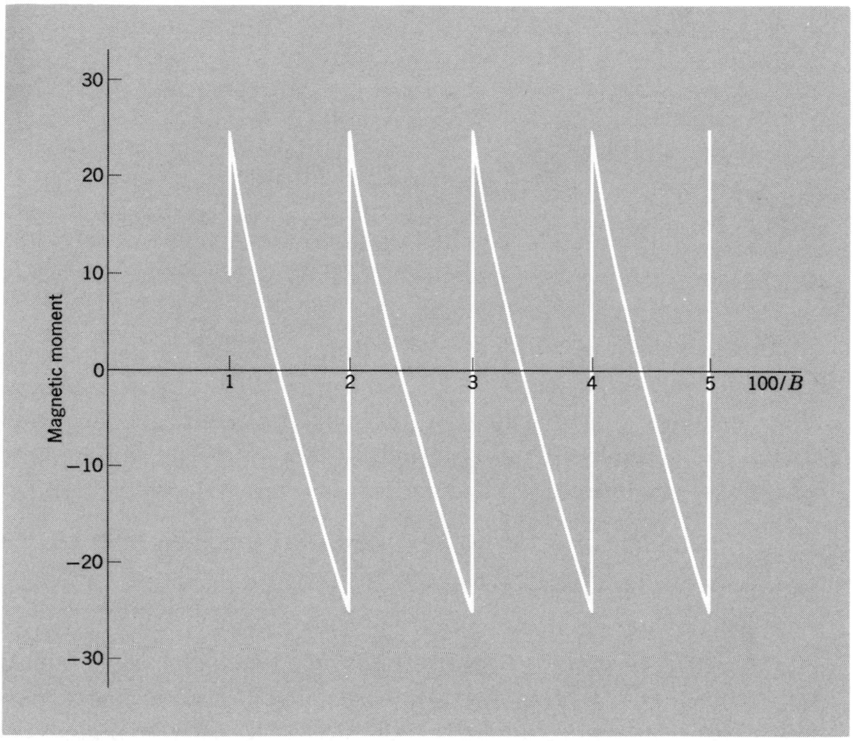

Figure 26 At absolute zero the magnetic moment is given by $-\partial U/\partial B$. The energy plotted in Fig. 25 leads to the magnetic moment shown here, an oscillatory function of $1/B$. In impure specimens the oscillations are smudged out in part because the energy levels are no longer sharply defined.

[4]For experimental methods, see D. Shoenberg, Proc. Phys. Soc. (London) **79,** 1–9 (1962).

Extremal orbits. One point in the interpretation of the dHvA effect is subtle. For a Fermi surface of general shape the sections at different values of k_B will have different periods. The response will be the sum of contributions from all sections or all orbits. *But the dominant response of the system comes from orbits whose periods are stationary with respect to small changes in* k_B. Such orbits are called **extremal orbits.** Thus in Fig. 27 the section AA' dominates the observed cyclotron period. The argument can be put in mathematical form, but we do not give the proof here.[5] Essentially it is a question of phase cancellation: the contributions of different nonextremal orbits cancel, but near extrema the phase varies only slowly and there is a net signal from these orbits. Both theory and experiment agree that sharp resonances are obtained even from complicated Fermi surfaces because the experiment selects the extremal orbits.

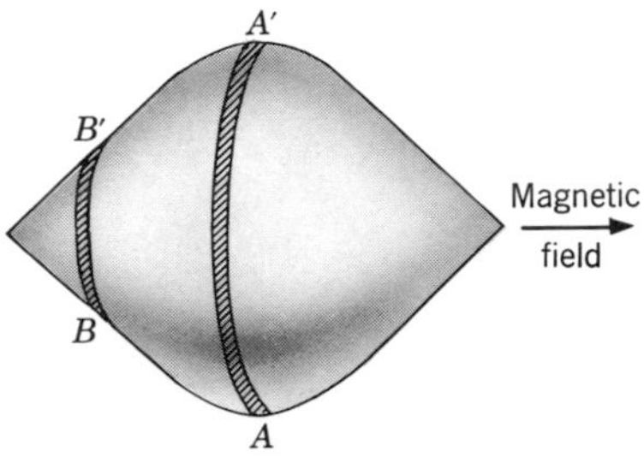

Figure 27 The orbits in the section AA' are extremal orbits: the cyclotron period is roughly constant over a reasonable section of the Fermi surface. Other orbits such as BB' vary in period over the section.

Fermi Surface of Copper. The Fermi surface of copper is distinctly nonspherical: eight necks make contact with the hexagonal faces of the first Brillouin zone of the fcc lattice. The electron concentration in a monovalent metal with an fcc structure is $n = 4/a^3$: there are four electrons in a cube of volume a^3. The radius of a free electron Fermi sphere is

$$k_F = (3\pi^2 n)^{1/3} = (12\pi^2/a^3)^{1/3} \cong (4.90/a) \ , \tag{35}$$

and the diameter is $9.80/a$. The shortest distance across the Brillouin zone (the distance between hexagonal faces) is $(2\pi/a)\,(3)^{\frac{1}{2}} = 10.90/a$, somewhat larger. The free electron sphere does not touch the zone boundary, but we know from Chapter 7 that the presence of a zone boundary tends to lower the band energy near the boundary. Thus it is plausible that the Fermi surface necks out to meet the hexagonal faces of the zone, Fig. 28. The square faces of the zone are more distant, with separation $12.57/a$.

[5]*QTS*, p. 223; Ziman, p. 322.

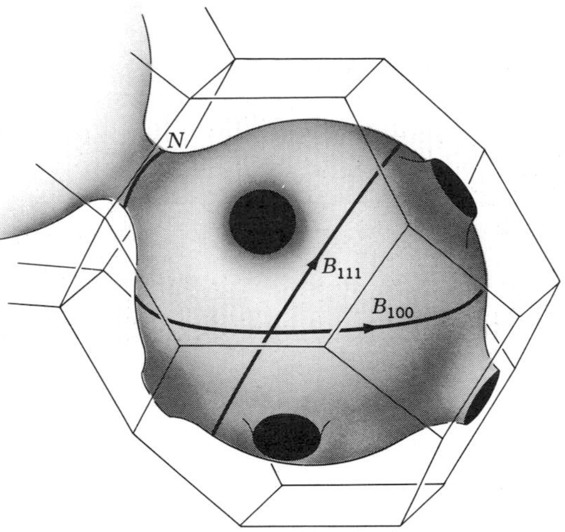

Figure 28 Fermi surface of copper, after Pippard. The Brillouin zone of the fcc structure is the truncated octahedron derived in Chapter 2. The Fermi surface makes contact with the boundary at the center of the hexagonal faces of the zone, in the [111] directions in **k** space. Two "belly" extremal orbits are shown, denoted by B; the extremal "neck" orbit is denoted by N.

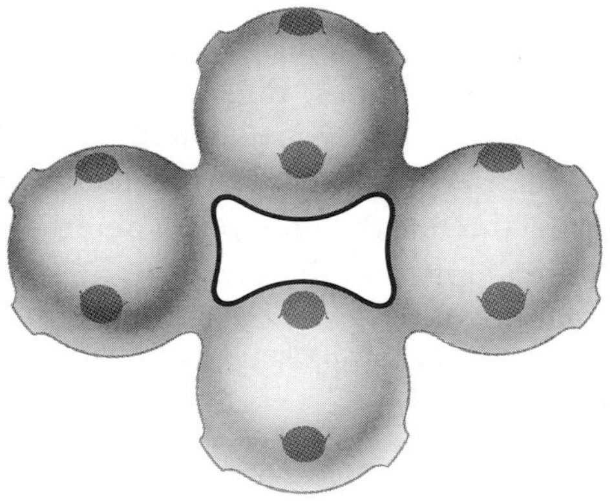

Figure 29 Dog's bone orbit of an electron on the Fermi surface of copper or gold in a magnetic field. This is a holelike orbit because the energy increases toward the interior of the orbit.

EXAMPLE: ***Fermi Surface of Gold.*** In gold for quite a wide range of field directions Shoenberg finds the magnetic moment has a period of 2×10^{-9} gauss^{-1}. This period corresponds to an extremal orbit of area

$$S = \frac{2\pi e/\hbar c}{\Delta(1/B)} \cong \frac{9.55 \times 10^7}{2 \times 10^{-9}} \cong 4.8 \times 10^{16} \text{ cm}^{-2} \ .$$

From Table 6.1, we have $k_F = 1.2 \times 10^8$ cm^{-1} for a free electron Fermi sphere for gold, or an extremal area of 4.5×10^{16} cm^{-2}, in general agreement with the experimental value. The actual periods reported by Shoenberg are 2.05×10^{-9} gauss^{-1} for the orbit B_{111} of Fig. 28 and 1.95×10^{-9} gauss^{-1} for B_{100}. In the [111] direction in Au a large period of 6×10^{-8} gauss^{-1} is also found; the corresponding orbital area is 1.6×10^{15} cm^{-2}. This is the "neck" orbit N. Another extremal orbit, the "dog's bone," is shown in Fig. 29; its area in Au is about 0.4 of the belly area. Experimental results are shown in Fig. 30. To do the example in SI, drop c from the relation for S and use as the period 2×10^{-5} tesla^{-1}.

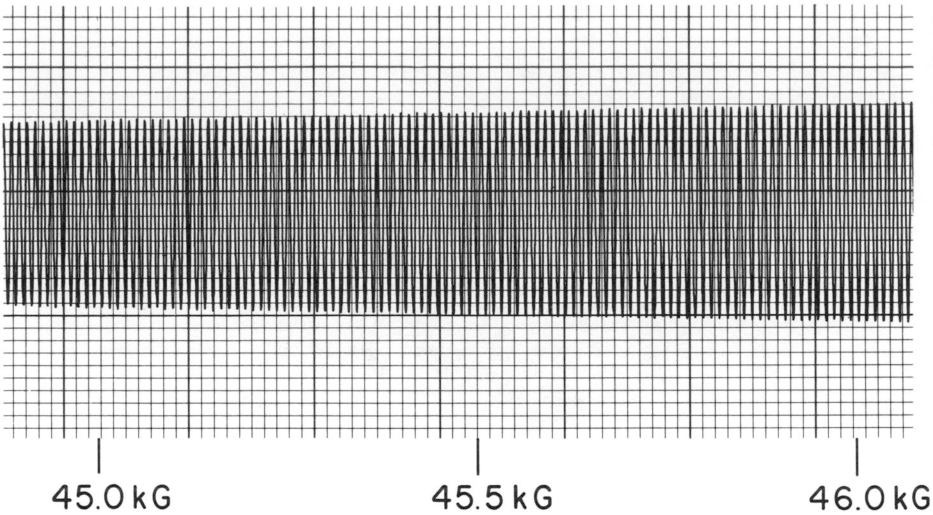

Figure 30 De Haas-van Alphen effect in gold with **B** ∥ [110]. The oscillation is from the dog's bone orbit of Fig. 30. The signal is related to the second derivative of the magnetic moment with respect to field. The results were obtained by a field modulation technique in a high-homogeneity superconducting solenoid at about 1.2 K. (Courtesy of I. M. Templeton.)

The free electron Fermi sphere of aluminum fills the first zone entirely and has a large overlap into the second and third zones, Fig. 1. The third zone Fermi surface is quite complicated, even though it is just made up of certain pieces of the surface of the free electron sphere. The free electron model also gives small pockets of holes in the fourth zone, but when the lattice potential is taken into account these empty out, the electrons being added to the third zone. The general features of the predicted Fermi surface of aluminum are quite well verified by experiment. Figure 31 shows part of the free electron Fermi surface of magnesium.

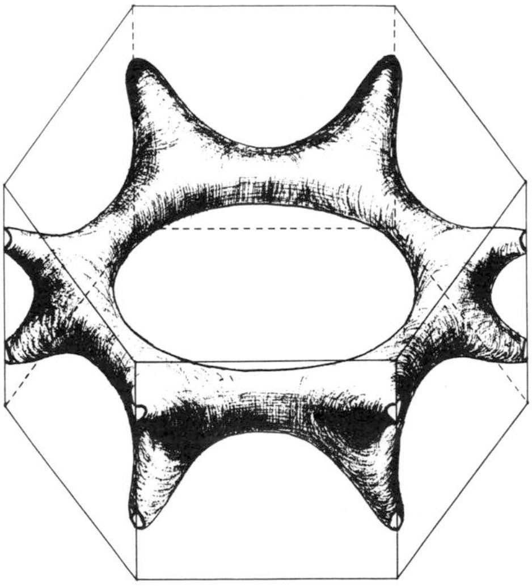

Figure 31 Multiply connected hole surface of magnesium in bands 1 and 2, according to L. M. Falicov. (Drawing by Marta Puebla.)

Magnetic Breakdown. Electrons in sufficiently large magnetic fields may move in free particle orbits, the circular cyclotron orbits of Fig. 32a. Here the magnetic forces are dominant, and the lattice potential is a slight perturbation. In this limit the classification of the orbitals into bands may have little importance. However, we know that at low magnetic fields the motion is described by (8.7) with the band structure $\epsilon_{\mathbf{k}}$ that obtains in the absence of a magnetic field. The eventual breakdown of this description as the magnetic field is increased is called magnetic breakdown.[6] The passage to strong magnetic fields may drastically change the connectivity of the orbits, as in the figure. The onset of magnetic breakdown will be revealed by physical properties such as magnetoresistance that depend sensitively on the connectivity.

The condition for magnetic breakdown is that $\hbar\omega_c\epsilon_F > E_g^2$, approximately. Here ϵ_F is the free electron Fermi energy and E_g is the energy gap. This condition is much milder than the naïve condition that the magnetic splitting $\hbar\omega_c$ exceed the gap, especially in metals with small gaps. Small gaps may be found in hcp metals where the gap across the hexagonal face of the zone would be zero except for a small splitting introduced by the spin-orbit interaction. In Mg the splitting is of the order of 10^{-3} eV; for this gap and $\epsilon_F \sim 10$ eV, the breakdown condition is $\hbar\omega_c > 10^{-5}$ eV, or $B > 1000$ G.

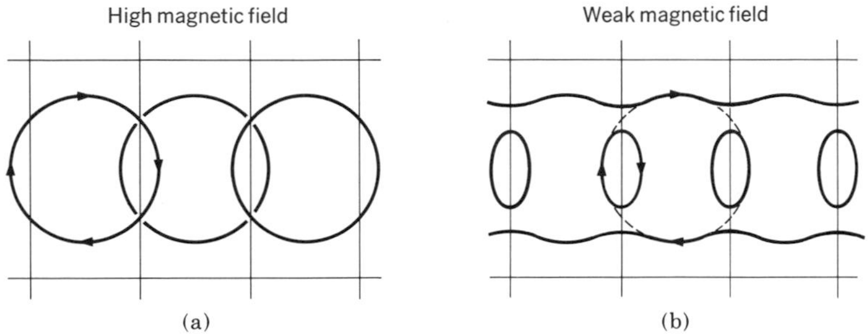

Figure 32 Breakdown of band structure by a strong magnetic field. The free electron orbits (a) in a strong field change connectivity in a weak field (b) to become open orbits in the first band and electron orbits in the second band. Both bands are mapped together.

[6]M. H. Cohen and L. M. Falicov, Phys. Rev. Letters **7**, 231 (1961).

ELECTRICAL RESISTIVITY

The relation between the periodic zone scheme and the reduced zone scheme is well illustrated by the umklapp scattering of electrons. Umklapp scattering accounts for most of the electrical resistivity of metals at low temperatures. It is seen especially well in potassium. Consider a section perpendicular to [100] through two adjacent Brillouin zones in bcc potassium, with the equivalent Fermi spheres inscribed within each (Fig. 33). The lower half of the figure shows the normal electron-phonon collision $\mathbf{k}' = \mathbf{k} + \mathbf{q}$, while the upper half shows a possible scattering process $\mathbf{k}' = \mathbf{k} + \mathbf{q} + \mathbf{G}$ involving the same phonon and terminating outside the first Brillouin zone, at the point A. This point is exactly equivalent to the point A' inside the original zone, where AA' is a reciprocal lattice vector $\mathbf{G}$. This scattering is an umklapp process, in analogy to phonons. Such collisions are strong scatterers because the scattering angle can be close to π, and a single collision can practically restore the electron to its ground orbital. When the Fermi surface does not intersect the zone boundary, there is some minimum phonon wavevector q_0 for umklapp scattering. At low enough temperatures the number of phonons available for umklapp scattering falls as $\exp(-\theta_U/T)$, where θ_U is a characteristic temperature calculable from the geometry of the Fermi surface inside the Brillouin zone. For a spherical Fermi surface with one electron orbital per atom inside the bcc Brillouin zone, one can show by geometry that $q_0 = 0.267\ k_F$. The experimental data for potassium have the expected exponential form with $\theta_U = 23$ K compared with the Debye $\theta = 91$ K. At the very lowest temperatures (below about 2 K in potassium) the number of umklapp processes is negligible and the lattice resistivity is then caused only by small angle scattering, the normal scattering. Bloch obtained an analytic result for the normal scattering, with $\rho_L \propto T^5/\theta^6$ at very low temperatures. This is a classic limiting result. These normal processes contribute to the resistivity in all metals, but they have not yet been clearly isolated for any metal because of the large competing effects of imperfection scattering and of umklapp scattering.

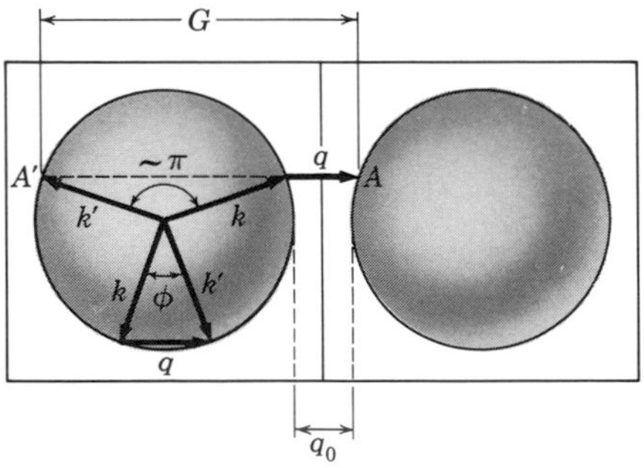

Figure 33 Two Fermi spheres in adjacent zones: a construction to show the role of umklapp processes in electrical resistivity.

SUMMARY

1. A Fermi surface is the surface in $\mathbf{k}$ space of constant energy equal to ϵ_F. The Fermi surface separates filled states from empty states at absolute zero. The form of the Fermi surface is usually exhibited best in the reduced zone scheme, but the connectivity of the surfaces is clearest in the periodic zone scheme.

2. An energy band is a single branch of the $\epsilon_{\mathbf{k}}$ versus $\mathbf{k}$ surface.

3. The cohesion of simple metals is accounted for by the lowering of energy of the $\mathbf{k} = 0$ conduction band orbital when the boundary conditions on the wavefunction are changed from Schrödinger to Wigner-Seitz.

4. The periodicity in the de Haas-van Alphen effect measures the extremal cross-section area S in $\mathbf{k}$ space of the Fermi surface, the cross section being taken perpendicular to $\mathbf{B}$:

$$\Delta\left(\frac{1}{B}\right) = \frac{2\pi e}{\hbar c S} .$$

PROBLEMS

1. ***Brillouin zones of rectangular lattice.*** Make a plot of the first two Brillouin zones of a primitive rectangular two-dimensional lattice with axes a, $b = 3a$.

2. ***Brillouin zone, rectangular lattice.*** A two-dimensional metal has one atom of valency one in a simple rectangular primitive cell $a = 2$ Å; $b = 4$ Å. (a) Draw the first Brillouin zone. Give its dimensions, in cm^{-1}. (b) Calculate the radius of the free electron Fermi sphere, in cm^{-1}. (c) Draw this sphere to scale on a drawing of the first Brillouin zone. Make another sketch to show the first few periods of the free electron band in the periodic zone scheme, for both the first and second energy bands. Assume there is a small energy gap at the zone boundary.

3. ***Hexagonal close-packed structure.*** Consider the first Brillouin zone of a crystal with a simple hexagonal lattice in three dimensions with lattice constants a and c. Let $\mathbf{G}_c$ denote the shortest reciprocal lattice vector parallel to the $\mathbf{c}$ axis of the crystal lattice. (a) Show that for a hexagonal close-packed crystal structure the Fourier component $U(\mathbf{G}_c)$ of the crystal potential $U(\mathbf{r})$ is zero. (b) Is $U(2\mathbf{G}_c)$ also zero? (c) Why is it possible in principle to obtain an insulator made up of divalent atoms at the lattice points of a simple hexagonal lattice? (d) Why is it not possible to obtain an insulator made up of monovalent atoms in a hexagonal-close-packed structure?

4. ***Brillouin zones of two-dimensional divalent metal.*** A two-dimensional metal in the form of a square lattice has two conduction electrons per atom. In the almost free electron approximation, sketch carefully the electron and hole energy surfaces. For the electrons choose a zone scheme such that the Fermi surface is shown as closed.

5. ***Open orbits.*** An open orbit in a monovalent tetragonal metal connects opposite faces of the boundary of a Brillouin zone. The faces are separated by $G = 2 \times 10^8\ \mathrm{cm}^{-1}$. A magnetic field $B = 10^3$ gauss $= 10^{-1}$ tesla is normal to the plane of the open orbit. (a) What is the order of magnitude of the period of the motion in **k** space? Take $v \approx 10^8$ cm/sec. (b) Describe in real space the motion of an electron on this orbit in the presence of the magnetic field.

6. ***Cohesive energy for a square well potential.*** (a) Find an expression for the binding energy of an electron in one dimension in a single square well of depth U_0 and width a. (This is the standard first problem in elementary quantum mechanics). Assume that the solution is symmetric about the midpoint of the well. (b) Find a numerical result for the binding energy in terms of U_0 for the special case $|U_0| = 2\hbar^2/ma^2$ and compare with the appropriate limit of Fig. 20. In this limit of widely separated wells the band width goes to zero, so the energy for $k = 0$ is the same as the energy for any other k in the lowest energy band. Other bands are formed from the excited states of the well, in this limit.

7. ***De Haas-van Alphen period of potassium.*** (a) Calculate the period $\Delta(1/B)$ expected for potassium on the free electron model. (b) What is the area in real space of the extremal orbit, for $B = 10$ kG $= 1$ T? The same period applies to oscillations in the electrical resistivity, known as the Schubnikow-de Haas effect.

*8. ***Band edge structure on k · p perturbation theory.*** Consider a nondegenerate orbital $\psi_{n\mathbf{k}}$ at $\mathbf{k} = 0$ in the band n of a cubic crystal. Use second-order perturbation theory to find the result

$$\epsilon_n(\mathbf{k}) = \epsilon_n(0) + \frac{\hbar^2 k^2}{2m} + \frac{\hbar^2}{m^2} \sum_j{}' \frac{|\langle n0|\mathbf{k}\cdot\mathbf{p}|j0\rangle|^2}{\epsilon_n(0) - \epsilon_j(0)} , \tag{36}$$

where the sum is over all other orbitals $\psi_{j\mathbf{k}}$ at $\mathbf{k} = 0$. The effective mass at this point is

$$\frac{m}{m^*} = 1 + \frac{2}{m} \sum_j{}' \frac{|\langle n0|\mathbf{p}|j0\rangle|^2}{\epsilon_n(0) - \epsilon_j(0)} . \tag{37}$$

The mass at the conduction band edge in a narrow gap semiconductor is often dominated by the effect of the valence band edge, whence

$$\frac{m}{m^*} \approx \frac{2}{mE_g} \sum_v |\langle c|\mathbf{p}|v\rangle|^2 , \tag{38}$$

where the sum is over the valence bands; E_g is the energy gap. For given matrix elements, small gaps lead to small masses.

9. ***Wannier functions.*** The Wannier functions of a band are defined in terms of the Bloch functions of the same band by

$$w(\mathbf{r} - \mathbf{r}_n) = N^{-\frac{1}{2}} \sum_{\mathbf{k}} e^{-i\mathbf{k}\cdot\mathbf{r}_n}\, \psi_{\mathbf{k}}(\mathbf{r}) , \tag{39}$$

*This problem requires familiarity with perturbation theory.

where $\mathbf{r}_n$ is a lattice point. (a) Prove that Wannier functions about different lattice points n,m are orthogonal:

$$\int dV\, w^*(\mathbf{r}-\mathbf{r}_n)\,w(\mathbf{r}-\mathbf{r}_m)=0\ , \qquad n\neq m\ . \tag{40}$$

This orthogonality property makes the functions often of greater use than atomic orbitals centered on different lattice sites, because the latter are not generally orthogonal. (b) The Wannier functions are peaked around the lattice sites. Show that for $\psi_k = N^{-\frac{1}{2}}\, e^{ikx} u_0(x)$ the Wannier function is

$$w(x-x_n)=u_0(x)\,\frac{\sin \pi(x-x_n)/a}{\pi(x-x_n)/a}\ ,$$

for N atoms on a line of lattice constant a.

10. ***Open orbits and magnetoresistance.*** We considered the transverse magnetoresistance of free electrons in Problem 6.9 and of electrons and holes in Problem 8.6. In some crystals the magnetoresistance saturates except in special crystal orientations. An open orbit carries current only in a single direction in the plane normal to the magnetic field; such carriers are not deflected by the field. In the arrangement of Fig. 6.10, let the open orbits be parallel to k_x; in real space these orbits carry current parallel to the y axis. Let $\sigma_{yy} = s\sigma_0$ be the conductivity of the open orbits; this defines the constant s. The magnetoconductivity tensor in the high field limit $\omega_c\tau \gg 1$ is

$$\sigma_0\begin{pmatrix} Q^{-2} & -Q^{-1} & 0 \\ Q^{-1} & s & 0 \\ 0 & 0 & 1 \end{pmatrix},$$

with $Q \equiv \omega_c\tau$. (a) Show that the Hall field is $E_y = -E_x/sQ$. (b) Show that the effective resistivity in the x direction is $\rho = (Q^2/\sigma_0)(s/s+1)$, so that the resistivity does not saturate, but increases as B^2.

References

J. Callaway, *Quantum theory of the solid state*, Academic Press, 1974.

J. F. Cochran and R. R. Haering, eds., *Electrons in metals*, Gordon and Breach, 1968.

T. E. Faber, *Introduction to the theory of liquid metals*, Cambridge, 1972.

G. C. Fletcher, *Electron band theory of solids*, North-Holland, 1971.

W. A. Harrison, *Solid state theory*, McGraw-Hill, 1970.

J. C. Slater, *Quantum theory of molecules and solids:* Vol. 1, *Electronic structure of molecules* (1963); Vol. 2, *Symmetry and energy bands in crystals* (1965); Vol. 3, *Insulators, semiconductors and solids* (1967); and Vol. 4, *The self-consistent field for molecules and solids* (1974). McGraw-Hill.

J. Ziman, ed., *Electrons*, Vol. 1 of *Physics of metals*, Cambridge, 1969.

Fermi Surfaces

A. P. Cracknell and K. C. Wong, *Fermi surface*, Oxford, 1973. Excellent general reference.

L. M. Falicov, "Fermi surface studies," *Electrons in crystalline solids*, IAEA, Vienna, 1973.

W. A. Harrison and M. B. Webb, eds., *Fermi surface*, Wiley, 1960.

I. M. Lifshitz, M. Ya. Azbel, and M. I. Kaganov, *Electronic theory of metals*, Consultants Bureau, 1973.

W. Mercouroff, *La surface de Fermi des metaux*, Masson, Paris, 1967.

A. B. Pippard, *Dynamics of conduction electrons*, Gordon and Breach, 1965.

P. B. Visscher and L. M. Falicov, "Fermi surface properties of metals," phys. stat. sol. (**b**) **54**, 9 (1972). An index to Fermi surfaces.

Classic Texts

P. W. Anderson, *Concepts in solids*, Benjamin, 1963.

N. F. Mott and H. Jones, *Theory of the properties of metals and alloys*, Oxford, 1936. (Dover reprint.)

R. Peierls, *Quantum theory of solids*, Oxford, 1955.

F. Seitz, *Modern theory of solids*, McGraw-Hill, 1940.

J. M. Ziman, *Principles of the theory of solids*, 2nd ed., Cambridge, 1972.

Pseudopotentials

Solid state physics **24** (1970) contains three articles on pseudopotentials: V. Heine, "The pseudopotential concept," p. 1; M. L. Cohen and V. Heine "Fitting of pseudopotentials to experimental data and their subsequent application," p. 37; V. Heine and D. Weaire, "Pseudopotential theory of cohesion and structure," p. 249.

Bibliography of Energy Band Calculations to 1972

J. C. Slater, in *A.I.P. Handbook*, 3rd ed., McGraw-Hill, 1972, pp. **9**–31 to **9**–38.

10 PLASMONS, POLARITONS, AND POLARONS

The text and problems of this chapter assume real facility in the use of electromagnetic theory at the level of a good senior course.

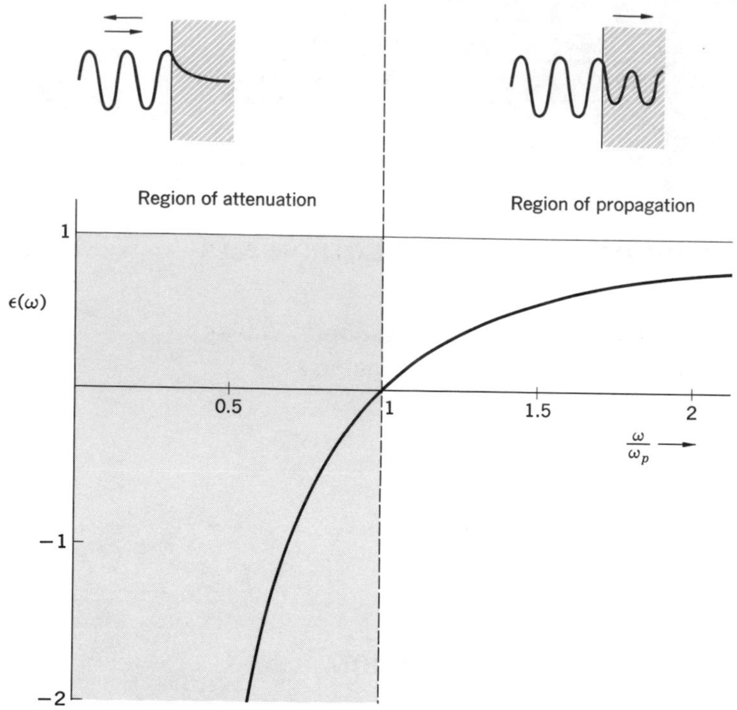

Figure 1 Dielectric function $\epsilon(\omega)$ of a free electron gas versus frequency in units of the plasma frequency ω_p. Electromagnetic waves propagate only when ϵ is positive. Electromagnetic waves are totally reflected from the medium when ϵ is negative.

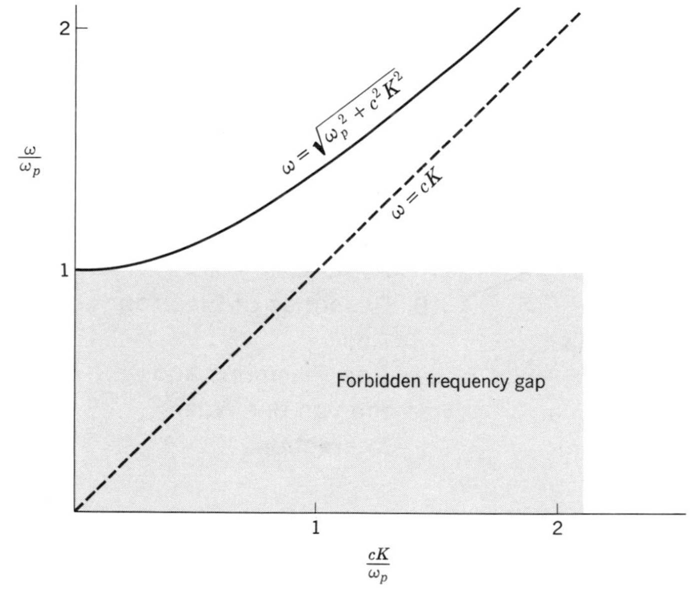

Figure 2 Dispersion relation for transverse electromagnetic waves in a plasma. The group velocity $v_g = d\omega/dK$ is the slope of the dispersion curve. Although the dielectric function is between zero and one, the group velocity is less than the velocity of light in vacuum.

10 PLASMONS, POLARITONS, AND POLARONS

DIELECTRIC FUNCTION OF THE ELECTRON GAS

The dielectric function $\epsilon(\omega,\mathbf{K})$ of the electron gas, with its strong dependence on frequency and wavevector, has significant consequences for the physical properties of solids. In one limit, $\epsilon(\omega,0)$ describes the collective excitations of the Fermi sea—the volume and surface plasmons. In another limit, $\epsilon(0,\mathbf{K})$ describes the electrostatic screening of the electron-electron, electron-lattice, and electron-impurity interactions in crystals. We use the dielectric function of an ionic crystal to derive the polariton spectrum. Later we discuss the properties of polarons. But first we are concerned with the electron gas in metals.

Definition of the Dielectric Function. The dielectric constant ϵ of electrostatics is defined in terms of the electric field $\mathbf{E}$ and the polarization $\mathbf{P}$, the dipole moment density:

$$\text{(CGS)}\quad \mathbf{D} = \mathbf{E} + 4\pi\mathbf{P} = \epsilon\mathbf{E} \; ; \qquad \text{(SI)}\quad \mathbf{D} = \epsilon_0\mathbf{E} + \mathbf{P} = \epsilon\epsilon_0\mathbf{E} \; . \tag{1}$$

Thus defined, ϵ is also known as the relative permittivity. The introduction of the displacement $\mathbf{D}$ is motivated by the usefulness of a vector related to the external applied charge density ρ_{ext} in the same way as $\mathbf{E}$ is related to the total charge density $\rho = \rho_{ext} + \rho_{ind}$, where ρ_{ind} is the charge density induced in the system by ρ_{ext}. Thus the divergence relation is

(CGS)	(SI)	
$\mathrm{div}\,\mathbf{D} = \mathrm{div}\,\epsilon\mathbf{E} = 4\pi\rho_{ext}$	$\mathrm{div}\,\mathbf{D} = \mathrm{div}\,\epsilon\epsilon_0\mathbf{E} = \rho_{ext}$	(2)
$\mathrm{div}\,\mathbf{E} = 4\pi\rho = 4\pi(\rho_{ext} + \rho_{ind})$	$\mathrm{div}\,\mathbf{E} = \rho/\epsilon_0 = (\rho_{ext} + \rho_{ind})/\epsilon_0$	(3)

Parts of this chapter will be written in CGS; to obtain results in SI, write $1/\epsilon_0$ for 4π.

We shall need relations between the Fourier components of **D**, **E**, ρ, and the electrostatic potential φ. For brevity we do not exhibit here the frequency dependence. Define $\epsilon(\mathbf{K})$ such that

$$\mathbf{D}(\mathbf{K}) = \epsilon(\mathbf{K})\mathbf{E}(\mathbf{K}) \ ; \tag{3a}$$

then (3) becomes

$$\text{div } \mathbf{E} = \text{div } \Sigma \ \mathbf{E}(\mathbf{K}) \ e^{i\mathbf{K}\cdot\mathbf{r}} = 4\pi \ \Sigma \ \rho(\mathbf{K}) \ e^{i\mathbf{K}\cdot\mathbf{r}} \ , \tag{3b}$$

and (2) becomes

$$\text{div } \mathbf{D} = \text{div } \Sigma \ \epsilon(\mathbf{K})\mathbf{E}(\mathbf{K}) \ e^{i\mathbf{K}\cdot\mathbf{r}} = 4\pi \ \Sigma \ \rho_{\text{ext}}(\mathbf{K}) \ e^{i\mathbf{K}\cdot\mathbf{r}} \ . \tag{3c}$$

Each of the equations must be satisfied term by term; we divide one by the other to obtain

$$\epsilon(\mathbf{K}) = \frac{\rho_{\text{ext}}(\mathbf{K})}{\rho(\mathbf{K})} = 1 - \frac{\rho_{\text{ind}}(\mathbf{K})}{\rho(\mathbf{K})} \ . \tag{3d}$$

The electrostatic potential φ_{ext} defined by $-\nabla\varphi_{\text{ext}} = \mathbf{D}$ satisfies the Poisson equation $\nabla^2\varphi_{\text{ext}} = -4\pi\rho_{\text{ext}}$; and the potential φ defined by $-\nabla\varphi = \mathbf{E}$ satisfies $\nabla^2\varphi = -4\pi\rho$. Their Fourier components must therefore satisfy

$$\frac{\varphi_{\text{ext}}(\mathbf{K})}{\varphi(\mathbf{K})} = \frac{\rho_{\text{ext}}(\mathbf{K})}{\rho(\mathbf{K})} = \epsilon(\mathbf{K}) \ , \tag{3e}$$

by (3d). We use this relation in the treatment of the screened coulomb potential.

Plasma Optics

The long wavelength dielectric response $\epsilon(\omega,0)$ or $\epsilon(\omega)$ of an electron gas is obtained from the equation of motion of a free electron in an electric field:

$$m \frac{d^2x}{dt^2} = -eE \ . \tag{4}$$

If x and E have the time dependence $e^{-i\omega t}$, then

$$-\omega^2 mx = -eE \ ; \qquad x = eE/m\omega^2 \ . \tag{5}$$

The dipole moment of one electron is $-ex = -e^2E/m\omega^2$, and the polarization, defined as the dipole moment per unit volume, is

$$P = -nex = -\frac{ne^2}{m\omega^2}E \ , \tag{6}$$

where n is the electron concentration.

The dielectric function at frequency ω is

(CGS) $$\epsilon(\omega) \equiv \frac{D(\omega)}{E(\omega)} \equiv 1 + 4\pi\frac{P(\omega)}{E(\omega)} \ ;$$

(SI) $$\epsilon(\omega) = \frac{D(\omega)}{\epsilon_0 E(\omega)} = 1 + \frac{P(\omega)}{\epsilon_0 E(\omega)} \ .$$

The dielectric function of the free electron gas follows:

(CGS) $\epsilon(\omega) = 1 - \dfrac{4\pi ne^2}{m\omega^2}$; (SI) $\epsilon(\omega) = 1 - \dfrac{ne^2}{\epsilon_0 m\omega^2}$. (8)

The **plasma frequency** is defined by the relation

(CGS) $\omega_p{}^2 \equiv 4\pi ne^2/m$; (SI) $\omega_p{}^2 \equiv ne^2/\epsilon_0 m$. (9)

A plasma is a medium with equal concentration of positive and negative charges, of which at least one charge type is mobile. In a solid the negative charges of the conduction electrons are balanced by an equal concentration of positive charge of the ion cores. We write the dielectric function as

$$\boxed{\epsilon(\omega) = 1 - \frac{\omega_p{}^2}{\omega^2}} \ , \tag{10}$$

plotted in Fig. 1. If the positive ion background has a dielectric constant labeled $\epsilon(\infty)$ essentially constant up to frequencies well above ω_p, then (8) becomes

$$\epsilon(\omega) = \epsilon(\infty) - 4\pi ne^2/m\omega^2 = \epsilon(\infty)[1 - \tilde{\omega}_p{}^2/\omega^2] \ , \tag{11}$$

where $\tilde{\omega}_p$ is defined as

$$\tilde{\omega}_p{}^2 = 4\pi ne^2/\epsilon(\infty)m \ . \tag{12}$$

Notice that $\epsilon = 0$ at $\omega = \tilde{\omega}_p$.

Dispersion Relation for Electromagnetic Waves

In a nonmagnetic isotropic medium the electromagnetic wave equation is

$$\text{(CGS)}\quad \partial^2\mathbf{D}/\partial t^2 = c^2\nabla^2\mathbf{E}\ ; \qquad \text{(SI)}\quad \mu_0\,\partial^2\mathbf{D}/\partial t^2 = \nabla^2\mathbf{E}\ . \tag{13}$$

We look for a solution with $\mathbf{E} \propto e^{-i\omega t}\,e^{i\mathbf{K}\cdot\mathbf{r}}$ and $\mathbf{D} = \epsilon(\omega,\mathbf{K})\mathbf{E}$; then we have the dispersion relation for electromagnetic waves:

$$\text{(CGS)}\quad \epsilon(\omega,\mathbf{K})\omega^2 = c^2K^2\ ; \qquad \text{(SI)}\quad \epsilon(\omega,\mathbf{K})\epsilon_0\mu_0\omega^2 = K^2\ . \tag{14}$$

This relation tells us a great deal. Consider

1. $\epsilon > 0$. For ω real, K is real and a transverse electromagnetic wave propagates with the phase velocity $c/\epsilon^{\frac{1}{2}}$.
2. $\epsilon < 0$. For ω real, K is imaginary and the wave is damped with a characteristic length $1/|K|$.
3. $\epsilon = \infty$. This means the system has a finite response in the absence of an applied force; thus the poles of $\epsilon(\omega,\mathbf{K})$ define the frequencies of the free oscillations of the medium.
4. $\epsilon = 0$. We shall see that longitudinally polarized waves are possible only at the zeros of ϵ.
5. ϵ complex. For ω real, $\mathbf{K}$ is complex at the zeros and poles of ϵ, and the waves are damped in space.

Transverse Optical Modes in a Plasma

The dispersion relation (14) becomes, for low K,

$$\text{(CGS)}\qquad \epsilon(\omega)\omega^2 = \epsilon(\infty)(\omega^2 - \tilde{\omega}_p{}^2) = c^2K^2\ , \tag{15}$$

using (11). For $\omega < \tilde{\omega}_p$ we have $K^2 < 0$, so that K is imaginary. The solutions are of the form $e^{-|K|x}$ in the frequency region $0 < \omega \leq \tilde{\omega}_p$. Waves incident on the medium in this frequency region will be totally reflected. An electron gas is transparent only when $\omega > \tilde{\omega}_p$, for here the dielectric function is positive. The dispersion relation in this region may be written as

$$\text{(CGS)}\qquad \omega^2 = \tilde{\omega}_p{}^2 + c^2K^2/\epsilon(\infty)\ ; \tag{16}$$

this describes transverse electromagnetic waves in a plasma (Fig. 2). Values of the plasma frequency ω_p and of the free space wavelength $\lambda_p \equiv 2\pi c/\omega_p$ for electron concentrations of interest are given below. A wave will propagate if its free space wavelength is less than λ_p; otherwise the wave is reflected.

n, electrons/cm^3	10^{22}	10^{18}	10^{14}	10^{10}
ω_p, s^{-1}	5.7×10^{15}	5.7×10^{13}	5.7×10^{11}	5.7×10^{9}
λ_p, cm	3.3×10^{-5}	3.3×10^{-3}	0.33	33

Transparency of Alkali Metals in the Ultraviolet. From the preceding discussion of the dielectric function we conclude that simple metals should reflect light in the visible region and be transparent to ultraviolet light. The effect was discovered by Wood[1] and explained by Zener. A comparison of calculated and observed cutoff wavelengths is given in Table 1. The reflection of light from a metal is entirely similar to the reflection of radio waves from the ionosphere, for the free electrons in the ionosphere make the dielectric constant negative at low frequencies. Experimental results for InSb with $n = 4 \times 10^{18}$ cm^{-3} are shown in Fig. 3. The plasma frequency is near 0.09 eV.

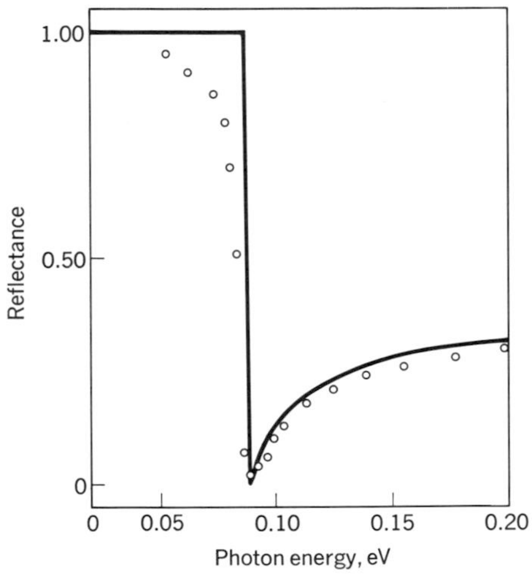

Figure 3 Reflectance of indium antimonide with $n = 4 \times 10^{18}$ cm^{-3}. [After W. G. Spitzer and H. Y. Fan, Phys. Rev. **106**, 882 (1957), from J. N. Hodgson.]

TABLE 1 Ultraviolet Transmission Limits of Alkali Metals, in Å

	Li	Na	K	Rb	Cs
λ_p, calculated, mass m	1550	2090	2870	3220	3620
λ_p, observed	1550	2100	3150	3400	–

[1]R. W. Wood, Phys. Rev. **44**, 353 (1933); C. Zener, Nature **132**, 968 (1933).

Longitudinal Plasma Oscillations

The zeros of the dielectric function determine the frequencies of the longitudinal modes. That is, the condition

$$\epsilon(\omega_L) = 0 \tag{17}$$

determines the longitudinal frequency ω_L near $K = 0$. By the geometry of a longitudinal polarization wave there is a depolarization field $\mathbf{E} = -4\pi\mathbf{P}$, discussed below. Thus $\mathbf{D} \equiv \mathbf{E} + 4\pi\mathbf{P} = 0$ for a longitudinal wave in a plasma or more generally in a crystal. In SI units, $\mathbf{D} = \epsilon_0\mathbf{E} + \mathbf{P} = 0$. At the zero of the dielectric function of an electron gas

$$\epsilon(\omega_L) = 1 - \omega_p^2/\omega_L^2 = 0 \ , \tag{18}$$

whence $\omega_L = \omega_p$. Thus there is a free longitudinal oscillation mode (Fig. 4) of an electron gas at the plasma frequency described earlier as the cutoff of transverse electromagnetic waves, Eq. (15).

A longitudinal plasma oscillation with $K = 0$ is shown in Fig. 5 as a uniform displacement of an electron gas in a thin metallic slab. The electron gas is moved as a whole with respect to the positive ion background. The displacement u creates an electric field $E = 4\pi neu$ that acts as a restoring force. The equation of motion of a unit volume of the electron gas is

$$\text{(CGS)} \qquad nm\frac{d^2u}{dt^2} = -neE = -4\pi n^2e^2u \ , \tag{19}$$

or

$$\text{(CGS)} \qquad \frac{d^2u}{dt^2} + \omega_p^2 u = 0 \ ; \qquad \omega_p = \left(\frac{4\pi ne^2}{m}\right)^{\frac{1}{2}} . \tag{20}$$

This is the equation of motion of a simple harmonic oscillator of frequency ω_p, the plasma frequency. The expression for ω_p is identical with (9), which arose in a different connection. In SI, the displacement u creates the electric field $E = neu/\epsilon_0$, whence $\omega_p = (ne^2/\epsilon_0 m)^{\frac{1}{2}}$.

A plasma oscillation of small wavevector has approximately the frequency ω_p. The wavevector dependence of the dispersion relation for longitudinal oscillations in a Fermi gas is given by

$$\omega \cong \omega_p\left(1 + \frac{3k^2v_F^2}{10\omega_p^2} + \cdots\right) , \tag{21}$$

where v_F is the electron velocity at the Fermi energy.

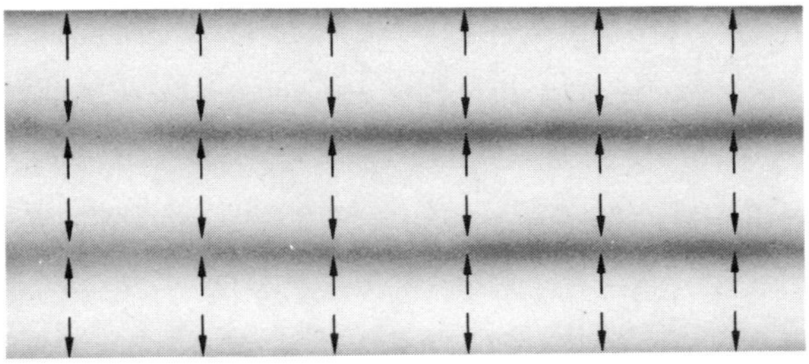

Figure 4 A plasma oscillation. The arrows indicate the direction of displacement of the electrons.

PLASMONS

A plasma oscillation in a metal is a collective longitudinal excitation of the conduction electron gas. A **plasmon** is a quantum of a plasma oscillation; we may excite a plasmon by passing an electron through a thin metallic film (Figs. 6 and 7) or by reflecting an electron or a photon from a film. The charge of the electron couples with the electrostatic field fluctuations of the plasma oscillations. The reflected or transmitted electron will show an energy loss equal to integral multiples of the plasmon energy. Experimental excitation spectra for Al and Mg are shown in Fig. 8. A comparison of observed and calculated values of plasmon energies is given in Table 2; further data are given in the reviews by Raether and by Daniels. It is equally possible to excite collective plasma oscillations in dielectric films; results for several dielectrics are included. The calculated plasma energies of Si, Ge, and InSb are based on four valence electrons per atom. In a dielectric the plasma oscillation is physically the same as in a metal: the entire valence electron sea oscillates back and forth with respect to the ion cores.

TABLE 2 Volume Plasmon Energies, in eV

		Calculated	
Material	Observed	$\hbar\omega_p$	$\hbar\tilde{\omega}_p$
Metals			
Li	7.12	8.02	7.96
Na	5.71	5.95	5.58
K	3.72	4.29	3.86
Mg	10.6	10.9	
Al	15.3	15.8	
Dielectrics			
Si	16.4–16.9	16.0	
Ge	16.0–16.4	16.0	
InSb	12.0–13.0	12.0	

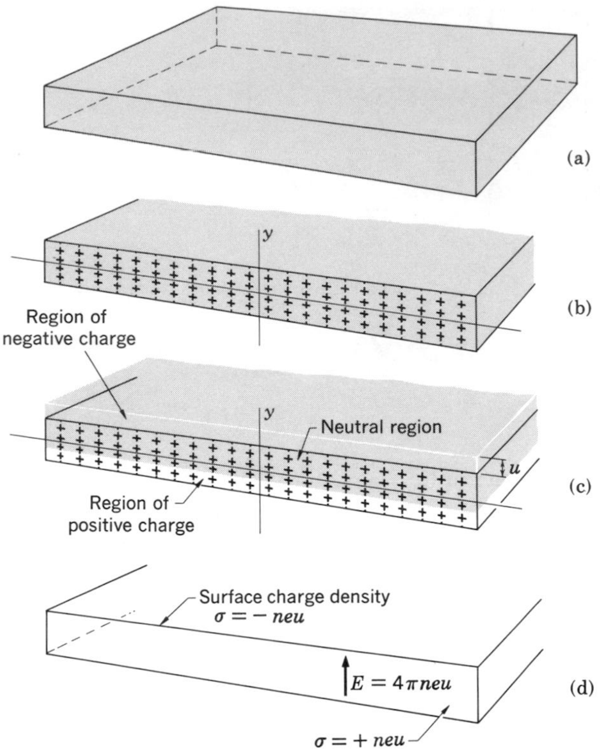

Figure 5 In (a) is shown a thin slab or film of a metal. A cross section is shown in (b), with the positive ion cores indicated by + signs and the electron sea indicated by the gray background. The slab is electrically neutral. In (c) the negative charge has been displaced upward uniformly by a small distance u, shown exaggerated in the figure. As in (d), this displacement establishes a surface charge density $-neu$ on the upper surface of the slab and $+neu$ on the lower surface, where n is the electron concentration. An electric field $E = 4\pi neu$ is produced inside the slab. This field tends to restore the electron sea to its equilibrium position (b). In SI units, $E = neu/\epsilon_0$.

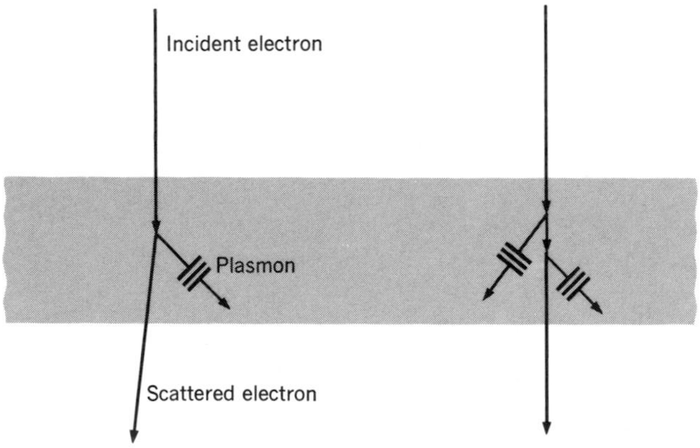

Figure 6 Creation of a plasmon in a metal film by inelastic scattering of an electron. The incident electron typically has an energy 1 to 10 keV; the plasmon energy may be of the order of 10 eV. An event is also shown in which two plasmons are created.

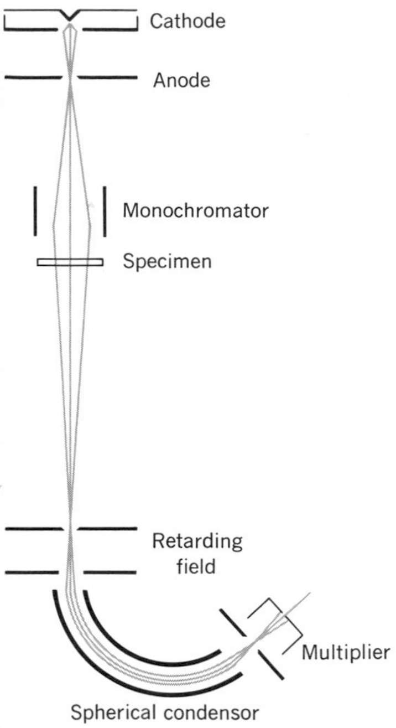

Figure 7 A spectrometer with electrostatic analyzer for the study of plasmon excitation by electrons. [After J. Daniels et al., *Springer tracts in modern physics* **54,** 77 (1970).]

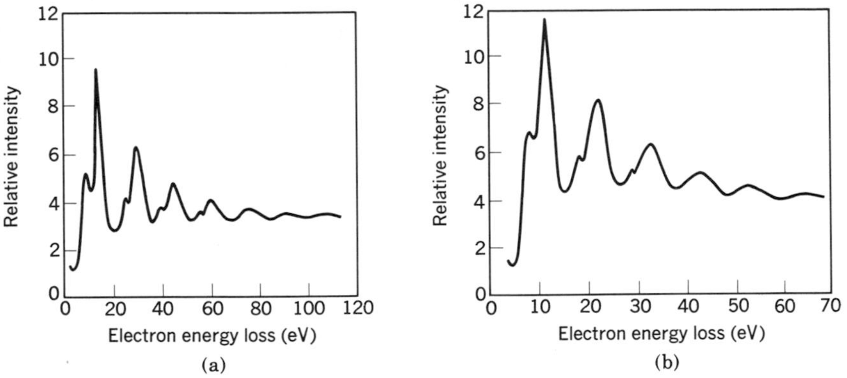

Figure 8 Energy loss spectra for electrons reflected from films of (a) aluminum and (b) magnesium, for primary electron energies of 2020 eV. The 12 loss peaks observed in Al are made up of combinations of 10.3 and 15.3 eV losses, where the 10.3 eV loss is due to surface plasmons and the 15.3 eV loss is due to volume plasmons. The 10 loss peaks observed in Mg are made up of combinations of 7.1 eV surface plasmons and 10.6 eV volume plasmons. Surface plasmons are the subject of Problem 1. [After C. J. Powell and J. B. Swan, Phys. Rev. **115,** 869 (1959); **116,** 81 (1959).]

ELECTROSTATIC SCREENING

The important screening behavior of an electron gas is described by the wavevector dependence of the static dielectric function $\epsilon(0,\mathbf{K})$. We consider the response of the electrons to an applied external electrostatic field. We start with a uniform gas of electrons of charge concentration $-n_0e$ superimposed on a background of positive charge of concentration n_0e. The positive charge background is now deformed mechanically to produce a sinusoidal variation of charge density in the x direction:

$$\rho^+(x) = n_0e + \rho_{\text{ext}}(K) \sin Kx \ . \tag{22}$$

The term $\rho_{\text{ext}}(K) \sin Kx$ gives rise to an electrostatic field that we call the external field applied to the electron gas.

The electrostatic potential φ of a charge distribution is found from the Poisson equation $\nabla^2\varphi = -4\pi\rho$, by (3) with $\mathbf{E} = -\nabla\varphi$. With

$$\varphi = \varphi_{\text{ext}}(K) \sin Kx \ ; \qquad \rho = \rho_{\text{ext}}(K) \sin Kx \ , \tag{23}$$

the Poisson equation gives the relation

$$K^2\varphi_{\text{ext}}(K) = 4\pi\rho_{\text{ext}}(K) \ . \tag{24}$$

The electron gas will be deformed by the combined influences of the electrostatic potential $\varphi_{\text{ext}}(K)$ of the positive charge distribution and of the as yet unknown induced electrostatic potential $\varphi_{\text{ind}}(K) \sin Kx$ of the deformation of the electron gas itself. The electron charge density is

$$\rho^-(x) = -n_0e + \rho_{\text{ind}}(K) \sin Kx \ , \tag{25}$$

where $\rho_{\text{ind}}(K)$ is the amplitude of the charge density variation induced in the electron gas. We want to find $\rho_{\text{ind}}(K)$ in terms of $\rho_{\text{ext}}(K)$.

The amplitude of the total electrostatic potential $\varphi(K) = \varphi_{\text{ext}}(K) + \varphi_{\text{ind}}(K)$ of the positive and negative charge distributions is related to the total charge density variation $\rho(K) = \rho_{\text{ext}}(K) + \rho_{\text{ind}}(K)$ by the Poisson equation. Then, as in Eq. (24),

$$K^2\varphi(K) = 4\pi\rho(K) \ . \tag{26}$$

To go further we need another equation that relates the electrostatic potential to the electron concentration. We now develop the Thomas-Fermi approximation. The electrochemical potential of the electron gas must be constant in equilibrium, independent of position. In a region where there is no electrostatic potential the chemical potential μ is related to the uniform concentration n_0 by

$$\mu = \epsilon_F{}^0 = \frac{\hbar^2}{2m}(3\pi^2 n_0)^{2/3} \tag{27}$$

at absolute zero, according to (6.17). In a region where the electrostatic potential is $\varphi(x)$, the electrochemical potential (Fig. 9) is constant and

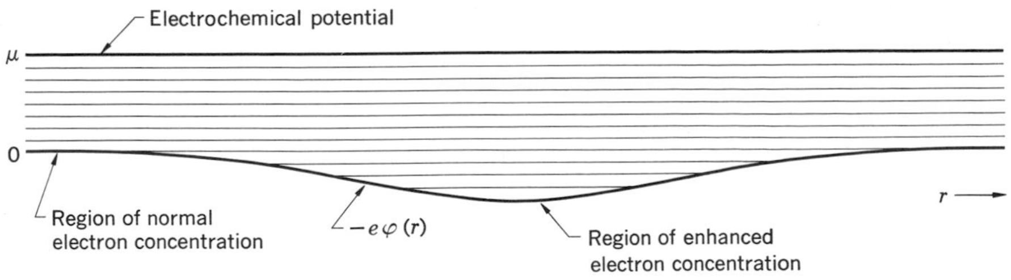

Figure 9 In thermal and diffusive equilibrium the electrochemical potential is constant; to maintain it constant we increase the electron concentration in regions of space where the potential energy is low, and we decrease the concentration where the potential is high.

equal to

$$\mu = \epsilon_F(x) - e\varphi(x) \cong \frac{\hbar^2}{2m}\,[3\pi^2 n(x)]^{2/3} - e\varphi(x) \cong \frac{\hbar^2}{2m}\,[3\pi^2 n_0]^{2/3} \; . \tag{28}$$

It is the Thomas-Fermi approximation that the electrochemical potential can be written in this way; the expression is valid for electrostatic potentials that vary slowly in comparison with the wavelength of an electron. By a Taylor series expansion of ϵ_F, Eq. (28) may be written as

$$\frac{d\epsilon_F}{dn_0}\,[n(x) - n_0] \cong e\varphi(x) \; . \tag{29}$$

From (27) we have $d\epsilon_F/dn_0 = 2\epsilon_F/3n_0$, whence

$$n(x) - n_0 \cong \frac{3}{2}\, n_0\, \frac{e\varphi(x)}{\epsilon_F} \; . \tag{30}$$

The left-hand side is the induced part of the concentration; thus the Fourier components of this equation are

$$\rho_{\mathrm{ind}}(K) = -(3n_0 e^2/2\epsilon_F)\varphi(K) \; . \tag{31}$$

By (26) this becomes

$$\rho_{\mathrm{ind}}(K) = -(6\pi n_0 e^2/\epsilon_F K^2)\rho(K) \; , \tag{32}$$

and by (3d) we have

$$\epsilon(0,K) = 1 - \frac{\rho_{\mathrm{ind}}(K)}{\rho(K)} = 1 + k_s^2/K^2 \; ; \tag{33}$$

here, after some rearrangement,

$$k_s^2 \equiv 6\pi n_0 e^2/\epsilon_F = 4(3/\pi)^{1/3}\, n_0^{1/3}/a_0 = 4\pi e^2 \mathfrak{D}(\epsilon_F) \; , \tag{34}$$

where a_0 is the Bohr radius and $\mathfrak{D}(\epsilon_F)$ is the density of states for a free electron gas. The approximation (33) for $\epsilon(0,K)$ is called the Thomas-Fermi dielectric function, and $1/k_s$ is the Thomas-Fermi screening length, as in (40) below. For copper with $n_0 = 8.5 \times 10^{22}$ cm^{-3}, the screening length is 0.55 Å.

We have derived two limiting expressions for the dielectric function of an electron gas:

$$\epsilon(0,K) = 1 + \frac{k_s^2}{K^2} \; ; \qquad \epsilon(\omega,0) = 1 - \frac{\omega_p^2}{\omega^2} \; . \tag{35}$$

We notice that $\epsilon(0,K)$ as $K \to 0$ does not approach the same limit as $\epsilon(\omega,0)$ as $\omega \to 0$. This means that great care must be taken with the dielectric function near the origin of the ω-K plane. The full theory for the general function $\epsilon(\omega,K)$ is due to Lindhard.[2]

Screened Coulomb Potential. We consider a point charge q placed in a sea of conduction electrons. The Poisson equation for the unscreened coulomb potential is

$$\nabla^2\varphi_0 = -4\pi q\delta(\mathbf{r}) \; , \tag{36}$$

and we know that $\varphi_0 = q/r$. Let us write

$$\varphi_0(\mathbf{r}) = (2\pi)^{-3}\int d\mathbf{K}\; \varphi_0(\mathbf{K})\; e^{i\mathbf{K}\cdot\mathbf{r}} \; . \tag{37}$$

We use the Fourier representation of the delta function:

$$\delta(\mathbf{r}) = (2\pi)^{-3}\int d\mathbf{K}\; e^{i\mathbf{K}\cdot\mathbf{r}} \; , \tag{38}$$

whence $K^2\varphi_0(K) = 4\pi q$. By (3e),

$$\varphi_0(K)/\varphi(K) = \epsilon(K) \; ,$$

where $\varphi(K)$ is the total or screened potential. We use $\epsilon(K)$ in the Thomas-Fermi form (33) to find

$$\varphi(\mathbf{K}) = \frac{4\pi q}{K^2 + k_s^2} \; . \tag{39}$$

The screened coulomb potential is the transform of $\varphi(\mathbf{K})$:

$$\begin{aligned}\varphi(r) &= \frac{4\pi q}{(2\pi)^3}\int_0^\infty dK\, \frac{2\pi K^2}{K^2 + k_s^2}\int_{-1}^{1} d(\cos\theta)\; e^{iKr\cos\theta} \\ &= \frac{2q}{\pi r}\int_0^\infty dK\, \frac{K\sin Kr}{K^2 + k_s^2} = \frac{q}{r}\, e^{-k_s r} \; ,\end{aligned} \tag{40}$$

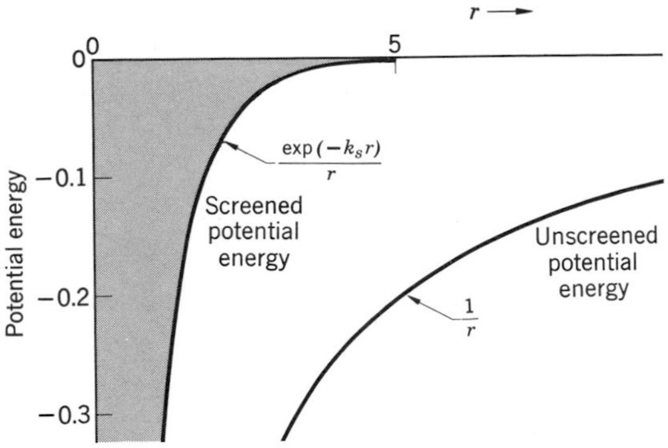

Figure 10 Comparison of screened and unscreened coulomb potentials of a unit positive charge as seen by an electron. The screening length k_s^{-1} has been taken as equal to unity.

as in Fig. 10. The screening parameter k_s is defined by (34). The exponential factor reduces the range of the coulomb potential. The bare potential q/r is obtained on letting the charge concentration $n_0 \rightarrow 0$, for then $k_s \rightarrow 0$. In the vacuum limit $\varphi(K) = 4\pi q/K^2$.

One application of the screened interaction is to the resistivity of certain alloys. The atoms of the series Cu, Zn, Ga, Ge, As have valences 1, 2, 3, 4, 5. An atom of Zn, Ga, Ge, or As added substitutionally to metallic Cu has an excess charge, referred to Cu, of 1, 2, 3, or 4 if all the valence electrons join the conduction band of the host metal. The foreign atom scatters the conduction electrons, with an interaction given by the screened coulomb potential. This scattering contributes to the residual electrical resistivity, and calculations by Mott[3] of the resistivity increase are in fair agreement with experiment.

[2]A good discussion of the Lindhard dielectric function is given by J. Ziman, *Principles of the theory of solids,* 2nd ed., Cambridge, 1972, Chap. 5. The algebraic steps in the evaluation of Ziman's equation (5.16) are given in detail by C. Kittel, *Solid state physics* **22**, 1 (1968), Section 6.
[3]N. F. Mott, Proc. Camb. Phil. Soc. **32**, 281 (1936).

Mott Metal-Insulator Transition

According to the independent-electron model, a crystal with an odd number of electrons per primitive cell is always a metal. The model fails for many transition metal oxides. For example, CoO is a semiconductor and not a metal, although the number of electrons is odd. An expanded space lattice of hydrogenlike atoms may not necessarily form a metal, but an insulator or a semiconductor. There is a further criterion for a metal: A lattice of hydrogen atoms at absolute zero will be a metal if the lattice constant is smaller than some critical value a_c, where on Mott's early estimate[4] $a_c \simeq 4.5a_0$, with $a_0 = \hbar^2/me^2$ as the radius of the first Bohr orbit of a hydrogen atom. At values of the lattice constant $a > a_c$, the crystal will be an insulator. There is strong experimental support of such a metal-insulator transition.

To approach this question, imagine that we start in the metallic state and expand a lattice of hydrogen atoms until the solid eventually becomes an insulator. The change of state begins when a conduction electron of the metal can form a bound and therefore insulating state with the ion, here the proton. In the metallic state the ion potential is screened: at high electron concentration the screening prevents the formation of a bound state, but as the concentration is lowered the screening length increases and finally a bound state is possible.

The screened potential energy of an electron-proton pair is

$$U(R) = -\frac{e^2}{r}\, e^{-k_s r} \ , \tag{41}$$

where $k_s^2 = 3.939\ n_0^{1/3}/a_0$, as in (34). The potential is known[5] to have a bound state for electrons in the field of a fixed positive charge e only if $k_s < 1.19/a_0$, or

$$\frac{3.939\ n_0^{1/3}}{a_0} < \frac{1.42}{a_0^2} \ .$$

With $n_0 = 1/a^3$ for an sc lattice, we have an insulator if

$$a > 2.78\ a_0 \ . \tag{42}$$

The condition derived in this way is close to the Mott result, derived differently.

[4] N. F. Mott, *Metal-insulator transitions*, Taylor & Francis, London, 1974. The possibility of a metallic form of hydrogen is reviewed by W. D. Metz, "Metallic hydrogen: simulating Jupiter in the laboratory," Science **180**, 398 (1973).

[5] F. J. Rogers, H. C. Graboske, Jr., and D. J. Harwood, Phys. Rev. **A1**, 1577 (1970).

Screening and Phonons in Metals

An interesting application of our two limiting forms of the dielectric function is to longitudinal acoustic phonons in metals. For longitudinal modes the total dielectric function, ions plus electrons, must be zero. Provided the sound velocity is less than the Fermi velocity of the electrons, we may use for the electrons the Thomas-Fermi dielectric function

$$\epsilon_{el}(\omega,K) = 1 + k_s^2/K^2 \ . \tag{43}$$

Provided also that the ions are well-spaced and move independently, we may use for them the $\epsilon(\omega,0)$ limit with the appropriate mass M. The total dielectric function, lattice plus electrons, but without the electronic polarizability of the ion cores, is

$$\epsilon(\omega,K) = 1 - \frac{4\pi ne^2}{M\omega^2} + \frac{k_s^2}{K^2} \ . \tag{44}$$

At low K and ω we neglect the term 1. At a zero of $\epsilon(\omega,K)$ we have, with $\epsilon_F \equiv \frac{1}{2}mv_F^2$,

$$\omega^2 = \frac{4\pi ne^2}{Mk_s^2}K^2 = \frac{4\pi ne^2}{M} \cdot \frac{\epsilon_F}{6\pi ne^2}K^2 = \frac{m}{3M}v_F^2K^2 \ , \tag{45}$$

or

$$\omega = vK \ ; \qquad v = \sqrt{\frac{m}{3M}}\, v_F \ . \tag{46}$$

This describes long wavelength acoustic phonons and in the alkali metals is in quite good agreement with the observed longitudinal wave velocity. For potassium we calculate $v = 1.8 \times 10^5$ cm s^{-1}; the observed velocity at 4 K in the [100] direction is 2.2×10^5 cm s^{-1}.

There is another zero of $\epsilon(\omega,K)$ for positive ions imbedded in an electron sea. For high frequencies we use the dielectric contribution $-\omega_p^2/\omega^2$ of the electron gas:

$$\epsilon(\omega,0) = 1 - \frac{4\pi ne^2}{M\omega^2} - \frac{4\pi ne^2}{m\omega^2} \ , \tag{47}$$

and this function has a zero when

$$\omega^2 = \frac{4\pi ne^2}{\mu} \ ; \qquad \frac{1}{\mu} = \frac{1}{M} + \frac{1}{m} \ . \tag{48}$$

This is the electron plasma frequency, but with the reduced mass correction for the motion of the positive ions.

POLARITONS

Longitudinal optical phonons and transverse optical phonons were discussed in Chapter 4, but we deferred treatment of the interaction of transverse optical phonons with transverse electromagnetic waves. At resonance the phonon-photon coupling entirely changes the character of the propagation, and a forbidden band is established for reasons that have nothing to do with the periodicity of the lattice. By resonance we mean a condition in which the frequencies and wavevectors of both particles are approximately equal. The region of the crossover of the two dashed curves in Fig. 11 is the resonance region; the two dashed curves are the dispersion relations for photons and transverse optical phonons in the absence of any coupling between them. In reality, however, there always is coupling implicit in Maxwell's equations and expressed by the dielectric function. The quantum of the coupled phonon-photon field is called a **polariton.**

In this section we see how the coupling is responsible for the dispersion relations shown as solid curves in the figure. All of this takes place at very low values of the wavevector in comparison with a zone boundary, because at crossover $\omega(\text{photon}) = ck(\text{photon}) = \omega(\text{phonon}) \approx 10^{13}\ \text{s}^{-1}$; thus $k \approx 300\ \text{cm}^{-1}$. An early warning: Although the symbol ω_L will necessarily arise in the theory, the effects do not concern longitudinal optical phonons. Longitudinal phonons do not couple to transverse photons in the bulk of a crystal.

The coupling of the electric field E of the photon with the dielectric polarization P of the TO phonon is described by the electromagnetic wave equation:

(CGS)
$$c^2K^2E = \omega^2(E + 4\pi P) \tag{49}$$

At low wavevectors the TO phonon frequency ω_T is independent of K. The polarization is proportional to the displacement of the positive ions relative to the negative ions, so that the equation of motion of the polarization is like that of an oscillator and may be written as, with $P = Nqu$,

$$-\omega^2P + \omega_T^2P = (Nq^2/M)E \ , \tag{50}$$

where there are N ion pairs of effective charge q and reduced mass M, per unit volume. The equations (49) and (50) have a solution when

$$\begin{vmatrix} \omega^2 - c^2K^2 & 4\pi\omega^2 \\ Nq^2/M & \omega^2 - \omega_T^2 \end{vmatrix} = 0 \ . \tag{51}$$

This gives the polariton dispersion relation plotted in Figs. 11 and 12. At $K = 0$ there are two roots, $\omega = 0$ for the photon and

$$\omega^2 = \omega_T^2 + 4\pi Nq^2/M = \omega_L^2 \tag{52}$$

for the polariton. Here ω_T is the TO phonon frequency in the absence of

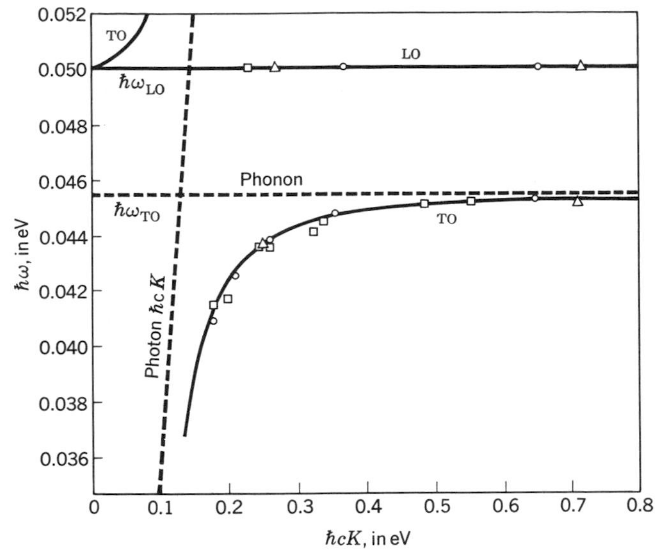

Figure 11 A plot of the observed energies and wavevectors of the polaritons and of the LO phonons in GaP. The theoretical dispersion curves are shown by the solid lines. The dispersion curves for the uncoupled phonons and photons are shown by the short, dashed lines. [After C. H. Henry and J. J. Hopfield, Phys. Rev. Letters **15**, 964 (1965).]

coupling. This relation defines ω_L, but we must establish the motivation for the notation.

The dielectric function obtained from (50) is:

$$\epsilon(\omega) = 1 + 4\pi P/E = 1 + \frac{4\pi Nq^2/M}{\omega_T^2 - \omega^2} \; . \tag{53}$$

If there is an optical electronic contribution to the polarization from the ion cores, this should be counted in as well. In the frequency range from zero up through the infrared, we write

$$\epsilon(\omega) = \epsilon(\infty) + \frac{4\pi Ne^2/M}{\omega_T^2 - \omega^2} \; , \tag{54}$$

in accord with the definition of $\epsilon(\infty)$ as the optical dielectric constant, obtained as the square of the optical refractive index.

We set $\omega = 0$ to obtain the static dielectric function:

$$\epsilon(0) = \epsilon(\infty) + 4\pi Ne^2/M\omega_T^2 \; , \tag{55}$$

which may be combined with (54) to obtain $\epsilon(\omega)$ in terms of accessible parameters, with ω_L given by (52):

$$\epsilon(\omega) = \epsilon(\infty) + [\epsilon(0) - \epsilon(\infty)] \frac{\omega_T^2}{\omega_T^2 - \omega^2} = \epsilon(\infty) \left(\frac{\omega_L^2 - \omega^2}{\omega_T^2 - \omega^2} \right) \; . \tag{56}$$

The zero of $\epsilon(\omega)$ is at ω_L and the pole is at ω_T.

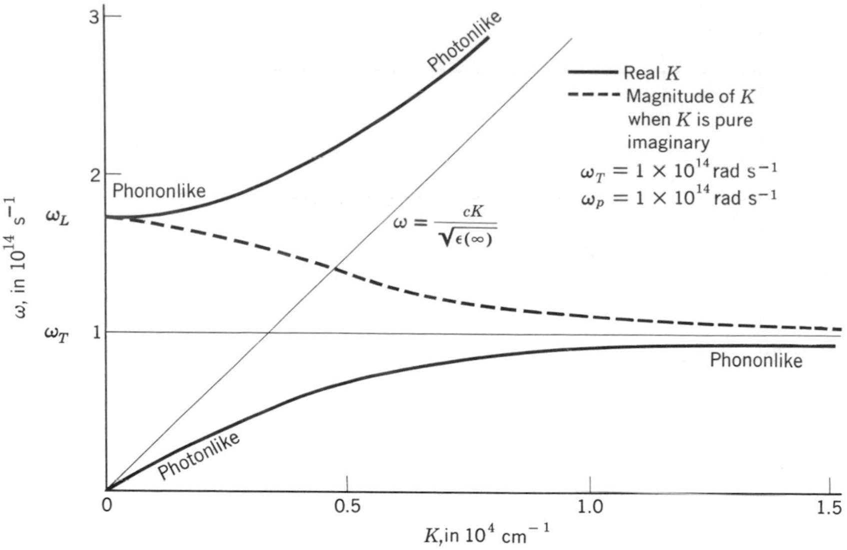

Figure 12 Coupled modes of photons and transverse optical phonons in an ionic crystal. The fine horizontal line represents oscillators of frequency ω_T in the absence of coupling to the electromagnetic field, and the fine line labeled $\omega = cK/\sqrt{\epsilon(\infty)}$ corresponds to electromagnetic waves in the crystal, but uncoupled to the lattice oscillators ω_T. The heavy lines are the dispersion relations in the presence of coupling between the lattice oscillators and the electromagnetic wave. One effect of the coupling is to create the frequency gap between ω_L and ω_T: within this gap the wavevector is pure imaginary of magnitude given by the broken line in the figure. In the gap the wave attenuates as $e^{-|K|x}$, and we see from the plot that the attenuation is much stronger near ω_T than near ω_L. The character of the branches varies with K; there is a region of mixed electric-mechanical aspects near the nominal crossover. Note, finally, it is intuitively obvious that the group velocity of light in the medium is always $< c$, because the slope $\partial\omega/\partial K$ for the actual dispersion relations (heavy lines) is everywhere less than the slope c for the uncoupled photon in free space.

Waves do not propagate[6] in the frequency region for which $\epsilon(\omega)$ is negative, between its pole at $\omega = \omega_T$ and its zero at $\omega = \omega_L$:

$$0 = \epsilon(\infty) + [\epsilon(0) - \epsilon(\infty)] \frac{\omega_T^2}{\omega_T^2 - \omega_L^2} , \tag{57}$$

as in Fig. 13. The zero of $\epsilon(\omega)$, by our earlier argument, is the LO frequency at low K, Fig. 14. Just as with ω_p, the frequency ω_L has two meanings, one as the LO frequency at low K and the other as the upper cutoff frequency of the forbidden band for propagation of an electromagnetic wave. The value of ω_L is identical at both frequencies.

[6]For negative ϵ, waves do not propagate because then K is imaginary for real ω, and $\exp(iKx) \rightarrow \exp(-|K|x)$, damped in space.

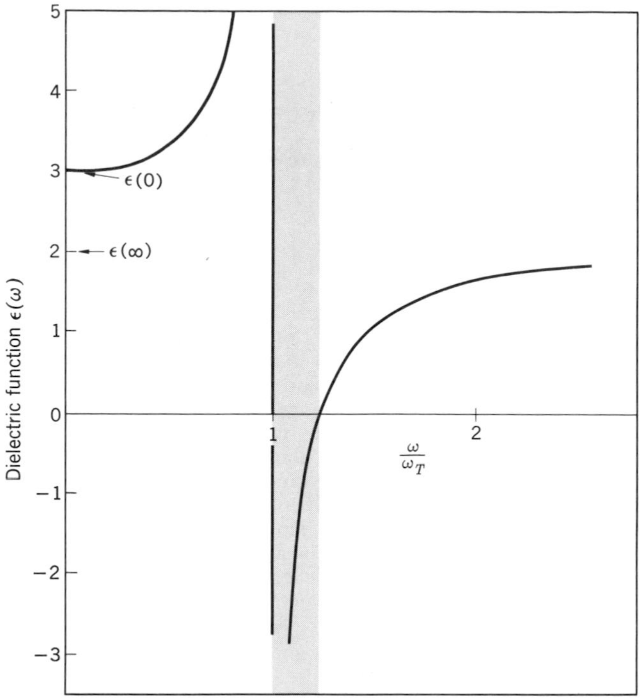

(a)

Figure 13a Plot of $\epsilon(\omega)$ from (56) for $\epsilon(\infty) = 2$ and $\epsilon(0) = 3$. The dielectric constant is negative between $\omega = \omega_T = \omega_L = (3/2)^{1/2}\omega_T$; that is, between the pole (infinity) of $\epsilon(\omega)$ and the zero of $\epsilon(\omega)$. Incident electromagnetic waves of frequencies $\omega_T < \omega < \omega_L$ will not propagate in the medium, but will be reflected at the boundary.

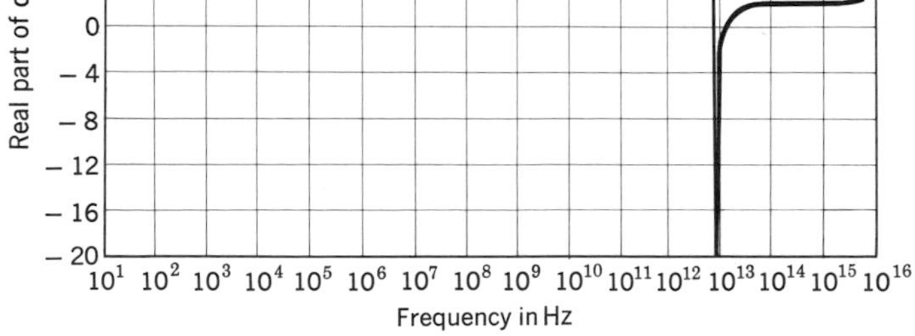

(b)

Figure 13b Dielectric function (real part) of SrF_2 measured over a wide frequency range, exhibiting the decrease of the ionic polarizability at high frequencies. (Courtesy of A. von Hippel.)

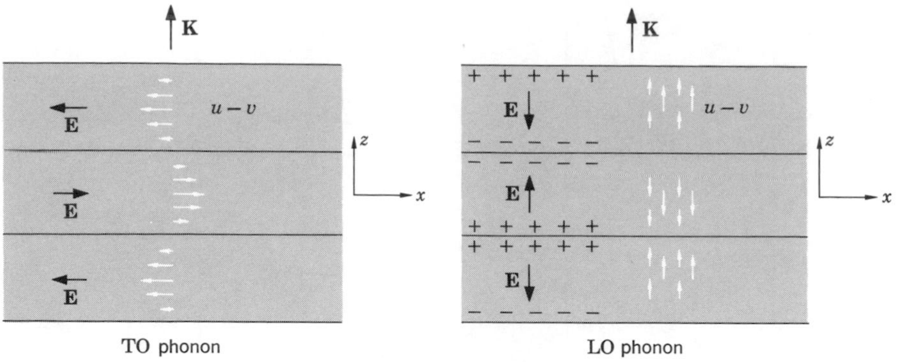

Figure 14 Relative displacements of the positive and negative ions at one instant of time for a wave in an optical mode traveling along the z axis. The planes of nodes (zero displacement) are shown; for long wavelength phonons the nodal planes are separated by many planes of atoms. In the transverse optical phonon mode the particle displacement is perpendicular to the wavevector **K**; the macroscopic electric field in an infinite medium will lie only in the $\pm x$ direction for the mode shown, and by the symmetry of the problem $\partial E_x/\partial x = 0$. It follows that div $\mathbf{E} = 0$ for a TO phonon. In the longitudinal optical phonon mode the particle displacements and hence the dielectric polarization **P** are parallel to the wavevector. The macroscopic electric field **E** satisfies $\mathbf{D} = \mathbf{E} + 4\pi\mathbf{P} = 0$ in CGS or $\epsilon_0\mathbf{E} + \mathbf{P} = 0$ in SI; by symmetry **E** and **P** are parallel to the z axis, and $\partial E_z/\partial z \neq 0$. Thus div $\mathbf{E} \neq 0$ for an LO phonon, and $\epsilon(\omega)$ div **E** is zero only if $\epsilon(\omega) = 0$.

LST Relation. Equation (57) can be rewritten as

$$\boxed{\frac{\omega_L^2}{\omega_T^2} = \frac{\epsilon(0)}{\epsilon(\infty)}} \ , \tag{58}$$

where $\epsilon(0)$ is the static dielectric constant and $\epsilon(\infty)$ is the high-frequency limit of the dielectric function, defined to include the core electron contribution. This result is the Lyddane-Sachs-Teller relation. The derivation assumed a cubic crystal with two atoms per primitive cell.[7] For soft modes with $\omega_T \rightarrow 0$ we see that $\epsilon(0) \rightarrow \infty$, a characteristic of ferroelectricity.

Undamped electromagnetic waves with frequencies within the gap cannot propagate in a thick crystal. The reflectivity of a crystal surface is expected to be high in this frequency region, as in Fig. 15. For films of thickness less than a wavelength the situation is changed. Because for frequencies in the gap the wave attenuates as $e^{-|K|x}$, it is possible for the radiation to be transmitted through a film for the small values of $|K|$ near ω_L, but for the large values of $|K|$ near ω_T the wave will be reflected. By reflection at nonnormal incidence the frequency ω_L of longitudinal optical phonons can be observed, as in Fig. 16.

[7]For the extension to complex structures, see A. S. Barker, Jr., "Infrared dielectric behavior of ferroelectric crystals," in *Ferroelectricity*, E. F. Weller, ed., Elsevier, 1967. Barker gives an interesting derivation of the LST relation by a causality argument.

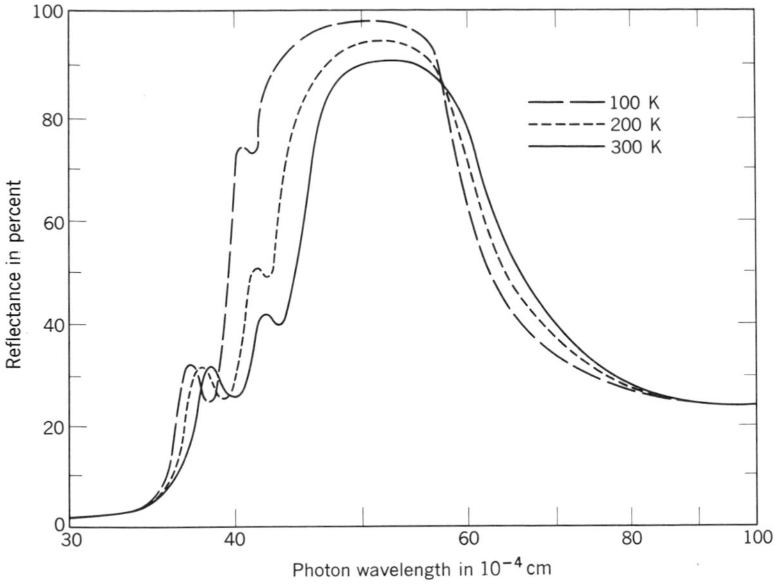

Figure 15 Reflectance of a crystal of NaCl at several temperatures, versus wavelength. The nominal values of ω_L and ω_T at room temperature correspond to wavelengths of 38 and 61×10^{-4} cm, respectively. [After A. Mitsuishi et al., J. Opt. Soc. Am. **52,** 14 (1962).]

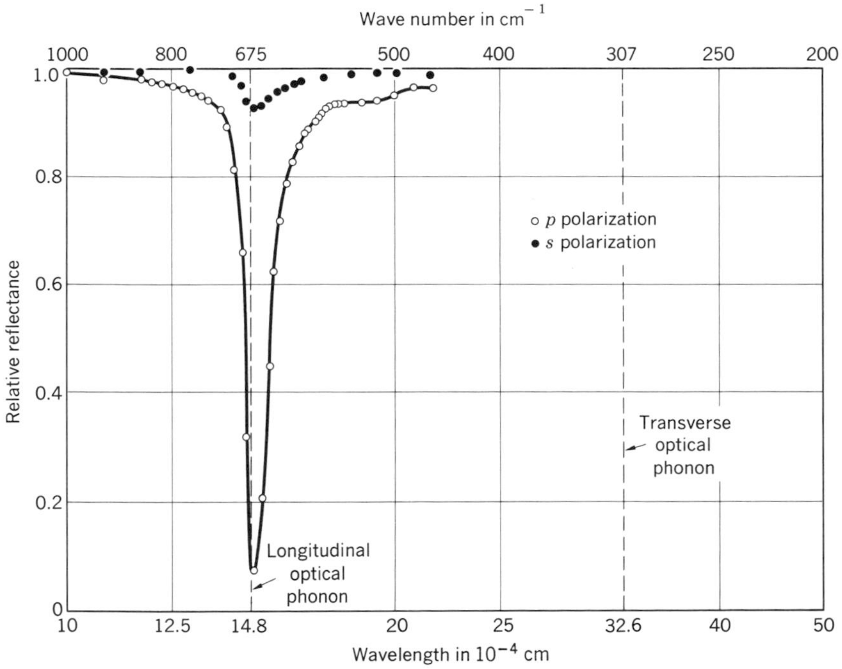

Figure 16 Reflectance versus wavelength of a LiF film backed by silver, for radiation incident near 30°. The longitudinal optical phonon absorbs strongly the radiation polarized (p) in the plane normal to the film, but absorbs hardly at all the radiation polarized (s) parallel to the film. [After D. W. Berreman, Phys. Rev. **130,** 2193 (1963).]

Experimental values of $\epsilon(0)$, $\epsilon(\infty)$, and ω_T are given in Table 3, with values of ω_L calculated using the LST relation, Eq. (58). Values of ω_T at a temperature of 4 K have been given by Jones.[8] We compare values of ω_L/ω_T obtained by inelastic neutron scattering[9] with experimental values of $[\epsilon(0)/\epsilon(\infty)]^{\frac{1}{2}}$ obtained by dielectric measurements:

	NaI	KBr	GaAs
ω_L/ω_T	1.44 ± 0.05	1.39 ± 0.02	1.07 ± 0.02
$[\epsilon(0)/\epsilon(\infty)]^{\frac{1}{2}}$	1.45 ± 0.03	1.38 ± 0.03	1.08

The agreement with the LST relation is excellent. The relation may often be used in the form $\epsilon(0) = (n\omega_L/\omega_T)^2$, where $n = [\epsilon(\infty)]^{\frac{1}{2}}$ is the optical refractive index. This is used to find a value of $\epsilon(0)$ intrinsic to the pure crystal when actual measurements of $\epsilon(0)$ are distorted by ionic or electronic conductivity, or when a single crystal of suitable form is not available for electrical measurements. The LST relation is also useful where $\epsilon(0)$ is very high. Thus in the semiconductor PbTe at 1.2 K we measure $\omega_T = 18\ \text{cm}^{-1}$, $\omega_L = 114\ \text{cm}^{-1}$, and $\epsilon(\infty) = 32$, whence by the LST relation $\epsilon(0) = 1300$. Here the direct measurement of $\epsilon(0)$ is difficult because of the inevitable presence of some free carriers in PbTe, never sufficiently pure.

ELECTRON-ELECTRON INTERACTION

Fermi Liquid. Because of the interaction of the conduction electrons with each other through their electrostatic interaction, the electrons suffer collisions. Also, a moving electron causes an inertial reaction in the surrounding electron gas, thereby increasing the effective mass of the electron. The effects of electron-electron interactions are usually described within the framework of the Landau theory of a Fermi liquid.[10] The object of the theory is to give a unified account of the effect of interactions. A Fermi gas is a system of noninteracting fermions; the same system with interactions is a Fermi liquid.

[8]G. O. Jones et al., Proc. Roy. Soc. (London) A**261**, 10 (1961).
[9]A. D. B. Woods et al., Phys. Rev. **131**, 1025 (1963); J. L. T. Waugh and G. Dolling, Phys. Rev. **132**, 2410 (1963).
[10]L. Landau, "Theory of a Fermi liquid," Soviet Physics JETP **3**, 920 (1957); L. Landau, "Oscillations in a Fermi liquid," Soviet Physics JETP **5**, 10 (1957); see also D. Pines and P. Nozieres, *Theory of quantum liquids*, Benjamin, 1966, Vol. I.

TABLE 3 Lattice Frequencies, Chiefly at 300 K

Crystal	Static Dielectric Constant $\epsilon(0)$	Optical Dielectric Constant $\epsilon(\infty)$	ω_T, in 10^{13} s^{-1} Experimental	ω_L, in 10^{13} s^{-1} LST Relation
LiH	12.9	3.6	11.	21.
LiF	8.9	1.9	5.8	12.
LiCl	12.0	2.7	3.6	7.5
LiBr	13.2	3.2	3.0	6.1
NaF	5.1	1.7	4.5	7.8
NaCl	5.9	2.25	3.1	5.0
NaBr	6.4	2.6	2.5	3.9
KF	5.5	1.5	3.6	6.1
KCl	4.85	2.1	2.7	4.0
KI	5.1	2.7	1.9	2.6
RbF	6.5	1.9	2.9	5.4
RbI	5.5	2.6	1.4	1.9
CsCl	7.2	2.6	1.9	3.1
CsI	5.65	3.0	1.2	1.6
TlCl	31.9	5.1	1.2	3.0
TlBr	29.8	5.4	0.81	1.9
AgCl	12.3	4.0	1.9	3.4
AgBr	13.1	4.6	1.5	2.5
MgO	9.8	2.95	7.5	14.
GaP	10.7	8.5	6.9	7.6
GaAs	12.9	10.9	5.1	5.5
GaSb	16.1	14.4	4.3	4.6
InP	12.4	9.6	5.7	6.5
InAs	14.9	12.3	4.1	4.5
InSb	17.7	15.6	3.5	3.7
SiC	9.6	6.7	14.9	17.9
C	5.5	5.5	25.1	25.1
Si	11.7	11.7	9.9	9.9
Ge	15.8	15.8	5.7	5.7

Data largely due to E. Burstein.

Landau's theory gives a good account of the low-lying single particle excitations of the system of interacting electrons. These single particle excitations are called **quasiparticles;** they have a one-to-one correspondence with the single particle excitations of the free electron gas. A quasiparticle may be thought of as a single particle accompanied by a distortion cloud in the electron gas. One effect of the coulomb interactions between electrons is to change the effective mass of the electron; in the alkali metals the increase is roughly of the order of 25 percent.

Electron-Electron Collisions. It is an astonishing property of metals that conduction electrons, although crowded together only 2 Å apart, travel long distances between collisions with each other. The mean free paths for electron-electron collisions are longer than 10^4 Å at room temperature and longer than 10 cm at 1 K. Two factors are responsible for these long mean free paths, without which the free electron model of metals would have little value: the most powerful factor is the exclusion principle (Fig. 17), and the second factor is the screening of the coulomb interaction between two electrons. We show how the exclusion principle reduces the collision frequency of an electron that has a low excitation energy ϵ_1 outside a filled Fermi sphere (Fig. 18). We estimate the effect of the exclusion principle on the two-body collision $\mathbf{1} + \mathbf{2} \rightarrow \mathbf{3} + \mathbf{4}$ between an electron in the excited orbital **1** and an electron in the filled orbital **2** in the Fermi sea. It is convenient to refer all energies to the Fermi level μ taken as the zero of energy; thus ϵ_1 will be positive and ϵ_2 will be negative. Because of the exclusion principle the orbitals **3** and **4** of the electrons after collision must lie outside the Fermi sphere, all orbitals within the sphere being already occupied; thus both energies ϵ_3, ϵ_4 must be positive.

The conservation of energy requires that $|\epsilon_2| < \epsilon_1$, for otherwise $\epsilon_3 + \epsilon_4 = \epsilon_1 + \epsilon_2$ could not be positive. This means that collisions are possible only if the orbital **2** lies within a shell of thickness ϵ_1 within the Fermi surface, as in Fig. 18a. Thus the fraction $\approx \epsilon_1/\epsilon_F$ of the electrons in filled orbitals provides a suitable target for electron **1.** But even if the target electron **2** is in the suitable energy shell, only a small fraction of the final orbitals compatible with conservation of energy and momentum are allowed by the exclusion principle. This gives a second factor of ϵ_1/ϵ_F. In Fig. 18c we show a small sphere on which all pairs of orbitals **3, 4** at opposite ends of a diameter satisfy the conservation laws, but collisions can occur only if both orbitals **3, 4** lie outside the Fermi sea. The product of the two fractions is $(\epsilon_1/\epsilon_F)^2$. If ϵ_1 corresponds to 1 K and ϵ_F to 5×10^4 K, we have $(\epsilon_1/\epsilon_F)^2 \approx 4 \times 10^{-10}$, the factor by which the exclusion principle reduces the collision rate.

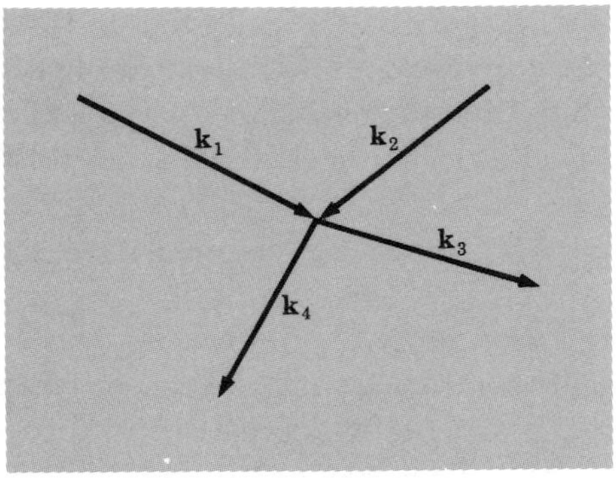

Figure 17 A collision between two electrons of wavevector $\mathbf{k}_1$ and $\mathbf{k}_2$. After the collision the particles have wavevector $\mathbf{k}_3$ and $\mathbf{k}_4$. The Pauli exclusion principle allows collisions only to final states $\mathbf{k}_3$, $\mathbf{k}_4$ which were vacant before the collision.

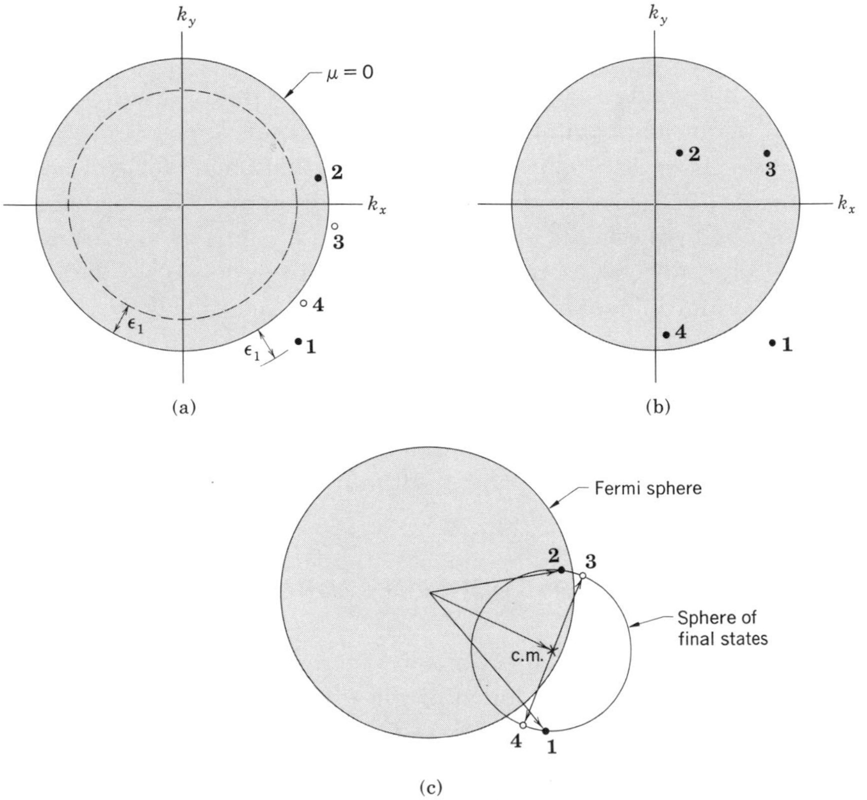

Figure 18 In (a) the electrons in initial orbitals **1** and **2** collide. If the orbitals **3** and **4** are initially vacant, the electrons **1** and **2** can occupy orbitals **3** and **4** after the collision. Energy and momentum are conserved. In (b) the electrons in initial orbitals **1** and **2** have no vacant final orbitals available that allow energy to be conserved in the collision. Orbitals such as **3** and **4** would conserve energy and momentum, but they are already filled with other electrons. In (c) we have denoted with × the wavevector of the center of mass **1** and **2**. All pairs of orbitals **3** and **4** conserve momentum and energy if they lie at opposite ends of a diameter of the small sphere. The small sphere was drawn from the center of mass to pass through **1** and **2**. But not all pairs of points **3, 4** are allowed by the exclusion principle, for both **3, 4** must lie outside the Fermi sphere: the fraction allowed is $\approx \epsilon_1/\epsilon_F$.

The argument is not changed for a thermal distribution of electrons at a low temperature such that $k_BT \ll \epsilon_F$. We replace ϵ_1 by the thermal energy $\approx k_BT$, and now the rate at which electron-electron collisions take place is reduced below the classical value by $(k_BT/\epsilon_F)^2$, so that the effective collision cross section is

$$\sigma \approx (k_BT/\epsilon_F)^2\sigma_0 \ , \tag{59}$$

where σ_0 is the cross section for the screened coulomb interaction. The interaction of an electron with another electron has a range of the order of the screening length $1/k_s$. From numerical calculations the effective cross section with screening for collisions between electrons is of the order of 10^{-15} cm^2 or 10 Å^2 in typical metals. The effect of screening in electron-electron collisions is to reduce σ_0 below the value expected from the Rutherford scattering equation for the unscreened coulomb potential, but much the greatest reduction in σ is caused by the Pauli factor $(k_BT/\epsilon_F)^2$.

At room temperature in a typical metal k_BT/ϵ_F is $\sim 10^{-2}$, so that $\sigma \sim 10^{-4}\sigma_0 \sim 10^{-19}$ cm^2. The mean free path for electron-electron collisions is $\ell \approx 1/n\sigma \sim 10^{-4}$ cm at room temperature. This is longer than the mean free path due to electron-phonon collisions by at least a factor of 10, so that at room temperature collisions with phonons are dominant. At liquid helium temperatures a contribution proportional to T^2 has been reported[11] in the resistivity of indium and aluminum, consistent with the electron-electron scattering cross section (59). The mean free path in indium at 2 K is of the order of 30 cm, as expected. Thus the Pauli principle explains one of the central problems of the theory of metals: how do the electrons travel long distances without colliding with each other?

ELECTRON-PHONON INTERACTION: POLARONS

The most common effect of the electron-phonon interaction is seen in the temperature dependence of the electrical resistivity, which for pure copper is 1.55 microhm-cm at 0 °C and 2.28 microhm-cm at 100 °C. The electrons are scattered by the phonons, and the higher the temperature, the more phonons there are and hence more scattering. Above the Debye temperature (70 °C for copper) the number of thermal phonons is roughly proportional to the absolute temperature, and we find that the resistivity increases as the absolute temperature in any reasonably pure metal in this temperature region. A simple calculation is found in the second edition of this text, pp. 302–304.

[11] J. C. Garland and R. Bowers, Phys. Rev. Letters **21**, 1007 (1968).

A more subtle effect of the electron-phonon interaction is the apparent increase in electron mass in metals and insulators because the electron drags the heavy ion cores along with it. In an insulator the combination of the electron and its strain field is known as a **polaron,**[12] Fig. 19. The effect is large in ionic crystals because of the strong coulomb interaction between ions and electrons. In covalent crystals the effect is weak because neutral atoms have only a weak interaction with electrons.

The strength of the electron-lattice interaction is measured by the dimensionless coupling constant α given by

$$\frac{1}{2}\alpha = \frac{\text{deformation energy}}{\hbar\omega_L} , \tag{60}$$

where ω_L is the longitudinal optical phonon frequency near zero wavevector. We view $\frac{1}{2}\alpha$ as "the number of phonons which surround a slow-moving electron in a crystal." Values deduced from diverse experiments and theory are given in Table 4 after F. C. Brown. The values of α are high in ionic crystals and low in covalent crystals. The values of the effective mass m^*_{pol} of the polaron are from cyclotron resonance experiments. The values given for the band effective mass m^* were calculated from m^*_{pol}. The last row in the table gives the factor m^*_{pol}/m^* by which the band mass is increased by the deformation of the lattice. Theory relates the effective mass of the polaron m^*_{pol} to the effective band mass m^* of the electron in the undeformed lattice by the relation[13]

$$m^*_{\text{pol}} \cong m^* \left(\frac{1 - 0.0008\alpha^2}{1 - \frac{1}{6}\alpha + 0.0034\alpha^2} \right) ; \tag{61}$$

for $\alpha \ll 1$ this is approximately $m^*(1 + \frac{1}{6}\alpha)$. Because the coupling constant α is always positive, the polaron mass is greater than the bare mass, as we expect from the inertia of the ions.

It is common to speak of large and small polarons. The electron associated with a large polaron moves in a band, but the mass is slightly enhanced; these are the polarons we have discussed above. The electron associated with a small polaron[14] spends most of its time trapped on a single ion. At high temperatures the electron moves from site to site by thermally activated hopping; at low temperatures the electron tunnels slowly through the crystal, as if in a band of large effective mass.

[12] See *QTS*, Chap. 7; also C. G. Kuper and G. D. Whitfield, ed., *Polarons and excitons*, Plenum Press, New York, 1963, particularly the article "Experiments on the polaron" by F. C. Brown, I. G. Austen, and N. F. Mott, "Polarons in crystalline and non-crystalline materials," Advances in Physics **18**, 41 (1969).
[13] D. C. Langreth, Phys. Rev. **159**, 717 (1967); see also *QTS*, p. 141.
[14] The theory of small polarons is treated by T. Holstein, Annals of Physics **8**, 343 (1959); see also references given by J. Appel, *Solid state physics* **21**, 193 (1968).

TABLE 4 Polaron Coupling Constants α, Masses m^*_{pol}, and Band Masses m^* for Electrons in the Conduction Band

Crystal	KCl	KBr	AgCl	AgBr	ZnO	PbS	InSb	GaAs
α	3.97	3.52	2.00	1.69	0.85	0.16	0.014	0.06
m^*_{pol}/m	1.25	0.93	0.51	0.33	—	—	0.014	—
m^*/m	0.50	0.43	0.35	0.24	—	—	0.014	—
m^*_{pol}/m^*	2.5	2.2	1.5	1.4	—	—	1.0	—

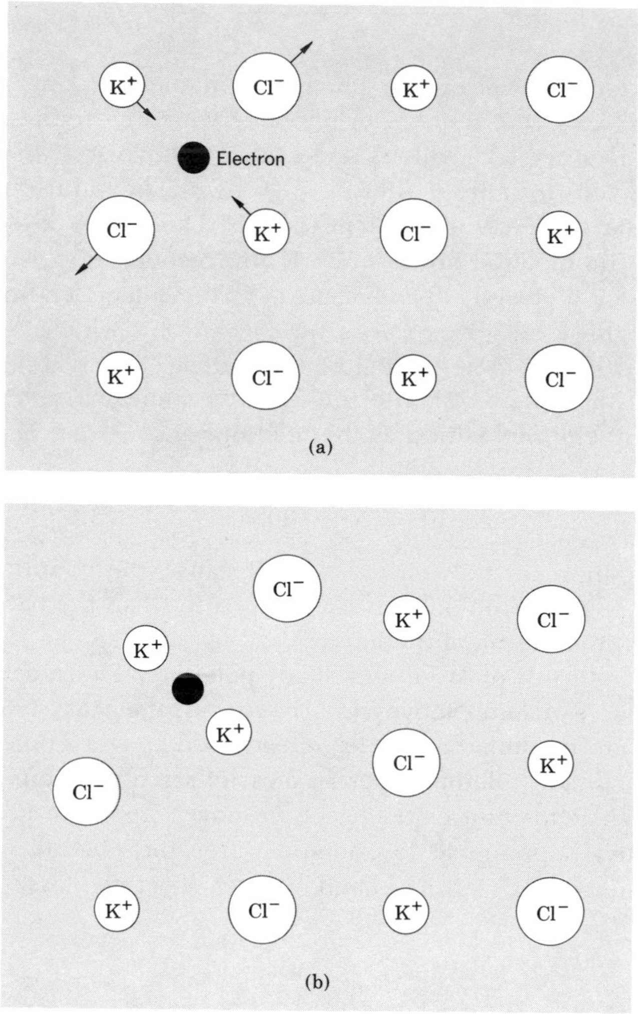

Figure 19 The formation of a polaron. (a) A conduction electron is shown in a rigid lattice of an ionic crystal, KCl. The forces on the ions adjacent to the electron are shown. (b) The electron is shown in an elastic or deformable lattice. The electron plus the associated strain field is called a polaron. The displacement of the ions increases the effective inertia and, hence, the effective mass of the electron; in KCl the mass is increased by a factor of 2.5 with respect to the band theory mass in a rigid lattice. In extreme situations, often with holes, the particle can become self-trapped (localized) in the lattice. In covalent crystals the forces on the atoms from the electron are weaker than in ionic crystals, so that polaron deformations are small in covalent crystals.

Holes or electrons can become self-trapped by inducing an asymmetric local deformation of the lattice. This is most likely to occur when the band edge is degenerate and the crystal is polar (such as an alkali halide or silver halide), with strong coupling of the particle to the lattice. The valence band edge is more often degenerate than the conduction band edge. Holes appear to be self-trapped in all the alkali and silver halides.

Ionic solids at room temperature generally have very low conductivities for the motion of ions through the crystal, less than 10^{-6} (ohm-cm)$^{-1}$, but a family of compounds has been reported[15] with conductivities of 0.2 (ohm-cm)$^{-1}$ at 20°C. The compounds have the composition MAg_4I_5, where M denotes K, Rb, or NH_4. The Ag^+ ions occupy only a fraction of the equivalent lattice sites available, and the ionic conductivity proceeds by the hopping of a silver ion from one site to a nearby vacant site. The crystal structures also have parallel open channels.

PEIERLS INSTABILITY OF LINEAR METALS

Consider a one-dimensional metal with an electron gas filling all conduction band orbitals out to the wavevector k_F, at absolute zero of temperature. Peierls suggested that such a linear metal is unstable with respect to a static lattice deformation of wavevector $G = 2k_F$. Such a deformation creates an energy gap at the Fermi surface, thereby lowering the energy of electrons below the energy gap, Fig. 20. The deformation proceeds until limited by the increase of elastic energy: the equilibrium deformation Δ is given by the root of

$$\frac{d}{d\Delta}\left(E_{\text{electronic}} + E_{\text{elastic}}\right) = 0 \ . \tag{62}$$

Consider the elastic strain $\Delta \cos 2k_F x$. The spatial-average elastic energy per unit length is $E_{\text{elastic}} = \frac{1}{2}C\Delta^2\langle\cos^2 2k_F x\rangle = \frac{1}{4}C\Delta^2$, where C is the force constant of the linear metal. We next calculate $E_{\text{electronic}}$. Suppose that the lattice potential seen by a conduction electron is proportional to the deformation: $U(x) = 2A\Delta \cos 2k_F x$. From (7.41) we have

$$\epsilon_K = (\hbar^2/2m)(k_F^2 + K^2) \pm [4(\hbar^2 k_F^2/2m)(\hbar^2 K^2/2m) + A^2\Delta^2]^{\frac{1}{2}} \ . \tag{63}$$

It is convenient to define

$$x_K \equiv \hbar^2 K^2/m \ ; \qquad x_F \equiv \hbar^2 k_F^2/m \ ; \qquad x \equiv \hbar^2 K k_F/m \ .$$

We form

$$\frac{d\epsilon_K}{d\Delta} = \frac{-A^2\Delta}{(x_F x_K + A^2\Delta^2)^{\frac{1}{2}}} \ ,$$

[15] See B. B. Owens and G. R. Argue, Science **157**, 308 (1967); S. Geller, Science **157**, 310 (1967).

whence, with $1/\pi$ as the number of orbitals per unit length,

$$\frac{dE_{\text{electronic}}}{d\Delta} = \frac{2}{\pi}\int_0^{k_F} dK\,\frac{d\epsilon_K}{d\Delta} = -(2A^2\Delta/\pi)\int_0^{k_F}\frac{dK}{(x_F x_K + A^2\Delta^2)^{\frac{1}{2}}}$$

$$= -(2A^2\Delta/\pi)\,(k_F/x_F)\int_0^{x_F}\frac{dx}{(x^2 + A^2\Delta^2)^{\frac{1}{2}}} = -\,(2A^2\Delta/\pi)\,(k_F/x_F)\,\sinh^{-1}(x_F/A\Delta)\ .$$

We put it all together. The equilibrium deformation is the root of

$$\tfrac{1}{2}C\Delta - (2A^2 m\Delta/\pi\hbar^2 k_F)\,\sinh^{-1}(\hbar^2 k_F{}^2/mA\Delta) = 0\ .$$

The root Δ that corresponds to the minimum energy is given by

$$\hbar^2 k_F{}^2/mA\Delta = \sinh(-\hbar^2 k_F \pi C/4mA^2)\ , \tag{64}$$

whence

$$|A|\,\Delta \simeq (2\hbar^2 k_F{}^2/m)\,\exp(-\hbar^2 k_F \pi C/4mA^2)\ , \tag{65}$$

if the argument of the sinh in (64) is $\gg 1$.

This result is of the form of the energy gap equation in the BCS theory of superconductivity, Chapter 12. The deformation Δ is a collective effect of all the electrons. If we set $W = \hbar^2 k_F{}^2/2m$ = conduction band width; $N(0) = 2m/\pi\hbar^2 k_F$ = density of orbitals at Fermi level; $V = 2A^2/C$ = effective electron-electron interaction energy, then we can write (65) as

$$|A|\,\Delta \simeq 4W\exp[-1/N(0)V]\ , \tag{66}$$

which is analogous to the BCS energy gap equation.[16]

---

[16]R. E. Merrifield, "Exciton analog of the Peierls instability of one-dimensional metals", J. Chem. Phys. **44,** 4005 (1966). The result (66) is attributed to D. J. Scalapino. The competition between superconductivity and ferroelectricity is discussed for a one-dimensional metal with an array of polarizable side groups. See, however, R. A. Harris and L. M. Falicov, "Self-consistent theory of bond alternation in polyenes: normal state, charge-density waves, and spin-density waves," J. Chem. Phys. **51,** 5034 (1969). A review of charge density waves and superlattices in metallic layered structures is given by J. A. Wilson, F. J. Di Salvo and S. Mahajan, Advances in Physics **24,** 117 (1975); this article contains some of the most beautiful diffraction photographs in solid state physics.

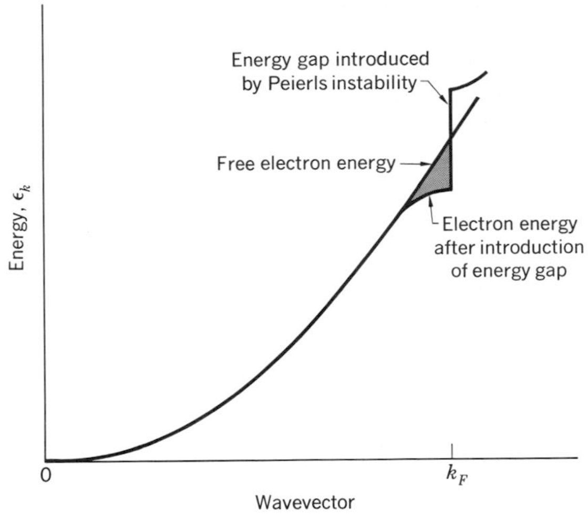

Figure 20 Peierls instability. Electrons with wavevectors near the Fermi surface have their energy lowered by a lattice deformation.

SUMMARY

1. The dielectric function may be defined as

$$\epsilon(\omega,\mathbf{K}) = \frac{\rho_{\text{ext}}(\omega,\mathbf{K})}{\rho_{\text{ext}}(\omega,\mathbf{K}) + \rho_{\text{ind}}(\omega,\mathbf{K})} ,$$

in terms of the applied and induced charge density components at $\omega,\mathbf{K}$.

2. The plasma frequency $\tilde{\omega}_p = [4\pi ne^2/\epsilon(\infty)m]^{\frac{1}{2}}$ is the frequency of the uniform collective longitudinal oscillation of the electron gas against a background of fixed positive ions. It is also the low cutoff for propagation of transverse electromagnetic waves in the plasma.

3. The poles of the dielectric function define ω_T and the zeros define ω_L.

4. In a plasma the coulomb interaction is screened; it becomes $(q/r)\exp(-k_s r)$, where the screening length $1/k_s = (\epsilon_F/6\pi n_0 e^2)^{\frac{1}{2}}$.

5. A metal-insulator transition may occur when the nearest-neighbor separation a becomes of the order of $4a_0$, where a_0 is the radius of the first Bohr orbit in the insulator. The metallic phase exists at smaller values of a.

6. A polariton is a quantum of the coupled TO phonon-photon fields. The coupling is assured by the Maxwell equations. The spectral region $\omega_T < \omega < \omega_L$ is forbidden to electromagnetic wave propagation.

7. The Lyddane-Sachs-Teller relation is $\omega_L{}^2/\omega_T{}^2 = \epsilon(0)/\epsilon(\infty)$.

PROBLEMS

1. ***Surface plasmons.*** Consider a semi-infinite plasma on the positive side of the plane $z = 0$. A solution of Laplace's equation $\nabla^2\varphi = 0$ in the plasma is $\varphi_i(x,z) = A \cos kx\, e^{-kz}$, whence $E_{zi} = kA \cos kx\, e^{-kz}$; $E_{xi} = kA \sin kx\, e^{-kz}$. (a) Show that in the vacuum $\varphi_0(x,z) = A \cos kx\, e^{kz}$ for $z < 0$ satisfies the boundary condition that the tangential component of **E** be continuous at the boundary; that is, find E_{xo}. (b) Note that $\mathbf{D}_i = \epsilon(\omega)\mathbf{E}_i$; $\mathbf{D}_o = \mathbf{E}_o$. Show that the boundary condition that the normal component of **D** be continuous at the boundary requires that $\epsilon(\omega) = -1$, whence from (10) we have the Stern-Ferrell result:

$$\omega_s^2 = \tfrac{1}{2}\omega_p^2 \tag{67}$$

for the frequency ω_s of a surface plasma oscillation.

2. ***Interface plasmons.*** We consider the plane interface $z = 0$ between metal 1 at $z > 0$ and metal 2 at $z < 0$. Metal 1 has bulk plasmon frequency ω_{p1}; metal 2 has ω_{p2}. The dielectric constants in both metals are those of free electron gases. Show that surface plasmons associated with the interface have the frequency

$$\omega = [\tfrac{1}{2}(\omega_{p1}^2 + \omega_{p2}^2)]^{\frac{1}{2}} \ .$$

3. ***Alfvén waves.*** Consider a solid with an equal concentration n of electrons of mass m_e and holes of mass m_h. This situation may arise in a semimetal or in a compensated semiconductor. Place the solid in a uniform magnetic field $\mathbf{B} = B\hat{\mathbf{z}}$. Introduce the coordinate $\xi = x + iy$ appropriate for circularly polarized motion, with ξ having time dependence $e^{-i\omega t}$. Let $\omega_e \equiv eB/m_ec$ and $\omega_h \equiv eB/m_hc$. (a) In CGS units, show that $\xi_e = eE^+/m_e\omega(\omega + \omega_e)$; $\xi_h = -eE^+/m_h\omega(\omega - \omega_h)$ are the displacements of the electrons and holes in the electric field $E^+ e^{-i\omega t} = (E_x + iE_y)\, e^{-i\omega t}$. (b) Show that the dielectric polarization $P^+ = ne(\xi_h - \xi_e)$ in the regime $\omega \ll \omega_e$, ω_h may be written as $P^+ = nc^2(m_h + m_e)E^+/B^2$, and the dielectric function $\epsilon(\omega) \equiv \epsilon_l + 4\pi P^+/E^+ = \epsilon_l + 4\pi c^2\rho/B^2$, where ϵ_l is the dielectric constant of the host lattice and $\rho \equiv n(m_e + m_n)$ is the mass density of the carriers. If ϵ_l may be neglected, the dispersion relation $\omega^2\epsilon(\omega) = c^2K^2$ becomes, for electromagnetic waves propagating in the z direction, $\omega^2 = (B^2/4\pi\rho)K^2$. Such waves are known as Alfvén waves; they propagate with the constant velocity $B/(4\pi\rho)^{\frac{1}{2}}$. If $B = 10$ kG; $n = 10^{18}$ cm^{-3}; $m = 10^{-27}$ g, the velocity is $\sim 10^8$ cm s^{-1}. Alfvén waves have been observed in semimetals and in electron-hole drops in germanium (Chapter 11).

4. ***Helicon waves.*** (a) Employ the method of Problem 3 to treat a specimen with only one carrier type, say holes in concentration p, and in the limit $\omega \ll \omega_h = eB/m_hC$. Show that $\epsilon(\omega) \cong 4\pi pe^2/m_h\omega\omega_h$, where $D^+(\omega) = \epsilon(\omega)E^+(\omega)$. The term ϵ_l in ϵ has been neglected. (b) Show further that the dispersion relation becomes $\omega = (Bc/4\pi pe)K^2$, the helicon dispersion relation, in CGS. For $K = 1$ cm^{-1} and $B = 1000$ G, estimate the helicon frequency in sodium metal. (The frequency is negative; with circular-polarized modes the sign of the frequency refers to the sense of the rotation.) For experiments, see R. Bowers *et al.*, Phys. Rev. Letters **7**, 339 (1961).

5. ***Plasmon mode of a sphere.*** The frequency of the uniform plasmon mode of a sphere is determined by the depolarization field $\mathbf{E} = -4\pi\mathbf{P}/3$ of a sphere, where the polarization $\mathbf{P} = -ne\mathbf{r}$, with $\mathbf{r}$ as the average displacement of the electrons of concentration n. Show from $\mathbf{F} = m\mathbf{a}$ that the resonance frequency of the electron gas is $\omega_0^2 = 4\pi ne^2/3m$. Because all electrons participate in the oscillation, such an excitation is called a collective excitation or collective mode of the electron gas.

6. ***Magnetoplasma frequency.*** Use the method of Problem 5 to find the frequency of the uniform plasmon mode of a sphere placed in a constant uniform magnetic field $\mathbf{B}$. Let $\mathbf{B}$ be along the z axis. The solution should go to the cyclotron frequency $-\omega_c = -eB/mc$ in one limit and to $\omega_0 = (4\pi ne^2/3m)^{\frac{1}{2}}$ in another limit. Take the motion in the $x - y$ plane.

7. ***Photon branch at low wavevector.*** (a) Find what (52) becomes when $\epsilon(\infty)$ is taken into account. (b) Show that there is a solution of (51) which at low wavevector is $\omega = cK/\sqrt{\epsilon(0)}$, which is what you expect for a photon in a crystal of refractive index $n^2 = \epsilon$.

8. ***Plasma frequency and electrical conductivity.*** An organic conductor has recently been found by optical studies to have $\omega_p = 1.80 \times 10^{15}$ s^{-1} for the plasma frequency, and $\tau = 2.83 \times 10^{-15}$ s for the electron relaxation time at room temperature. (a) Calculate the electrical conductivity from these data. The carrier mass is not known and is not needed here. Take $\epsilon(\infty) = 1$. Convert the result to units $(\Omega\ \text{cm})^{-1}$. (b) From the crystal and chemical structure, the conduction electron concentration is 4.7×10^{21} cm^{-3}. Calculate the electron effective mass m^*.

9. ***Bulk modulus of the Fermi gas.*** Show that the contribution of the kinetic energy to the bulk modulus of the electron gas at absolute zero is $B = \frac{1}{3}nmv_F^2$. It is convenient to use (3.18). We can use our result for B to find the velocity of sound, which in a compressible fluid is $v = (B/\rho)^{\frac{1}{2}}$, whence $v = (m/3M)^{\frac{1}{2}}v_F$, in agreement with (46). These estimates neglect attractive interactions.

10. ***Response of electron gas.*** It is sometimes stated erroneously in books on electromagnetism that the static conductivity σ, which in gaussian units has the dimensions of a frequency, measures the response frequency of a metal to an electric field suddenly applied. Criticize this statement as it might apply to copper at room temperature. The resistivity is ~ 1 μohm-cm; the electron concentration is 8×10^{22} cm^{-3}; the mean free path is ~ 400 Å; the Fermi velocity is 1.6×10^8 cm s^{-1}. You will not necessarily need all these data. Give the order of magnitude of the three frequencies σ, ω_p, and $1/\tau$ that might be relevant in the problem. Set up and solve the problem of the response $x(t)$ of the system to an electric field $E(t < 0) = 0$; $E(t > 0) = 1$. The system is a sheet of copper; the field is applied normal to the sheet. Include the damping. Solve the differential equation by elementary methods.

*11. ***Gap plasmons and the van der Waals interaction.*** Consider two semi-infinite media with plane surfaces $z = 0, d$. The dielectric function of the identical media is $\epsilon(\omega)$. Show that for surface plasmons symmetrical with respect to the gap the frequency must satisfy $\epsilon(\omega) = -\tanh(Kd/2)$, where $K^2 = k_x^2 + k_y^2$. The

*This problem is somewhat difficult.

electric potential will have the form

$$\varphi = f(z)\,\exp(ik_x x + ik_y y - i\omega t)\ .$$

Look for nonretarded solutions—that is, solutions of the Laplace equation rather than of the wave equation. The sum of the zero-point energy of all gap modes is the nonretarded part of the van der Waals attraction between the two specimens—see N. G. van Kampen, B. R. A. Nijboer, and K. Schram, Physics Letters **26A,** 307 (1968).

References

M. P. Givens, "Optical properties of metals," *Solid state physics* **6,** 313 (1958).

J. N. Hodgson, *Optical absorption and dispersion in solids,* Chapman and Hall, 1970. Good introductory survey.

E. A. Kaner and V. G. Skobov, "Electromagnetic waves in metals in a magnetic field," Advances in Physics **17,** 605 (1968).

D. Pines, "Electron interaction in metals," *Solid state physics* **1,** 367 (1955).

A. C. Smith, J. F. Janak, and R. B. Adler, *Electronic conduction in solids,* McGraw-Hill, 1967.

F. Stern, "Elementary theory of the optical properties of metals," *Solid state physics* **15,** 300 (1963).

H. G. Schuster, ed., *One-dimensional conductors,* Springer, 1975.

Solid State Plasmas

A. C. Baynham and A. D. Boardman, *Plasma effects in semiconductors: helicon and Alfvén waves,* Taylor & Francis, 1971.

J. Daniels, C. v. Festenberg, H. Raether, and K. Zeppenfeld, "Optical constants of solids by electron spectroscopy," *Springer tracts in modern physics* **54,** 78 (1970).

E. N. Economou and K. L. Ngai, "Surface plasma oscillations and related surface effects in solids," *Advances in chemical physics* **27,** 265 (1974).

M. F. Hoyaux, *Solid state plasmas,* Pion, 1970.

P. M. Platzman and P. A. Wolff, *Waves and interactions in solid state plasmas,* Academic Press, 1973. Good general reference; also contains account of Fermi liquid theory and experiments.

H. Raether, "Solid state excitations by electrons; plasma oscillations and single electron transitions," *Springer tracts in modern physics* **38,** 84 (1965).

Polarons

J. T. Devreese, ed., *Polarons in ionic crystals and polar semiconductors,* North-Holland, 1972.

11 OPTICAL PROCESSES AND EXCITONS

NOTE TO THE READER: This chapter treats selected topics from a large field of current interest. Many of the topics may be passed over in a first reading, but the section on excitons is of basic interest.

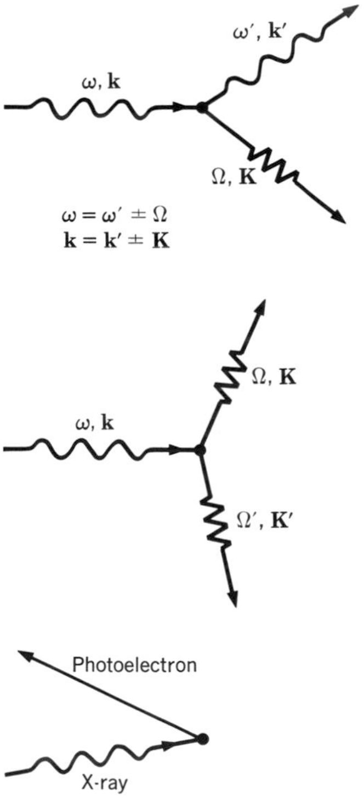

Raman scattering (generic term):
Brillouin scattering when acoustic phonon is involved; polariton scattering when optical phonon is involved.

Signs { + for phonon emission (Stokes process); − for phonon absorption (anti-Stokes) }

Two phonon infrared absorption:
the absorption is strongest for phonons near the zone boundary where the density of modes is a maximum.

Electron spectroscopy with x-rays (XPS):
incident x-ray photon ejects valence or core electron from solid.

Figure 1 There are many types of experiments in which light interacts with wavelike excitations in a crystal. Several processes are illustrated here.

11 OPTICAL PROCESSES AND EXCITONS

The dielectric function $\epsilon(\omega,\mathbf{K})$ was introduced in the preceding chapter to describe the response of a crystal to an electromagnetic field. The dielectric function depends sensitively on the electronic band structure of a crystal, and studies of the dielectric function by optical spectroscopy are very useful in the determination of the overall band structure of a crystal. Indeed, in the last decade optical spectroscopy developed into the single most important experimental tool for band structure determination.

In the infrared, visible, and ultraviolet spectral regions the wavevector K of the radiation is very small compared with the shortest reciprocal lattice vector, and therefore K may usually be taken as zero. We are concerned then with the real ϵ' and imaginary ϵ'' parts of the dielectric function $\epsilon(\omega) = \epsilon'(\omega) + i\epsilon''(\omega)$, also written as $\epsilon_1(\omega) + i\epsilon_2(\omega)$. However, the dielectric function is not directly accessible experimentally from optical measurements: directly accessible are the reflectance power $R(\omega)$, the refractive index $n(\omega)$, and the extinction coefficient $K(\omega)$. Our first objective is to derive relations between the experimentally observable quantities and the real and imaginary parts of the dielectric function.

OPTICAL REFLECTANCE

The optical measurements that give the fullest information on the electronic system are measurements of the reflectivity of light at normal incidence on single crystals. The **reflectivity coefficient** $r(\omega)$ is a complex function defined at the crystal surface as the ratio of the reflected electric field $E(\text{refl})$ to the incident electric field $E(\text{inc})$:

$$E(\text{refl})/E(\text{inc}) \equiv r(\omega) \equiv \rho(\omega)\, e^{i\theta(\omega)} , \qquad (1)$$

where we have separated the amplitude $\rho(\omega)$ and phase $\theta(\omega)$ components of the reflectivity coefficient. The **refractive index** $n(\omega)$ and the **extinction coefficient** $K(\omega)$ in the crystal are related to the reflectivity at normal incidence by

$$r(\omega) = \frac{n + iK - 1}{n + iK + 1} , \qquad (2)$$

as derived in Problem 3 by elementary consideration of the continuity of the components of **E** and **B** parallel to the crystal surface. By definition $n(\omega)$ and $K(\omega)$ are related to the dielectric function $\epsilon(\omega)$ by

$$\sqrt{\epsilon(\omega)} \equiv n(\omega) + iK(\omega) \equiv N(\omega) \ , \tag{3}$$

where $N(\omega)$ is the complex refractive index. Do not confuse $K(\omega)$ as used here with a wavevector. If the incident traveling wave has the wavevector k:

$$E_y(\text{inc}) = E(\text{inc})\ e^{i(kx-\omega t)} \ , \tag{4}$$

the transmitted wave in the medium is attenuated because, by the dispersion relation for electromagnetic waves, the wavevector in the medium is related to the incident k in vacuum by $(n + iK)k$:

$$E_y(\text{trans}) \propto e^{i[(n+iK)kx-\omega t]} = e^{-Kx}\ e^{i(nkx-\omega t)} \ . \tag{5}$$

One quantity measured in experiments is the **reflectance**, which is the ratio of the reflected intensity to the incident intensity:

$$R = E^*\ (\text{refl})E(\text{refl})/E^*\ (\text{inc})E(\text{inc}) = r^*r = \rho^2 \ . \tag{6}$$

It is difficult to measure the phase $\theta(\omega)$ of the reflected wave, but we show below that it can be calculated from the measured reflectance $R(\omega)$ if this is known at all frequencies. Then we know both $R(\omega)$, $\theta(\omega)$, and we can proceed by (2) to obtain $n(\omega)$ and $K(\omega)$. We use these in (3) to obtain $\epsilon(\omega) = \epsilon'(\omega) + i\epsilon''(\omega)$, where $\epsilon'(\omega)$ and $\epsilon''(\omega)$ are the real and imaginary parts of the dielectric function. The inversion of (3) gives

$$\epsilon'(\omega) = n^2 - K^2 \ ; \qquad \epsilon''(\omega) = 2nK \ . \tag{7}$$

We now show how to find the phase $\theta(\omega)$ as an integral over the reflectance $R(\omega)$; by a similar method we relate the real and imaginary parts of the dielectric function. In this way we can find everything from the experimental $R(\omega)$.

Kramers-Kronig Relations

The Kramers-Kronig relations enable us to find the real part of the response of a linear passive system if we know the imaginary part of the response at all frequencies, and vice versa. They are central to the analysis of optical experiments on solids. The response of any linear passive system can be represented as the superposition of the responses of a collection of damped harmonic oscillators. Let the response function $\alpha(\omega) = \alpha'(\omega) + i\alpha''(\omega)$ of the collection of oscillators be defined by

$$x_\omega = \alpha(\omega)\ F_\omega \ , \tag{8}$$

where the applied force F is the real part of $F_\omega \exp(-i\omega t)$ and the total displacement x is the real part of $x_\omega \exp(-i\omega t)$. From the equation of motion,

$$\sum_j M_j(d^2/dt^2 + \rho_j d/dt + \omega_j^2)x = F \ ,$$

we have the complex response function of the oscillator system:

$$\alpha(\omega) = \sum_j \frac{f_j}{\omega_j^2 - \omega^2 - i\omega\rho_j} = \sum f_j \frac{\omega_j^2 - \omega^2 + i\omega\rho_j}{(\omega_j^2 - \omega^2)^2 + \omega^2\rho_j^2} \ , \tag{9}$$

where the constants f_j and relaxation frequencies ρ_j are all positive for a passive system. In a mechanical system with masses M_j, we have $f_j = 1/M_j$. If $\alpha(\omega)$ is the dielectric polarizability of atoms in concentration n, then f has the form of an oscillator strength times ne^2/m; such a dielectric response function is said to be of the Kramers-Heisenberg form. The relations we develop also apply to the electrical conductivity $\sigma(\omega)$ in Ohm's law, $j_\omega = \sigma(\omega)E_\omega$.

We need not assume the specific form (9), but we make use of three properties of the response function viewed as a function of the complex variable ω. Any function with the following properties will satisfy the Kramers-Kronig relations (11):

(a) The poles of $\alpha(\omega)$ are all below the real axis.
(b) The integral of $\alpha(\omega)/\omega$ vanishes when taken around an infinite semicircle in the upper half of the complex ω-plane. It suffices that $\alpha(\omega) \to 0$ uniformly as $|\omega| \to \infty$.
(c) The function $\alpha'(\omega)$ is even and $\alpha''(\omega)$ is odd with respect to real ω.

Consider the Cauchy integral in the form

$$\alpha(\omega) = \frac{1}{\pi i} \mathrm{P} \int_{-\infty}^{\infty} \frac{\alpha(s)}{s - \omega} ds \ , \tag{10}$$

where P denotes the principal part of the integral, as discussed in the mathematical note that follows. The right-hand side is to be completed by an integral over the semicircle at infinity in the upper half-plane; but we have seen in (b) that this integral vanishes. We equate the real parts of (10) to obtain

$$\alpha'(\omega) = \frac{1}{\pi} \mathrm{P} \int_{-\infty}^{\infty} \frac{\alpha''(s)}{s - \omega} ds = \frac{1}{\pi} \mathrm{P} \left[\int_0^{\infty} \frac{\alpha''(\omega)}{s - \omega} ds + \int_{-\infty}^{0} \frac{\alpha''(p)}{p - \omega} dp \right] .$$

In the last integral we substitute s for $-p$ and use property (c) that $\alpha''(-s) = -\alpha''(s)$; this integral then becomes

$$\int_0^{\infty} \frac{\alpha''(s)}{s + \omega} ds,$$

and we have, with

$$\frac{1}{s - \omega} + \frac{1}{s + \omega} = \frac{2s}{s^2 - \omega^2} \ ,$$

the result

$$\boxed{\alpha'(\omega) = \frac{2}{\pi} \text{P} \int_0^\infty \frac{s\alpha''(s)}{s^2 - \omega^2} ds \ .} \tag{11a}$$

This is one of the Kramers-Kronig relations. The other relation follows on equating the imaginary parts of Eq. (10):

$$\alpha''(\omega) = -\frac{1}{\pi} \text{P} \int_{-\infty}^\infty \frac{\alpha'(s)}{s - \omega} ds = -\frac{1}{\pi} \text{P} \left[\int_0^\infty \frac{\alpha'(s)}{s - \omega} ds - \int_0^\infty \frac{\alpha'(s)}{s + \omega} ds \right] ,$$

whence

$$\boxed{\alpha''(\omega) = -\frac{2\omega}{\pi} \text{P} \int_0^\infty \frac{\alpha'(s)}{s^2 - \omega^2} ds \ .} \tag{11b}$$

These relations are applied below to the analysis of optical reflectance data; this is their most important application.

Let us apply the Kramers-Kronig relations to $r(\omega)$ viewed as a response function between the incident and reflected waves in Eq. (1). We apply (11) to

$$\ln r(\omega) = \ln R^{\frac{1}{2}}(\omega) + i\theta(\omega) \tag{12}$$

to obtain the phase in terms of the reflectance:

$$\theta(\omega) = -\frac{\omega}{\pi} \text{P} \int_0^\infty \frac{\ln R(s)}{s^2 - \omega^2} ds \ . \tag{13}$$

We integrate by parts to obtain a form that gives insight into the contributions to the phase angle:

$$\theta(\omega) = -\frac{1}{2\pi} \int_0^\infty \ln \left| \frac{s + \omega}{s - \omega} \right| \frac{d \ln R(s)}{ds} ds \ . \tag{14}$$

Spectral regions in which the reflectance is constant do not contribute to the integral; further, spectral regions $s \gg \omega$ and $s \ll \omega$ do not contribute much because the function $\ln |(s + \omega)/(s - \omega)|$ is small in these regions.

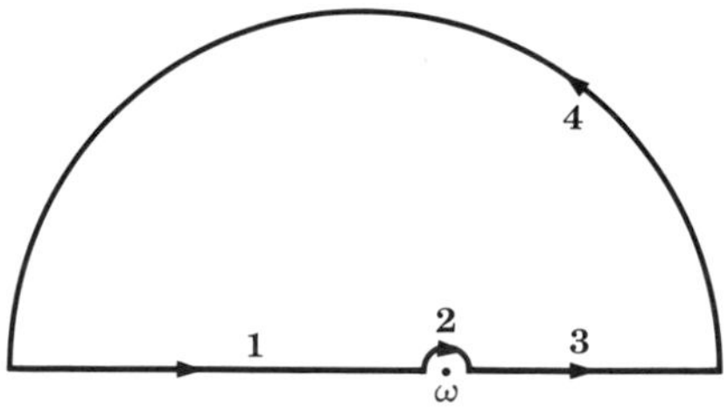

Figure 2 Contour for the Cauchy principal value integral.

Mathematical Note. To obtain the Cauchy integral (10) we take the integral $\int \alpha(s)(s-\omega)^{-1}ds$ over the contour[1] in Fig. 2. The function $\alpha(s)$ is analytic in the upper half-plane, so that the value of the integral is zero. The contribution of segment **4** to the integral vanishes if the integrand $\alpha(s)/s \to 0$ faster than $|s|^{-1}$ as $|s| \to \infty$. For the response function (9) the integrand $\to 0$ as $|s|^{-3}$; and for the conductivity $\sigma(s)$ the integrand $\to 0$ as $|s|^{-2}$. The segment **2** contributes, in the limit as $u \to 0$,

$$\int_{(2)} \frac{\alpha(s)}{s-\omega}\, ds \to \alpha(\omega) \int_{\pi}^{0} \frac{iu\, e^{i\theta}\, d\theta}{u\, e^{i\theta}} = -\pi i \alpha(\omega)$$

to the integral, where $s = \omega + u\, e^{i\theta}$. The segments **1** and **3** are by definition the principal part of the integral between $-\infty$ and ∞. Because the integral over **1 + 2 + 3 + 4** must vanish,

$$\int_{(1)} + \int_{(3)} \equiv \mathrm{P} \int_{-\infty}^{\infty} \frac{\alpha(s)}{s-\omega}\, ds = \pi i \alpha(\omega) \ .$$

EXAMPLE: ***Conductivity of collisionless electron gas.*** Consider a gas of free electrons in the limit as the collision frequency goes to zero. From (9) the response function is, with $f = 1/m$,

$$\alpha(\omega) = -\frac{1}{m\omega} \lim_{\rho \to 0} \frac{1}{\omega + i\rho} = -\frac{1}{m\omega}\left[\frac{1}{\omega} - i\pi\delta(\omega)\right] , \tag{15}$$

by the Dirac identity. We confirm that the delta function in (15) satisfies the Kramers-Kronig relation (11a), by which

$$\alpha'(\omega) = \frac{2}{m}\int_0^{\infty} \frac{\delta(s)}{s^2 - \omega^2}\, ds = -\frac{1}{m\omega^2} , \tag{16}$$

in agreement with (12).

We obtain the electrical conductivity $\sigma(\omega)$ from the dielectric function

$$\epsilon(\omega) - 1 = 4\pi P_\omega/E_\omega = -4\pi n e x_\omega/E_\omega = 4\pi n e^2 \alpha(\omega) , \tag{17}$$

where $\alpha(\omega) = x_\omega/(-e)E_\omega$ is the response function. We use the equivalence

$$\sigma(\omega) = (-i\omega/4\pi)[\epsilon(\omega) - 1] , \tag{18}$$

for the Maxwell equation can be written either as c curl $\mathbf{H} = 4\pi\sigma(\omega)\mathbf{E} - i\omega\mathbf{E}$ or as c curl $\mathbf{H} = -i\omega\epsilon(\omega)\mathbf{E}$. We combine (15), (17), and (18) to find the conductivity of a collisionless electron gas:

$$\sigma'(\omega) + i\sigma''(\omega) = \frac{ne^2}{m}\left[\pi\delta(\omega) + \frac{i}{\omega}\right] . \tag{19}$$

For collisionless electrons the real part of the conductivity has a delta function at $\omega = 0$.

---

[1]See E. T. Whittaker and G. N. Watson, *Modern analysis*, Cambridge, 1935, p. 117.

In Fig. 3 we show the experimental reflectance spectrum of a crystal of CdS, taken with E perpendicular to the c axis of the crystal. By a Kramers-Kronig inversion of this spectrum, we obtain the refractive index $n(\omega)$, extinction coefficient $K(\omega)$, and the real part $\epsilon'(\omega)$ and imaginary part $\epsilon''(\omega)$ of the dielectric function, as shown in Figs. 4, 5, 6, and 7, all credited to M. Balkanski. There is also plotted in Fig. 7 the energy loss function, defined as the imaginary part of $-\epsilon^{-1}$:

$$\frac{1}{\epsilon}=\frac{\epsilon'-i\epsilon''}{(\epsilon')^2+(\epsilon'')^2}\ ; \qquad -\mathrm{Im}\{\epsilon^{-1}\}=\frac{\epsilon''}{(\epsilon')^2+(\epsilon'')^2}=\frac{2nK}{(n^2+K^2)^2}\ . \tag{20}$$

The spectrum of the energy loss function describes the loss of energy of a charged particle that passes through a solid, as discussed below.

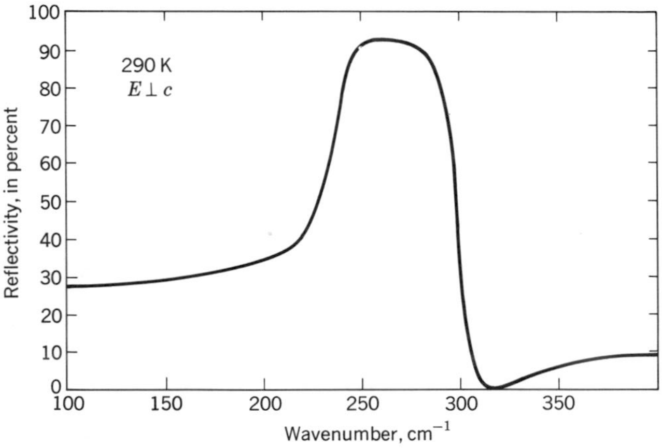

Figure 3 Reflectance spectrum for CdS single crystals. (After Balkanski.)

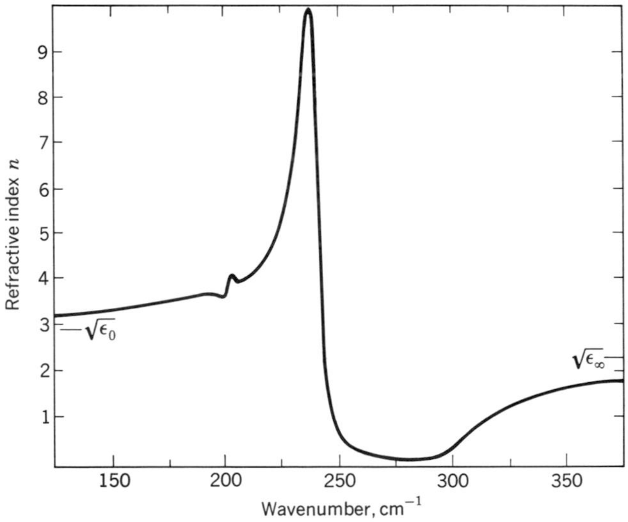

Figure 4 Refractive index for CdS.

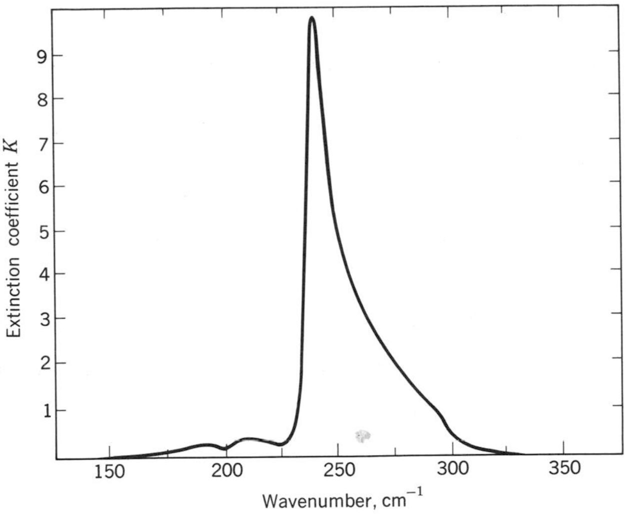

Figure 5 Extinction coefficient for CdS.

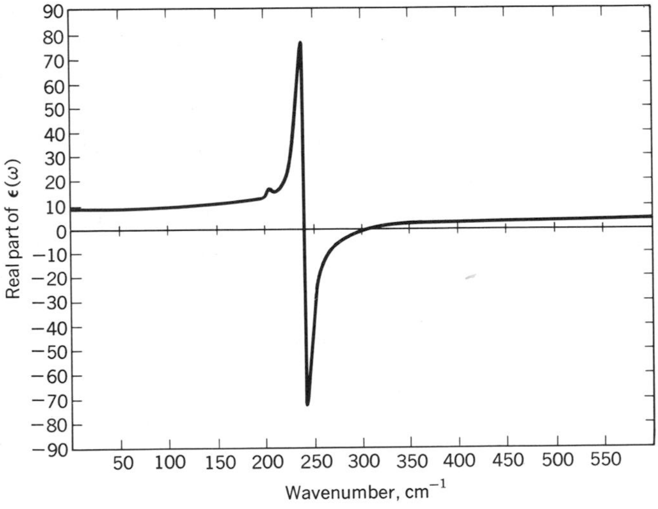

Figure 6 Real part of the dielectric constant of CdS.

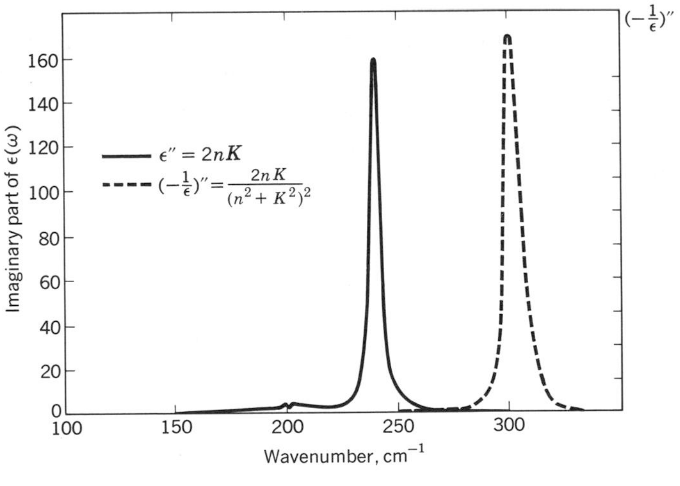

Figure 7 Imaginary part of the dielectric constant and the function $(-1/\epsilon)'' = 2nK/(n^2 + K^2)^2$ as a function of the frequency, for CdS.

Electronic Interband Transitions

It came as a surprise that optical spectroscopy developed as an important experimental tool for the determination of band structure. It was a surprise for two reasons. First, the absorption and reflection bands of crystals are broad and apparently featureless functions of the photon energy when this is greater than the band gap. Second, direct interband absorption of a photon $\hbar\omega$ will occur at all points in the Brillouin zone for which energy is conserved:

$$\hbar\omega = \epsilon_c(\mathbf{k}) - \epsilon_v(\mathbf{k}) \ , \tag{21}$$

where c is an empty band and v is a filled band. The total absorption at given ω is an integral over all transitions in the zone that satisfy (21). There seemed, therefore, little reason to expect one photon spectroscopy to be the key to band structure.

Three factors unraveled the spectra:

1. The broad bands are not like a spectral line greatly broadened by damping, but the bands convey much intelligence which emerges when derivatives are taken of the reflectance (Figs. 8 and 9): derivatives with respect to wavelength, electric field, temperature, pressure, or uniaxial stress, for example. The spectroscopy of derivatives is called modulation spectroscopy.
2. The relation (21) does not exclude spectral structure in a crystal, because transitions accumulate at frequencies for which the bands c, v are parallel—that is, at frequencies where

$$\nabla_{\mathbf{k}}[\epsilon_c(\mathbf{k}) - \epsilon_v(\mathbf{k})] = 0 \ . \tag{22}$$

   At these critical points in $\mathbf{k}$ space the joint density of states $\mathcal{D}_c(\epsilon_v + \hbar\omega)\mathcal{D}_v(\epsilon_v)$ is singular, according to the same argument we used in (5.26) to show that the density of phonon modes $\mathcal{D}(\omega)$ is singular when $\nabla_{\mathbf{k}}\omega$ is zero.
3. The empirical pseudopotential method for calculating energy bands is helpful for identifying the positions in the Brillouin zone of the critical points found in modulation spectra. Band-band energy differences can be calculated with an accuracy as good as 0.1 eV. The experimental results can then be fed back to give improvements in the pseudopotential calculations. Most modern energy band calculations are adjusted or reconciled to results of optical studies of interband electronic transitions.

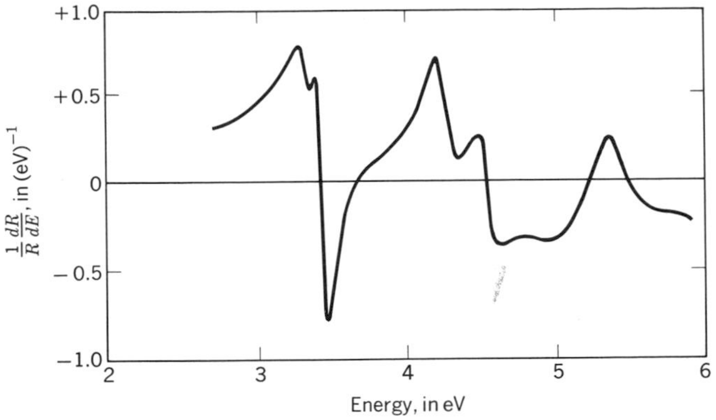

Figure 8 Wavelength modulation reflectance spectra of silicon at 5 K. [After R. R. L. Zucca, J. P. Walter, Y. R. Shen, and M. L. Cohen, Solid State Commun. **8**, 627 (1970).]

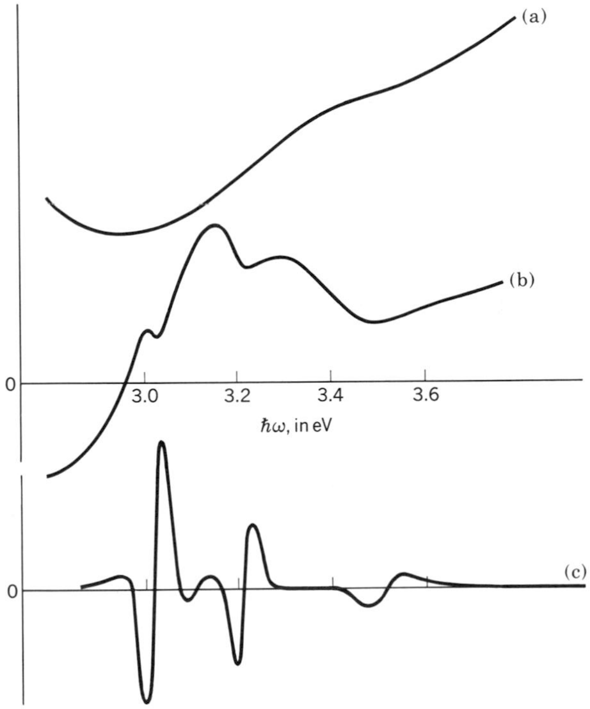

Figure 9 Comparison of (a) reflectance, (b) wavelength derivative reflectance (first derivative), and (c) electroreflectance (third derivative), of the spectral region in germanium between 3.0 and 3.6 eV. (After data by D. D. Sell, E. O. Kane, and D. E. Aspnes.)

EXCITONS

Reflectance and absorption spectra often show structure for photon energies just below the energy gap, where we might expect the crystal to be transparent. This structure is caused by the absorption of a photon with the creation of an exciton, either in a direct or indirect process. An electron and a hole may be bound together by their attractive electrostatic interaction, just as an electron is bound to a proton. The bound electron-hole is called an exciton, Fig. 10. It can move through the crystal transporting excitation energy, but it is electrically neutral. Excitons can be formed in every insulating crystal, although some types of excitons are intrinsically unstable with respect to decay into a free electron and free hole. All excitons are unstable against the ultimate recombination process in which the electron drops into the hole.

We have seen that a free electron and free hole are created whenever a photon of energy greater than the energy gap is absorbed in a crystal. The threshold for this process is $\hbar\omega > E_g$ in a direct process. In the indirect phonon-assisted process of Chapter 8 the threshold is lower by the phonon energy $\hbar\Omega$. But in the formation of excitons the energy is lowered with respect to these thresholds by the binding energy of the exciton,[2] which may be in the range 1 meV to 1 eV, as in Table 1. Excitons can be formed by photon absorption at any critical point (22), for if $\nabla_{\mathbf{k}}\epsilon_v = \nabla_{\mathbf{k}}\epsilon_c$ the group velocities of electron and hole are equal and the particles may be bound by their coulomb attraction. Transitions leading to the formation of excitons below the energy gap are indicated in Figs. 11 and 12.

The binding energy of the exciton can be measured in three ways:

1. In optical transitions from the valence band, by the difference between the energy required to create an exciton and the energy to create a free electron and free hole, Fig. 13.
2. In recombination luminescence, by comparison of the energy of the free electron-hole recombination line with the energy of the exciton recombination line.
3. By photo-ionization of excitons, to form free carriers. This is a difficult experiment that requires a high concentration of excitons; it has been carried out in Ge by a Moscow group using mm wave radiation.

We discuss excitons in two different limiting approximations, one by Frenkel in which the exciton is small and tightly bound, and the other by Mott and Wannier in which the exciton is weakly bound, with an electron-hole separation large in comparison with a lattice constant.

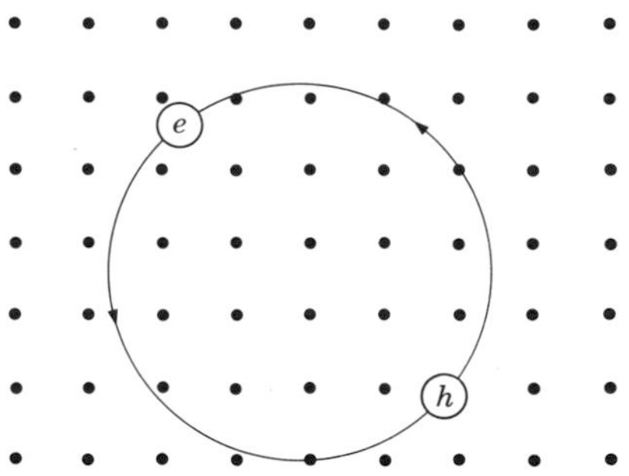

Figure 10a An exciton is a bound electron-hole pair, usually free to move together through the crystal. In some respects it is similar to an atom of positronium, formed from a positron and an electron. The exciton shown is a Mott exciton: it is weakly bound, with an average electron-hole distance large in comparison with a lattice constant.

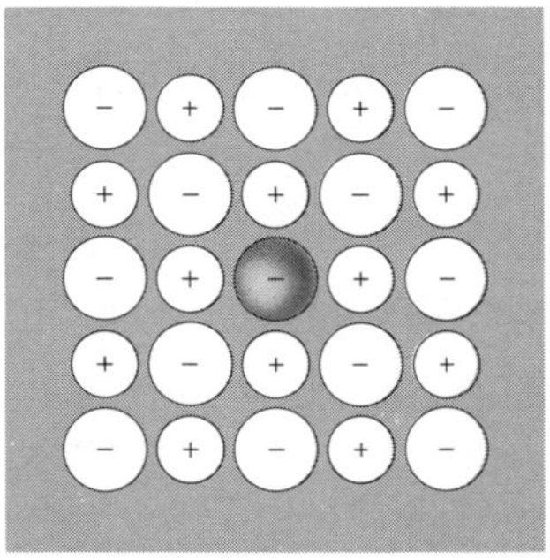

Figure 10b A tightly bound or Frenkel exciton shown localized on one atom in an alkali halide crystal. An ideal Frenkel exciton will travel as a wave throughout the crystal, but the electron is always close to the hole.

TABLE 1 Binding Energy of Excitons, in meV

Si	14.7	BaO	56.	RbCl	440.
Ge	3.8–4.1	InP	4.0	LiF	(1000)
GaAs	4.2	InSb	(0.4)	AgBr	20.
GaP	3.5	KI	480.	AgCl	30.
CdS	29.	KCl	400.	TlCl	11.
CdSe	15.	KBr	400.	TlBr	6.

Data assembled by Frederick C. Brown and Arnold Schmidt.

[2]Important early papers on excitons include J. Frenkel, Phys. Rev. **37**, 17, 1276 (1931); Physik. Z. Sowjetunion **9**, 158 (1936); R. Peierls, Ann. Physik **13**, 905 (1932); J. C. Slater and W. Shockley, Phys. Rev. **50**, 705 (1936); G. H. Wannier, Phys. Rev. **52**, 191 (1937); W. R. Heller and A. Marcus, Phys. Rev. **84**, 809 (1951); N. F. Mott, Trans. Faraday Soc. **34**, 500 (1938); D. L. Dexter and W. R. Heller, Phys. Rev. **84**, 377 (1951); A. S. Davydov, J.E.T.P. **18**, 210 (1948).

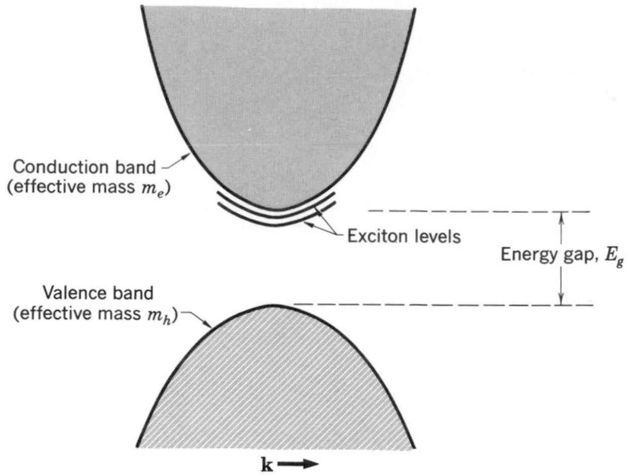

Figure 11 Exciton levels in relation to the conduction band edge, for a simple band structure with both conduction and valence band edges at $k = 0$. An exciton can have translational kinetic energy. Excitons are unstable with respect to radiative recombination in which the electron drops into the hole in the valence band, accompanied by the emission of a photon or phonons.

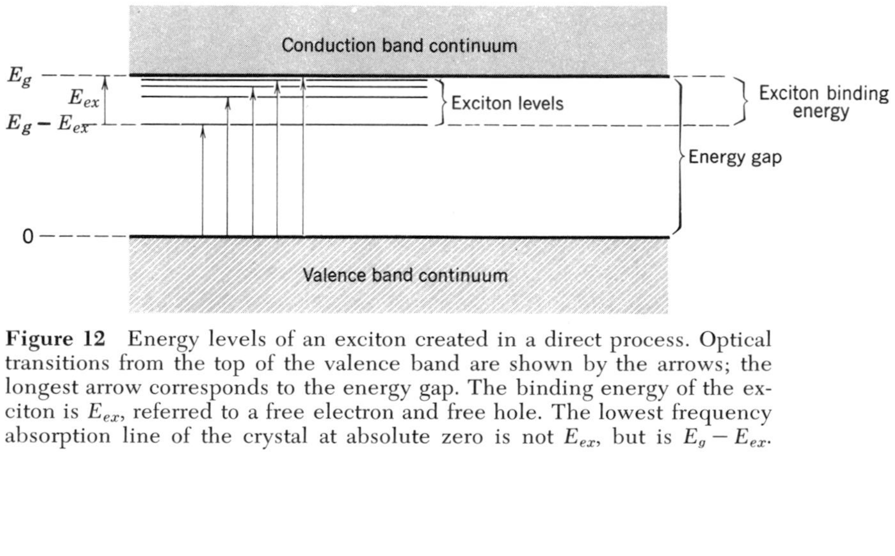

Figure 12 Energy levels of an exciton created in a direct process. Optical transitions from the top of the valence band are shown by the arrows; the longest arrow corresponds to the energy gap. The binding energy of the exciton is E_{ex}, referred to a free electron and free hole. The lowest frequency absorption line of the crystal at absolute zero is not E_{ex}, but is $E_g - E_{ex}$.

Frenkel Excitons

In a tightly bound exciton the excitation is localized on or near a single atom: the hole is usually on the same atom as the electron although the pair may be anywhere in the crystal. A Frenkel exciton is essentially an excited state of a single atom, but the excitation can hop from one atom to another by virtue of the coupling between neighbors. The excitation wave travels through the crystal much as the reversed spin of a magnon travels through the crystal. The crystalline inert gases have excitons which in their ground states correspond to the Frenkel model. Atomic krypton has its lowest strong atomic transition at 9.99 eV. The corresponding transition in the crystal is closely equal and is at 10.17 eV, Fig. 14. The energy gap in the crystal is 11.7 eV, so the exciton ground state energy is $11.7 - 10.17 = 1.5$ eV, referred to a free electron and free hole separated and at rest in the crystal.

The states of Frenkel excitons have the form of propagating waves. Consider a crystal of N atoms on a line or ring. If u_j is the ground state of atom j, the ground state of the crystal is

$$\psi_g = u_1 u_2 \cdots u_{N-1} u_N \ , \tag{23}$$

if interactions between the atoms are neglected. If a single atom j is in an excited state v_j, the system is described by

$$\varphi_j = u_1 u_2 \cdots u_{j-1} v_j u_{j+1} \cdots u_N \ . \tag{24}$$

This function has the same energy as the function φ_l with any other atom l excited. If there is an interaction between an excited atom and a nearby unexcited atom, the excitation energy will be transferred from atom to atom. The functions φ that describe a definite single excited atom (and $N-1$ unexcited atoms) are not stationary states of the problem.

The eigenfunctions are easily found. We can describe the effect of the interaction. When the hamiltonian of the system operates on the function φ_j with the jth atom excited, we obtain

$$\mathcal{H}\varphi_j = \epsilon\varphi_j + T(\varphi_{j-1} + \varphi_{j+1}) \ , \tag{25}$$

where ϵ is the free atom excitation energy; the interaction T measures the rate of transfer of the excitation from j to its nearest neighbors, $j-1$ and $j+1$. The solutions of (25) are waves of the Bloch form:

$$\psi_k = \sum_j \exp(ijka)\ \varphi_j \ . \tag{26}$$

To see this we let $\mathcal{H}$ operate on ψ_k:

$$\mathcal{H}\psi_k = \sum_j e^{ijka}\ \mathcal{H}\varphi_j = \sum_j e^{ijka}[\epsilon\varphi_j + T(\varphi_{j-1} + \varphi_{j+1})] \ , \tag{27}$$

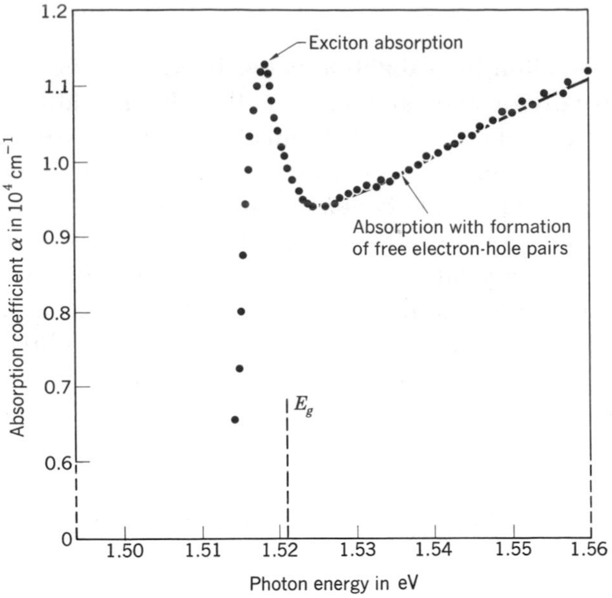

Figure 13 Effect of an exciton level on the optical absorption of a semiconductor for photons of energy near the band gap E_g in gallium arsenide at 21 K. The vertical scale is the intensity absorption coefficient, as in $I(x) = I_0 \exp(-\alpha x)$. The energy gap and exciton binding energy are deduced from the shape of the absorption curve: the gap E_g is 1.521 eV and the exciton binding energy is 0.0034 eV. [After M. D. Sturge, Phys. Rev. **127,** 768 (1962).]

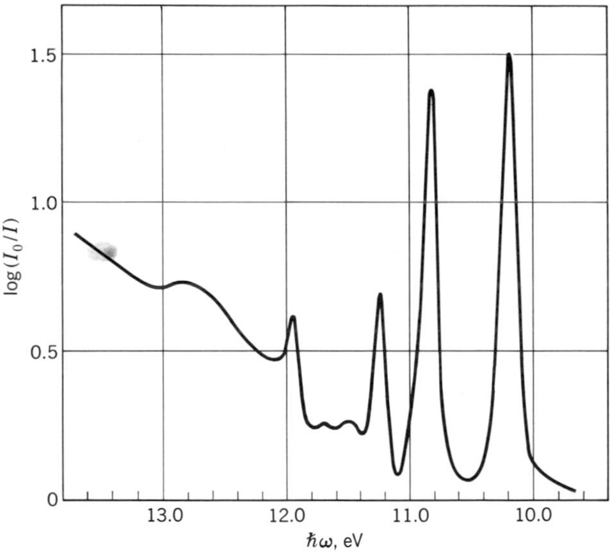

Figure 14 Absorption spectrum of solid krypton at 20 K. [After G. Baldini, Phys. Rev. **128,** 1562 (1962). This paper also has results for argon and xenon.]

from (25). We rearrange the right-hand side to obtain

$$\mathcal{H}\psi_k = \sum_j e^{ijka}[\epsilon + T(e^{ika} + e^{-ika})]\varphi_j = (\epsilon + 2T \cos ka)\psi_k \ , \qquad (28)$$

so that the energy eigenvalues of the problem are

$$E_k = \epsilon + 2T \cos ka \ , \qquad (29)$$

as in Fig. 15. The application of periodic boundary conditions determines the allowed values of the wavevector k:

$$k = 2\pi s/Na; \qquad s = -\tfrac{1}{2}N, -\tfrac{1}{2}N + 1, \cdots, \tfrac{1}{2}N - 1 \ . \qquad (30)$$

Alkali Halides. In alkali halide crystals (Fig. 10b) the excitons of lowest energy are localized on the negative halogen ions: the negative ions have lower electronic excitation levels than the positive ions. Alkali halide crystals when pure are transparent in the visible region of the spectrum, but in the vacuum ultraviolet region they show a considerable structure in spectral absorption from the creation of excitons. A doublet structure is particularly evident in sodium bromide in Fig. 16. This structure is similar to the doublet structure of the lowest excited state of the krypton atom, which has the same number of electrons as the Br^- ion. The resemblance of the doublet structures supports the identification of the lowest energy absorption in the alkali bromides as arising from the creation of an exciton centered on the Br^- ions. The splitting is caused by the spin-orbit interaction.

Molecular Crystals. In molecular crystals the covalent binding within a molecule is strong in comparison with the van der Waals binding between molecules. Electronic excitation lines of an individual molecule appear in the crystalline solid as an exciton, often with little shift in frequency. At low temperatures the lines in the solid are quite sharp, although there may be more structure to the lines in the solid than in the molecule because of the Davydov splitting, as discussed in Problem 7.

Considerable work has been done on excitons in crystals of anthracene, Fig. 17. The low-lying exciton levels of anthracene are indicated in Fig. 18. The optical transition between the ground state (singlet) and the lowest electronic excited state (triplet) is not an allowed electric dipole transition, but the transition probability is nonzero and a significant concentration of triplet excitons can be produced by irradiation of the crystal with an intense laser beam of photons of energy 1.79 eV. Two triplet excitons can combine[3] to form one singlet exciton of energy 3.15 eV, the excess energy being carried off by phonons. The 3.15 eV exciton is connected with the ground state by an allowed transition, and the emitted photon at 3.15 eV is detected in the experiments (Fig. 19).

[3] R. G. Kepler, J. C. Caris, P. Avakian, and E. Abramson, Phys. Rev. Letters **10**, 400 (1963); P. Avakian and R. E. Merrifield, Phys. Rev. Letters **13**, 541 (1964).

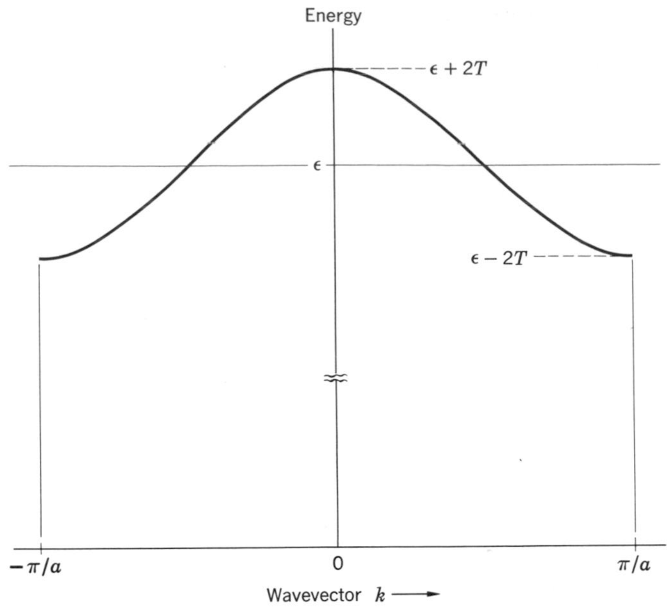

Figure 15 Energy versus wavevector for a Frenkel exciton, calculated with positive nearest-neighbor transfer interaction T.

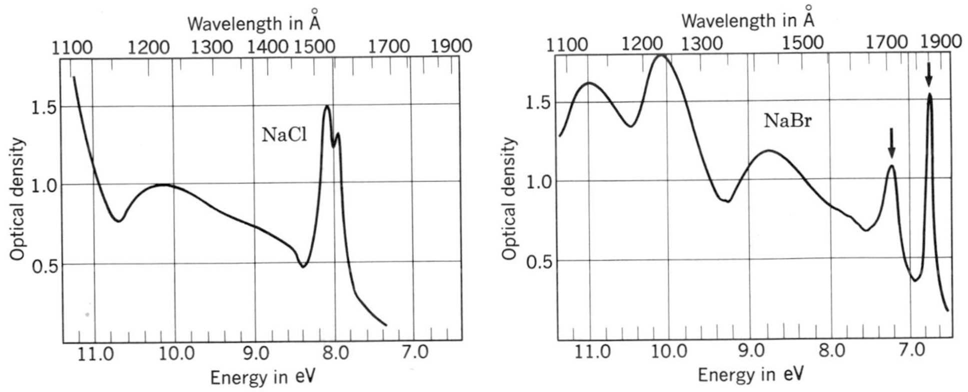

Figure 16 Optical absorption spectra of thin films of sodium halides at 80 K, after J. E. Eby, K. J. Teegarden, and D. B. Dutton, Phys. Rev. **116,** 1099 (1959). A doublet structure in the bromides is marked by arrows. The lowest excitations of the halogen ions are doublets: in Br^- the ground state is a $^1S_0$ state arising from the electron configuration $4p^6$; the lowest excited states arise from the configuration $4p^55s$. The $4p^5$ core may have total angular momentum $J = \frac{3}{2}$ or $\frac{1}{2}$, split by about 0.5 eV; the observed doublet structure results from this splitting.

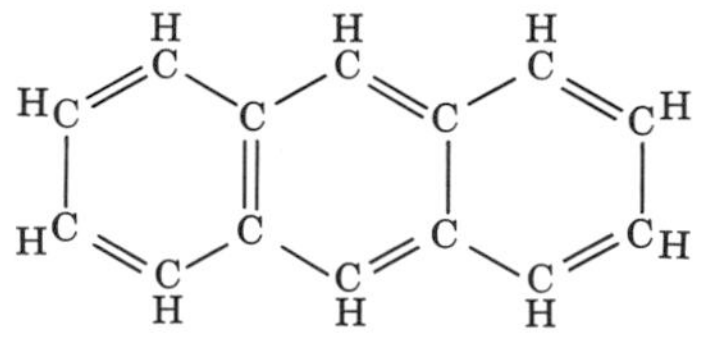

Figure 17 Chemical structure of anthracene.

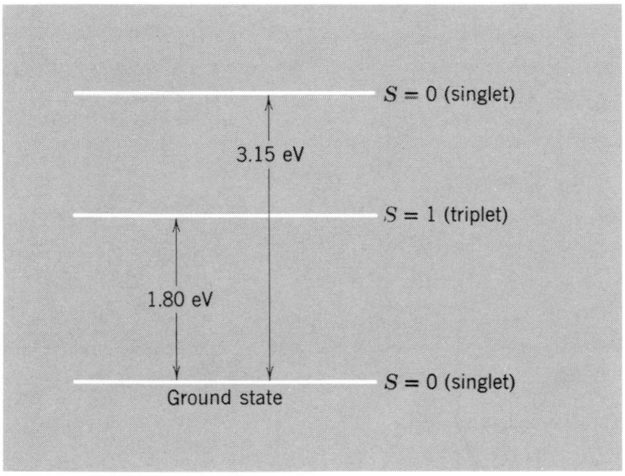

Figure 18 Low-lying exciton states of anthracene. The electron spin is denoted by S.

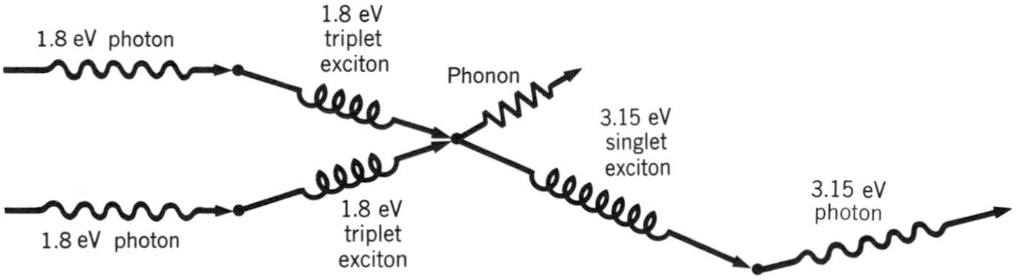

Figure 19 Detection of excitons in anthracene. (After Avakian, Kepler et al.)

Weakly Bound Excitons

Consider an electron in the conduction band and a hole in the valence band. The electron and hole attract each other by the coulomb potential

(CGS) $$U(r) = -e^2/\epsilon r \ , \tag{31}$$

where r is the distance between the particles and ϵ is the appropriate dielectric constant. (The lattice polarization contribution to the dielectric constant should not be included if the frequency of motion of the exciton is higher than the optical phonon frequencies.) There will be bound states of the exciton system having total energies lower than the bottom of the conduction band. The problem is the hydrogen atom problem if the energy surfaces for the electron and hole are spherical and non-degenerate. The energy levels referred to the top of the valence band are given by a modified Rydberg equation

(CGS) $$E_n = E_g - \frac{\mu e^4}{2\hbar^2\epsilon^2 n^2} \ . \tag{32}$$

Here n is the principal quantum number and μ is the reduced mass:

$$\frac{1}{\mu} = \frac{1}{m_e} + \frac{1}{m_h} \ , \tag{33}$$

formed from the effective masses m_e, m_h of the electron and hole.

The exciton ground state energy is obtained on setting $n = 1$ in (32); this is the ionization energy of the exciton. Gross[4] and others have studied optical absorption lines in cuprous oxide, Cu_2O, at low temperatures, with results for the spacing of the exciton levels in good agreement with the Rydberg equation, particularly for levels with $n > 2$. The results are shown in Fig. 20. An empirical fit to the lines is obtained with the relation $\nu(\mathrm{cm}^{-1}) = 17{,}508 - (800/n^2)$. Taking $\epsilon = 10$, we find $\mu \simeq 0.7\ m$ from the coefficient of $1/n^2$. The constant term 17,508 cm^{-1} corresponds to an energy gap $E_g = 2.17$ eV.

[4]E. F. Gross, B. P. Zakharchenya, and N. M. Reinov, Doklady Akad. Nauk S.S.S.R. **92**, 265 (1953); **97**, 57, 221 (1954); **99**, 231, 527 (1954); S. Nikitine, G. Perny, and M. Sieskind, Compt. rend. (Paris) **238**, 67 (1954); J. H. Apfel and L. N. Hadley, Phys. Rev. **100**, 1689 (1955); A. G. Samoilovich and L. L. Kornblit, Doklady Akad. Nauk S.S.S.R. **100**, 43 (1955); S. Nikitine, Helv. Phys. Acta **28**, 307 (1955). For further details of the exciton series in Cu_2O, see articles by R. J. Elliott and by M. Grosmann in *Polarons and excitons*, Plenum, New York, 1963.

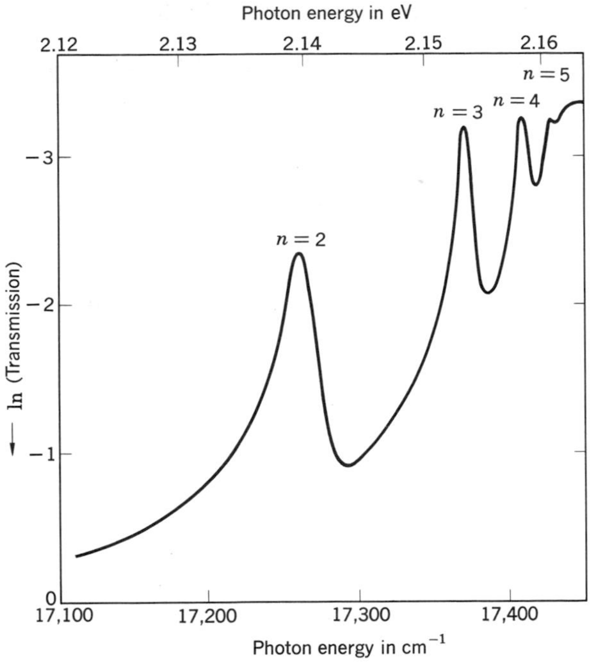

Figure 20 Logarithm of the optical transmission versus photon energy in cuprous oxide at 77 K, showing a series of exciton lines. Note that on the vertical axis the logarithm is plotted decreasing upward; thus a peak corresponds to absorption. The band gap E_g is 2.17 eV. [After P. W. Baumeister, Phys. Rev. **121,** 359 (1961).]

Exciton Condensation into Electron-Hole Drops (EHD)

A condensed phase of an electron-hole plasma forms in Ge and Si when maintained at a low temperature and irradiated by light. The following sequence of events takes place when an electron-hole drop (EHD) is formed in Ge: The absorption of a photon of energy $\hbar\omega > E_g$ produces a free electron and free hole, with high efficiency. These combine rapidly, perhaps in 1 ns, to form an exciton. The exciton may decay with annihilation of the e-h pair with a lifetime of 8 μs. But if the exciton concentration is sufficiently high—over 10^{13} cm^{-3} at 2 K—most of the excitons will condense into a drop. The drop lifetime is 40 μs, but in strained Ge may be as long as 600 μs. Within the drop the excitons dissolve into a degenerate Fermi gas of electrons and holes, with metallic properties: this state was predicted by L. V. Keldysh. The binding energy is 2 meV with respect to free excitons, and the concentration $n = p = 2 \times 10^{17}$ cm^{-3}. Experimental studies[5] of the condensed EHD phase have utilized the methods of recombination (luminescence) radiation, light scattering, plasma resonance, Alfvén wave resonance, and $p - n$ junction noise. Figure 21a shows the recombination radiation in Ge from free excitons (714 meV) and from the EHD phase (709 meV). The width of the 714 meV line is accounted for by Doppler broadening, and the width of the 709 meV line is compatible with the kinetic energy distribution of electrons and holes in a Fermi gas of concentration 2×10^{17} cm^{-3}. Figure 21b is a photograph of a large EHD.

[5]An early review is by Y. E. Pokrovskii, physica status solidi. (a) **11,** 385 (1972); see further C. D. Jeffries, Science **189,** 955 (1975).

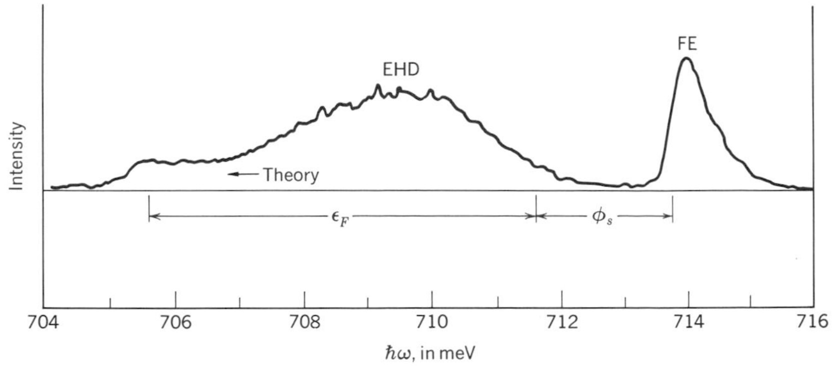

(a)

Figure 21a Recombination radiation of free electrons and of electron-hole drops in Ge at 3.04 K. The Fermi energy in the drop is ϵ_F and the cohesive energy of the drop with respect to a free exciton is ϕ_s. [After T. K. Lo, Solid State Comm. **15**, 1231 (1974).]

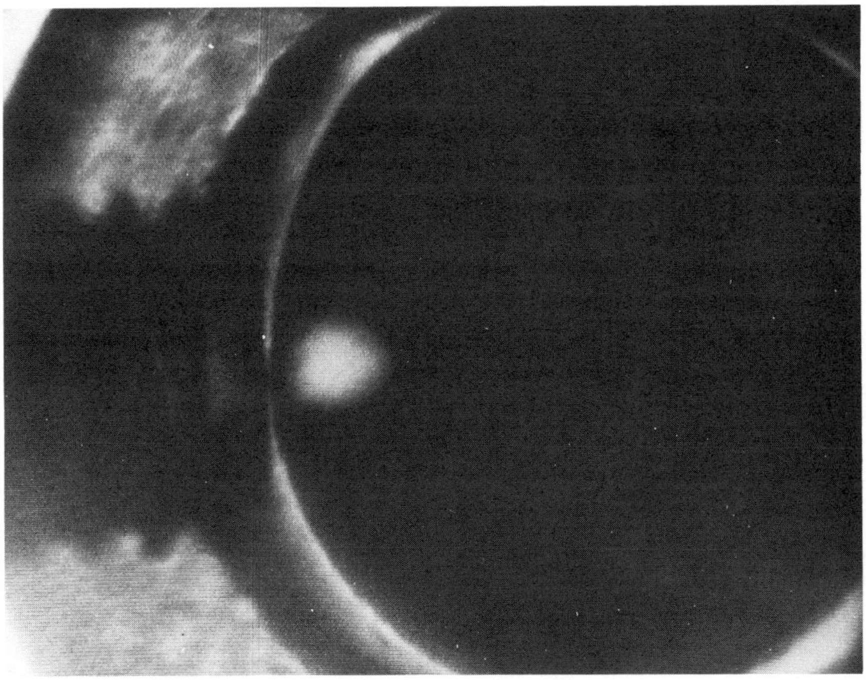

(b)

Figure 21b Photograph of an electron-hole drop in a 4 mm disk of pure germanium. The drop is the intense spot adjacent to the set screw on the left of the disk. The photograph is the image of the drop obtained by focusing its electron-hole recombination luminescence onto the surface of an infrared-sensitive videcon image tube. [After J. P. Wolfe, W. L. Hansen, E. E. Haller, R. S. Markiewicz, C. Kittel, and C. D. Jeffries, Phys. Rev. Letters **34**, 1292 (1975).]

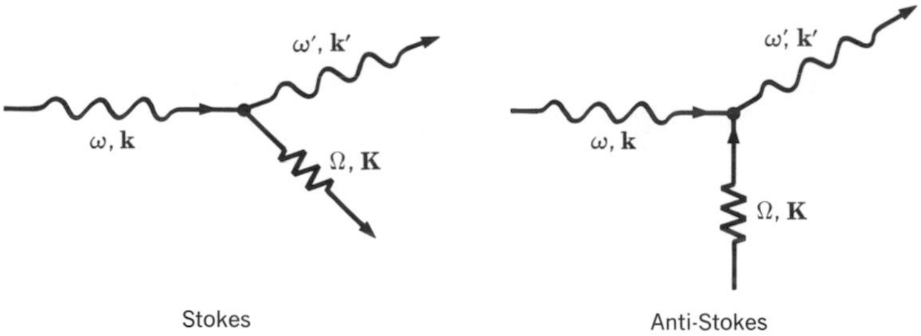

Figure 22 Raman scattering of a photon with emission or absorption of a phonon. The process is called Brillouin scattering when an acoustic phonon is involved and polariton scattering when an optical phonon is involved. Similar processes occur with magnons (spin waves).

RAMAN EFFECT IN CRYSTALS

Raman scattering involves two photons—one in, one out—and is one step more complex than the one photon processes treated earlier in this chapter. In the Raman effect a photon is scattered inelastically by a crystal, with creation or annihilation of a phonon or magnon (Fig. 22). The process is identical to the inelastic scattering of x-rays and it is similar to the inelastic scattering of neutrons by a crystal. The selection rules for the first-order Raman effect are

$$\omega = \omega' \pm \Omega \ ; \qquad \mathbf{k} = \mathbf{k}' \pm \mathbf{K} \ , \tag{34}$$

where ω, $\mathbf{k}$ refer to the incident photon; ω', $\mathbf{k}'$ refer to the scattered photon, and Ω, $\mathbf{K}$ refer to the phonon created or destroyed in the scattering event. In the second-order Raman effect, two phonons are involved in the inelastic scattering of the photon.

The Raman effect is made possible by the strain-dependence of the electronic polarizability. We suppose that the polarizability associated with a phonon mode may be written as a power series in the phonon amplitude u:

$$\alpha = \alpha_0 + \alpha_1 u + \alpha_2 u^2 + \cdots . \tag{35}$$

If $u(t) = u_0 \cos \Omega t$ and the incident electric field is $E(t) = E_0 \cos \omega t$, then the induced electric dipole moment has a component

$$\alpha_1 E_0 u_0 \cos \omega t \cos \Omega t = \tfrac{1}{2}\alpha_1 E_0 u_0 [\cos(\omega + \Omega)t + \cos(\omega - \Omega)t] . \tag{36}$$

Thus photons at frequencies $\omega + \Omega$ and $\omega - \Omega$ can be emitted, accompanied by absorption or emission of a phonon of frequency Ω.

The photon at $\omega - \Omega$ is called the Stokes line and that at $\omega + \Omega$ is the anti-Stokes line. The intensity of the Stokes line involves the matrix element for phonon creation, which are just the matrix elements for the harmonic oscillator, as in elementary quantum mechanics texts:

$$I(\omega - \Omega) \propto |\langle n_{\mathbf{K}} + 1|u|n_{\mathbf{K}}\rangle|^2 \propto n_{\mathbf{K}} + 1 , \tag{37}$$

where $n_{\mathbf{K}}$ is the initial population of phonon mode **K.** The anti-Stokes line involves phonon annihilation, with a photon intensity proportional to

$$I(\omega + \Omega) \propto |\langle n_{\mathbf{K}} - 1|u|n_{\mathbf{K}}\rangle|^2 \propto n_{\mathbf{K}} . \tag{38}$$

If the phonon population is initially in thermal equilibrium at temperature T, the intensity ratio of the two lines is

$$\frac{I(\omega + \Omega)}{I(\omega - \Omega)} = \frac{\langle n_{\mathbf{K}}\rangle}{\langle n_{\mathbf{K}}\rangle + 1} = e^{-\hbar\Omega/k_BT} , \tag{39}$$

with $\langle n_{\mathbf{K}}\rangle$ given by the Planck distribution function $1/(e^{\hbar\Omega/k_BT} - 1)$. We see that the relative intensity of the anti-Stokes lines vanishes as $T \rightarrow 0$, because here there are no thermal phonons available to be annihilated.

Observations on the $\mathbf{K} = 0$ optical phonon in silicon are shown in Figs. 23 and 24. Silicon has two identical atoms in the primitive cell, and there is no electric dipole moment associated with the primitive cell in the absence of deformation by phonons. But $\alpha_1 u$ does not vanish for silicon at $\mathbf{K} = 0$, so that we can observe the mode by first-order Raman scattering of light.

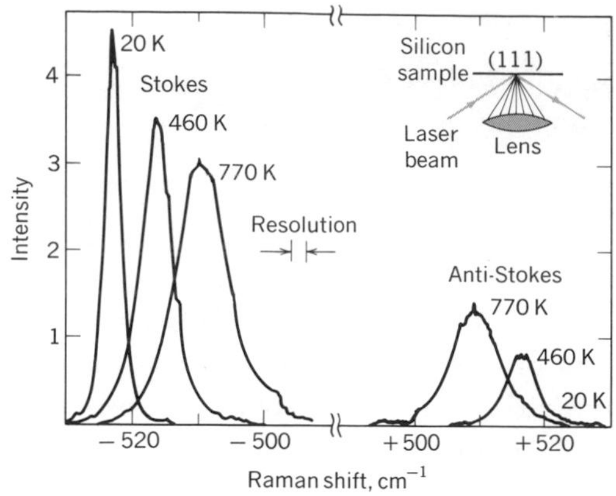

Figure 23 First-order Raman spectra of the $K \simeq 0$ optical mode of a silicon crystal observed at three temperatures. The incident photon has a wavelength of 5145 Å. The optical phonon frequency is equal to the frequency shift; it depends slightly on the temperature. [After T. R. Hart, R. L. Aggarwal, and B. Lax, Phys. Rev. **B1**, 638 (1970).]

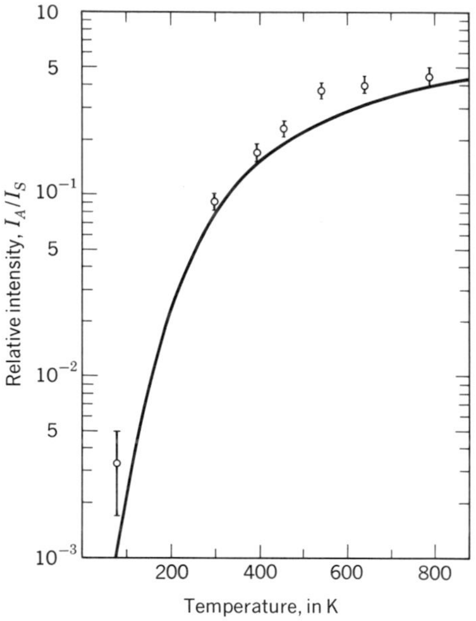

Figure 24 Intensity ratio anti-Stokes to Stokes lines as a function of temperature, for the observations of Fig. 23 on the optical mode of silicon. The observed temperature dependence is in good agreement with the prediction of Eq. (39): the solid curve is of the function $\exp(-\hbar\Omega/k_BT)$.

The second-order Raman effect arises from the term $\alpha_2 u^2$ in the polarizability. Inelastic scattering of light in this order is accompanied by the creation of two phonons, or the absorption of two phonons, or the creation of one and the absorption of another phonon. The phonons may have different frequencies. The intensity distribution in the scattered photon spectrum may be quite complicated if there are several atoms in the primitive cell because of the corresponding large number of optical phonon modes. Such second-order Raman spectra have been observed and analyzed in numerous crystals. Measurements on gallium phosphide (GaP) are shown in Fig. 25.

The selection rules in the second-order Raman effect are:

$$\omega = \omega' \pm \Omega \pm \Omega' \ ; \qquad \mathbf{k} = \mathbf{k}' \pm \mathbf{K} \pm \mathbf{K}' + \mathbf{G} \ , \tag{40}$$

where now we have included the possibility of a reciprocal lattice vector **G**. In the first-order Raman effect in the optical region it is not possible energetically to have a reciprocal lattice vector in (34) because the photon wavevectors **k**, **k**′ and their difference are much smaller than the shortest **G**, except for photons in the x-ray and γ-ray regions. But in the second-order effect it is possible for the difference of the phonon wavevectors **K**, **K**′ to be a reciprocal lattice vector.

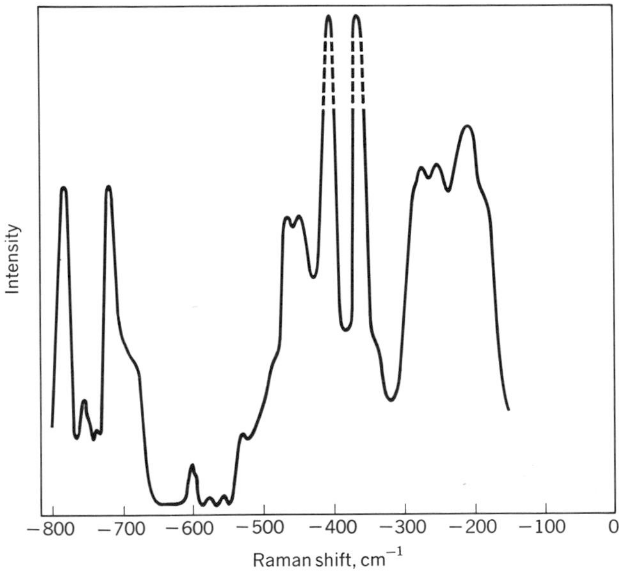

Figure 25 Raman spectrum of GaP at 20 K. The two highest peaks are the first-order Raman lines associated with the excitation of an *LO* phonon at 404 cm^{-1} and a *TO* phonon at 366 cm^{-1}. All the other peaks involve two phonons. [After M. V. Hobden and J. P. Russell, Physics Letters **13**, 39 (1964).]

Electron Spectroscopy with X-Rays

The next degree of complexity involves a photon in and an electron out of the solid. The important techniques of x-ray photoemission from solids (XPS) and ultraviolet photoemission (UPS) have recently been developed. In solid state physics they are used in band structure studies and surface physics, including catalysis and adsorption. XPS and UPS spectra can be compared directly with valence band densities of states $\mathfrak{D}(\epsilon)$. The specimen is irradiated with highly monochromatized x-rays or ultraviolet photons. The photon is absorbed, with the emission of a photoelectron whose kinetic energy is equal to the energy of the incident photon minus the binding energy of the electron in the solid. The electrons come from a thin layer near the surface, typically 50 Å in depth. The resolution of the ESCA spectrometer systems is of the order of 0.6 eV, which permits refined studies of band structure, particularly of core electrons. ESCA stands for electron spectroscopy for chemical analysis.

The valence band structure of silver is shown in Fig. 26. The zero of energy has been set at the Fermi level: thermal excitation is visible above the Fermi level. Below the Fermi level the electrons in the first 3 eV come from the 5*s* conduction band, and below 3 eV the strong peak with structure is from the 4*d* valence band electrons. Excitations are also observed from deeper levels, and these excitations may be accompanied by the inelastic excitation of plasmons: thus in silicon the 2*p* electron at a binding energy close to 99.2 eV is observed in replica also at 117 eV and 134.7 eV—the first is with single plasmon excitation and the second is with two plasmon excitation.

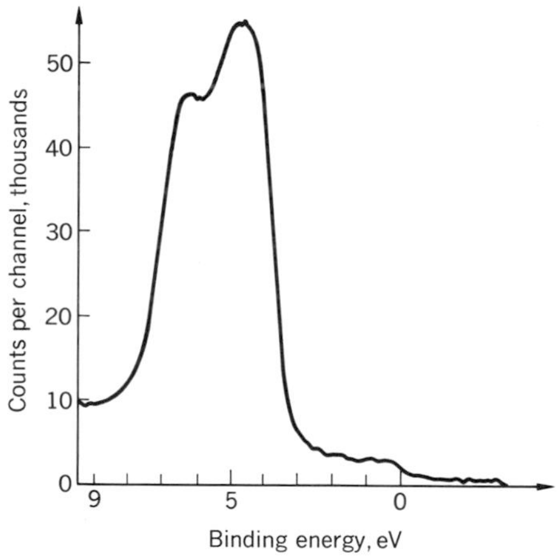

Figure 26 Valence band electron emission from silver, after Siegbahn and co-workers, Science **176**, 245 (1972). For studies on gold at high resolution, see D. A. Shirley, Phys. Rev. B**5**, 4709 (1972).

ENERGY LOSS OF FAST PARTICLES IN A SOLID

So far we have used photons as probes of the electronic structure of solids. We can also use electron beams for the same purpose. The results also involve the dielectric function, now through the imaginary part of $1/\epsilon(\omega)$. The dielectric function enters as $\mathrm{Im}\{\epsilon(\omega)\}$ into the energy loss by an electromagnetic wave in a solid, but as $-\mathrm{Im}\{1/\epsilon(\omega)\}$ into the energy loss by a charged particle that penetrates a solid. Consider this difference. The general result from electromagnetic theory for the power dissipation density by dielectric losses is

(CGS) $$\mathcal{P} = \frac{1}{4\pi}\,\mathbf{E}\cdot(\partial\mathbf{D}/\partial t)\ ,$$

per unit volume. With a transverse electromagnetic wave $Ee^{-i\omega t}$ in the crystal, we have $dD/dt = -i\omega\,\epsilon(\omega)Ee^{-i\omega t}$, whence the time-average power is

$$\begin{aligned}\mathcal{P} &= \frac{1}{4\pi}\,\langle \mathrm{Re}\{Ee^{-i\omega t}\}\mathrm{Re}\{-i\omega\epsilon(\omega)Ee^{-i\omega t}\}\rangle \\ &= \frac{1}{4\pi}\,\omega E^2\,\langle(\epsilon''\cos\omega t - \epsilon'\sin\omega t)\cos\omega t\rangle = \frac{1}{8\pi}\,\omega\epsilon''(\omega)E^2\ ,\end{aligned} \tag{41}$$

proportional to $\epsilon''(\omega)$. The tangential component of E is continuous across the boundary of the solid.

If a particle of charge e and velocity $\mathbf{v}$ enters a crystal, the dielectric displacement is

(CGS) $$\mathbf{D}(\mathbf{r},t) = -\mathrm{grad}\,\frac{e}{|\mathbf{r}-\mathbf{v}t|}\ , \tag{42}$$

because by the Poisson equation it is $\mathbf{D}$, and not $\mathbf{E}$, that is related to the free charge. In an isotropic medium the Fourier component $E(\omega,\mathbf{k})$ is related to the Fourier component $D(\omega,\mathbf{k})$ of $D(\mathbf{r},t)$ by $E(\omega,\mathbf{k}) = D(\omega,\mathbf{k})/\epsilon(\omega,\mathbf{k})$. The time-average power dissipation associated with this Fourier component is

$$\begin{aligned}\mathcal{P}(\omega,\mathbf{k}) &= \frac{1}{4\pi}\,\langle \mathrm{Re}\{\epsilon^{-1}(\omega,\mathbf{k})D(\omega,\mathbf{k})\,e^{-i\omega t}\}\mathrm{Re}\{-i\omega\,D(\omega,\mathbf{k})\,e^{-i\omega t}\}\rangle \\ &= \frac{1}{4\pi}\,\omega\,D^2(\omega,\mathbf{k})\left\langle\left[\left(\frac{1}{\epsilon}\right)'\cos\omega t + \left(\frac{1}{\epsilon}\right)''\sin\omega t\right][-\sin\omega t]\right\rangle\ ,\end{aligned}$$

whence

$$\mathcal{P}(\omega,\mathbf{k}) = -\frac{1}{8\pi}\,\omega\left(\frac{1}{\epsilon}\right)''D^2(\omega,\mathbf{k}) = \frac{1}{8\pi}\,\omega\,\frac{\epsilon_2(\omega)}{|\epsilon|^2}\,D^2(\omega,\mathbf{k})\ . \tag{43}$$

The result[6] is the motivation for the introduction of the **energy loss function** $-\mathrm{Im}\{1/\epsilon(\omega,\mathbf{k})\}$ and it is also the motivation[7] for many of the experiments on energy losses by fast electrons in thin films.

If the dielectric function is independent of **k**, the power loss is

$$\mathcal{P}(\omega) = -\frac{2}{\pi}\frac{e^2}{\hbar v}\,\mathrm{Im}\{1/\epsilon(\omega)\}\ln(k_0 v/\omega) \ , \tag{44}$$

where $\hbar k_0$ is the maximum possible momentum transfer from the primary particle to an electron of the crystal. Figure 27 shows the excellent experimental agreement between values of $\epsilon''(\omega)$ deduced from optical reflectivity measurements with values deduced from electron energy loss measurements.

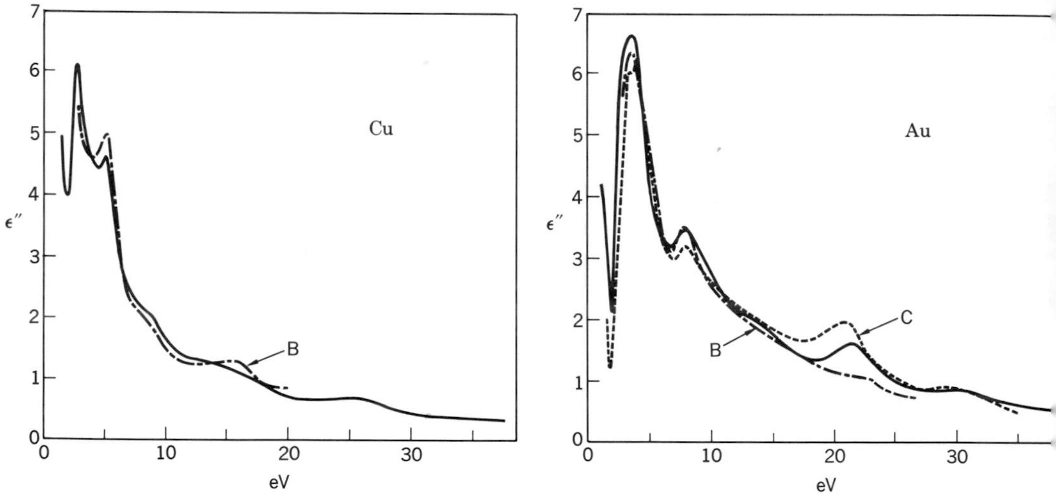

Figure 27 $\epsilon''(\omega)$ for Cu and Au; the bold lines are from energy loss measurements by J. Daniels, and the other lines were calculated from optical measurements by D. Beaglehole and L. R. Canfield et al.

---

[6]The result is due to H. Fröhlich and H. Pelzer, Proc. Phys. Soc. (London) **A68,** 525 (1955); the complete calculation is given by H. Fröhlich, "Phenomenological theory of the energy loss of fast particles in solids," in W. Frank, ed., *Max-Planck-Festschrift,* VEB Deutscher Verlag, Berlin, 1958, p. 277.

[7]For a review see H. Raether, "Solid state excitations by electrons," *Springer tracts in modern physics* **38,** 84 (1965); J. Daniels, C. v. Festenberg, H. Raether, and K. Zeppenfeld, "Optical constants of solids by electron spectroscopy," *Springer tracts in modern physics* **54,** 77 (1970).

SUMMARY

1. The Kramers-Kronig relations connect the real and imaginary parts of a response function:

$$\alpha'(\omega) = \frac{2}{\pi} \mathrm{P} \int_0^\infty \frac{s\alpha''(s)}{s^2 - \omega^2}\, ds \ ; \qquad \alpha''(\omega) = -\frac{2\omega}{\pi} \mathrm{P} \int_0^\infty \frac{\alpha'(s)}{s^2 - \omega^2}\, ds \ .$$

2. The complex refractive index $N(\omega) = n(\omega) + iK(\omega)$, where n is the refractive index and K is the extinction coefficient; further, $\epsilon(\omega) = N^2(\omega)$, whence $\epsilon'(\omega) = n^2 - K^2$ and $\epsilon''(\omega) = 2nK$.

3. The reflectance at normal incidence is

$$R = \frac{(n-1)^2 + K^2}{(n+1)^2 + K^2} \ .$$

4. The energy loss function $-\mathrm{Im}\{1/\epsilon(\omega)\}$ gives the energy loss by a charged particle moving in a solid.

PROBLEMS

1. ***Causality and the response function.*** The Kramers-Kronig relations are consistent with the principle that an effect not precede its cause. Consider a delta-function force applied at time $t = 0$:

$$F(t) = \delta(t) = \frac{1}{2\pi} \int_{-\infty}^{\infty} e^{-i\omega t}\, d\omega \ ,$$

whence $F_\omega = 1/2\pi$. (a) Show by direct integration or by use of the KK relations that the oscillator response function

$$\alpha(\omega) = (\omega_0{}^2 - \omega^2 - i\omega\rho)^{-1}$$

gives zero displacement, $x(t) = 0$, for $t < 0$ under the above force. For $t < 0$ the contour integral may be completed by a semicircle in the upper half-plane. (b) Evaluate $x(t)$ for $t > 0$. Note that $\alpha(\omega)$ has poles at $\pm(\omega_0{}^2 - \frac{1}{4}\rho^2)^{1/2} - \frac{1}{2}i\rho$, both in the lower half-plane.

2. ***Dissipation sum rule.*** By comparison of $\alpha'(\omega)$ from (9) and from (11a) in the limit $\omega \to \infty$, show that the following sum rule for the oscillator strengths must hold:

$$\Sigma f_j = \frac{2}{\pi} \int_0^\infty s\alpha''(s)\, ds$$

3. ***Reflection at normal incidence.*** Consider an electromagnetic wave in vacuum, with field components of the form

$$E_y(\text{inc}) = B_z(\text{inc}) = Ae^{i(kx-\omega t)} \ .$$

Let the wave be incident upon a medium of dielectric constant ϵ and permeability $\mu = 1$ that fills the half-space $x > 0$. Show that the reflectivity coefficient $r(\omega)$ as defined by $E(\text{refl}) = r(\omega)E(\text{inc})$ is given by

$$r(\omega) = \frac{n + iK - 1}{n + iK + 1} \ ,$$

where $n + iK \equiv \epsilon^{\frac{1}{2}}$, with n and K real. Show further that the reflectance is

$$R(\omega) = \frac{(n-1)^2 + K^2}{(n+1)^2 + K^2} \ .$$

*4. ***Conductivity sum rule and superconductivity.*** We write the electrical conductivity as $\sigma(\omega) = \sigma'(\omega) + i\sigma''(\omega)$, where σ', σ'' are real. (a) Show by a Kramers-Kronig relation that

$$\lim_{\omega \to \infty} \omega \, \sigma''(\omega) = \frac{2}{\pi} \int_0^\infty \sigma'(s) \, ds \ .$$

This result is used in the theory of superconductivity. If at very high frequencies (such as x-ray frequencies) $\sigma''(\omega)$ is identical for the superconducting and normal states, then we must have

$$\int_0^\infty \sigma_s'(\omega) \, d\omega = \int_0^\infty \sigma_n'(\omega) \, d\omega \ .$$

But at frequencies $0 < \omega < \omega_g$ within the superconducting energy gap the real part of the conductivity of a superconductor vanishes, so that in this region the integral on the left-hand side is lower by $\approx \sigma_n'\omega_g$. There must be an additional contribution to σ_s' to balance this deficiency. (b) Show that if $\sigma_s'(\omega < \omega_g) < \sigma_n'(\omega < \omega_g)$, as is observed experimentally, then $\sigma_s'(\omega)$ can have a delta function contribution at $\omega = 0$, and from the delta function there is a contribution $\sigma_s''(\omega) \approx \sigma_n'\omega_g/\omega$. The delta function corresponds to infinite conductivity at zero frequency. (c) By elementary consideration of the classical motion of conduction electrons at very high frequencies, show that

$$\int_0^\infty \sigma'(\omega) \, d\omega = \pi n e^2/2m \ ,$$

a result found by Ferrell and Glover.

5. ***Dielectric constant and the semiconductor energy gap.*** The effect on $\epsilon''(\omega)$ of an energy gap ω_g in a semiconductor may be approximated very roughly by substituting $\frac{1}{2}\delta(\omega - \omega_g)$ for $\delta(\omega)$ in the response function (15); that is, we take $\epsilon''(\omega) = (2\pi n e^2/m\omega)\pi\delta(\omega - \omega_g)$. This is crude because it puts all the absorption at the gap frequency. The factor 1/2 enters as soon as we move the delta function away from the origin, because the integral in the sum rule of Problem 2 starts at the origin. Show that the real part of the dielectric constant on this

*This problem is somewhat difficult.

model is

$$\epsilon'(\omega) = 1 + \omega_p{}^2/(\omega_g{}^2 - \omega^2), \qquad \omega_p{}^2 \equiv 4\pi ne^2/m \ .$$

It follows that the static dielectric constant is $\epsilon'(0) = 1 + \omega_p{}^2/\omega_g{}^2$, widely used as a rule of thumb.

6. ***Hagen-Rubens relation for infrared reflectivity of metals.*** The complex refractive index $n + iK$ of a metal for $\omega\tau \ll 1$ is given by

(CGS) $$\epsilon(\omega) \equiv (n + iK)^2 = 1 + 4\pi i\sigma_0/\omega \ ,$$

where σ_0 is the conductivity for static fields. We assume here that intraband currents are dominant; interband transitions are neglected. Using the result of Problem 3 for the reflection coefficient at normal incidence, show that

(CGS) $$R \simeq 1 - (2\omega/\pi\sigma_0)^{1/2} \ ,$$

provided that $\sigma_0 \gg \omega$. This is the Hagen-Rubens relation. For sodium at room temperature, $\sigma_0 \simeq 2.1 \times 10^{17}\ \mathrm{s}^{-1}$ in CGS and $\tau \simeq 3.1 \times 10^{-14}$ s, as deduced from $\tau = \sigma_0 m/ne^2$. Radiation of 10 μm has $\omega = 1.88 \times 10^{14}\ \mathrm{s}^{-1}$, so that the Hagen-Rubens result should apply: $R = 0.976$. The result calculated from experimental values of n and K is 0.987. Hint: If $\sigma_0 \gg \omega$, then $n^2 \simeq K^2$. This simplifies the algebra.

*7. ***Davydov splitting of exciton lines.*** The Frenkel exciton band of Fig. 15 is doubled when there are two atoms A, B in a primitive cell. Extend the theory of Eqs. (25) to (29) to a linear crystal AB.AB.AB.AB. with transfer integrals T_1 between AB and T_2 between B.A. Find an equation for the two bands as functions of the wavevector. The splitting between the bands at $k = 0$ is called the Davydov splitting.

*This problem is somewhat difficult.

References

(See also references in Chapter 10.)

M. Balkanski, ed., *Light scattering in solids,* Flammarion, Paris, 1971.
N. Bloembergen, *Nonlinear optics,* Benjamin, 1965.
M. Cardona, *Modulation spectroscopy,* Academic Press, 1969.
C. B. Duke and N. Holonyak, Jr., "Advances in light-emitting diodes," Physics Today, December 1973, p. 23.
D. L. Greenway and G. Harbeke, *Optical properties and band structure of semiconductors,* Pergamon, 1968.
J. N. Hodgson, *Optical absorption and dispersion in solids,* Chapman and Hall, 1971. Excellent short monograph.
T. S. Moss, G. J. Burrell, and B. Ellis, *Semiconductor optoelectronics,* Wiley, 1973.
J. C. Phillips, "Fundamental optical spectra of solids," *Solid state physics* **18,** 56 (1966).
D. A. Shirley, ed., *Electron spectroscopy,* North-Holland, 1972.
W. E. Spicer, "Ultraviolet photoemission spectroscopy," Comments on Solid State Physics **5,** 105 (1973).
J. Tauc, ed., *Optical properties of solids,* Academic Press, 1966.
G. Turrell, *Infrared and Raman spectra of crystals,* Academic Press, 1972.

Excitons

A. S. Davydov, *Theory of molecular excitons,* McGraw-Hill, 1962.
D. Dexter and R. S. Knox, *Excitons,* Wiley, 1965.
D. Fox, M. M. Davis, and A. Weissburger, *Physics and chemistry of the organic solid state,* Wiley, Vol. I (1963), Vol. II (1965), Vol. III (1967).
L. V. Keldysh, *Excitons in semiconductors,* Nauka (Moscow), 1974, In Russian.
C. C. Klick and J. S. Schulman, "Luminescence in solids," *Solid state physics* **5,** 97 (1957).
R. S. Knox, *Theory of excitons,* Academic Press, 1963.
C. G. Kuper and G. D. Whitfield, eds., *Polarons and excitons,* Plenum, 1963.
D. S. McClure, "Electronic spectra of molecules and ions in crystals," *Solid state physics* **8,** 1 (1959).
A. Rose, *Concepts in photoconductivity,* Interscience, 1963.
J. Tauc, ed., *Optical properties of solids,* Academic Press, 1966.
R. F. Wallis, ed., *Localized excitations in solids,* Plenum, 1968.
H. C. Wolf, "The electronic spectra of aromatic molecular crystals," *Solid state physics* **9,** 1 (1959).

12 SUPERCONDUCTIVITY

NOTATION: In this chapter, B_a denotes the applied magnetic field. In the CGS system the critical value B_{ac} of the applied field will be denoted by the symbol H_c in accordance with the custom of workers in superconductivity. Values of B_{ac} are given in gauss in CGS units and in teslas in SI units. In SI we have $B_{ac} = \mu_0 H_c$.

TABLE 1 Superconductivity Parameters of the Elements

An asterisk denotes an element superconducting only in thin films or under high pressure in a crystal modification not normally stable. Data courtesy of B. T. Matthias.

Key: Transition temperature in K / Critical magnetic field at absolute zero in gauss (10^{-4} tesla)

Li	Be											B	C	N	O	F	Ne
Na	Mg											Al 1.140 105	Si*	P*	S	Cl	Ar
K	Ca	Sc	Ti 0.39 100	V 5.38 1420	Cr*	Mn	Fe	Co	Ni	Cu	Zn 0.875 53	Ga 1.091 51	Ge*	As*	Se*	Br	Kr
Rb	Sr	Y*	Zr 0.546 47	Nb 9.50 1980	Mo 0.92 95	Tc 7.77 1410	Ru 0.51 70	Rh	Pd	Ag	Cd 0.56 30	In 3.4035 293	Sn (w) 3.722 309	Sb*	Te*	I	Xe
Cs*	Ba*	La fcc 6.00 1100	Hf 0.12	Ta 4.483 830	W 0.012 1.07	Re 1.4 198	Os 0.655 65	Ir 0.14 19	Pt	Au	Hg (α) 4.153 412	Tl 2.39 171	Pb 7.193 803	Bi*	Po	At	Rn
Fr	Ra	Ac															

Ce*	Pr	Nd	Pm	Sm	Eu	Gd	Tb	Dy	Ho	Er	Tm	Yb	Lu 0.1
Th 1.368 1.62	Pa 1.4	U*(α)	Np	Pu	Am	Cm	Bk	Cf	Es	Fm	Md	No	Lr

12 SUPERCONDUCTIVITY

The electrical resistivity of many metals and alloys drops suddenly to zero when the specimen is cooled to a sufficiently low temperature, often a temperature in the liquid helium range. This phenomenon was observed first by Kamerlingh Onnes[1] in Leiden in 1911, three years after he first liquified helium. At a critical temperature T_c the specimen undergoes a phase transition from a state of normal electrical resistivity to a superconducting state, Fig. 1.

EXPERIMENTAL SURVEY

In the superconducting state the dc electrical resistivity is zero, or so close to zero that persistent electrical currents have been observed to flow without attenuation in superconducting rings for more than a year, until at last the experimentalist wearied of the experiment. The decay of supercurrents in a solenoid was studied by File and Mills[2] using precision nuclear magnetic resonance methods to measure the magnetic field associated with the supercurrent. They concluded that the decay time of the supercurrent is not less than 100,000 years. We estimate the decay time below. In some superconducting materials, particularly those used for superconducting magnets, finite decay times are observed because of an irreversible redistribution of magnetic flux in the material.

The magnetic properties exhibited by superconductors are as dramatic as their electrical properties. The magnetic properties cannot be accounted for by the assumption that the superconducting state is characterized properly by zero electrical resistivity. It is an experimental fact that a

[1]H. Kamerlingh Onnes, Akad. van Wetenschappen (Amsterdam) **14,** 113, 818 (1911): "The value of the mercury resistance used was 172.7 ohms in the liquid condition at 0°C; extrapolation from the melting point to 0°C by means of the temperature coefficient of solid mercury gives a resistance corresponding to this of 39.7 ohms in the solid state. At 4.3 K this had sunk to 0.084 ohms, that is, to 0.0021 times the resistance which the solid mercury would have at 0°C. At 3 K the resistance was found to have fallen below 3×10^{-6} ohms, that is to one ten-millionth of the value which it would have at 0°C. As the temperature sank further to 1.5 K this value remained the upper limit of the resistance." Historical references are given by C. J. Gorter, Rev. Mod. Phys. **36,** 1 (1964).
[2]J. File and R. G. Mills, Phys. Rev. Letters **10,** 93 (1963).

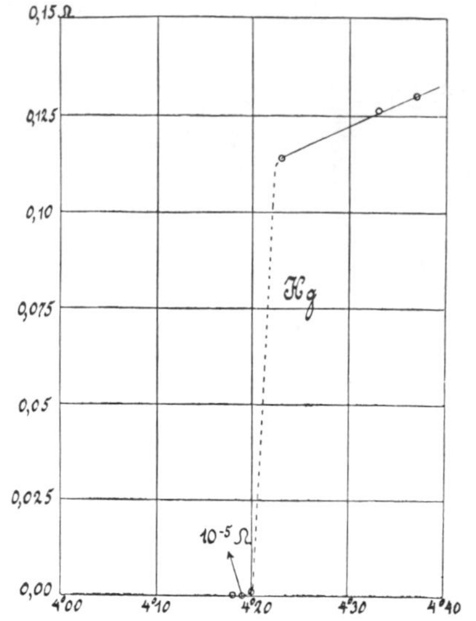

Figure 1 Resistance in ohms of a specimen of mercury versus absolute temperature. This plot by Kamerlingh Onnes marked the discovery of superconductivity.

bulk superconductor in a weak magnetic field will act as a perfect diamagnet, with zero magnetic induction in the interior. When a specimen is placed in a magnetic field and is then cooled through the transition temperature for superconductivity, the magnetic flux originally present is ejected from the specimen. This is called the **Meissner effect.** The sequence of events is shown in Fig. 2. The unique magnetic properties of superconductors are of central importance to the characterization of the superconducting state.

The superconducting state is known to be an ordered state of the conduction electrons of the metal. The order is in the formation of loosely associated pairs of electrons. The electrons are ordered at temperatures below the transition temperature, and they are disordered above the transition temperature. The nature and origin of the ordering was explained by Bardeen, Cooper, and Schrieffer.[3] In the present chapter we develop as far as we can in an elementary way the physics of the superconducting state. We shall also discuss the basic physics of the materials used for superconducting magnets, but not their technology.

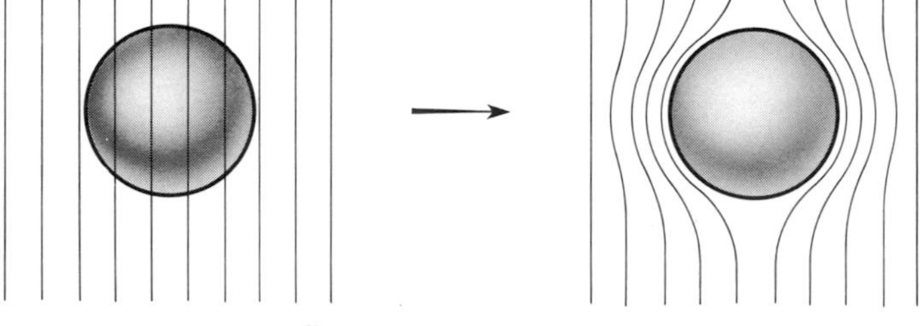

Figure 2 Meissner effect in a superconducting sphere cooled in a constant applied magnetic field; on passing below the transition temperature the lines of induction **B** are ejected from the sphere.

Occurrence of Superconductivity

Superconductivity occurs[4] in many metallic elements of the periodic system and also in alloys, intermetallic compounds, and semiconductors. The range of transition temperatures at present extends from 23.2 K for the alloy Nb_3Ge to 0.01 K for some semiconductors. In many metals superconductivity has not been found down to the lowest temperatures at which the metal was examined, usually well below 1 K. Thus Li, Na, and K have been investigated for superconductivity down to 0.08 K, 0.09 K, and 0.08 K, respectively, where they were still normal conductors. Similarly, Cu, Ag, and Au have been investigated down to 0.05 K, 0.35 K, and 0.05 K, where they are still normal conductors. It has been predicted[5] by theoretical calculation that if sodium and potassium superconduct at all, the transition temperature will be much less than 10^{-5} K. These statements refer to atmospheric pressure. Cesium becomes a superconductor ($T_c = 1.5$ K) at a pressure of 110 kbar, after several phase transformations. The elements known to be superconducting are displayed in Table 1.

Will every nonmagnetic metallic element become a superconductor at sufficiently low temperatures? We do not know. In experimental searches for superconductors with ultralow transition temperatures it is important to eliminate from the specimen even trace quantities of foreign paramagnetic elements, because they can lower the transition temperature severely. A few parts per million of Fe will destroy the superconductivity of Mo, which when pure has $T_c = 0.92$ K; and 1 at. percent of gadolinium lowers the transition temperature of lanthanum from 5.6 K to 0.6 K. Nonmagnetic impurities have no very marked effect on the transition temperature. The transition temperatures of a number of interesting superconducting compounds are listed in Table 2.

TABLE 2 Superconductivity of Selected Compounds

Compound	T_c, in K	Compound	T_c, in K
Nb_3Sn	18.05	V_3Ga	16.5
Nb_3Ge	23.2	V_3Si	17.1
Nb_3Al	17.5	$Pb_1Mo_{5.1}S_6$	14.4
NbN	16.0	Ti_2Co	3.44
$(SN)_x$ polymer	0.26	La_3In	10.4

[3]J. Bardeen, L. N. Cooper, and J. R. Schrieffer, Phys. Rev. **106,** 162 (1957); **108,** 1175 (1957).

[4]A review of data on the occurrence of superconductivity is given in the *A.I.P. Handbook*, 3rd ed. For Nb_3Ge, see the review in Science **183,** 293 (1974) and also J. R. Gavaler, Appl. Phys. Letters **23,** 480 (1973).

[5]J. P. Carbotte and R. C. Dynes, Phys. Rev. **172,** 476 (1968); for the experiments on cesium, see J. Wittig, Phys. Rev. Letters **24,** 812 (1970).

Destruction of Superconductivity by Magnetic Fields

A sufficiently strong magnetic field will destroy superconductivity. The threshold or critical value of the applied magnetic field for the destruction of superconductivity is denoted by $H_c(T)$ and is a function of the temperature. At the critical temperature the critical field is zero: $H_c(T_c) = 0$. The variation of the critical field with temperature for several superconducting elements is shown in Fig. 3. The threshold curves separate the superconducting state in the lower left of the figure from the normal state in the upper right. Note: We should denote the critical value of the applied magnetic field as B_{ac}, but this is not common practice among workers in superconductivity. In the CGS system we shall always understand that $H_c \equiv B_{ac}$, and in the SI we have $H_c \equiv B_{ac}/\mu_0$. The symbol B_a denotes the applied magnetic field.

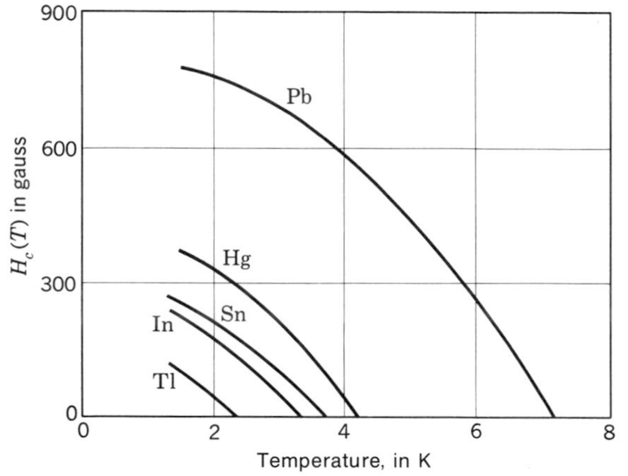

Figure 3 Threshold curves of the critical field $H_c(T)$ versus temperature for several superconductors. A specimen is superconducting below the curve and normal above the curve.

Meissner Effect

Meissner and Ochsenfeld[6] found that if a superconductor is cooled in a magnetic field to below the transition temperature, then at the transition the lines of induction B are pushed out (Fig. 2). This phenomenon is called the Meissner effect. It shows that a bulk superconductor behaves in an applied external field B_a as if inside the specimen $B = 0$. We obtain a particularly useful form of this result if we limit ourselves to long thin specimens with long axis parallel to B_a; now the demagnetizing field contribution (see Chapter 13) to B will be negligible, whence:[7]

$$\text{(CGS)} \qquad B = B_a + 4\pi M = 0 \ ; \qquad \text{or} \qquad \frac{M}{B_a} = -\frac{1}{4\pi} \ ; \qquad (1)$$

$$\text{(SI)} \qquad B = B_a + \mu_0 M = 0 \ ; \qquad \text{or} \qquad \frac{M}{B_a} = -\frac{1}{\mu_0} \ .$$

This result cannot be derived from the characterization of a superconductor as a medium of zero resistivity. From Ohm's law, $\mathbf{E} = \rho\mathbf{j}$, we see that if the resistivity ρ goes to zero while $\mathbf{j}$ is held finite, then $\mathbf{E}$ must be zero. By a Maxwell equation $d\mathbf{B}/dt$ is proportional to curl $\mathbf{E}$, so that zero resistivity implies $d\mathbf{B}/dt = 0$. This argument is not entirely transparent, but the result predicts that the flux through the metal cannot change on cooling through the transition. The Meissner effect contradicts this result and suggests that perfect diamagnetism is an essential property of the superconducting state. We expect another difference between a superconductor and a perfect conductor, defined as a conductor in which there is nothing to scatter the electrons. When the problem is solved in detail, it turns out that a perfect conductor when placed in a magnetic field cannot produce a permanent eddy current screen: the field will penetrate about 1 cm in an hour.[8]

The magnetization curve expected for a superconductor under the conditions of the Meissner-Ochsenfeld experiment is sketched in Fig. 4a. This applies quantitatively to a specimen in the form of a long solid cylinder[9] placed in a longitudinal magnetic field. Pure specimens of many materials exhibit this behavior; they are called **type I superconductors** or, formerly, soft superconductors. The values of H_c are always too low for type I superconductors to have any useful technical application in coils for superconducting magnets.

[6] W. Meissner and R. Ochsenfeld, Naturwiss. **21**, 787 (1933).
[7] Diamagnetism, the magnetization M, and the magnetic susceptibility are defined in Chapter 14. The magnitude of the apparent diamagnetic susceptibility of bulk superconductors is very much larger than in typical diamagnetic substances. In (1), M is the magnetization equivalent to the superconducting currents in the specimen.
[8] A. B. Pippard, *Dynamics of conduction electrons*, Gordon and Breach, 1965.
[9] In other geometries the field may not be homogeneous around the specimen and the field may penetrate below H_c; for a sphere the field penetrates at $\frac{2}{3}H_c$ (see Problem 7).

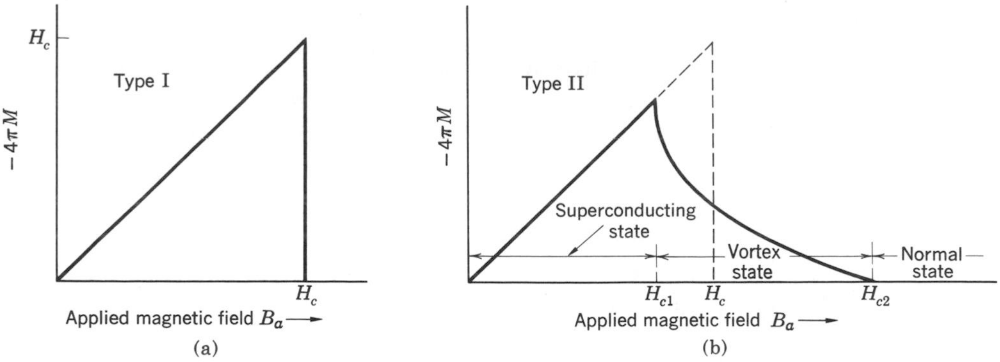

Figure 4 (a) Magnetization versus applied magnetic field for a bulk superconductor exhibiting a complete Meissner effect (perfect diamagnetism). A superconductor with this behavior is called a type I superconductor. Above the critical field H_c the specimen is a normal conductor and the magnetization is too small to be seen on this scale. Note that minus $4\pi M$ is plotted on the vertical scale: the negative value of M corresponds to diamagnetism. (b) Superconducting magnetization curve of a type II superconductor. The flux starts to penetrate the specimen at a field H_{c1} lower than the thermodynamic critical field H_c. The specimen is in a vortex state between H_{c1} and H_{c2}, and it has superconducting electrical properties up to H_{c2}. Above H_{c2} the specimen is a normal conductor in every respect, except for possible surface effects. For given H_c the area under the magnetization curve is the same for a type II superconductor as for a type I. (CGS units in all parts of this figure.)

Other materials exhibit a magnetization curve of the form of Fig. 4b and are known as **type II superconductors.** They tend to be alloys (as in Fig. 5) or transition metals with high values of the electrical resistivity in the normal state: that is, the electronic mean free path in the normal state is short. We shall see later why the mean free path is involved in the "magnetization" of superconductors. Type II superconductors have superconducting electrical properties up to a field denoted by H_{c2}. Between the lower critical field H_{c1} and the upper critical field H_{c2} the flux density $B \neq 0$ and the Meissner effect is said to be incomplete. The value of H_{c2} may be 100 times or more higher than the value of the critical field H_c calculated from the thermodynamics of the transition. In the region between H_{c1} and H_{c2} the superconductor is threaded by flux lines and is said to be in the **vortex state.** A field H_{c2} of 410 kG (41 teslas) has been attained[10] in an alloy of Nb, Al, and Ge at the boiling point of helium (Fig. 6) and 510 kG has been reported for $Pb_1Mo_{5.1}S_6$. Commercial solenoids wound with a hard superconductor produce high steady fields, some over 100 kG. A hard superconductor is a type II superconductor with a large amount of magnetic hysteresis or flux pinning induced by mechanical treatment.

[10]For detailed data on high H_{c2} materials, see T. G. Berlincourt and R. R. Hake, Phys. Rev. **131,** 140 (1963); Y. Shapira and L. J. Neuringer, Phys. Rev. **140,** A1638 (1965); S. Foner, E. J. McNiff, Jr., B. T. Matthias, T. H. Geballe, R. H. Willens and E. Corenzwit, Physics Letters **31A,** 349 (1970). For lead molybdenum polysulfide, see S. Foner, E. J. McNiff, Jr., and E. Alexander, Phys. Letters **49A,** 269 (1974); O. Fischer, H. Jones, G. Bongi, M. Sergent, and R. Chevrel, J. Phys. C. **7,** L450 (1974).

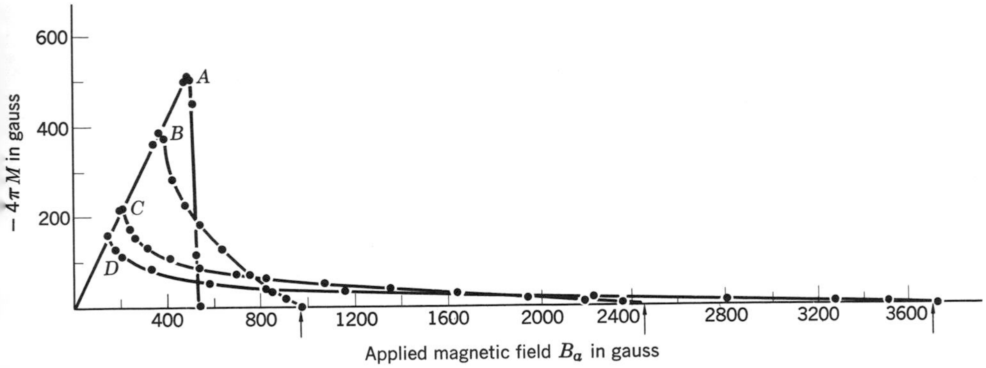

Figure 5 Superconducting magnetization curves of annealed polycrystalline lead and lead-indium alloys at 4.2 K. (*A*) lead; (*B*) lead–2.08 wt. percent indium; (*C*) lead–8.23 wt. percent indium; (*D*) lead–20.4 wt. percent indium. (After Livingston.)

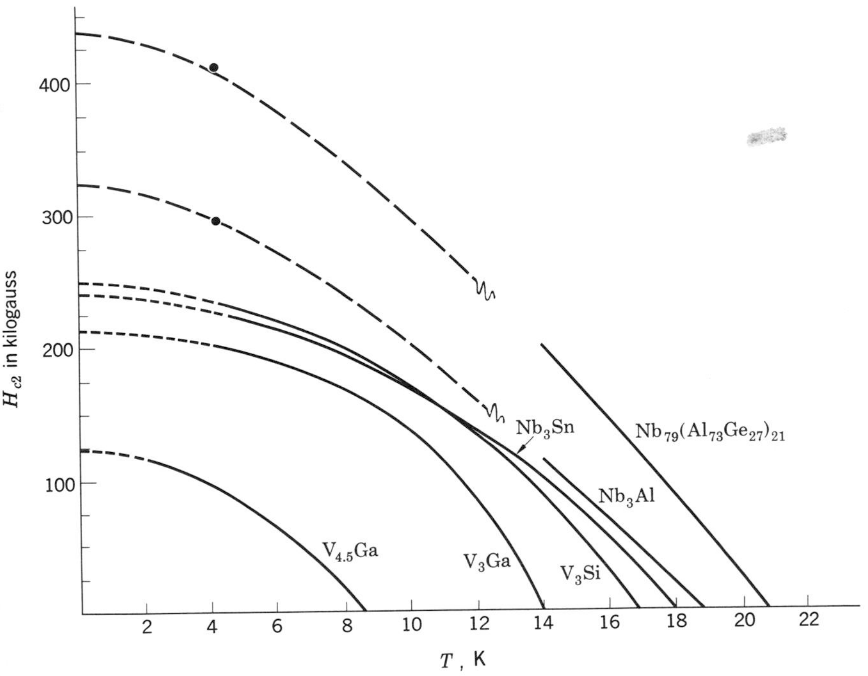

Figure 6 Upper critical field versus temperature, for various high field superconductors. (After S. Foner, E. J. McNiff, Jr., B. T. Matthias, T. H. Geballe, R. H. Willens, and E. Corenzwit.)

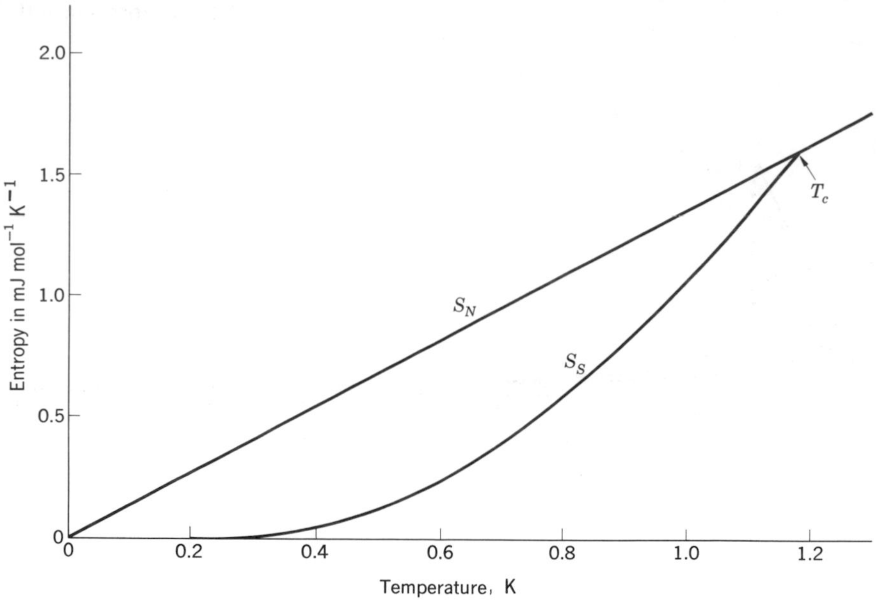

Figure 7a Entropy S of aluminum in the normal and superconducting states as a function of the temperature. The entropy is lower in the superconducting state because the electrons are more ordered here than in the normal state. At any temperature below the critical temperature T_c the specimen can be put in the normal state by application of a magnetic field stronger than the critical field.

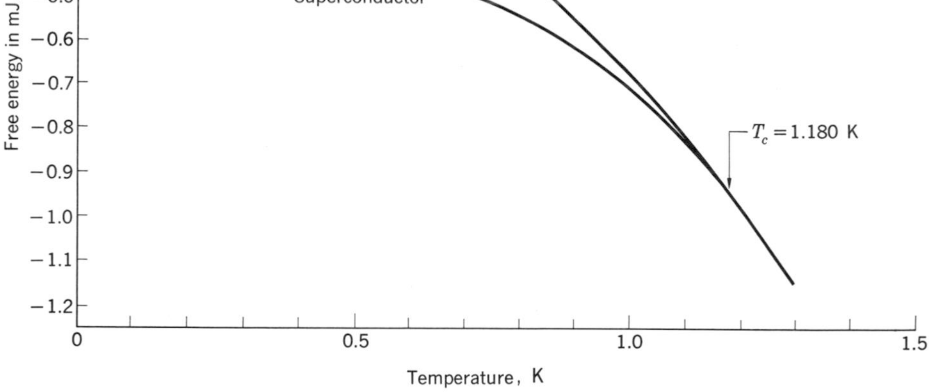

Figure 7b Experimental values of the free energy as a function of temperature for aluminum in the superconducting state and in the normal state. Below the transition temperature $T_c = 1.180$ K the free energy is lower in the superconducting state. The two curves merge at the transition temperature, so that the phase transition is second order (there is no latent heat of transition at T_c). The curve F_S is measured in zero magnetic field, and F_N is measured in a magnetic field sufficient to put the specimen in the normal state. (Courtesy of N. E. Phillips.)

Heat Capacity

In all superconductors the entropy decreases markedly on cooling below the critical temperature T_c. Measurements for aluminum are plotted in Fig. 7a. The decrease in entropy between the normal state and the superconducting state tells us that the superconducting state is more ordered than the normal state, for the entropy is a measure of the disorder of a system. Some or all of the electrons thermally excited in the normal state are ordered in the superconducting state. The change in entropy is small, in aluminum of the order of $10^{-4}\ k_B$ per atom.

The heat capacity of gallium is plotted in Fig. 8: (a) compares the normal and superconducting states; (b) shows that the electronic contribution to the heat capacity in the superconducting state is an exponential form with an argument proportional to $-1/T$, suggestive of excitation of electrons across an energy gap. An energy gap (Fig. 9) is a characteristic, but not universal, feature of the superconducting state. The gap is accounted for by the Bardeen-Cooper-Schrieffer (BCS) theory of superconductivity.

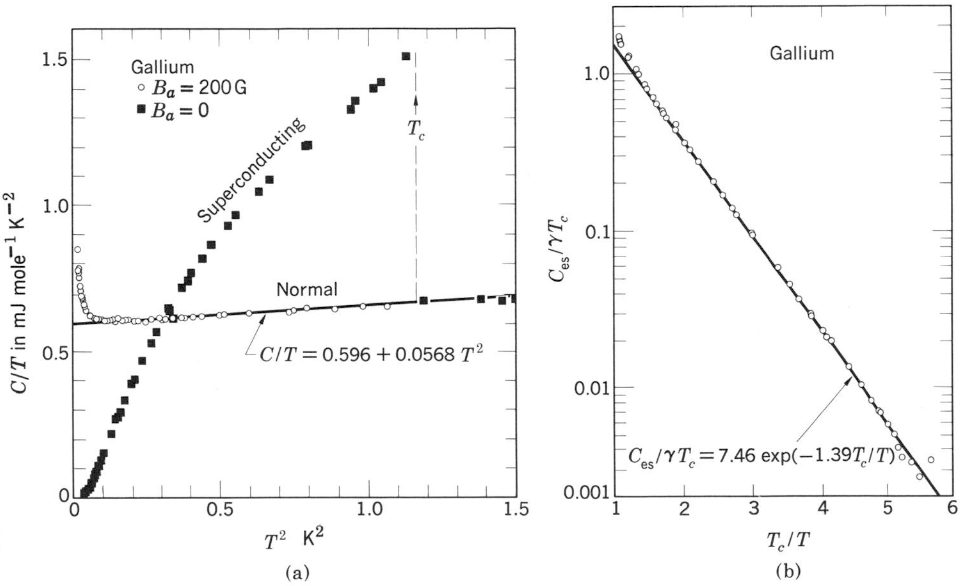

Figure 8 (a) The heat capacity of gallium in the normal and superconducting states. The normal state (which is restored by a 200 G field) has electronic, lattice, and (at low temperatures) nuclear quadrupole contributions. In (b) the electronic part C_{es} of the heat capacity in the superconducting state is plotted on a log scale versus T_c/T: the exponential dependence on $1/T$ is evident. Here $\gamma = 0.60$ mJ mol^{-1} deg^{-2}. [After N. E. Phillips, Phys. Rev. **134**, 385 (1964).]

Energy Gap

The energy gap in superconductors is of an entirely different nature than the energy gap in insulators. In an insulator the gap is tied to the lattice; in a superconductor the gap is tied to the Fermi gas. The argument of the exponential in the heat capacity is found to be one-half of the gap, from comparison with optical and other determinations of the gap. Values of the energy gaps[11] in several superconductors are given in Table 3; the values were obtained by the electron tunneling method to be described.

The transition in zero magnetic field from the superconducting state to the normal state is observed to be a second-order phase transition. At a second-order transition there is no latent heat, but there is a discontinuity in the heat capacity, evident in Fig. 8a. Further, the energy gap decreases continuously to zero as the temperature is increased to the transition temperature T_c, as in Fig. 10. A first-order transition would be characterized by a latent heat and by a discontinuity in the energy gap.

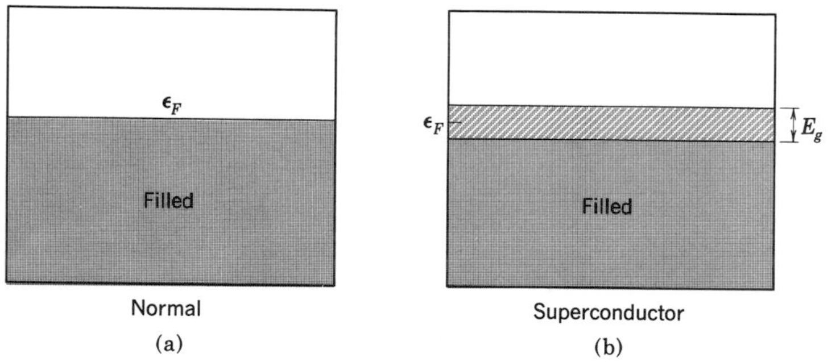

Figure 9 (a) Conduction band in the normal state; (b) energy gap at the Fermi level in the superconducting state. Electrons in excited states above the gap behave as normal electrons in rf fields: they cause resistance; at dc they are shorted out by the superconducting electrons. At absolute zero there are no electrons above the gap. The gap E_g is exaggerated in the figure: typically $E_g \sim 10^{-4}\ \epsilon_F$.

[11]For a review, see D. H. Douglass, Jr., and L. M. Falicov, *Progress in low temperature physics* **4**, 97–193 (1964).

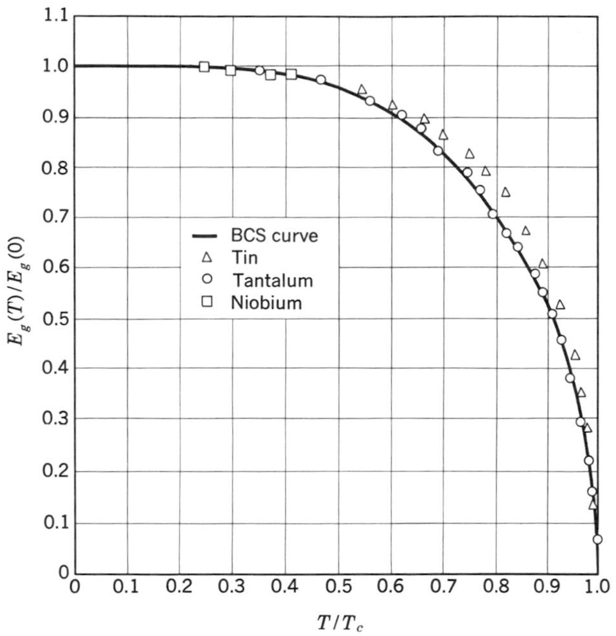

Figure 10 Reduced values of the observed energy gap $E_g(T)/E_g(0)$ as a function of the reduced temperature T/T_c, after Townsend and Sutton. The solid curve is drawn for the BCS theory.

TABLE 3 Energy Gaps in Superconductors, at $T = 0$

$E_g(0)$ in 10^{-4}eV.
$E_g(0)/k_BT_c$.

										Al 3.4 3,3	**Si**
Sc	**Ti**	**V** 16. 3.4	**Cr**	**Mn**	**Fe**	**Co**	**Ni**	**Cu**	**Zn** 2.4 3.2	**Ga** 3.3 3.5	**Ge**
Y	**Zr**	**Nb** 30.5 3.80	**Mo** 2.7 3.4	**Tc**	**Ru**	**Rh**	**Pd**	**Ag**	**Cd** 1.5 3.2	**In** 10.5 3.6	**Sn** (w) 11.5 3.5
La fcc 19. 3.7	**Hf**	**Ta** 14. 3.60	**W**	**Re**	**Os**	**Ir**	**Pt**	**Au**	**Hg** (α) 16.5 4.6	**Tl** 7.35 3.57	**Pb** 27.3 4.38

Microwave and Infrared Properties

The existence of an energy gap in superconductors means that photons of energy less than the gap energy are not absorbed. Nearly all the photons incident are reflected as for any metal because of the impedance mismatch at the boundary between vacuum and metal, but for a very thin (~20 Å) film more photons are transmitted in the superconducting state than in the normal state. For photon energies less than the energy gap, the resistivity of a superconductor vanishes at absolute zero. At $T \ll T_c$ the resistance in the superconducting state has a sharp threshold at the gap energy. Photons of lower energy see a resistanceless surface. Photons of higher energy see a resistance that approaches that of the normal state because such photons cause transitions to unoccupied normal energy levels above the gap. As the temperature is increased not only does the gap decrease in energy, but the resistivity for photons below the gap energy no longer vanishes, except at zero frequency. At zero frequency the superconducting electrons short-circuit any normal electrons that have been thermally excited above the gap. At finite frequencies the inertia of the superconducting electrons prevents them from completely screening the electric field, so that thermally excited normal electrons now can absorb energy (Problem 3).

Isotope Effect

It has been observed that the critical temperature of superconductors varies with isotopic mass. In mercury T_c varies from 4.185 K to 4.146 K as the average atomic mass M varies from 199.5 to 203.4 atomic mass units. The transition temperature changes smoothly when we mix different isotopes of the same element. The experimental results within each series of isotopes may be fitted by a relation of the form

$$M^\alpha T_c = \text{constant} \ . \tag{2}$$

Observed values of α are given in Table 4. From the dependence of T_c on the isotopic mass we learn that lattice vibrations and hence electron-lattice interactions are deeply involved in superconductivity.[12] There is no other reason for the superconducting transition temperature to depend on the number of neutrons in the nucleus. The original BCS model gave the result $T_c \propto \theta_{\text{Debye}} \propto M^{-\frac{1}{2}}$, so that $\alpha = \frac{1}{2}$ in (2), but the inclusion of coulomb interactions between the electrons changes the relation. Nothing is sacred about $\alpha = \frac{1}{2}$.

[12]The absence of an isotope effect in Ru and Os has been accounted for in terms of the band structure of these metals. J. W. Garland, Jr., Phys. Rev. Letters **11,** 114 (1963); see also W. I. McMillan, Phys. Rev. **167,** 331 (1968).

TABLE 4 Isotope Effect in Superconductors

Experimental values of α in $M^{\alpha}T_c = \text{constant}$, where M is the isotopic mass.

Substance	α	Substance	α
Zn	0.45 ± 0.05	Ru	0.00 ± 0.05
Cd	0.32 ± 0.07	Os	0.15 ± 0.05
Sn	0.47 ± 0.02	Mo	0.33
Hg	0.50 ± 0.03	Nb_3Sn	0.08 ± 0.02
Pb	0.49 ± 0.02	Mo_3Ir	0.33 ± 0.03
Tl	0.61 ± 0.10	Zr	0.00 ± 0.05

After a tabulation by J. W. Garland, Jr., Phys. Rev. Letters **11**, 114 (1963), with revisions suggested by Dr. V. Compton.

THEORETICAL SURVEY

A theoretical understanding of the phenomena associated with superconductivity has been reached in several ways. Certain results follow directly from thermodynamics. Many important results can be described by phenomenological equations: the London equations and the Landau-Ginzburg equations. A successful quantum theory of superconductivity was given by Bardeen, Cooper, and Schrieffer, and has provided the basis for advanced work. Josephson and Anderson discovered the importance of the phase of the superconducting wavefunction.

Thermodynamics of the Superconducting Transition

The transition between the normal and superconducting states is thermodynamically reversible, just as the transition between liquid and vapor phases of a substance is reversible. Thus we may apply thermodynamics to the transition, and we thereby obtain an expression for the entropy difference between normal and superconducting states in terms of the critical field curve H_c versus T. This is analogous to the vapor pressure equation for the liquid-gas coexistence curve (*TP*, Chapter 20). We treat a type I superconductor with a complete Meissner effect, so that $B = 0$ inside the superconductor. We shall see that the critical field H_c is a quantitative measure of the energy difference between the superconducting and normal states at absolute zero. The symbol H_c will always refer to a bulk specimen, never to a thin film. For type II superconductors, H_c is understood to be the thermodynamic critical field related to the stabilization energy.

The stabilization energy of the superconducting state with respect to the normal state can be determined calorimetrically or magnetically. Direct measurements of the heat capacity are made for the normal metal and for the superconductor in zero magnetic field. From their difference we compute the energy difference at absolute zero, the **stabilization energy** of the superconducting state. It is also possible to obtain the stabilization energy and the free energy simply from the value of the applied magnetic field that will destroy the superconducting state and will carry the specimen into the normal state. The argument follows. Consider the work done (Fig. 11) on a superconductor when it is brought from a position at infinity where the applied field is zero to a position **r** in the field of a permanent magnet:

$$W = -\int_0^{B_a} \mathbf{M} \cdot d\mathbf{B}_a \tag{3}$$

per unit volume of specimen. This appears in the energy of the magnetic field. The thermodynamic identity for the process is, as in *TP*, Chapter 23,

$$dU = T\,dS - \mathbf{M} \cdot d\mathbf{B}_a \ . \tag{4}$$

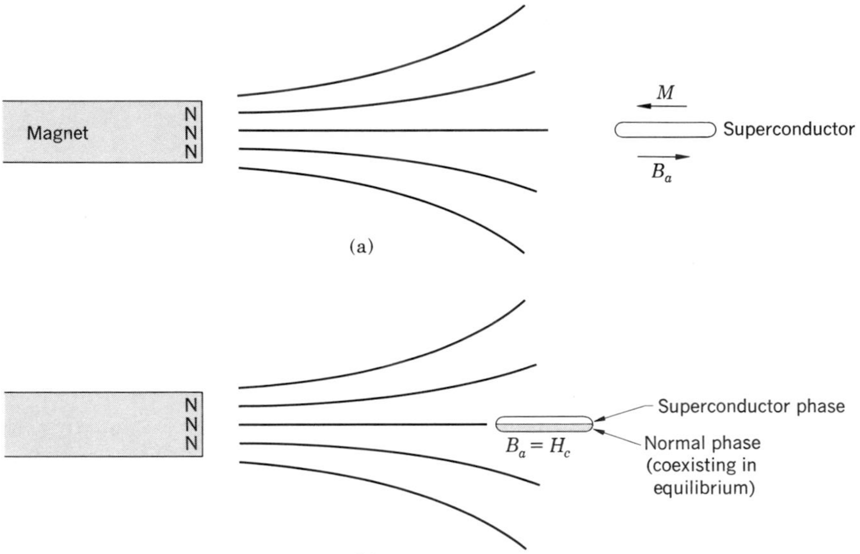

Figure 11 (a) A superconductor in which the Meissner effect is complete has $B = 0$, as if the magnetization were $M = -B_a/4\pi$, in CGS units. The work done on the superconductor in a displacement from infinity to the position where the field of the permanent magnet is B_a is

(CGS) $$W = -\int \mathbf{M} \cdot d\mathbf{B}_a = B_a^2/8\pi$$

per unit volume. (b) When the applied field reaches the value B_{ac}, the normal state can coexist in equilibrium with the superconducting state. In coexistence the free energy densities are equal: $F_N(T, B_{ac}) = F_S(T, B_{ac})$.

For a superconductor with **M** related to $\mathbf{B}_a$ by (1) we have

(CGS) $$dU_S = T\,dS + \frac{1}{4\pi} B_a\,dB_a \ ; \qquad (5)$$

(SI) $$dU_S = T\,dS + \frac{1}{\mu_0} B_a\,dB_a \ .$$

Thus at absolute zero, where $T\,dS = 0$, the increase in the energy density of the superconductor is

(CGS) $$U_S(B_a) - U_S(0) = B_a^2/8\pi \ ; \qquad (6)$$

(SI) $$U_S(B_a) - U_S(0) = B_a^2/2\mu_0 \ ,$$

on being brought from a position where the applied field is zero to a position where the applied field is B_a.

Now consider a normal nonmagnetic metal. If we neglect the small susceptibility[13] of a metal in the normal state, then $M = 0$ and the energy of the normal metal is independent of field. At the critical field we have

$$U_N(B_{ac}) = U_N(0) \ . \qquad (7)$$

The results (6) and (7) are all we need to determine the stabilization energy of the superconducting state at absolute zero. At the critical value B_{ac} of the applied magnetic field the energies are equal in the normal and superconducting states:

(CGS) $$U_N(B_{ac}) = U_S(B_{ac}) = U_S(0) + B_{ac}^2/8\pi \ , \qquad (8)$$

(SI) $$U_N(B_{ac}) = U_S(B_{ac}) = U_S(0) + B_{ac}^2/2\mu_0 \ .$$

In SI units $H_c \equiv B_{ac}/\mu_0$, whereas in CGS units $H_c \equiv B_{ac}$. The specimen is stable in either state when the applied field is equal to the critical field. Now by (7) it follows that

(CGS) $$\Delta U \equiv U_N(0) - U_S(0) = B_{ac}^2/8\pi \ , \qquad (9)$$

where ΔU is the stabilization energy density of the superconducting state at absolute zero. For aluminum, B_{ac} at absolute zero is 105 gauss, so that $\Delta U = (105)^2/8\pi = 439$ erg cm^{-3}, in excellent agreement with the result of thermal measurements, 430 erg cm^{-3}.

[13]This is an adequate assumption for type I superconductors. In type II superconductors $H_c \ll H_{c2}$ and in high fields the change in spin paramagnetism of the conduction electrons lowers the energy of the normal phase significantly. In some type II superconductors the upper critical field is limited by this effect. Clogston has suggested that $H_{c2}(\max) = 18{,}400\ T_c$, where H_{c2} is in gauss and T_c in K. See A. M. Clogston, Phys. Rev. Letters **9**, 266 (1962); B. Chandrasekhar, Appl. Phys. Letters **1**, 7 (1962).

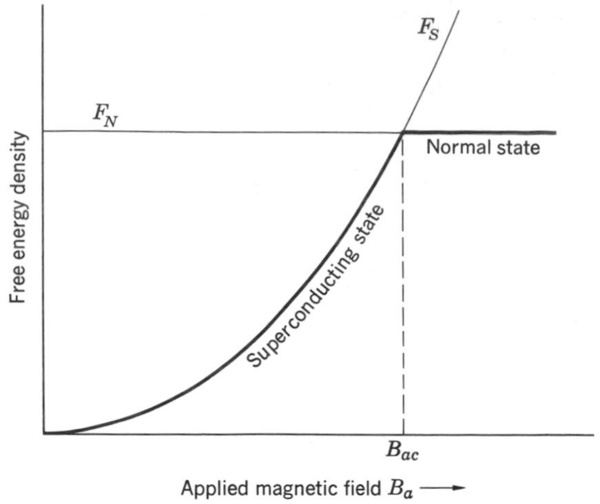

Figure 12 The free energy density F_N of a nonmagnetic normal metal is approximately independent of the intensity of the applied magnetic field B_a. At a temperature $T < T_c$ the metal is a superconductor in zero magnetic field, so that $F_S(T, 0)$ is lower than $F_N(T, 0)$. An applied magnetic field increases F_S by $B_a^2/8\pi$, in CGS units, so that $F_S(T, B_a) = F_S(T, 0) + B_a^2/8\pi$. If B_a is larger than the critical field B_{ac} the free energy density is lower in the normal state than in the superconducting state, and now the normal state is the stable state. The origin of the vertical scale in the drawing is at $F_S(T, 0)$. The figure equally applies to U_S and U_N at $T = 0$.

At a finite temperature the normal and superconducting phases are in equilibrium when their free energies $F = U - TS$ are equal. The free energies of the two phases are sketched in Fig. 12 as a function of the magnetic field. Experimental curves of the free energies of the two phases in aluminum as a function of the temperature are given in Fig. 7b. The slopes dF/dT are equal at the transition so that there is no latent heat.

London Equation

We saw that the Meissner effect implies a magnetic susceptibility $\chi = -1/4\pi$ in CGS in the superconducting state or, in SI, $\chi = -1$. This sweeping assumption tends to cut off further discussion, and it does not account for the flux penetration observed in thin films. Can we modify a constitutive equation of electrodynamics (such as Ohm's law) in some way to obtain the Meissner effect? We do not want to modify the Maxwell equations themselves.

Electrical conduction in the normal state of a metal is described by Ohm's law $\mathbf{j} = \sigma\mathbf{E}$. We need to modify this drastically to describe conduction and the Meissner effect in the superconducting state. Let us make a *postulate* and see what happens. We postulate that in the superconducting state the current density is directly proportional to the vector potential $\mathbf{A}$ of the local magnetic field, where $\mathbf{B} = \text{curl } \mathbf{A}$. In CGS units we write the constant of proportionality as $-c/4\pi\lambda_L^2$ for reasons that will become clear. Here c is the speed of light and λ_L is a constant with the dimensions of length. In SI units we write $-1/\mu_0\lambda_L^2$. Thus

$$\text{(CGS)} \quad \mathbf{j} = -\frac{c}{4\pi\lambda_L^2}\mathbf{A} \; ; \qquad \text{(SI)} \quad \mathbf{j} = -\frac{1}{\mu_0\lambda_L^2}\mathbf{A} \; . \tag{10}$$

This is the London equation.[14] We express it another way by taking the curl of both sides to obtain

$$\text{(CGS)} \quad \text{curl}\,\mathbf{j} = -\frac{c}{4\pi\lambda_L^2}\mathbf{B} \; ; \qquad \text{(SI)} \quad \text{curl}\,\mathbf{j} = -\frac{1}{\mu_0\lambda_L^2}\mathbf{B} \; . \tag{11}$$

First we show that the London equation leads to the Meissner effect. By a Maxwell equation we know that

$$\text{(CGS)} \quad \text{curl}\,\mathbf{B} = \frac{4\pi}{c}\mathbf{j} \; ; \qquad \text{(SI)} \quad \text{curl}\,\mathbf{B} = \mu_0\mathbf{j}_0 \; ; \tag{12}$$

under static conditions. We take the curl of both sides to obtain

$$\text{(CGS)} \qquad \text{curl curl}\,\mathbf{B} = -\nabla^2\mathbf{B} = \frac{4\pi}{c}\,\text{curl}\,\mathbf{j} \; ;$$

$$\text{(SI)} \qquad \text{curl curl B} = -\nabla^2\mathbf{B} = \mu_0\,\text{curl}\,\mathbf{j} \; ,$$

which may be combined with (11) to give for a superconductor

$$\nabla^2\mathbf{B} = \mathbf{B}/\lambda_L^2 \; . \tag{13}$$

This equation accounts for the Meissner effect because it does not allow a solution uniform in space, so that a uniform magnetic field cannot exist in a superconductor. That is, $\mathbf{B}(\mathbf{r}) = \mathbf{B}_0 =$ constant is not a solution of (13) unless the constant field $\mathbf{B}_0$ is identically zero. The result follows because $\nabla^2\mathbf{B}_0$ is always zero, but $\mathbf{B}_0/\lambda_L^2$ is not zero unless $\mathbf{B}_0$ is zero. Note that (12) ensures that $\mathbf{j} = 0$ wherever $\mathbf{B} = 0$.

[14]F. London and H. London, Proc. Roy. Soc. (London) **A149,** 72 (1935); Physica **2,** 341 (1935). The London equation (10) is understood to be written with the vector potential in the London gauge in which div $\mathbf{A} = 0$, and $\mathbf{A}_n = 0$ on any external surface through which no external current is fed. The subscript n denotes the component normal to the surface. Thus div $\mathbf{j} = 0$ and $\mathbf{j}_n = 0$, the actual physical boundary conditions. The form (10) applies to a simply connected superconductor; additional terms may be present in a ring or cylinder, but (11) holds true independent of geometry.

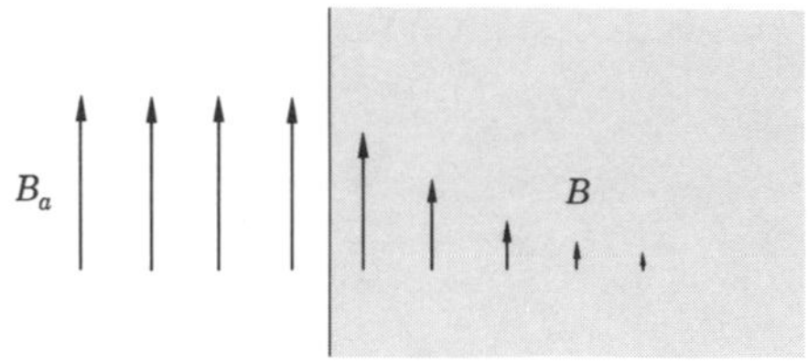

Figure 13 Penetration of an applied magnetic field into a semi-infinite superconductor. The penetration depth λ is defined as the distance in which the field decreases by the factor e^{-1}. Typically, $\lambda \approx 500$ Å in a pure superconductor.

In the pure superconducting state the only field allowed is exponentially damped as we go in from an external surface. Let a semi-infinite superconductor occupy the space on the positive side of the x axis, as in Fig. 13. If $B(0)$ is the field at the plane boundary, then the field inside is

$$B(x) = B(0)\exp(-x/\lambda_L) \ , \tag{14}$$

for this is a solution of (13). In this example the magnetic field is assumed to be parallel to the boundary. Thus we see λ_L measures the depth of penetration of the magnetic field; it is known as the **London penetration depth.** Actual penetration depths are not described precisely by λ_L alone, for the London equation is now known to be somewhat oversimplified. It is shown by comparison of (22) with (11) that

$$\lambda_L = (mc^2/4\pi nq^2)^{1/2} \ , \tag{14a}$$

for particles of charge q and mass m in concentration n. Values are given in Table 5 below.

An applied magnetic field B_a will penetrate a thin film fairly uniformly if the thickness is much less than λ_L; thus in a thin film the Meissner effect is not complete. In this situation the induced field is much less than B_a, and there is little effect of B_a on the energy density of the superconducting state, so that (6) does not apply. From our thermodynamic argument, it follows that the critical field H_c of thin films in parallel magnetic fields should be very high.

Coherence Length

The London penetration depth λ_L is a fundamental length that characterizes a superconductor. Another independent length of equal importance is the **coherence length** ξ. The coherence length is a measure of the distance within which the gap parameter cannot change drastically in a spatially-varying magnetic field. The London equation is a *local* equation: it relates the current density at a point **r** to the vector potential at the same point. So long as $\mathbf{j}(\mathbf{r})$ is given as a constant times $\mathbf{A}(\mathbf{r})$, the current is required to follow exactly any variation in the vector potential. But the coherence length ξ is a measure of the range over which we should average **A** to obtain **j**. It is also a measure of the minimum spatial extent of a transition layer between normal and superconductor. Actually, the two coherence lengths are different, but we shall not go into this.

Any spatial variation in the state of an electronic system requires extra kinetic energy. A modulation of an eigenfunction increases the kinetic energy because the modulation will increase the integral of $d^2\varphi/dx^2$. It is reasonable to restrict the spatial variation of $\mathbf{j}(\mathbf{r})$ in such a way that the extra energy is less than the stabilization energy of the superconducting state. A suggestive argument (based on the uncertainty principle) for the coherence length at absolute zero follows. We compare the plane wave $\psi(x) = e^{ikx}$ with the strongly modulated wavefunction:

$$\varphi(x) = 2^{-\frac{1}{2}}(e^{i(k+q)x} + e^{ikx}) \ . \tag{15}$$

The probability density associated with the plane wave is uniform in space: $\psi^*\psi = e^{-ikx}\, e^{ikx} = 1$, whereas $\varphi^*\varphi$ is modulated with the wavevector q:

$$\begin{aligned} \varphi^*\varphi &= \tfrac{1}{2}(e^{-i(k+q)x} + e^{-ikx})(e^{i(k+q)x} + e^{ikx}) \\ &= \tfrac{1}{2}(2 + e^{iqx} + e^{-iqx}) = 1 + \cos qx . \end{aligned}$$

The kinetic energy of the wave $\psi(x)$ is $\epsilon = \hbar^2k^2/2m$; the kinetic energy of the modulated density distribution is higher, for

$$\int dx\ \varphi^* \left(-\frac{\hbar^2}{2m}\frac{d^2}{dx^2}\right)\varphi = \frac{1}{2}\left(\frac{\hbar^2}{2m}\right)[(k+q)^2 + k^2] \cong \frac{\hbar^2}{2m}k^2 + \frac{\hbar^2}{2m}kq \ ,$$

where we neglect q^2 for $q \ll k$.

The increase of energy required to modulate is $\hbar^2kq/2m$. If this increase exceeds the energy gap E_g, superconductivity will be destroyed. The critical value q_0 of the modulation wavevector is given by

$$\frac{\hbar^2}{2m}k_Fq_0 = E_g \ .$$

We define an **intrinsic coherence length** ξ_0 related to the critical modulation by $\xi_0 = 1/q_0$. We have

$$\xi_0 = \hbar^2k_F/2mE_g = \hbar v_F/2E_g \ , \tag{16}$$

where v_F is the electron velocity at the Fermi surface. On the BCS theory[15] a similar result is found:

$$\boxed{\xi_0 = 2\hbar v_F/\pi E_g} \ . \qquad (17)$$

Calculated values of ξ_0 from (17) are given in Table 5. The intrinsic coherence length ξ_0 is characteristic of a pure superconductor. In impure materials and in alloys the coherence length ξ is shorter than ξ_0. This may be understood qualitatively: in impure material the electron eigenfunctions already have wiggles in them: we can construct a given localized variation of current density with less energy from wavefunctions with wiggles than from smooth wavefunctions.

The coherence length ξ first appeared in the solutions of a pair of phenomenological equations known as the Landau-Ginzburg equations; these equations also follow from the BCS theory. They describe the structure of the transition layer between normal and superconducting phases in contact. The coherence length and the actual penetration depth λ depend on the mean free path ℓ of the electrons measured in the normal state; the relationships are indicated in Fig. 14. When the superconductor is very impure, with a very small ℓ, then $\xi \approx (\xi_0 \ell)^{1/2}$ and $\lambda \approx \lambda_L \ (\xi_0/\ell)^{1/2}$.

TABLE 5 Calculated Intrinsic Coherence Length and London Penetration Depth, at Absolute Zero

Metal	Intrinsic Pippard Coherence Length ξ_0, in 10^{-6} cm	London Penetration Depth λ_L, in 10^{-6} cm	$\frac{\xi_0}{\lambda_L}$
Sn	23.	3.4	6.2
Al	160.	1.6	100.
Pb	8.3	3.7	2.2
Cd	76.	11.0	6.9
Nb	3.8	3.9	0.98

After R. Meservey and B. B. Schwartz.

[15]See *QTS*, pp. 173–174.

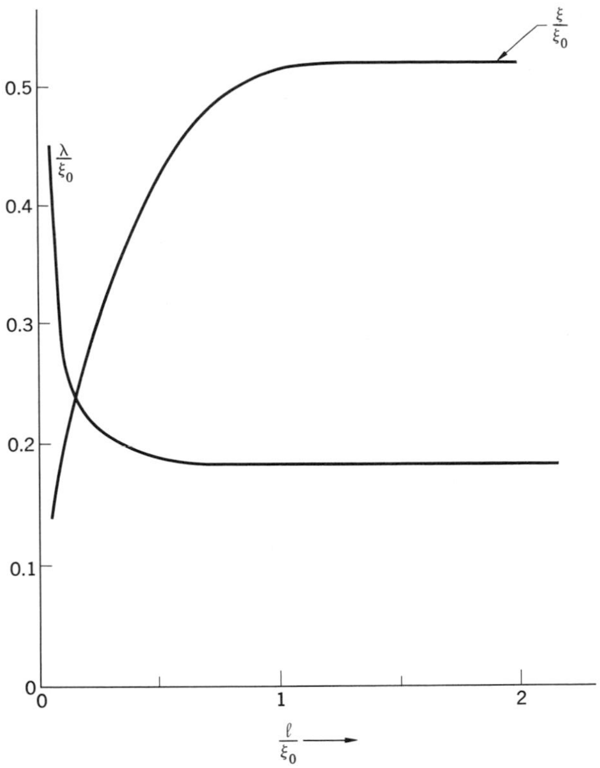

Figure 14 Penetration depth λ and the coherence length ξ as functions of the mean free path ℓ of the conduction electrons in the normal state. All lengths are in units of ξ_0, the intrinsic coherence length. The curves are sketched for $\xi_0 = 10\lambda_L$. For short mean free paths the coherence length becomes shorter and the penetration depth becomes longer. The increase in the ratio λ/ξ favors type II superconductivity.

BCS Theory of Superconductivity

The basis of a quantum theory of superconductivity was laid by the classic 1957 papers of Bardeen, Cooper, and Schrieffer.[16] The accomplishments of the BCS theory include:

1. An attractive interaction between electrons can lead to a ground state separated from excited states by an energy gap. The critical field, the thermal properties, and most of the electromagnetic properties are consequences of the energy gap. In special circumstances, superconductivity may occur without an actual energy gap.
2. The electron-lattice-electron interaction leads to an energy gap of the observed magnitude. The indirect interaction proceeds when one electron interacts with the lattice and deforms it; a second electron sees the deformed lattice and adjusts itself to take advantage of the deformation to lower its energy. Thus the second electron interacts with the first electron via the lattice deformation.
3. The penetration depth and the coherence length emerge as natural consequences of the BCS theory. The London equation is obtained for magnetic fields that vary slowly in space. Thus the central phenomenon in superconductivity, the Meissner effect, is obtained in a natural way.
4. The criterion for the transition temperature of an element or alloy involves the electron density of orbitals $\mathfrak{D}(\epsilon_F)$ at the Fermi level and the electron-lattice interaction U, which can be estimated from the electrical resistivity. For $U\mathfrak{D}(\epsilon_F) \ll 1$ the BCS theory predicts

$$T_c = 1.14\theta \exp[-1/U\mathfrak{D}(\epsilon_F)] \ , \tag{18}$$

   where θ is the Debye temperature and U is an attractive interaction. The result for T_c is satisfied at least qualitatively by the experimental data. There is an interesting apparent paradox: the higher the resistivity at room temperature,[17] the more likely it is that a metal will be a superconductor when cooled.
5. Magnetic flux through a superconducting ring is quantized and the effective unit of charge is $2e$ rather than e. The BCS ground state involves pairs of electrons; thus flux quantization in terms of the pair charge $2e$ is a consequence of the theory.

[16] J. Bardeen, L. N. Cooper, and J. R. Schrieffer, Phys. Rev. **106,** 162 (1957); **108,** 1175 (1957).

[17] Because the resistivity at room temperature is a measure of the electron-phonon interaction. In the BCS relation quoted for T_c we should take $\mathfrak{D}(\epsilon_F)$ as the density of orbitals of one spin.

BCS Ground State

The filled Fermi sea is the ground state of a Fermi gas of noninteracting electrons. This state allows arbitrarily small excitations—we can form an excited state by taking an electron from the Fermi surface and raising it just above the Fermi surface. The BCS theory shows that with an appropriate attractive interaction between electrons the new ground state is superconducting and is separated by a finite energy E_g from its lowest excited state.

The formation of the BCS ground state is suggested by Fig. 15. The BCS state in (b) contains admixtures of one-electron orbitals from above the Fermi energy ϵ_F. At first sight the BCS state appears to have a higher energy than the Fermi state: the comparison of (b) with (a) shows that the kinetic energy of the BCS state is higher than that of the Fermi state. But the attractive potential energy of the BCS state, although not represented in the figure, acts to lower the total energy of the BCS state with respect to the Fermi state. When the BCS ground state of a many-electron system is described in terms of the occupancy of one-particle orbitals, those near ϵ_F are filled somewhat like a Fermi-Dirac distribution for some finite temperature. However, the central feature of the BCS state is that the one-particle orbitals are occupied in pairs: if an orbital with wavevector **k** and spin up is occupied, then the orbital with wavevector $-\mathbf{k}$ and spin down is also occupied. If $\mathbf{k}_{1\uparrow}$ is vacant, then $-\mathbf{k}_{1\downarrow}$ is also vacant. The pairs are called **Cooper pairs.** They have spin zero and have many attributes of bosons.[18]

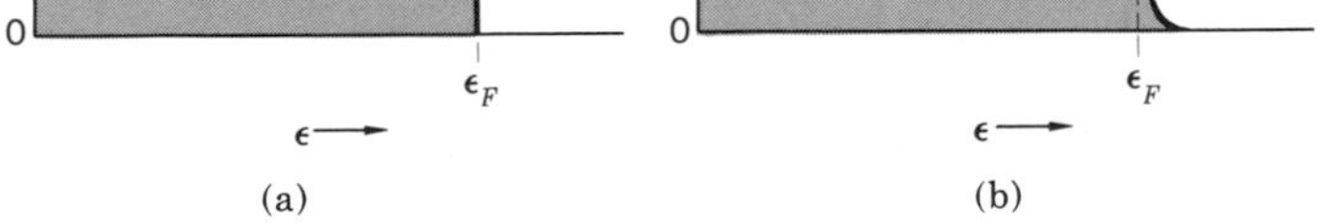

Figure 15 (a) Probability P that an orbital of kinetic energy ϵ is occupied in the ground state of the noninteracting Fermi gas; (b) the BCS ground state differs from the Fermi state in a region of width of the order of the energy gap E_g. Both curves are for absolute zero.

[18]The boson condensation temperature calculated for metallic concentrations is of the order of the Fermi temperature ($10^4 - 10^5$ K). The superconducting transition temperature is much lower and takes place when the electron pairs break up into two fermions. The model of a superconductor as composed of noninteracting bosons cannot be taken too literally, for there are about 10^6 electrons in the volume occupied by a single Cooper pair.

Flux Quantization in a Superconducting Ring

We prove that the total magnetic flux that passes through a superconducting ring may assume only quantized values, integral multiples of the flux quantum $2\pi\hbar c/q$, where by experiment $q = 2e$, the charge of an electron pair. Flux quantization is a beautiful example of a long-range quantum effect in which the coherence of the superconducting state extends over a ring or solenoid. Let us first consider the electromagnetic field as an example of a similar boson field. The electric field intensity $E(\mathbf{r})$ acts qualitatively as a probability field amplitude. When the total number of photons is large, the energy density may be written as

$$E^*(\mathbf{r})E(\mathbf{r})/4\pi \cong n(\mathbf{r})\hbar\omega \ ,$$

where $n(\mathbf{r})$ is the number density of photons of frequency ω. Then we may write the electric field in a semiclassical approximation as

$$E(\mathbf{r}) \cong (4\pi\hbar\omega)^{\frac{1}{2}} n(\mathbf{r})^{\frac{1}{2}}\, e^{i\theta(\mathbf{r})} \qquad E^*(\mathbf{r}) \cong (4\pi\hbar\omega)^{\frac{1}{2}} n(\mathbf{r})^{\frac{1}{2}}\, e^{-i\theta(\mathbf{r})} \ ,$$

where $\theta(\mathbf{r})$ is the phase of the field. A similar probability amplitude describes Cooper pairs.

The arguments that follow apply to a boson gas with a large number of bosons in the same orbital. We then can treat the boson probability amplitude as a classical quantity, just as the electromagnetic field is used for photons. Both amplitude and phase are then meaningful and observable. The arguments do not apply to a metal in the normal state because an electron in the normal state acts as a single unpaired fermion that cannot be treated classically. We first show that a charged boson gas obeys the London equation. Let $\psi(\mathbf{r})$ be the particle probability amplitude. We suppose that the pair concentration $n = \psi^*\psi = \text{constant}$. At absolute zero n is one-half of the concentration of electrons in the conduction band, for n refers to pairs. Then we may write

$$\psi = n^{\frac{1}{2}}\, e^{i\theta(\mathbf{r})} \ ; \qquad \psi^* = n^{\frac{1}{2}}\, e^{-i\theta(\mathbf{r})} \ . \tag{19}$$

The phase $\theta(\mathbf{r})$ is important for what follows. In SI units, set c = 1 in the equations that follow.

The velocity of a particle is, from the Hamilton equations of mechanics,

$$\mathbf{v} = \frac{1}{m}\left(\mathbf{p} - \frac{q}{c}\mathbf{A}\right) = \frac{1}{m}\left(-i\hbar\nabla - \frac{q}{c}\mathbf{A}\right) .$$

The particle flux is given by

$$\psi^*\mathbf{v}\psi = \frac{n}{m}\left(\hbar\nabla\theta - \frac{q}{c}\mathbf{A}\right) , \tag{20}$$

so that the electric current density is

$$\mathbf{j} = q\psi^*\mathbf{v}\psi = \frac{nq}{m}\left(\hbar\nabla\theta - \frac{q}{c}\mathbf{A}\right) . \tag{21}$$

We may take the curl of both sides to obtain the London equation:

$$\text{curl } \mathbf{j} = -\frac{nq^2}{mc}\mathbf{B} \ , \tag{22}$$

with use of the fact that the curl of the gradient of a scalar is identically zero. The constant that multiplies **B** agrees with (14a). We recall that the Meissner effect is a consequence of the London equation, which we have here derived.

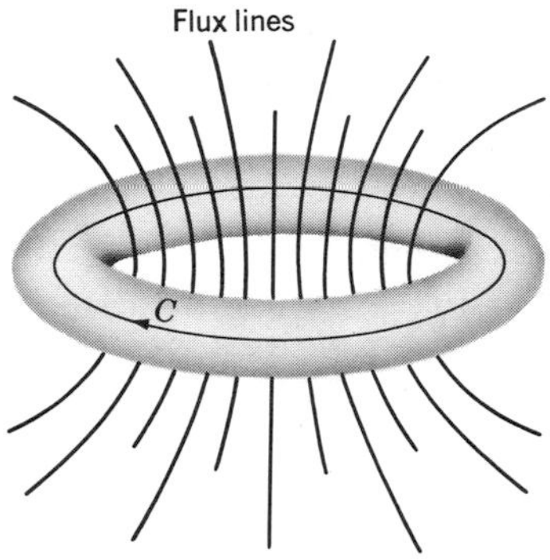

Figure 16 Path of integration C through the interior of a superconducting ring. The flux through the ring is the sum of the flux Φ_{ext} from external sources and the flux Φ_{sc} from the superconducting currents which flow in the surface of the ring: $\Phi = \Phi_{ext} + \Phi_{sc}$. The flux Φ is quantized. There is normally no quantization condition on the flux from external sources, so that Φ_{sc} must adjust itself appropriately in order that Φ assume a quantized value.

Quantization of the magnetic flux through a ring is a dramatic consequence of Eq. (21). Let us take a closed path C through the interior of the superconducting material, well away from the surface (Fig. 16). The Meissner effect tells us that **B** and **j** are zero in the interior. Now (21) is zero if

$$\hbar c \nabla \theta = q\mathbf{A} \ . \tag{23}$$

We form

$$\oint_C \nabla\theta \cdot dl = \theta_2 - \theta_1$$

for the change of phase on going once around the ring. The probability amplitude is measurable in the classical approximation, so that ψ must be single-valued and

$$\theta_2 - \theta_1 = 2\pi s \ , \tag{24}$$

where s is an integer. By the Stokes theorem,

$$\oint_C \mathbf{A} \cdot dl = \int_C (\text{curl } \mathbf{A}) \cdot d\boldsymbol{\sigma} = \int_C \mathbf{B} \cdot d\boldsymbol{\sigma} = \Phi \ , \tag{25}$$

where $d\sigma$ is an element of area on a surface bounded by the curve C, and Φ is the magnetic flux through C. From (23), (24), and (25) we have $2\pi\hbar cs = q\Phi$, or

$$\Phi = (2\pi\hbar c/q)s \ . \tag{26}$$

Thus the flux through the ring is quantized[19] in integral multiples of $2\pi\hbar c/q$. By experiment $q = -2e$ as appropriate for electron pairs, so that the quantum of flux in a superconductor is

$$\Phi_0 = 2\pi\hbar c/2e \cong 2.0678 \times 10^{-7} \text{ gauss-cm}^2 \ . \tag{27}$$

This unit of flux is called a **fluxoid.** We remark that the simple result (26) does not hold if the flux penetrates the ring itself, as when the material of the ring is thin.

The flux through the ring is the sum of the flux Φ_{ext} from external sources and the flux Φ_{sc} from the superconducting currents which flow in the surface of the ring: $\Phi = \Phi_{ext} + \Phi_{sc}$. The flux Φ is quantized. There is normally no quantization condition on the flux from external sources, so that Φ_{sc} must adjust itself appropriately in order that Φ assume a quantized value.

Suppose a magnetic monopole of strength g is situated just below the center of a superconducting ring. The magnetic flux through the ring is $(g/r^2)(2\pi r^2) = 2\pi g$, and by (27) this must equal an integral multiple of $\pi\hbar c/e$. Thus the minimum permissible value of g is $\hbar c/2e$, the famous Dirac result.

Duration of Persistent Currents

Consider a persistent current that flows in a ring of a type I superconductor of wire of length L and cross-sectional area A. The persistent current maintains a flux through the ring of some integral number of fluxoids (27). A fluxoid cannot leak out of the ring and thereby reduce the persistent current unless by a thermal fluctuation a minimum volume of the superconducting ring is momentarily in the normal state. The probability per unit time that a fluxoid will leak out is the product

$$P = (\text{attempt frequency})(\text{activation barrier factor}) \ . \tag{28}$$

The activation barrier factor is $\exp(-\Delta F/k_B T)$, where the free energy of the barrier is

$$\Delta F \approx (\text{minimum volume})(\text{excess free energy density of normal state}) \ .$$

[19]The effect was discovered by B. S. Deaver, Jr., and W. M. Fairbank, Phys. Rev. Letters **7**, 43 (1961), and R. Doll and M. Näbauer, Phys. Rev. Letters **7**, 51 (1961).

The minimum volume of the ring that must turn normal to allow a fluxoid to escape is of the order of $R\xi^2$, where ξ is the coherence length of the superconductor and R the wire thickness. The excess free energy density of the normal state is $H_c{}^2/8\pi$, whence the barrier free energy is

$$\Delta F \approx R\xi^2 H_c{}^2/8\pi \ . \tag{29}$$

Let the wire thickness be 10^{-4} cm, the coherence length $= 10^{-4}$ cm, and $H_c = 10^3$ G; then $\Delta F \approx 10^{-7}$ erg. As we approach the transition temperature from below, ΔF will decrease toward zero, but the value given is a fair estimate between absolute zero and 0.8 T_c. Thus the activation barrier factor is

$$\exp(-\Delta F/k_B T) \approx \exp(-10^8) \approx 10^{-(4.34\times 10^7)} \ .$$

The characteristic frequency with which the minimum volume can attempt to change its state must be of order of $E_g/\hbar$. If $E_g = 10^{-15}$ erg, the attempt frequency is $\approx 10^{-15}/10^{-27} \approx 10^{12}$ s^{-1}. The leakage probability (28) becomes

$$P \approx 10^{12} 10^{-4.34\times 10^7} \ \text{s}^{-1} \approx 10^{-4.34\times 10^7} \ \text{s}^{-1} \ .$$

The reciprocal of this is a measure of the time required for a fluxoid to leak out, $T = 1/P = 10^{4.34\times 10^7}$ s. The age of the universe is only 10^{18} s, so that a fluxoid will never leak out in the age of the universe, under our assumed conditions. Accordingly, the current is maintained.

There are two circumstances in which the activation energy is much lower and a fluxoid can be observed to leak out of a ring—either very close to the critical temperature, where H_c is very small, or when the material of the ring is a type II superconductor and already has fluxoids embedded in it. These special situations are discussed in the literature under the subject of fluctuations in superconductors.

Type II Superconductors

There is no difference in the mechanism of superconductivity in type I and type II superconductors. Both types have similar thermal properties at the superconductor-normal transition in zero magnetic field. But the Meissner effect is entirely different (Fig. 5). A good type I superconductor excludes a magnetic field until superconductivity is destroyed suddenly, and then the field penetrates completely. A good type II superconductor excludes the field completely, up to a field H_{c1}. Above H_{c1} the field is partially excluded, but the specimen remains electrically superconducting. At a much higher field, H_{c2}, the flux penetrates completely and superconductivity vanishes. (An outer surface layer of the specimen may remain superconducting up to a still higher field H_{c3}.) An important difference in a type I and a type II superconductor is in the mean free path of the conduction electrons in the normal state. If the coherence length is longer than the penetration depth, the superconductor will be type I. Most pure metals are type I.

But if the mean free path is short, the coherence length is short and the penetration depth is great (Fig. 14); then the superconductor will be type II. We can change some metals from type I to type II by a modest addition of an alloying element. In Figure 5 the addition of 2 wt. percent of indium changes lead from type I to type II, although the transition temperature is scarcely changed at all. Nothing fundamental has been done to the electronic structure of lead by this amount of alloying, but the magnetic behavior has changed drastically. The theory of type II superconductors was developed by Ginzburg, Landau, Abrikosov, and Gorkov. Later Kunzler and co-workers observed that Nb_3Sn wires can carry large supercurrents in fields approaching 100 kG; this led to the commercial development of strong field superconducting magnets.

Consider the interface between a region in the superconducting state and a region in the normal state. The interface has a surface energy that may be positive or negative and that decreases as the applied magnetic field is increased. A superconductor is type I if the surface energy is always positive, and type II if the surface energy becomes negative as the magnetic field is increased. The sign of the surface energy has no importance for the transition temperature.

The energy of a bulk superconductor is increased when the magnetic field is expelled. However, a parallel field can penetrate a very thin film nearly uniformly (Fig. 17), only a part of the flux is expelled, and the

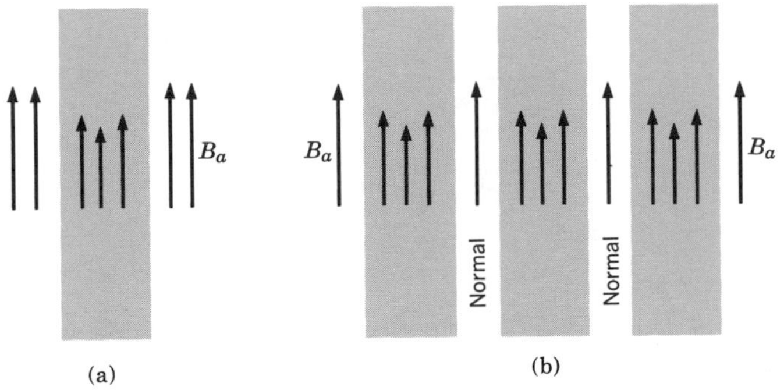

Figure 17 (a) Magnetic field penetration into a thin film of thickness equal to the penetration depth λ. The arrows indicate the intensity of the magnetic field. (b) Magnetic field penetration in a homogeneous bulk structure in the mixed or vortex state, with alternate layers in normal and superconducting states. The superconducting layers are thin in comparison with λ. The laminar structure is shown for convenience; the actual structure consists of rods of the normal state surrounded by the superconducting state. The N regions in the vortex state are not exactly normal, but are described by low values of the stabilization energy density.

energy of the superconducting film will increase only slowly as the external magnetic field is increased. This causes a large increase in the field intensity required for the destruction of superconductivity. The film has the usual energy gap and will be resistanceless. A thin film is not a type II superconductor, but the film results show that under suitable conditions superconductivity can exist in high magnetic fields.

Vortex State. The results for thin films suggest the question: Are there stable configurations of a superconductor in a magnetic field with regions in the form of thin rods (or plates) in the normal state, each normal region surrounded by a superconducting region? In such a mixed state the external magnetic field will penetrate the thin normal regions uniformly, and the field will also penetrate somewhat into the surrounding superconducting material, as in Fig. 18. The term vortex state describes the circulation of superconducting currents in vortices throughout the bulk specimen, as in Fig. 19 below. There is no chemical or crystallographic difference between the normal and the superconducting regions in the vortex state. The vortex state is stable because the penetration of the applied field into the superconducting material causes the surface energy to become negative. A type II superconductor is characterized by a vortex state stable over a certain range of magnetic field strength; namely, between H_{c1} and H_{c2}.

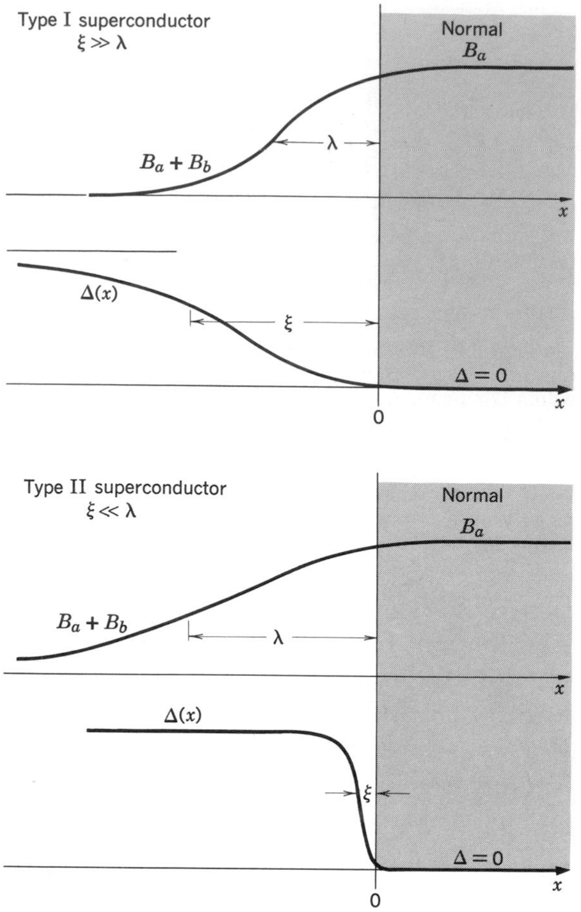

Figure 18 Variation of the magnetic field and energy gap parameter $\Delta(x)$ at the interface of superconducting and normal regions, for type I and type II superconductors. The energy gap parameter is a measure of the stabilization energy density of the superconducting state.

Estimation of H_{c1} and H_{c2}. We can estimate H_{c1} from the penetration depth λ. The field in the normal core of the fluxoid will be H_{c1} when the applied field is H_{c1}. The field will extend out from the normal core a distance λ into the superconducting environment. The flux thus associated with a single core is $\pi\lambda^2 H_{c1}$, and this must be equal to the flux quantum Φ_0 defined by (27). Thus

$$H_{c1} \approx \Phi_0/\pi\lambda^2 \ . \tag{30}$$

This is the field for nucleation of a single fluxoid.

At H_{c2} the fluxoids are packed together as tightly as possible, consistent with the preservation of the superconducting state. This means as densely as the coherence length will allow. The external field penetrates the specimen almost uniformly, with small ripples on the scale of the fluxoid lat-

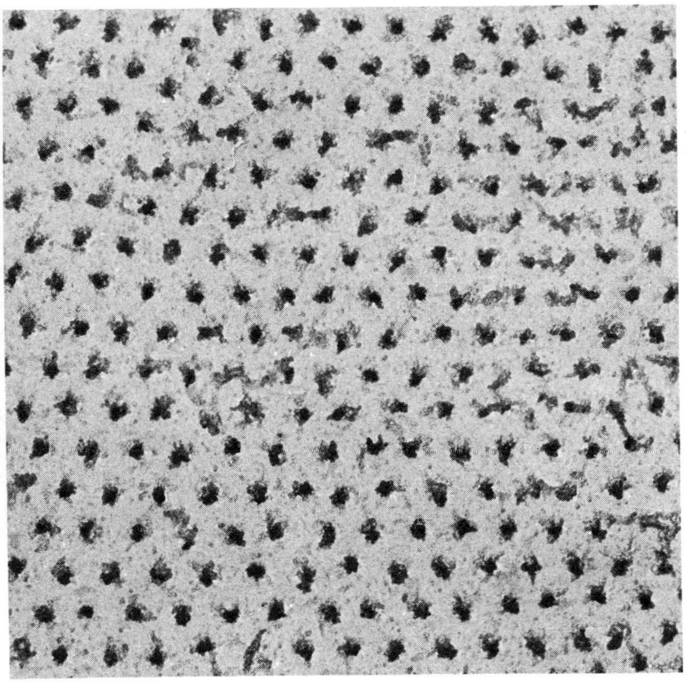

Figure 19 Triangular lattice of flux lines through top surface of a superconducting cylinder. The points of exit of the flux lines are decorated with fine ferromagnetic particles. The electron microscope image is at a magnification of 8300, by U. Essmann and H. Träuble. Neutron diffraction is another powerful means of studying the flux distribution; for a review, see U. Ullmaier, Comments in Solid State Physics **5,** 81 (1973).

tice. Each core is responsible for carrying a flux of the order of $\pi\xi^2 H_{c2}$, which also is quantized to Φ_0. Thus

$$H_{c2} \approx \Phi_0/\pi\xi^2 \tag{31}$$

gives the upper critical field. The larger the ratio λ/ξ, the larger is the ratio of H_{c2} to H_{c1}.

It remains for us to find a relation between these critical fields and the thermodynamic critical field H_c that measures the stabilization energy density of the superconducting state, which is known by (9) to be $H_c^2/8\pi$. In a type II superconductor we can determine H_c only indirectly by calorimetric measurement of the stabilization energy. To estimate H_{c1} in terms of H_c, we consider the stability of the vortex state at absolute zero in the impure limit $\xi < \lambda$; here the coherence length is short in comparison with the penetration depth. We estimate in the vortex state the stabilization energy of a fluxoid core as a normal metal cylinder which carries an average magnetic field B_a. The radius is of the order of the coherence length, the thickness of the boundary between N and S phases. The energy of the normal core referred to the energy of a pure superconductor is given by the product of the stabilization energy times the area of the core:

$$\text{(CGS)} \qquad f_{\text{core}} \approx \frac{1}{8\pi} H_c^2 \times \pi\xi^2 \ , \qquad (32)$$

per unit length. But there is also a decrease in magnetic energy because of the penetration of the applied field B_a into the superconducting material around the core:

$$\text{(CGS)} \qquad f_{\text{mag}} \approx -\frac{1}{8\pi} B_a^2 \times \pi\lambda^2 \ . \qquad (33)$$

For a single fluxoid we add these two contributions to obtain

$$\text{(CGS)} \qquad f = f_{\text{core}} + f_{\text{mag}} \approx \tfrac{1}{8}(H_c^2\xi^2 - B_a^2\lambda^2) \ . \qquad (34)$$

The core is stable if $f < 0$. The threshold field for a stable fluxoid is at $f = 0$, or, with H_{c1} written for B_a,

$$H_{c1}/H_c \approx \xi/\lambda \ . \qquad (35)$$

The threshold field divides the region of positive surface energy from the region of negative surface energy.

We can combine (30) and (35) to obtain a relation for H_c:

$$\pi\xi\lambda H_c \approx \Phi_0 \ . \qquad (36)$$

We can combine (30), (31), and (35) to obtain

$$(H_{c1}H_{c2})^{1/2} \approx H_c \ . \qquad (37)$$

Single Particle Tunneling

Consider two metals separated by an insulator, as in Fig. 20. The insulator normally acts as a barrier to the flow of conduction electrons from one metal to the other. If the barrier is sufficiently thin (less than 10 or 20 Å) there is a significant probability that an electron which impinges on the barrier will pass from one metal to the other: this is called **tunneling.** The concept that particles can tunnel through potential barriers is as old as quantum mechanics. In many experiments the insulating layer is simply a thin oxide layer formed on one of two evaporated metal films, as in Fig. 21.

When both metals are normal conductors, the current-voltage relation of the sandwich or tunneling junction is ohmic at low voltages,[20] with the current directly proportional to the applied voltage. Giaever[21] discovered that if one of the metals becomes superconducting the current-voltage characteristic changes from the straight line of Fig. 19a to the curve shown in Fig. 22b. Figure 23a contrasts the electron density of orbitals in the superconductor with that in the normal metal. In the superconductor there is an energy gap centered at the Fermi level. At absolute zero no

[20]See, for example, I. Giaever and K. Megerle, Phys. Rev. **122,** 1101 (1961); J. Bardeen, Phys. Rev. Letters **6,** 57 (1961).
[21]I. Giaever, Phys. Rev. Letters **5,** 147, 464 (1960).

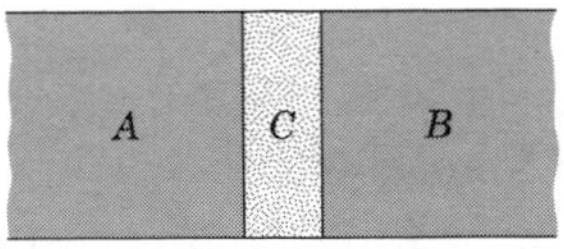

Figure 20 Two metals, A and B, separated by a thin layer of an insulator C.

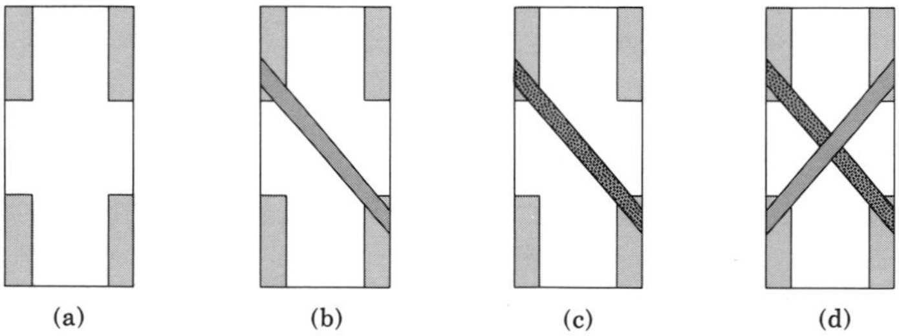

Figure 21 Preparation of an $Al/Al_2O_3/Sn$ sandwich. (a) Glass slide with indium contacts. (b) An aluminum strip 1 mm wide and 1000 to 3000 Å thick has been deposited across the contacts. (c) The aluminum strip has been oxidized to form an Al_2O_3 layer 10 to 20 Å in thickness. (d) A tin film has been deposited across the aluminum film, forming an $Al/Al_2O_3/Sn$ sandwich. The external leads are connected to the indium contacts; two contacts are used for the current measurement and two for the voltage measurement. (After Giaever and Megerle.)

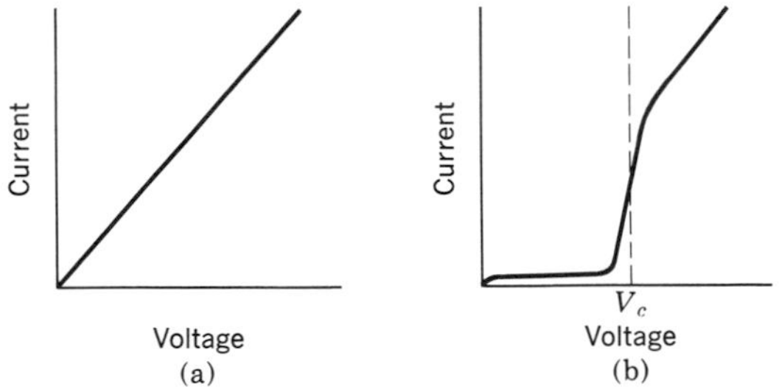

Figure 22 (a) Linear current-voltage relation for junction of normal metals separated by oxide layer; (b) current-voltage relation with one metal normal and the other metal superconducting.

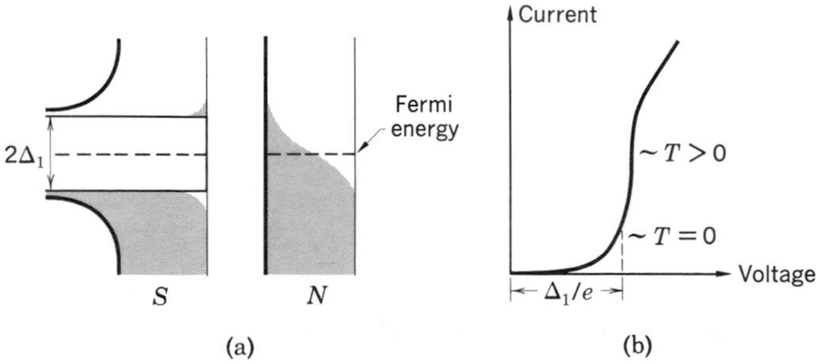

Figure 23 The density of orbitals and the current-voltage characteristic for a tunneling junction. In (a) the energy is plotted on the vertical scale and the density of orbitals on the horizontal scale. One metal is in the normal state and one in the superconducting state. (b) I versus V; the dashes indicate the expected break at $T = 0$. (After Giaever and Megerle.)

current can flow until the applied voltage is $V = E_g/2e = \Delta/e$. The gap E_g corresponds to the break-up of a pair of electrons in the superconducting state, with the formation of two electrons, or an electron and a hole, in the normal state. The current starts when $eV = \Delta$. At finite temperatures there is a small current flow even at low voltages, because of electrons in the superconductor that are thermally excited across the energy gap.

Josephson Superconductor Tunneling

Under suitable conditions we observe remarkable effects associated with the tunneling of superconducting electron pairs from a superconductor through a layer of an insulator into another superconductor. Such a junction is called a weak link. The effects of pair tunneling include:

Dc Josephson effect. A dc current flows across the junction in the absence of any electric or magnetic field.

Ac Josephson effect. A dc voltage applied across the junction causes rf current oscillations across the junction. This effect has been utilized in a precision determination of the value of $\hbar/e$. Further, an rf voltage applied with the dc voltage can then cause a dc current across the junction.

Macroscopic long-range quantum interference. A dc magnetic field applied through a superconducting circuit containing two junctions causes the maximum supercurrent to show interference effects as a function of magnetic field intensity. This effect can be utilized in sensitive magnetometers.

Dc Josephson Effect. Our discussion of Josephson junction phenomena follows the discussion of flux quantization. Let ψ_1 be the probability amplitude of electron pairs on one side of a junction, and let ψ_2 be the amplitude on the other side. For simplicity, let both superconductors be identical. For the present we suppose that they are both at zero potential. The time-dependent Schrödinger equation $i\hbar\partial\psi/\partial t = \mathcal{H}\psi$ applied to the two amplitudes gives

$$i\hbar \frac{\partial\psi_1}{\partial t} = \hbar T\psi_2 \; ; \qquad i\hbar \frac{\partial\psi_2}{\partial t} = \hbar T\psi_1 \; . \tag{38}$$

Here $\hbar T$ represents the effect of the electron-pair coupling or transfer interaction across the insulator; T has the dimensions of a rate or frequency. It is a measure of the leakage of ψ_1 into the region 2, and of ψ_2 into the region 1. If the insulator is very thick, T is zero and there is no pair tunneling.

Let $\psi_1 = n_1^{\frac{1}{2}}\, e^{i\theta_1}$ and $\psi_2 = n_2^{\frac{1}{2}}\, e^{i\theta_2}$. Then

$$\frac{\partial\psi_1}{\partial t} = \tfrac{1}{2} n_1^{-\frac{1}{2}}\, e^{i\theta_1} \frac{\partial n_1}{\partial t} + i\psi_1 \frac{\partial\theta_1}{\partial t} = -iT\psi_2 \; ; \tag{39}$$

$$\frac{\partial\psi_2}{\partial t} = \tfrac{1}{2} n_2^{-\frac{1}{2}}\, e^{i\theta_2} \frac{\partial n_2}{\partial t} + i\psi_2 \frac{\partial\theta_2}{\partial t} = -iT\psi_1 \; . \tag{40}$$

We multiply (39) by $n_1^{\frac{1}{2}}\, e^{-i\theta_1}$ to obtain, with $\delta \equiv \theta_2 - \theta_1$,

$$\frac{1}{2}\frac{\partial n_1}{\partial t}+in_1\frac{\partial \theta_1}{\partial t}=-iT(n_1n_2)^{\frac{1}{2}}\,e^{i\delta}\ . \tag{41}$$

We multiply (40) by $n_2^{\frac{1}{2}}\,e^{-i\theta_2}$ to obtain

$$\frac{1}{2}\frac{\partial n_2}{\partial t}+in_2\frac{\partial \theta_2}{\partial t}=-iT(n_1n_2)^{\frac{1}{2}}\,e^{-i\delta}\ . \tag{42}$$

Now equate the real and imaginary parts of (41) and similarly of (42):

$$\frac{\partial n_1}{\partial t}=2T(n_1n_2)^{\frac{1}{2}}\sin\delta\ ;\qquad \frac{\partial n_2}{\partial t}=-2T(n_1n_2)^{\frac{1}{2}}\sin\delta\ ; \tag{43}$$

$$\frac{\partial \theta_1}{\partial t}=-T\left(\frac{n_2}{n_1}\right)^{\frac{1}{2}}\cos\delta\ ;\qquad \frac{\partial \theta_2}{\partial t}=-T\left(\frac{n_1}{n_2}\right)^{\frac{1}{2}}\cos\delta\ . \tag{44}$$

If $n_1 \cong n_2$ as for identical superconductors 1 and 2, we have from (44) that

$$\frac{\partial \theta_1}{\partial t}=\frac{\partial \theta_2}{\partial t}\ ;\qquad \frac{\partial}{\partial t}(\theta_2-\theta_1)=0\ . \tag{45}$$

From (43) we see that

$$\frac{\partial n_2}{\partial t}=-\frac{\partial n_1}{\partial t}\ . \tag{46}$$

The current flow from (1) to (2) is proportional to $\partial n_2/\partial t$ or, the same thing, $-\partial n_1/\partial t$. We therefore conclude from (43) that the current J of superconductor pairs across the junction depends on the phase difference δ as

$$\boxed{J=J_0\sin\delta=J_0\sin(\theta_2-\theta_1)}\ , \tag{47}$$

where J_0 is proportional to the transfer interaction T. The current J_0 is the maximum zero-voltage current that can be passed by the junction. With no applied voltage a dc current will flow across the junction (Fig. 24), with a value between J_0 and $-J_0$ according to the value of the phase difference $\theta_2-\theta_1$. This is the **dc Josephson effect.**

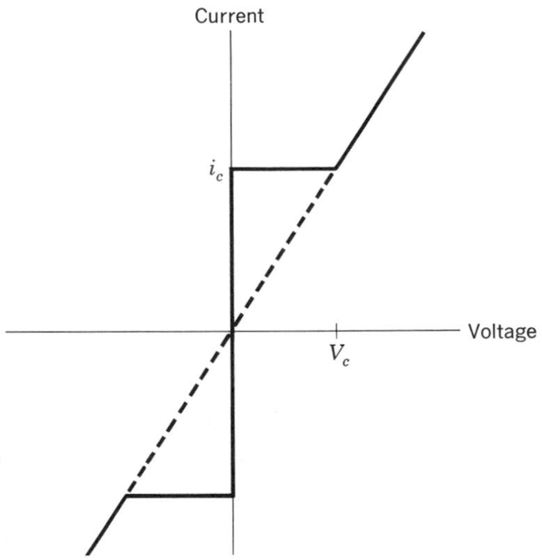

Figure 24 Current-voltage characteristic of a Josephson junction. Dc currents flow under zero applied voltage up to a critical current i_c: this is the dc Josephson effect. At voltages above V_c the junction has a finite resistance, but the current has an oscillatory component of frequency $\omega = 2eV/\hbar$: this is the ac Josephson effect.

Ac Josephson Effect. Let a voltage V be applied across the junction. We can do this because the junction is an insulator. An electron pair experiences a potential energy difference qV on passing across the junction, where $q = -2e$. We can say that a pair on one side is at potential energy $-eV$ and a pair on the other side is at eV. The equations of motion that replace (38) are

$$i\hbar\, \partial\psi_1/\partial t = \hbar T\psi_2 - eV\psi_1 \ ; \qquad i\hbar\, \partial\psi_2/\partial t = \hbar T\psi_1 + eV\psi_2 \ . \tag{48}$$

We proceed as above to find in place of (41) the equation

$$\frac{1}{2}\frac{\partial n_1}{\partial t} + in_1\frac{\partial \theta_1}{\partial t} = ieVn_1\hbar^{-1} - iT(n_1n_2)^{\frac{1}{2}}\, e^{i\delta} \ . \tag{49}$$

This equation breaks up into the real part

$$\partial n_1/\partial t = 2T(n_1n_2)^{\frac{1}{2}} \sin\delta \ , \tag{50}$$

exactly as without the voltage V, and the imaginary part

$$\partial\theta_1/\partial t = (eV/\hbar) - T(n_2/n_1)^{\frac{1}{2}} \cos\delta \ , \tag{51}$$

which differs from (44) by the term $eV/\hbar$.

Further, by extension of (42),

$$\frac{1}{2}\frac{\partial n_2}{\partial t} + in_2\frac{\partial \theta_2}{\partial t} = -i\, eVn_2\hbar^{-1} - iT(n_1n_2)^{\frac{1}{2}}\, e^{-i\delta} \ , \tag{52}$$

whence

$$\partial n_2/\partial t = -2T(n_1n_2)^{\frac{1}{2}} \sin\delta \ ; \tag{53}$$

$$\partial\theta_2/\partial t = -(eV/\hbar) - T(n_1/n_2)^{\frac{1}{2}} \cos\delta \ . \tag{54}$$

From (51) and (54) with $n_1 \cong n_2$, we have

$$\partial(\theta_2 - \theta_1)/\partial t = \partial\delta/\partial t = -2eV/\hbar \ . \tag{55}$$

We see by integration of (55) that with a dc voltage across the junction the relative phase of the probability amplitudes vary as

$$\delta(t) = \delta(0) - (2eVt/\hbar). \tag{56}$$

The current is now given by (47) with (56) for the phase:

$$\boxed{J = J_0 \sin\,[\delta(0) - (2eVt/\hbar)]\ .} \tag{57}$$

The current oscillates with frequency

$$\omega = 2eV/\hbar \ . \tag{58}$$

This is the **ac Josephson effect.** A dc voltage of 1 μV produces a frequency of 483.6 MHz. The relation (58) says that a photon of energy $\hbar\omega = 2eV$ is emitted or absorbed when an electron pair crosses the barrier. By measuring the voltage and the frequency it is possible to obtain a very precise value[22] of $e/\hbar$.

[22]W. H. Parker, B. N. Taylor, and D. N. Langenberg, Phys. Rev. Letters **18,** 287 (1967).

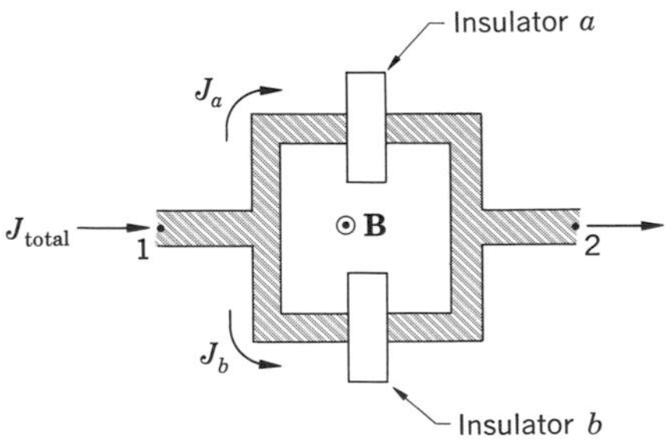

Figure 25 The arrangement for experiment on macroscopic quantum interference. A magnetic flux Φ passes through the interior of the loop.

Macroscopic Quantum Interference. We saw in (24) and (26) that the phase difference $\theta_2 - \theta_1$ around a closed circuit which encompasses a total magnetic flux Φ is given by

$$\theta_2 - \theta_1 = (2e/\hbar c)\Phi \ . \tag{59}$$

The flux is the sum of that due to external fields and that due to currents in the circuit itself. We consider two Josephson junctions in parallel, as in Fig. 25. No voltage is applied. Let the phase difference between points 1 and 2 taken on a path through junction a be δ_a. When taken on a path through junction b, the phase difference is δ_b. In the absence of a magnetic field these two phases must be equal. Now let the flux Φ pass through the interior of the circuit. We do this with a straight solenoid normal to the plane of the paper and lying inside the circuit. By (59), $\delta_b - \delta_a = (2e/\hbar c)\Phi$, or

$$\delta_b = \delta_0 + \frac{e}{\hbar c}\Phi \ ; \qquad \delta_a = \delta_0 - \frac{e}{\hbar c}\Phi \ . \tag{60}$$

The total current is the sum of J_a and J_b. The current through each junction is of the form (47), so that

$$J_{\text{Total}} = J_0 \left\{ \sin\left(\delta_0 + \frac{e}{\hbar c}\Phi\right) + \sin\left(\delta_0 - \frac{e}{\hbar c}\Phi\right) \right\} = 2(J_0 \sin \delta_0) \cos \frac{e\Phi}{\hbar c} \ . \tag{61}$$

The current varies with Φ. The magnitude of the current has maxima when

$$e\Phi/\hbar c = s\pi \ , \qquad s = \text{integer} \ . \tag{62}$$

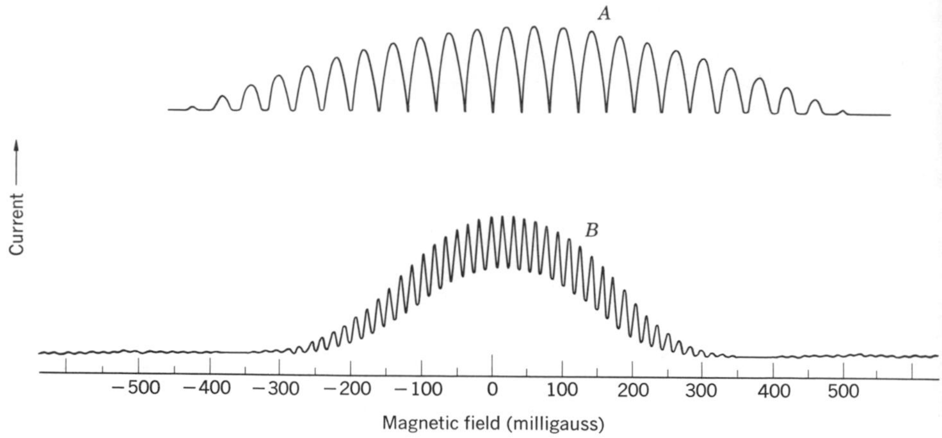

Figure 26 Experimental trace of J_{max} versus magnetic field showing interference and diffraction effects for two junctions *A* and *B*. The field periodicity is 39.5 and 16 mG for *A* and *B*, respectively. Approximate maximum currents are 1 mA (*A*) and 0.5 mA (*B*). The junction separation is 3 mm and junction width 0.5 mm for both cases. The zero offset of *A* is due to a background magnetic field. [After R. C. Jaklevic, J. Lambe, J. E. Mercereau and A. H. Silver, Phys. Rev. **140,** A1628 (1965).]

The periodicity of the current is shown in Fig. 26. The short period variation is the interference effect of the two junctions predicted by (61) and (62). The longer variation is a diffraction effect and arises from the finite dimensions of each junction, which causes Φ to depend on the particular path we integrate over. The diffraction effect was responsible for the failure to observe pair tunneling in early experiments on single electron tunneling. The single electrons are not bosons and are not liable to diffraction effects, but the contribution of the pairs is washed out by diffraction unless special care is taken in construction and magnetic shielding.

SUMMARY (In CGS units)

1. A superconductor exhibits infinite conductivity.

2. A bulk specimen of metal in the superconducting state exhibits perfect diamagnetism, with the magnetic induction $\mathbf{B} = 0$. This is the Meissner effect. The external magnetic field will penetrate the surface of the specimen over a distance determined by the penetration depth λ.

3. There are two types of superconductors, I and II. In a bulk specimen of type I superconductor the superconducting state is destroyed and the normal state is restored by application of an external magnetic field in excess of a critical value H_c. A type II superconductor has two critical fields, $H_{c1} < H_c < H_{c2}$; a vortex state exists in the range between H_{c1} and H_{c2}. The stabilization energy density of the pure superconducting state is $H_c^2/8\pi$ in both type I and II superconductors.

4. In the superconducting state an energy gap $E_g \approx 4k_BT_c$ separates superconducting electrons below from normal electrons above the gap. The gap is detected in experiments on heat capacity, infrared absorption, and tunneling.

5. The London equation

$$\mathbf{j} = -\frac{c}{4\pi\lambda_L^2}\,\mathbf{A} \qquad \text{or} \qquad \text{curl}\,\mathbf{j} = -\frac{c}{4\pi\lambda_L^2}\,\mathbf{B}$$

leads to the Meissner effect through the penetration equation $\nabla^2 B = B/\lambda_L^2$, where $\lambda_L \approx (mc^2/4\pi ne^2)^{\frac{1}{2}}$ is the London penetration depth.

6. In the London equation $\mathbf{A}$ or $\mathbf{B}$ should be a weighted average over the coherence length ξ. The intrinsic coherence length $\xi_0 = 2\hbar v_F/\pi E_g$.

7. The BCS theory accounts for a superconducting state formed from pairs of electrons $\mathbf{k}\uparrow$ and $-\mathbf{k}\downarrow$.

8. Three important lengths enter the theory of superconductivity: the London penetration depth λ_L; the intrinsic coherence length ξ_0; and the normal electron mean free path ℓ.

9. Type II superconductors have $\xi < \lambda$. The critical fields are related by $H_{c1} \approx (\xi/\lambda)H_c$ and $H_{c2} \approx (\lambda/\xi)H_c$. Values of H_{c2} are as high as 500 kG = 50 T.

PROBLEMS

1. ***Magnetic field penetration in a plate.*** The penetration equation may be written as $\lambda^2\nabla^2 B = B$, where λ is the penetration depth. (a) Show that $B(x)$ inside a superconducting plate perpendicular to the x axis and of thickness δ is given by

$$B(x) = B_a \frac{\cosh(x/\lambda)}{\cosh(\delta/2\lambda)} ,$$

where B_a is the field outside the plate and parallel to it; here $x = 0$ is at the center of the plate. (b) The effective magnetization $M(x)$ in the plate is defined by $B(x) - B_a = 4\pi M(x)$. Show that, in CGS, $4\pi M(x) = -B_a(1/8\lambda^2)(\delta^2 - 4x^2)$, for $\delta \ll \lambda$. In SI we replace the 4π by μ_0.

2. ***Critical field of thin films.*** (a) Using the result of Problem 1b, show that the energy density at $T = 0$ K within a superconducting film of thickness δ in an external magnetic field B_a is given by, for $\delta \ll \lambda$,

(CGS) $$U_S(x, B_a) = U_S(0) + (\delta^2 - 4x^2)B_a^2/64\pi\lambda^2 .$$

In SI the factor π is replaced by $\frac{1}{4}\mu_0$. We neglect a kinetic energy contribution to the problem. (b) Show that the magnetic contribution to U_S when averaged over the thickness of the film is $B_a^2(\delta/\lambda)^2/96\pi$. (c) Show that the critical field of the thin film is proportional to $(\lambda/\delta)H_c$, where H_c is the bulk critical field, if we consider only the magnetic contribution to U_S.

3. ***Two-fluid model of a superconductor.*** On the two-fluid model of a superconductor we assume that at temperatures $0 < T < T_c$ the current density may be written as the sum of the contributions of normal and superconducting electrons: $\mathbf{j} = \mathbf{j}_N + \mathbf{j}_S$, where $\mathbf{j}_N = \sigma_0\mathbf{E}$ and $\mathbf{j}_S$ is given by the London equation. Here σ_0 is an ordinary normal conductivity, decreased by the reduction in the number of normal electrons at temperature T as compared to the normal state. Neglect inertial effects on both j_N and j_S. (a) Show from the Maxwell equations that the dispersion relation connecting wavevector $\mathbf{k}$ and frequency ω for electromagnetic waves in the superconductor is

(CGS) $$k^2c^2 = 4\pi\sigma_0\omega i - c^2\lambda_L^{-2} + \omega^2 ;$$ or

(SI) $$k^2c^2 = (\sigma_0/\epsilon_0)\,\omega i - c^2\lambda_L^{-2} + \omega^2 ,$$

where λ_L^2 is given by (14a) with n replaced by n_S. Recall that curl curl $\mathbf{B} = -\nabla^2\mathbf{B}$. (b) If τ is the relaxation time of the normal electrons and n_N is their concentration, show by use of the expression $\sigma_0 = n_N e^2\tau/m$ that at frequencies $\omega \ll 1/\tau$ the dispersion relation does not involve the normal electrons in an important way, so that the motion of the electrons is described by the London equation alone. The supercurrent short-circuits the normal electrons. The London equation itself only holds true if $\hbar\omega$ is small in comparison with the energy gap. Note: The frequencies of interest are such that $\omega \ll \omega_p$, where ω_p is the plasma frequency.

*4. ***Structure of a vortex.*** (a) Find a solution to the London equation that has cylindrical symmetry and applies outside a line core. In cylindrical polar coordinates, we want a solution of

$$B - \lambda^2 \nabla^2 B = 0$$

that is singular at the origin and for which the total flux is the flux quantum:

$$2\pi \int_0^\infty d\rho \; \rho B(\rho) = \Phi_0 \; .$$

The equation is in fact valid only outside the normal core of radius ξ. (b) Show that the solution has the limits

$$B(\rho) \simeq (\Phi_0/2\pi\lambda^2)\ln(\lambda/\rho) \; , \qquad (\xi \ll \rho \ll \lambda)$$

$$B(\rho) \simeq (\Phi_0/2\pi\lambda^2)(\pi\lambda/2\rho)^{1/2} \exp(-\rho/\lambda) \; . \qquad (\rho \gg \lambda)$$

5. ***London penetration depth.*** (a) Take the time derivative of the London equation (10) to show that $\partial \mathbf{j}/\partial t = (c^2/4\pi\lambda_L{}^2)\mathbf{E}$. (b) If $m d\mathbf{v}/dt = q\mathbf{E}$, as for free carriers of charge q and mass m, show that $\lambda_L{}^2 = mc^2/4\pi nq^2$.

*6. ***Diffraction effect of Josephson junction.*** Consider a junction of rectangular cross section with a magnetic field B applied in the plane of the junction, normal to an edge of width w. Let the thickness of the junction be T. Assume for convenience that the phase difference of the two superconductors is $\pi/2$ when $B = 0$. Show that the dc current in the presence of the magnetic field is

$$J = J_0 \frac{\sin(wTBe/\hbar c)}{(wTBe/\hbar c)} \; .$$

7. ***Meissner effect in sphere.*** Consider a sphere of a type I superconductor with critical field H_c. (a) Show that in the Meissner regime the effective magnetization M within the sphere is given by $-8\pi M/3 = B_a$, the uniform applied magnetic field. (b) Show that the magnetic field at the surface of the sphere in the equatorial plane is $3B_a/2$. (It follows that the applied field at which the Meissner effect starts to break down is $2H_c/3$.) Reminder: The demagnetization field of a uniformly magnetized sphere is $-4\pi M/3$.

*This problem is somewhat difficult.

References

J. M. Blatt, *Theory of superconductivity*, Academic Press, 1964.

C. G. Kuper, *Introduction to the theory of superconductivity*, Oxford, 1968.

E. A. Lynton, *Superconductivity*, 3rd ed., Halsted, 1971. A short introduction.

V. L. Newhouse, ed., *Applied Superconductivity, 2 vols.*, Academic Press, New York, 1975.

P. Nozières, "Superfluidity in Bose and Fermi fluids," D. F. Brewer, ed., *Quantum fluids*, North-Holland, 1966. Especially good discussion of the Meissner effect on pp. 1–9.

R. D. Parks, ed., *Superconductivity*, Dekker, 1969. Thorough, useful collection of review articles; 1412 pp.

G. Rickayzen, *Theory of superconductivity*, Interscience, 1965.

J. R. Schrieffer, *Theory of superconductivity*, Benjamin, 1964.

M. Tinkham, *Introduction to superconductivity*, McGraw-Hill, 1975.

P. R. Wallace, ed., *Superconductivity*, Gordon and Breach, 1969, 2 vols.

Type II Superconductivity

P. G. de Gennes, *Superconductivity of metals and alloys*, Benjamin, 1966.

D. Saint-James, G. Sarma, E. J. Thomas, *Type II superconductivity*, Pergamon, 1969.

Josephson Effects

P. W. Anderson, in *Lectures on the many-body problem*, E. R. Caianiello, ed., Academic Press, 1964; Vol. 2, pp. 113–135. (For physical insight.)

J. Clarke, "The Josephson effect and $e/\hbar$," Amer. J. Phys. **38,** 1071 (1970). Elementary review.

R. Feynman, *Lectures on physics*, Addison-Wesley, 1963, Vol. 3, Sec. 21/9. Simple discussion.

R. C. Jaklevic, J. Lambe, J. E. Mercereau, and A. H. Silver, "Macroscopic quantum interference in superconductors," Phys. Rev. **140,** A1628 (1965).

I. O. Kulik and I. K. Yanson, *Josephson effect in superconductive tunneling structures*, Halsted, 1972.

P. L. Richards, S. Shapiro, and C. C. Grimes, "Student laboratory demonstration of flux quantization and the Josephson effect in superconductors," Amer. J. Phys. **36,** 690 (1968).

L. Solymar, *Superconductive tunnelling and applications*, Halsted, 1972.

Conference Proceedings

International Conferences on Low Temperature Physics: LT 12, Kyoto, 1970; LT 13, Boulder, 1972; LT 14, Helsinki, 1975.

Historical

F. London, *Superfluids*, Vol. I., Wiley, 1950.

K. Mendelssohn, *Quest for absolute zero*, McGraw-Hill, 1966, paperback.

13 DIELECTRICS AND FERROELECTRICS

NOTATION: $\epsilon_0 = 10^7/4\pi c^2$;

(CGS) $D = E + 4\pi P = \epsilon E = (1 + 4\pi\chi)E$; $\alpha = p/E_{\mathrm{local}}$;

(SI) $D = \epsilon_0 E + P = \epsilon\epsilon_0 E = (1 + \chi)\epsilon_0 E)$; $\alpha = p/E_{\mathrm{local}}$;

$\epsilon_{\mathrm{CGS}} = \epsilon_{\mathrm{SI}}$; $4\pi\chi_{\mathrm{CGS}} = \chi_{\mathrm{SI}}$; $\alpha_{\mathrm{SI}} = 4\pi\epsilon_0\alpha_{\mathrm{CGS}}$.

Maxwell Equations

(CGS)	(SI)
$\mathrm{curl}\ \mathbf{H} = \dfrac{4\pi}{c}\mathbf{j} + \dfrac{1}{c}\dfrac{\partial}{\partial t}(\mathbf{E} + 4\pi\mathbf{P})$;	$\mathrm{curl}\ \mathbf{H} = \mathbf{j} + \dfrac{\partial}{\partial t}(\epsilon_0\mathbf{E} + \mathbf{P})$;
$\mathrm{curl}\ \mathbf{E} = -\dfrac{1}{c}\dfrac{\partial \mathbf{B}}{\partial t}$;	$\mathrm{curl}\ \mathbf{E} = -\dfrac{\partial \mathbf{B}}{\partial t}$;
$\mathrm{div}(\mathbf{E}) = 4\pi\rho$;	$\mathrm{div}\ \epsilon_0\mathbf{E} = \rho$;
$\mathrm{div}\ \mathbf{B} = 0$;	$\mathrm{div}\ \mathbf{B} = 0$.

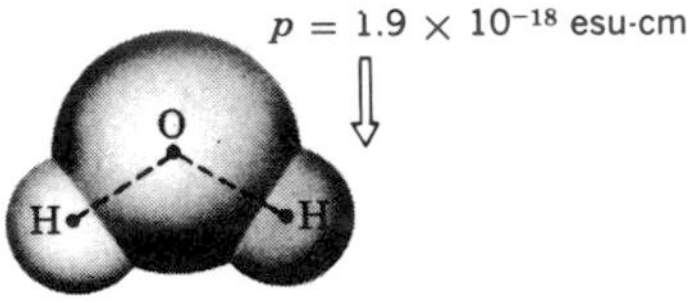

Figure 1 The permanent dipole moment of a molecule of water has the magnitude 1.9×10^{-18} esu-cm and is directed from O^{--} ion toward the midpoint of the line connecting the H^+ ions. (To convert to SI units, multiply by $\frac{1}{3} \times 10^{11}$.)

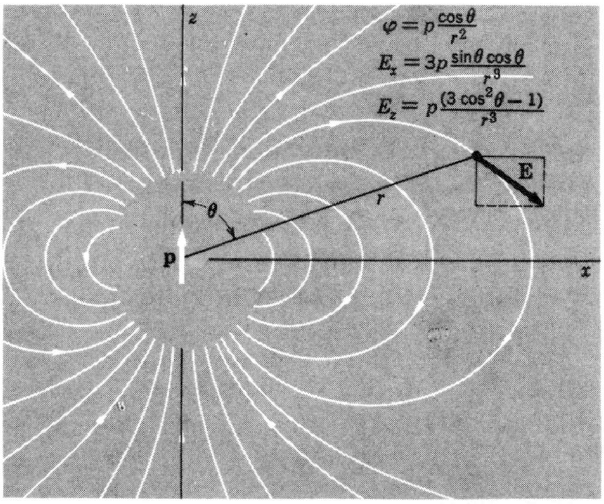

Figure 2 Electrostatic potential and field components in CGS at position r, θ for a dipole $\mathbf{p}$ directed along the z axis. For $\theta = 0$, we have $E_x = E_y = 0$ and $E_z = 2p/r^3$; for $\theta = \pi/2$ we have $E_x = E_y = 0$ and $E_z = -pr^3$. To convert to SI, replace p by $p/4\pi\epsilon_0$. (After E. M. Purcell.)

13 DIELECTRICS AND FERROELECTRICS

First we relate the applied electric field to the internal electric field in a dielectric crystal. The study of the electric field within dielectric matter arises when we ask:

1. What is the relation in the material between the dielectric polarization **P** and the macroscopic electric field **E** in the Maxwell equations?
2. What is the relation between the dielectric polarization and the local electric field which acts at the site of an atom in the lattice? The local field determines the dipole moment of the atom.

Polarization

The **polarization P** is defined as the **dipole moment per unit volume,** averaged over the volume of a cell. The total dipole moment is defined as

$$\mathbf{p} = \Sigma q_n \mathbf{r}_n \ , \qquad (1)$$

where $\mathbf{r}_n$ is the position vector of the charge q_n. The value of the sum will be independent of the origin chosen for the position vectors, provided that the system is neutral. The dipole moment of a water molecule is shown in Fig. 1. The electric field at a point **r** from a point of moment **p** is given by a standard result of elementary electrostatics:

$$\text{(CGS)} \quad \mathbf{E}(\mathbf{r}) = \frac{3(\mathbf{p}\cdot\mathbf{r})\mathbf{r} - r^2\mathbf{p}}{r^5} \ ; \qquad \text{(SI)} \quad \mathbf{E}(\mathbf{r}) = \frac{3(\mathbf{p}\cdot\mathbf{r})\mathbf{r} - r^2\mathbf{p}}{4\pi\epsilon_0 r^5} \ . \qquad (2)$$

The lines of force of a dipole pointing along the z axis are shown in Fig. 2.

MACROSCOPIC ELECTRIC FIELD

One contribution to the electric field inside a body is that of the applied electric field, defined as

$$\boxed{\mathbf{E}_0 \equiv \text{field produced by fixed charges external to the body} \ .} \qquad (3)$$

The other contribution to the electric field is the sum of the fields of all charges that constitute the body. If the body is neutral, the contribution to the average field may be expressed in terms of the sum of the fields of atomic dipoles.

We define the average electric field $\mathbf{E}(\mathbf{r}_0)$ as the **average field over the volume of the crystal cell** that contains the lattice point $\mathbf{r}_0$:

$$\mathbf{E}(\mathbf{r}_0) = \frac{1}{V_c} \int dV \, \mathbf{e}(\mathbf{r}) \ , \tag{4}$$

where $\mathbf{e}(\mathbf{r})$ is the microscopic electric field at the point $\mathbf{r}$. The field $\mathbf{E}$ is a much smoother quantity than the microscopic field $\mathbf{e}$. We could well have written the dipole field (2) as $\mathbf{e}(\mathbf{r})$ because it is a microscopic unsmoothed field.

We call $\mathbf{E}$ the **macroscopic electric field;** it is adequate for all problems in the electrodynamics of crystals provided that we know the connection between $\mathbf{E}$, the polarization $\mathbf{P}$, and the current density $\mathbf{j}$, and provided that the wavelengths of interest are long in comparison with the lattice spacing.[1]

To find the contribution of the polarization to the macroscopic field, we can simplify the sum over all the dipoles in the specimen. By a famous theorem of electrostatics[2] the macroscopic electric field caused by a uniform polarization is equal to the electric field in vacuum of a fictitious surface charge density $\sigma = \hat{\mathbf{n}} \cdot \mathbf{P}$ on the surface of the body. Here $\hat{\mathbf{n}}$ is the unit normal to the surface, drawn outward from the polarized matter. We apply the result to a thin dielectric slab (Fig. 3a) with a uniform volume polarization $\mathbf{P}$. The electric field $\mathbf{E}_1(\mathbf{r})$ produced by the polarization is equal to the field produced by the fictitious surface charge density $\sigma = \hat{\mathbf{n}} \cdot \mathbf{P}$ on the surface of the slab. On the upper boundary the unit vector $\hat{\mathbf{n}}$ is directed upward and on the lower boundary $\hat{\mathbf{n}}$ is directed downward. The upper boundary bears the fictitious charge $\sigma = \hat{\mathbf{n}} \cdot \mathbf{P} = P$ per unit area, and the lower boundary bears $-P$ per unit area.

The electric field $\mathbf{E}_1$ due to these charges has a simple form at any point between the plates, but comfortably removed from their edges. By Gauss's law

$$\text{(CGS)} \quad E_1 = -4\pi|\sigma| = -4\pi P \ ; \qquad \text{(SI)} \quad E_1 = -\frac{|\sigma|}{\epsilon_0} = -\frac{P}{\epsilon_0} \ . \tag{4a}$$

We add $\mathbf{E}_1$ to the applied field $\mathbf{E}_0$ to obtain the total macroscopic field inside the slab, with $\hat{\mathbf{z}}$ the unit vector normal to the plane of the slab:

$$\text{(CGS)} \qquad \mathbf{E} = \mathbf{E}_0 + \mathbf{E}_1 = \mathbf{E}_0 - 4\pi P\hat{\mathbf{z}} \ ; \tag{5}$$

$$\text{(SI)} \qquad \mathbf{E} = \mathbf{E}_0 + \mathbf{E}_1 = \mathbf{E}_0 - \frac{P}{\epsilon_0}\hat{\mathbf{z}} \ .$$

We define

$$\boxed{\mathbf{E}_1 \equiv \text{field of the surface charge density } \hat{\mathbf{n}} \cdot \mathbf{P} \text{ on the boundary} \ .} \tag{6}$$

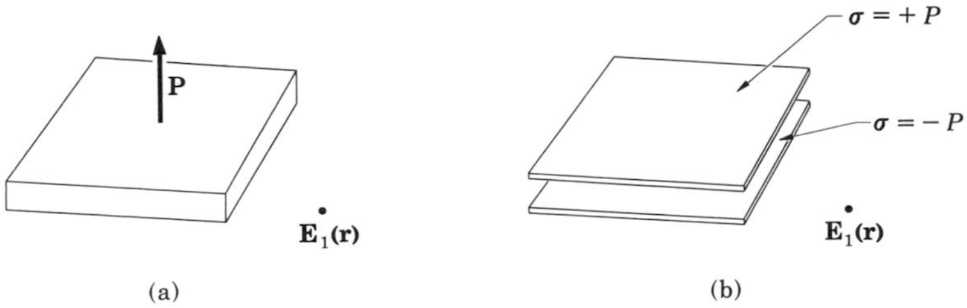

Figure 3 (a) A uniformly polarized dielectric slab, with the polarization vector **P** normal to the plane of the slab. (b) A pair of uniformly charged parallel plates which give rise to the identical electric field $\mathbf{E}_1$ as in (a). The upper plate has the surface charge density $\sigma = +P$, and the lower plate has $\sigma = -P$.

This field is smoothly varying in space inside and outside the body and satisfies the Maxwell equations as written for the macroscopic field **E**. The reason $\mathbf{E}_1$ is a smooth function when viewed on an atomic scale is that we have replaced the discrete lattice of dipoles $\mathbf{p}_j$ with the smoothed polarization **P**.

[1]A detailed derivation of the Maxwell equations for the macroscopic fields **E** and **B**, starting from the Maxwell equations in terms of the microscopic fields **e** and **h**, is given by E. M. Purcell, *Electricity and magnetism*, McGraw-Hill, 1965, Chaps. 9 and 10.
[2]The electrostatic potential in CGS units of a dipole **p** is $\varphi(\mathbf{r}) = \mathbf{p} \cdot \operatorname{grad}(1/r)$. For a volume distribution of polarization **P** we have

$$\varphi(\mathbf{r}) = \int dV \left(\mathbf{P} \cdot \operatorname{grad} \frac{1}{r}\right) ,$$

which by a vector identity becomes

$$\varphi(\mathbf{r}) = \int dV \left(-\frac{1}{r} \operatorname{div} \mathbf{P} + \operatorname{div} \frac{\mathbf{P}}{r}\right) .$$

If **P** is constant, then $\operatorname{div} \mathbf{P} = 0$ and by the Gauss theorem we have

$$\varphi(\mathbf{r}) = \int dS \frac{P_n}{r} = \int dS \frac{\sigma}{r} ,$$

where dS is an element of area on the surface of the body. This completes the proof.

Depolarization Field, $\mathbf{E}_1$

The geometry in many of our problems is such that the polarization is uniform within the body, and then the only contributions to the macroscopic field are from $\mathbf{E}_0$ and $\mathbf{E}_1$:

$$\boxed{\mathbf{E} = \mathbf{E}_0 + \mathbf{E}_1 \ .} \tag{7}$$

Here $\mathbf{E}_0$ is the applied field and $\mathbf{E}_1$ is the field due to the uniform polarization.

The field $\mathbf{E}_1$ is called the **depolarization field,** for within the body it tends to oppose the applied field $\mathbf{E}_0$ as in Fig. 4. Specimens in the shape of ellipsoids, a class that includes spheres, cylinders, and discs as limiting forms, have an advantageous property: a uniform polarization produces a uniform depolarization field. This is a famous mathematical result demonstrated in classic texts on electricity and magnetism.[3] If P_x, P_y, P_z are the components of the polarization $\mathbf{P}$ referred to the principal axes of an ellipsoid, then the components of the depolarization field are written

$$\text{(CGS)} \qquad E_{1x} = -N_x P_x \ ; \qquad E_{1y} = -N_y P_y \ ; \qquad E_{1z} = -N_z P_z \ ; \tag{8}$$

$$\text{(SI)} \qquad E_{1x} = -\frac{N_x P_x}{\epsilon_0} \ ; \qquad E_{1y} = -\frac{N_y P_y}{\epsilon_0} \ ; \qquad E_{1z} = -\frac{N_z P_z}{\epsilon_0} \ .$$

Here N_x, N_y, N_z are the **depolarization factors;** their values depend on the ratios of the principal axes of the ellipsoid. The N's are positive and satisfy the sum rule $N_x + N_y + N_z = 4\pi$ in CGS, and $N_x + N_y + N_z = 1$ in SI.

Values of N parallel to the figure axis of ellipsoids of revolution are plotted in Fig. 5; additional cases have been calculated by Osborn[4] and by Stoner. In limiting cases N has the values:

Shape	*Axis*	*N* (CGS)	*N* (SI)
Sphere	any	$4\pi/3$	1/3
Thin slab	normal	4π	1
Thin slab	in plane	0	0
Long circular cylinder	longitudinal	0	0
Long circular cylinder	transverse	2π	1/2

We can reduce the depolarization field to zero in two ways, either by working with a long fine specimen or by making an electrical connection between electrodes deposited on the opposite surfaces of a thin slab.

[3] R. Becker, *Electromagnetic fields and interactions,* Blaisdell, 1964, pp. 102–107.
[4] J. A. Osborn, Phys. Rev. **67**, 351 (1945); E. C. Stoner, Philosophical Magazine **36,** 803 (1945).

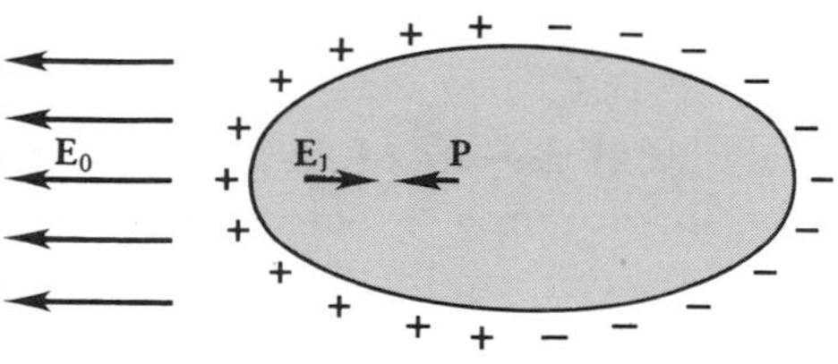

Figure 4 The depolarization field $\mathbf{E}_1$ is opposite to **P**. The fictitious surface charges are indicated: the field of these charges is $\mathbf{E}_1$ within the ellipsoid.

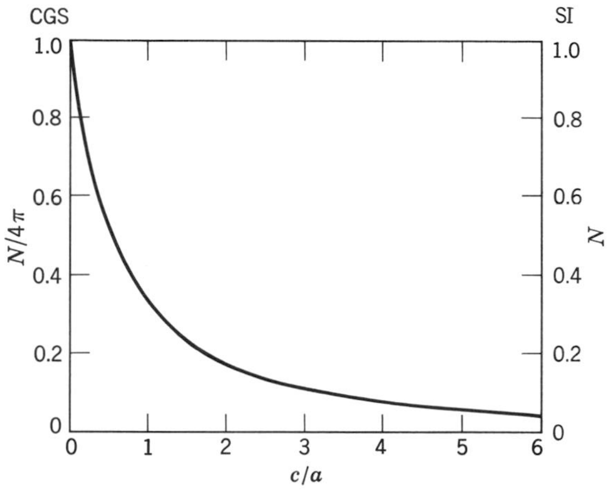

Figure 5 Depolarization factor N parallel to the figure axis of ellipsoids of revolution, as a function of the axial ratio c/a.

A uniform applied field $\mathbf{E}_0$ will induce uniform polarization in an ellipsoid. We introduce the **dielectric susceptibility** χ such that

$$\text{(CGS)}\quad \mathbf{P} = \chi\mathbf{E}\ ; \qquad \text{(SI)}\quad \mathbf{P} = \epsilon_0\chi\mathbf{E}\ , \tag{9}$$

connect the macroscopic field **E** inside the ellipsoid with the polarization **P**. If $\mathbf{E}_0$ is uniform and parallel to a principal axis of the ellipsoid, then

$$\text{(CGS)}\quad E = E_0 + E_1 = E_0 - NP\ ; \qquad \text{(SI)}\quad E = E_0 - \frac{NP}{\epsilon_0}\ , \tag{10}$$

by (8), whence

$$\text{(CGS)}\qquad P = \chi(E_0 - NP)\ ; \qquad P = \frac{\chi}{1 + N\chi}E_0\ ; \tag{11}$$

$$\text{(SI)}\qquad P = \chi(\epsilon_0 E_0 - NP)\ ; \qquad P = \frac{\chi\epsilon_0}{1 + N\chi}E_0\ .$$

The value of the polarization depends on the depolarization factor N.

LOCAL ELECTRIC FIELD AT AN ATOM

The value of the local electric field that acts at the site of an atom is significantly different from the value of the macroscopic electric field. We can convince ourselves of this by consideration of the local field at a site with a cubic arrangement of neighbors[5] in a crystal of spherical shape. The macroscopic electric field in a sphere is

$$\text{(CGS)} \qquad \mathbf{E} = \mathbf{E}_0 + \mathbf{E}_1 = \mathbf{E}_0 - \frac{4\pi}{3}\mathbf{P} \; ; \tag{12}$$

$$\text{(SI)} \qquad \mathbf{E} = \mathbf{E}_0 + \mathbf{E}_1 = \mathbf{E}_0 - \frac{1}{3\epsilon_0}\mathbf{P} \; ,$$

by (9). But consider the field that acts on the atom at the center of the sphere (this atom is not unrepresentative). If all dipoles are parallel to the z axis and have magnitude p, the z component of the field at the center due to all other dipoles is, from (2),

$$\text{(CGS)} \qquad E_{\text{dipole}} = p \sum_i \frac{3z_i^2 - r_i^2}{r_i^5} = p \sum_i \frac{2z_i^2 - x_i^2 - y_i^2}{r_i^5} \; . \tag{13}$$

In SI we replace p by $p/4\pi\epsilon_0$. The x, y, z directions are equivalent because of the symmetry of the lattice and of the sphere; thus

$$\sum_i \frac{z_i^2}{r_i^5} = \sum_i \frac{x_i^2}{r_i^5} = \sum_i \frac{y_i^2}{r_i^5} \; ,$$

whence $E_{\text{dipole}} = 0$. The correct local field is just equal to the applied field, $\mathbf{E}_{\text{local}} = \mathbf{E}_0$, for an atom site with a cubic environment in a spherical specimen. Thus the local field is not the same as the macroscopic average field.

We now develop an expression for the local field at a general lattice site, not necessarily of cubic symmetry. The local field at an atom is the sum of the electric field $\mathbf{E}_0$ from external sources and of the field from the dipoles within the specimen. It is convenient to decompose the dipole field so that part of the summation over dipoles may be replaced by integration. We write

$$\mathbf{E}_{\text{local}} = \mathbf{E}_0 + \mathbf{E}_1 + \mathbf{E}_2 + \mathbf{E}_3 \; . \tag{14}$$

Here

$\mathbf{E}_0$ = field produced by fixed charges external to the body;

$\mathbf{E}_1$ = depolarization field, from a surface charge density $\hat{\mathbf{n}} \cdot \mathbf{P}$ on the outer surface of the specimen;

$\mathbf{E}_2$ = Lorentz cavity field: field from polarization charges on inside of a spherical cavity cut (as a mathematical fiction) out of the specimen with the reference atom as center, as in Fig. 6; $\mathbf{E}_1 + \mathbf{E}_2$ is field due to uniform polarization of the body in which a hole has been created;

$\mathbf{E}_3$ = field of atoms inside cavity.

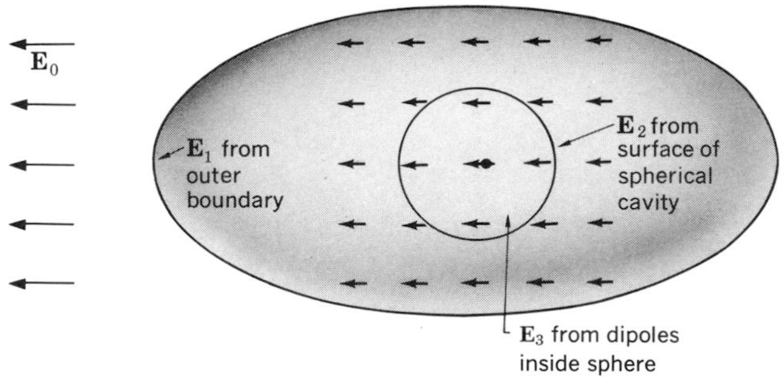

Figure 6 The internal electric field on an atom in a crystal is the sum of the external applied field $\mathbf{E}_0$ and of the field due to the other atoms in the crystal. The standard method of summing the dipole fields of the other atoms is first to sum individually over a moderate number of neighboring atoms inside an imaginary sphere concentric with the reference atom: this defines the field $\mathbf{E}_3$, which vanishes at a reference site with cubic symmetry. The atoms outside the sphere can be treated as a uniformly polarized dielectric. Their contribution to the field at the reference point is $\mathbf{E}_1 + \mathbf{E}_2$, where $\mathbf{E}_1$ is the depolarization field associated with the other boundary and $\mathbf{E}_2$ is the field associated with the surface of the spherical cavity.

The contribution $\mathbf{E}_1 + \mathbf{E}_2 + \mathbf{E}_3$ to the local field is the total field at one atom caused by the dipole moments of all the other atoms in the specimen:

$$\text{(CGS)} \qquad \mathbf{E}_1 + \mathbf{E}_2 + \mathbf{E}_3 = \sum_i \frac{3(\mathbf{p}_i \cdot \mathbf{r}_i)\mathbf{r}_i - r_i^2\mathbf{p}_i}{r_i^5} , \qquad (15)$$

and in SI we replace $\mathbf{p}_i$ by $\mathbf{p}_i/4\pi\epsilon_0$. Dipoles at distances greater than perhaps ten lattice constants from the reference site make a smoothly varying contribution to this sum, a contribution which may be replaced by two surface integrals. One surface integral is taken over the outer surface of the ellipsoidal specimen and defines $\mathbf{E}_1$, as in Eq. (6). The second surface integral defines $\mathbf{E}_2$ and may be taken over any interior surface that is a suitable distance (say 50 Å) from the reference site. We count in $\mathbf{E}_3$ any dipoles not included in the volume bounded by the inner and outer surfaces. It is convenient to let the interior surface be spherical.

[5]Atom sites in a cubic crystal do not necessarily have cubic symmetry: thus the O^{2-} sites in the barium titanate structure of Fig. 10 do not have a cubic environment. However, the Na^+ and Cl^- sites in the NaCl structure and the Cs^+ and Cl^- sites in the CsCl structure have cubic symmetry.

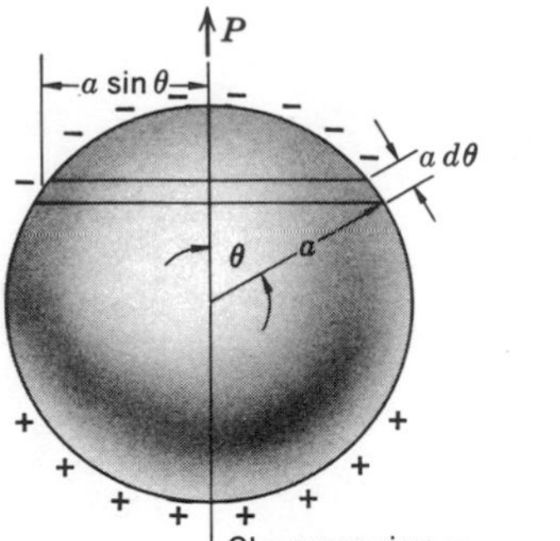

Figure 7 Calculation of the field in a spherical cavity in a uniformly polarized medium.

Lorentz Field, E_2

The field $\mathbf{E}_2$ due to the polarization charges on the surface of the fictitious cavity was calculated by Lorentz. If θ is the polar angle (Fig. 7) referred to the polarization direction, the surface charge density on the surface of the cavity is $-P\cos\theta$. The electric field at the center of the spherical cavity of radius a is

$$\text{(CGS)} \quad \mathbf{E}_2 = \int_0^{\pi} (a^{-2})(2\pi a \sin\theta)(a\,d\theta)(\mathbf{P}\cos\theta)(\cos\theta) = \frac{4\pi}{3}\mathbf{P} \; ; \qquad (16)$$

$$\text{(SI)} \qquad \mathbf{E}_2 = \frac{1}{3\epsilon_0}\mathbf{P} \; .$$

This is the negative of the field in a polarized sphere.

Field of Dipoles Inside Cavity, E_3

The field $\mathbf{E}_3$ due to the dipoles within the spherical cavity is the only term that depends on the crystal structure. We showed for a reference site with cubic surroundings in a sphere that $\mathbf{E}_3 = 0$ if all the atoms may be replaced by point dipoles parallel to each other.

The total local field at a cubic site is, from (14) and (16),

$$\text{(CGS)} \qquad \mathbf{E}_{\text{local}} = \mathbf{E}_0 + \mathbf{E}_1 + \frac{4\pi}{3}\mathbf{P} = \mathbf{E} + \frac{4\pi}{3}\mathbf{P} \; ; \qquad (17)$$

$$\text{(SI)} \qquad \mathbf{E}_{\text{local}} = \mathbf{E} + \frac{1}{3\epsilon_0}\mathbf{P} \; .$$

This is the **Lorentz relation:** the field acting at an atom in a cubic site is the macroscopic field $\mathbf{E}$ of Eq. (7) plus $4\pi\mathbf{P}/3$ or $\mathbf{P}/3\epsilon_0$ from the polarization of the other atoms in the specimen. Experimental data for cubic ionic crystals[6] support the Lorentz relation.

[6] J. Tessman, A. Kahn, and W. Shockley, Phys. Rev. **92**, 890 (1953); E. Kartheuser and J. Deltour, Physics Letters **19**, 548 (1965).

DIELECTRIC CONSTANT AND POLARIZABILITY

The **dielectric constant** ϵ of an isotropic or cubic medium relative to vacuum is defined in terms of the macroscopic field E:

(CGS) $$\epsilon \equiv \frac{E + 4\pi P}{E} = 1 + 4\pi\chi \ ; \qquad (18)$$

(SI) $$\epsilon = \frac{\epsilon_0 E + P}{\epsilon_0 E} = 1 + \chi \ .$$

The susceptibility (9) is related to the dielectric constant by

(CGS) $$\chi = \frac{P}{E} = \frac{\epsilon - 1}{4\pi} \ ;$$ (SI) $$\chi = \frac{P}{\epsilon_0 E} = \epsilon - 1 \ . \qquad (19)$$

In a noncubic crystal the dielectric response is described by the components of the susceptibility tensor or of the dielectric constant tensor:

(CGS) $$P_\mu = \chi_{\mu\nu} E_\nu \ ; \qquad \epsilon_{\mu\nu} = 1 + 4\pi\chi_{\mu\nu} \ . \qquad (20)$$

(SI) $$P_\mu = \chi_{\mu\nu} \epsilon_0 E_\nu \ ; \qquad \epsilon_{\mu\nu} = 1 + \chi_{\mu\nu} \ .$$

The **polarizability** α of an atom is defined in terms of the local electric field at the atom:

$$p = \alpha E_{\text{local}} \ , \qquad (21)$$

where p is the dipole moment. This definition applies in CGS and in SI. The polarizability is an atomic property, but the dielectric constant will depend on the manner in which the atoms are assembled to form a crystal. The polarization of a crystal may be expressed approximately as the product of the polarizabilities of the atoms times the local electric field:

$$P = \sum_j N_j p_j = \sum_j N_j \alpha_j E_{\text{loc}}(j) \ , \qquad (22)$$

where N_j is the concentration and α_j the polarizability of atoms j, and $E_{\text{loc}}(j)$ is the local field at atom sites j. We want to relate the dielectric constant to the polarizabilities; the result will depend on the relation that holds between the macroscopic electric field and the local electric field. We give the derivation in CGS units and state the result in both systems of units.

If the local field is given by the Lorentz relation (17), then

(CGS) $$P = (\Sigma N_j \alpha_j) \left(E + \frac{4\pi}{3} P \right) \ ;$$

and we solve for P to find the susceptibility

(CGS) $$\chi = \frac{P}{E} = \frac{\Sigma N_j\alpha_j}{1 - \dfrac{4\pi}{3}\Sigma N_j\alpha_j} \ . \qquad (23)$$

By definition $\epsilon = 1 + 4\pi\chi$ in CGS; we may rearrange (23) to obtain

(CGS) $$\frac{\epsilon - 1}{\epsilon + 2} = \frac{4\pi}{3}\Sigma N_j\alpha_j \ ;$$ (SI) $$\frac{\epsilon - 1}{\epsilon + 2} = \frac{1}{3\epsilon_0}\Sigma N_j\alpha_j \ , \qquad (24)$$

the **Clausius-Mossotti relation.** This relates the dielectric constant to the electronic polarizability, but only for crystal structures for which the Lorentz local field obtains.

Electronic Polarizability

The total polarizability may usually be separated into three parts: electronic, ionic, and dipolar, as in Fig. 8. The electronic contribution arises from the displacement of the electron shell relative to a nucleus. The ionic contribution comes from the displacement of a charged ion with respect to other ions. The dipolar polarizability arises from molecules with a permanent electric dipole moment that can change orientation in an applied electric field. In heterogeneous materials there is usually also an interfacial polarization arising from the accumulation of charge at structural interfaces. This is of little fundamental interest, but it is of considerable practical interest because commercial insulating materials are usually heterogeneous.

The dielectric constant at optical frequencies arises almost entirely from the electronic polarizability. The dipolar and ionic contributions are small at high frequencies because of the inertia of the molecules and ions. In the optical range (24) reduces to

(CGS) $$\frac{n^2 - 1}{n^2 + 2} = \frac{4\pi}{3}\Sigma N_j\alpha_j(\text{electronic}) \ ; \qquad (25)$$

here we have used the relation $n^2 = \epsilon$, where n is the refractive index. By applying (25) to large numbers of crystals we determine in Table 1 empirical values of the electronic polarizabilities that are reasonably consistent with the observed values of the refractive index. The scheme is not entirely self-consistent, because the electronic polarizability of an ion depends somewhat on the environment in which it is placed. The negative ions are highly polarizable because they are large.

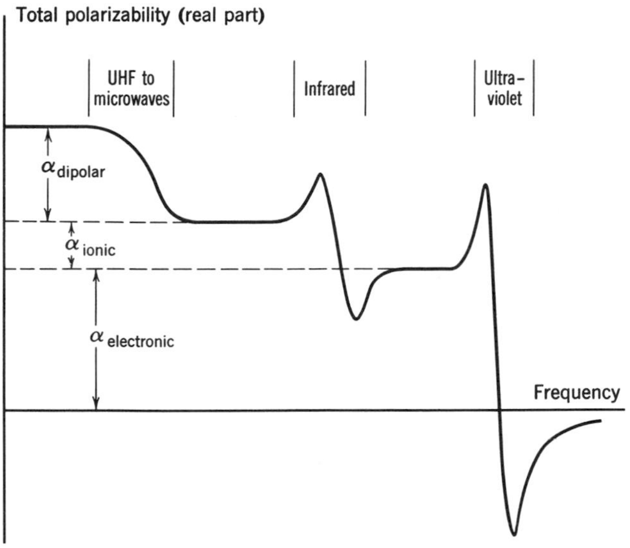

Figure 8 Frequency dependence of the several contributions to the polarizability.

TABLE 1 Electronic Polarizabilities of Ions in 10^{-24} cm^3

			He	Li^+	Be^{2+}	B^{3+}	C^{4+}
Pauling			0.201	0.029	0.008	0.003	0.0013
JS				0.029			
	O^{2-}	F^-	Ne	Na^+	Mg^{2+}	Al^{3+}	Si^{4+}
Pauling	3.88	1.04	0.390	0.179	0.094	0.052	0.0165
JS-(TKS)	(2.4)	0.858		0.290			
	S^{2-}	Cl^-	Ar	K^+	Ca^{2+}	Sc^{3+}	Ti^{4+}
Pauling	10.2	3.66	1.62	0.83	0.47	0.286	0.185
JS-(TKS)	(5.5)	2.947		1.133	(1.1)		(0.19)
	Se^{2-}	Br^-	Kr	Rb^+	Sr^{2+}	Y^{3+}	Zr^{4+}
Pauling	10.5	4.77	2.46	1.40	0.86	0.55	0.37
JS-(TKS)	(7.)	4.091		1.679	(1.6)		
	Te^{2-}	I^-	Xe	Cs^+	Ba^{2+}	La^{3+}	Ce^{4+}
Pauling	14.0	7.10	3.99	2.42	1.55	1.04	0.73
JS-(TKS)	(9.)	6.116		2.743	(2.5)		

Values from L. Pauling, Proc. Roy. Soc. (London) **A114**, 181 (1927); S. S. Jaswal and T. P. Sharma, J. Phys. Chem. Solids **34**, 509 (1973); and J. Tessman, A. Kahn, and W. Shockley, Phys. Rev. **92**, 890 (1953). The TKS polarizabilities are at the frequency of the D lines of sodium. The values are in CGS; to convert to SI, multiply by $\frac{1}{9} \times 10^{-15}$.

Classical Theory of Electronic Polarizability. An electron bound harmonically to an atom will show resonance absorption at a frequency $\omega_0 = (\beta/m)^{\frac{1}{2}}$, where β is the force constant. The displacement x of the electron occasioned by the application of a field E_{loc} is given by

$$-eE_{\text{loc}} = \beta x = m\omega_0^2 x \ , \tag{26}$$

so that the static electronic polarizability is

$$\alpha(\text{electronic}) = p/E_{\text{loc}} = -ex/E_{\text{loc}} = e^2/m\omega_0^2 \ . \tag{27}$$

The electronic polarizability will depend on frequency, and it is shown in the following example that for frequency ω

$$\alpha(\text{electronic}) = \frac{e^2/m}{\omega_0^2 - \omega^2} \ , \tag{28}$$

but in the visible region the frequency dependence (dispersion) is not usually very important in most transparent materials.

EXAMPLE: ***Frequency dependence.*** Find the frequency dependence of the electronic polarizability of an electron having the resonance frequency ω_0, treating the system as a simple harmonic oscillator.

The equation of motion in the local electric field $E_{\text{loc}} \sin \omega t$ is

$$m \frac{d^2x}{dt^2} + m\omega_0^2 x = -eE_{\text{loc}} \sin \omega t \ ,$$

so that, for $x = x_0 \sin \omega t$,

$$m(-\omega^2 + \omega_0^2)x_0 = -eE_{\text{loc}} \ .$$

The dipole moment has the amplitude

$$p_0 = -ex_0 = \frac{e^2 E_{\text{loc}}}{m(\omega_0^2 - \omega^2)} \ ,$$

from which (28) follows.

In quantum theory the expression corresponding to (28) is

$$\alpha(\text{electronic}) = \frac{e^2}{m} \sum_j \frac{f_{ij}}{\omega_{ij}^2 - \omega^2} \ , \tag{29}$$

where f_{ij} is called the **oscillator strength** of the electric dipole transition between the atomic states i and j.

FERROELECTRIC CRYSTALS

A ferroelectric crystal[7] exhibits an electric dipole moment even in the absence of an external electric field. In the **ferroelectric state** the center of positive charge of the crystal does not coincide with the center of negative charge. A typical plot of polarization versus electric field for the ferroelectric state is shown in Fig. 9. The loop is called a **hysteresis loop;** it is a sign of a ferroelectric state. A crystal in a normal dielectric state usually does not show perceptible hysteresis when the electric field is increased and reversed slowly. In some crystals the ferroelectric dipole moment may not be changed by an electric field of the maximum intensity which it is possible to apply without causing electrical breakdown of the crystal. In these crystals we are often able to observe a change in the spontaneous moment when they are heated: changing the temperature changes the value of the dipole moment. Such crystals are called **pyroelectric,** whereas crystals such that the direction of the spontaneous moment can be altered by an electric field are called ferroelectric. Lithium niobate, $LiNbO_3$, is pyroelectric at room temperature; it has a high transition temperature ($T_c = 1470$ K) and a very high saturation polarization. It can only be given a remanent polarization by an electric field applied over 1000°C.

Ferroelectricity usually disappears above a certain temperature called the **transition temperature;** above the transition the crystal is said to be in a **paraelectric state.** The term paraelectric suggests an analogy with paramagnetism and implies a rapid decrease in the dielectric constant as the temperature increases.

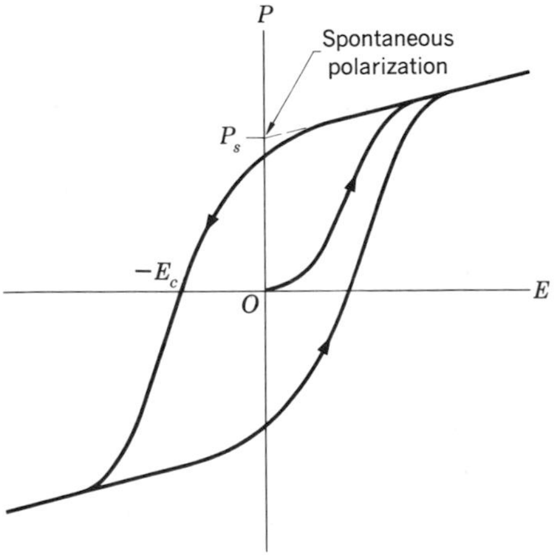

Figure 9 Ferroelectric hysteresis loop; the coercive force is E_c.

[7]The early history of ferroelectrics is recalled by Joseph Valasek, who discovered the first ferroelectric crystal (Rochelle salt): Ferroelectrics **2,** 239 (1971); the original reference is Phys. Rev. **15,** 537 (1920).

Classification of Ferroelectric Crystals

We list in Table 2 some of the crystals commonly considered to be ferroelectric, along with the transition temperature or Curie point T_c at which the crystal changes from the low temperature polarized state to the high temperature unpolarized state. Thermal motion tends to destroy the ferroelectric order. Some ferroelectric crystals have no Curie point because they melt before leaving the ferroelectric phase. The table also includes values of the spontaneous polarization P_s. Ferroelectric crystals may be classified into two main groups, order-disorder or displacive, according to whether the transition is associated with the individual ordering of ions or is associated with the displacement of a whole sublattice of ions of one type relative to another sublattice.

The order-disorder class of ferroelectrics includes crystals with hydrogen bonds in which the motion of the protons is related to the ferroelectric properties, as in potassium dihydrogen phosphate (KH_2PO_4) and isomorphous salts. The behavior of crystals in which the hydrogen has been replaced by deuterium is interesting:

	KH_2PO_4	KD_2PO_4	KH_2AsO_4	KD_2AsO_4
Curie temperature	123 K	213 K	96 K	162 K

The substitution of deuterons for protons nearly doubles T_c, although the fractional change in the molecular weight of the compound is less than 2 percent. This extraordinarily large isotope shift is believed to be a quantum effect[8] involving the mass-dependence of the de Broglie wavelength. Neutron diffraction data[9] show that above the Curie temperature the proton distribution along the hydrogen bond is symmetrically elongated. Below the Curie temperature the distribution is more concentrated and asymmetric with respect to neighboring ions, so that one end of the hydrogen bond is preferred by the proton over the other end.

The displacive class of ferroelectrics includes ionic crystal structures closely related to the perovskite and ilmenite structures. The simplest ferroelectric crystal[10] is GeTe with the sodium chloride structure. We shall devote ourselves primarily to crystals with the perovskite structure, Fig. 10. Representative examples of displacive structural phase transitions, other than ferroelectric, are given in Table 3.

[8]J. Pirenne, Physica **15,** 1019 (1949); J. C. Slater, J. Chem. Phys. **9,** 16 (1941); T. Nagamiya, Prog. Theor. Phys. **7,** 275 (1952); H. B. Silsbee, E. A. Uehling, and V. H. Schmidt, Phys. Rev. **133,** A165 (1964); C. W. Fairall and W. Reese, Phys. Rev. **B6,** 193 (1972).

[9]B. C. Frazer and R. Pepinsky, Acta Cryst. **6,** 273 (1953); R. S. Pease and G. E. Bacon, Proc. R. Soc. (London) **A220,** 397 (1953); **A236,** 359 (1955).

[10]G. S. Pawley, W. Cochran, R. A. Cowley, and G. Dolling, Phys. Rev. Letters **17,** 753 (1966).

TABLE 2 Ferroelectric Crystals[a]

To obtain P_s in the SI unit of C m^{-2}, divide the value given in CGS by 3×10^5. To obtain P_s in μcoul/cm^2, divide the value in CGS by 3×10^3.

		T_c, in K	P_s in esu cm^{-2}, at T K	
KDP type	KH_2PO_4	123	16,000	[96]
	KD_2PO_4	213	13,500	—
	RbH_2PO_4	147	16,800	[90]
	RbH_2AsO_4	111	—	—
	KH_2AsO_4	96	15,000	[80]
	GeTe	670	—	—
TGS type	Tri-glycine sulfate	322	8,400	[293]
	Tri-glycine selenate	295	9,600	[273]
Perovskites	$BaTiO_3$	393	78,000	[296]
	$SrTiO_3$	~0	(9,000)	[4]
	$KNbO_3$	712	90,000	[523]
	$PbTiO_3$	763	>150,000	[300]
	$LiTaO_3$	890	70,000	[720]
	$LiNbO_3$	1470	900,000	—

[a] A compilation of data on ferroelectric and antiferroelectric materials is given by E. C. Subbarao, Ferroelectrics **5**, 267 (1973). Valuable bibliographies by K. Toyoda appear semiannually in this journal.

Consider the order of magnitude of the ferroelectric effects in barium titanate: The observed saturation polarization P_s at room temperature (Fig. 11) is 8×10^4 esu cm^{-2}. The volume of a cell is $(4 \times 10^{-8})^3 = 64 \times 10^{-24}$ cm^3, so that the dipole moment of a cell is

$$\text{(CGS)} \quad p \cong (8 \times 10^4 \text{ esu cm}^{-2})(64 \times 10^{-24} \text{ cm}^3) \cong 5 \times 10^{-18} \text{ esu cm} \ ;$$

$$\text{(SI)} \quad p \cong (3 \times 10^{-1} \text{ C m}^{-2})(64 \times 10^{-30} \text{ m}^3) \cong 2 \times 10^{-29} \text{ C m} \ .$$

If the positive ions Ba^{2+} and Ti^{4+} were moved[11] by $\delta = 0.1$ Å with respect to the negative O^{2-} ions, the dipole moment of a cell would be $6e\delta \cong 3 \times 10^{-18}$ esu cm.

[11] For a discussion of the x-ray and neutron diffraction studies of the ionic displacements, see J. Harada, T. Pedersen, and Z. Barnea, Acta Crystall. **A26**, 336 (1970). In $LiNbO_3$ the displacements are considerably larger, being 0.9 Å and 0.5 Å for the lithium and niobum ions respectively.

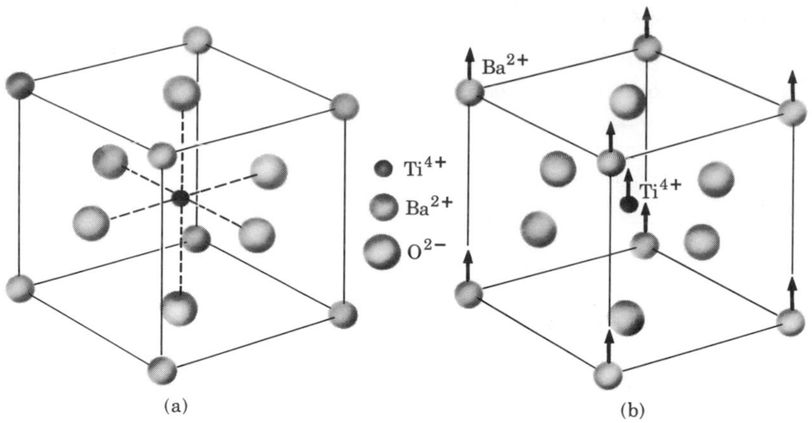

Figure 10 (a) The crystal structure of barium titanate. The prototype crystal is calcium titanate (perovskite). The structure is cubic, with Ba^{2+} ions at the cube corners, O^{2-} ions at the face centers, and a Ti^{4+} ion at the body center. (b) Below the Curie temperature the structure is slightly deformed, with Ba^{++} and Ti^{4+} ions displaced relative to the O^{2-} ions, thereby developing a dipole moment. The upper and lower oxygen ions may move downward slightly.

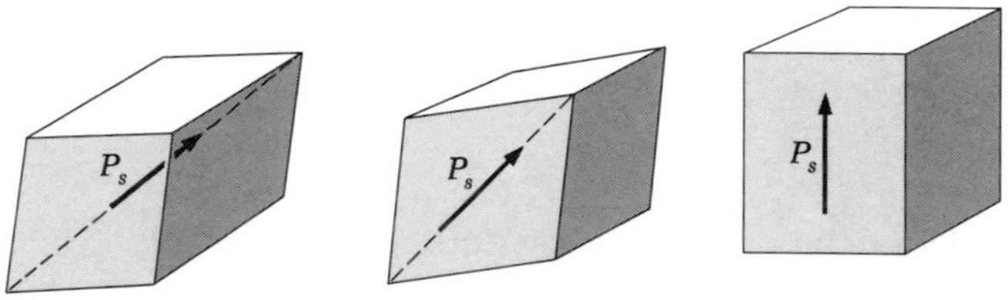

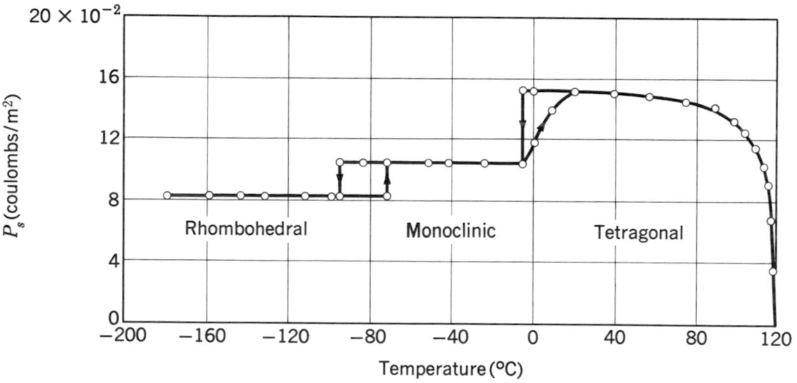

Figure 11 Spontaneous polarization projected on cube edge of barium titanate, as a function of temperature. [After W. J. Merz, Phys. Rev. **76,** 1221 (1949).]

TABLE 3 Nonferroelectric Displacive Structural Phase Transformations

	Structure	Type	T_c, in K
Quartz	Rhombohedral	Nonpolar	850
$SrTiO_3$	Cubic	Antiferroelectric	105
$KMnF_3$	Perovskite	Antiferroelectric	185
$LaAlO_3$	Perovskite	Antiferroelectric	810
$Gd_2(MoO_4)_3$	Tetragonal	Ferroelectric	430
V_3Si	Cubic	Elastic	21
Nb_3Sn	β-W	$(C_{11} - C_{12}) \rightarrow 0$	46

After J. D. Axe and G. Shirane, Physics Today, September 1973, p. 32.

POLARIZATION CATASTROPHE

Two related viewpoints contribute to an understanding of the occurrence of ferroelectricity. We may speak in terms of a polarization catastrophe in which for some critical condition the polarization becomes very large; or we may speak in terms of a transverse optical phonon of very low frequency.[12] In a polarization catastrophe the local electric field caused by the polarization increases faster than the elastic restoring force on an ion in the crystal, thereby leading to an asymmetrical shift in ionic positions. The shift is limited to a finite displacement by higher order restoring forces. The occurrence of ferroelectricity in an appreciable number of crystals with the perovskite structure suggests that this structure is favorably disposed to a polarizability catastrophe. Calculations of local fields[13] have made clear the physical reason for the favored position of the perovskite structure.

We gave first the simple form of the catastrophe theory, supposing that the local field at all atoms is equal to $\mathbf{E} + 4\pi\mathbf{P}/3$ in CGS or $\mathbf{E} + \mathbf{P}/3\epsilon_0$ in SI. The theory given now leads to a second-order transition; the physical ideas can be carried over to a first-order transition. In a second-order transition there is no latent heat; the order parameter (in this instance, the polarization) is not discontinuous at the transition temperature. In a first-order transition there is a latent heat; the order parameter changes discontinuously at the transition temperature.

We rewrite (24) for the dielectric constant in the form

$$\text{(CGS)} \qquad \epsilon = \frac{1 + \frac{8\pi}{3}\Sigma N_i\alpha_i}{1 - \frac{4\pi}{3}\Sigma N_i\alpha_i} , \qquad (30)$$

[12]P. W. Anderson, Moscow dielectric conference, 1960; W. Cochran, Adv. in Physics **9**, 387 (1960); V. L. Ginzburg, Sov. Physics (Solid State) **2**, 1824 (1960).

[13]J. C. Slater, Phys. Rev. **78**, 748 (1950). We note that the O^{2-} ions in the perovskite structure do not have cubic surroundings. The local field factors at these ions turn out to be large.

where α_i is the electronic plus ionic polarizability of an ion of type i and N_i is the number of ions i per unit volume. The dielectric constant becomes infinite and permits a finite polarization in zero applied field when

$$\text{(CGS)} \qquad \Sigma N_i\alpha_i = 3/4\pi \ . \qquad (31)$$

This is the condition for a polarization catastrophe. The value of ϵ in (30) is sensitive to small departures of $\Sigma N_i\alpha_i$ from the critical value $3/4\pi$. If we write

$$\text{(CGS)} \qquad (4\pi/3)\ \Sigma N_i\alpha_i = 1 - 3s \ , \qquad (32)$$

where $s \ll 1$, the dielectric constant in (30) becomes

$$\epsilon \simeq 1/s \ . \qquad (33)$$

Suppose near the critical temperature s varies linearly with temperature:

$$s \simeq (T - T_c)/\xi \ , \qquad (34)$$

where ξ is a constant. Such a variation of s or $\Sigma N_i\alpha_i$ might come from normal thermal expansion of the lattice. The dielectric constant has the form

$$\epsilon \simeq \frac{\xi}{T - T_c} \ , \qquad (35)$$

close to the observed temperature variation in the paraelectric state, Fig. 12.

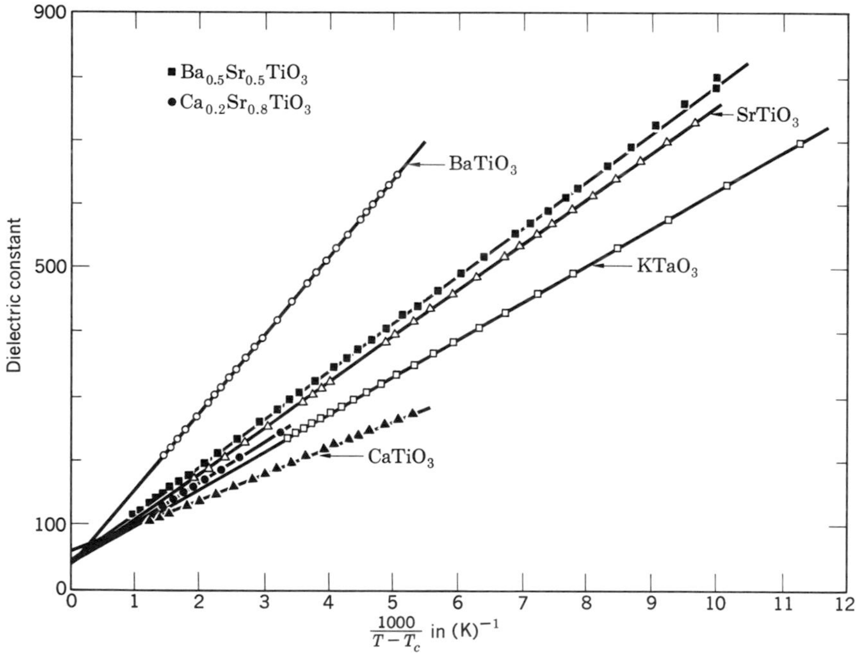

Figure 12 Dielectric constant versus $1/(T - T_c)$ in the paraelectric state ($T > T_c$) of perovskites, after G. Rupprecht and R. O. Bell, Phys. Rev. **135**, A748 (1964).

Landau Theory of the Phase Transition

A ferroelectric with a first-order phase transition between the ferroelectric and the paraelectric state is distinguished by a discontinuous change of the saturation polarization at the transition temperature. The transition between the normal and superconducting states is a second-order transition, as is the transition between the ferromagnetic and paramagnetic states. In these transitions the degree of order goes to zero without a discontinuous change as the temperature is increased.

We can obtain a consistent formal thermodynamic theory of the behavior of a ferroelectric crystal by considering the form of the expansion of the energy as a function of the polarization P. We assume that the Landau[14] free energy density $\hat{F}$ in one dimension may be expanded formally as

$$\hat{F}(P;T,E) = -EP + g_0 + \tfrac{1}{2}g_2P^2 + \tfrac{1}{4}g_4P^4 + \tfrac{1}{6}g_6P^6 + \cdots \ , \tag{36}$$

where the coefficients g_n depend on the temperature. The series does not contain terms in odd powers of P if the unpolarized crystal has a center of inversion symmetry, but crystals are known in which odd powers are important. Power series expansions of the free energy do not always exist, for nonanalytic terms are known to occur. For example, the transition in KH_2PO_4 appears to have a logarithmic singularity in the heat capacity at the transition, which is not classifiable as either first or second order. The value of P in thermal equilibrium is given by the minimum of $\hat{F}$ as a function of P; the value of $\hat{F}$ at this minimum defines the Helmholtz free energy $F(T,E)$. The equilibrium polarization in an electric field E satisfies

$$\frac{\partial \hat{F}}{\partial P} = 0 = -E + g_2P + g_4P^3 + g_6P^5 + \cdots \ . \tag{37}$$

In this section we suppose that the specimen is a long rod with E parallel to the long axis; E denotes the external applied field.

To obtain a ferroelectric state we must suppose that the coefficient of the term in P^2 in (36) passes through zero at some temperature T_0:

$$g_2 = \gamma(T - T_0) \ , \tag{38}$$

where γ is taken as a positive constant and T_0 may be equal to or lower than the transition temperature. A small positive value of g_2 means that the lattice is "soft" and is close to instability. A negative value of g_2 means that the unpolarized lattice is unstable. The variation of g_2 with temperature can be accounted for by thermal expansion and other effects of anharmonic lattice interactions.

[14]In *TP*, see pp. 300 to 304 for a discussion of the Landau function and p. 361 for the term $-EP$. The Helmholtz free energy $F(T, E)$ is the A scheme of *TP*, Chapter 22.

Second-Order Transition

If g_4 is positive, nothing new is added by the term in g_6, and this may then be neglected. The polarization for zero applied electric field is found from (37):

$$\gamma(T - T_0)P_s + g_4 P_s^3 = 0 \ , \tag{39}$$

so that either $P_s = 0$ or $P_s^2 = (\gamma/g_4)(T_0 - T)$. For $T \geq T_0$ the only real root of (39) is at $P_s = 0$, because γ and g_4 are positive. Thus T_0 is the Curie temperature. For $T < T_0$ the minimum of the Landau free energy in zero applied field is at

$$|P_s| = (\gamma/g_4)^{\frac{1}{2}}(T_0 - T)^{\frac{1}{2}} \ , \tag{40}$$

as plotted in Fig. 13. The phase transition is a second-order transition because the polarization goes continuously to zero at the transition temperature. The transition in triglycine sulfate is an example (Fig. 14) of a second-order transition; the observed exponent[15] of $(T_0 - T)$ in the saturation polarization P_s is 0.51 ± 0.05.

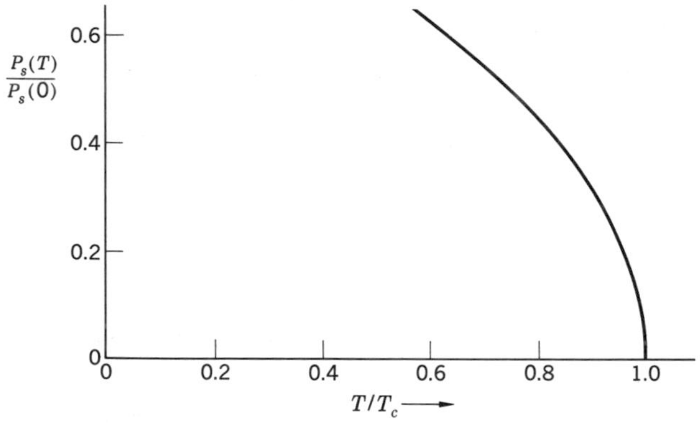

Figure 13 Spontaneous polarization versus temperature, for a second-order phase transition.

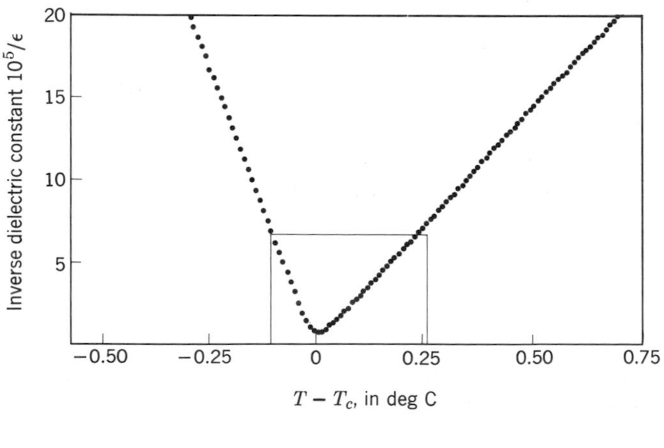

Figure 14 Plot of inverse dielectric constant of triglycine sulfate versus temperature difference $\Delta T = T - T_c$ from the transition temperature, 49.92°C. The linearity is as predicted for a second-order phase transition, and the difference in slopes is also accounted for by the theory. (After J. A. Gonzalo.)

First-Order Transition

The transition is first order if g_4 is negative. We must now retain g_6 and take it positive in order to restrain $\hat{F}$ from going to minus infinity (Fig. 15). The equilibrium condition for $E = 0$ is given by (37):

$$\gamma(T - T_0)P_s - |g_4|P_s^3 + g_6 P_s^5 = 0 \ , \qquad (41)$$

so that either $P_s = 0$ or

$$\gamma(T - T_0) - |g_4|P_s^2 + g_6 P_s^4 = 0 \ . \qquad (42)$$

At the transition temperature T_c the free energies of the paraelectric and ferroelectric phases will be equal. That is, the value of F for $P_s = 0$ will be equal to the value of F at the minimum given by (42). In Fig. 16 we show the characteristic variation with temperature of P_s for a first-order phase transition; contrast this with the variation shown in Fig. 13 for a second-order phase transition. The transition in $BaTiO_3$ is first order.

The dielectric constant is calculated from the equilibrium polarization in an applied electric field E and is found from (37). In equilibrium at temperatures over the transition the terms in P^4 and P^6 may be neglected; thus $E = \gamma(T - T_0)P$, or

$$\text{(CGS)} \qquad \epsilon(T > T_c) = 1 + 4\pi P/E = 1 + 4\pi/\gamma(T - T_0) \ . \qquad (43)$$

of the form of (35). The result applies whether the transition is of the first or second order, but if second order we have $T_0 = T_c$; if first order, then $T_0 < T_c$. Equation (38) defines T_0, but T_c is the transition temperature.

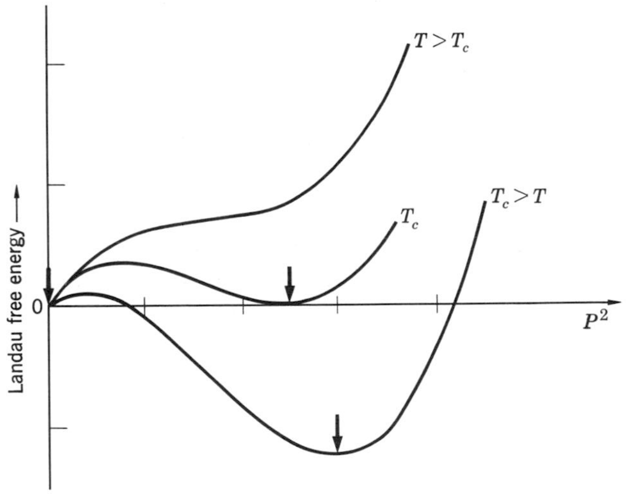

Figure 15 Landau free energy function versus (polarization)^2 in a first-order transition, at representative temperatures. At T_c the Landau function has equal minima at $P = 0$ and at a finite P as shown. For T below T_c the absolute minimum is at larger values of P; as T passes through T_c there is a discontinuous change in the position of the absolute minimum. The arrows mark the minima.

[15] J. A. Gonzalo, Phys. Rev. **144,** 662 (1966); P. P. Craig, Phys. Letters **20,** 140 (1966).

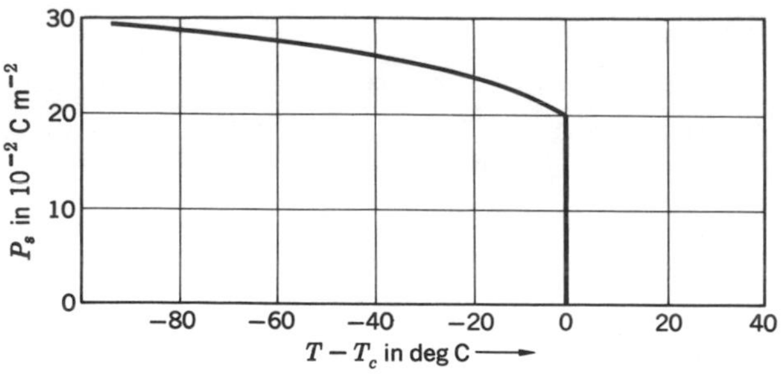

Figure 16 Calculated values of the spontaneous polarization as a function of temperature, with parameters as for barium titanate. (After W. Cochran.)

SOFT OPTICAL PHONONS

The Lyddane-Sachs-Teller relation (Chapter 10) is

$$\omega_T^2/\omega_L^2 = \epsilon(\infty)/\epsilon(0) \quad . \tag{44}$$

It is important for ferroelectricity that the static dielectric constant increases as when the transverse optical phonon frequency decreases. When the static dielectric constant $\epsilon(0)$ has a high value, such as 100 to 10,000, when ω_T has a low value. When $\omega_T = 0$ the crystal is unstable because there is no effective restoring force. The ferroelectric $BaTiO_3$ at 24°C has a TO mode at 12 cm^{-1}, a low frequency for an optical mode. If the transition to a ferroelectric state is first order, we do not find $\omega_T = 0$ or $\epsilon(0) = \infty$ at the transition. The LST relation suggests only that $\epsilon(0)$ extrapolates to a singularity at a temperature T_0 below T_c. In disordered ("dirty") ferroelectrics the static dielectric constant can be larger than suggested by the LST relations.[15a]

The association of a high static dielectric constant with a low frequency optical mode is supported by experiments on strontium titanate, $SrTiO_3$. According to the LST relation, if the reciprocal of the static dielectric constant has a temperature dependence $1/\epsilon(0) \propto (T - T_0)$, then the square of the optical mode frequency will have a similar temperature dependence: $\omega_T^2 \propto (T - T_0)$, if ω_L is independent of temperature. The result for ω_T^2 is very well confirmed by Fig. 17. Measurements of ω_T versus T for another ferroelectric crystal, SbSI, are shown in Fig. 18.

[15a] G. Burns and E. Burstein, Ferroelectrics **7**, 297 (1974).

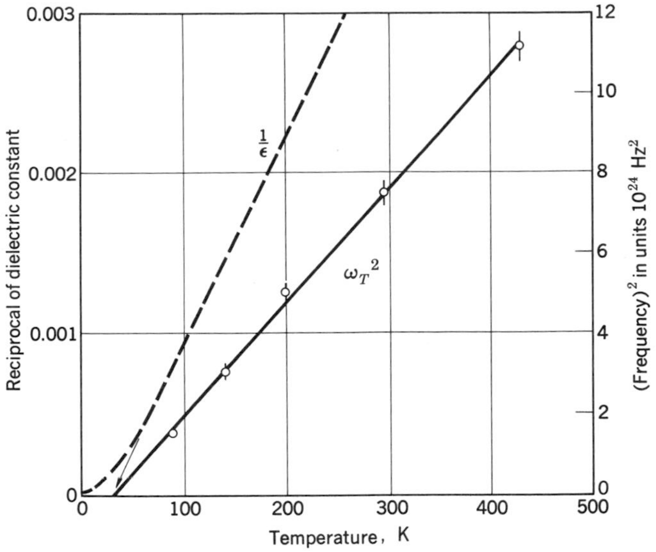

Figure 17 Plot of the square of the frequency of the zero wavevector transverse optic mode against temperature, for $SrTiO_3$, as observed in neutron diffraction experiments by Cowley. The broken line is the reciprocal of the dielectric constant from the measurements of Mitsui and Westphal.

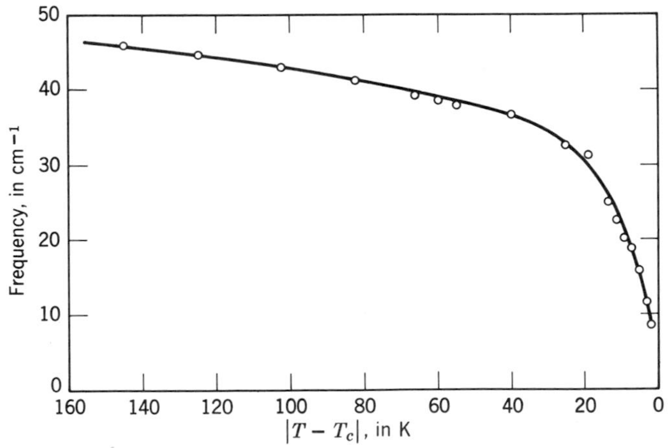

Figure 18 Decrease of a transverse phonon frequency as the Curie temperature is approached from below, in the ferroelectric crystal antimony sulphoiodide (SbSI). [After Raman scattering experiments by C. H. Perry and D. K. Agrawal, Solid State Comm. **8**, 225 (1970).]

Antiferroelectricity

A ferroelectric displacement is not the only type of instability that may develop in a dielectric crystal. In the perovskite structure other deformations occur. These deformations, even if they do not give a spontaneous polarization, may be accompanied by changes in the dielectric constant. One type of deformation is called **antiferroelectric** and has neighboring lines of ions displaced in opposite senses, as in Fig. 19. The perovskite structure appears to be susceptible to many types of deformation, often with little difference in energy between them. The phase diagrams of mixed perovskite systems, such as the $PbZrO_3$–$PbTiO_3$ system, show transitions between para-, ferro-, and antiferroelectric states.

Ordered antiferroelectric arrangements of permanent electric dipole moments occur at low temperatures in ammonium salts and in hydrogen halides. Several crystals believed to have an ordered nonpolar state are listed in Table 4.

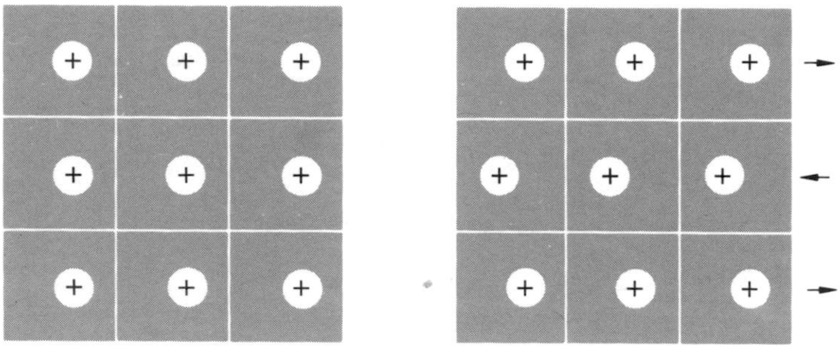

(a) (b)

Figure 19 Comparison of (a) ferroelectric and (b) antiferroelectric deformations.

TABLE 4 Antiferroelectric Crystals

Crystal	Transition Temperature to Antiferroelectric State, in K
WO_3	1010
$NaNbO_3$	793, 911
$PbZrO_3$	506
$PbHfO_3$	488
$NH_4H_2PO_4$	148
$ND_4D_2PO_4$	242
$NH_4H_2AsO_4$	216
$ND_4D_2AsO_4$	304
$(NH_4)_2H_3IO_6$	254

From a compilation by Walter J. Merz.

Ferroelectric Domains

Consider a ferroelectric crystal (such as barium titanate in the tetragonal phase) in which the spontaneous polarization may be either up or down the c axis of the crystal. A ferroelectric crystal generally consists of regions called **domains** within each of which the polarization is in the same direction, but in adjacent domains the polarization is in different directions. In Fig. 20 the polarization is in opposite directions. The net polarization depends on the difference in the volumes of the upward- and downward-directed domains. The crystal as a whole will appear to be unpolarized, as measured by the charge on electrodes covering the ends, when the volumes of domains in opposite senses are equal. The total dipole moment of the crystal may be changed by the movement of the walls between domains or by the nucleation of new domains.

Figure 21 is a series of photomicrographs of a single crystal of barium titanate in an electric field normal to the plane of the photographs and parallel to the tetragonal axis. The closed curves are boundaries between domains polarized into and out of the plane of the photographs. The domain boundaries change size and shape when the intensity of the electric field is altered. The motion of domain walls in ferroelectrics is not simple: it is known[16] that in an electric field a 180° wall[17] in $BaTiO_3$ appears to move by the repeated nucleation[18] of steps by thermal fluctuations along the parent wall.

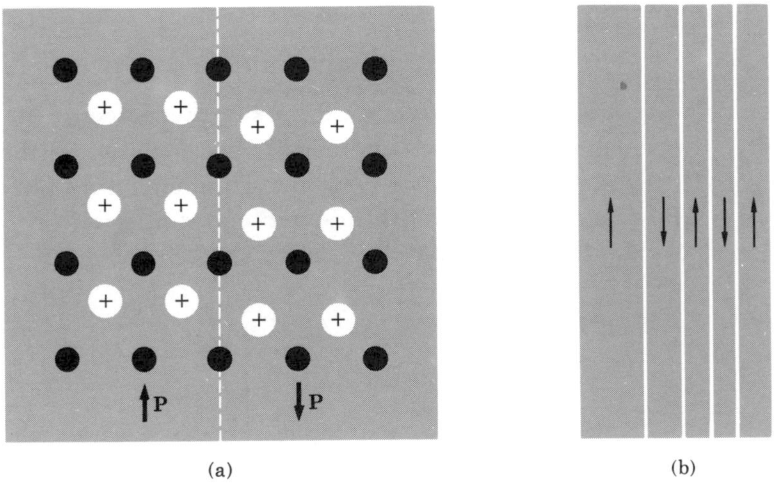

Figure 20 (a) Schematic drawing of atomic displacements on either side of a boundary between domains polarized in opposite directions in a ferroelectric crystal; (b) view of a domain structure, showing 180° boundaries between domains polarized in opposite directions.

[16]W. J. Merz, Phys. Rev. **95,** 690 (1954); R. Landauer, J. Appl. Phys. **28,** 227 (1957); R. C. Miller and A. Savage, Phys. Rev. **115,** 1176 (1959).
[17]A "180° wall" is the boundary between regions having opposite polarization directions.
[18]R. C. Miller and G. Weinreich, Phys. Rev. **117,** 1460 (1960).

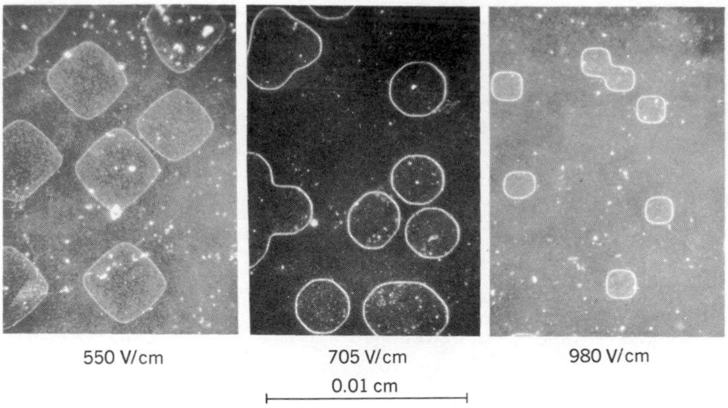

Figure 21 Ferroelectric domains on the face of a single crystal of barium titanate. The face is normal to the tetragonal or c axis. The net polarization of the crystal as judged by domain volumes is increased markedly as the electric field intensity parallel to the axis is increased from 550 volts/cm to 980 V/cm. The domain boundaries are made visible by etching the crystal in a weak acid solution. (Courtesy of R. C. Miller.)

Piezoelectricity

All crystals in a ferroelectric state are also piezoelectric: a stress Z applied to the crystal will change the electric polarization (Fig. 22). Similarly, an electric field E applied to the crystal will cause the crystal to become strained. In schematic one-dimensional notation, the piezoelectric equations are

$$\text{(CGS)} \qquad P = Zd + E\chi \ ; \qquad e = Zs + Ed \ , \tag{45}$$

where P is the polarization, Z the stress, d the **piezoelectric strain constant,** E the electric field, χ the dielectric susceptibility, e the elastic strain, and s the elastic compliance constant. To obtain (45) in SI, replace χ by $\epsilon_0\chi$. These relations exhibit the development of polarization by an applied stress and the development of elastic strain by an applied electric field.

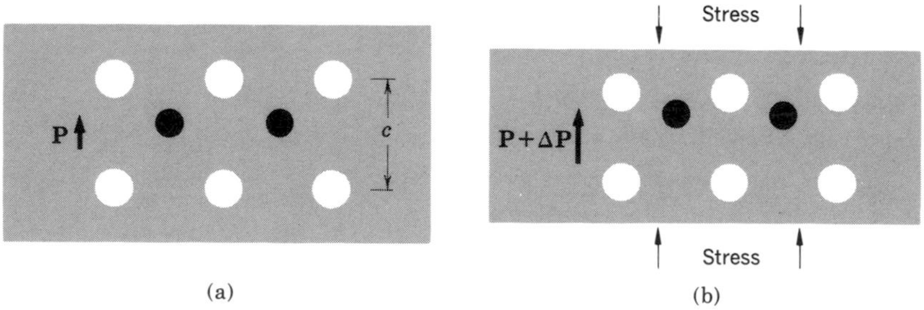

Figure 22 (a) Unstressed ferroelectric crystal and (b) stressed ferroelectric crystal. The stress changes the polarization by $\Delta\mathbf{P}$, the induced piezoelectric polarization.

A crystal may be piezoelectric without being ferroelectric: a schematic example of such a structure is given in Fig. 23. Quartz is piezoelectric, but not ferroelectric; barium titanate is both. For order of magnitude, in quartz $d \approx 10^{-7}$ cm/statvolt and in barium titanate $d \approx 10^{-5}$ cm/statvolt. The general definition[19] of the piezoelectric strain constants is

$$d_{ik} = (\partial e_k/\partial E_i)_Z \ , \tag{46}$$

where $i \equiv x, y, z$ and $k \equiv xx, yy, zz, yz, zx, xy$. To convert to cm/stat V from values of d_{ik} given in m/V, multiply by 3×10^4.

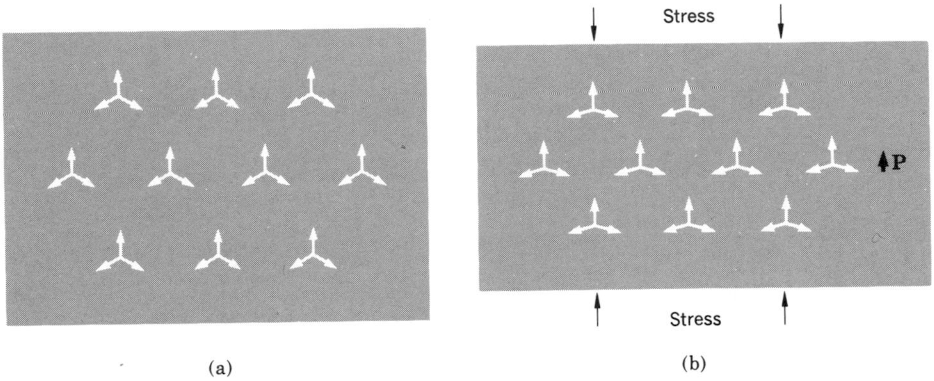

Figure 23 (a) The unstressed crystal has a threefold symmetry axis. The arrows represent dipole moments, each set of three arrows represents a planar group of ions denoted by $A_3^+B^{3-}$, with a B^{3-} ion at each vertex. The sum of the three dipole moments at each vertex is zero. (b) The crystal when stressed develops a polarization in the direction indicated. The sum of the dipole moments about each vertex is no longer zero.

Ferroelasticity

A crystal is ferroelastic if it has two or more stable orientation states in the absence of mechanical stress and if it can be reproducibly transformed from one to another of these states by application of mechanical stress.[20] Realizations are few: gadolinium molybdate $Gd_2(MoO_4)_3$ has a combined ferroelectric-ferroelastic transition; lead phosphate $Pb_3(PO_4)_2$ has a pure ferroelastic transition; several crystals have combined ferromagnetic-ferroelastic transitions.

[19]The piezoelectric strain constants form a third-rank tensor. A concise treatment of the effect of symmetries on the tensor properties of crystals is given in the *A.I.P. Handbook*, 3rd ed., pp. 9–10 to 9–15.
[20]K. Aizu, Phys. Rev. **B2**, 754 (1970). A table of displacive phase transitions is given by G. A. Samara, T. Sakudo, and K. Yoshimitsu, Phys. Rev. Letters **35**, 1767 (1975).

SUMMARY (In CGS units)

1. The electric field averaged over the volume of the specimen defines the macroscopic electric field $\mathbf{E}$ of the Maxwell equations.

2. The electric field that acts at the site $\mathbf{r}_j$ of an atom j is the local electric field, $\mathbf{E}_{loc}$. It is a sum over all charges, grouped in terms as $\mathbf{E}_{loc}(\mathbf{r}_j) = \mathbf{E}_0 + \mathbf{E}_1 + \mathbf{E}_2 + \mathbf{E}_3(\mathbf{r}_j)$, where only $\mathbf{E}_3$ varies rapidly within a cell. Here:

   $\mathbf{E}_0$ = external electric field;
   $\mathbf{E}_1$ = depolarization field associated with the boundary of the specimen;
   $\mathbf{E}_2$ = field from polarization outside a sphere centered about $\mathbf{r}_j$;
   $\mathbf{E}_3(\mathbf{r}_j)$ = field at $\mathbf{r}_j$ due to all atoms inside the sphere.

3. The macroscopic field $\mathbf{E}$ of the Maxwell equations is equal to $\mathbf{E}_0 + \mathbf{E}_1$, which, in general, is not equal to $\mathbf{E}_{loc}(\mathbf{r}_j)$.

4. The depolarization field in an ellipsoid is $E_{1\mu} = -N_{\mu\nu}P_\nu$, where $N_{\mu\nu}$ is the depolarization tensor; the polarization $\mathbf{P}$ is the dipole moment per unit volume. In a sphere $N = 4\pi/3$.

5. The Lorentz field is $\mathbf{E}_2 = 4\pi\mathbf{P}/3$.

6. The polarizability α of an atom is defined in terms of the local electric field as $\mathbf{p} = \alpha\mathbf{E}_{loc}$.

7. The dielectric susceptibility χ and dielectric constant ϵ are defined in terms of the macroscopic electric field $\mathbf{E}$ as $\mathbf{D} = \mathbf{E} + 4\pi\mathbf{P} = \epsilon\mathbf{E} = (1 + 4\pi\chi)\mathbf{E}$, or $\chi = P/E$. In SI, we have $\chi = P/\epsilon_0 E$.

8. An atom at a site with cubic symmetry has $\mathbf{E}_{loc} = \mathbf{E} + (4\pi/3)\mathbf{P}$ and satisfies the Clausius-Mossotti relation (24).

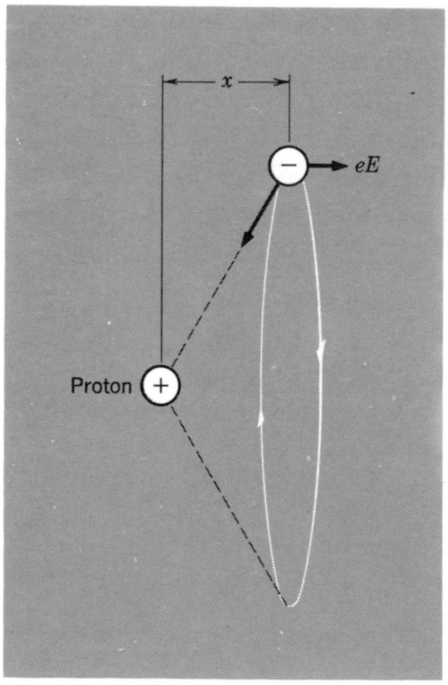

Figure 24 An electron in a circular orbit of radius a_H is displaced a distance x on application of an electric field E in the $-x$ direction. The force on the electron due to the nucleus is e^2/a_H^2 in CGS or $e^2/4\pi\epsilon_0 a_H^2$ in SI.

PROBLEMS

1. ***Polarizability of atomic hydrogen.*** Consider a semiclassical model of the ground state of the hydrogen atom in an electric field normal to the plane of the orbit (Fig. 24), and show that for this model $\alpha = a_H{}^3$, where a_H is the radius of the unperturbed orbit. *Note:* If the applied field is in the x direction, then the x component of the field of the nucleus at the displaced position of the electron orbit must be equal to the applied field. The correct quantum-mechanical result is larger than this by the factor $\frac{9}{2}$. (We are speaking of α_0 in the expansion $\alpha = \alpha_0 + \alpha_1 E + \cdots$.) We assume $x < a_H$.

2. ***Polarizability of conducting sphere.*** Show that the polarizability of a conducting metallic sphere of radius a is $\alpha = a^3$. This result is most easily obtained by noting that $E = 0$ inside the sphere and then using the depolarization factor $4\pi/3$ for a sphere (Fig. 25). The result gives values of α of the order

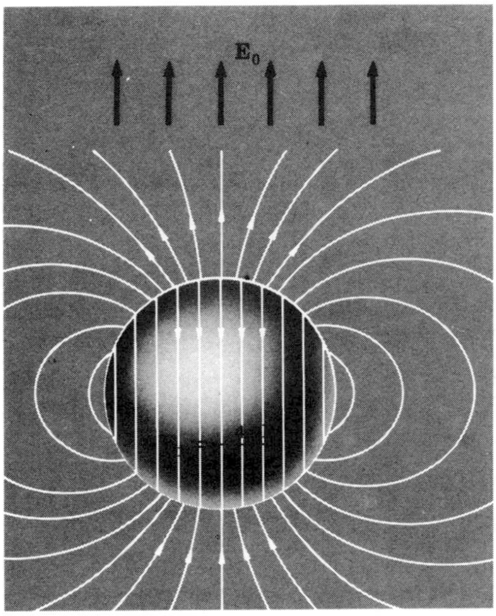

Figure 25 The total field inside a conducting sphere is zero. If a field $\mathbf{E}_0$ is applied externally, then the field $\mathbf{E}_1$ due to surface charges on the sphere must just cancel $\mathbf{E}_0$, so that $\mathbf{E}_0 + \mathbf{E}_1 = 0$ within the sphere. But $\mathbf{E}_1$ can be simulated by the depolarization field $-4\pi\mathbf{P}/3$ of a uniformly polarized sphere of polarization $\mathbf{P}$. Relate $\mathbf{P}$ to $\mathbf{E}_0$ and calculate the dipole moment $\mathbf{p}$ of the sphere. In SI the depolarization field is $-\mathbf{P}/3\epsilon_0$.

of magnitude of the observed polarizabilities of atoms. A lattice of N conducting spheres per unit volume has dielectric constant $\epsilon = 1 + 4\pi Na^3$; for $Na^3 \ll 1$. The suggested proportionality of α to the cube of the ionic radius is satisfied quite well for alkali and halogen ions. To do the problem in SI, use $\frac{1}{3}$ as the depolarization factor; the result is the same.

3. ***Effect of air gap.*** Discuss the effect of an air gap (Fig. 26) between condenser plates and dielectric on the measurement of high dielectric constants. What is

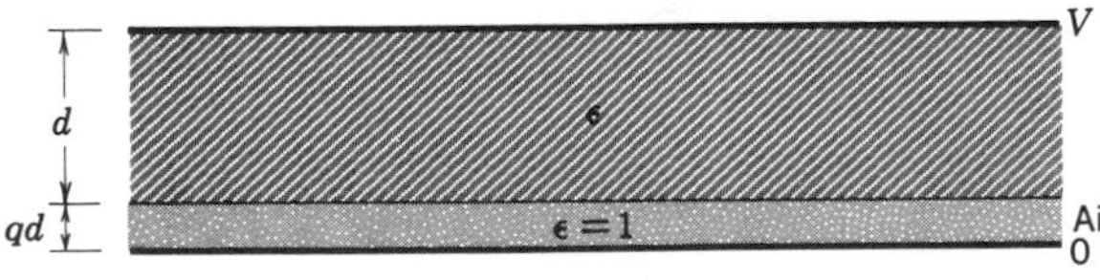

Figure 26 An air gap of thickness qd is in series in a capacitor with a dielectric slab of thickness d.

the highest apparent dielectric constant possible if the air gap thickness is 10^{-3} of the total thickness? The presence of air gaps can seriously distort the measurement of high dielectric constants.

4. ***Interfacial polarization.*** Show that a parallel-plate capacitor made up of two parallel layers of material—one layer with dielectric constant ϵ, zero conductivity, and thickness d, and the other layer with $\epsilon = 0$ for convenience, finite conductivity σ, and thickness qd—behaves as if the space between the condenser plates were filled with a homogeneous dielectric with dielectric constant

$$\text{(CGS)} \qquad \epsilon_{\text{eff}} = \frac{\epsilon(1+q)}{1-(i\epsilon\omega q/4\pi\sigma)} ,$$

where ω is the angular frequency. Values of ϵ_{eff} as high as 10^4 or 10^5 caused largely by this Maxwell-Wagner mechanism are sometimes found, but the high values are always accompanied by large losses.

5. ***Polarization of sphere.*** A sphere of dielectric constant ϵ is placed in a uniform external electric field E_0. (a) What is the volume average electric field E in the sphere? (b) Show that the polarization in the sphere is $P = \chi E_0/[1 + (4\pi\chi/3)]$, where $\chi = (\epsilon - 1)/4\pi$. Hint: You do not need to calculate E_{loc} in this problem; in fact it is confusing to do so, because ϵ and χ are defined so that $P = \chi E$. We require E_0 to be unchanged by insertion of the sphere. We can produce a fixed E_0 by placing positive charges on one thin plate of an insulator and negative charges on an opposite plate. If the plates are always far from the sphere, the field of the plates will remain unchanged when the sphere is inserted between them. The results above are in CGS.

6. ***Ferroelectric criterion for atoms.*** Consider a system of two neutral atoms separated by a fixed distance a, each atom having a polarizability α. Find the relation between a and α for such a system to be ferroelectric. Hint: The dipolar field is strongest along the axis of the dipole.

7. ***Saturation polarization at Curie point.*** In a first-order transition the equilibrium condition (42) with T set equal T_c gives one equation for the polarization $P_s(T_c)$. A further condition at the Curie point is that $\hat{F}(P_s, T_c) = \hat{F}(0, T_c)$. (a) Combining these two conditions, show that $P_s^2(T_c) = 3|g_4|/4g_6$. (b) Using this result, show that $T_c = T_0 + 3g_4^2/16\gamma g_6$.

8. ***Dielectric constant below transition temperature.*** In terms of the parameters in the Landau free energy expansion, show that for a second-order phase transition the dielectric constant below the transition temperature is

$$\epsilon = 1 + 4\pi\Delta P/E = 1 + 2\pi/\gamma(T_c - T) .$$

This result may be compared with (43) above the transition.

9. ***Soft modes and lattice transformations.*** Sketch a monatomic linear lattice of lattice constant a. (a) Add to each of six atoms a vector to indicate the direction of the displacement at a given time caused by a longitudinal phonon with wavevector at the zone boundary. (b) Sketch the crystal structure that results if this zone boundary phonon becomes unstable ($\omega \rightarrow 0$) as the crystal is cooled through T_c. (c) Sketch on one graph the essential aspects of the longitudinal phonon dispersion relation for the monatomic lattice at T well above T_c and at $T = T_c$. Add to the graph the same information for phonons in the new structure at T well below T_c.

10. ***Ferroelectric linear array.*** Consider a line of atoms of polarizability α and separation a. Show that the array can polarize spontaneously if $\alpha \geq a^3/4 \, \Sigma \, n^{-3}$, where the sum is over all positive integers and is given in tables as 1.202. . . .

References

Dielectrics

H. Fröhlich, *Theory of dielectrics: dielectric constant and dielectric loss,* Oxford, 1958, 2nd ed.
J. H. Van Vleck, *Theory of electric and magnetic susceptibilities,* Oxford, 1932.
W. F. Brown, Jr., "Dielectrics," *Encyclo. of physics* **17,** 1–263 (1956).
V. Daniel, *Dielectric relaxation,* Academic Press, 1967.
I. S. Zheludev, *Physics of crystalline dielectrics,* Plenum, 1971.
Digest of literature on dielectrics, National Academy of Sciences. A series.
J. B. Birks, ed., *Progress in dielectrics,* Chemical Rubber. A series.

Ferroelectrics

R. Blinc and B. Zeks, *Soft modes in ferroelectrics and antiferroelectrics,* North-Holland, 1974.
J. C. Burfoot, *Ferroelectrics, an introduction to the physical principles,* Van Nostrand, 1967.
E. Fatuzzo and W. J. Merz, *Ferroelectricity,* North-Holland, 1967.
F. Jona and G. Shirane, *Ferroelectric crystals,* Pergamon, 1962.
W. Cochran, "Crystal stability and the theory of ferroelectricity," Advances in Physics **9,** 387 (1960); **10,** 401–420 (1961).
W. Känzig, "Ferroelectrics and antiferroelectrics," *Solid state physics* **4,** 1–197 (1957).
P. W. Forsbergh, "Piezoelectricity, electrostriction, and ferroelectricity," *Encyclo. of physics* **17,** 264–392 (1956).
J. F. Scott, "Soft mode spectroscopy," Rev. Mod. Phys. **48,** 83 (1975).
Proceedings of the Second International Conference on Ferroelectricity, Kyoto, 1969. J. Phys. Soc. Japan **28,** Suppl. (1970).
Ferroelectrics. A journal.

14 DIAMAGNETISM AND PARAMAGNETISM

NOTATION: In the problems treated in this chapter the magnetic field B is always closely equal to the applied field B_a, so that we write B for B_a in most instances.

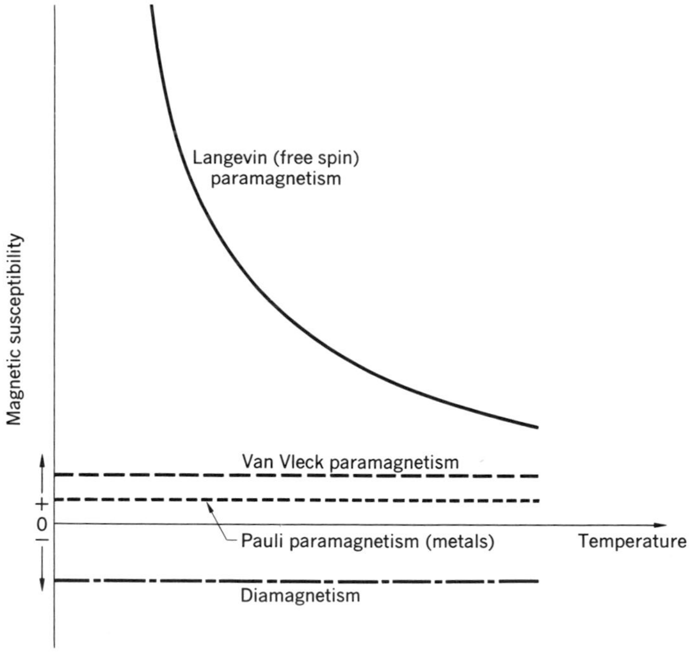

Figure 1 Characteristic magnetic susceptibilities of diamagnetic and paramagnetic substances.

14 DIAMAGNETISM AND PARAMAGNETISM

Magnetism is inseparable from quantum mechanics, for a strictly classical system in thermal equilibrium can display no magnetic moment, even in a magnetic field. The magnetic moment of a free atom has three principal sources: the spin with which electrons are endowed; their orbital angular momentum about the nucleus; and the change in the orbital moment induced by an applied magnetic field. The first two effects give paramagnetic contributions to the magnetization, and the third gives a diamagnetic contribution. In the ground $1s$ state of the hydrogen atom the orbital moment is zero, and the magnetic moment is that of the electron spin along with a small induced diamagnetic moment. In the $1s^2$ state of helium the spin and orbital moments are both zero, and there is only an induced moment. Atoms with filled electron shells have zero spin and zero orbital moment: these moments are associated with unfilled shells.

The magnetization M is defined as the magnetic moment per unit volume. The **magnetic susceptibility** per unit volume is defined as

$$\text{(CGS)} \quad \chi = \frac{M}{B} \; ; \qquad \text{(SI)} \quad \chi = \frac{\mu_0 M}{B} \; , \tag{1}$$

where B is the macroscopic magnetic field intensity. In both systems of units χ is dimensionless. We shall sometimes for convenience refer to M/B as the susceptibility without specifying the system of units. Quite frequently a susceptibility is defined referred to unit mass or to a mole of the substance. The molar susceptibility is written as χ_M; the magnetic moment per gram is sometimes written as σ. Substances with a negative magnetic susceptibility are called **diamagnetic.** Substances with a positive susceptibility are called **paramagnetic,** as in Fig. 1. Ordered arrays of magnetic moments are discussed in Chapter 15; the arrays may be ferromagnetic, ferrimagnetic, antiferromagnetic, helical, or more complex in form. Nuclear magnetic moments give rise to **nuclear paramagnetism.** Magnetic moments of nuclei are of the order of 10^{-3} times smaller than the magnetic moment of the electron.

LANGEVIN DIAMAGNETISM EQUATION

Diamagnetism is associated with the tendency of electrical charges partially to shield the interior of a body from an applied magnetic field. In electromagnetism we are familiar with Lenz's law: when the flux through an electrical circuit is changed, an induced current is set up in such a direction as to oppose the flux change. In a superconductor or in an electron orbit within an atom, the induced current persists as long as the field is present. The magnetic field of the induced current is opposite to the applied field, and the magnetic moment associated with the current is a diamagnetic moment. Even in a normal metal there is a diamagnetic contribution from the conduction electrons, and this diamagnetism is not destroyed by collisions of the electrons.

The usual treatment of the diamagnetism of atoms and ions employs the Larmor theorem: in a magnetic field the motion of the electrons around a central nucleus is, to the first order in B, the same as a possible motion in the absence of B except for the superposition of a precession of the electrons with angular frequency

$$\text{(CGS)} \quad \omega = eB/2mc \ ; \qquad\qquad \text{(SI)} \quad \omega = eB/2m \ . \tag{2}$$

If the field is applied slowly, the motion in the rotating reference system will be the same as the original motion in the rest system before the application of the field. If the average electron current around the nucleus is zero initially, the application of the magnetic field will cause a finite current around the nucleus. The current is equivalent to a magnetic moment opposite to the applied field. It is assumed that the Larmor frequency (2) is much lower than the frequency of the original motion in the central field. This condition is not satisfied in free carrier cyclotron resonance, and the cyclotron frequency is twice the frequency (2).

The Larmor precession of Z electrons is equivalent to an electric current

$$\text{(SI)} \quad I = \text{(charge)(revolutions per unit time)} = (-Ze)\left(\frac{1}{2\pi} \cdot \frac{eB}{2m}\right) . \tag{3}$$

The magnetic moment μ of a current loop is given by the product (current) $\times$ (area of the loop). The area of a loop of radius ρ is $\pi\rho^2$. We have

$$\text{(SI)} \quad \mu = -\frac{Ze^2B}{4m}\langle\rho^2\rangle \ ; \qquad\qquad \text{(CGS)} \quad \mu = -\frac{Ze^2B}{4mc^2}\langle\rho^2\rangle \ . \tag{4}$$

Here $\langle\rho^2\rangle = \langle x^2\rangle + \langle y^2\rangle$ is the mean square of the perpendicular distance of the electron from the field axis through the nucleus. The mean square distance of the electrons from the nucleus is $\langle r^2\rangle = \langle x^2\rangle + \langle y^2\rangle + \langle z^2\rangle$. For a spherically symmetrical distribution of charge we have $\langle x^2 \rangle = \langle y^2 \rangle = \langle z^2 \rangle$, so that $\langle r^2 \rangle = \frac{3}{2}\langle \rho^2 \rangle$. From (4) the diamagnetic susceptibility per unit volume is, if N is the number of atoms per unit volume,

$$\text{(CGS)} \qquad \chi = \frac{N\mu}{B} = -\frac{NZe^2}{6mc^2}\langle r^2 \rangle \; ; \qquad (5)$$

$$\text{(SI)} \qquad \chi = \frac{\mu_0 N\mu}{B} = -\frac{\mu_0 NZe^2}{6m}\langle r^2 \rangle \; .$$

This is the classical Langevin result.

The problem of calculating the diamagnetic susceptibility of an isolated atom is reduced to the calculation of $\langle r^2 \rangle$ for the electron distribution within the atom. The distribution can be calculated by quantum mechanics. Experimental values for neutral atoms are most easily obtained for the inert gases. Typical experimental values of the molar susceptibilities are the following:

	He	Ne	Ar	Kr	Xe
χ_M in CGS in 10^{-6} cm³/mole:	−1.9	−7.2	−19.4	−28.0	−43.0

In dielectric solids the diamagnetic contribution of the ion cores is described roughly by the Langevin result. The contribution of conduction electrons is more complicated, as is evident from the de Haas–van Alphen effect discussed in Chapter 9.

PARAMAGNETISM

Electronic paramagnetism (positive contribution to χ) is found in:

1. Atoms, molecules, and lattice defects possessing an odd number of electrons, as here the total spin of the system cannot be zero. Examples: free sodium atoms; gaseous nitric oxide (NO); organic free radicals such as triphenylmethyl, $C(C_6H_5)_3$; *F* centers in alkali halides.
2. Free atoms and ions with a partly filled inner shell: transition elements; ions isoelectronic with transition elements; rare earth and actinide elements. Examples: Mn^{2+}, Gd^{3+}, U^{4+}. Paramagnetism is exhibited by many of these ions when incorporated into solids, but not invariably.
3. A few compounds with an even number of electrons, including molecular oxygen and organic biradicals.
4. Metals.

We shall consider classes (2) and (4).

QUANTUM THEORY OF PARAMAGNETISM

The magnetic moment of an atom or ion in free space is given by

$$\boxed{\boldsymbol{\mu} = \gamma\hbar\mathbf{J} = -g\mu_B\mathbf{J}}\ , \tag{6}$$

where the total angular momentum $\hbar\mathbf{J}$ is the sum of the orbital $\hbar\mathbf{L}$ and spin $\hbar\mathbf{S}$ angular momenta. The constant γ is the ratio of the magnetic moment to the angular momentum; γ is called the **gyromagnetic ratio** or **magnetogyric ratio.** For electronic systems a quantity g called the g factor or the spectroscopic splitting factor is defined by

$$g\mu_B \equiv -\gamma\hbar\ . \tag{7}$$

For an electron spin $g = 2.0023$, usually taken as 2.00. For a free atom the g factor is given by the Landé equation

$$g = 1 + \frac{J(J+1) + S(S+1) - L(L+1)}{2J(J+1)}\ . \tag{8}$$

The **Bohr magneton** μ_B is defined as $e\hbar/2mc$ in CGS and $e\hbar/2m$ in SI. It is closely equal to the spin magnetic moment of a free electron.

The energy levels of the system in a magnetic field are

$$U = -\boldsymbol{\mu}\cdot\mathbf{B} = m_J g\mu_B B\ , \tag{9}$$

where m_J is the azimuthal quantum number and has the values J, $J-1, \ldots, -J$. For a single spin with no orbital moment we have $m_J = \pm\frac{1}{2}$ and $g = 2$, whence $U = \pm\mu_B B$. This splitting is shown in Fig. 2.

If a system has only two levels the equilibrium populations are, with $\tau \equiv k_B T$,

$$\frac{N_1}{N} = \frac{\exp(\mu B/\tau)}{\exp(\mu B/\tau) + \exp(-\mu B/\tau)}\ ; \tag{10}$$

$$\frac{N_2}{N} = \frac{\exp(-\mu B/\tau)}{\exp(\mu B/\tau) + \exp(-\mu B/\tau)}\ ; \tag{11}$$

here N_1, N_2 are the populations of the lower and upper levels, and $N = N_1 + N_2$ is the total number of atoms. The fractional populations are plotted in Fig. 3.

The projection of the magnetic moment of the upper state along the field direction is $-\mu$ and of the lower state is μ. The resultant magnetization for N atoms per unit volume is, with $x \equiv \mu B/k_B T$,

$$M = (N_1 - N_2)\mu = N\mu \cdot \frac{e^x - e^{-x}}{e^x + e^{-x}} = N\mu \tanh x\ . \tag{12}$$

For $x \ll 1$, $\tanh x \cong x$, and we have

$$M \cong N\mu(\mu B/k_B T)\ . \tag{13}$$

In a magnetic field an atom with angular momentum quantum number J has $2J + 1$ equally spaced energy levels. The magnetization (Fig. 4) is

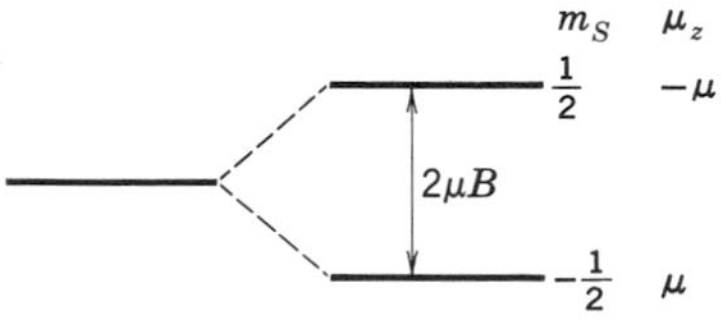

Figure 2 Energy level splitting for one electron in a magnetic field B directed along the positive z axis. For an electron the magnetic moment μ is opposite in sign to the spin S, so that $\mu = -g\mu_B S$. In the low energy state the magnetic moment is parallel to the magnetic field.

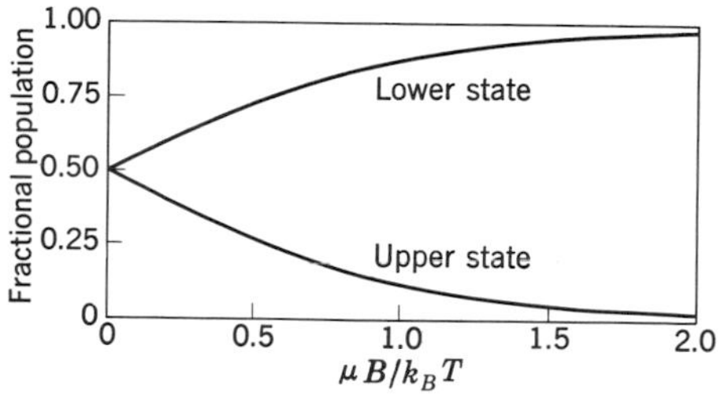

Figure 3 Fractional populations of a two-level system in thermal equilibrium at temperature T in a magnetic field B. The magnetic moment is proportional to the difference between the two curves.

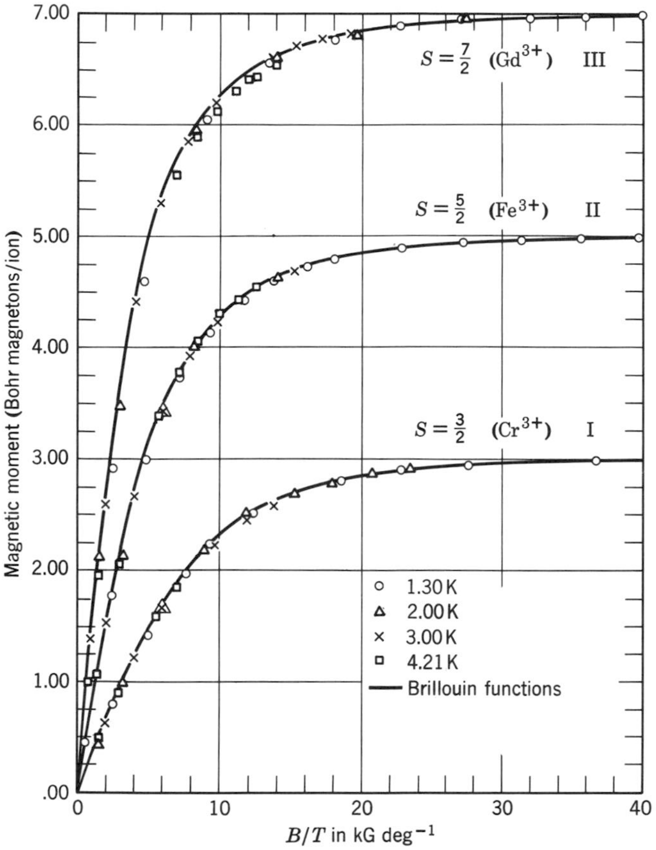

Figure 4 Plot of magnetic moment *versus* B/T for spherical samples of (I) potassium chromium alum, (II) ferric ammonium alum, and (III) gadolinium sulfate octahydrate. Over 99.5% magnetic saturation is achieved at 1.3 K and about 50,000 gauss. [After W. E. Henry, Phys. Rev. **88**, 559 (1952).]

given by

$$M = NgJ\mu_B B_J(x) \ , \qquad (x \equiv gJ\mu_B B/k_B T) \ , \tag{14}$$

where the **Brillouin function** B_J is defined by

$$B_J(x) = \frac{2J+1}{2J} \operatorname{ctnh}\left(\frac{(2J+1)x}{2J}\right) - \frac{1}{2J} \operatorname{ctnh}\left(\frac{x}{2J}\right) \ . \tag{15}$$

Equation (12) is a special case of (15) for $J = \frac{1}{2}$. For $x \ll 1$, we have

$$\operatorname{ctnh} x = \frac{1}{x} + \frac{x}{3} - \frac{x^3}{45} + \cdots \ , \tag{16}$$

and the susceptibility is

$$\frac{M}{B} \cong \frac{NJ(J+1)g^2\mu_B{}^2}{3k_BT} = \frac{Np^2\mu_B{}^2}{3k_BT} = \frac{C}{T} \ . \tag{17}$$

Here p is the **effective number of Bohr magnetons,** defined as

$$p \equiv g[J(J+1)]^{\frac{1}{2}} \ . \tag{18}$$

The constant C is known as the **Curie constant.** The form (17) is known as the **Curie law.** Results for the paramagnetic ions in a gadolinium salt are shown in Fig. 5.

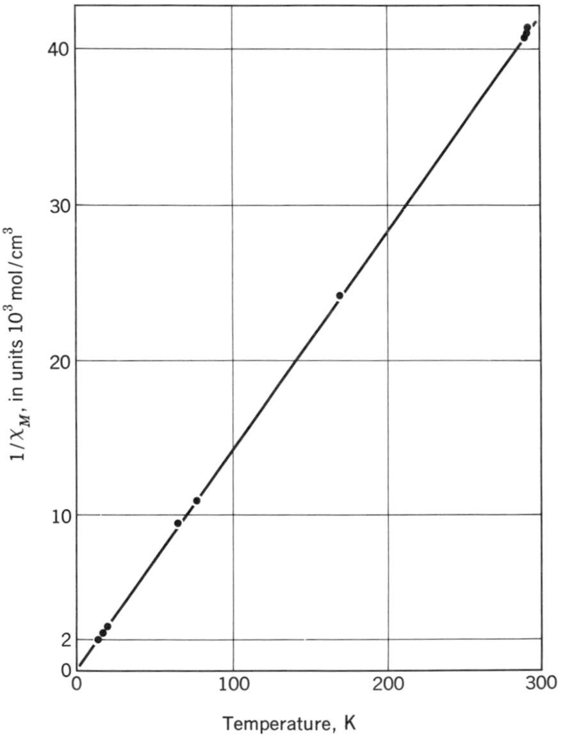

Figure 5 Plot of $1/\chi$ *vs* T for a gadolinium salt, $Gd(C_2H_5SO_4)_3 \cdot 9H_2O$. The straight line is the Curie law. (After L. C. Jackson and H. Kamerlingh Onnes, Leiden Communications 168a.)

Rare Earth Ions

The ions of the rare-earth elements have closely similar chemical properties, and their chemical separation in tolerably pure form was accomplished only long after their discovery. Their magnetic properties are fascinating: the ions exhibit a systematic variety and intelligible complexity. The chemical properties of the trivalent ions are similar because the outermost electron shells are identically in the $5s^2\ 5p^6$ configuration, like neutral xenon. In lanthanum, just before the rare earth group begins, the $4f$ shell is empty; at cerium there is one $4f$ electron, and the number of $4f$ electrons increases steadily through the group until we have $4f^{13}$ at ytterbium and the filled shell $4f^{14}$ at lutecium. The radii of the trivalent ions contract fairly smoothly as we go through the group from 1.11 Å at cerium to 0.94 Å at ytterbium. This is the famous "lanthanide contraction." What distinguishes the magnetic behavior of one ion species from another is the number of $4f$ electrons compacted in the inner shell with a radius of perhaps 0.3 Å. Even in the metals the $4f$ core retains its integrity and its atomic properties: no other group of elements in the periodic table is as interesting.

The preceding discussion of paramagnetism applies to atoms that have a $(2J + 1)$-fold degenerate ground state, the degeneracy being lifted by a magnetic field. The influence of all higher energy states of the system is neglected. These assumptions appear to be satisfied by a number of rare-earth ions, Table 1. The calculated magneton numbers are obtained with g values from the Landé result (8) and the ground-state level assignment predicted by the Hund theory of spectral terms. The discrepancy between the experimental magneton numbers and those calculated on these assumptions is quite marked for Eu^{3+} and Sm^{3+} ions. For these ions it is necessary to consider the influence of the high states of the L–S multiplet, as the intervals between successive states of the multiplet are not large compared to k_BT at room temperature. A multiplet is the set of levels of different J values arising out of a given L and S. The levels of a multiplet are split by the spin-orbit interaction.

TABLE 1 Effective Magneton Numbers *p* for Trivalent Lanthanide Group Ions

(Near room temperature)

Ion	Configuration	Basic Level	$p(\text{calc}) = g[J(J+1)]^{\frac{1}{2}}$	$p(\text{exp})$, Approximate
Ce^{3+}	$4f^1 5s^2 p^6$	$^2F_{5/2}$	2.54	2.4
Pr^{3+}	$4f^2 5s^2 p^6$	$^3H_4$	3.58	3.5
Nd^{3+}	$4f^3 5s^2 p^6$	$^4I_{9/2}$	3.62	3.5
Pm^{3+}	$4f^4 5s^2 p^6$	$^5I_4$	2.68	—
Sm^{3+}	$4f^5 5s^2 p^6$	$^6H_{5/2}$	0.84	1.5
Eu^{3+}	$4f^6 5s^2 p^6$	$^7F_0$	0	3.4
Gd^{3+}	$4f^7 5s^2 p^6$	$^8S_{7/2}$	7.94	8.0
Tb^{3+}	$4f^8 5s^2 p^6$	$^7F_6$	9.72	9.5
Dy^{3+}	$4f^9 5s^2 p^6$	$^6H_{15/2}$	10.63	10.6
Ho^{3+}	$4f^{10} 5s^2 p^6$	$^5I_8$	10.60	10.4
Er^{3+}	$4f^{11} 5s^2 p^6$	$^4I_{15/2}$	9.59	9.5
Tm^{3+}	$4f^{12} 5s^2 p^6$	$^3H_6$	7.57	7.3
Yb^{3+}	$4f^{13} 5s^2 p^6$	$^2F_{7/2}$	4.54	4.5

Hund Rules

The Hund rules as applied to electrons in a given shell of an atom affirm that electrons will occupy orbitals in such a way that the ground state is characterized by the following:

1. The maximum value of the total spin S allowed by the exclusion principle;
2. The maximum value of the orbital angular momentum L consistent with this value of S;
3. The value of the total angular momentum J is equal to $|L - S|$ when the shell is less than half full and to $L + S$ when the shell is more than half full. When the shell is just half full, the application of the first rule gives $L = 0$, so that $J = S$.

The first Hund rule has its origin in the exclusion principle and the coulomb repulsion between electrons. The exclusion principle prevents two electrons of the same spin from being at the same place at the same time. Thus electrons of the same spin are kept apart, further apart than electrons of opposite spin. Because of the coulomb interaction the energy of electrons of the same spin is lower—the average potential energy is less positive for parallel spin than for antiparallel spin. A good example is the ion Mn^{2+}. This ion has five electrons in the $3d$ shell, which is therefore half-filled. The spins can all be parallel if each electron enters a different orbital, and there are exactly five different orbitals available, characterized by the orbital quantum numbers $m_L = 2, 1, 0, -1, -2$. These will be occupied by one electron each. We expect $S = \frac{5}{2}$, and because $\Sigma m_L = 0$ the only possible value of L is 0, as observed.

The second Hund rule is best approached by model calculations. Pauling and Wilson,[1] for example, give a calculation of the spectral terms that arise from the configuration p^2. The third Hund rule is a consequence of the sign of the spin-orbit interaction: For a single electron the energy is lowest when the spin is antiparallel to the orbital angular momentum. But the low energy pairs m_L, m_S are progressively used up as we add electrons to the shell; by the exclusion principle when the shell is more than half full the state of lowest energy necessarily has the spin parallel to the orbit.

Consider two examples of the Hund rules: The ion Ce^{3+} has a single f electron; an f electron has $l = 3$ and $s = \frac{1}{2}$. Because the f shell is less than half full, the J value by the preceding rule is $|L - S| = L - \frac{1}{2} = \frac{5}{2}$. The ion Pr^{3+} has two f electrons: one of the rules tells us that the spins add to give $S = 1$. Both f electrons cannot have $m_l = 3$ without violating the Pauli exclusion principle, so that the maximum L consistent with the Pauli principle is not 6, but 5. The J value is $|L - S| = 5 - 1 = 4$.

Iron Group Ions

Table 2 shows that the experimental magneton numbers for salts of the iron transition group of the periodic table are in poor agreement with (18). The values often agree quite well with magneton numbers $p = 2[S(S+1)]^{\frac{1}{2}}$ calculated as if the orbital moment were not there at all. We say that the orbital moments are **quenched.**

TABLE 2 Effective Magneton Numbers for Iron Group Ions

Ion	Configuration	Basic Level	$p(\text{calc}) = g[J(J+1)]^{\frac{1}{2}}$	$p(\text{calc}) = 2[S(S+1)]^{\frac{1}{2}}$	$p(\text{exp})$[a]
Ti^{3+}, V^{4+}	$3d^1$	$^2D_{3/2}$	1.55	1.73	1.8
V^{3+}	$3d^2$	$^3F_2$	1.63	2.83	2.8
Cr^{3+}, V^{2+}	$3d^3$	$^4F_{3/2}$	0.77	3.87	3.8
Mn^{3+}, Cr^{2+}	$3d^4$	$^5D_0$	0	4.90	4.9
Fe^{3+}, Mn^{2+}	$3d^5$	$^6S_{5/2}$	5.92	5.92	5.9
Fe^{2+}	$3d^6$	$^5D_4$	6.70	4.90	5.4
Co^{2+}	$3d^7$	$^4F_{9/2}$	6.63	3.87	4.8
Ni^{2+}	$3d^8$	$^3F_4$	5.59	2.83	3.2
Cu^{2+}	$3d^9$	$^2D_{5/2}$	3.55	1.73	1.9

[a]Representative values.

[1]L. Pauling and E. B. Wilson, *Introduction to quantum mechanics*, McGraw-Hill, 1935, pp. 239–246.

Crystal Field Splitting

The difference in behavior of the rare earth and the iron group salts is that the $4f$ shell responsible for paramagnetism in the rare earth ions lies deep inside the ions, within the $5s$ and $5p$ shells, whereas in the iron group ions the $3d$ shell responsible for paramagnetism is the outermost shell. The $3d$ shell experiences the intense inhomogeneous electric field produced by neighboring ions. This inhomogeneous electric field is called the **crystal field.** The interaction of the paramagnetic ions with the crystal field has two major effects: the coupling of $\mathbf{L}$ and $\mathbf{S}$ vectors is largely broken up, so that the states are no longer specified by their J values; further, the $2L + 1$ sublevels belonging to a given L which are degenerate in the free ion may now be split by the crystal field, as in Fig. 6. This splitting diminishes the contribution of the orbital motion to the magnetic moment.

Quenching of the Orbital Angular Momentum

In an electric field directed toward a fixed nucleus, the plane of a classical orbit is fixed in space, so that all the orbital angular momentum components L_x, L_y, L_z are constant. In quantum theory one angular momentum component, usually taken as L_z, and the square of the total orbital angular momentum L^2 are constant in a central field. In a noncentral field the plane of the orbit will move about; the angular momentum components are no longer constant and may average to zero. In a crystal L_z will no longer be a constant of the motion, although to a good approximation L^2 may continue to be constant. When L_z averages to zero, the orbital angular momentum is said to be quenched. The magnetic moment of a state is given by the average value of the magnetic moment operator $\mu_B(\mathbf{L} + 2\mathbf{S})$. In a magnetic field along the z direction the orbital contribution to the magnetic moment is proportional to the quantum expectation value of L_z; the orbital magnetic moment is quenched if the mechanical moment L_z is quenched.

When the spin-orbit interaction energy is introduced, the spin may drag some orbital moment along with it. If the sign of the interaction favors parallel orientation of the spin and orbital magnetic moments, the total magnetic moment will be larger than for the spin alone, and the g value will be larger than 2. The experimental results are in agreement with the known variation of sign of the spin-orbit interaction: $g > 2$ when the $3d$ shell is more than half full, $g = 2$ when the shell is half full, and $g < 2$ when the shell is less than half full.

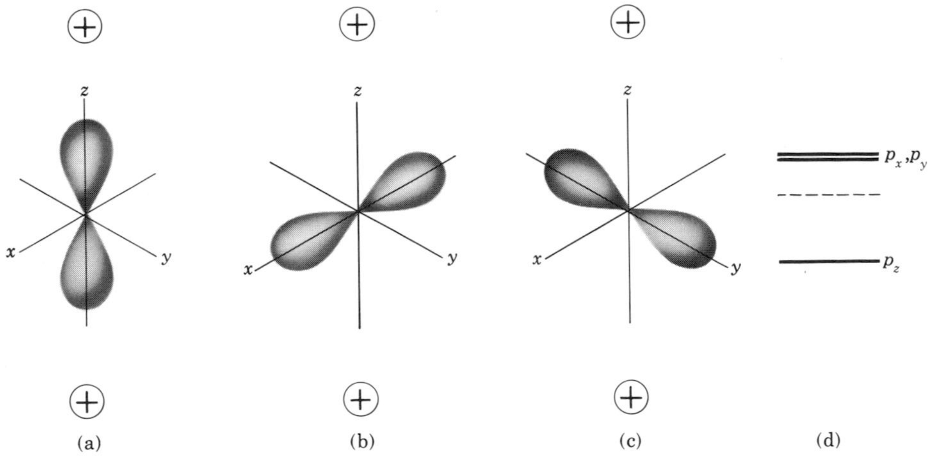

Figure 6 Consider an atom with orbital angular momentum $L = 1$ placed in the uniaxial crystalline electric field of the two positive ions along the z axis. In the free atom the states $m_L = \pm 1$, 0 have identical energies—they are degenerate. In the crystal the atom has a lower energy when the electron cloud is close to positive ions as in (a) than when it is oriented midway between them, as in (b) and (c). The wavefunctions that give rise to these charge densities are of the form $zf(r)$, $xf(r)$ and $yf(r)$ and are called the p_z, p_x, p_y orbitals, respectively. In an axially symmetric field, as shown, the p_x and p_y orbitals are degenerate. The energy levels referred to the free atom (dotted line) are shown in (d). If the electric field does not have axial symmetry, all three states will have different energies.

COOLING BY ADIABATIC DEMAGNETIZATION OF A PARAMAGNETIC SALT

The first method for attaining temperatures much below 1 K was that of adiabatic demagnetization.[2] By its use temperatures of 10^{-3} K and lower have been reached. The method rests on the fact that at a fixed temperature the entropy of a system of magnetic moments is lowered by the application of a magnetic field. The entropy is a measure of the disorder of a system: the greater the disorder, the higher is the entropy. In the magnetic field the moments will be partly lined up (partly ordered), so that the entropy is lowered by the field. The entropy is also lowered if the temperature is lowered, as more of the moments line up.

If the magnetic field can then be removed without changing the entropy of the spin system, the order of the spin system will look like a lower temperature than the same degree of order in the presence of the field. When the specimen is demagnetized adiabatically, entropy can flow into the spin system only from the system of lattice vibrations, as in Fig. 7. At the temperatures of interest the entropy of the lattice vibrations is usually negligible; thus the entropy of the spin system will be essentially constant during adiabatic demagnetization of the specimen. Magnetic cooling is a one-shot operation, not cyclic.

We first find an expression for the spin entropy of a system of N ions, each of spin S, at a temperature sufficiently high that the spin system is entirely disordered. That is, T is supposed to be much higher than some temperature Δ which characterizes the energy of the interactions ($E_{\text{int}} \equiv k_B\Delta$) tending to orient the spins preferentially. Some of these interactions are discussed in Chapter 16. The definition of the entropy σ of a system of G accessible states is $\sigma = k_B \ln G$. At a temperature so high that all of the $2S + 1$ states of each ion are nearly equally populated, G is the number of ways of arranging N spins in $2S + 1$ states. Thus $G = (2S + 1)^N$, whence the spin entropy σ_S is:

$$\sigma_S = k_B \ln (2S + 1)^N = Nk_B \ln (2S + 1) \ . \tag{19}$$

It is this spin entropy which is reduced by a magnetic field. The field separates the $2S + 1$ states in energy, and the lower levels gain in population.

The steps carried out in the cooling process are shown in Fig. 8. The field is applied at temperature T_1 with the specimen in good thermal contact with the surroundings, giving the isothermal path ab. The specimen is then insulated ($\Delta\sigma = 0$) and the field removed; the specimen follows the constant entropy path bc, ending up at temperature T_2. The thermal contact at T_1 is provided by helium gas, and the thermal contact is broken by removing the gas with a pump.

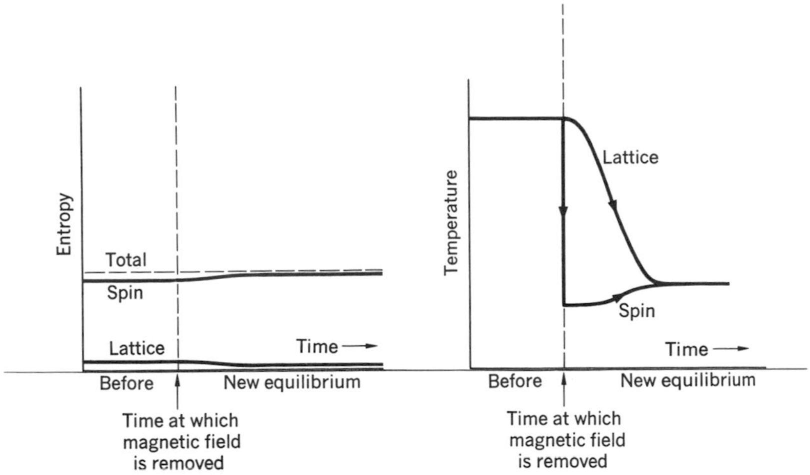

Figure 7 During adiabatic demagnetization the total entropy of the specimen is constant. The initial entropy of the lattice should be small in comparison with the entropy of the spin system.

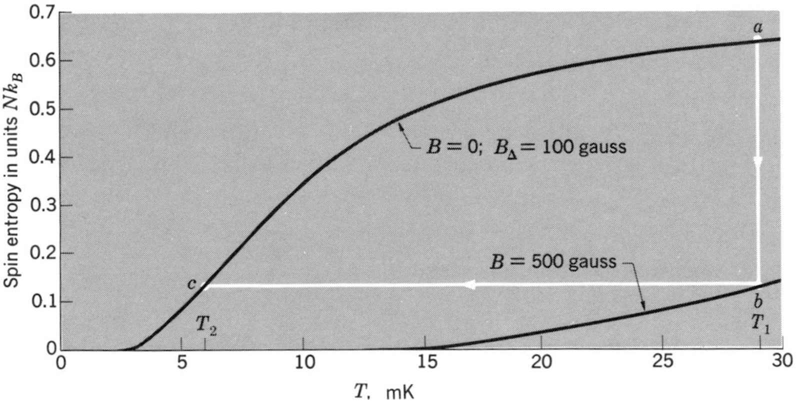

Figure 8 Entropy for a spin $\frac{1}{2}$ system as a function of temperature, assuming an internal random magnetic field B_Δ of 100 gauss. The specimen is magnetized isothermally along *ab*, and is then insulated thermally. The external magnetic field is turned off along *bc*. In order to keep the figure on a reasonable scale the initial temperature T_1 is lower than would be used in practice, as is also the external magnetic field.

[2]The method was suggested by P. Debye, Ann. Physik **81,** 1154 (1926); and W. F. Giauque, J. Am. Chem. Soc. **49,** 1864 (1927). For many purposes the method has been supplanted by the He^3-He^4 dilution refrigerator which operates in a continuous cycle. The He^3 atoms in solution in liquid He^4 play the role of atoms in a gas, and cooling is effected by "vaporization" of He^3. References are given by J. C. Wheatley, Amer. J. Phys. **36,** 181 (1968); see especially pp. 198–210.

Nuclear Demagnetization

The population of a magnetic sublevel is a function only of $\mu B/k_B T$, hence of B/T. The spin-system entropy is a function only of the population distribution; hence the spin entropy is a function only of B/T. If B_Δ is the effective field that corresponds to the local interactions, the final temperature T_2 reached in an adiabatic demagnetization experiment is

$$\boxed{T_2 = T_1(B_\Delta/B)} \ , \qquad (20)$$

where B is the initial field and T_1 the initial temperature.

Because nuclear magnetic moments are weak, nuclear magnetic interactions are much weaker than similar electronic interactions. We expect to reach a temperature 100 times lower with a nuclear paramagnet than with an electron paramagnet. The initial temperature T_1 of the nuclear stage in a nuclear spin-cooling experiment must be lower than in an electron spin-cooling experiment. If we start at $B = 50$ kG and $T_1 = 0.01$ K, then $\mu B/k_B T_1 \approx 0.5$, and the entropy decrease on magnetization is over 10 percent of the maximum spin entropy. This is sufficient to overwhelm the lattice and from (20) we estimate a final temperature $T_2 \approx 10^{-7}$ K. The first[3] nuclear cooling experiment was carried out on Cu nuclei in the metal, starting from a first stage at about 0.02 K as attained by electronic cooling. The lowest temperature reached was 1.2×10^{-6} K. The results in Fig. 9 fit a line of the form of (20): $T_2 = T_1(3.1/B)$ with B in gauss, so that $B_\Delta = 3.1$ gauss. This is the effective interaction field of the magnetic moments of the Cu nuclei. The motivation for using nuclei in a metal is that conduction electrons help ensure rapid thermal contact of lattice and nuclei at the temperature of the first stage.

---

[3]N. Kurti, F. N. H. Robinson, F. E. Simon, and D. A. Spohr, Nature **178**, 450 (1956); R. E. Walstedt, E. L. Hahn, C. Froidevaux, and E. Geissler, Proc. Roy. Soc. (London) **A284**, 499 (1965); for reviews see N. Kurti, Cryogenics **1**, 2 (1960); Adv. in Cryogenic Engineering **8**, 1 (1963).

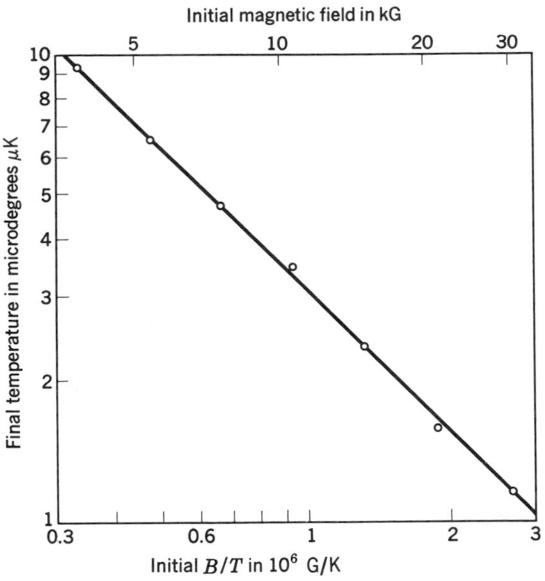

Figure 9 Nuclear demagnetizations of copper nuclei in the metal, starting from 0.012 K and various fields. [After M. V. Hobden and N. Kurti, Phil. Mag. **4**, 1092 (1959).]

PARAMAGNETIC SUSCEPTIBILITY OF CONDUCTION ELECTRONS

We are going to try to show how on the basis of these statistics the fact that many metals are diamagnetic or only weakly paramagnetic, can be brought into agreement with the existence of a magnetic moment of the electrons.

W. Pauli, 1927

Classical free electron theory gives an unsatisfactory account of the paramagnetic susceptibility of the conduction electrons. An electron has associated with it a magnetic moment of one Bohr magneton, μ_B. One might expect that the conduction electrons would make a Curie-type paramagnetic contribution (17) to the magnetization of the metal: $M = N\mu_B{}^2B/k_BT$. Instead it is observed that the magnetization of most normal nonferromagnetic metals is independent of temperature.

Pauli[4] showed that the application of the Fermi-Dirac distribution (Chapter 6) would correct the theory as required. We first give a qualitative explanation of the situation. The result (13) tells us that the probability an atom will be lined up parallel to the field B exceeds the probability of the antiparallel orientation by roughly $\mu B/k_BT$. For N atoms per unit volume, this gives a net magnetization $\approx N\mu^2B/k_BT$, the standard result.

[4]W. Pauli, Z. Physik **41**, 81 (1927).

Most conduction electrons in a metal, however, have no possibility of turning over when a field is applied, because most orbitals in the Fermi sea with parallel spin are already occupied. Only the electrons within a range k_BT of the top of the Fermi distribution have a chance to turn over in the field; thus only the fraction T/T_F of the total number of electrons contribute to the susceptibility. Hence

$$M \approx \frac{N\mu^2 B}{k_B T} \cdot \frac{T}{T_F} = \frac{N\mu^2}{kT_F} B \ ,$$

which is independent of temperature and of the observed order of magnitude.

We now calculate the expression for the paramagnetic susceptibility of a free electron gas at $T \ll T_F$. We follow the method of calculation suggested by Fig. 10. An alternate derivation is the subject of Problem 5. The concentration of electrons with magnetic moments parallel to the magnetic field is

$$N_+ = \frac{1}{2}\int_{-\mu}^{\epsilon_F} d\epsilon\, f(\epsilon)\ \mathfrak{D}(\epsilon + \mu B) \cong \frac{1}{2}\int_0^{\epsilon_F} d\epsilon\, f(\epsilon)\ \mathfrak{D}(\epsilon) + \frac{1}{2}\mu B\ \mathfrak{D}(\epsilon_F) \ ,$$

where $f(\epsilon)$ is the Fermi-Dirac distribution function and $\frac{1}{2}\mathfrak{D}(\epsilon + \mu B)$ is the density of orbitals of one spin orientation, with allowance for the downward shift of energy by $-\mu B$. (Make a drawing for yourself of the integrands in the expressions for N_+ and N_-.) The approximation is written for $k_BT \ll \epsilon_F$. The concentration of electrons with magnetic moments antiparallel to the magnetic field is

$$N_- = \frac{1}{2}\int_{\mu_B}^{\epsilon_F} d\epsilon\, f(\epsilon)\ \mathfrak{D}(\epsilon - \mu B) \cong \frac{1}{2}\int_0^{\epsilon_F} d\epsilon\, f(\epsilon)\ \mathfrak{D}(\epsilon) - \frac{1}{2}\mu B\ \mathfrak{D}(\epsilon_F) \ .$$

The magnetization is given by $M = \mu(N_+ - N_-)$, so that

$$M = \mu^2\ \mathfrak{D}(\epsilon_F) B = \frac{3N\mu^2}{2k_BT_F} B \ , \qquad (21)$$

with $\mathfrak{D}(\epsilon_F) = 3N/2\epsilon_F = 3N/2k_BT_F$ from Chapter 6. The result (21) gives the **Pauli spin magnetization** of the conduction electrons.

In deriving the paramagnetic susceptibility, we have supposed that the spatial motion of the electrons is not affected by the magnetic field. But the wavefunctions are modified by the magnetic field; Landau has shown

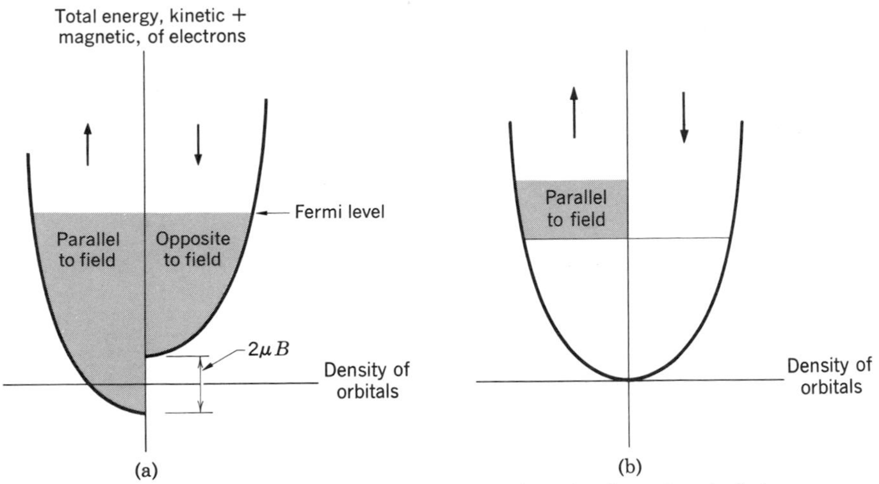

Figure 10 Pauli paramagnetism at absolute zero; the orbitals in the shaded regions in (a) are occupied. The numbers of electrons in the "up" and "down" band will adjust to make the energies equal at the Fermi level. The chemical potential (Fermi level) of the spin up electrons is equal to that of the spin down electrons. In (b) we show the excess of spin up electrons in the magnetic field.

that for free electrons this causes a diamagnetic moment equal to $-\frac{1}{3}$ of the paramagnetic moment. Thus the total magnetization of a free electron gas is

$$M = \frac{N\mu_B{}^2}{k_B T_F} B \ . \tag{22}$$

Before comparing (22) with the experiment we must take account of the diamagnetism of the ionic cores, of band effects, and of electron-electron interactions. In sodium the interaction effects increase the spin susceptibility by perhaps 75 percent. The magnetic susceptibility is considerably higher for most transition metals (with unfilled inner electron shells) than for the alkali metals (Fig. 11). The high values suggest that the density of orbitals is unusually high for transition metals, in agreement with measurements of the electronic heat capacity. We saw in Chapter 9 how this arises from band theory.

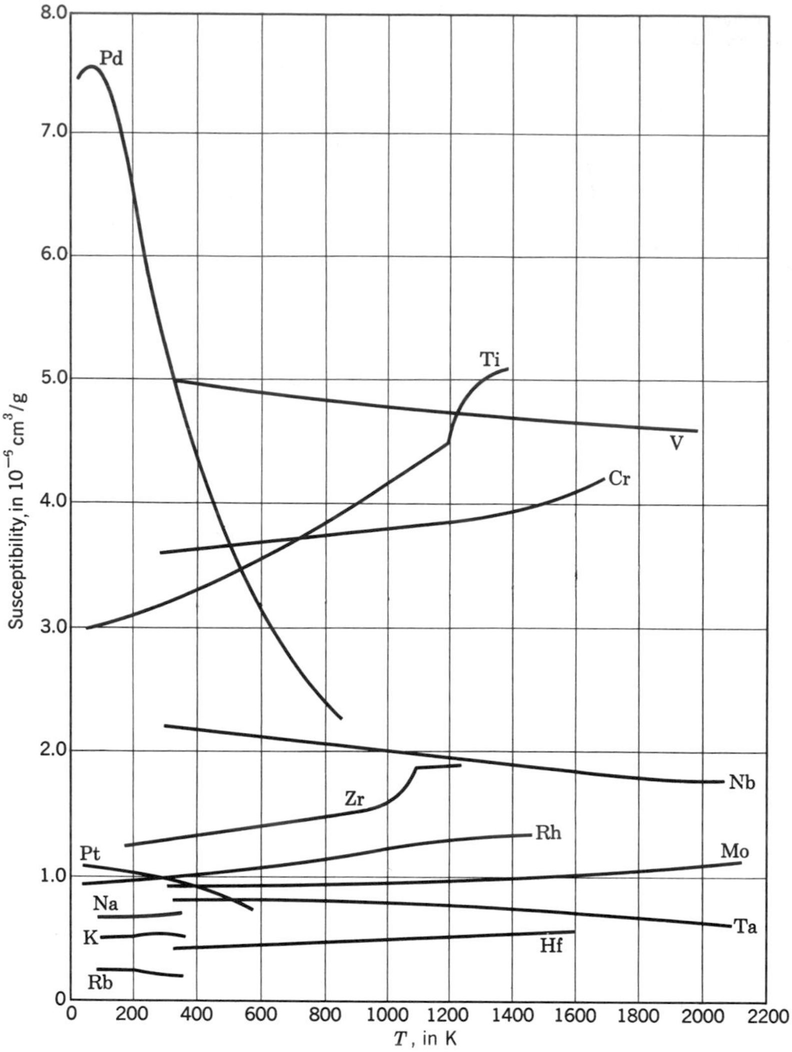

Figure 11 Temperature dependence of the magnetic susceptibility of metals. (Courtesy of C. J. Kriessman.)

SUMMARY (CGS units)

1. The diamagnetic susceptibility of N atoms of atomic number Z is $\chi = -Ze^2N\langle r^2\rangle/6mc^2$, where $\langle r^2\rangle$ is the mean square atomic radius. (Langevin)

2. Atoms with a permanent magnetic moment $\boldsymbol{\mu}$ have a paramagnetic susceptibility $\chi = N\mu^2/3k_BT$, for $\mu B \ll k_BT$. (Curie-Langevin)

3. For a system of spins $S = \frac{1}{2}$, the exact magnetization is $M = N\mu \tanh(\mu B/k_BT)$, where $\mu = \frac{1}{2}g\mu_B$. (Brillouin)

4. The ground state of electrons in the same shell have the maximum value of S allowed by the Pauli principle and the maximum L consistent with this S. The J value is $L + S$ if the shell is more than half full and $|L - S|$ if the shell is less than half full.

5. A cooling process operates by demagnetization of a paramagnetic salt at constant entropy. The final temperature reached is of the order of $(B_\Delta/B)T_{\text{initial}}$, where B_Δ is the effective local field and B is the initial applied magnetic field.

6. The paramagnetic susceptibility of a Fermi gas of conduction electrons is $\chi = 3N\mu^2/2\epsilon_F$, independent of temperature for $k_BT \ll \epsilon_F$. (Pauli)

PROBLEMS

1. ***Diamagnetic susceptibility of atomic hydrogen.*** The wave function of the hydrogen atom in its ground state (1s) is $\psi = (\pi a_0^3)^{-\frac{1}{2}} e^{-r/a_0}$, where $a_0 = \hbar^2/me^2 = 0.529 \times 10^{-8}$ cm. The charge density is $\rho(x, y, z) = -e|\psi|^2$, according to the statistical interpretation of the wave function. Show that for this state $\langle r^2 \rangle = 3a_0^2$, and calculate the molar diamagnetic susceptibility of atomic hydrogen (-2.36×10^{-6} cm^3/mole).

2. ***Hund rules.*** Apply the Hund rules to find the ground state (the basic level in the notation of Table 1) of (a) Eu^{++}, in the configuration $4f^7\, 5s^2p^6$; (b) Yb^{3+}; (c) Tb^{3+}. The results for (b) and (c) are in Table 1, but you should give the separate steps in applying the rules.

3. ***Triplet excited states.*** Some organic molecules have a triplet ($S = 1$) excited state at an energy $k_B\Delta$ above a singlet ($S = 0$) ground state. (a) Find an expression for the magnetic moment $\langle \mu \rangle$ in a field B. (b) Show that the susceptibility for $T \gg \Delta$ is approximately independent of Δ. (c) With the help of a diagram of energy levels versus field and a rough sketch of entropy versus field, explain how this system might be cooled by adiabatic magnetization (not demagnetization).

4. ***Heat capacity from internal degrees of freedom.*** (a) Consider a two-level system with an energy splitting $k_B\Delta$ between upper and lower states; the splitting may arise from a magnetic field or in other ways. Show that the heat capacity per system is

$$C = \left(\frac{\partial U}{\partial T}\right)_\Delta = k_B \frac{(\Delta/T)^2\, e^{\Delta/T}}{(1 + e^{\Delta/T})^2} \;.$$

The function is plotted in Fig. 12. Peaks of this type in the heat capacity are often known as Schottky anomalies. The maximum heat capacity is quite high, but for $T \ll \Delta$ and for $T \gg \Delta$ the heat capacity is low. (b) Show that for $T \gg \Delta$ we have $C \cong k_B(\Delta/2T)^2 + \cdots$. The hyperfine interaction between nuclear and electronic magnetic moments in paramagnetic salts (and in systems having electron spin order) causes splittings with $\Delta \approx 1$ to 100 mK. These splittings are often detected experimentally by the presence of a term in $1/T^2$ in the heat capacity in the region $T \gg \Delta$. Nuclear electric quadrupole interactions (see Chapter 16) with crystal fields also cause splittings, as in Fig. 13.

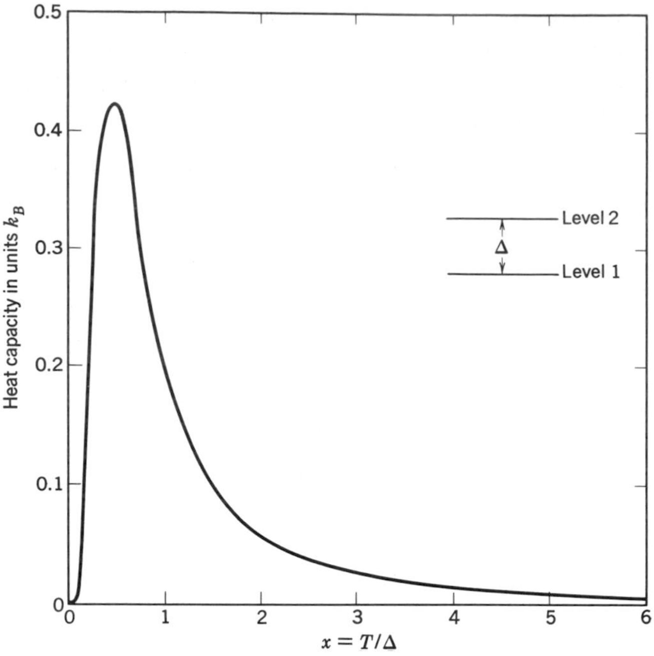

Figure 12 Heat capacity of a two-level system as a function of T/Δ, where Δ is the level splitting. The Schottky anomaly is a very useful tool for determining energy level splittings of ions in rare-earth and transition-group metals, compounds, and alloys.

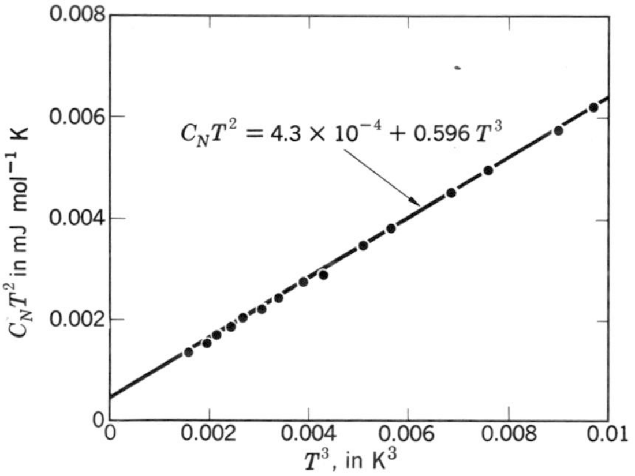

Figure 13 The normal-state heat capacity of gallium at $T < 0.21$ K. The nuclear quadrupole ($C \propto T^{-2}$) and conduction electron ($C \propto T$) contributions dominate the heat capacity at very low temperatures. [After N. E. Phillips, Phys. Rev. **134,** 385 (1964).]

5. *Pauli spin susceptibility.* The spin susceptibility of a conduction electron gas at absolute zero may be discussed by another method. Let

$$N^+ = \tfrac{1}{2}N(1+\zeta) \ ; \qquad N^- = \tfrac{1}{2}N(1-\zeta)$$

be the concentrations of spin-up and spin-down electrons. (a) Show that in a magnetic field B the total energy of the spin-up band in a free electron gas is

$$E^+ = E_0(1+\zeta)^{\frac{5}{3}} - \tfrac{1}{2}N\mu B(1+\zeta) \ ,$$

where $E_0 = \frac{3}{10}N\epsilon_F$, in terms of the Fermi energy ϵ_F in zero magnetic field. Find a similar expression for E^-. (b) Minimize $E_{total} = E^+ + E^-$ with respect to ζ and solve for the equilibrium value of ζ in the approximation $\zeta \ll 1$. Go on to show that the magnetization is $M = 3N\mu^2B/2\epsilon_F$, in agreement with Eq. (21).

6. ***Conduction electron ferromagnetism.*** We approximate the effect of exchange interactions among the conduction electrons if we assume that electrons with parallel spins interact with each other with energy $-V$, and V is positive, while electrons with antiparallel spins do not interact with each other.
(a) Show with the help of Problem 5 that the total energy of the spin-up band is

$$E^+ = E_0(1+\zeta)^{\frac{5}{3}} - \tfrac{1}{8}VN^2(1+\zeta)^2 - \tfrac{1}{2}N\mu B(1+\zeta) \; ;$$

find a similar expression for E^-. (b) Minimize the total energy and solve for ζ in the limit $\zeta \ll 1$. Show that the magnetization is

$$M = \frac{3N\mu^2}{2\epsilon_F - \frac{3}{2}VN} B \; ,$$

so that the exchange interaction enhances the susceptibility. (c) Show that with $B = 0$ the total energy is unstable at $\zeta = 0$ when $V > 4\epsilon_F/3N$. If this is satisfied a ferromagnetic state ($\zeta \neq 0$) will have a lower energy than the paramagnetic state. Because of the assumption $\zeta \ll 1$, this is a sufficient condition for ferromagnetism, but it may not be a necessary condition.

7. ***Two-level system.*** The result of Problem 4 is often seen in another form. (a) If the two energy levels are at Δ and $-\Delta$, show that the energy and heat capacity are

$$U = -\Delta \tanh(\Delta/k_BT) \; ; \qquad C = k_B(\Delta/k_BT)^2 \operatorname{sech}^2(\Delta/k_BT) \; .$$

(b) If the system has a random composition such that all values of Δ are equally likely up to some limit Δ_0, show that the heat capacity is linearly proportional to the temperature, provided $k_BT \ll \Delta_0$. This result was applied to the heat capacity of dilute magnetic alloys by W. Marshall, Phys. Rev. **118**, 1519 (1960). It is also used in the theory of glasses.

8. ***Paramagnetism of $S = 1$ system.*** (a) Find the magnetization as a function of magnetic field and temperature for a system of spins with $S = 1$, moment μ, and concentration n. (b) Show that in the limit $\mu B \ll kT$ the result is $M \cong (2n\mu^2/3kT)B$.

References

A. Abragam and B. Bleaney, *Electron paramagnetic resonance of transition ions*, Oxford, 1970.

L. F. Bates, *Modern magnetism*, 4th ed., Cambridge, 1961.

H. B. G. Casimir, *Magnetism and very low temperatures*, Cambridge, 1940. A classic.

M. I. Darby and K. N. R. Taylor, *Physics of rare earth solids*, Halsted, 1972.

A. J. Freeman, *The actinides: electronic structure and related properties*, Academic, 1974.

R. D. Hudson, *Principles and applications of magnetic cooling*, Elsevier, 1972.

W. J. Huiskamp and O. V. Lounasmaa, "Ultralow temperatures—how and why," Rep. Prog. Phys. **36**, 423 (1973).

H. Knoepfel, *Pulsed high magnetic fields*, North-Holland, 1970.

W. A. Little, "Magnetic cooling," *Prog. in cryogenics* **4**, 101 (1964).

O. V. Lounasmaa, *Experimental principles and methods below 1 K*, Academic Press, 1974.

L. Orgel, *Introduction to transition metal chemistry*, 2nd ed., Wiley, 1966.

R. L. Sanford and I. L. Cooter, "Basic magnetic quantities and the measurement of the magnetic properties of materials," NBS Monograph **47** (1962).

J. H. Van Vleck, *The theory of electric and magnetic susceptibilities*, Oxford, 1932. Superb derivations of basic theorems.

G. K. White, *Experimental techniques in low temperature physics*, 2nd ed., Oxford, 1968.

R. M. White, *Quantum theory of magnetism*, McGraw-Hill, 1970.

15 FERROMAGNETISM AND ANTIFERROMAGNETISM

NOTATION: (CGS) $B = H + 4\pi M$; (SI) $B = \mu_0(H + M)$. We call B_a the applied magnetic field in both systems of units: in CGS we have $B_a = H_a$ and in SI we have $B_a = \mu_0 H_a$. The susceptibility is $\chi = M/B_a$ in CGS and $\chi = M/H_a = \mu_0 M/B_a$ in SI.

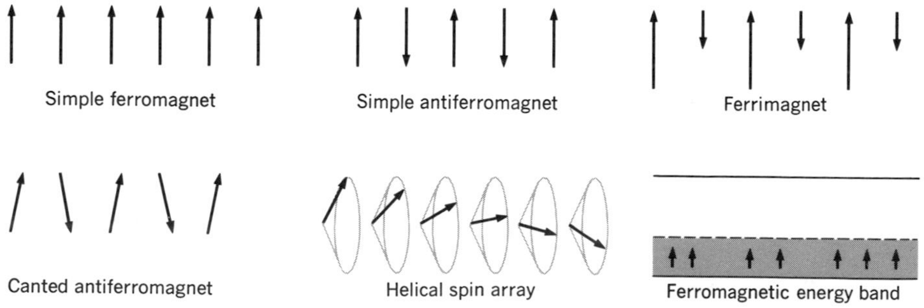

Figure 1 Ordered arrangements of electron spins.

15 FERROMAGNETISM AND ANTIFERROMAGNETISM

FERROMAGNETIC ORDER

A ferromagnet has a spontaneous magnetic moment—a magnetic moment even in zero applied magnetic field. A spontaneous moment suggests that electron spins and magnetic moments are arranged in a regular manner. The order need not be simple: all of the spin arrangements sketched in Fig. 1 except the simple antiferromagnet have a spontaneous magnetic moment, called the **saturation moment.**

Curie Point and the Exchange Integral

Consider a paramagnet with a concentration of N ions of spin S. Given an internal interaction tending to line up the magnetic moments parallel to each other, we shall have a ferromagnet. Let us postulate such an interaction and call it the **exchange field.**[1] The orienting effect of the exchange field is opposed by thermal agitation, and at elevated temperatures the spin order is destroyed.

We treat the exchange field as equivalent to a magnetic field $\mathbf{B}_E$. The magnitude of the exchange field may be as high as 10^7 gauss. We assume that $\mathbf{B}_E$ is proportional to the magnetization $\mathbf{M}$. The magnetization is defined as the magnetic moment per unit volume; unless otherwise specified it is understood to be the value in thermal equilibrium at the temperature T. If domains (regions magnetized in different directions) are present, the magnetization refers to the value within a domain. In the **mean field approximation** we assume each magnetic atom experiences a field proportional to the magnetization:

$$\mathbf{B}_E = \lambda \mathbf{M} \ , \tag{1}$$

[1]Also called the molecular field or the Weiss field, after Pierre Weiss who was the first to imagine such a field. The exchange field B_E simulates a real magnetic field in the expressions for the energy $-\boldsymbol{\mu} \cdot \mathbf{B}_E$ and the torque $\boldsymbol{\mu} \times \mathbf{B}_E$ on a magnetic moment μ. But B_E is not really a magnetic field and therefore does not enter into the Maxwell equations: for example, there is no current density $\mathbf{j}$ related to $\mathbf{B}_E$ by curl $\mathbf{H} = 4\pi\mathbf{j}/c$. The magnitude of B_E is typically 10^4 larger than the average magnetic field of the magnetic dipoles of the ferromagnet.

where λ is a constant, independent of temperature. According to (1), each spin sees the average magnetization of all the other spins. In truth, it may see only near neighbors, but our simplification is good for a first look at the problem.

The **Curie temperature** T_c is the temperature above which the spontaneous magnetization vanishes; it separates the disordered paramagnetic phase at $T > T_c$ from the ordered ferromagnetic phase at $T < T_c$. We can find T_c in terms of λ. Consider the paramagnetic phase: an applied field B_a will cause a finite magnetization and this in turn will cause a finite exchange field B_E. If χ_p is the paramagnetic susceptibility,

$$\text{(CGS)} \quad M = \chi_p(B_a + B_E) \ ; \qquad \text{(SI)} \quad \mu_0 M = \chi_p(B_a + B_E) \ . \tag{2}$$

We saw in Chapter 14 that we can write the magnetization as a constant susceptibility times a field only if the fractional alignment is small: this is where the assumption enters that the specimen is in the paramagnetic phase. The paramagnetic susceptibility is given by the Curie law $\chi_p = C/T$, where C is the Curie constant. Substitute (1) in (2); we find $MT = C(B_a + \lambda M)$ and

$$\text{(CGS)} \qquad \chi = \frac{M}{B_a} = \frac{C}{(T - C\lambda)} \ . \tag{3}$$

The susceptibility has a singularity at $T = C\lambda$. At this temperature (and below) there exists a spontaneous magnetization, because if χ is infinite we can have a finite M for zero B_a. From (3) we have the **Curie-Weiss law**

$$\text{(CGS)} \qquad \chi = \frac{C}{T - T_c} \ ; \qquad T_c = C\lambda \ . \tag{4}$$

This expression describes fairly well the observed susceptibility variation in the paramagnetic region above the Curie point. Detailed calculations[2] predict

$$\chi \propto \frac{1}{(T - T_c)^{1.33}}$$

at temperatures close to T_c, in general agreement with the experimental data summarized in Table 1. The reciprocal susceptibility of nickel is plotted in Fig. 2.

From (4) and the definition (14.17) of the Curie constant C we may determine the value of the mean field constant λ:

$$\text{(CGS)} \qquad \lambda = \frac{T_c}{C} = \frac{3k_B T_c}{Ng^2 S(S+1)\mu_B^2} \ . \tag{5}$$

For iron $T_c \approx 1000$ K, $g \approx 2$, and $S \approx 1$; from (5) we have $\lambda \approx 5000$. With $M_s \approx 1700$ we have $B_E \approx \lambda M \approx (5000)(1700) \approx 10^7$ G. The exchange field in iron is very much stronger than the real magnetic field due to the other magnetic ions in the crystal: a magnetic ion produces a field $\approx \mu_B/a^3$ or about 10^3 G at a neighboring lattice point.

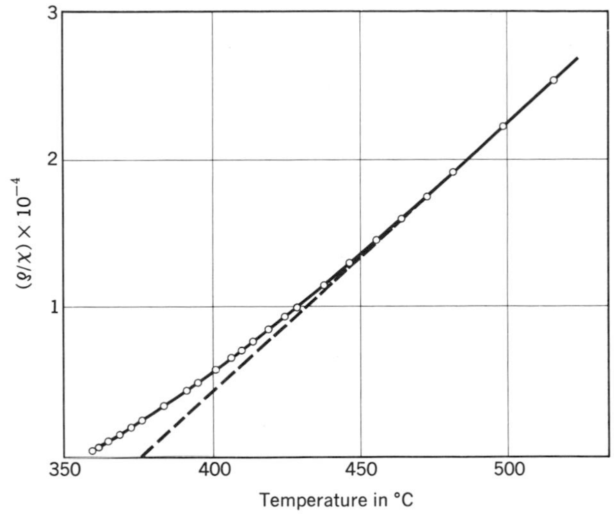

Figure 2 Reciprocal of the susceptibility per gram of nickel in the neighborhood of the Curie temperature (358°C). The density is ρ. The dashed line is a linear extrapolation from high temperatures. (After P. Weiss and R. Forrer.) An analysis of the curve near T_c has been given by J. S. Kouvel and M. E. Fisher, Phys. Rev. **136,** A1626 (1964).

TABLE 1 Critical Point Exponents for Ferromagnets

As $T \rightarrow T_c$ from above, the susceptibility χ becomes proportional to $(T - T_c)^{-\gamma}$; as $T \rightarrow T_c$ from below, the magnetization M_s becomes proportional to $(T_c - T)^{\beta}$. In the mean field approximation, $\gamma = 1$ and $\beta = \frac{1}{2}$.

	γ	β	T_c, in K
Fe	1.33 ± 0.015	0.34 ± 0.04	1043
Co	1.21 ± 0.04	—	1388
Ni	1.35 ± 0.02	0.42 ± 0.07	627.2
Gd	1.3 ± 0.1	—	292.5
CrO_2	1.63 ± 0.02	—	386.5
$CrBr_3$	1.215 ± 0.02	0.368 ± 0.005	32.56
EuS	—	0.33 ± 0.015	16.50

Experimental data collected by H. E. Stanley.

[2]Experimentally the susceptibility for $T \gg T_c$ is given quite accurately by $C/(T - \theta)$, where θ is appreciably greater than the actual transition temperature T_c. See the review by C. Domb in *Magnetism*, Vol. 2A, G. T. Rado and H. Suhl, eds., Academic Press, 1965.

The exchange field gives an approximate representation of the quantum-mechanical exchange interaction. On certain assumptions it can be shown[3] that the energy of interaction of atoms i, j bearing spins $\mathbf{S}_i$, $\mathbf{S}_j$ contains a term

$$U = -2J\mathbf{S}_i \cdot \mathbf{S}_j \ , \tag{6}$$

where J is the exchange integral and is related to the overlap of the charge distributions of the atoms i, j. Equation (6) is called the **Heisenberg model.**

The charge distribution of a system of two spins depends on whether the spins are parallel or antiparallel,[4] for the Pauli principle excludes two electrons of the same spin from being at the same place at the same time. It does not exclude two electrons of opposite spin. Thus the electrostatic energy of a system will depend on the relative orientation of the spins: the difference in energy defines the **exchange energy.** The exchange energy of two electrons may be written in the form $-2J\mathbf{s}_1 \cdot \mathbf{s}_2$ as in (6), just as if there were a direct coupling between the directions of the two spins. For many purposes in ferromagnetism it is a good approximation to treat the spins as classical angular momentum vectors.

We can establish an approximate connection between the exchange integral J in (6) and the exchange constant λ in (1). We suppose that the atom under consideration has z nearest neighbors, each connected with the central atom by the interaction J. For more distant neighbors we take J as zero. The mean field theory result is

$$J = \frac{3k_BT_c}{2zS(S+1)} \ . \tag{7}$$

Better approximations give somewhat different results. For the sc, bcc, and fcc structures with $S = \frac{1}{2}$, Rushbrooke and Wood give $k_BT_c/zJ = 0.28$; 0.325; and 0.346, respectively, as compared with 0.500 from (7) for all three structures. If iron is represented by the Heisenberg model with $S = 1$, then the observed Curie temperature corresponds to $J = 11.9$ meV.

[3]See most texts on quantum theory; also J. H. Van Vleck, Revs. Modern Phys. **17,** 27 (1945). The origin of exchange in insulators is reviewed by P. W. Anderson in Rado and Suhl, Vol. **I,** 25 (1963); in metals by C. Herring in Vol. IV (1966).

[4]If two spins are antiparallel, the wavefunctions of the two electrons must be symmetric, as in the combination $u(\mathbf{r}_1)v(\mathbf{r}_2) + u(\mathbf{r}_2)v(\mathbf{r}_1)$. If the two spins are parallel, the Pauli principle requires that the orbital part of the wavefunction be antisymmetric, as in $u(\mathbf{r}_1)v(\mathbf{r}_2) - u(\mathbf{r}_2)v(\mathbf{r}_1)$, for here if we interchange the coordinates $\mathbf{r}_1$, $\mathbf{r}_2$ the wavefunction changes sign. If we set the positions equal so that $\mathbf{r}_1 = \mathbf{r}_2$, then the antisymmetric function vanishes: for parallel spins there is zero probability of finding the two electrons at the same position.

Temperature Dependence of the Saturation Magnetization

We can also use the mean field approximation below the Curie temperature to find the magnetization as a function of temperature. We proceed as before, but instead of the Curie law we use the complete Brillouin expression for the magnetization. For spin $\frac{1}{2}$ this is $M = N\mu \tanh(\mu B/k_B T)$. If we omit the applied magnetic field and replace B by the molecular field $B_E = \lambda M$, then

$$M = N\mu \tanh(\mu\lambda M/k_B T) \ . \tag{8}$$

We shall see that solutions of this equation with nonzero M exist in the temperature range between 0 and T_c. To solve (8) we write it in terms of the reduced magnetization $m \equiv M/N\mu$ and the reduced temperature $t \equiv k_B T/N\mu^2\lambda$, whence

$$m = \tanh(m/t) \ . \tag{9}$$

We then plot the right and left sides of this equation separately as functions of m, as in Fig. 3. The intercept of the two curves gives the value of m at the temperature of interest. The critical temperature is $t = 1$, or $T_c = N\mu^2\lambda/k_B$. The curves of M versus T obtained in this way reproduce roughly the features of the experimental results, as shown in Fig. 4 for nickel. As T increases the magnetization decreases smoothly to zero at $T = T_c$. This behavior classifies the usual ferromagnetic/paramagnetic transition as second order.

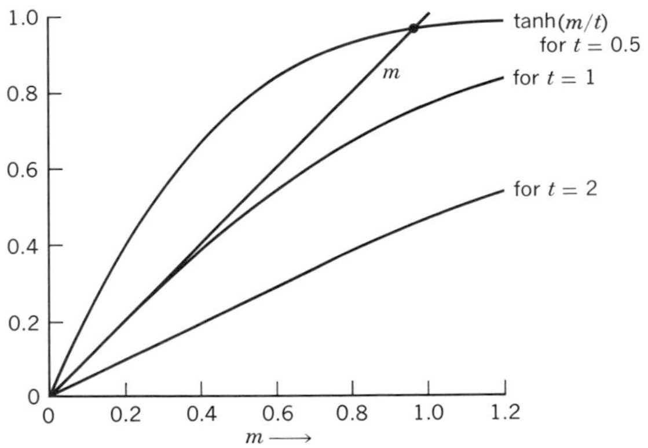

Figure 3 Graphical solution of Eq. (9) for the reduced magnetization m as a function of temperature. The reduced magnetization is defined as $m = M/N\mu$. The left-hand side of Eq. (9) is plotted as a straight line m with unit slope. The right-hand side is $\tanh(m/t)$ and is plotted vs. m for three different values of the reduced temperature $t = k_B T/N\mu^2\lambda = T/T_c$. The three curves correspond to the temperatures $2T_c$, T_c, and $0.5T_c$. The curve for $t = 2$ intersects the straight line m only at $m = 0$, as appropriate for the paramagnetic region (there is no external applied magnetic field). The curve for $t = 1$ (or $T = T_c$) is tangent to the straight line m at the origin; this temperature marks the onset of ferromagnetism. The curve for $t = 0.5$ is in the ferromagnetic region and intersects the straight line m at about $m = 0.94N\mu$. As $t \to 0$ the intercept moves up to $m = 1$, so that all magnetic moments are lined up at absolute zero.

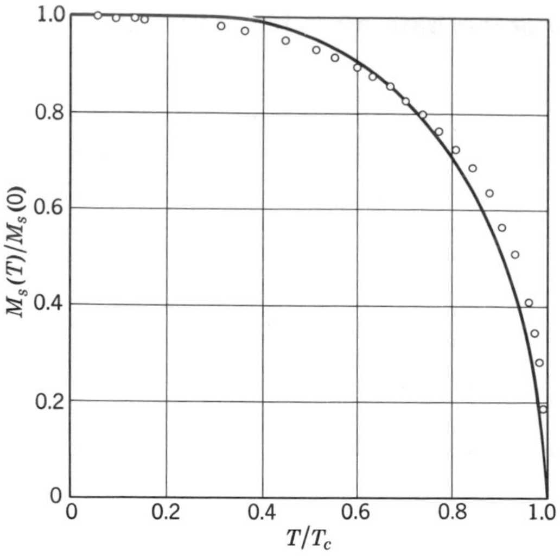

Figure 4 Saturation magnetization of nickel as a function of temperature, together with the theoretical curve for $S = \frac{1}{2}$ on the mean field theory. Experimental values by P. Weiss and R. Forrer, Ann. Phys. **5,** 153 (1926).

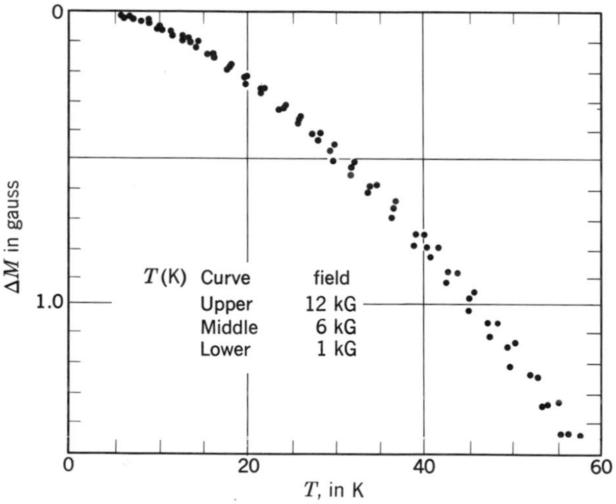

Figure 5 Decrease in magnetization of nickel with temperature, after Argyle, Charap, and Pugh. In the plot $\Delta M \equiv 0$ at 4.2 K.

The mean field theory does not give a good description of the variation of M at low temperatures. For $T \ll T_c$ the argument of tanh in (9) is large, and

$$\tanh \xi \cong 1 - 2e^{-2\xi} \ldots .$$

To lowest order the magnetization deviation $\Delta M \equiv M(0) - M(T)$ is

$$\Delta M \cong 2N\mu \exp(-2\lambda N\mu^2/k_BT) . \qquad (10)$$

The argument of the exponential is equal to $-2T_c/T$. For $T = 0.1T_c$ we have $\Delta M/N\mu \cong 4 \times 10^{-9}$. The experimental results show a much more rapid dependence of ΔM on temperature at low temperatures. At $T = 0.1T_c$ we have $\Delta M/M \cong 2 \times 10^{-3}$ from the data of Fig. 5. The leading term in ΔM is observed from experiment to have the form

$$\frac{\Delta M}{M(0)} = AT^{3/2} , \qquad (11)$$

where the constant A has the experimental value $(7.5 \pm 0.2) \times 10^{-6}$ deg$^{-3/2}$ for Ni and $(3.4 \pm 0.2) \times 10^{-6}$ deg$^{-3/2}$ for Fe. The result (11) finds a natural explanation in terms of spin wave theory, as discussed below.

TABLE 2 Ferromagnetic Crystals

	Magnetization M_s, in gauss			Curie
Substance	Room Temperature	0 K	n_B(0 K), per Formula Unit	Temperature, in K
Fe	1707	1740	2.22	1043
Co	1400	1446	1.72	1388
Ni	485	510	0.606	627
Gd	—	2060	7.63	292
Dy	—	2920	10.2	88
MnAs	670	870	3.4	318
MnBi	620	680	3.52	630
MnSb	710	—	3.5	587
CrO_2	515	—	2.03	386
$MnOFe_2O_3$	410	—	5.0	573
$FeOFe_2O_3$	480	—	4.1	858
$NiOFe_2O_3$	270	—	2.4	858
$CuOFe_2O_3$	135	—	1.3	728
$MgOFe_2O_3$	110	—	1.1	713
EuO	—	1920	6.8	69
$Y_3Fe_5O_{12}$	130	200	5.0	560

Saturation Magnetization at Absolute Zero

Table 2 gives representative values of the saturation magnetization M_s, the ferromagnetic Curie temperature, and the effective magneton number defined by $M_s(0) = n_B N \mu_B$, where N is the number of formula units per unit volume. Do not confuse n_B with the paramagnetic effective magneton number p defined by (14.18). Observed values of n_B are often nonintegral. There are many possible causes. One is the spin-orbit interaction which adds or subtracts some orbital magnetic moment. Another cause in ferromagnetic metals is the conduction electron magnetization induced locally about a paramagnetic ion core. A third cause is suggested by the drawing in Fig. 1 of the spin arrangement in a ferrimagnet: if there is one atom of spin projection $-S$ for every two atoms $+S$, the average spin is $\frac{1}{3}S$. Are there in fact any simple ferromagnetic insulators, with all ionic spins parallel in the ground state? The few simple ferromagnets known at present include $CrBr_3$, EuO, and EuS.

A band model[5] is the most likely model for the ferromagnetism of the transition metals Fe, Co, Ni. The approach is indicated in Figs. 6 and 7. The relationship of $4s$ and $3d$ bands is shown in Fig. 6 for copper, which is not ferromagnetic. If we remove one electron from copper, we obtain nickel which has the possibility of a vacant state in the $3d$ band. In the band structure of nickel shown in Fig. 7a for $T > T_c$ we have taken $2 \times 0.27 = 0.54$ of an electron away from the $3d$ band and 0.46 away from the $4s$ band, as compared with copper. The band structure of nickel at absolute zero is shown in Fig. 7b. Nickel is ferromagnetic, and at absolute zero $n_B = 0.60$ Bohr magnetons per atom. After allowance for the magnetic moment contribution of orbital electronic motion, nickel has an excess of 0.54 electron per atom having spin preferentially oriented in one direction. The exchange enhancement of the susceptibility of metals was the subject of Problem 14.6.

[5]C. Herring in Rado and Suhl, Vol. IV (1966).

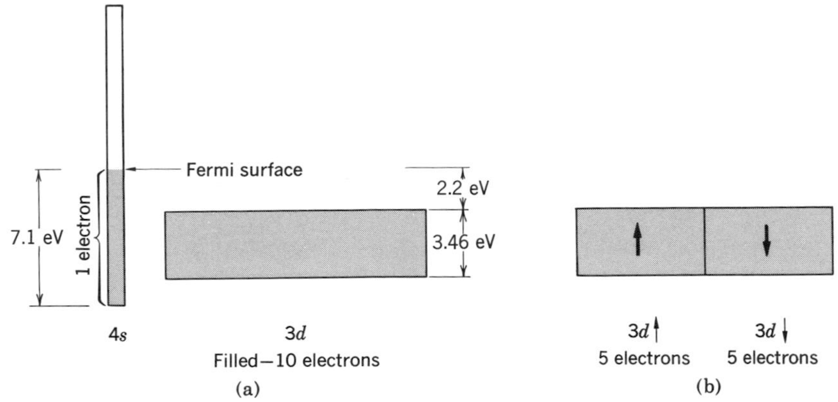

Figure 6a Schematic relationship of 4s and 3d bands in metallic copper. The 3d band holds 10 electrons per atom and is filled in copper. The 4s band can hold two electrons per atom; it is shown half-filled, as copper has one valence electron outside the filled 3d shell.

Figure 6b The filled 3d band of copper shown as two separate sub-bands of opposite electron spin orientation, each band holding five electrons. With both sub-bands filled as shown, the net spin (and hence the net magnetization) of the d band is zero.

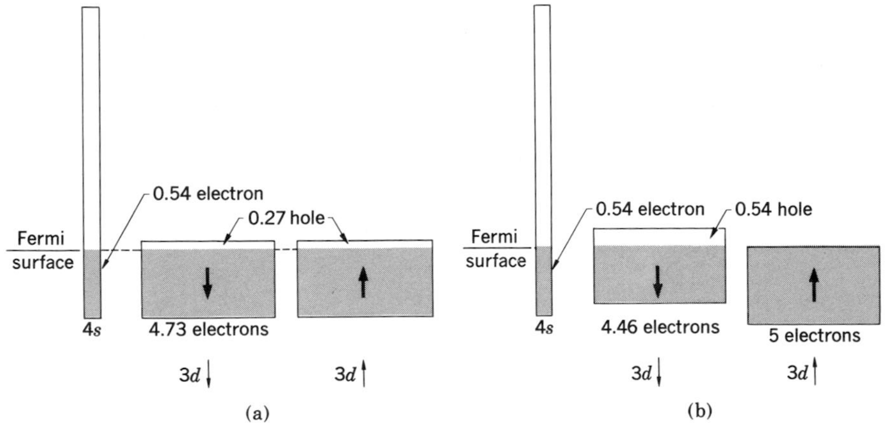

Figure 7a Band relationships in nickel above the Curie temperature. The net magnetic moment is zero, as there are equal numbers of holes in both $3d\downarrow$ and $3d\uparrow$ bands.

Figure 7b Schematic relationship of bands in nickel at absolute zero. The energies of the $3d\uparrow$ and $3d\downarrow$ sub-bands are separated by an exchange interaction. The $3d\uparrow$ band is filled; the $3d\downarrow$ band contains 4.46 electrons and 0.54 hole. The 4s band is usually thought to contain approximately equal numbers of electrons in both spin directions, and so we have not troubled to divide it into sub-bands. The net magnetic moment of 0.54 μ_B per atom arises from the excess population of the $3d\uparrow$ band over the $3d\downarrow$ band. It is often convenient to speak of the magnetization as arising from the 0.54 hole in the $3d\downarrow$ band.

MAGNONS

A magnon is a quantized spin wave. We use a classical argument, just as we did for phonons, to find the magnon dispersion relation. We then quantize the magnon energy and interpret the quantization in terms of spin reversal. The ground state of a simple ferromagnet has all spins parallel, as in Fig. 8a. Consider N spins each of magnitude S on a line or a ring, with nearest neighbor spins coupled by the Heisenberg interaction:

$$U = -2J \sum_{p=1}^{N} \mathbf{S}_p \cdot \mathbf{S}_{p+1} \ . \tag{12}$$

Here J is the exchange integral and $\hbar \mathbf{S}_p$ is the angular momentum of the spin at site p. If we treat the spins $\mathbf{S}_p$ as classical vectors, then in the ground state $\mathbf{S}_p \cdot \mathbf{S}_{p+1} = S^2$ and the exchange energy of the system is $U_0 = -2NJS^2$. What is the energy of the first excited state? Consider an excited state with one particular spin reversed, as in Fig. 8b. We see from (12) that this increases the energy by $8JS^2$, so that $U_1 = U_0 + 8JS^2$.

We can form an excitation of much lower energy if we let all the spins share the reversal, as in Fig. 8c. The elementary excitations of a spin system have a wavelike form and are called magnons (Fig. 9). These are analogous to lattice vibrations or phonons. Spin waves are oscillations in the relative orientations of spins on a lattice; lattice vibrations are oscillations in the relative positions of atoms on a lattice.

We now give a classical derivation of the magnon dispersion relation. The terms in (12) which involve the pth spin are

$$-2J\mathbf{S}_p \cdot (\mathbf{S}_{p-1} + \mathbf{S}_{p+1}) \ . \tag{13}$$

We write the magnetic moment at site p as $\boldsymbol{\mu}_p = -g\mu_B \mathbf{S}_p$, as in (14.6). Then (13) becomes

$$-\boldsymbol{\mu}_p \cdot [(-2J/g\mu_B)(\mathbf{S}_{p-1} + \mathbf{S}_{p+1})] \ , \tag{14}$$

which is of the form $-\boldsymbol{\mu}_p \cdot \mathbf{B}_p$, where the effective magnetic field or exchange field that acts on the pth spin is

$$\mathbf{B}_p = (-2J/g\mu_B)(\mathbf{S}_{p-1} + \mathbf{S}_{p+1}) \ . \tag{15}$$

From mechanics the rate of change of the angular momentum $\hbar \mathbf{S}_p$ is equal to the torque $\boldsymbol{\mu}_p \times \mathbf{B}_p$ which acts on the spin: $\hbar \ d\mathbf{S}_p/dt = \boldsymbol{\mu}_p \times \mathbf{B}_p$, or

$$d\mathbf{S}_p/dt = (-g\mu_B/\hbar) \ \mathbf{S}_p \times \mathbf{B}_p = (2J/\hbar)(\mathbf{S}_p \times \mathbf{S}_{p-1} + \mathbf{S}_p \times \mathbf{S}_{p+1}) \ . \tag{16}$$

In Cartesian components

$$dS_p^x/dt = (2J/\hbar)[S_p^y(S_{p-1}^z + S_{p+1}^z) - S_p^z(S_{p-1}^y + S_{p+1}^y)] \ , \tag{17}$$

and similarly for dS_p^y/dt and dS_p^z/dt. These equations involve products of spin components and are nonlinear.

If the amplitude of the excitation is small (if $S_p^x, S_p^y \ll S$), we may obtain an approximate set of linear equations by taking all $S_p^z = S$, and by

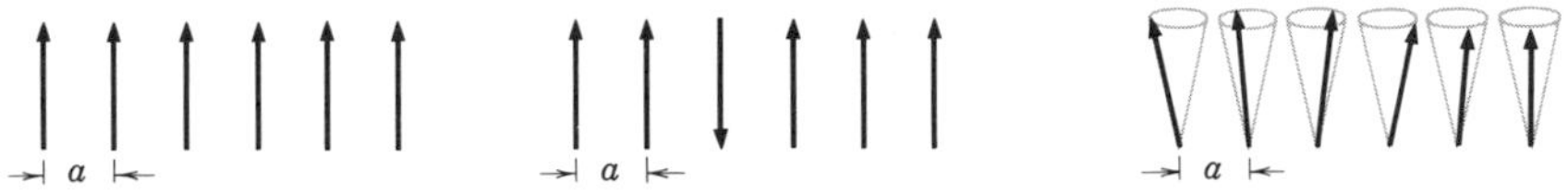

Figure 8 (a) Classical picture of the ground state of a simple ferromagnet; all spins are parallel. (b) A possible excitation; one spin is reversed. (c) The low-lying elementary excitations are spin waves. The ends of the spin vectors precess on the surfaces of cones, with successive spins advanced in phase by a constant angle.

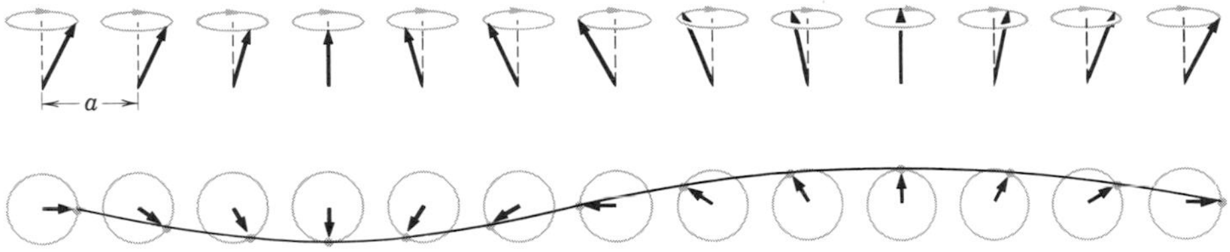

Figure 9 A spin wave on a line of spins. (a) The spins viewed in perspective. (b) Spins viewed from above, showing one wavelength. The wave is drawn through the ends of the spin vectors.

neglecting terms in the product of S^x and S^y which appear in the equation for dS^z/dt. The linearized equations are

$$dS_p^x/dt = (2JS/\hbar)(2S_p^y - S_{p-1}^y - S_{p+1}^y) \ ; \qquad (18a)$$

$$dS_p^y/dt = -(2JS/\hbar)(2S_p^x - S_{p-1}^x - S_{p+1}^x) \ ; \qquad (18b)$$

$$dS_p^z/dt = 0 \ . \qquad (19)$$

By analogy with the phonon problems in Chapter 4 we look for traveling wave solutions of (18) of the form

$$S_p^x = u \exp[i(pka - \omega t)] \ ; \qquad S_p^y = v \exp[i(pka - \omega t)] \ , \qquad (20)$$

where u, v are constants, p is an integer, and a is the lattice constant. On substitution into (18) we have

$$-i\omega u = (2JS/\hbar)(2 - e^{-ika} - e^{ika})v = (4JS/\hbar)(1 - \cos ka)v \ ;$$

$$-i\omega v = -(2JS/\hbar)(2 - e^{-ika} - e^{ika})u = -(4JS/\hbar)(1 - \cos ka)u \ .$$

These equations have a solution for u and v if the determinant of the coefficients is equal to zero:

$$\begin{vmatrix} i\omega & (4JS/\hbar)(1 - \cos ka) \\ -(4JS/\hbar)(1 - \cos ka) & i\omega \end{vmatrix} = 0 \ , \qquad (21)$$

whence

$$\hbar\omega = 4JS(1 - \cos ka) \ . \qquad (22)$$

This result is plotted in Fig. 10. With this solution we find that $v = -iu$, corresponding to circular precession[6] of each spin about the z axis. Equation (22) is the dispersion relation for spin waves in one dimension with

[6]We see this on taking real parts of (20), with v set equal to $-iu$. Then

$$S_p^x = u \cos(pka - \omega t); \qquad S_p^y = u \sin(pka - \omega t) \ .$$

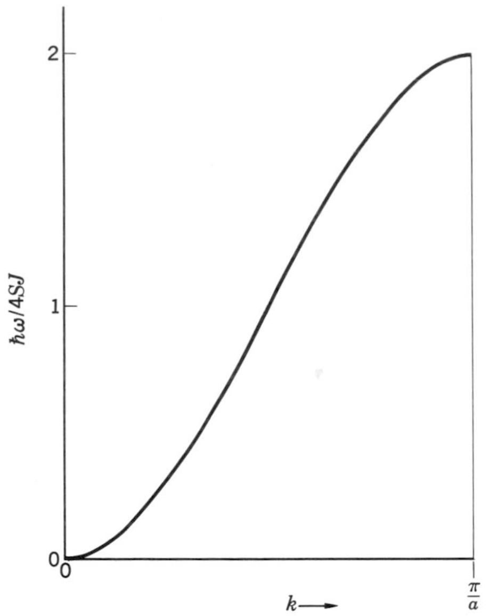

Figure 10 Dispersion relation for magnons in a ferromagnet in one dimension with nearest-neighbor interactions.

nearest-neighbor interactions.[7] At long wavelengths $ka \ll 1$, so that $(1 - \cos ka) \cong \frac{1}{2}(ka)^2$ and

$$\hbar\omega \cong (2JSa^2)k^2 \ . \tag{23}$$

The frequency is proportional to k^2; in the same limit the frequency of a phonon is proportional to k.

The dispersion relation for a ferromagnetic cubic lattice with nearest-neighbor interactions

$$\hbar\omega = 2JS\left[z - \sum_{\delta} \cos(\mathbf{k} \cdot \boldsymbol{\delta})\right] \ , \tag{24}$$

where the summation is over the z vectors denoted by $\boldsymbol{\delta}$ which join the central atom to its nearest neighbors. For $ka \ll 1$,

$$\hbar\omega = (2JSa^2)k^2 \tag{25}$$

for all three cubic lattices, where a is the lattice constant. The coefficient of k^2 often may be determined accurately by spin wave resonance in thin films, Chapter 16. By neutron diffraction G. Shirane and coworkers find, in the equation $\hbar\omega = Dk^2$, the values 281, 500, and 364 meV Å^2 for D at 295 K in Fe, Co, Ni, respectively.

Quantization of Spin Waves. The quantization of spin waves proceeds exactly as for photons and phonons. The energy of a mode of frequency ω_k with n_k magnons is given by

$$\epsilon_k = (n_k + \tfrac{1}{2})\hbar\omega_k \ . \tag{26}$$

The excitation of a magnon corresponds to the reversal of one spin $\frac{1}{2}$.

[7]Precisely the same result is obtained from the quantum-mechanical solution; see *QTS*, Chap. 4.

Thermal Excitation of Magnons

In thermal equilibrium the average value of $n_{\mathbf{k}}$ is given by the Planck distribution:[8]

$$\langle n_{\mathbf{k}}\rangle = \frac{1}{\exp(\hbar\omega_{\mathbf{k}}/k_BT) - 1} \; . \tag{27}$$

The total number of magnons excited at a temperature T is

$$\sum_{\mathbf{k}} n_{\mathbf{k}} = \int d\omega \; \mathcal{D}(\omega)\langle n(\omega)\rangle \; , \tag{28}$$

where $\mathcal{D}(\omega)$ is the number of magnon modes per unit frequency range. The integral is taken over the allowed range of **k**, which is the first Brillouin zone. At sufficiently low temperatures we may carry the integral between 0 and ∞ because $\langle n(\omega)\rangle \rightarrow 0$ exponentially as $\omega \rightarrow \infty$.

Magnons have a single polarization for each value of **k**. In three dimensions the number of modes of wavevector less than k is $(1/2\pi)^3(4\pi k^3/3)$ per unit volume, whence the number of magnons $\mathcal{D}(\omega)d\omega$ with frequency in $d\omega$ at ω is $(1/2\pi)^3(4\pi k^2)(dk/d\omega)\;d\omega$. In the approximation (25),

$$\frac{d\omega}{dk} = \frac{4JSa^2k}{\hbar} = 2\left(\frac{2JSa^2}{h}\right)^{\frac{1}{2}} \omega^{\frac{1}{2}} \; .$$

Thus the density of modes for magnons is

$$\mathcal{D}(\omega) = \frac{1}{4\pi^2}\left(\frac{\hbar}{2JSa^2}\right)^{3/2} \omega^{\frac{1}{2}} \; , \tag{29}$$

so that (28) becomes

$$\sum_{\mathbf{k}} n_{\mathbf{k}} = \frac{1}{4\pi^2}\left(\frac{\hbar}{2JSa^2}\right)^{3/2} \int_0^\infty d\omega \, \frac{\omega^{\frac{1}{2}}}{e^{\beta\hbar\omega} - 1} = \frac{1}{4\pi^2}\left(\frac{k_BT}{2JSa^2}\right)^{3/2} \int_0^\infty dx \, \frac{x^{\frac{1}{2}}}{e^x - 1} \; .$$

The definite integral is found in tables and has the value $(0.0587)(4\pi^2)$.

The number N of atoms per unit volume is Q/a^3, where $Q = 1, 2, 4$ for sc, bcc, fcc lattices, respectively. Now $(\Sigma n_{\mathbf{k}})/NS$ is equal to the fractional change of magnetization $\Delta M/M(0)$, whence

$$\boxed{\frac{\Delta M}{M(0)} = \frac{0.0587}{SQ} \cdot \left(\frac{k_BT}{2JS}\right)^{3/2}} \; . \tag{30}$$

This result due to Felix Bloch is known as the **Bloch $\mathbf{T^{3/2}}$ law**; it has the form found experimentally. In neutron scattering experiments spin waves have been observed up to temperatures near the Curie temperature and even above the Curie temperature.

[8]The argument is exactly as for phonons or photons. The Planck distribution follows for any problem where the energy levels are identical with those of a harmonic oscillator or collection of harmonic oscillators.

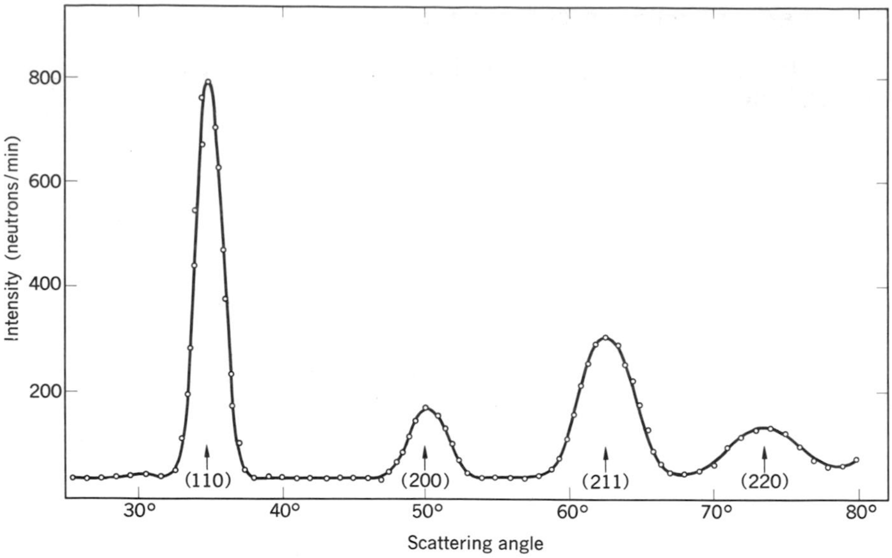

Figure 11 Neutron diffraction pattern for iron. [After C. G. Shull, E. O. Wollan, and W. C. Koehler, Phys. Rev. **84**, 912 (1951).]

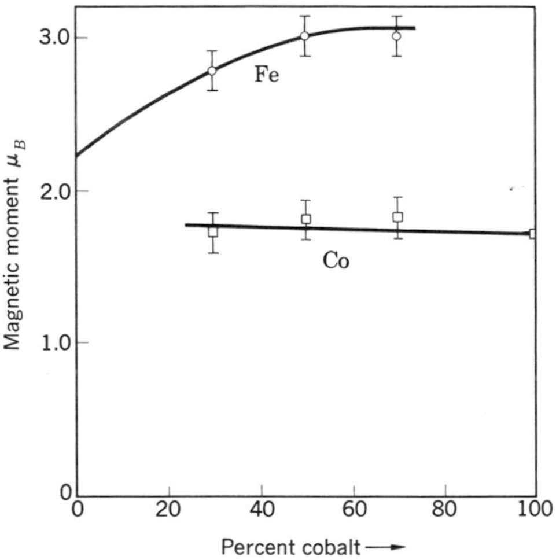

Figure 12 Moments attributable to $3d$ electrons in Fe-Co alloys as a function of composition, after M. F. Collins and J. B. Forsyth. Phil. Mag. **8**, 401 (1963).

NEUTRON MAGNETIC SCATTERING

An x-ray photon sees the spatial distribution of electronic charge, whether or not the charge density is magnetized or unmagnetized. A neutron sees two aspects of a crystal: the distribution of nuclei and the distribution of electronic magnetization. The neutron diffraction pattern for iron is shown in Fig. 11. The magnetic moment of the neutron interacts with the magnetic moment of the electron. The cross section for the neutron-electron interaction is of the same order of magnitude as for the neutron-nuclear interaction. Diffraction of neutrons by a magnetic crystal allows the determination of the distribution, direction, and order of the magnetic moments. Further, a neutron can be inelastically scattered by the magnetic structure, with the creation or annihilation of a magnon; such events make possible the experimental determination of magnon spectra.

The magnetic moments associated with particular components of alloys may be investigated by neutron diffraction. Results for the Fe-Co binary alloy system (which is ferromagnetic) are shown in Fig. 12. Notice that the magnetic moment on the cobalt atom does not appear to be affected by alloying, but that on the iron atom increases to about $3\mu_B$ as the cobalt concentration increases. The magnetizations are shown in Fig. 13, a famous plot.

In an inelastic scattering event a neutron may create or destroy a magnon (Fig. 14). If the incident neutron has wavevector $\mathbf{k}_n$ and is scattered to $\mathbf{k}_n'$ with the creation of a magnon of wavevector $\mathbf{k}$, then by conservation of crystal momentum $\mathbf{k}_n = \mathbf{k}_n' + \mathbf{k} + \mathbf{G}$, where $\mathbf{G}$ is a reciprocal lattice vector. By conservation of energy

$$\frac{\hbar^2 k_n^2}{2M_n} = \frac{\hbar^2 k_n'^2}{2M_n} + \hbar\omega_{\mathbf{k}} \ , \qquad (31)$$

where $\hbar\omega_{\mathbf{k}}$ is the energy of the magnon created in the process. The observed magnon spectrum for $MnPt_3$ is shown in Fig. 15.

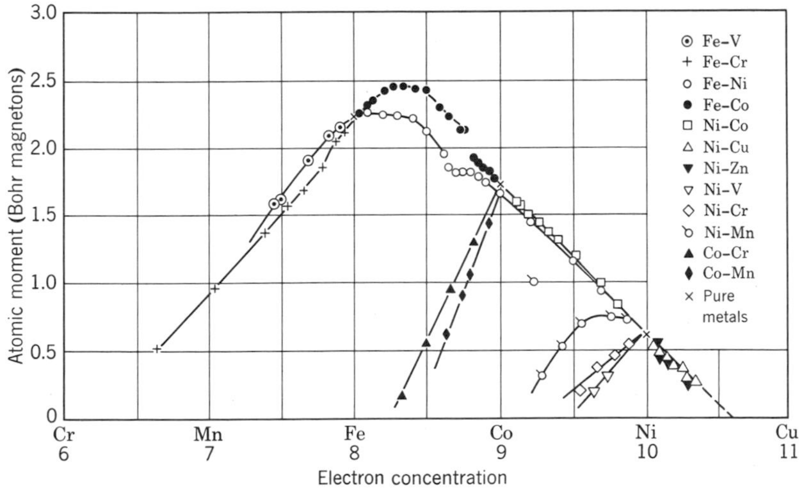

Figure 13 Average atomic moments of binary alloys of the elements in the iron group. (After Bozorth.)

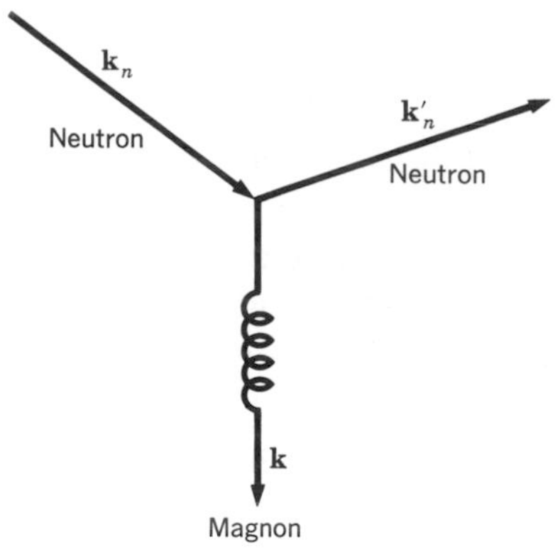

Figure 14 Scattering of a neutron with creation of a magnon.

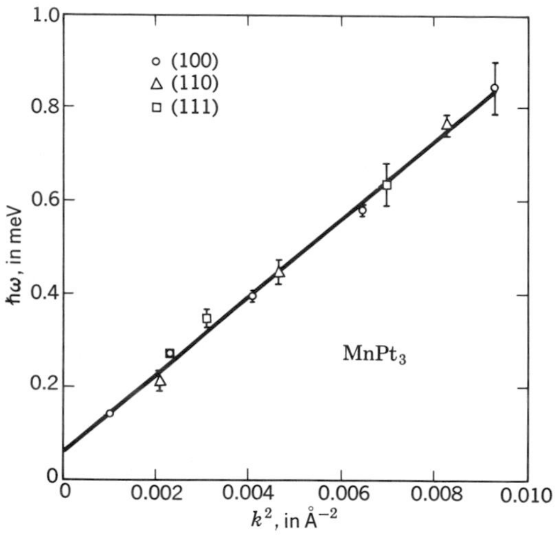

Figure 15 Magnon energy as a function of the square of the wavevector, for the ferromagnet $MnPt_3$. [After B. Antonini and V. J. Minkiewicz, Solid State Comm. **10**, 203 (1972).]

FERRIMAGNETIC ORDER

In many ferromagnetic crystals the saturation magnetization at $T = 0$ K does not correspond to parallel alignment of the magnetic moments of the constituent paramagnetic ions, even in crystals for which there is strong evidence that the individual paramagnetic ions have their normal magnetic moments. The most familiar example is magnetite, Fe_3O_4 or $FeO \cdot Fe_2O_3$. From Table 14.2 we see that ferric (Fe^{3+}) ions are in a state with spin $S = \frac{5}{2}$ and zero orbital moment. Thus each ion should contribute $5\mu_B$ to the saturation moment. The ferrous (Fe^{2+}) ions have a spin of 2 and should contribute $4\mu_B$, apart from any residual orbital moment contribution. Thus the effective number of Bohr magnetons per Fe_3O_4 formula unit should be about $2 \times 5 + 4 = 14$ if all spins were parallel. The observed value (Table 2) is 4.1. The discrepancy is accounted for if the moments of the Fe^{3+} ions are antiparallel to each other: then the observed moment arises only from the Fe^{2+} ion, as in Fig. 16. Neutron diffraction results agree with this model.

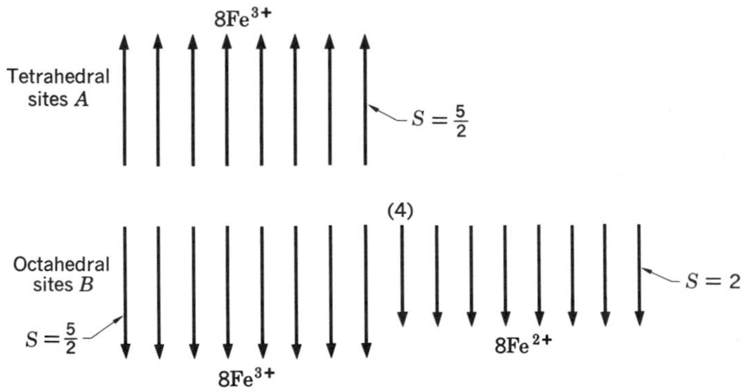

Figure 16 Spin arrangements in magnetite, $FeO \cdot Fe_2O_3$, showing that the moments of the Fe^{3+} ions cancel out, leaving only the moments of the Fe^{2+} ions.

A systematic discussion of the consequences of this type of spin order was given by L. Néel with reference to an important class of magnetic oxides known as ferrites. The usual chemical formula of a ferrite is $MO \cdot Fe_2O_3$, where M is a divalent cation, often Zn, Cd, Fe, Ni, Cu, Co, or Mg. The term **ferrimagnetic** was coined originally to describe the ferrite-type ferromagnetic spin order such as Fig. 16, and by extension the term covers almost any compound in which some ions have a moment antiparallel to other ions. Many ferrimagnets are poor conductors of electricity, a quality exploited in device applications.

The cubic ferrites have the **spinel** crystal structure shown in Fig. 17. There are eight occupied tetrahedral (or A) sites and 16 occupied octahedral (or B) sites in a unit cube. The lattice constant is about 8 Å. A remarkable feature of the spinels is that all exchange interactions (AA, AB, and BB) favor *antiparallel* alignment of the spins connected by the interaction. But the AB interaction is the strongest, so that the A spins are parallel to each other and the B spins are parallel to each other, just in order that the A spins may be antiparallel to the B spins. We believe that all exchange integrals J_{AA}, J_{AB}, and J_{BB} are negative. If J in $U = -2J\mathbf{S}_i \cdot \mathbf{S}_j$ is positive, we say that the exchange integral is ferromagnetic; if J is negative, the exchange integral is antiferromagnetic.

We now prove that three antiferromagnetic interactions can result in ferrimagnetism. The mean exchange fields acting on the A and B spin lattices may be written

$$\mathbf{B}_A = -\lambda\mathbf{M}_A - \mu\mathbf{M}_B \ ; \qquad \mathbf{B}_B = -\mu\mathbf{M}_A - \nu\mathbf{M}_B \ ; \tag{32}$$

taking all constants λ, μ, ν to be positive. The minus sign then corresponds to an antiparallel interaction. The interaction energy density is

$$U = -\tfrac{1}{2}(\mathbf{B}_A \cdot \mathbf{M}_A + \mathbf{B}_B \cdot \mathbf{M}_B) = \tfrac{1}{2}\lambda M_A^2 + \mu\mathbf{M}_A \cdot \mathbf{M}_B + \tfrac{1}{2}\nu M_B^2 \ ; \tag{33}$$

this is lower when $\mathbf{M}_A$ is antiparallel to $\mathbf{M}_B$ than when $\mathbf{M}_A$ is parallel to $\mathbf{M}_B$. The energy when antiparallel should be compared with zero, because

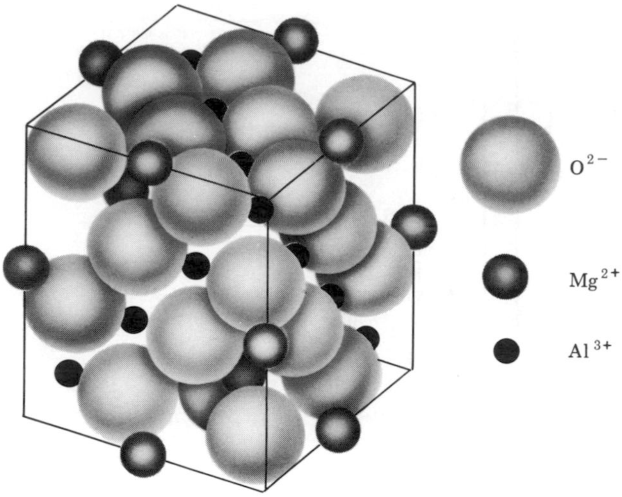

Figure 17 Crystal structure of the mineral spinel $MgAl_2O_4$; the Mg^{2+} ions occupy tetrahedral sites, each surrounded by four oxygen ions; the Al^{3+} occupy octahedral sites, each surrounded by six oxygen ions. This is a **normal spinel** arrangement: the divalent metal ions occupy the tetrahedral sites. In the **inverse spinel** arrangement the tetrahedral sites are occupied by trivalent metal ions, while the octahedral sites are occupied half by divalent and half by trivalent metal ions.

a possible solution is $M_A = M_B = 0$. Thus when

$$\mu M_A M_B > \tfrac{1}{2}(\lambda M_A{}^2 + \nu M_B{}^2) \ , \tag{34}$$

the ground state will have M_A directed oppositely to M_B. Under certain conditions there may be noncollinear spin arrays of still lower energy.

Curie Temperature and Susceptibility of Ferrimagnets

We define separate Curie constants C_A and C_B for the A and B sites. We need separate C's for the two lattices because the numbers and types of paramagnetic ions will usually be different on the two lattices. Let all interactions be zero except for an antiparallel interaction between the A and B sites: $\mathbf{B}_A = -\mu \mathbf{M}_B$; $\mathbf{B}_B = -\mu \mathbf{M}_A$, where μ is positive. The same constant μ is involved in both expressions because of the form of (33). In analogy to the argument of (1) to (4) we have in the mean field approximation

$$\text{(CGS)} \qquad M_A T = C_A(B_a - \mu M_B) \ ; \qquad M_B T = C_B(B_a - \mu M_A) \ , \tag{35}$$

where B_a is the applied field. These equations have a nonzero solution for M_A and M_B in zero applied field if

$$\begin{vmatrix} T & \mu C_A \\ \mu C_B & T \end{vmatrix} = 0 \ , \tag{36}$$

so that the ferrimagnetic Curie temperature is given by $T_c = \mu(C_A C_B)^{\frac{1}{2}}$.

We solve (35) for M_A and M_B to obtain the susceptibility at $T > T_c$:

$$\text{(CGS)} \qquad \chi = \frac{M_A + M_B}{B_a} = \frac{(C_A + C_B)T - 2\mu C_A C_B}{T^2 - T_c^2} \;, \tag{37}$$

a result more complicated than (4). Experimental values for Fe_3O_4 are plotted in Fig. 18. The curvature of the plot of $1/\chi$ versus T is a characteristic feature of a ferrimagnet. We consider below the antiferromagnetic case $C_A = C_B$.

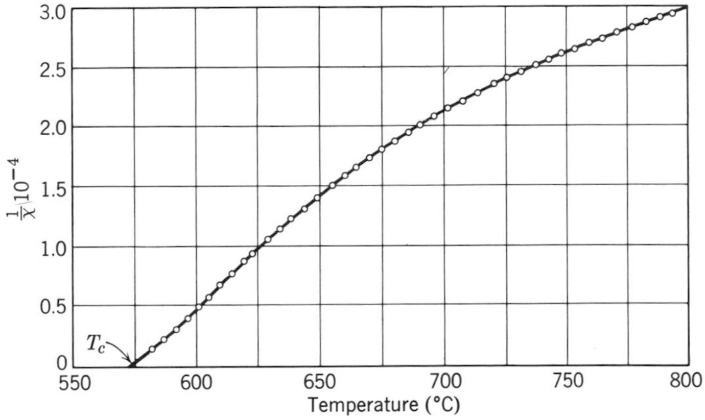

Figure 18 Reciprocal susceptibility of magnetite, $FeO \cdot Fe_2O_3$.

Iron Garnets. The iron garnets are cubic ferrimagnetic insulators with the general formula $M_3Fe_5O_{12}$, where M is a trivalent metal ion and the Fe is the trivalent ferric ion ($S = \frac{5}{2}$, $L = 0$). An example is yttrium iron garnet $Y_3Fe_5O_{12}$, known as YIG. Here Y^{3+} is diamagnetic. The net magnetization of YIG is due to the resultant of two oppositely magnetized lattices of Fe^{3+} ions. At absolute zero each ferric ion contributes $\pm 5\mu_B$ to the magnetization, but in each formula unit the three Fe^{3+} ions on sites denoted as d sites are magnetized in one sense and the two Fe^{3+} ions on a sites are magnetized in the opposite sense, giving a resultant of $5\mu_B$ per formula unit in good agreement with the measurements of Geller et al.[9] The mean field at an a site due to the ions on the d sites is $B_a = -(1.5 \times 10^4)M_d$. The observed Curie temperature 559 K of YIG is due to the a-d interaction. The only magnetic ions in YIG are the ferric ions. Because these are in an $L = 0$ state with a spherical charge distribution their interaction with lattice deformations and phonons is weak. As a result YIG is characterized by very narrow linewidths in ferromagnetic resonance experiments.

In the rare-earth iron garnets[10] the ions M^{3+} are paramagnetic trivalent rare-earth ions. Magnetization curves are given in Fig. 19. The rare-earth

[9]S. Geller, H. J. Williams, R. C. Sherwood, J. P. Remeika, and G. P. Espinosa, Phys. Rev. **131**, 1080 (1963).

[10]A review of their properties is given by L. Néel, R. Pauthenet, and B. Dreyfus, *Prog. in low temperature physics* **4**, 344–383 (1964).

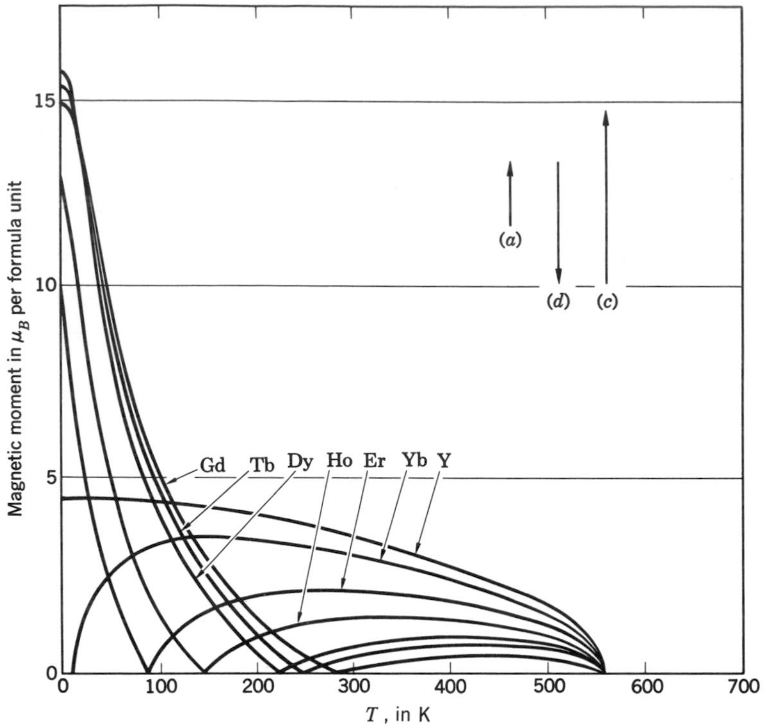

Figure 19 Experimental values of the saturation magnetization versus temperature of various iron garnets, after R. Pauthenet. The formula unit is $M_3Fe_5O_{12}$, where M is a trivalent metal ion. The temperature at which the magnetization crosses zero is called the compensation temperature; here the magnetization of the M sublattice is equal and opposite to the net magnetization of the ferric ion sublattices. Per formula unit there are 3 Fe^{3+} ions on tetrahedral sides *d*; 2 Fe^{3+} ions on octahedral sites *a*; and 3 M^{3+} ions on sites denoted by *c*. The ferric ions contribute $(3-2)5\mu_B = 5\mu_B$ per formula unit. The ferric ion coupling is strong and determines the Curie temperature. If the M^{3+} ions are rare-earth ions they are magnetized opposite to the resultant of the Fe^{3+} ions. The M^{3+} contribution drops rapidly with increasing temperature because the M-Fe coupling is weak. Measurements on single crystal specimens are reported by Geller et al., Phys. Rev. **137**, 1034 (1965).

ions occupy sites labeled *c*; the magnetization M_c of the ions on the *c* lattice is opposite to the net magnetization of the ferric ions on the $a + d$ sites. At low temperatures the combined moments of the three rare-earth ions in a formula unit may dominate the net moment of the Fe^{3+} ions, but because of the weak *c*-*a* and *c*-*d* coupling the rare-earth lattice loses its magnetization rapidly with increasing temperature. The total moment can pass through zero and then increase again as the Fe^{3+} moment starts to be dominant.

ANTIFERROMAGNETIC ORDER

A classical example of magnetic structure determination by neutrons is shown in Fig. 20 for MnO, which has the NaCl structure. At 80 K there are extra neutron reflections not present at 293 K. The reflections at 80 K may be classified in terms of a cubic unit cell of lattice constant 8.85 Å. At 293 K the reflections correspond to an fcc unit cell of lattice constant 4.43 Å. But the lattice constant determined by x-ray reflection is 4.43 Å at *both* temperatures, 80 K and 293 K. We conclude that the chemical unit cell has the 4.43 Å lattice parameter, but that at 80 K the electronic magnetic moments of the Mn^{2+} ions are ordered in some nonferromagnetic arrangement. If the ordering were ferromagnetic, the chemical and magnetic cells would give the same reflections. The spin arrangement shown in Fig. 21 is consistent with the neutron diffraction results and with magnetic measurements. The spins in a single [111] plane are parallel, but adjacent [111] planes are antiparallel. Thus MnO is an antiferromagnet, as in Fig. 22.

In an **antiferromagnet** the spins are ordered in an antiparallel arrangement with zero net moment at temperatures below the ordering or **Néel temperature.** The susceptibility of an antiferromagnet is not infinite at $T = T_N$, but has a weak cusp, as in Fig. 23. An antiferromagnet is a special case of a ferrimagnet for which both sublattices A and B have equal saturation magnetizations. Thus $C_A = C_B$ in (37), and the Néel temperature in the mean field approximation is given by

$$T_N = \mu C \ , \tag{38}$$

where C refers to a single sublattice. The susceptibility in the paramagnetic region $T > T_N$ is obtained from (37):

$$\chi = \frac{2CT - 2\mu C^2}{T^2 - (\mu C)^2} = \frac{2C}{T + \mu C} = \frac{2C}{T + T_N} \ . \tag{39}$$

The experimental results at $T > T_N$ are of the form

(CGS) $$\chi = \frac{2C}{T + \theta} \ . \tag{40}$$

Experimental values of θ / T_N listed in Table 3 often differ substantially from the value unity expected from (39). Values of θ / T_N of the observed magnitude may be obtained when next-nearest-neighbor interactions are provided for, and when sublattice arrangements are considered. If a mean field constant $-\epsilon$ is introduced to describe interactions within a sublattice, then $\theta / T_N = (\mu + \epsilon)/(\mu - \epsilon)$.

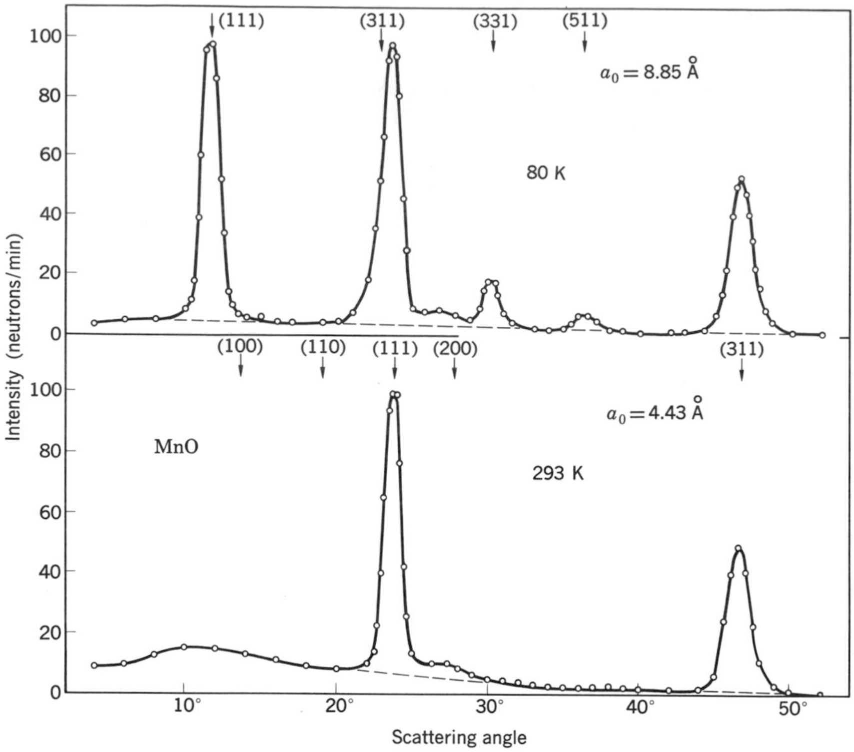

Figure 20 Neutron diffraction patterns for MnO below and above the spin-ordering temperature of 120 K, after C. G. Shull, W. A. Strauser, and E. O. Wollan. Phys. Rev. **83,** 333 (1951). The reflection indices are based on an 8.85 Å cell at 80 K and on a 4.43 Å cell at 293 K. At the higher temperature the Mn^{2+} ions are still magnetic, but they are no longer ordered.

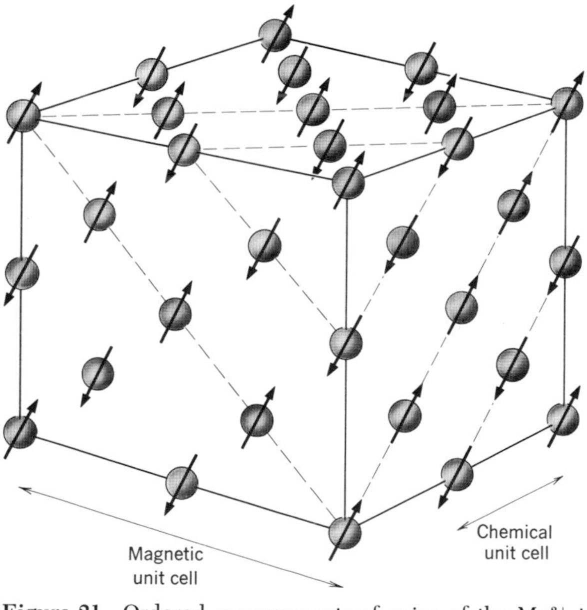

Figure 21 Ordered arrangements of spins of the Mn^{2+} ions in manganese oxide. MnO, as determined by neutron diffraction. The O^{2-} ions are not shown.

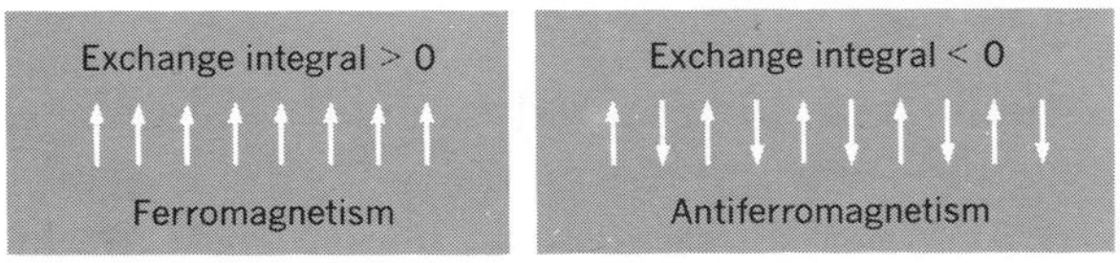

Figure 22 Spin ordering in ferromagnets ($J > 0$) and antiferromagnets ($J < 0$).

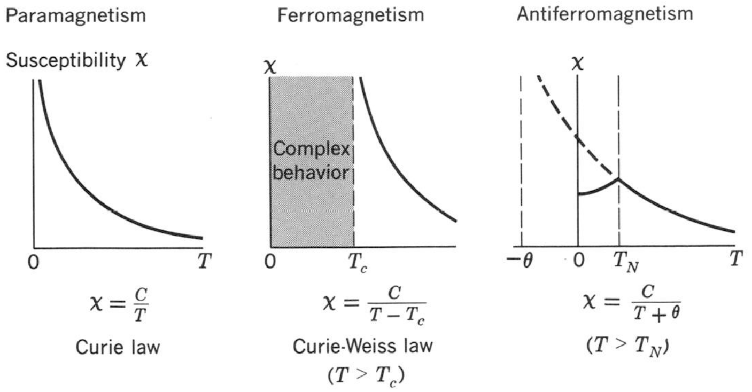

Figure 23 Temperature dependence of the magnetic susceptibility in paramagnets, ferromagnets, and antiferromagnets. Below the Néel temperature of an antiferromagnet the spins have antiparallel orientations; the susceptibility attains its maximum value at T_N where there is a well-defined kink in the curve of χ versus T. The transition is also marked by peaks in the heat capacity and the thermal expansion coefficient.

TABLE 3 Antiferromagnetic Crystals

Substance	Paramagnetic Ion Lattice	Transition Temperature, T_N, in K	Curie-Weiss θ, in K	$\frac{\theta}{T_N}$	$\frac{\chi(0)}{\chi(T_N)}$
MnO	fcc	116	610	5.3	$\frac{2}{3}$
MnS	fcc	160	528	3.3	0.82
MnTe	hex. layer	307	690	2.25	
MnF_2	bc tetr	67	82	1.24	0.76
FeF_2	bc tetr	79	117	1.48	0.72
$FeCl_2$	hex. layer	24	48	2.0	<0.2
FeO	fcc	198	570	2.9	0.8
$CoCl_2$	hex. layer	25	38.1	1.53	
CoO	fcc	291	330	1.14	
$NiCl_2$	hex. layer	50	68.2	1.37	
NiO	fcc	525	~2000	~4	
Cr	bcc	308			

Susceptibility Below the Néel Temperature

There are two situations: with the applied magnetic field perpendicular to the axis of the spins; and with the field parallel to the axis of the spins. At the Néel temperature the susceptibility is nearly independent of the direction of the field relative to the spin axis. For $\mathbf{B}_a$ perpendicular to the axis of the spins we can calculate the susceptibility by elementary considerations. The energy density in the presence of the field is, with $M = |M_A| = |M_B|$,

$$U = \mu \mathbf{M}_A \cdot \mathbf{M}_B - \mathbf{B}_a \cdot (\mathbf{M}_A + \mathbf{M}_B) \cong -\mu M^2[1 - \tfrac{1}{2}(2\varphi)^2] - 2B_a M\varphi \ , \quad (41)$$

where 2φ is the angle the spins make with each other (Fig. 24a). The energy is a minimum when

$$dU/d\varphi = 0 = 4\mu M^2 \varphi - 2B_a M \ ; \qquad \varphi = B_a/2\mu M \ , \quad (42)$$

so that

(CGS) $$\chi_\perp = 2M\varphi/B_a = 1/\mu \ . \quad (43)$$

In the parallel orientation (Fig. 24b) the magnetic energy is not changed if the spin systems A and B make equal angles with the field. Thus the susceptibility at $T = 0$ K is zero:

$$\chi_\parallel(0) = 0 \ . \quad (44)$$

The parallel susceptibility increases smoothly with temperature up to T_N. Measurements on MnF_2 are shown in Fig. 25. In very strong fields the spin systems will turn discontinuously from the parallel orientation to the perpendicular orientation where the energy is lower.

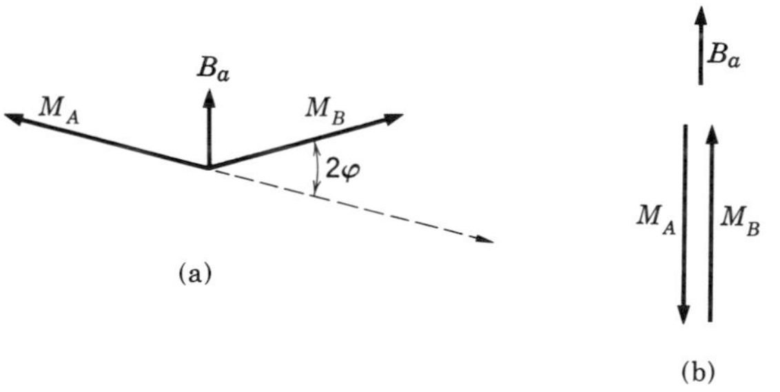

Figure 24 Calculation of (a) perpendicular and (b) parallel susceptibilities at 0 K, in the mean field approximation.

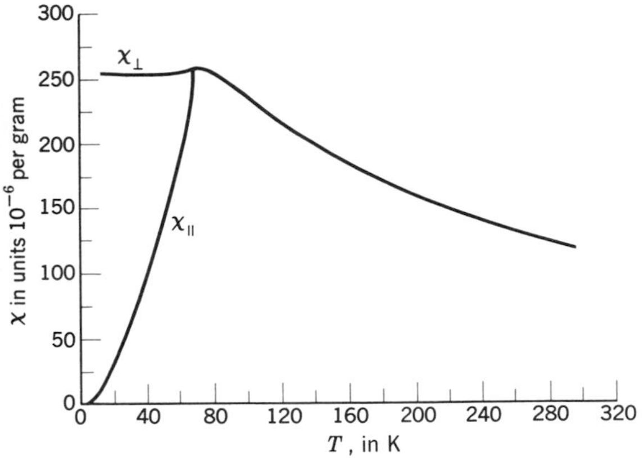

Figure 25 Magnetic susceptibility of manganese fluoride, MnF_2, parallel and perpendicular to the tetragonal axis. (After S. Foner.)

Antiferromagnetic Magnons

We obtain simply an expression for the dispersion relation of magnons in a one-dimensional antiferromagnet by making the appropriate substitutions in the treatment of the ferromagnetic line. Let spins with even indices $2p$ compose sublattice A, that with spins up ($S_z = S$); and let spins with odd indices $2p+1$ compose sublattice B, that with spins down ($S_z = -S$). We consider only nearest-neighbor interactions, with J negative. Then (18) written for A becomes, with a careful look at (17),

$$dS^x_{2p}/dt = (2JS/\hbar)(-2S^y_{2p} - S^y_{2p-1} - S^y_{2p+1}) \; ; \qquad (45a)$$

$$dS^y_{2p}/dt = -(2JS/\hbar)(-2S^x_{2p} - S^x_{2p-1} - S^x_{2p+1}) \; . \qquad (45b)$$

The corresponding equations for a spin on B are

$$dS^x_{2p+1}/dt = (2JS/\hbar)(2S^y_{2p+1} + S^y_{2p} + S^y_{2p+2}) \; ; \qquad (46a)$$

$$dS^y_{2p+1}/dt = -(2JS/\hbar)(2S^x_{2p+1} + S^x_{2p} + S^x_{2p+2}) \; . \qquad (46b)$$

We form $S^+ = S_x + iS_y$; then

$$dS^+_{2p}/dt = (2iJS/\hbar)(2S^+_{2p} + S^+_{2p-1} + S^+_{2p+1}) \; ; \qquad (47)$$

$$dS^+_{2p+1}/dt = -(2iJS/\hbar)(2S^+_{2p+1} + S^+_{2p} + S^+_{2p+2}) \; . \qquad (48)$$

We look for solutions of the form

$$S^+_{2p} = u \exp(ipka - i\omega t) \; ; \qquad S^+_{2p+1} = v \exp(ipka - i\omega t) \; , \qquad (49)$$

so that (47) and (48) become, with $\omega_{ex} \equiv -4JS/\hbar = 4|J|S/\hbar$,

$$\omega u = \tfrac{1}{2}\omega_{ex}(2u + ve^{-ika} + ve^{ika}) \; ; \qquad (50a)$$

$$-\omega v = \tfrac{1}{2}\omega_{ex}(2v + ue^{-ika} + ue^{ika}) \; . \qquad (50b)$$

Equations (50) have a solution if

$$\begin{vmatrix} \omega_{ex} - \omega & \omega_{ex}\cos ka \\ \omega_{ex}\cos ka & \omega_{ex} + \omega \end{vmatrix} = 0 \; ; \qquad (51)$$

thus

$$\omega^2 = \omega_{ex}^2(1 - \cos^2 ka) \; ; \qquad \omega = \omega_{ex}|\sin ka| \; . \qquad (52)$$

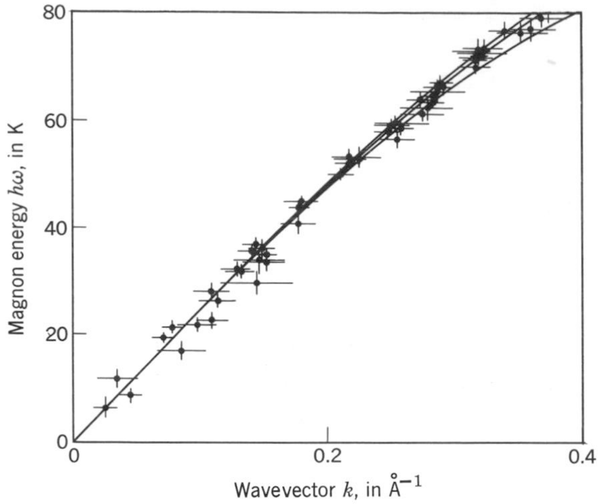

Figure 26 Magnon dispersion relation in the simple cubic antiferromagnet $RbMnF_3$ as determined at 4.2 K by inelastic neutron scattering. [After C. G. Windsor and R. W. H. Stevenson, Proc. Phys. Soc. (London) **87**, 501 (1966).]

The dispersion relation for magnons in an antiferromagnet is quite different from (22) for magnons in a ferromagnet. For $ka \ll 1$ we see that (52) is linear[11] in k: $\omega \cong \omega_{ex}|ka|$. The magnon spectrum of $RbMnF_3$ is shown in Fig. 26, as determined by inelastic neutron scattering experiments. There is a large region in which the magnon frequency is directly proportional to the wavevector. Well-resolved magnons have been observed in MnF_2 at specimen temperatures up to 0.93 of the Néel temperature. Thus even at high temperatures the magnon approximation is useful. Further details concerning antiferromagnetic magnons are given in *QTS*, Chap. 4.

FERROMAGNETIC DOMAINS

At temperatures well below the Curie point the electronic magnetic moments of a ferromagnet are essentially all lined up when regarded on a microscopic scale. Yet, looking at a specimen as a whole, the magnetic moment may be very much less than the saturation moment, and the application of an external magnetic field may be required to saturate the specimen. The behavior observed in polycrystalline specimens is similar to that in single crystals. Actual specimens are composed of small regions called domains, within each of which the local magnetization is saturated. The directions of magnetization of different domains need not be parallel. An arrangement of domains with approximately zero resultant magnetic moment is shown in Fig. 27. Domains form also in antiferromagnetics, ferroelectrics, antiferroelectrics, ferroelastics, superconductors, and sometimes in metals under conditions of a strong de Haas-van Alphen effect.

[11]A physical discussion of the difference between the dispersion relations for ferromagnetic and antiferromagnetic magnons is given by F. Keffer, H. Kaplan, and Y. Yafet, Amer. J. Phys. **21**, 250 (1953).

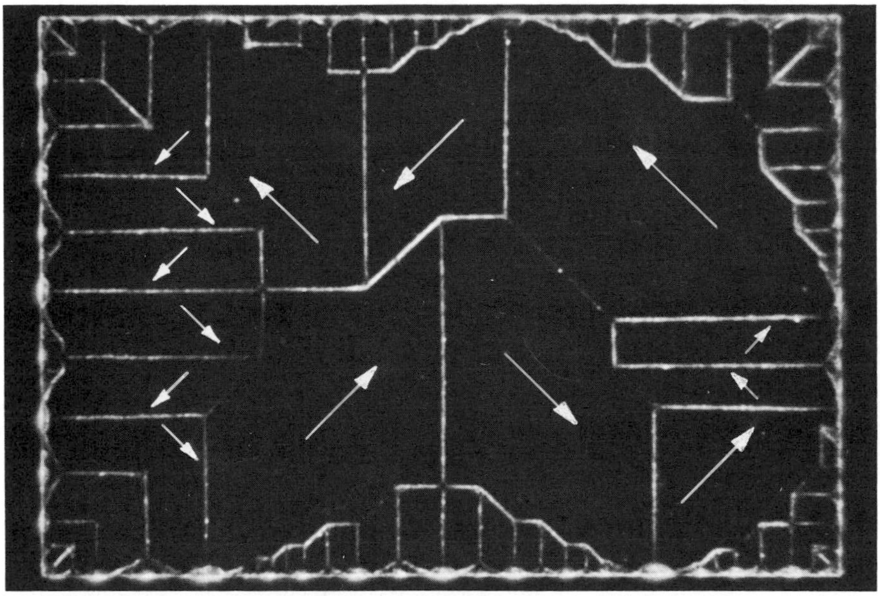

Figure 27 Ferromagnetic domain pattern on a single crystal platelet of nickel. The domain boundaries are made visible by the Bitter technique. The direction of magnetization within a domain is determined by observing growth or contraction of the domain in a magnetic field. (After R. W. De Blois.)

The increase in the magnetic moment of the specimen under the action of an applied magnetic field takes place by two independent processes:

1. In weak applied fields the volume of domains (Fig. 28) favorably oriented with respect to the field increases at the expense of unfavorably oriented domains;
2. In strong applied fields the magnetization rotates toward the direction of the field.

Technical terms defined by the hysteresis loop are shown in Fig. 29. The domain structure of ferromagnetic materials affects their practical properties. In a transformer core we want a high permeability; in a permanent magnet we want a high coercive force.[12] By making the material pure, homogeneous, and well oriented, we facilitate boundary displacement and thereby attain high permeability; values of the relative permeability up to 3.8×10^6 have been reported. By suppressing the possibility of boundary displacement, we may achieve a high coercivity; the suppression may be accomplished by using very fine particles or by precipitating a second metallurgical phase so that the specimen is heterogeneous on a very fine scale.

[12]The **coercive force** is defined as the reverse field needed to reduce the induction B or the magnetization M to zero, starting in a saturated condition. Usually the definition is understood to refer to B, except in theoretical work. When referred to M, one writes ${}_IH_c$ or ${}_MH_c$.

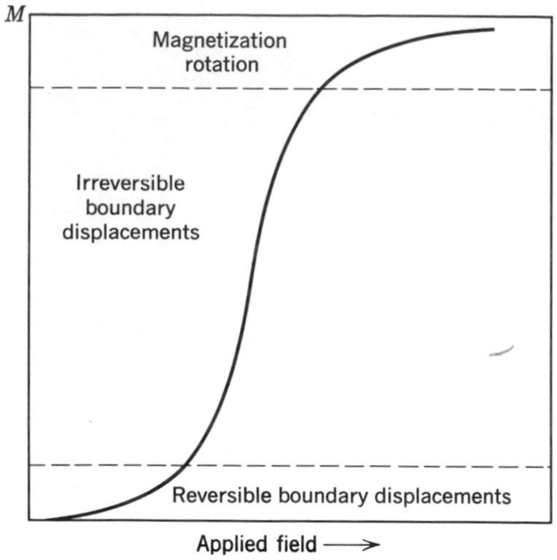

Figure 28 Representative magnetization curve, showing the dominant magnetization processes in the different regions of the curve.

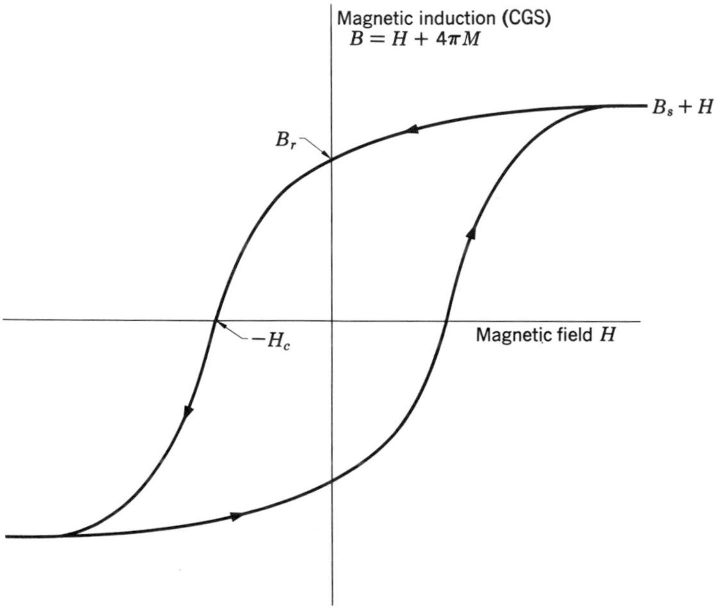

Figure 29 The technical magnetization curve. The **coercive force** H_c is the reverse field necessary to bring the induction B to zero; the **remanence** B_r is the value of B at $H = 0$; the **saturation induction** B_s is defined as the limiting value of $(B - H)$ for large H. The **saturation magnetization** M_s is given by $B_s/4\pi$. In SI the vertical axis is $B = \mu_0(H + M)$. (The vertical scale has been greatly compressed.)

Anisotropy Energy

There is an energy in a ferromagnetic crystal which directs the magnetization along certain definite crystallographic axes called directions of easy magnetization. This energy is called the **magnetocrystalline** or **anisotropy energy.** It does not come about from the pure isotropic exchange interaction considered thus far. Cobalt is a hexagonal crystal. The hexagonal axis is the direction of easy magnetization at room temperature, as shown in Fig. 30. One origin of the anisotropy energy is illustrated by Fig. 31. The magnetization of the crystal sees the crystal lattice through orbital overlap of the electrons: the spin interacts with the orbital motion by means of the spin-orbit coupling.

In cobalt the anisotropy energy density is given by

$$U_K = K_1' \sin^2 \theta + K_2' \sin^4 \theta \ , \tag{53}$$

where θ is the angle the magnetization makes with the hexagonal axis. At room temperature $K_1' = 4.1 \times 10^6$ erg/cm^3; $K_2' = 1.0 \times 10^6$ erg/cm^3. Iron is a cubic crystal, and the cube edges are the directions of easy magnetization. To represent the anisotropy energy of iron magnetized in an arbitrary direction with direction cosines α_1, α_2, α_3 referred to the cube edges, we are guided by cubic symmetry. The expression for the anisotropy energy must be an even power of each α_i, provided opposite ends of a crystal axis are equivalent magnetically, and it must be invariant under interchanges of the α_i among themselves. The lowest order combination satisfying the symmetry requirements is $\alpha_1^2 + \alpha_2^2 + \alpha_3^2$, but this is identically equal to unity and does not describe anisotropy effects. The next combination is of the fourth degree: $\alpha_1^2\alpha_2^2 + \alpha_1^2\alpha_3^2 + \alpha_3^2\alpha_2^2$, and then of the sixth degree: $\alpha_1^2\alpha_2^2\alpha_3^2$. Thus

$$U_K = K_1(\alpha_1^2\alpha_2^2 + \alpha_2^2\alpha_3^2 + \alpha_3^2\alpha_1^2) + K_2\alpha_1^2\alpha_2^2\alpha_3^2 \ ; \tag{54}$$

at room temperature $K_1 = 4.2 \times 10^5$ erg/cm^3 and $K_2 = 1.5 \times 10^5$ erg/cm^3.

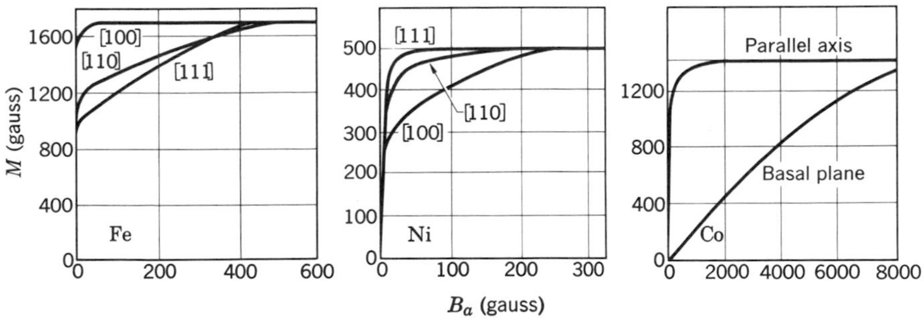

Figure 30 Magnetization curves for single crystals of iron, nickel, and cobalt. From the curves for iron we see that the [100] directions are easy directions of magnetization and the [111] directions are hard directions. The applied field is B_a. (After Honda and Kaya.)

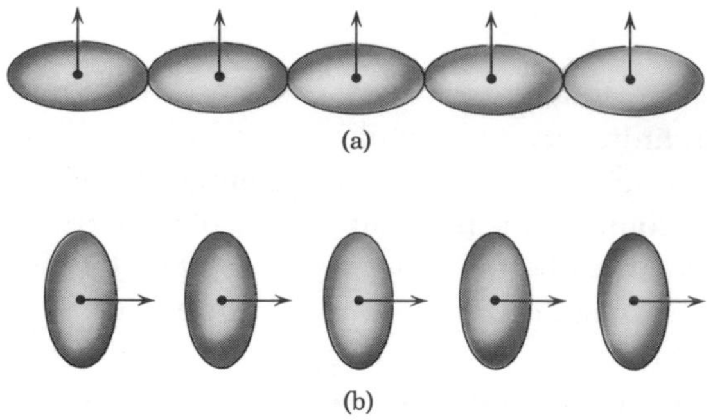

Figure 31 Asymmetry of the overlap of electron distributions on neighboring ions provides one mechanism of magnetocrystalline anisotropy. Because of spin-orbit interaction the charge distribution is spheroidal and not spherical. The asymmetry is tied to the direction of the spin, so that a rotation of the spin directions relative to the crystal axes changes the exchange energy and also changes the electrostatic interaction energy of the charge distributions on pairs of atoms. Both effects give rise to an anisotropy energy. The energy of (a) is not the same as the energy of (b).

Transition Region Between Domains

A **Bloch wall** in a crystal is the transition layer that separates adjacent regions (domains) magnetized in different directions. The entire change in spin direction between domains does not occur in one discontinuous jump across a single atomic plane, but takes place in a gradual way over many atomic planes (Fig. 32). The exchange energy is lower when the change is distributed over many spins. This behavior may be understood by interpreting the Heisenberg equation (6) classically; we also replace $\cos\varphi$ by $1-\frac{1}{2}\varphi^2$. Then $w_{ex}=JS^2\varphi^2$ is the exchange energy between two spins making a small angle φ with each other. Here J is the exchange integral and S is the spin quantum number; w_{ex} is referred to the energy for parallel spins. If a total change of π occurs in N equal steps, the angle between neighboring spins is π/N, and the exchange energy per pair of neighboring atoms is $w_{ex}=JS^2(\pi/N)^2$. The total exchange energy of a line of $N+1$ atoms is

$$Nw_{ex}=JS^2\pi^2/N \ . \tag{55}$$

The wall would thicken without limit were it not for the anisotropy energy, which acts to limit the width of the transition layer. The spins contained within the wall are largely directed away from the axes of easy magnetization, so there is an anisotropy energy associated with the wall, roughly proportional to the wall thickness.

Let us consider a wall parallel to the cube face of a simple cubic lattice and separating domains magnetized in opposite directions. We wish to de-

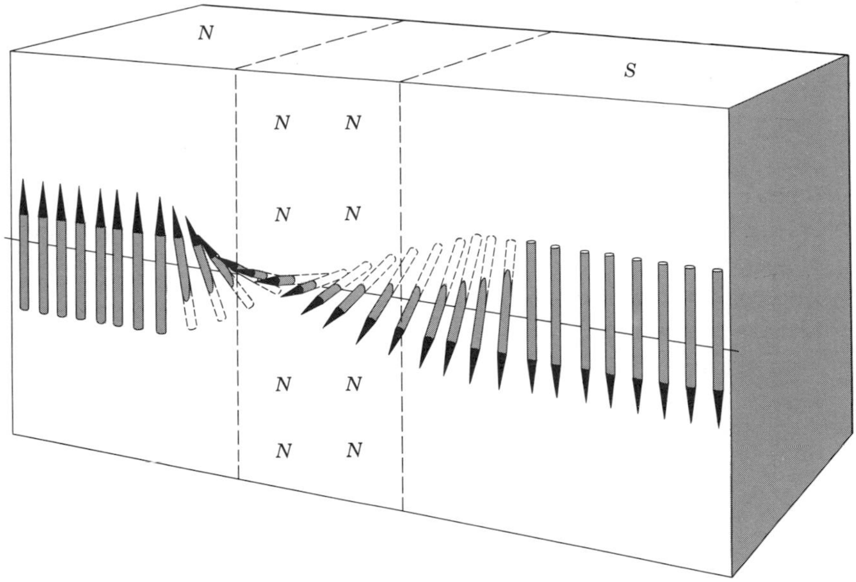

Figure 32 The structure of the Bloch wall separating domains. In iron the thickness of the transition region is about 300 lattice constants.

termine the number N of atomic planes contained within the wall. The energy per unit area of wall is the sum of contributions from exchange and anisotropy energies: $\sigma_w = \sigma_{ex} + \sigma_{\text{anis}}$. The exchange energy is given approximately by (55) for each line of atoms normal to the plane of the wall. There are $1/a^2$ such lines per unit area, where a is the lattice constant. Thus $\sigma_{ex} = \pi^2 JS^2/Na^2$. The anisotropy energy is of the order of the anisotropy constant times the thickness Na, or $\sigma_{\text{anis}} \approx KNa$; therefore

$$\sigma_w \approx (\pi^2 JS^2/Na^2) + KNa \ . \tag{56}$$

This is a minimum with respect to N when

$$\partial\sigma_w/\partial N = 0 = -(\pi^2 JS^2/N^2a^2) + Ka \ ; \tag{57}$$

or

$$N = (\pi^2 JS^2/Ka^3)^{\frac{1}{2}} \ . \tag{58}$$

For order of magnitude, $N \approx 300$ in iron. The total wall energy per unit area on our model is

$$\sigma_w = 2\pi(KJS^2/a)^{\frac{1}{2}} \ ; \tag{59}$$

in iron $\sigma_w \approx 1$ erg/cm^2. Accurate calculation for a 180° wall in a (100) plane gives $\sigma_w = 2(2K_1JS^2/a)^{\frac{1}{2}}$.

Origin of Domains

Landau and Lifshitz showed that domain structure is a natural consequence of the various contributions to the energy—exchange, anisotropy, and magnetic—of a ferromagnetic body. Direct evidence of domain structure is furnished by photomicrographs of domain boundaries obtained by the technique of magnetic powder patterns and by optical studies using Faraday rotation. The powder pattern method developed by F. Bitter consists in placing a drop of a colloidal suspension of finely divided ferromagnetic material, such as magnetite, on the surface of the ferromagnetic crystal. The colloid particles in the suspension concentrate strongly about the boundaries between domains where strong local magnetic fields exist which attract the magnetic particles. The discovery of transparent ferromagnetic compounds has encouraged the use also of optical rotation for domain studies.

We may understand the origin of domains by considering the structures shown in Fig. 33, each representing a cross section through a ferromagnetic single crystal. In (a) we have a single domain; as a consequence of the magnetic "poles" formed on the surfaces of the crystal this configuration will have a high value of the magnetic energy $(1/8\pi)\int B^2\, dV$. The magnetic energy density for the configuration shown will be of the order of $M_s^2 \approx 10^6$ erg/cm^3; here M_s denotes the saturation magnetization, and the units are CGS.

In (b) the magnetic energy is reduced by roughly one-half by dividing the crystal into two domains magnetized in opposite directions. In (c) with N domains the magnetic energy is reduced to approximately $1/N$ of the magnetic energy of (a), because of the reduced spatial extension of the field. In domain arrangements such as (d) and (e) the magnetic energy is zero. Here the boundaries of the triangular prism domains near the end faces of the crystal make equal angles (45°) with the magnetization in the rectangular domains and with the magnetization in the domains of closure. The component of magnetization normal to the boundary is continuous across the boundary and there is no magnetic field associated with the magnetization. The flux circuit is completed with the crystal—thus giving rise to the phrase **domains of closure** for surface domains that complete the flux circuit, as in Fig. 34.

Domain structures are often more complicated than our simple examples, but *domain structure always has its origin in the possibility of lowering the energy of a system by going from a saturated configuration with high magnetic energy to a domain configuration with a lower energy.*

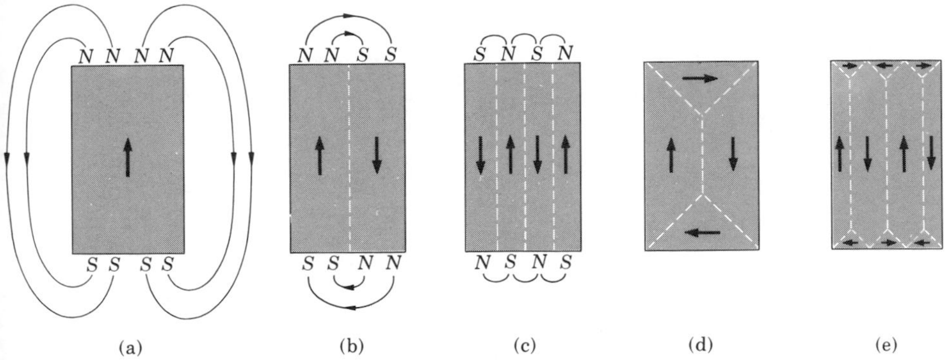

Figure 33 The origin of domains.

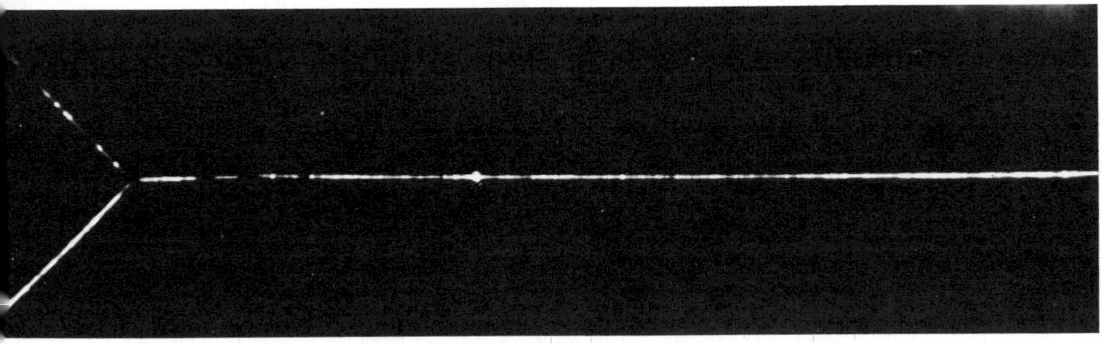

Figure 34 Domain of closure at the end of a single crystal iron whisker. The face is a (100) plane; the whisker axis is [001]. (Courtesy of R. V. Coleman, C. G. Scott, and A. Isin.)

Coercive Force and Hysteresis

The coercive force is the field H_c required to reduce the induction B to zero (Fig. 29). It is the most sensitive property of ferromagnetic materials which is subject to control. The coercive force may range from the value of 600 G in a loudspeaker permanent magnet (Alnico V) and 20,000 G in a special high stability magnet (Fe-Pt) to 0.5 G in a commercial power transformer (Si-Fe) or 0.004 G in a pulse transformer (Supermalloy). In a transformer low hysteresis is desired; this means a low coercive force. The coercive force decreases as the impurity content decreases

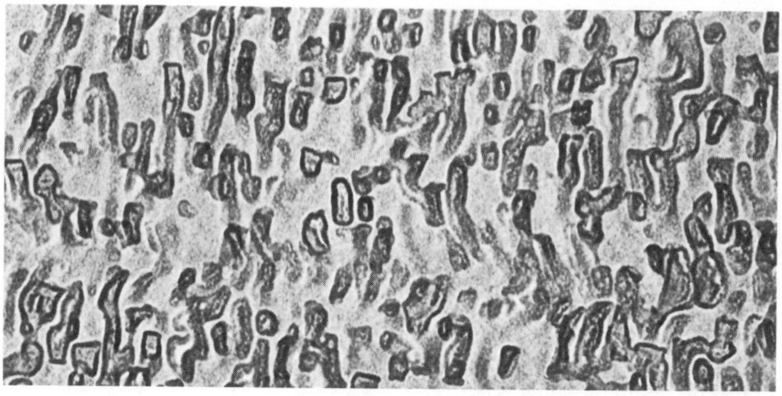

Figure 35 Microstructure of Alnico V in its optimum state as a permanent magnet. The composition of Alnico V is, by weight, 8 Al, 14 Ni, 24 Co, 3 Cu, 51 Fe. As a permanent magnet it is a two-phase system, with fine particles of one phase embedded in the other phase. The precipitation is carried out in a magnetic field, and the particles are oriented with their long axis parallel to the field direction. The width shown is 1.1 μm. (Courtesy of F. E. Luborsky.)

and also as internal strains are removed by annealing (slow cooling). Alloys that contain a precipitated phase may have a high coercivity, as in Fig. 35. The high coercivity of materials composed of very small grains or fine powders is well understood. A sufficiently small particle, with diameter less than 10^{-5} or 10^{-6} cm, is always magnetized to saturation as a single domain because the formation of a flux-closure configuration is energetically unfavorable.

In a sufficiently small single domain particle it will not be possible for magnetization reversal to take place by means of the process of boundary displacement, which usually requires relatively weak fields. Instead the magnetization of the particle must rotate as a whole, a process that may require large fields depending on the anisotropy energy of the material and the shape of the particle. The coercive force of fine iron particles is expected theoretically to be about 500 gauss on the basis of rotation opposed by the crystalline anisotropy energy, and this is of the order of the value reported by several observers. Higher coercivities have been reported for elongated iron particles, the rotation here being opposed by the shape anisotropy of the demagnetization energy. A promising permanent magnet material[13] is Co_5Sm: an anisotropy energy equivalent to a magnetic field of 290 kG or 29 T has been observed in a cobalt-samarium single crystal.

[13]A review of rare-earth-compound permanent magnet materials is given by J. J. Becker, J. Appl. Phys. **41,** 1055 (1970).

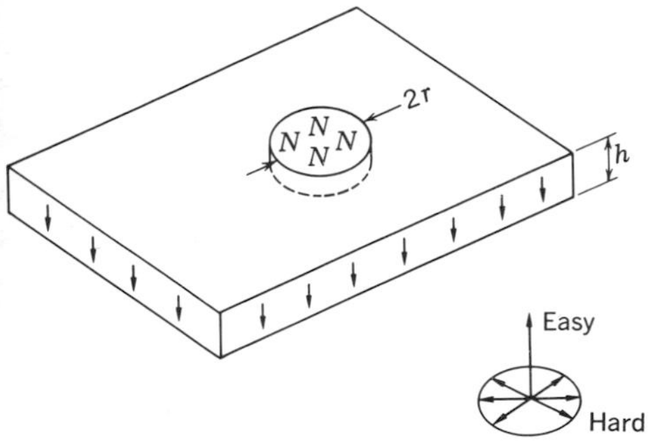

Figure 36 A magnetic bubble in a platelet of ferromagnetic material. The platelet is magnetized down and the bubble up.

Magnetic Bubble Domains

Small cylindrical magnetic domains can be stabilized in a thin crystal of a uniaxial magnetic material by a bias magnetic field. Such magnetic bubbles (Fig. 36) are of interest for high density memory storage.[14] The bubble diameter may be $\approx 10\ \mu\text{m}$, and the optimum thickness of the crystal platelet is comparable. The easy axis of magnetization and a bias field B must be normal to the platelet. There are three contributions to the energy of a bubble domain of radius r and thickness h:

1. The domain wall energy $U_w = 2\pi rh\sigma$.
2. The magnetostatic energy of the bubble; this is a complicated quantity[15] of the order of $-4\pi M_s^2$ times the volume, or $U_M \approx -4\pi r^2 hM_s^2$, for thin bubbles with $r \approx h$. This contribution is referred to the uniformly magnetized platelet with no bubble present.

Thus the total energy of a single bubble is given by Bobeck. It

$$U = 2\pi rh\sigma - 4\pi r^2 hM_s^2 + 2\pi r^2 hM_s B, \tag{60}$$

is approximately a minimum when

$$B = \frac{\sigma}{2M^2 r_0} - \frac{4\pi Mh}{h + 1.45r_0} \tag{61}$$

Now σ is characteristically $\approx 1\ \text{erg/cm}^2$, and $M_s \approx 10$ in the rare earth orthoferrites, such as $YFeO_3$, that are favorable materials for bubble domains. Thus the length $\sigma/2M_s^2 \approx 4 \times 10^{-3}$ cm is characteristic for the radius of a domain. The bias field B has an important role that is not brought out by the analysis above: The circular domain is not stable if B is below a critical value, which is 33 G in $YFeO_3$. Below this field the bubble is unstable with respect to a serpentine domain pattern.

[14] A. H. Bobeck and H. E. D. Scovil, Scientific American **224**, 78 (June 1971).
[15] A. H. Bobeck, Bell Syst. Tech Jour. **46**, 1901 (1967).

SUMMARY (In CGS units)

1. The susceptibility of a ferromagnet above the Curie temperature has the form $\chi = C/(T - T_c)$ in the mean field approximation.

2. In the mean field approximation the effective magnetic field seen by a magnetic moment in a ferromagnet is $\mathbf{B}_a + \lambda\mathbf{M}$, when $\lambda = T_c/C$ and $\mathbf{B}_a$ is the applied magnetic field.

3. The elementary excitations in a ferromagnet are magnons. Their dispersion relation for $ka \ll 1$ has the form $\hbar\omega \approx Jk^2a^2$ in zero external magnetic field. The thermal excitation of magnons leads at low temperatures to a heat capacity and to a fractional magnetization change both proportional to $T^{3/2}$.

4. In an antiferromagnet two spin lattices are equal, but antiparallel. In a ferrimagnet two lattices are antiparallel, but the magnetic moment of one is larger than the magnetic moment of the other.

5. In an antiferromagnet the susceptibility above the Néel temperature has the form $\chi = 2C/(T + \theta)$.

6. The magnon dispersion relation in an antiferromagnet has the form $\hbar\omega \approx Jka$. The thermal excitation of magnons leads at low temperatures to a term in T^3 in the heat capacity, in addition to the phonon term in T^3.

7. A Bloch wall separates domains magnetized in different directions. The thickness of a wall is $\approx (J/Ka^3)^{\frac{1}{2}}$ lattice constants, and the energy per unit area is $\approx (KJ/a)^{\frac{1}{2}}$, where K is the anisotropy energy density.

PROBLEMS

1. ***Magnon dispersion relation.*** Derive the magnon dispersion relation (24) for a spin S on a simple cubic lattice, $z = 6$. Hint: Show first that (18a) is replaced by

$$dS_\rho{}^x/dt = (2JS/\hbar)(6S_\rho{}^y - \sum_\delta S^y_{\rho+\delta}) \ ,$$

where the central atom is at $\boldsymbol{\rho}$ and the six nearest neighbors are connected to it by six vectors $\boldsymbol{\delta}$. Look for solutions of the equations for $dS_\rho{}^x/dt$ and $dS_\rho{}^y/dt$ of the form $\exp(i\mathbf{k}\cdot\boldsymbol{\rho} - i\omega t)$.

2. ***Heat capacity of magnons.*** Use the approximate magnon dispersion relation $\omega = Ak^2$ to find the leading term in the heat capacity of a three-dimensional ferromagnet at low temperatures $k_BT \ll J$. The result is $0.113\, k_B(k_BT/\hbar A)^{3/2}$, per unit volume. The zeta function that enters the result may be estimated numerically; it is tabulated in Jahnke-Emde. Heat capacity measurements on YIG are reviewed by A. J. Henderson, Jr., D. G. Onn, H. Meyer, and J. P. Remeika, Phys. Rev. **185**, 1218 (1969).

3. ***Néel temperature.*** Taking the effective fields on the two-sublattice model of an antiferromagnetic as

$$B_A = B_a - \mu M_B - \epsilon M_A \ ; \qquad B_B = B_a - \mu M_A - \epsilon M_B \ ,$$

show that

$$\frac{\theta}{T_N} = \frac{\mu + \epsilon}{\mu - \epsilon} \ .$$

4. ***Magnetoelastic coupling.*** In a cubic crystal the elastic energy density in terms of the usual strain components e_{ij} is

$$U_{el} = \tfrac{1}{2}C_{11}(e_{xx}^2 + e_{yy}^2 + e_{zz}^2) + \tfrac{1}{2}C_{44}(e_{xy}^2 + e_{yz}^2 + e_{zx}^2) + C_{12}(e_{yy}e_{zz} + e_{xx}e_{zz} + e_{xx}e_{yy}) \ ,$$

and the leading term in the magnetic anisotropy energy density is, from (54),

$$U_K = K_1(\alpha_1^2\alpha_2^2 + \alpha_2^2\alpha_3^2 + \alpha_3^2\alpha_1^2) \ .$$

Coupling between elastic strain and magnetization direction may be taken formally into account by including in the total energy density a term

$$U_c = B_1(\alpha_1^2 e_{xx} + \alpha_2^2 e_{yy} + \alpha_3^2 e_{zz}) + B_2(\alpha_1\alpha_2 e_{xy} + \alpha_2\alpha_3 e_{yz} + \alpha_3\alpha_1 e_{zx})$$

arising from the strain dependence of U_K; here B_1 and B_2 are called magnetoelastic coupling constants. Show that the total energy is a minimum when

$$e_{ii} = \frac{B_1[C_{12} - \alpha_i^2(C_{11} + 2C_{12})]}{[(C_{11} - C_{12})(C_{11} + 2C_{12})]} \ ; \qquad e_{ij} = -\frac{B_2\alpha_i\alpha_j}{C_{44}} \qquad (i \neq j) \ .$$

This explains the origin of magnetostriction, the change of length on magnetization.

5. ***Coercive force of a small particle.*** (a) Consider a small spherical single-domain particle of a uniaxial ferromagnet. Show that the reverse field along the axis required to reverse the magnetization is $B_a = 2K/M_s$, in CGS units. The coercive force of single-domain particles is observed to be of this magnitude. Take $U_K = K \sin^2 \theta$ as the anisotropy energy density and $U_M = -B_a M \cos \theta$ as the interaction energy density with the external field H; here θ is the angle between $\mathbf{B}_a$ and $\mathbf{M}$. Hint: Expand the energies for small angles about $\theta = \pi$, and find the value of B_a for which $U_K + U_M$ does not have a minimum near $\theta = \pi$. (b) Show that the magnetic energy of a saturated sphere of diameter d is $\approx M_s^2 d^3$. An arrangement with appreciably less magnetic energy has a single wall in an equatorial plane. The domain wall energy will be $\pi\sigma_w d^2/4$, where σ_w is the wall energy per unit area. *Estimate* for cobalt the critical radius below which the particles are stable as single domains, taking the value of JS^2/a as for iron.

6. ***Saturation magnetization near T_c.*** Show that in the mean field approximation the saturation magnetization just below the Curie temperature has the dominant temperature dependence $(T_c - T)^{\frac{1}{2}}$. Assume the spin is $\frac{1}{2}$. The result is the same as that for a second-order transition in a ferroelectric crystal, as discussed in Chapter 13. The experimental data for ferromagnets (Table 1) suggest that the exponent is closer to 0.33.

References

A. E. Berkowitz and E. Kneller, eds., *Magnetism and metallurgy*, Academic Press, 1970.
R. M. Bozorth, *Ferromagnetism*, Van Nostrand, 1951.
S. Chikazumi, *Physics of magnetism*, Wiley, 1964. Elementary account of ferromagnetism.
D. J. Craik, ed., *Magnetic oxides*, Wiley, 1975.
D. J. Craik and R. S. Tebble, *Ferromagnetism and ferromagnetic domains*, North-Holland, 1966.
E. Della Torre and A. H. Bobeck, *Magnetic bubbles*, North-Holland, 1974.
R. J. Elliott, ed., *Magnetic properties of rare earth metals*, Plenum, 1972.
A. Herpin, *Théorie du magnétisme*, Presses universitaires, 1968.
F. Keffer, "Spin waves," *Encyclo. of physics* **18**/2 (1966). Excellent review.
C. Kittel and J. K. Galt, "Ferromagnetic domains," *Solid state physics* **3**, 437 (1956).
E. Kneller, *Ferromagnetismus*, Springer, 1962.
D. C. Mattis, *Theory of magnetism*, Harper & Row, 1965.
A. H. Morrish, *Physical principles of magnetism*, Wiley, 1965.
T. H. O'Dell, *Magnetic bubbles*, Halsted, 1974.
P. M. Platzman and P. A. Wolff, *Waves and interactions in solid state plasmas*, Academic Press, 1973. For spin waves in paramagnetic materials.
G. T. Rado and H. Suhl, eds., *Magnetism*, Academic Press, 1963–, several volumes. An encyclopedic work.
K. J. Standley, *Oxide magnetic materials*, 2nd ed., Oxford, 1972.
R. S. Tebble, *Magnetic domains*, Halsted, 1969.
S. V. Vonsovskii, *Magnetism*, Halsted, 1975, 2 vol.
H. J. Zeiger, *Magnetic interactions in solids*, Oxford, 1973.
Magnetism and magnetic materials, proceedings of the annual U.S. magnetism conference, published by the American Institute of Physics.

16 MAGNETIC RESONANCE AND MASERS

NOTATION: In this chapter the symbols B_a and B_0 refer to the applied field, and B_i is the applied field plus the demagnetizing field. In particular we write $\mathbf{B}_a = B_0\hat{\mathbf{z}}$. For CGS readers it may be simpler to read H for B whenever it occurs in this chapter.

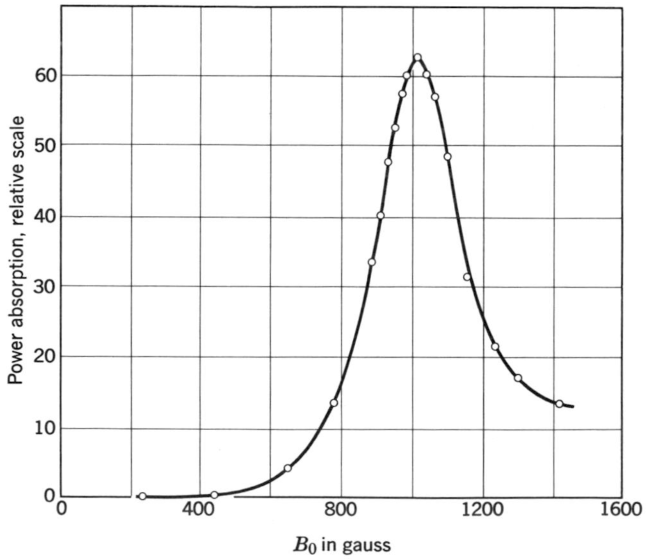

Figure 1 Electron spin resonance absorption in $MnSO_4$ at 298 K at 2.75 GHz, after Zavoisky.

16 MAGNETIC RESONANCE AND MASERS

In this chapter we discuss dynamical magnetic effects associated with the spin angular momentum of nuclei and of electrons. The principal phenomena are often identified in the literature by their initial letters, such as

NMR: nuclear magnetic resonance
NQR: nuclear quadrupole resonance
EPR: electron paramagnetic resonance (electron spin resonance)
FMR: ferromagnetic resonance
SWR: spin wave resonance (ferromagnetic films)
AFMR: antiferromagnetic resonance
CESR: conduction electron spin resonance

Zavoisky[1] performed the earliest magnetic resonance experiments in a solid. He observed strong electron spin resonance absorption (Fig. 1) in several paramagnetic salts. Spin resonance experiments on nuclei in liquids and solids were carried out first by Purcell, Torrey, and Pound[2] and by Bloch, Hansen, and Packard.[3]

The kinds of information that can be obtained about solids by resonance studies may in part be categorized as follows:

1. Electronic structure of single defects, as revealed by the fine structure of the absorption.
2. Motion of the spin or of the surroundings, as revealed by changes in the line width.
3. Internal magnetic fields sampled by the spin, as revealed by the position of the resonance line.
4. Collective spin excitations.

It is best to discuss NMR as a basis for a brief account of the other resonance experiments. However, the greatest impact of NMR has not been in

[1]E. Zavoisky, J. Phys. USSR **9**, 211, 245, 447 (1945).
[2]E. M. Purcell, H. C. Torrey, and R. V. Pound, Phys. Rev. **69**, 37 (1946).
[3]F. Bloch, W. W. Hansen, and M. Packard, Phys. Rev. **69**, 127 (1946).

TABLE 1 Nuclear Magnetic Resonance Data

For every element the most abundant magnetic isotope is shown.
After Varian Associates NMR Table, 4th ed., 1964.

Most abundant isotope with nonzero nuclear spin
Nuclear spin; in units of $\hbar$
Natural abundance of isotope, in percent
Nuclear magnetic moment, in units of $e\hbar/2M_pc$

H^{1} 1/2 99.98 2.792																	He^{3} 1/2 10^{-6} −2.127
Li^{7} 3/2 92.57 3.256	Be^{9} 3/2 100. −1.177											B^{11} 3/2 81.17 2.688	C^{13} 1/2 1.108 0.702	N^{14} 1 99.64 0.404	O^{17} 5/2 0.04 −1.893	F^{19} 1/2 100. 2.627	Ne^{21} 3/2 0.257 −0.662
Na^{23} 3/2 100. 2.216	Mg^{25} 5/2 10.05 0.855											Al^{27} 5/2 100. 3.639	Si^{29} 1/2 4.70 0.555	P^{31} 1/2 100. 1.131	S^{33} 3/2 0.74 0.643	Cl^{35} 3/2 75.4 0.821	Ar
K^{39} 3/2 93.08 0.391	Ca^{43} 7/2 0.13 −1.315	Sc^{45} 7/2 100. 4.749	Ti^{47} 5/2 7.75 0.787	V^{51} 7/2 ~100. 5.139	Cr^{53} 3/2 9.54 0.474	Mn^{55} 5/2 100. 3.461	Fe^{57} 1/2 2.245 0.090	Co^{59} 7/2 100. 4.639	Ni^{61} 3/2 1.25 0.746	Cu^{63} 3/2 69.09 2.221	Zn^{67} 5/2 4.12 0.874	Ga^{69} 3/2 60.2 2.011	Ge^{73} 9/2 7.61 0.877	As^{75} 3/2 100. 1.435	Se^{77} 1/2 7.50 0.533	Br^{79} 3/2 50.57 2.099	Kr^{83} 9/2 11.55 −0.967
Rb^{85} 5/2 72.8 1.348	Sr^{87} 9/2 7.02 1.089	Y^{89} 1/2 100. 0.137	Zr^{91} 5/2 11.23 1.298	Nb^{93} 9/2 100. 6.144	Mo^{95} 5/2 15.78 0.910	Tc	Ru^{101} 5/2 16.98 −0.69	Rh^{103} 1/2 100. 0.088	Pd^{105} 5/2 22.23 −0.57	Ag^{107} 1/2 51.35 −0.113	Cd^{111} 1/2 12.86 −0.592	In^{115} 9/2 95.84 5.507	Sn^{119} 1/2 8.68 −1.041	Sb^{121} 5/2 57.25 3.342	Te^{125} 1/2 7.03 −0.882	I^{127} 5/2 100. 2.794	Xe^{129} 1/2 26.24 −0.773
Cs^{133} 7/2 100. 2.564	Ba^{137} 3/2 11.32 0.931	La^{139} 7/2 99.9 2.761	Hf^{177} 7/2 18.39 0.61	Ta^{181} 7/2 100. 2.340	W^{183} 1/2 14.28 0.115	Re^{187} 5/2 62.93 3.176	Os^{189} 3/2 16.1 0.651	Ir^{193} 3/2 61.5 0.17	Pt^{195} 1/2 33.7 0.600	Au^{197} 3/2 100. 0.144	Hg^{199} 1/2 16.86 0.498	Tl^{205} 1/2 70.48 1.612	Pb^{207} 1/2 21.11 0.584	Bi^{209} 9/2 100. 4.039	Po	At	Rn
Fr	Ra	Ac															

Ce^{141}* 7/2 — 0.16	Pr^{141} 5/2 100. 3.92	Nd^{143} 7/2 12.20 −1.25	Pm	Sm^{147} 7/2 15.07 −0.68	Eu^{153} 5/2 52.23 1.521	Gd^{157} 3/2 15.64 −0.34	Tb^{159} 3/2 100. 1.52	Dy^{163} 5/2 24.97 −0.53	Ho^{165} 7/2 100. 3.31	Er^{167} 7/2 22.82 0.48	Tm^{169} 1/2 100. −0.20	Yb^{173} 5/2 16.08 −0.677	Lu^{175} 7/2 97.40 2.9
Th	Pa	U	Np	Pu	Am	Cm	Bk	Cf	Es	Fm	Md	No	Lr

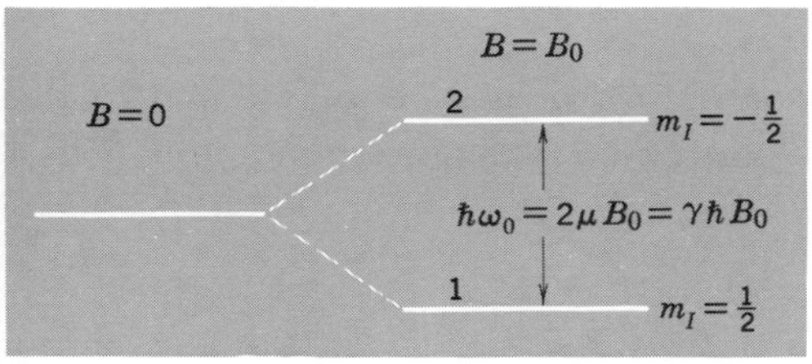

Figure 2 Energy level splitting of a nucleus of spin $I=\frac{1}{2}$ in a static magnetic field B_0.

solids, but in organic chemistry, where NMR provides a powerful tool for the identification and the structure determination of complex molecules. This success is due to the extremely high resolution attainable in diamagnetic liquids.

NUCLEAR MAGNETIC RESONANCE

We consider a nucleus that possesses a magnetic moment $\boldsymbol{\mu}$ and an angular momentum $\hbar\mathbf{I}$. The two quantities are parallel, and we may write

$$\boldsymbol{\mu} = \gamma\hbar\mathbf{I} \; ; \tag{1}$$

the magnetogyric ratio γ is constant. By convention $\mathbf{I}$ denotes the nuclear angular momentum measured in units of $\hbar$. The energy of interaction with the applied magnetic field is

$$U = -\boldsymbol{\mu}\cdot\mathbf{B}_a \; ; \tag{2}$$

if $\mathbf{B}_a = B_0\hat{\mathbf{z}}$, then

$$U = -\mu_z B_0 = -\gamma\hbar B_0 I_z \; . \tag{3}$$

The allowed values of I_z are $m_I = I, I-1, \ldots, -I$, and $U = -m_I\gamma\hbar B_0$.

In a magnetic field a nucleus with $I=\frac{1}{2}$ has two energy levels corresponding to $m_I = \pm\frac{1}{2}$, as in Fig. 2. If $\hbar\omega_0$ denotes the energy difference between the two levels, then $\hbar\omega_0 = \gamma\hbar B_0$ or

$$\omega_0 = \gamma B_0 \; . \tag{4}$$

This is the fundamental condition for magnetic resonance absorption. For the proton[4] $\gamma = 2.675\times10^4$ s^{-1} gauss^{-1} $= 2.675\times10^8$ s^{-1} tesla^{-1}, so that

$$\boxed{\nu(\mathrm{MHz}) = 4.258\, B_0(\mathrm{kilogauss}) = 42.58\, B_0(\mathrm{tesla}) \; ,} \tag{4a}$$

where ν is the frequency. One tesla is precisely 10^4 gauss. For the electron spin,

$$\boxed{\nu(\mathrm{GHz}) = 2.80\, B_0(\mathrm{kilogauss}) = 28.0\, B_0(\mathrm{tesla}) \; .} \tag{4b}$$

Magnetic data for selected nuclei are given in Table 1.

[4]The magnetic moment μ_p of the proton is 1.4106×10^{-23} erg G^{-1} or 1.4106×10^{-26} JT^{-1}, and $\gamma \equiv 2\mu_p/\hbar$. The **nuclear magneton** μ_n is defined as $e\hbar/2M_pc$ and is equal to 5.0509×10^{-24} erg G^{-1} or 5.0509×10^{-27} JT^{-1}; thus $\mu_p = 2.793$ nuclear magnetons.

Equations of Motion

The rate of change of angular momentum of a system is equal to the torque that acts on the system. The torque on a magnetic moment $\boldsymbol{\mu}$ in a magnetic field $\mathbf{B}$ is $\boldsymbol{\mu} \times \mathbf{B}$, so that we have the gyroscopic equation

$$\hbar d\mathbf{I}/dt = \boldsymbol{\mu} \times \mathbf{B}_a \ ; \tag{5}$$

or

$$d\boldsymbol{\mu}/dt = \gamma\boldsymbol{\mu} \times \mathbf{B}_a \ . \tag{6}$$

The nuclear magnetization $\mathbf{M}$ is the sum $\Sigma\boldsymbol{\mu}_i$ over all the nuclei in a unit volume. If only a single isotope is important, we consider only a single value of γ, so that

$$d\mathbf{M}/dt = \gamma\mathbf{M} \times \mathbf{B}_a \ . \tag{7}$$

We place the nuclei in a static field $\mathbf{B}_a = B_0\hat{\mathbf{z}}$. In thermal equilibrium at temperature T the magnetization will be along $\hat{\mathbf{z}}$:

$$M_x = 0 \ ; \qquad M_y = 0 \ ; \qquad M_z = M_0 = \chi_0 B_0 = CB_0/T \ , \tag{8}$$

where the Curie constant $C = N\mu^2/3k_B$. The magnetization of a system of spins with $I = \frac{1}{2}$ is related to the population difference $N_1 - N_2$ of the lower and upper levels in Fig. 2: $M_z = (N_1 - N_2)\mu$, where the N's refer to a unit volume. The population ratio in thermal equilibrium is just given by the Boltzmann factor for the energy difference $2\mu B_0$:

$$(N_2/N_1)_0 = \exp(-2\mu B_0/k_BT) \ . \tag{9}$$

The equilibrium magnetization is $M = N\mu \tanh(\mu B/k_BT)$.

When the magnetization component M_z is not in thermal equilibrium, we suppose that it approaches equilibrium at a rate proportional to the departure from the equilibrium value M_0:

$$\frac{dM_z}{dt} = \frac{M_0 - M_z}{T_1} \ . \tag{10}$$

In the standard notation T_1 is called the **longitudinal relaxation time** or the **spin-lattice relaxation time.** If at $t = 0$ an unmagnetized specimen is placed in a magnetic field $B_0\hat{\mathbf{z}}$, the magnetization will increase from the initial value $M_z = 0$ to a final value $M_z = M_0$. Before and just after the specimen is placed in the field the population N_1 will be equal to N_2, as appropriate to thermal equilibrium in zero magnetic field. It is necessary

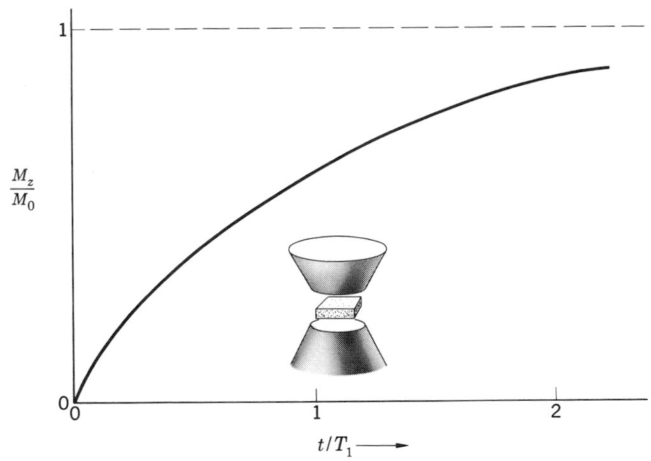

Figure 3 At time $t = 0$ an unmagnetized specimen $M_z(0) = 0$ is placed in a static magnetic field B_0. The magnetization increases with time and approaches the new equilibrium value $M_0 = \chi_0 B_0$. This experiment defines the longitudinal relaxation time T_1. The magnetic energy density $-\mathbf{M} \cdot \mathbf{B}$ decreases as part of the spin population moves into the lower level. The asymptotic value at $t \gg T_1$ is $-M_0 B_0$. The energy flows from the spin system to the system of lattice vibrations; thus T_1 is also called the spin-lattice relaxation time.

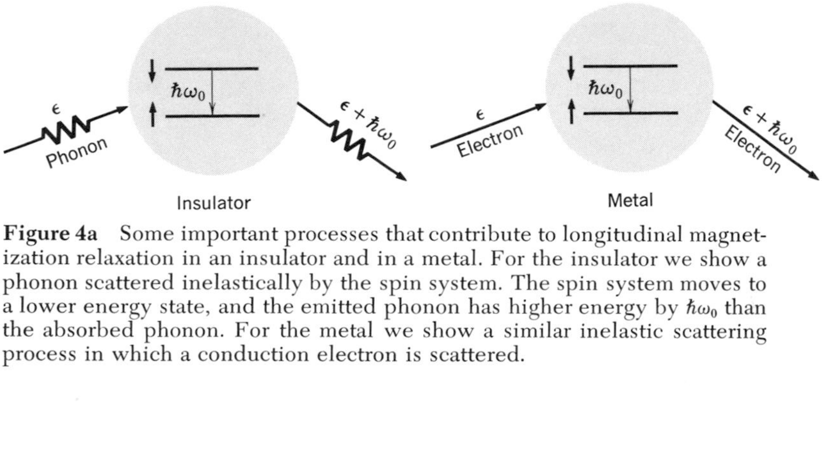

Figure 4a Some important processes that contribute to longitudinal magnetization relaxation in an insulator and in a metal. For the insulator we show a phonon scattered inelastically by the spin system. The spin system moves to a lower energy state, and the emitted phonon has higher energy by $\hbar\omega_0$ than the absorbed phonon. For the metal we show a similar inelastic scattering process in which a conduction electron is scattered.

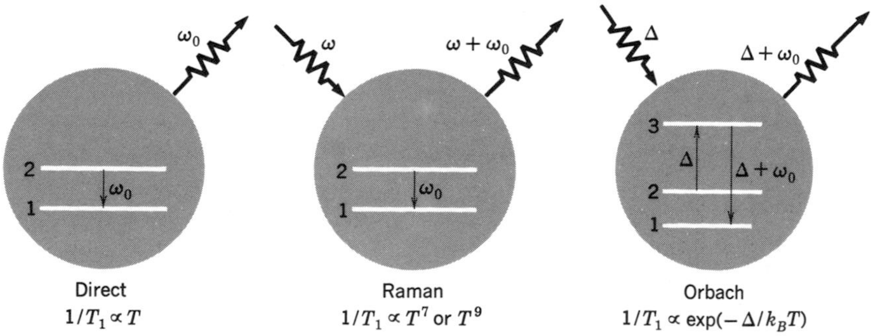

Figure 4b Spin relaxation from $2 \rightarrow 1$ by phonon emission, phonon scattering, and a two-stage phonon process. The temperature dependence of the longitudinal relaxation time T_1 is shown for the several processes.

to reverse some spins to establish the new equilibrium distribution in the field B_0. On integrating (10):

$$\int_0^{M_z} \frac{dM_z}{M_0 - M_z} = \frac{1}{T_1} \int_0^t dt \ , \tag{11}$$

or

$$\log \frac{M_0}{M_0 - M_z} = \frac{t}{T_1} \ ; \qquad M_z(t) = M_0(1 - e^{-t/T_1}) \ , \tag{12}$$

as in Fig. 3. The magnetic energy $-\mathbf{M}\cdot\mathbf{B}_a$ decreases as M_z approaches its new equilibrium value.

Typical processes whereby the magnetization approaches equilibrium are indicated in Fig. 4. The dominant spin-lattice interaction of paramagnetic ions in crystals is by the phonon modulation of the crystalline electric field. Relaxation proceeds by three principal processes (Fig. 4b): direct (emission or absorption of a phonon); Raman (scattering of a phonon); and Orbach[5] (intervention of a third state). A thorough experimental analysis of spin-lattice relaxation in several rare-earth salts at helium temperatures has been given by Scott and Jeffries;[6] they discuss evidence for the three processes and give useful references to the literature of paramagnetic relaxation.

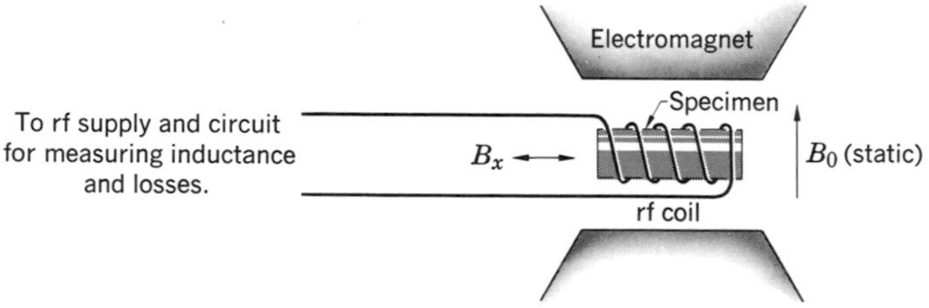

Figure 5 Schematic arrangement for magnetic resonance experiments.

[5]C. B. P. Finn, R. Orbach, and W. P. Wolf, Proc. Phys. Soc. (London) **77,** 261 (1961).
[6]P. L. Scott and C. D. Jeffries, Phys. Rev. **127,** 32 (1962).

Taking account of (10), the z component of the equation of motion (7) becomes

$$\frac{dM_z}{dt} = \gamma(\mathbf{M} \times \mathbf{B}_a)_z + \frac{M_0 - M_z}{T_1} \ , \tag{13a}$$

where $(M_0 - M_z)/T_1$ is an extra term in the equation of motion, arising from interactions not included in the magnetic field $\mathbf{B}_a$. That is, besides precessing about the magnetic field, $\mathbf{M}$ will relax to the equilibrium value $\mathbf{M}_0$. If in a static field $B_0\hat{\mathbf{z}}$ the transverse magnetization component M_x is not zero, then M_x will decay to zero, and similarly for M_y. The decay occurs because in thermal equilibrium the transverse components are zero. We can provide for transverse relaxation:

$$dM_x/dt = \gamma(\mathbf{M} \times \mathbf{B}_a)_x - M_x/T_2 \ ; \tag{13b}$$

$$dM_y/dt = \gamma(\mathbf{M} \times \mathbf{B}_a)_y - M_y/T_2 \ , \tag{13c}$$

where T_2 is called the **transverse relaxation time.** The magnetic energy $-\mathbf{M}\cdot\mathbf{B}_a$ does not change as M_x or M_y changes, provided that $\mathbf{B}_a$ is along $\hat{\mathbf{z}}$. No energy need flow out of the spin system for relaxation of M_x or M_y, so that the conditions that determine T_2 may be less strict than for T_1. Sometimes the two times are nearly equal, and sometimes $T_1 \gg T_2$, depending on local conditions. The time T_2 is a measure of the time during which the individual moments that contribute to M_x, M_y remain in phase with each other. Different local magnetic fields at the different spins will cause them to precess at different frequencies. If initially the spins have a common phase, the phases will become random in the course of time and the values of M_x, M_y will become zero. We can think of T_2 as a dephasing time.

The set of equations (13) are called the **Bloch equations.** They are not symmetrical in x, y, and z because we have biased the system with a static magnetic field along $\hat{\mathbf{z}}$. In the experiments an rf magnetic field is usually applied along the $\hat{\mathbf{x}}$ or $\hat{\mathbf{y}}$ axes. Our main interest is in the behavior of the magnetization in the combined rf and static fields, as in Fig. 5. The Bloch equations are plausible, but not exact; they do not describe all spin phenomena, particularly in solids.

We determine the frequency of free precession of the spin system in a static field $\mathbf{B}_a = B_0\hat{\mathbf{z}}$ and with $M_z = M_0$. The Bloch equations reduce to

$$\frac{dM_x}{dt} = \gamma B_0 M_y - \frac{M_x}{T_2} \ ; \qquad \frac{dM_y}{dt} = -\gamma B_0 M_x - \frac{M_y}{T_2} \ ; \qquad \frac{dM_z}{dt} = 0 \ . \tag{14}$$

We look for damped oscillatory solutions of the form

$$M_x = me^{-t/T'} \cos \omega t \; ; \qquad M_y = -me^{-t/T'} \sin \omega t \; ; \tag{15}$$

on substitution in (14) we have for the left-hand equation

$$-\omega \sin \omega t - \frac{1}{T'} \cos \omega t = -\gamma B_0 \sin \omega t - \frac{1}{T_2} \cos \omega t \; , \tag{16}$$

so that the free precession is characterized by

$$\omega_0 = \gamma B_0 \; ; \qquad T' = T_2 \; . \tag{17}$$

The motion is similar to that of a damped harmonic oscillator in two dimensions. The analogy suggests correctly that the spin system will show resonance absorption of energy from a driving field near the frequency $\omega_0 = \gamma B_0$, and the frequency width of the response of the system to the driving field will be $\Delta\omega \approx 1/T_2$. Figure 6 shows the resonance of protons in water.

The Bloch equations may be solved to give the power absorption from a rotating magnetic field of amplitude B_1:

$$B_x = B_1 \cos \omega t \; ; \qquad B_y = -B_1 \sin \omega t \; . \tag{18}$$

After a routine calculation one finds that the power absorption is

(CGS)
$$\mathcal{P}(\omega) = \frac{\omega\gamma M_z T_2}{1 + (\omega_0 - \omega)^2 T_2^2} B_1^2 \; . \tag{19}$$

The half-width of the resonance at half-maximum power is

$$(\Delta\omega)_{\frac{1}{2}} = 1/T_2 \; . \tag{20}$$

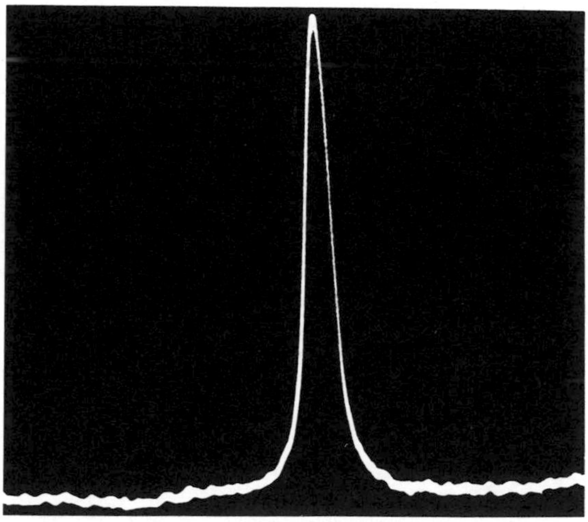

Figure 6 Proton resonance absorption in water. (Courtesy of E. L. Hahn.)

LINE WIDTH

The magnetic dipolar interaction is usually the most important cause of line broadening in a rigid lattice of magnetic dipoles. The magnetic field $\Delta\mathbf{B}$ seen by a magnetic dipole $\boldsymbol{\mu}_1$ due to a magnetic dipole $\boldsymbol{\mu}_2$ at a point $\mathbf{r}_{12}$ from the first dipole is

$$\text{(CGS)} \qquad \Delta\mathbf{B} = \frac{3(\boldsymbol{\mu}_2 \cdot \mathbf{r}_{12})\mathbf{r}_{12} - \boldsymbol{\mu}_2 r_{12}^2}{r_{12}^5} \ , \qquad (21)$$

by a fundamental result of magnetostatics. The order of magnitude of the interaction is, with B_i written for ΔB,

$$\text{(CGS)} \qquad B_i \approx \mu/r^3 \ . \qquad (22)$$

The strong dependence on r suggests that close neighbor interactions will be dominant, so that

$$\text{(CGS)} \qquad B_i \approx \mu/a^3 \ , \qquad (23)$$

where a is the separation of nearest neighbors. This result gives us a measure of the width of the spin resonance line, assuming random orientation of the neighbors. For protons at 2Å separation,

$$B_i \approx \frac{1.4 \times 10^{-23}\ \text{G cm}^3}{8 \times 10^{-24}\ \text{cm}^3} \approx 2\ \text{gauss} = 2 \times 10^{-4}\ \text{tesla} \ . \qquad (24)$$

To express (21), (22), and (23) in SI, multiply the right-hand sides by $\mu_0/4\pi$.

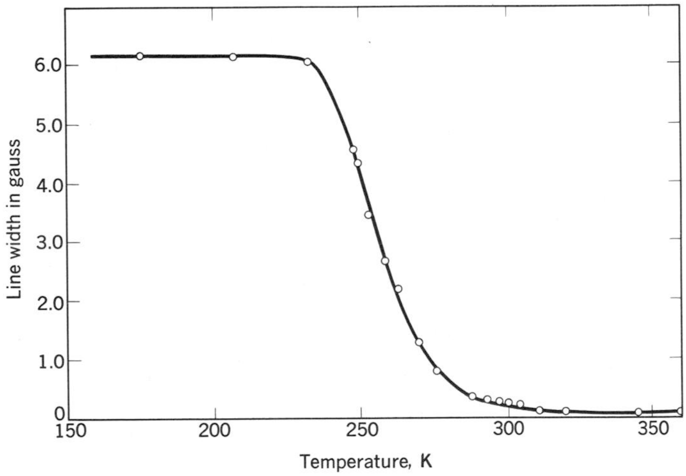

Figure 7 Effect of diffusion of nuclei on the Li^7 NMR line width in metallic lithium. At low temperatures the width agrees with the theoretical value for a rigid lattice. As the temperature increases the diffusion rate increases and the line width decreases. The abrupt decrease in line width above T = 230 K occurs when the diffusion hopping time τ becomes shorter than $1/\gamma B_i$. Thus the experiment gives a direct measure of the hopping time for an atom to change lattice sites. [After H. S. Gutowsky and B. R. McGarvey, J. Chem. Phys. **20**, 1472 (1952).]

Motional Narrowing

The line width decreases for nuclei in rapid relative motion. The effect in solids is illustrated by Fig. 7: diffusion resembles a random walk as atoms jump from one crystal site to another. An atom remains in one site for an average time τ that decreases markedly as the temperature increases. The motional effects on the line width are even more spectacular in liquids, because atoms are much more mobile in a liquid. The width of the proton resonance line in water is only 10^{-5} of the width expected for water molecules frozen in position.

The effects of nuclear motion on T_2 and on the line width can be understood by an elementary argument. We know from the Bloch equations that T_2 is a measure of the time in which an individual spin becomes dephased by one radian because of a local perturbation in the magnetic field intensity. Let $(\Delta\omega)_0 \approx \gamma B_i$ denote the local frequency deviation due to a perturbation B_i. The local field may be caused by dipolar interactions with other spins. If the atoms are in rapid relative motion, the local field B_i seen by a given spin will fluctuate rapidly in time. We suppose that the local field has a value $+B_i$ for an average time τ and then changes to $-B_i$, as in Fig. 8b. Such a random change could be caused by a change of the angle between $\boldsymbol{\mu}$ and $\mathbf{r}$ in (21). In the time τ the spin will precess by an extra phase angle $\delta\varphi = \pm\gamma B_i\tau$ relative to the phase angle of the steady precession in the applied field B_0. The motional narrowing effect arises for short τ such that $\delta\varphi \ll 1$. After n intervals of duration τ the mean square dephasing angle in the field B_0 will be

$$\langle\varphi^2\rangle = n(\delta\varphi)^2 = n\gamma^2 B_i^2\tau^2 \ , \tag{25}$$

by analogy with a random walk process: the mean square displacement from the initial position after n steps of length ℓ in random directions is $\langle r^2\rangle = n\ell^2$.

The average number of steps necessary to dephase a spin by one radian is $n = 1/\gamma^2 B_i^2\tau^2$. Spins dephased by much more than one radian do not contribute to the absorption signal. This number of steps takes place in a time

$$T_2 = n\tau = 1/\gamma^2 B_i^2\tau \ , \tag{26}$$

quite different from the rigid lattice result $T_2 \cong 1/\gamma B_i$. From (26) we obtain as the line width for rapid motion with a characteristic time τ:

$$\Delta\omega = 1/T_2 = (\gamma B_i)^2\tau \ , \tag{27}$$

or

$$\Delta\omega = 1/T_2 = (\Delta\omega)_0^2\tau \ , \tag{28}$$

where $(\Delta\omega)_0$ is the line width in the rigid lattice.

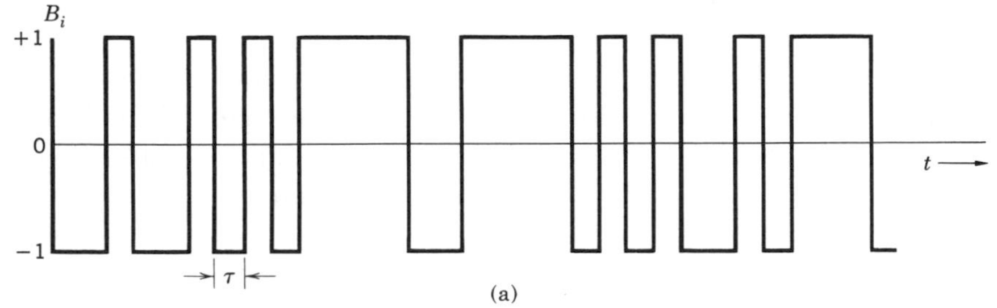

(a)

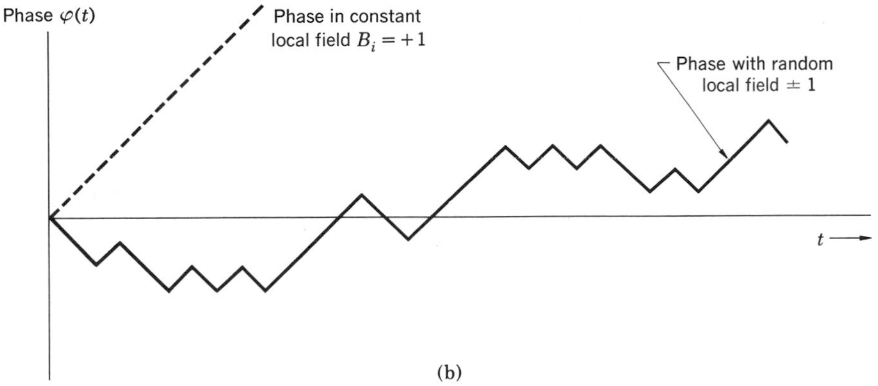

(b)

Figure 8 Phase of a spin in a constant local field, as compared with dephasing of a spin which after fixed time intervals τ hops at random among sites having local fields ± 1. The dephasing is measured relative to the phase of a spin in the applied field B_0.

The argument assumes that $(\Delta\omega)_0\tau \ll 1$, as otherwise $\delta\varphi$ will not be $\ll 1$. Thus $\Delta\omega \ll (\Delta\omega)_0$. The shorter is τ, the narrower is the resonance line! This remarkable effect is known as **motional narrowing.**[7] The rotational relaxation time of water molecules at room temperature is known from dielectric constant measurements to be of the order of 10^{-10} s; if $(\Delta\omega)_0 \approx 10^5\ \text{s}^{-1}$, then $(\Delta\omega)_0\tau \approx 10^{-5}$ and $\Delta\omega \approx (\Delta\omega)_0{}^2\tau \approx 1\ \text{s}^{-1}$. Thus the motion narrows the proton resonance line to about 10^{-5} of the static width.

[7]The physical ideas are due to N. Bloembergen, E. M. Purcell, and R. V. Pound, Phys. Rev. **73**, 679 (1948). The result differs from the theory of optical line width caused by *strong* collisions between atoms (as in a gas discharge), where a short τ gives a broad line. In the nuclear spin problem the collisions are weak. In most optical problems the collisions of atoms are strong enough to interrupt the phase of the oscillation. In nuclear resonance the phase may vary smoothly in a collision, although the frequency may vary suddenly from one value to another nearby value.

HYPERFINE SPLITTING

The hyperfine interaction is the magnetic interaction between the magnetic moment of a nucleus and the magnetic moment of an electron. To an observer stationed on the nucleus, the interaction is caused by the magnetic field produced by the motion of the electron about the nucleus. There is an electron current about the nucleus if the electron is in a state with orbital angular momentum about the nucleus. But even if the electron is in a state of zero orbital angular momentum, there is an electron spin current about the nucleus, and this current gives rise to the **contact hyperfine interaction,** of particular importance in solids. We can understand the origin of the contact by a qualitative physical argument, given in CGS.

The results of the Dirac theory of the electron suggest that the magnetic moment of $\mu_B = e\hbar/2mc$ of the electron arises from the circulation of an electron with velocity c in a current loop of radius approximately the electron Compton wavelength, $\lambda\!\!\!^{-}_e = \hbar/mc \sim 10^{-11}$ cm. The electric current associated with the circulation is

$$I \sim e \times (\text{turns per unit time}) \sim ec/\lambda\!\!\!^{-}_e \ , \tag{29}$$

and the magnetic field (Fig. 9) produced by the current is

$$\text{(CGS)} \qquad B \sim I/\lambda\!\!\!^{-}_e c \sim e/\lambda\!\!\!^{-}_e{}^2 \ . \tag{30}$$

The observer on the nucleus has the probability

$$P \approx |\psi(0)|^2 \lambda\!\!\!^{-}_e{}^3 \ . \tag{31}$$

of finding himself inside the electron, that is, within a sphere of volume $\lambda\!\!\!^{-}_e{}^3$ about the electron. Here $\psi(0)$ is the value of the electron wavefunction

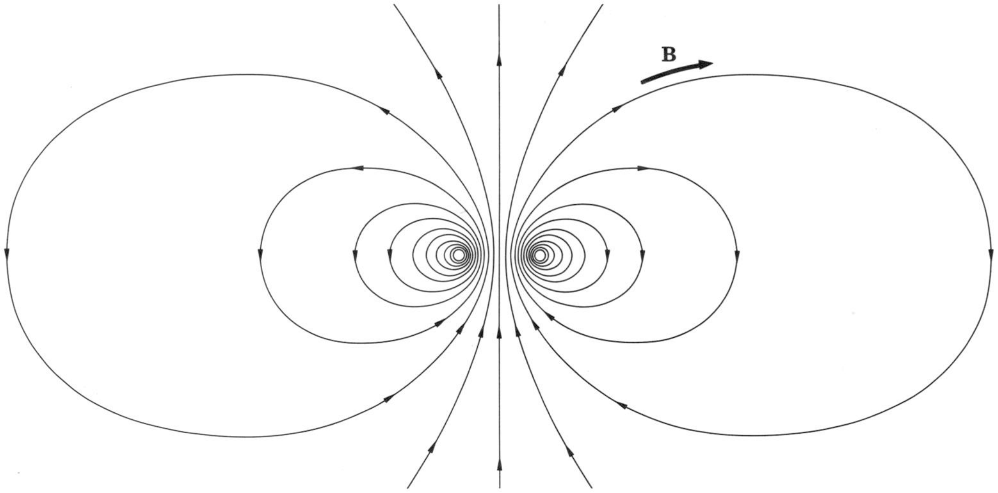

Figure 9 Magnetic field **B** produced by a charge moving in a circular loop. The contact part of the hyperfine interaction with a nuclear magnetic moment arises from the region within or near to the current loop. The field averaged over a spherical shell that encloses the loop gives zero. Thus for an s electron ($L = 0$) only the contact part contributes to the interaction.

at the nucleus. Thus the average value of the magnetic field seen by the nucleus is

$$\bar{B} \approx e|\psi(0)|^2 \lambda\!\!\!^{-}_e \approx \mu_B|\psi(0)|^2 \ , \qquad (32)$$

where $\mu_B = e\hbar/2mc = \frac{1}{2}e\lambda\!\!\!^{-}_e$ is the Bohr magneton. The contact part of the hyperfine interaction energy is

$$U = -\boldsymbol{\mu}_I \cdot \bar{\mathbf{B}} \approx -\boldsymbol{\mu}_I \cdot \boldsymbol{\mu}_B |\psi(0)|^2 \approx \gamma\hbar\mu_B|\psi(0)|^2 \mathbf{I} \cdot \mathbf{S} \ , \qquad (33)$$

where I is the nuclear spin in units of $\hbar$.

The contact interaction in an atom has the form

$$U = a\mathbf{I} \cdot \mathbf{S} \ . \qquad (34)$$

Values of the hyperfine constant a for the ground states of several free atoms are:

nucleus	H^1	Li^7	Na^{23}	K^{39}	K^{41}
I	$\frac{1}{2}$	$\frac{3}{2}$	$\frac{3}{2}$	$\frac{3}{2}$	$\frac{3}{2}$
a in gauss	507	144	310	83	85
a in MHz	1420	402	886	231	127

The value of a in gauss as seen by an electron spin is defined as $a/2\mu_B$.

In a strong magnetic field the energy level scheme of a free atom or ion is dominated by the Zeeman energy splitting of the electron levels; the hyperfine interaction gives an additional splitting that in strong fields is $U' \cong am_S m_I$, where m_S, m_I are the magnetic quantum numbers. For the energy level diagram of Fig. 10 the two electronic transitions have $\Delta m_S = \pm 1$, $\Delta m_I = 0$; the frequencies are $\omega = \gamma H_0 \pm a/2\hbar$. The nuclear transitions are not marked; they have $\Delta m_S = 0$, so that $\omega_{\text{nuc}} = a/2\hbar$. The frequency of the nuclear transition $1 \rightarrow 2$ is equal to that of $3 \rightarrow 4$. The hyperfine interaction in an atom may split the ground energy level. The splitting in hydrogen is 1420 MHz; and this is the radio frequency line of interstellar atomic hydrogen.

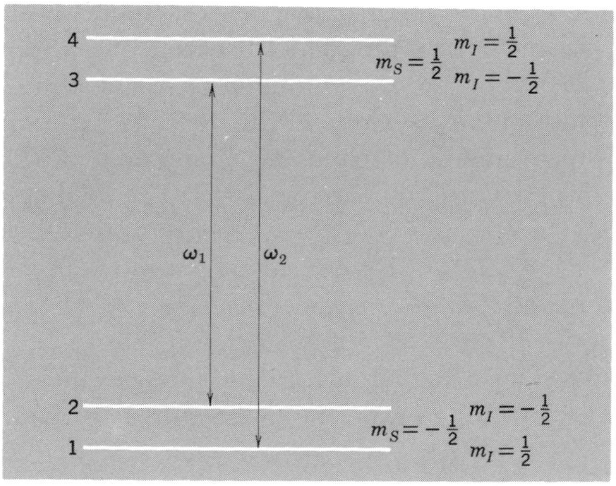

Figure 10 Energy levels in a magnetic field of a system with $S = \frac{1}{2}$, $I = \frac{1}{2}$. The diagram is drawn for the strong field approximation $\mu_B B \gg a$, where a is the hyperfine coupling constant, taken to be positive. The four levels are labeled by the magnetic quantum numbers m_S, m_I. The strong electronic transitions have $\Delta m_I = 0$, $\Delta m_S = \pm 1$.

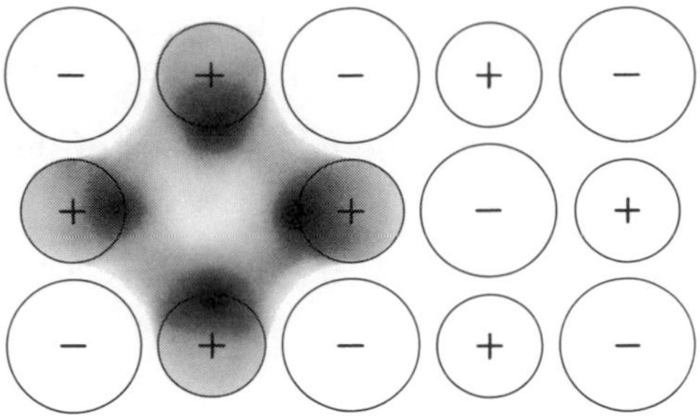

Figure 11 An F center is a negative ion vacancy with one excess electron bound at the vacancy. The distribution of the excess electron is largely on the positive metal ions adjacent to the vacant lattice site.

Examples: Paramagnetic Point Defects

The hyperfine splitting of the electron spin resonance furnishes valuable structural information about paramagnetic point defects, such as the F centers in alkali halide crystals and the donor atoms in semiconductor crystals.

F Centers in Alkali Halides. An F center is a negative ion vacancy with one excess electron bound at the vacancy (Fig. 11). The wavefunction of the trapped electron is shared chiefly among the six alkali ions adjacent to the vacant lattice site, with smaller amplitudes on the 12 halide ions that form the shell of second nearest neighbors. The counting applies to crystals with the NaCl structure. If $\varphi(\mathbf{r})$ is the wavefunction of the valence electron on a single alkali ion, then in the first approximation[8]

$$\psi(\mathbf{r}) = C \sum_p \varphi(\mathbf{r} - \mathbf{r}_p) \ , \tag{35}$$

where in the NaCl structure the six values of $\mathbf{r}_p$ mark the alkali ion sites that bound the lattice vacancy. The width of the electron spin resonance line of an F center is determined essentially by the hyperfine interaction of the trapped electron with the nuclear magnetic moments of the alkali ions adjacent to the vacant lattice site. The observed line width is evidence for the simple picture of the wavefunction of the electron. By line width we mean the width of the envelope of the possible hyperfine structure components. As an example, consider an F center in KCl. Natural potassium is 93 percent K^{39} with nuclear spin $I = \frac{3}{2}$. The total spin of the six potassium nuclei at the F center is $I_{\text{max}} = 6 \times \frac{3}{2} = 9$, so that the number of hyperfine components is $2I_{\text{max}} + 1 = 19$: this is the number of possible values of the quantum number m_I. There are $(2I + 1)^6 = 4^6 = 4096$ independent arrangements of the six spins distributed into the 19 components, as in Fig. 12. Often we observe only the envelope of the absorption line of an F center. A review of the experimental results is included in the survey by Markham.[9]

[8]Kip, Kittel, Levy and Portis, Phys. Rev. **91,** 1066 (1953); G. Feher, Phys. Rev. **105,** 1122 (1957).
[9]J. J. Markham, *F centers in alkali halides,* Academic Press, 1966.

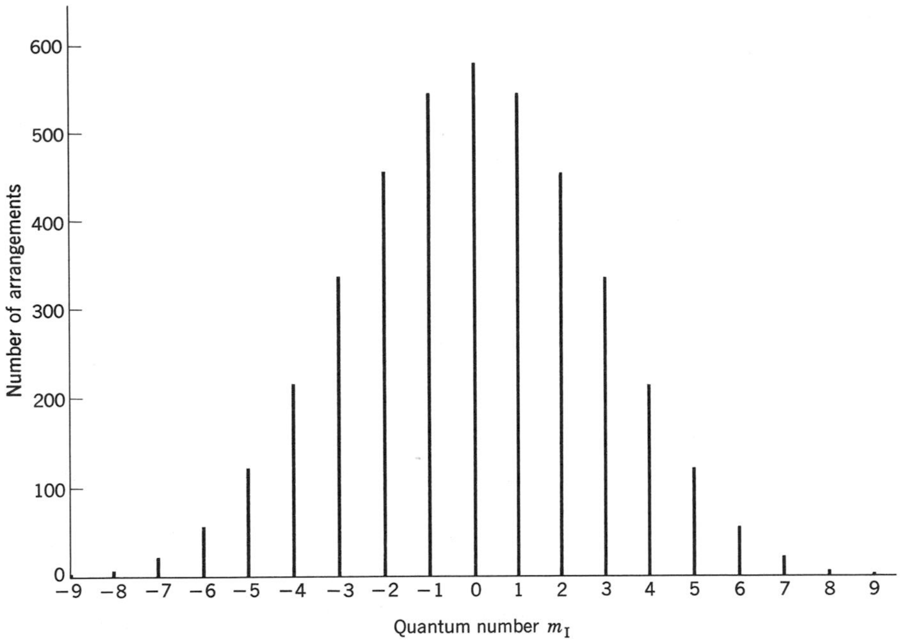

Figure 12 The 4096 arrangements of the six nuclear spins of K^{39} as distributed into 19 hyperfine components. Each component will be split further into a very large number of components by virtue of the residual hyperfine interaction with the 12 neighbor Cl nuclei, which may be Cl^{35} (75 percent) or Cl^{37} (25 percent). The envelope of the pattern is approximately gaussian in form.

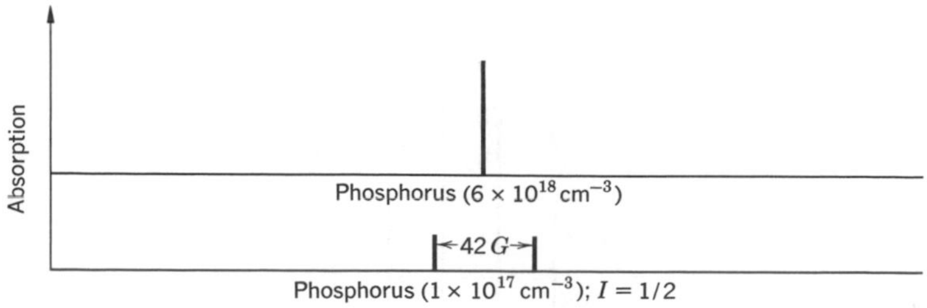

Figure 13 Electron spin resonance lines of P donor atoms in silicon. At the higher donor concentration a donor electron can hop from site to site so rapidly that the hyperfine structure is suppressed. [After R. C. Fletcher, W. A. Yager, G. L. Pearson, and F. R. Merritt, Phys. Rev. **95**, 844 (1954).]

Donor Atoms in Silicon. Phosphorus is a donor when present in silicon. Each donor atom has five electrons, of which four enter diamagnetically into the covalent bond network of the crystal, and the fifth bound electron acts as a paramagnetic center of spin $S = \frac{1}{2}$. The experimental hyperfine splitting in the strong field limit is shown in Fig. 13. When the concentration exceeds about 1×10^{18} donors cm^{-3}, the split line is replaced by a single narrow line. This is a motional narrowing effect (Eq. 28) of the rapid hopping of the donor electrons among many donor atoms. The rapid hopping averages out the hyperfine splitting. The hopping rate increases at the higher concentrations as the overlap of the donor electron wavefunctions is increased, a view supported by conductivity measurements.

The donor electron wavefunction extends not only over the central donor atom but also significantly over some hundreds of silicon atoms. The Si^{29} nuclear spins give additional hyperfine splittings first studied by Feher[10] with a powerful electron nuclear double resonance technique; the studies give information on the conduction electron wavefunctions and the location in the Brillouin zone of the conduction band edge in silicon.

Knight Shift

At a fixed frequency the resonance of a nuclear spin is observed at a slightly different magnetic field in a metal than in a diamagnetic solid. The effect is known as the **Knight shift** and is valuable as a tool for the study of conduction electrons. The interaction energy of a nucleus of spin

[10]G. Feher, Phys. Rev. **114**, 1219 (1959). The technique is known as ENDOR. The results are reviewed by W. Kohn, *Solid state physics* **5**, 257 (1957).

I and magnetogyric ratio γ_I is

$$U = (-\gamma_I \hbar B_0 + a\langle S_z \rangle) I_z \ , \tag{36}$$

where the first term is the interaction with the applied magnetic field B_0 and the second is the average hyperfine interaction of the nucleus with the conduction electrons. The average conduction electron spin $\langle S_z \rangle$ is related to the Pauli spin susceptibility χ_s of the conduction electrons: $M_z = gN\mu_B \langle S_z \rangle = \chi_s B_0$, whence the interaction may be written as

$$U = \left(-\gamma_I \hbar + \frac{a\chi_s}{gN\mu_B}\right) B_0 I_z = -\gamma_I \hbar B_0 \left(1 + \frac{\Delta B}{B_0}\right) I_z \ . \tag{37}$$

The Knight shift is defined as

$$K = -\frac{\Delta B}{B_0} = \frac{a\chi_s}{gN\mu_B \gamma_I \hbar} \ . \tag{38}$$

By the definition (34) of the hyperfine contact energy, the Knight shift is given approximately by $K \approx \chi_s \ |\psi(0)|^2/N$; that is, by the Pauli spin susceptibility increased in the ratio of the conduction electron concentration at the nucleus to the average conduction electron concentration. Experimental values are given in Table 2. The value of the hyperfine coupling constant a is somewhat different in the metal than in the free atom because the wave functions at the nucleus are different. From the Knight shift of metallic Li it is deduced that the value of $|\psi(0)|^2$ in the metal is 0.44 of the value in the free atom; a calculated value of the ratio is 0.49. It is only in rare instances that the absolute value of the spin contribution χ_s to the magnetic susceptibility can be determined, and then usually by very careful conduction electron spin resonance experiments. The Knight shift has been of value in the study of metals, alloys, soft and intermetallic superconductors, and unusual electronic systems such as Na_xWO_3. The field is reviewed by Drain.[11]

TABLE 2 Knight Shifts in NMR in Metallic Elements

(At room temperature)

Nucleus	Knight Shift in Percent	Nucleus	Knight Shift in Percent
Li^7	0.0261	Cu^{63}	0.237
Na^{23}	0.112	Rb^{87}	0.653
Al^{27}	0.162	Pd^{105}	−3.0
K^{39}	0.265	Pt^{195}	−3.533
V^{51}	0.580	Au^{197}	1.4
Cr^{53}	0.69	Pb^{207}	1.47

[11] L. E. Drain, "Nuclear magnetic resonance in metals," Metallurgical Reviews, Review 119 (1967).

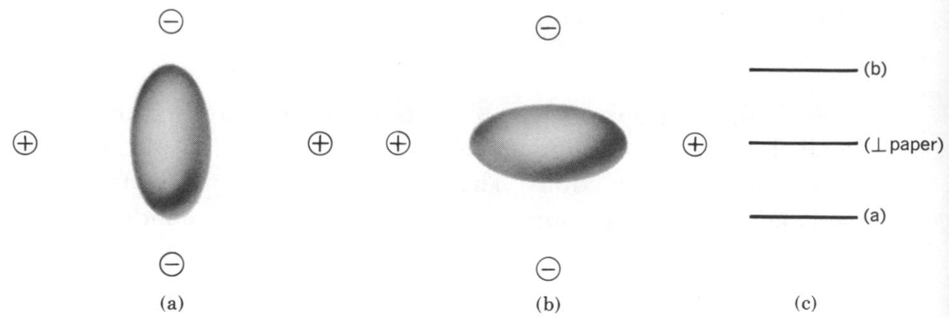

Figure 14 (a) Lowest energy orientation of a nuclear electric quadrupole moment ($Q > 0$) in the local electric field of the four ions shown. The electrons of the ion itself are not shown. (b) Highest energy orientation. (c) The energy level splitting for $I = 1$.

NUCLEAR QUADRUPOLE RESONANCE

Nuclei of spin $I \geq 1$ have an electric quadrupole moment. The quadrupole moment Q is a measure of the ellipticity of the distribution of charge in the nucleus. The quantity of interest is defined classically by

$$eQ = \tfrac{1}{2}\int (3z^2 - r^2)\rho(\mathbf{r})d^3x \ , \tag{39}$$

where $\rho(\mathbf{r})$ is the charge density. An egg-shaped nucleus has Q positive; a saucer-shaped nucleus has Q negative. The nucleus when placed in a crystal will see the electrostatic field of its environment, as in Fig. 14. If the symmetry of this field is lower than cubic, then the nuclear quadrupole moment will lead to a set of energy levels split by the interaction of the quadrupole moment with the local electric field.

The states that are split are the $2I + 1$ states of a spin I. The quadrupole splittings can often be observed directly because an rf magnetic field of the appropriate frequency can cause transitions between the levels. The term **nuclear quadrupole resonance** refers to observations of nuclear quadrupole splittings in the absence of a static magnetic field. The quadrupole splittings are particularly large in covalently bonded molecules such as Cl_2, Br_2, and I_2; the splittings are of the order 10^7 or 10^8 Hz.

FERROMAGNETIC RESONANCE

Spin resonance at microwave frequencies in ferromagnets[12] is similar in principle to nuclear spin resonance. The total magnetic moment of the specimen precesses about the direction of the static magnetic field, and energy is absorbed strongly from the rf transverse field when its frequency is equal to the precessional frequency. We may think of the macroscopic vector representing the total spin of the ferromagnet as quantized in the static magnetic field, with energy levels separated by the usual Zeeman frequencies; the selection rule $\Delta m_S = \pm 1$ allows transitions only between adjacent levels. The unusual features of ferromagnetic resonance include:

1. The transverse susceptibility components χ' and χ'' are very large because the magnetization of a ferromagnet in a given static field is very much larger than the magnetization of electronic or nuclear paramagnets in the same field.
2. The shape of the specimen plays an important role. Because the magnetization is large, the demagnetization field is large.
3. The strong exchange coupling between the ferromagnetic electrons tends to suppress the dipolar contribution to the line width, so that the ferromagnetic resonance lines can be quite sharp (<1 G) under favorable conditions.
4. Saturation effects occur at low rf power levels. It is not possible, as it is with nuclear spin systems, to drive a ferromagnetic spin system so hard that the magnetization M_z is reduced to zero or reversed. The ferromagnetic resonance excitation breaks down into spin wave modes before the magnetization vector can be rotated appreciably from its initial direction.

[12]Observed first by J. H. E. Griffiths, Nature **158**, 670 (1946).

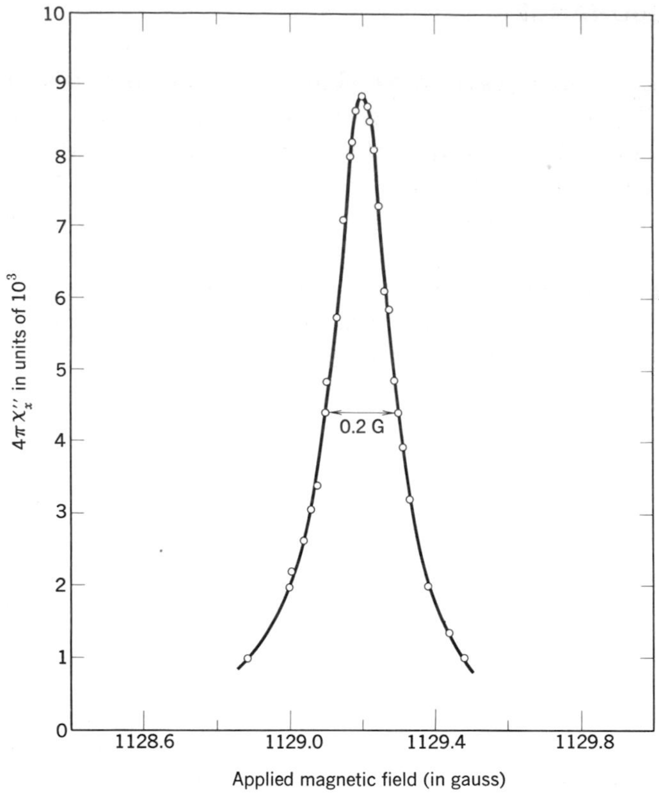

Figure 15 FMR in a polished sphere of yttrium iron garnet at 3.33 GHz and 300 K for $B_0 \parallel [111]$. The total line width at half-power is only 0.2 G. (After R. C. LeCraw and E. Spencer.)

Shape Effects in FMR

We treat the effects of specimen shape on the resonance frequency. Consider a specimen of a cubic ferromagnetic insulator in the form of an ellipsoid with principal axes parallel to x, y, z axes of a cartesian coordinate system. The **demagnetization factors** N_x, N_y, N_z are identical with the depolarization factors defined in Chapter 13. The components of the internal magnetic field $\mathbf{B}_i$ in the ellipsoid are related to the applied field by

$$B_x{}^i = B_x{}^0 - N_xM_x \ ; \qquad B_y{}^i = B_y{}^0 - N_yM_y \ ; \qquad B_z{}^i = B_z{}^0 - N_zM_z \ .$$

The Lorentz field $(4\pi/3)\mathbf{M}$ and the exchange field $\lambda\mathbf{M}$ do not contribute to the torque because their vector product with $\mathbf{M}$ vanishes identically. In SI we replace the components of $\mathbf{M}$ by $\mu_0\mathbf{M}$, with the appropriate redefinition of the N's. The components of the spin equation $\dot{\mathbf{M}} = \gamma(\mathbf{M} \times \mathbf{B}^i)$ become, for an applied static field $B_0\hat{\mathbf{z}}$,

$$\frac{dM_x}{dt} = \gamma(M_y B_z^i - M_z B_y^i) = \gamma[B_0 + (N_y - N_z)M]M_y \ ; \tag{40}$$

$$\frac{dM_y}{dt} = \gamma[M(-N_x M_x) - M_x(B_0 - N_z M)] = -\gamma[B_0 + (N_x - N_z)M]M_x \ .$$

To first order we may set $dM_z/dt = 0$ and $M_z = M$. Solutions of (40) with time dependence $e^{-i\omega t}$ exist if

$$\begin{vmatrix} i\omega & \gamma[B_0 + (N_y - N_z)M] \\ -\gamma[B_0 + (N_x - N_z)M] & i\omega \end{vmatrix} = 0 \ ,$$

so that the ferromagnetic resonance frequency in the applied field B_0 is

$$\text{(CGS)} \qquad \omega_0^2 = \gamma^2[B_0 + (N_y - N_z)M][B_0 + (N_x - N_z)M] \ ; \tag{41}$$

$$\text{(SI)} \qquad \omega_0^2 = \gamma^2[B_0 + (N_y - N_z)\mu_0 M][B_0 + (N_x - N_z)\mu_0 M] \ .$$

The frequency ω_0 is called the frequency of the **uniform mode,** in distinction to the frequencies of magnon and other nonuniform modes. In the uniform mode all the moments precess together in phase with the same amplitude.

For a sphere $N_x = N_y = N_z$, so that $\omega_0 = \gamma B_0$. A very sharp resonance line in this geometry is shown in Fig. 15. For a flat plate with B_0 perpendicular to the plate $N_x = N_y = 0$; $N_z = 4\pi$, whence

$$\text{(CGS)} \quad \omega_0 = \gamma(B_0 - 4\pi M) \ ; \qquad \text{(SI)} \quad \omega_0 = \gamma(B_0 - \mu_0 M) \ . \tag{42}$$

If B_0 is parallel to the plane of the plate, the xz plane, then $N_x = N_z = 0$; $N_y = 4\pi$, and

$$\text{(CGS)} \quad \omega_0 = \gamma[B_0(B_0 + 4\pi M)]^{\frac{1}{2}} \ ; \qquad \text{(SI)} \quad \omega_0 = \gamma[B_0(B_0 + \mu_0 M)]^{\frac{1}{2}} \ . \tag{43}$$

The experiments determine γ, which is related to the spectroscopic splitting factor g by $-\gamma \equiv g\mu_B/\hbar$. Values of g for metallic Fe, Co, Ni at room temperature are 2.10, 2.18, and 2.21, respectively.

Spin Wave Resonance

Uniform rf magnetic fields can excite long wavelength spin waves in thin ferromagnetic films[13] if the electron spins on the surfaces of the film see different anisotropy fields than the spins within the films. In effect, the surface spins may be pinned, as shown in Fig. 16. If the rf field is uniform, it can excite waves with an odd number of half-wavelengths within the thickness of the film. Waves with an even number of half-wavelengths have no net interaction energy with the field.

The condition for **spin wave resonance** (SWR) with the applied magnetic field normal to the film is obtained from (42) by adding to the right-hand side the exchange contribution to the frequency. The exchange contribution may be written as Dk^2, where D is the spin wave exchange constant. The assumption $ka \ll 1$ is valid for the SWR experiments. Thus in an applied field B_0:

$$\text{(CGS)} \qquad \omega_0 = \gamma(B_0 - 4\pi M) + Dk^2 = \gamma(B_0 - 4\pi M) + D(n/\pi L)^2 \ , \qquad (44)$$

where the wavevector for a mode of n half-wavelengths in a film of thickness L is $k = n\pi/L$. An experimental spectrum is shown in Fig. 17.

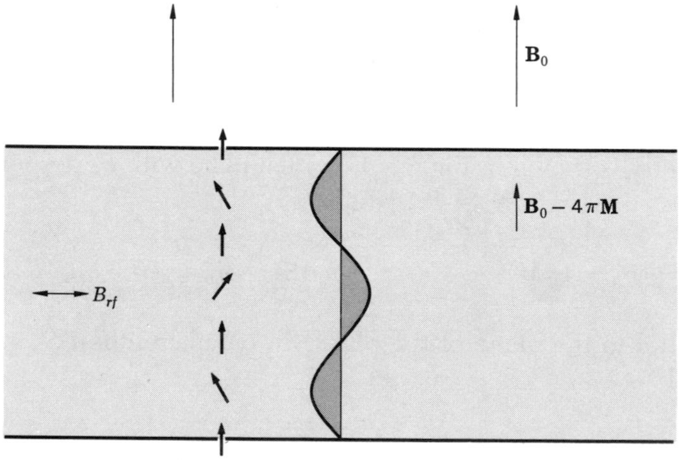

Figure 16 Spin wave resonance in a thin film. The plane of the film is normal to the applied magnetic field B_0. A cross section of the film is shown here. The internal magnetic field is $B_0 - 4\pi M$. The spins on the surfaces of the film are assumed to be held fixed in direction by surface anisotropy forces. A uniform rf field will excite spin wave modes having an odd number of half-wavelengths. The wave shown is for $n = 3$ half-wavelengths.

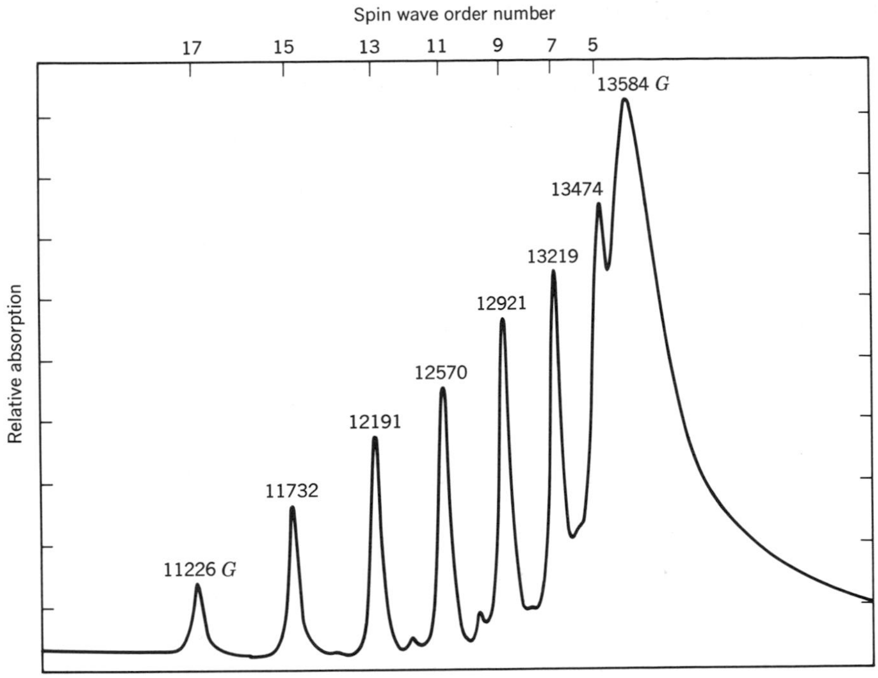

Figure 17 Spin wave resonance spectrum in a Permalloy (80Ni20Fe) film at 9 GHz. The order number is the number of half-wavelengths in the thickness of the film. (After R. Weber, IEEE Transactions on Magnetics, MAG-4:1, March 1968.)

[13]C. Kittel, Phys. Rev. **110,** 1295 (1958); M. H. Seavey, Jr., and P. E. Tannenwald, Phys. Rev. Letters **1,** 108 (1958); P. Pincus, Phys. Rev. **118,** 658 (1960); for a study of the angular dependence see M. Okochi, J. Phys. Soc. Japan **28,** 897 (1970).

ANTIFERROMAGNETIC RESONANCE

We consider a uniaxial antiferromagnetic with spins on two sublattices, 1 and 2. We suppose that the magnetization $\mathbf{M}_1$ on sublattice 1 is directed along the $+z$ direction by an anisotropy field $B_A\hat{\mathbf{z}}$; the anisotropy field (Chapter 15) results from an anisotropy energy density $U_K(\theta_1) = K \sin^2 \theta_1$. Here θ_1 is the angle between $\mathbf{M}_1$ and the z axis, whence $B_A = 2K/M$, with $M = |\mathbf{M}_1| = |\mathbf{M}_2|$. The magnetization $\mathbf{M}_2$ is directed along the $-z$ direction by an anisotropy field $-B_A\hat{\mathbf{z}}$. If $+z$ is an easy direction of magnetization, so is $-z$. If one sublattice is directed along $+z$, the other will be directed along $-z$.

The exchange interaction between $\mathbf{M}_1$ and $\mathbf{M}_2$ is treated in the mean field approximation. The exchange fields are

$$\mathbf{B}_1(\text{ex}) = -\lambda\mathbf{M}_2 \; ; \qquad \mathbf{B}_2(\text{ex}) = -\lambda\mathbf{M}_1 \; , \tag{45}$$

where λ is positive. Here $\mathbf{B}_1$ is the field that acts on the spins of sublattice 1, and $\mathbf{B}_2$ acts on sublattice 2. In the absence of an external magnetic field the total field acting on $\mathbf{M}_1$ is $\mathbf{B}_1 = -\lambda\mathbf{M}_2 + B_A\hat{\mathbf{z}}$; the total field on $\mathbf{M}_2$ is $\mathbf{B}_2 = -\lambda\mathbf{M}_1 - \mathbf{B}_A\hat{\mathbf{z}}$, as in Fig. 18. In what follows we set $M_1^z = M$; $M_2^z = -M$. The linearized equations of motion are

$$\begin{aligned}
dM_1^x/dt &= \gamma[M_1^y(\lambda M + B_A) - M(-\lambda M_2^y)] \; ; \\
dM_1^y/dt &= \gamma[M(-\lambda M_2^x) - M_1^x(\lambda M + B_A)] \; ; \qquad (46) \\
dM_2^x/dt &= \gamma[M_2^y(-\lambda M - B_A) - (-M)(-\lambda M_1^y)] \; ; \\
dM_2^y/dt &= \gamma[(-M)(-\lambda M_1^x) - M_2^x(-\lambda M - B_A)] \; . \qquad (47)
\end{aligned}$$

We define $M_1^+ = M_1^x + iM_1^y$; $M_2^+ = M_2^x + iM_2^y$. Then (46) and (47) become, for time dependence $\exp(-i\omega t)$,

$$\begin{aligned}
-i\omega M_1^+ &= -i\gamma[M_1^+(B_A + \lambda M) + M_2^+(\lambda M)] \; ; \\
-i\omega M_2^+ &= i\gamma[M_2^+(B_A + \lambda M) + M_1^+(\lambda M)] \; .
\end{aligned}$$

These equations have a solution if, with $B_E \equiv \lambda M$,

$$\begin{vmatrix} \gamma(B_A + B_E) - \omega & \gamma B_E \\ \gamma B_E & \gamma(B_A + B_E) + \omega \end{vmatrix} = 0 \; .$$

Thus the AFMR frequency[14] is given by

$$\omega_0^2 = \gamma^2 B_A(B_A + 2B_E) \; . \tag{48}$$

[14]T. Nagamiya, Prog. Theor. Phys. **6**, 342 (1951); C. Kittel, Phys. Rev. **82**, 565 (1951); F. Keffer and C. Kittel, Phys. Rev. **85**, 329 (1952).

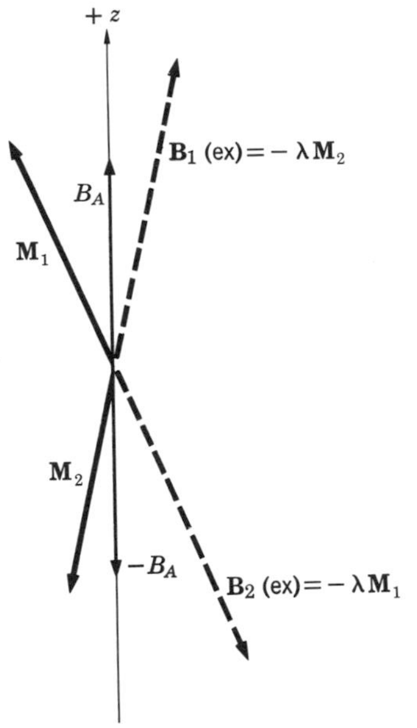

Figure 18 Effective fields in antiferromagnetic resonance. The magnetization $\mathbf{M}_1$ of sublattice 1 sees a field $-\lambda\mathbf{M}_2 + B_A\hat{\mathbf{z}}$; the magnetization $\mathbf{M}_2$ sees $-\lambda\mathbf{M}_1 - B_A\hat{\mathbf{z}}$. Both ends of the crystal axis are "easy axes" of magnetization.

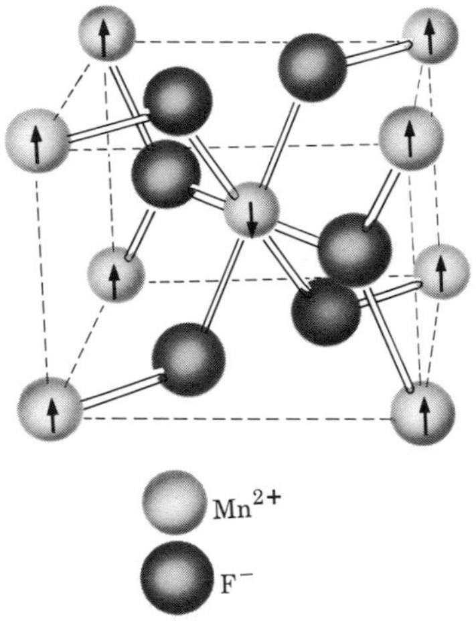

Figure 19 Chemical and magnetic structure of MnF_2. The arrows indicate the direction and arrangement of the magnetic moments assigned to the manganese atoms.

MnF_2 is an extensively studied antiferromagnet. The structure is shown in Fig. 19. The observed variation of ω_0 with temperature is shown in Fig. 20. Careful estimates were made by Keffer of B_A and B_E for MnF_2. He estimated $B_E = 540$ kG and $B_A = 8.8$ kG at 0 K, whence $(2B_AB_E)^{\frac{1}{2}} = 100$ kG. The observed value is 93 kG. Richards has made a compilation of AFMR frequencies as extrapolated to 0 K:

Crystal	CoF_2	NiF_2	MnF_2	FeF_2	MnO	NiO
Frequency in 10^{10} Hz	85.5	93.3	26.0	158.	82.8	109

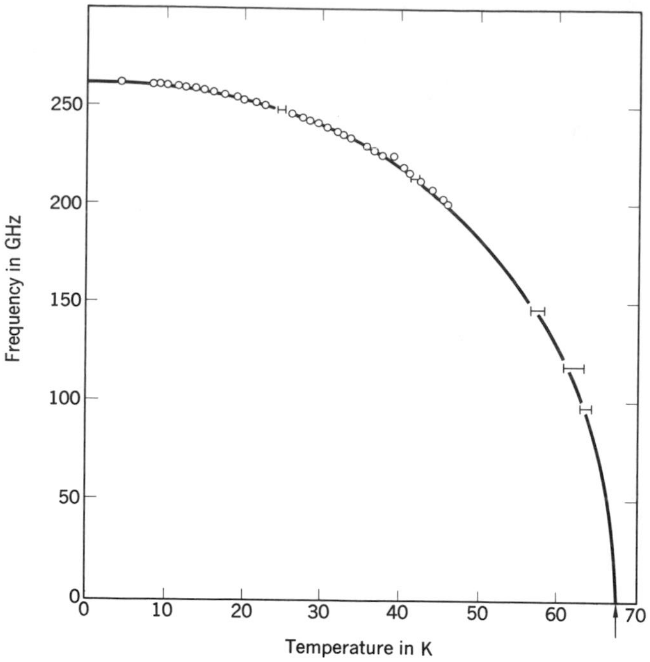

Figure 20 Antiferromagnetic resonance frequency for MnF_2 versus temperature. (After Johnson and Nethercot.)

ELECTRON PARAMAGNETIC RESONANCE

Electron spin resonance is now a vast area of research. We mention several topics of interest.

Exchange Narrowing

We consider a paramagnet with an exchange interaction J among nearest-neighbor electron spins. The temperature is assumed to be well above any spin-ordering temperature T_c. Under these conditions the width of the spin resonance line is usually much narrower than expected for the dipole-dipole interaction. The effect is called **exchange narrowing;** there is a close analogy with motional narrowing. We interpret the exchange frequency $\omega_{ex} \approx J/\hbar$ as a hopping frequency $1/\tau$. Then by generalization of the motional-narrowing result (28) we have for the width of the exchange-narrowed line:

$$\Delta\omega \approx (\Delta\omega)_0^2/\omega_{ex} , \tag{49}$$

where $(\Delta\omega)_0^2 = \gamma^2\langle B_i^2\rangle$ is the square of the static dipolar width in the absence of exchange. A useful and striking example of exchange narrowing is the paramagnetic organic crystal known as the g marker or DPPH.[15] This free radical has a 1.35 G half-width of the resonance line at half-power, only a few percent of the pure dipole width.

Zero-Field Splitting

A number of paramagnetic ions have crystal field splittings of their magnetic ground state energy levels in the range of $10^{10} - 10^{11}$ Hz, which is conveniently accessible by microwave techniques. Much of the early work is due to B. Bleaney and co-workers at Oxford. The Mn^{2+} ion is popular and has been studied in many crystals as an additive impurity. A ground state splitting in the range $10^7 - 10^9$ Hz is observed, according to the environment.

PRINCIPLE OF MASER ACTION

Crystals can be used as microwave and light amplifiers and as sources of coherent radiation. A **maser** amplifies microwaves by the stimulated emission of radiation; a **laser** amplifies light by the same method. The principle, due to Townes, may be understood from the two-level magnetic system of Fig. 21. There are n_u atoms in the upper state and n_l atoms in the lower state. We immerse the system in radiation at frequency ω; the

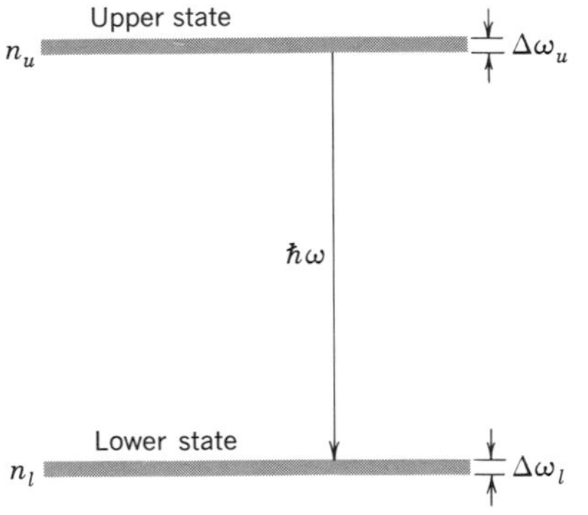

Figure 21 A two-level system, to explain maser operation. The populations of the upper and lower states are n_u and n_l, respectively. The frequency of the emitted radiation is ω; the combined width of the states is $\Delta\omega = \Delta\omega_u + \Delta\omega_l$.

[15]Diphenyl picryl hydrazyl. The g value is 2.0036 ± 0.0002. Because of the sharp line width it is often used for calibration of the magnetic field; see A. N. Holden et al., Phys. Rev. **77**, 147 (1950).

amplitude of the magnetic component of the radiation field is B_{rf}. The probability per atom per unit time of a transition between the upper and lower states is

$$P = \left(\frac{\mu B_{rf}}{\hbar}\right)^2 \frac{1}{\Delta\omega} \; ; \tag{50}$$

here μ is the magnetic moment, and $\Delta\omega$ is the combined width of the two levels. The result (50) is from a standard result of quantum mechanics, called Fermi's golden rule. The net energy emitted from atoms in both upper and lower states is

$$\mathcal{P} = \left(\frac{\mu B_{rf}}{\hbar}\right)^2 \frac{1}{\Delta\omega} \cdot \hbar\omega \cdot (n_u - n_l) \; , \tag{51}$$

per unit time. Here $\mathcal{P}$ denotes the power out; $\hbar\omega$ is the energy per photon; and $n_u - n_l$ is the excess of the number of atoms n_u initially able to emit a photon over the number of atoms n_l able to absorb a photon. In thermal equilibrium $n_u < n_l$, so we cannot get net emission of radiation, but in a nonequilibrium condition with $n_u > n_l$ there will be emission. In fact, if we start with $n_u > n_l$ and reflect the emitted radiation back onto the system, we increase B_{rf} and thereby stimulate a higher rate of emission. The enhanced stimulation continues until the population in the upper state decreases and becomes equal to the population in the lower state.

We can build up the intensity of the radiation field by placing the crystal in an electromagnetic cavity. There will be some power loss in the walls of the cavity: the rate of power loss is

$$\text{(CGS)} \quad \mathcal{P}_L = \frac{B_{rf}^2 V}{8\pi} \cdot \frac{\omega}{Q} \; ; \qquad \text{(SI)} \quad \mathcal{P}_L = \frac{B_{rf}^2 V}{2\mu_0} \cdot \frac{\omega}{Q} \; , \tag{52}$$

where V is the volume and Q is the Q factor of the cavity. We understand B_{rf}^2 to be a volume average. The condition for maser action is that the emitted power $\mathcal{P}$ exceed the power loss $\mathcal{P}_L$. Both quantities involve B_{rf}^2. The maser condition can be expressed in terms of the population excess in the upper state:

$$\text{(CGS)} \quad n_u - n_l > \frac{V\,\Delta B}{8\pi\mu Q} \; , \qquad \text{(SI)} \quad n_u - n_l > \frac{V\,\Delta B}{2\mu_0\mu Q} \; , \tag{53}$$

where μ is the magnetic moment and the line width ΔB is defined in terms of the combined line width $\Delta\omega$ of the upper and lower states as $\mu\Delta B = \hbar\Delta\omega$. The central problem of the maser or laser is to obtain a suitable excess population in the upper state. This is accomplished in various ways in various devices.

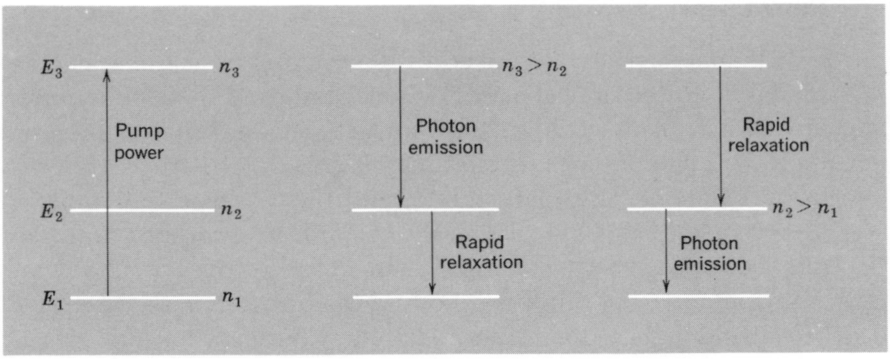

Figure 22 Three-level maser system. Two possible modes of operation are shown, starting from rf saturation of the states 3 and 1 to obtain $n_3 = n_1$.

Three-Level Maser

The three-level maser system (Fig. 22) is a clever solution to the excess population problem. Such a system may derive its energy levels from magnetic ions in a crystal.[16] Rf power is applied at the pump frequency $\hbar\omega_p = E_3 - E_1$ in sufficient intensity to maintain the population of level 3 substantially equal to the population of level 1. Now consider the rate of change of the population n_2 of level 2 owing to normal thermal relaxation processes. In terms of the indicated transition rates P,

$$dn_2/dt = -n_2P(2 \rightarrow 1) - n_2P(2 \rightarrow 3) + n_3P(3 \rightarrow 2) + n_1P(1 \rightarrow 2) \ . \quad (55)$$

In the steady state $dn_2/dt = 0$, and by virtue of the saturation rf power we have $n_3 = n_1$, whence

$$\frac{n_2}{n_1} = \frac{P(3 \rightarrow 2) + P(1 \rightarrow 2)}{P(2 \rightarrow 1) + P(2 \rightarrow 3)} \ . \quad (56)$$

The transition rates are affected by many details of the paramagnetic ion and its environment, but one can hardly lose with this system, for either $n_2 > n_1$ and we get maser action between levels 2 and 1, or else $n_2 < n_1 = n_3$, and we get maser action between levels 3 and 2.

Ruby is an excellent crystal for a three-level maser. Ruby is Al_2O_3 with Cr^{3+} impurity. The Cr^{3+} ions have spin $S = \frac{3}{2}$; the ground level splits into four states in a magnetic field. Three of the four levels are utilized in the maser. Good low noise amplifiers at microwave frequencies have been made from ruby and these find application in radio astronomy and space communication.

[16]N. Bloembergen, Phys. Rev. **104**, 329 (1956).

Ruby Laser

The same crystal, ruby, used in the microwave maser was also the first crystal to exhibit optical maser action,[17] but a different set of energy levels of Cr^{3+} are involved (Fig. 23). About 15,000 cm^{-1} above the ground state there lie a pair of states labeled ^2E, spaced 29 cm^{-1} apart. Above ^2E lie two broad bands of states, labeled $^4F_1$ and $^4F_2$. Because the bands are broad they can be populated efficiently by optical absorption from broadband light sources such as xenon flash lamps.

In operation of a ruby laser both of the broad ^4F bands are populated by broadband light. Atoms thus excited will decay in 10^{-7} sec by radiationless processes with the emission of phonons to the states ^2E. Photon emission from the lower of the states ^2E to the ground state occurs slowly, in about 5×10^{-3} sec, so that a large excited population can pile up in ^2E. For laser action this population must exceed that in the ground state. The stored energy in ruby is 10^8 erg cm^{-3} if 10^{20} Cr^{3+} ions cm^{-3} are in an excited state. The ruby laser can emit at a very high power level if all this stored energy comes out in a short burst. The overall efficiency of conversion of a ruby laser from input electrical energy to output laser light is about one percent. Another popular solid state laser is the neodymium glass laser, made of calcium tungstate glass doped with Nd^{3+} ions. This operates as a four level system (Fig. 24). Here it is not necessary to empty out the ground state before laser action can occur.

[17]T. H. Maiman, Phys. Rev. Letters **4**, 564 (1960); Phys. Rev. **123**, 1145 (1961); for the principles of infrared and optical lasers, see A. L. Schawlow and C. H. Townes, Phys. Rev. **112**, 1940 (1958).

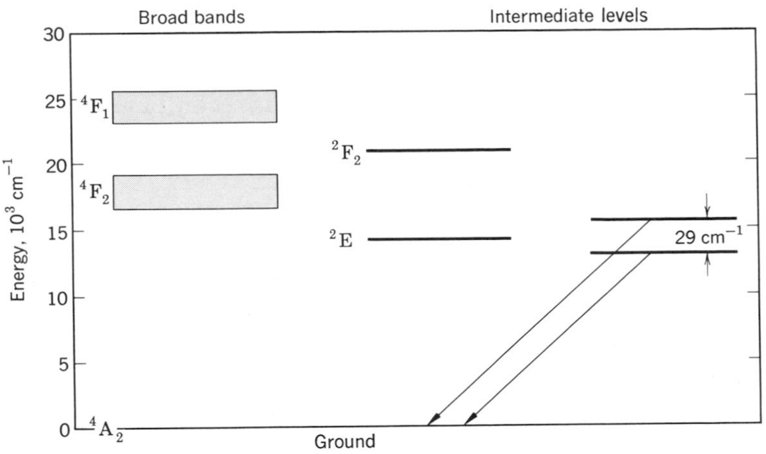

Figure 23 Energy level diagram of Cr^{3+} in ruby, as used in laser operation. The initial excitation takes place to the broad bands; they decay to the intermediate levels by the emission of phonons, and the intermediate levels radiate photons as the ion makes the transition to the ground level.

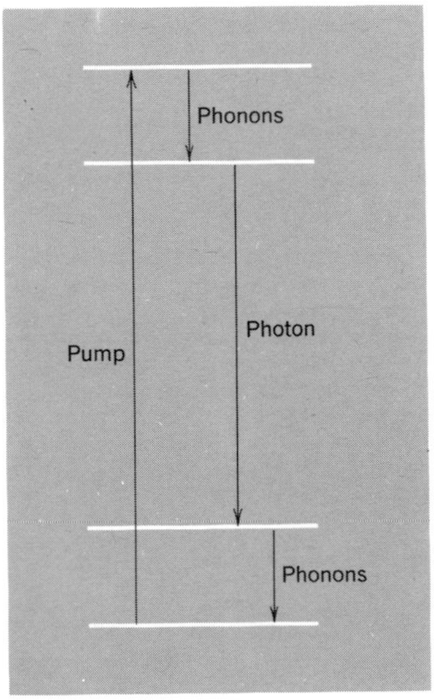

Figure 24 Four-level laser system, as in the neodymium glass laser.

Semiconductor Junction Lasers

Stimulated emission of radiation can occur at a *p-n* semiconductor junction from the radiation emitted when electrons recombine with holes (Fig. 25). The pumping action is provided by the dc voltage. The diode wafer provides its own electromagnetic cavity, for the reflectivity at the crystal-air interface is high. Crystals are usually polished or cleaved to provide two flat parallel surfaces; the radiation is emitted in the plane of the junction. Crystals with direct band gaps (both band edges at the same point in **k** space) have high recombination probabilities and are required normally for junction lasers. Indirect gaps involve phonons as well as photons; carriers recombine less efficiently because of competing processes. Gallium arsenide[18] was the first to work and has been widely studied; it emits in the near infrared at about 8383 Å (1.48 eV), with the exact wavelength tunable by variation of temperature or pressure. The gap is direct, with both conduction and valence band edges at $\mathbf{k} = 0$. The system is very efficient; the ratio of light energy output to dc electrical energy input is near 50 percent. Other semiconductor junction lasers include GaP ($\lambda = 0.65 \mu m$) and InSb ($\lambda = 5.3 \mu m$).

[18]For early experiments see R. N. Hall, G. E. Fenner, J. D. Kingsley, T. J. Soltys, and R. O. Carlson, Phys. Rev. Letters **9**, 366 (1962); M. I. Nathan, W. P. Dumke, G. Burns, F. H. Dill, Jr., and G. Lasher, Appl. Phys. Letters **1**, 62 (1962); T. M. Quist, R. H. Rediker, R. J. Keyes, W. E. Krag, B. Lax, A. L. McWhorter, and H. J. Zeiger, Appl. Phys. Letters **1**, 91 (1962).

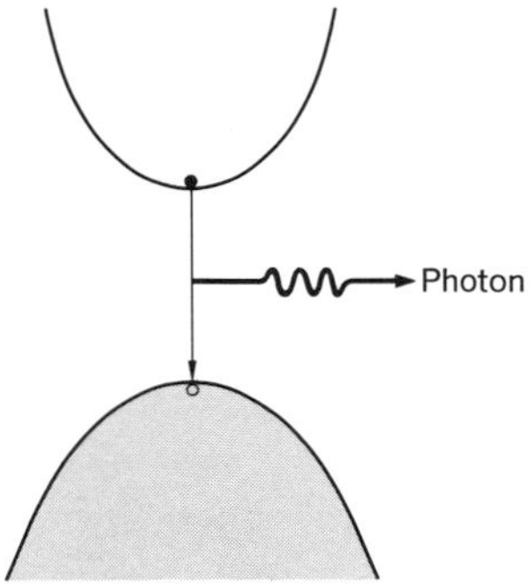

Figure 25a Direct photon emission from the recombination of an electron with a hole.

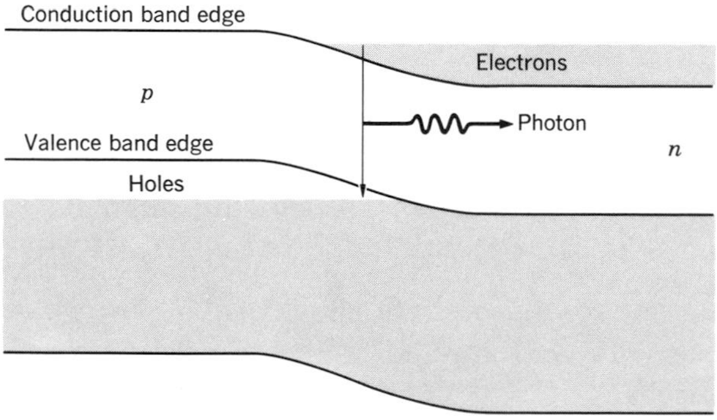

Figure 25b Biased semiconductor junction between a degenerate *p*-type region and a degenerate *n*-type region. Recombination radiation is obtained by biasing the *p*-*n* junction diode in the forward direction. This injection of minority carriers across the junction permits recombination, with the consequent emission of photons.

SUMMARY (In CGS Units)

1. The resonance frequency of a free spin is $\omega_0 = \gamma B_0$, where $\gamma = \mu/\hbar I$ is the magnetogyric ratio.

2. The Bloch equations are

$$dM_x/dt = \gamma(\mathbf{M} \times \mathbf{B})_x - M_x/T_2 \ ;$$
$$dM_y/dt = \gamma(\mathbf{M} \times \mathbf{B})_y - M_y/T_2 \ ;$$
$$dM_z/dt = \gamma(\mathbf{M} \times \mathbf{B})_z + (M_0 - M_z)/T_1 \ .$$

3. The half-width of the resonance at half-power is $(\Delta\omega)_{\frac{1}{2}} = 1/T_2$.

4. Saturation effects at high rf power enter when $\gamma^2 B_1^2 T_1 T_2$ exceeds unity.

5. The dipolar line width in a rigid lattice is $(\Delta B)_0 \approx \mu/a^3$.

6. If the magnetic moments are ambulatory, with a characteristic time $\tau \ll 1/(\Delta\omega)_0$, the line width is reduced by the factor $(\Delta\omega)_0\tau$. In this limit $1/T_1 \approx 1/T_2 \approx (\Delta\omega)_0^2\tau$. With exchange coupling in a paramagnet the line width becomes $\approx (\Delta\omega)_0^2/\omega_{ex}$.

7. The ferromagnetic resonance frequency in an ellipsoid of demagnetization factors N_x, N_y, N_z is $\omega_0^2 = \gamma^2[B_0 + (N_y - N_z)M][B_0 + (N_x - N_z)M]$.

8. The antiferromagnetic resonance frequency is $\omega_0^2 = \gamma^2 B_A(B_A + 2B_E)$, in a spherical specimen with zero applied field. Here B_A is the anisotropy field and B_E is the exchange field.

9. The condition for maser action is that

$$n_u - n_l > V\Delta B/8\pi\mu Q \ .$$

PROBLEMS

1. ***Equivalent electrical circuit.*** Consider an empty coil of inductance L_0 in a series with a resistance R_0; show if the coil is completely filled with a spin system characterized by the susceptibility components $\chi'(\omega)$ and $\chi''(\omega)$ that the inductance at frequency ω becomes $L = [1 + 4\pi\chi'(\omega)]L_o$, in series with an effective resistance $R = 4\pi\omega\chi''(\omega)L_0 + R_0$. In this problem $\chi = \chi' + i\chi''$ is defined for a linearly polarized rf field. Hint: Consider the impedance of the circuit. (CGS units.)

2. ***Rotating coordinate system.*** We define the vector $\mathbf{F}(t) = F_x(t)\hat{\mathbf{x}} + F_y(t)\hat{\mathbf{y}} + F_z(t)\hat{\mathbf{z}}$. Let the coordinate system of the unit vectors $\hat{\mathbf{x}}$, $\hat{\mathbf{y}}$, $\hat{\mathbf{z}}$ rotate with an instantaneous angular velocity Ω, so that $d\hat{\mathbf{x}}/dt = \Omega_y\hat{\mathbf{z}} - \Omega_z\hat{\mathbf{y}}$, etc. (a) Show that $d\mathbf{F}/dt = (d\mathbf{F}/dt)_R + \mathbf{\Omega} \times \mathbf{F}$, where $(d\mathbf{F}/dt)_R$ is the time derivative of $\mathbf{F}$ as viewed in the rotating frame R. (b) Show that (7) may be written $(d\mathbf{M}/dt)_R = \gamma\mathbf{M} \times (\mathbf{B}_a + \mathbf{\Omega}/\gamma)$. This is the equation of motion of $\mathbf{M}$ in a rotating coordinate system. The transformation to a rotating system is extraordinarily useful; it is exploited widely in the literature. (c) Let $\mathbf{\Omega} = -\gamma B_0\hat{\mathbf{z}}$; thus in the rotating frame there is no static magnetic field. Still in the rotating frame, we now apply a dc pulse $B_1\hat{\mathbf{x}}$ for a time t. If the magnetization is initially along $\hat{\mathbf{z}}$, find an expression for the pulse length t such that the magnetization will be directed

along $-\hat{\mathbf{z}}$ at the end of the pulse. (Neglect relaxation effects.) (d) Describe this pulse as viewed from the laboratory frame of reference.

3. ***Hyperfine effects on ESR in metals.*** We suppose that the electron spin of a conduction electron in a metal sees an effective magnetic field from the hyperfine interaction of the electron spin with the nuclear spin. Let the z component of the field seen by the conduction electron be written

$$B_i = \left(\frac{a}{N}\right) \sum_{j=1}^{N} I_j^z \ ,$$

where I_j^z is equally likely to be $\pm\frac{1}{2}$. (a) Show that $\langle B_i^2 \rangle = (a/2N)^2 N$. (b) Show that $\langle B_i^4 \rangle = 3(a/2N)^4 N^2$, for $N \gg 1$.

4. ***FMR in the anisotropy field.*** Consider a spherical specimen of a uniaxial ferromagnetic crystal with an anisotropy energy density of the form $U_K = K \sin^2 \theta$, where θ is the angle between the magnetization and the z axis. We assume that K is positive. Show that the ferromagnetic resonance frequency in an external magnetic field $B_0\hat{\mathbf{z}}$ is $\omega_0 = \gamma(B_0 + B_A)$, where $B_A \equiv 2K/M_s$.

5. ***Exchange frequency resonance.*** Consider a ferrimagnet with two sublattices A and B of magnetizations $\mathbf{M}_A$ and $\mathbf{M}_B$, where $\mathbf{M}_B$ is opposite to $\mathbf{M}_A$ when the spin system is at rest. The gyromagnetic ratios are γ_A, γ_B and the molecular fields are $\mathbf{B}_A = -\lambda \mathbf{M}_B$; $\mathbf{B}_B = -\lambda \mathbf{M}_A$. Show that there is a resonance at

$$\omega_0^2 = \lambda^2(\gamma_A |M_B| - \gamma_B |M_A|)^2 \ .$$

This is called the exchange frequency resonance.

6. ***Rf saturation.*** Given, at equilibrium for temperature T, a two-level spin system in a magnetic field $H_0\hat{\mathbf{z}}$, with populations N_1, N_2 and transition rates W_{12}, W_{21}. We apply an rf signal that gives a transition rate W_{rf}. (a) Derive the equation for dM_z/dt and show that in the steady state

$$M_z = M_0/(1 + 2W_{rf}T_1) \ ,$$

where $1/T_1 = W_{12} + W_{21}$. It will be helpful to write $N = N_1 + N_2$; $n = N_1 - N_2$; and $n_0 = N(W_{21} - W_{12})/(W_{21} + W_{12})$. We see that as long as $2W_{rf}T_1 \ll 1$ the absorption of energy from the rf field does not substantially alter the population distribution from its thermal equilibrium value. (b) Using the expression for n, write down the rate at which energy is absorbed from the rf field. What happens as W_{rf} approaches $1/2T_1$? This effect is called saturation, and its onset may be used to measure T_1.

References

Introduction

G. E. Pake and T. L. Estle, *Physical principles of paramagnetic resonance*, 2nd ed., Benjamin, 1973.

C. P. Slichter, *Principles of magnetic resonance, with examples from solid state physics*, Harper & Row, 1963. An excellent introduction.

Nuclear Magnetic Resonance

A. Abragam, *Nuclear magnetism*, Oxford, 1961. Definitive and comprehensive.

T. P. Das and E. L. Hahn, "Nuclear quadrupole resonance spectroscopy," *Solid state physics*, Supplement **1**, (1958).

V. Jaccarino, "Nuclear resonance in antiferromagnets," *Magnetism*[19] IIA.

A. M. Portis and R. H. Lindquist, "Nuclear resonance in ferromagnetic materials," *Magnetism* IIA, 357.

E. A. Turov and M. P. Petrov, *Nuclear magnetic resonance in ferro- and antiferromagnetics*, Halsted, 1972.

Electron Spin Resonance

A. Abragam and B. Bleaney, *Electron paramagnetic resonance of transition ions*, Oxford, 1970.

S. A. Al'tshuler and B. M. Kozyrev, *Electron paramagnetic resonance in compounds of transition elements*, 2nd ed., Halsted, 1974.

B. Bleaney and K. W. H. Stevens, "Paramagnetic resonance," Repts. Prog. Phys. **16**, 108 (1953).

K. D. Bowers and J. Owen, "Paramagnetic resonance II," Repts. Prog. Phys. **18**, 304 (1955).

J. W. Orton, "Paramagnetic resonance data," Repts. Prog. Phys, **22**, 204 (1959).

W. Low, "Paramagnetic resonance in solids," *Solid state physics*, Supplement **2**, 1960.

K. J. Standley and R. A. Vaughan, *Electron spin relaxation in solids*, Hilger, London, 1969.

Ferro- and Antiferromagnetic Resonance

M. Sparks, *Ferromagnetic relaxation theory*, McGraw-Hill, 1964.

S. Foner, "Antiferromagnetic and ferrimagnetic resonance," *Magnetism* I, 384.

C. W. Haas and H. B. Callen, "Ferromagnetic relaxation and resonance line widths," *Magnetism* I, 450.

R. W. Damon, "Ferromagnetic resonance at high power," *Magnetism* I, 552.

Quantum Electronics

G. J. F. Troup, *Masers and lasers*, 2nd ed., Halsted, 1973.

A. Yariv, *Quantum electronics*, Wiley, 1967.

D. Ross, *Lasers, light amplifiers, and oscillators*, Academic Press, 1969.

W. V. Smith and P. P. Sorokin, *The laser*, McGraw-Hill, 1966.

C. H. Gooch, ed., *Gallium arsenide lasers*, Wiley, 1969.

J. W. Orton, D. H. Paxman, and J. C. Walling, *Solid state maser*, Pergamon, 1970.

H. M. Nussenzveig, *Introduction to quantum optics*, Gordon and Breach, 1973.

[19] Edited by G. T. Rado and H. Suhl, Academic Press.

17 POINT DEFECTS AND ALLOYS

From *Go fly a kite, Charlie Brown,* by kind permission of C. M. Schulz, © 1960 by United Features Syndicate, Inc.

17 POINT DEFECTS AND ALLOYS

Any deviation in a crystal from a perfect periodic lattice or structure is an imperfection. The common point imperfections are chemical impurities, vacant lattice sites, and extra atoms not in regular lattice positions. Line imperfections are considered in the chapter on dislocations. The crystal surface is a planar imperfection, with its surface electron and phonon states. Real crystals are always imperfect in some respect. The nature of the common imperfections are fairly well understood for some solids. Much work is concerned with the alkali and silver halides, germanium, silicon, copper, and alloys in general. An alloy may represent a high concentration of point imperfections.

Many important properties are controlled as much by imperfections as by the nature of the host crystal, which may act only as a vehicle or solvent or matrix for the imperfections. The conductivity of some semiconductors is due entirely to trace amounts of chemical impurities. The color of many crystals arises from imperfections. The luminescence of crystals is nearly always connected with the presence of impurities. Diffusion of atoms may be accelerated enormously by impurities or imperfections. The mechanical and plastic properties are usually controlled by imperfections.

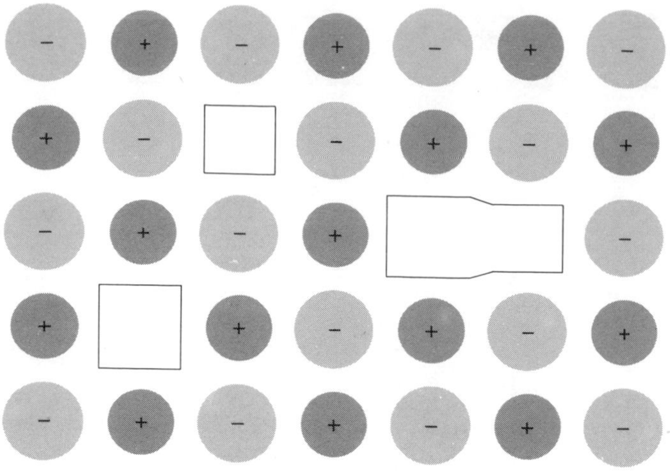

Figure 1 A plane of a pure alkali halide crystal, showing a vacant positive ion site, a vacant negative ion site, and a coupled pair of vacant sites of opposite sign.

LATTICE VACANCIES

The simplest imperfection is a **lattice vacancy,** which is a missing atom or ion, also known as a **Schottky defect.** A lattice vacancy is often indicated in illustrations and in chemical equations by a square (Fig. 1). We create a Schottky defect in a perfect crystal by transferring an atom from a lattice site in the interior to a lattice site on the surface of the crystal. In thermal equilibrium in an otherwise perfect crystal a certain number of lattice vacancies are always present, because the entropy is increased by the presence of disorder in the structure. In metals with close-packed structures the proportion of lattice sites vacant at temperatures just below the melting point is of the order of 10^{-3} to 10^{-4}. But in some alloys, in particular the very hard transition metal carbides such as TiC, the proportion of vacant sites of one component can be as high as 50 percent.

The probability that a given lattice site is vacant is proportional to the Boltzmann factor for thermal equilibrium: $P = \exp(-E_V/k_BT)$, where E_V is the energy required to take an atom from a lattice site inside the crystal to a lattice site on the surface. If there are N atoms, the equilibrium number n of vacancies is given by

$$\frac{n}{N-n} = \exp(-E_V/k_BT) \ . \tag{1}$$

If $n \ll N$, then

$$n/N \cong \exp(-E_V/k_BT) \ . \tag{2}$$

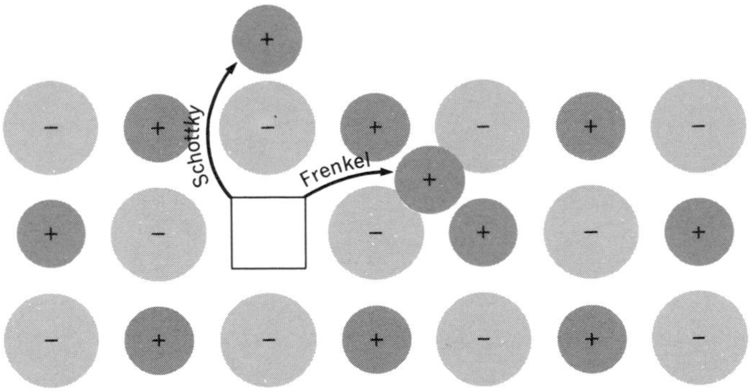

Figure 2 Schottky and Frenkel defects in an ionic crystal. The arrows indicate the displacement of the ions. In a Schottky defect the ion ends up on the surface of the crystal; in a Frenkel defect it is removed to an interstitial position.

If $E_V \approx 1$ eV and $T \approx 1000$ K, then $n/N \approx e^{-12} \approx 10^{-5}$. The equilibrium concentration of vacancies decreases as the temperature decreases. The actual concentration of vacancies will be higher than the equilibrium value if the crystal is grown at an elevated temperature and then cooled suddenly, thereby freezing in the vacancies (see the discussion of diffusion below).

In ionic crystals it is usually favorable energetically to form roughly equal numbers of positive and negative ion vacancies. The formation of pairs of vacancies keeps the crystal electrostatically neutral on a local scale. From a statistical calculation we obtain

$$n \cong N \exp(-E_p/2k_BT) \tag{3}$$

for the number of pairs, where E_p is the energy of formation of a pair.

Another vacancy defect is the **Frenkel defect** (Fig. 2) in which an atom is transferred from a lattice site to an **interstitial position,** a position not normally occupied by an atom. The calculation of the equilibrium number of Frenkel defects proceeds along the lines of Problem 1. If the number n of Frenkel defects is much smaller than the number of lattice sites N and the number of interstitial sites N', the result is

$$n \cong (NN')^{\frac{1}{2}} \exp(-E_I/2k_BT) \ , \tag{4}$$

where E_I is the energy necessary to remove an atom from a lattice site to an interstitial position. In pure alkali halides the most common lattice vacancies are Schottky defects; in pure silver halides the most common vacancies are Frenkel defects.

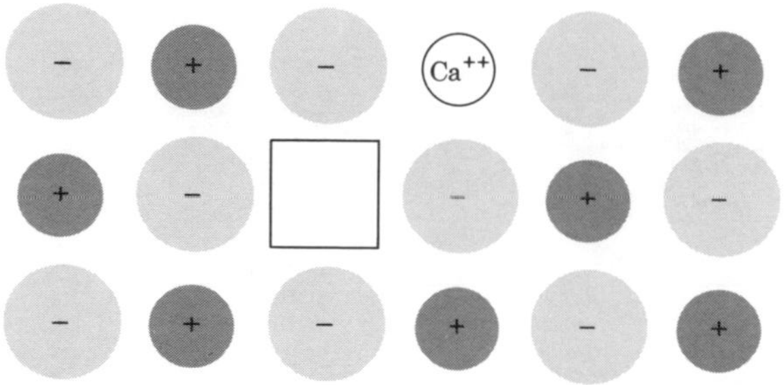

Figure 3 Production of lattice vacancy by the solution of $CaCl_2$ in KCl: to ensure electrical neutrality a positive ion vacancy is introduced into the lattice with each divalent cation Ca^{++}. The two Cl^- ions of $CaCl_2$ enter normal negative ion sites.

Lattice vacancies are present in alkali halides containing additions of divalent elements. If a crystal of KCl is grown with controlled amounts of $CaCl_2$, the density varies as if a K^+ lattice vacancy were formed for each Ca^{2+} ion in the crystal. The Ca^{2+} enters the lattice in a normal K^+ site and the two Cl^- ions enter two Cl^- sites in the KCl crystal (Fig. 3). Demands of charge neutrality results in a vacant metal ion site. The experimental results show that the addition of $CaCl_2$ to KCl lowers the density of the crystal. The density would increase if no vacancies were produced, because Ca^{2+} is a heavier and smaller ion than K^+.

The mechanism of electrical conductivity in alkali and silver halide crystals is usually the motion of ions and not the motion of electrons. This has been established by comparing the transport of charge with the transport of mass as measured by the material plated out on electrodes in contact with the crystal. The study of ionic conductivity is an important tool in the investigation of lattice defects. Work on alkali and silver halides containing known additions of divalent metal ions shows that at not too high temperatures the ionic conductivity is directly proportional to the amount of divalent addition. This is not because the divalent ions are intrinsically highly mobile, for it is predominantly the monovalent metal ion which deposits at the cathode. The lattice vacancies introduced with the divalent ions are responsible for the enhanced diffusion (Fig. 4c). The diffusion of a vacancy in one direction is equivalent to the diffusion of an atom in the opposite direction. When lattice defects are generated thermally, their energy of formation gives an extra contribution to the heat capacity of the crystal, as shown in Fig. 5.

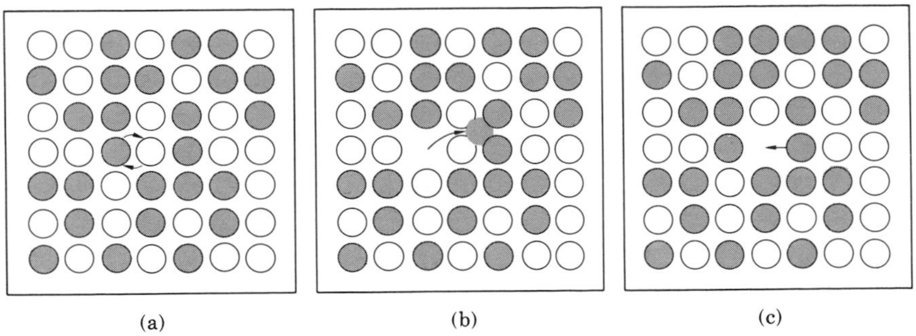

Figure 4 Three basic mechanisms of diffusion: (a) Interchange by rotation about a midpoint. More than two atoms may rotate together. (b) Migration through interstitial sites. (c) Atoms exchange position with vacant lattice sites. (From Seitz.)

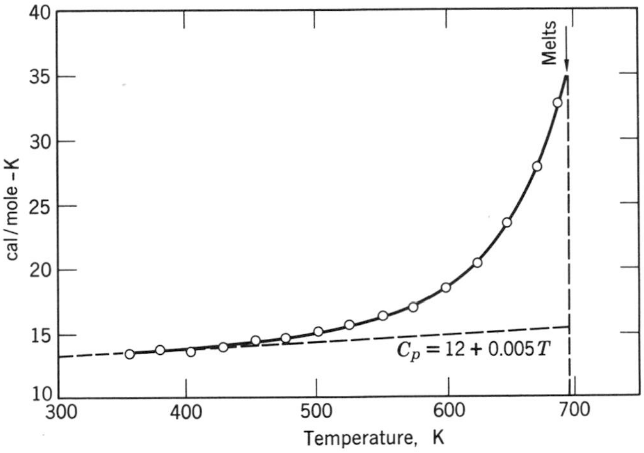

Figure 5 Heat capacity of silver bromide exhibiting the excess heat capacity from the formation of lattice defects. [After R. W. Christy and A. W. Lawson, J. Chem. Phys. **19**, 517 (1951).]

An associated pair of vacancies of opposite sign exhibits an electric dipole moment, with contributions to the dielectric constant and dielectric loss due to the motion of pairs of vacancies. The dielectric relaxation time is a measure of the time required for one of the vacant sites to jump by one atomic position with respect to the other. The dipole moment can change at low frequencies, but not at high. In sodium chloride the relaxation frequency is 1000 s^{-1} at 85°C.

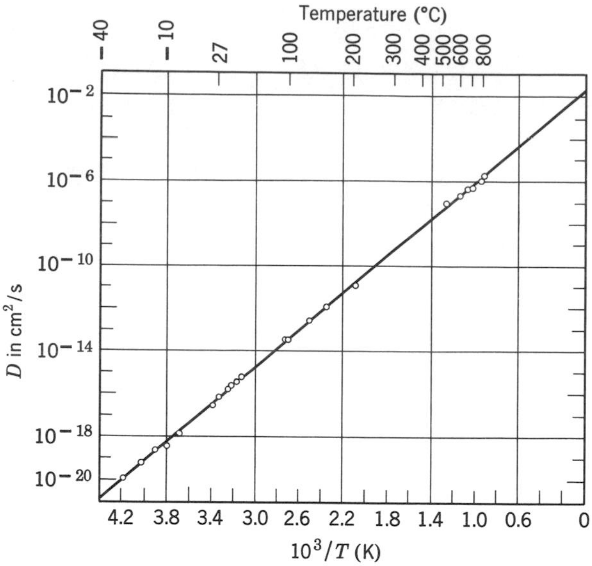

Figure 6 Diffusion coefficient of carbon in iron, after Wert. The logarithm of D is directly proportional to $1/T$. See also Fig. 15 of A. E. Lord, Jr. and D. N. Beshers, Acta Met. **14,** 1659 (1966).

DIFFUSION

When there is a concentration gradient of impurity atoms or vacancies in a solid, there will be a flux of these through the solid. In equilibrium the impurities or vacancies will be distributed uniformly. The net flux J_N of atoms of one species in a solid is related to the gradient of the concentration N of this species by a phenomenological relation called **Fick's law:**

$$\mathbf{J}_N = -D \text{ grad } N \ . \tag{5}$$

Here J_N is the number of atoms crossing unit area in unit time; the constant D is the **diffusion constant** or **diffusivity** and has the units cm²/s. The minus sign means that diffusion occurs away from regions of high concentration. The form (5) of the law of diffusion is often adequate, but rigorously the gradient of the chemical potential is the driving force for diffusion and not the concentration gradient alone. The diffusion constant is often found to vary with temperature as

$$D = D_0 \exp(-E/k_BT) \ ; \tag{6}$$

here E is the **activation energy** for the process. Experimental results on the diffusion of carbon in alpha iron are shown in Fig. 6. The data are represented by $E = 0.87$ eV, $D_0 = 0.020$ cm²/s. Representative values of D_0 and E are given in Table 1.

TABLE 1 Diffusion Constants and Activation Energies

Host Crystal	Atom	D_0 cm^2s^{-1}	E eV	Host Crystal	Atom	D_0 cm^2s^{-1}	E eV
Cu	Cu	0.20	2.04	Si	Al	8.0	3.47
Cu	Zn	0.34	1.98	Si	Ga	3.6	3.51
Ag	Ag	0.40	1.91	Si	In	16.0	3.90
Ag	Cu	1.2	2.00	Si	As	0.32	3.56
Ag	Au	0.26	1.98	Si	Sb	5.6	3.94
Ag	Pb	0.22	1.65	Si	Li	2×10^{-3}	0.66
Na	Na	0.24	0.45	Si	Au	1×10^{-3}	1.13
U	U	2×10^{-3}	1.20	Ge	Ge	10.0	3.1

In order to diffuse, an atom must surmount the potential energy barrier presented by its neighbors. We consider the diffusion of impurity atoms in interstitial sites; identical results apply to the diffusion of vacancies. If the barrier is of height E, the atom will have sufficient thermal energy to pass over the barrier only a fraction $\exp(-E/k_BT)$ of the time. Quantum tunneling through a barrier is usually important only for the lightest nuclei, because for a given energy the de Broglie wavelength increases as the mass of the particle decreases. If ν is a characteristic atomic vibrational frequency, then the probability p that sometime during one second the atom will have enough thermal energy to pass over the barrier is

$$p \approx \nu \exp(-E/k_BT) \ . \tag{7}$$

In unit time the atom makes ν passes at the barrier, with a probability $\exp(-E/k_BT)$ of surmounting the barrier on each try. The quantity p is called the *jump frequency*. Values of ν are of the order of 10^{14} s^{-1}.

We consider two parallel planes of impurity atoms in interstitial sites. The planes are separated by lattice constant a. There are S impurity atoms on one plane and $(S + a\ dS/dx)$ on the other. The net number of atoms crossing between the planes in one second is $\approx -pa\ dS/dx$. If N is the total concentration of impurity atoms, then $S = aN$ per unit area of a plane. The diffusion flux may now be written as

$$J_N \approx -pa^2(dN/dx) \ . \tag{8}$$

On comparison with (5) we have the result

$$\boxed{D = \nu a^2 \exp(-E/k_BT) \ ,} \tag{9}$$

of the form (6) with $D_0 = \nu a^2$.

TABLE 2 Activation Energy E_+ for Motion of a Positive Ion Vacancy

Values of the energy of formation of a vacancy pair, E_f, are also given. The numbers given in parentheses for the silver salts refer to interstitial silver ions.

Crystal	E_+(eV)	E_f(eV)	Workers
NaCl	0.86	2.02	Etzel and Maurer
LiF	0.65	2.68	Haven
LiCl	0.41	2.12	Haven
LiBr	0.31	1.80	Haven
LiI	0.38	1.34	Haven
KCl	0.89	2.1–2.4	Wagner; Kelting and Witt
AgCl	0.39(0.10)	1.4[a]	Teltow
AgBr	0.25(0.11)	1.1[a]	Compton

[a] For Frenkel defect.

If the impurities are charged, we may find the ionic mobility μ and the conductivity σ from the diffusivity by using the Einstein relation $k_BT\mu = qD$:

$$\mu = (q\nu a^2/k_BT)\exp(-E/k_BT) \; ; \tag{10}$$

$$\sigma = Nq\mu = (Nq^2\nu a^2/k_BT)\exp(-E/k_BT) \; , \tag{11}$$

where N is the concentration of impurity ions of charge q.

The proportion of vacancies is independent of temperature in the range in which the number of vacancies is determined by the number of divalent metal ions. Then the slope of a plot of $\ln\sigma$ versus $1/k_BT$ gives E_+, the barrier activation energy for the jumping of positive ion vacancies (Table 2). At room temperature the jump frequency is of the order of 1 s^{-1}, and at 100 K it is of the order of 10^{-25} s^{-1}. Diffusion is very slow at low temperatures. The proportion of vacancies in the temperature range in which the concentration of defects is determined by thermal generation is given by

$$f \cong \exp(-E_f/2k_BT) \; , \tag{12}$$

where E_f is the energy of formation of a vacancy pair, according to the theory of Schottky or Frenkel defects. Here the slope of a plot of $\ln\sigma$ versus $1/k_BT$ will be $E_+ + \frac{1}{2}E_f$, according to (10) and (12). From measurements in different temperature ranges we determine the energy of formation of a vacancy pair E_f and the jump activation energy E_+.

The diffusion constant can be measured by radioactive tracer techniques. The diffusion of a known initial distribution of radioactive ions is followed as a function of time or distance. Values of the diffusion constant thus determined may be compared with values from ionic conductivities. The two sets of values do not usually agree within the experimental accuracy, so that there may be present a diffusion mechanism that does not involve the transport of charge. For example, the diffusion of pairs of positive and negative ion vacancies does not involve the transport of charge.

Metals

Self-diffusion in monatomic metals most commonly proceeds by lattice vacancies. **Self-diffusion** means the diffusion of atoms of the metal itself, and not of impurities. The activation energy for self-diffusion in copper[1] is expected to be in the range 2.4 to 2.7 eV for diffusion through vacancies and 5.1 to 6.4 eV for diffusion through interstitial sites. Observed values of the activation energy are 1.7 to 2.1 eV.

Activation energies for diffusion in Li and Na can be determined from measurements of the temperature dependence of the nuclear resonance line width. As discussed in Chapter 16, the resonance line width narrows when the jump frequency of an atom between sites becomes rapid in comparison with the frequency corresponding to the static line width. The values 0.57 eV and 0.45 eV were determined by NMR for Li and Na.[2] Self-diffusion measurements for sodium also give 0.45 eV.

[1] Vacancy formation in Cu has been reviewed by W. M. Lomer, *Prog. metal physics* **8** (1959).

[2] D. F. Holcomb and R. E. Norberg, Phys. Rev. **93,** 919 (1954).

COLOR CENTERS

Pure alkali halide crystals are transparent throughout the visible region of the spectrum. The crystals may be colored in a number of ways:

1. by the introduction of chemical impurities;
2. by the introduction of an excess of the metal ion (we may heat the crystal in the vapor of the alkali metal and then cool it quickly—an NaCl crystal heated in the presence of sodium vapor becomes yellow; a KCl crystal heated in potassium vapor becomes magenta);
3. by x-ray, γ-ray, neutron, and electron bombardment; and
4. by electrolysis. A **color center** is a lattice defect that absorbs visible light. An ordinary lattice vacancy does not color alkali halide crystals, although it affects the absorption in the ultraviolet.

F Centers

The simplest color center is an *F* center. The name comes from the German word for color, *Farbe*. We usually produce *F* centers by heating the crystal in excess alkali vapor or by x-irradiation. The central absorption band (*F* band) associated with *F* centers in several alkali halides are shown in Fig. 7, and the quantum energies are listed in Table 3. Experimental properties of *F* centers have been investigated in detail, originally by Pohl. As discussed in Chapter 16, the *F* center has been identified by electron spin resonance as an electron bound at a negative ion vacancy (Fig. 8), in agreement with a model suggested by de Boer. When excess alkali atoms are added to an alkali halide crystal, a corresponding number of negative ion vacancies are created. The valence electron of the alkali atom is not bound to the atom; the electron migrates in the crystal and becomes bound to a vacant negative ion site. A negative ion vacancy in a perfect periodic lattice has the effect of an isolated positive charge: it attracts and binds an electron. We can simulate the electrostatic effect of a negative ion vacancy by adding a positive charge q to the normal charge $-q$ of an occupied negative ion site. The *F* center is the simplest trapped-electron center in alkali halide crystals. The optical absorption of an *F* center arises from an electric dipole transition to a bound excited state of the center.

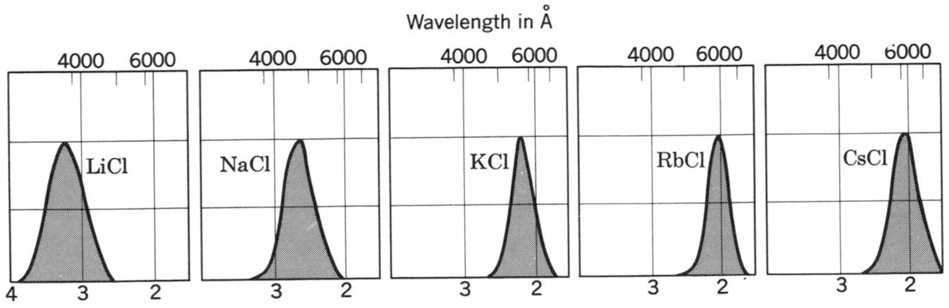

Figure 7 The F bands for several alkali halides: optical absorption versus wavelength for crystals that contain F centers.

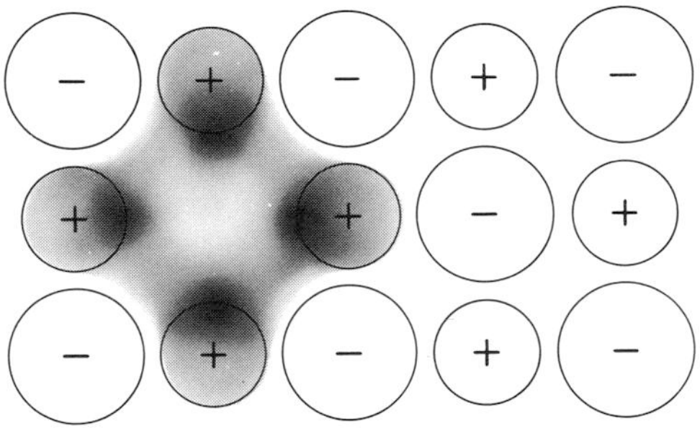

Figure 8 An F center is a negative ion vacancy with one excess electron bound at the vacancy. The distribution of the excess electron is largely on the positive metal ions adjacent to the vacant lattice site.

TABLE 3 Experimental *F* Center Absorption Energies in eV

LiCl	3.1	NaBr	2.3
NaCl	2.7	KBr	2.0
KCl	2.2	RbBr	1.8
RbCl	2.0	LiF	5.0
CsCl	2.0	NaF	3.6
LiBr	2.7	KF	2.7

The model is consistent with the experimental facts:

1. The F band optical absorption is characteristic of the crystal and not of the alkali metal used in the vapor; that is, the band in potassium chloride is almost exactly the same whether the crystal is heated in potassium or sodium vapor. The role of the alkali vapor is to produce an F center in the host crystal.
2. Crystals colored by heating in alkali vapor are found by chemical analysis to contain an excess of alkali metal atoms, typically 10^{16} to 10^{19} per cm^3. The integrated spectral absorption in the F band corresponds quantitatively to that expected from the known amount of excess alkali metal.
3. Colored crystals are less dense than uncolored crystals. The introduction of vacancies lowers the density.

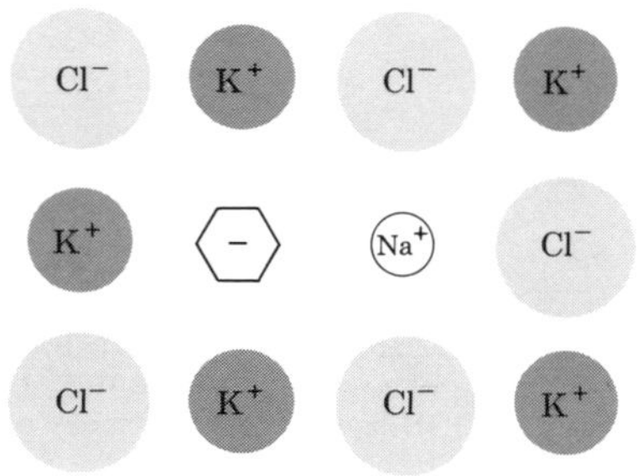

Figure 9 An F_A center in KCl: one of the six K^+ ions which bound an F center is replaced by another alkali ion, here Na^+.

Other Centers in Alkali Halides

In the F_A center one of the six nearest neighbors of the F center has been replaced by a different alkali ion, as in Fig. 9. Complex trapped-electron centers are formed by groups of F centers, as in Figs. 10 and 11. Two adjacent F centers form an M center; three F centers form an R center. These and other centers are usually identified by their optical properties.

Holes may be trapped also to form color centers. Hole centers differ from electron centers: a hole in the filled p^6 shell of a halogen ion leaves the halogen in a p^5 electron configuration, whereas an electron added to the filled p^6 shell of an alkali ion leaves the ion in a p^6s configuration. The chemistry of the two configurations is different. The antimorph to the F center is a hole trapped at a positive ion vacancy, but no such center has been identified in experiments. The best-known trapped-hole center is the V_K center (Fig. 12). It is believed that a hole may be trapped by any halogen ion in a perfect alkali halide crystal to form a V_K center. The structure of the V_K center is known from electron spin resonance experiments to resemble a negative halogen molecule ion: in KCl the V_K center is like a Cl_2^- ion.

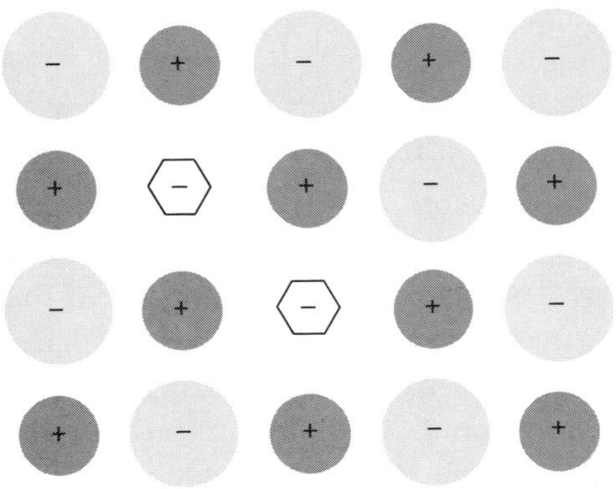

Figure 10 An M center consists of two adjacent F centers.

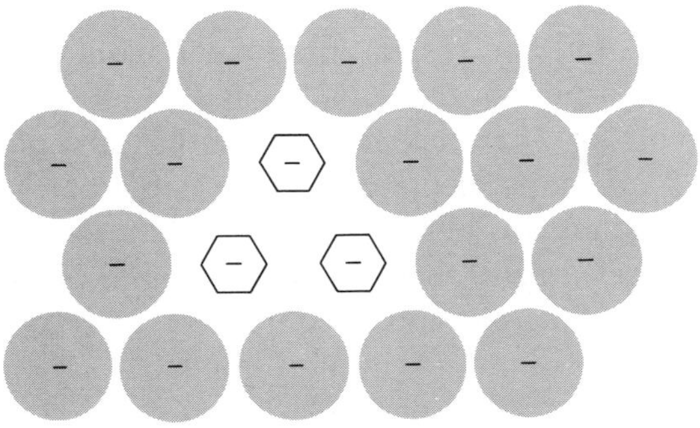

Figure 11 An R center consists of three adjacent F centers; that is, a group of three negative ion vacancies in a [111] plane of the NaCl structure, with three associated electrons.

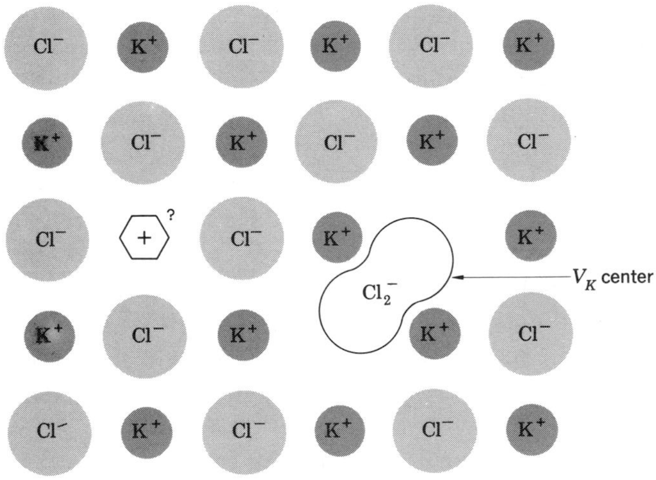

Figure 12 A V_K center formed when a hole is trapped by a pair of negative ions resembles a negative halogen molecule-ion, which is Cl_2^- in KCl. No lattice vacancies or extra atoms are involved in a V_K center. The center at the left of the figure probably is not stable: the hexagon represents a hole trapped near a positive ion vacancy; such a center would be the antimorph to an F center. Holes have a lower energy trapped in a V_K center than in an anti-F center.

ALLOYS

The band structure of solids developed in Chapters 7 and 9 is based on the assumption that the crystal is periodic under the translations of the primitive lattice. What happens to the band structure if the crystal is not periodic, but contains impurities that occupy lattice sites at random, or if the crystal is an alloy of two elements? The translational symmetry of the lattice is no longer perfect. Can we expect that all consequences of band theory, such as the Fermi surface and energy gaps, will no longer exist? Will insulators become conductors? Experiment and theory agree that the consequences of the destruction of the translational symmetry are much weaker than one expects at first sight. These questions can be discussed using the method of orthogonalized plane waves and effective potentials. A low concentration of impurity atoms cannot have much effect on the Fourier components U_G of the potential $U(\mathbf{r})$ responsible for the band gaps and for the behavior of the energy surfaces near the band gap: An impurity will introduce Fourier components of $U(\mathbf{r})$ at wavevectors that are not reciprocal lattice vectors, but such components are never large if the impurity atoms are located at random, by which we mean that regular lattice sites in the host crystal are occupied randomly by impurities. Thus one obtains sharp x-ray diffraction lines from random alloys.

If the impurity belongs to the same column of the periodic table as the host element it replaces, then the effects are particularly small, in part because the average number of valence electrons remains constant. One measure of the effect of alloying is the residual resistivity. One atomic percent of copper dissolved in silver increases the residual electrical resistivity by 0.07 μohm-cm, which corresponds to a scattering cross section of perhaps 0.03 of the area of the impurity atom. Similarly, the mobility of electrons in Si–Ge alloys is much higher than one would expect from the simple geometrical argument that a germanium atom (32 electrons) is very unlike a silicon atom (14 electrons) and therefore a Si atom in Ge or a Ge atom in Si should act as an efficient scattering center for charge carriers. The theory shows us that the effective scattering potential of an impurity can be very small. There is no experimental evidence for an intrinsic reduction in band gap due to the random aspects of alloying. For example, silicon and germanium form solid solutions over the entire composition range; the energies of the band edges in the alloys vary continuously with composition (Fig. 13). It is believed that the density of orbitals near the band edges are somewhat smudged out by alloying.

We now discuss substitutional solid solutions of one metal A in another B of different valence. We suppose that atoms A and B occupy equivalent

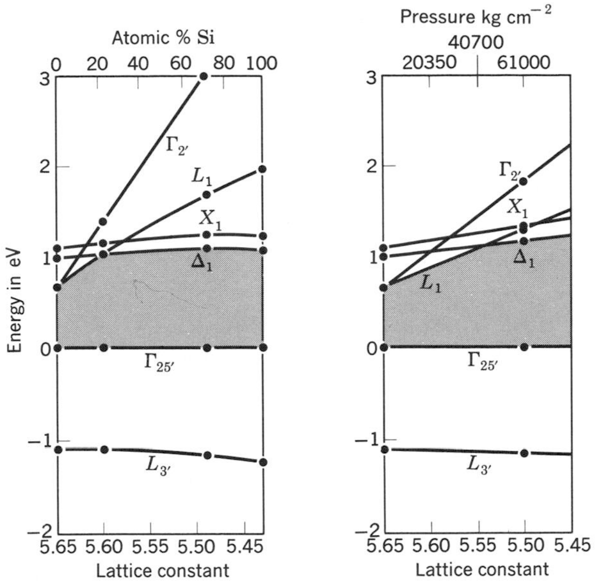

Figure 13 (a) Calculated variation of the major band edges in Ge-Si alloys as a function of the silicon concentration. The point $\Gamma_{25'}$ in **k** space is the valence band edge in both Ge and Si; L_1 is the conduction band edge in Ge. (Compare with Fig. 8.15.) The calculations are by F. Bassani and D. Brust, Phys. Rev. **131**, 1524 (1963), who give extensive references to the experimental data. (b) The calculated variation as a function of pressure for pure Ge. Qualitatively the variation of the bands with pressure is similar to the variation with alloying, as we see by comparing energy values at equal values of the lattice constant.

lattice positions at random. The distinct effects that occur when the occupancies are regular and not random are considered under the heading of the order-disorder transformation. Hume-Rothery has discussed empirical requirements for solid solutions to occur. One requirement is size. It is difficult to form solid solutions if the atomic diameters of A to B differ by more than 15 percent. The sizes are favorable in the Cu (2.55 Å)–Zn (2.66 Å) system: zinc dissolves in copper as a fcc solid solution up to 38 atomic percent zinc. The sizes are somewhat unfavorable in the Cu (2.55 Å)–Cd (2.97 Å) system: only 1.7 atomic percent cadmium is soluble in copper. The atomic diameters referred to copper are 1.04 for zinc and 1.165 for cadmium.

Although the sizes may be favorable, solid solutions will not form if there is a strong tendency for A and B to form stable compounds of definite chemical proportions. If A is strongly electronegative and B strongly electropositive, it is likely that compounds such as AB and A_2B will precipitate out of solution. Although the atomic diameter ratio is favorable (1.02) for As in Cu, only 6 percent As is soluble. The ratio is also favorable (1.09) for Sb in Mg, yet the solubility of Sb in Mg is very small.

Several aspects of the electronic structure of alloys can be discussed in terms of the average number of conduction electrons per atom, denoted by n. The value of n in the alloy 50 percent Cu–50 percent Zn is 1.50; in 50 Cu–50 Al, $n = 2.00$. Some of the principal effects of alloying elements of different valence come from the change in electron concentration. Hume-Rothery first drew attention to the importance of the average electron concentration as determining structural changes in certain alloy systems. The phase diagram of the copper-zinc system[3] is shown in Fig. 14. The fcc structure of pure copper ($n = 1$) persists on the addition of zinc ($n = 2$) until the electron concentration reaches 1.38. A bcc structure occurs at a minimum electron concentration of about 1.48. The γ phase exists for the approximate range of n between 1.58 and 1.66, and the hcp phase ϵ occurs near 1.75. The term **electron compound** denotes an intermediate phase (such as the β phase of CuZn) whose crystal structure is determined by a fairly well defined electron to atom ratio. The values of the ratio are called the **Hume-Rothery rules;** they are 1.50 for the β phase, 1.62 for the γ phase, and 1.75 for the ϵ phase. Representative experimental values are collected in Table 4, based on the usual chemical valence of 1 for Cu and Ag; 2 for Zn and Cd; 3 for Al and Ga; 4 for Si, Ge, and Sn.

The Hume-Rothery rules find a simple expression in terms of the band theory of nearly free electrons. The observed limit of the fcc phase occurs close to the electron concentration of 1.36 at which an inscribed Fermi sphere makes contact with the Brillouin zone boundary for the fcc lattice. The observed electron concentration of the bcc phase is close to the concentration 1.48 at which an inscribed Fermi sphere makes contact with the zone boundary for the bcc lattice. Contact of the Fermi sphere with the zone boundary for the γ phase is at the concentration 1.54. Contact for the hcp phase is at the concentration 1.69 for the ideal c/a ratio.

Why is there a connection between the electron concentration at which a new phase appears and the electron concentration at which the Fermi surface makes contact with the Brillouin zone boundary? It is costly in energy to add further electrons to an alloy once the filled states reach the zone boundary. Additional electrons can be accommodated only in states above the energy gap which occurs at the boundary or in the states of high energy near the corners of the lower zone. It may therefore be energetically favorable for the crystal structure to change to one which can contain a larger Fermi surface before contact. In this way the sequence fcc, bcc, γ, hcp was made plausible by H. Jones.

[3]The phases of interest are usually denoted by metallurgists by Greek characters: in the Cu-Zn system we have α (fcc), β (bcc), γ (complex cubic cell of 52 atoms), ϵ (hcp) and η (hcp); ϵ and η differ considerably in c/a ratio. The meaning of the characters depends on the alloy system.

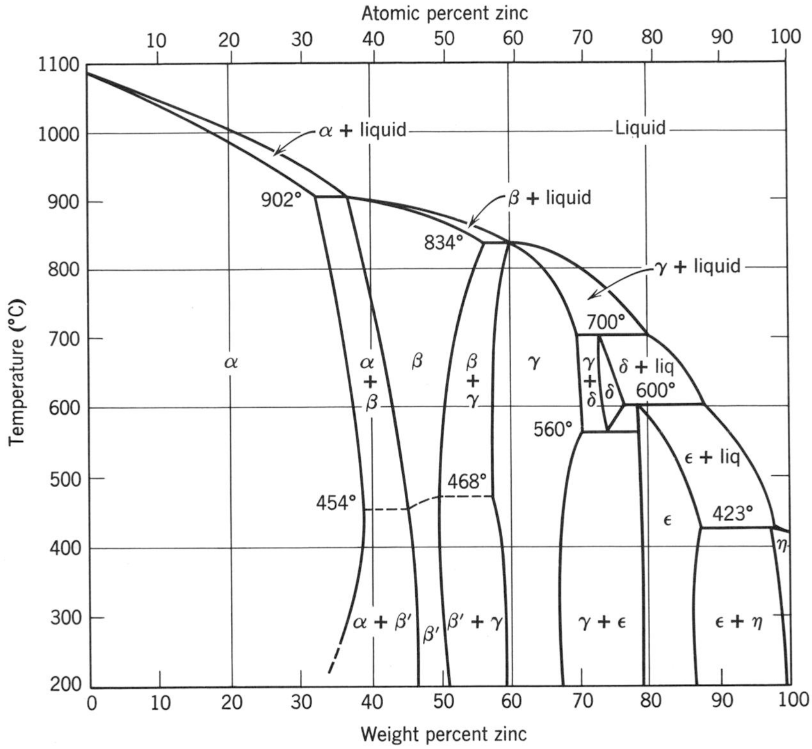

Figure 14 Equilibrium diagram of phases in the copper-zinc alloy system. The α phase is fcc; β and β' are bcc; γ is a complex structure; ϵ and η are both hcp, but ϵ has a c/a ratio near 1.56 and η (for pure Zn) has $c/a = 1.86$. The β' phase is ordered bcc, by which we mean that most of the Cu atoms occupy sites on one sc sublattice and most of the Zn atoms occupy sites on a second sc sublattice that interpenetrates the first sublattice. The β phase is disordered bcc: any site is equally likely to be occupied by a Cu or a Zn atom, almost irrespective of what atoms are in the neighboring sites.

TABLE 4 Electron/Atom Ratios of Electron Compounds

Alloy	fcc Phase Boundary	Minimum bcc Phase Boundary	γ-Phase Boundaries	hcp Phase Boundaries
Cu–Zn	1.38	1.48	1.58–1.66	1.78–1.87
Cu–Al	1.41	1.48	1.63–1.77	
Cu–Ga	1.41			
Cu–Si	1.42	1.49		
Cu–Ge	1.36			
Cu–Sn	1.27	1.49	1.60–1.63	1.73–1.75
Ag–Zn	1.38		1.58–1.63	1.67–1.90
Ag–Cd	1.42	1.50	1.59–1.63	1.65–1.82
Ag–Al	1.41			1.55–1.80

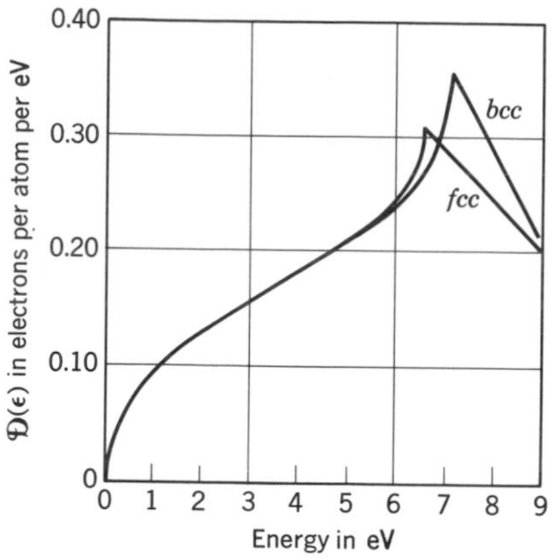

Figure 15 Number of orbitals per unit energy range for the first Brillouin zone of the fcc and bcc lattices, as a function of energy.

The transformation from fcc to bcc is illustrated by Fig. 15; this shows the number of orbitals per unit energy range as a function of energy, for the fcc and bcc structures. As the number of electrons is increased, a point is reached where it is easier to accommodate additional electrons in the Brillouin zone of the bcc lattice rather than in the Brillouin zone of the fcc lattice. The figure is drawn for copper.

Magnetic Alloys and the Kondo Effect

In dilute solid solutions of a magnetic ion in a nonmagnetic metal crystal (such as Mn in Cu) the exchange coupling between the ion and the conduction electrons has important consequences. The conduction electron gas is magnetized in the vicinity of the magnetic ion, with the spatial dependence shown in Fig. 16. This magnetization causes an indirect exchange interaction[4] between two magnetic ions, because a second ion perceives the magnetization induced by the first ion. The interaction, known as the RKKY interaction, also plays a role in the magnetic spin order of the rare-earth metals, where the spins of the $4f$ ion cores are coupled together by the magnetization induced in the conduction electron gas.

[4]A review of indirect exchange interactions in metals is given by C. Kittel, *Solid state physics* **22,** 1 (1968); a review of the Kondo effect is given by J. Kondo, "Theory of dilute magnetic alloys," *Solid state physics* **23,** 184 (1969) and A. J. Heeger, "Localized moments and nonmoments in metals: the Kondo effect," *Solid state physics* **23,** 248 (1969).

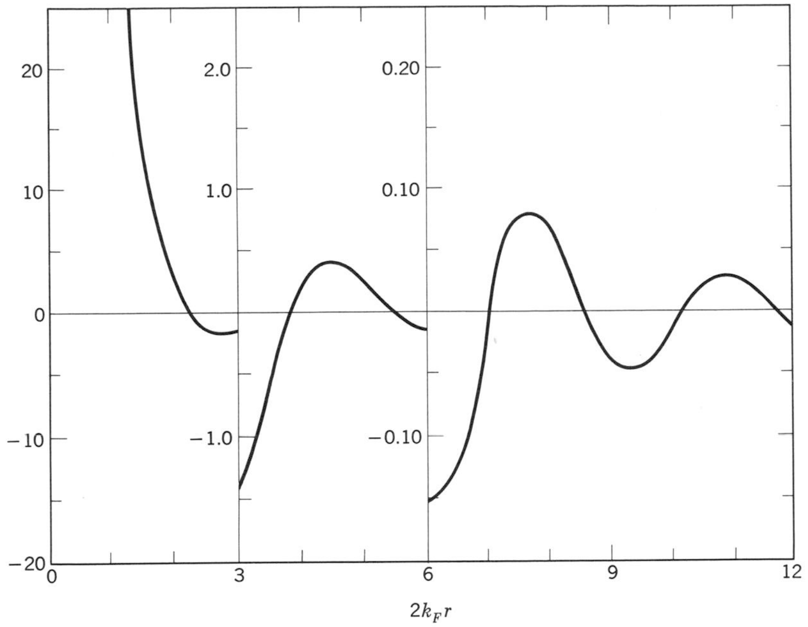

Figure 16 Magnetization of a free electron gas in neighborhood of a point magnetic moment at the origin $r = 0$, according to the RKKY theory. The horizontal axis is $2k_F r$, where k_F is the wave vector on the Fermi sphere. (After de Gennes.)

A dramatic consequence of the magnetic ion-conduction electron interaction is the **Kondo effect.** A minimum in the resistivity-temperature curve of dilute magnetic alloys at low temperatures has been observed in alloys of Cu, Ag, Au, Mg, Zn with Cr, Mn, and Fe as impurities, among others. The occurrence of a resistance minimum is connected with the existence of localized magnetic moments on the impurity atoms. Where a resistance minimum is found, there is inevitably a local moment. Kondo showed that the anomalously high scattering probability of magnetic ions at low temperatures is an esoteric consequence of the dynamic nature of the scattering by the exchange coupling and of the sharpness of the Fermi surface at low temperatures. The temperature region in which the Kondo effect is important is shown in Fig. 17. The central result is that the spin-dependent contribution to the resistivity is

$$\rho_{\text{spin}} = c\rho_M \left[1 + \frac{3zJ}{\epsilon_F} \ln T \right] , \tag{13}$$

where J is the exchange energy; z the number of nearest neighbors; c the concentration; and ρ_M is a measure of the strength of the exchange scattering. We see that the spin resistivity increases toward low temperatures if J is negative. If the phonon resistivity goes as T^5 in the region of interest and if the resistivities are additive, then the total resistivity has the form

$$\rho = aT^5 + c\rho_0 - c\rho_1 \ln T \ , \tag{14}$$

with a minimum at

$$d\rho/dT = 5aT^4 - c\rho_1/T = 0 \ , \tag{15}$$

whence

$$T_{\text{min}} = (c\rho_1/5a)^{1/5} \ . \tag{16}$$

The temperature at which the resistivity is a minimum varies as the one-fifth power of the concentration of the magnetic impurity, in agreement with experiment at least for Fe in Cu.

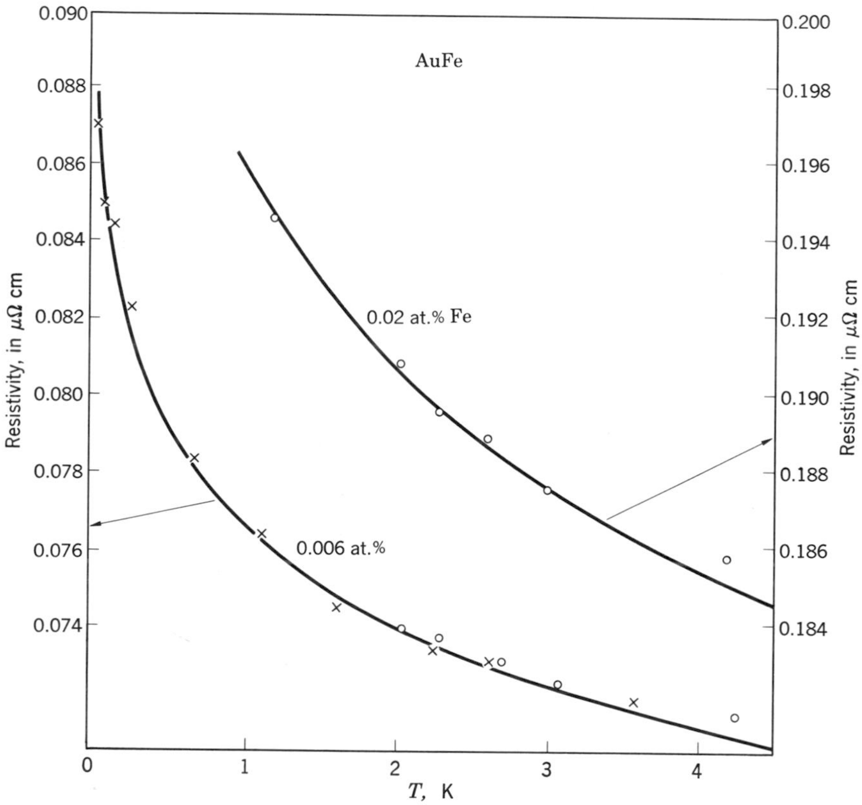

Figure 17 A comparison of experimental and theoretical results for the increase of electrical resistivity at low temperatures in dilute alloys of iron in gold. The resistance minimum lies to the right of the figure, for the resistivity increases at high temperatures because of scattering of electrons by thermal phonons. The experiments are due to D. K. C. MacDonald, W. B. Pearson, and I. M. Templeton, Proc. Roy. Soc. (London) **A266,** 161 (1962); the theory is by J. Kondo, Prog. Theo. Physics **32,** 37 (1964).

ORDER-DISORDER TRANSFORMATION

The dashed horizontal line in the β phase region of the phase diagram of the Cu–Zn (Fig. 14) represents the transition temperature between ordered and disordered states of the alloy. Consider an alloy composed of equal numbers of two types of metal atoms, A and B. The alloy is said to be **ordered** if the A and B atoms have a regular periodic arrangement with respect to one another, as in Fig. 18a. The alloy is **disordered** if the A and B are randomly arranged, as in Fig. 18b. Many properties of an alloy are sensitive to the degree of order. In the common ordered arrangement of an AB alloy with a bcc structure all the nearest-neighbor atoms of a B atom are A atoms, and vice versa. This arrangement results when the dominant interaction among the atoms is an attraction between A and B atoms. (If the dominant interaction is a repulsion between A and B atoms, a two-phase system is formed in which some crystallites are largely A and other crystallites are largely B.)

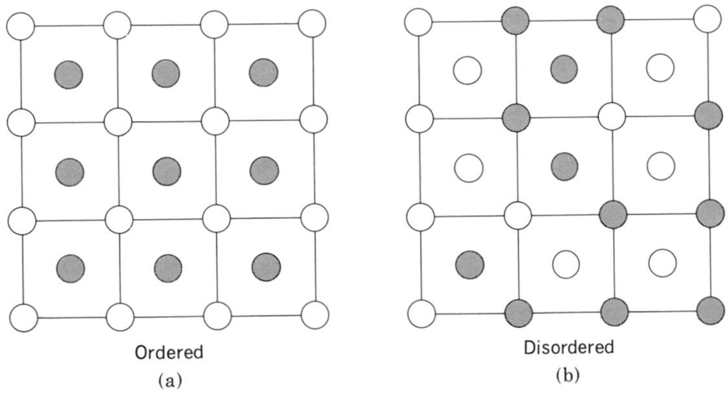

Figure 18 Ordered (a) and disordered (b) arrangements of AB ions in the alloy AB.

The alloy is completely ordered at absolute zero. It becomes less ordered as the temperature is increased, until a transition temperature is reached above which the structure is disordered. The transition temperature marks the disappearance of **long-range order,** which is order over many interatomic distances, but some **short-range order** or correlation among near neighbors may persist above the transition. The long-range order in an AB alloy is shown in Fig. 19a. Long- and short-range order for an alloy of composition AB_3 is given in Fig. 19b. The degree of order is defined below.

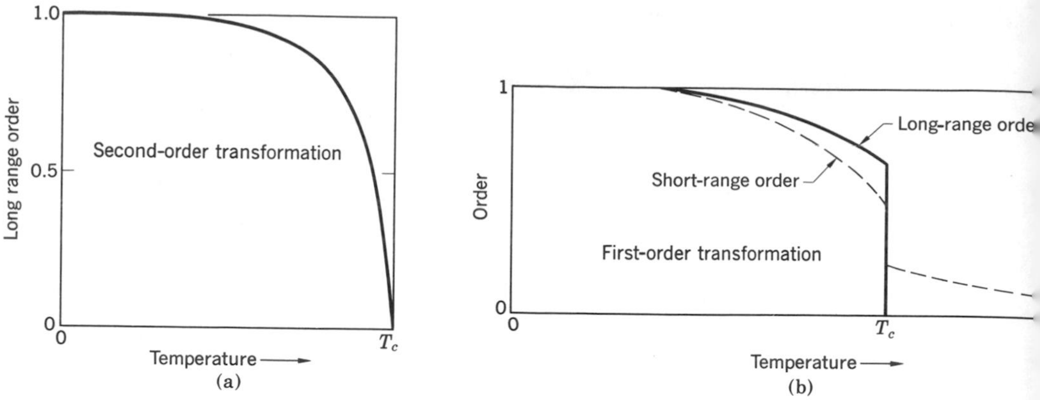

Figure 19 (a) Long-range order versus temperature for an *AB* alloy. The transformation is second order in the sense of Chapter 13. (b) Long-range and short-range order for an AB_3 alloy, after Nix and Shockley. The transformation for this composition is first order.

If an alloy is cooled rapidly from high temperatures to a temperature below the transition, a metastable condition may be produced in which a nonequilibrium disorder is frozen in the structure. The reverse effect occurs when an ordered specimen is disordered at constant temperature by heavy irradiation with nuclear particles. The degree of order may be investigated experimentally by x-ray diffraction. The disordered structure in Fig. 18 has diffraction lines at the same positions as if the lattice points were all occupied by only one type of atom, because the effective scattering power of each plane is equal to the average of the *A* and *B* scattering powers. The ordered structure has extra diffraction lines not possessed by the disordered structure. The extra lines are called **superstructure lines.**

The structure of the ordered CuZn alloy is the cesium chloride structure (Fig. 1.23). The space lattice is simple cubic, and the basis has one Cu atom at 000 and one Zn atom at $\frac{1}{2}\frac{1}{2}\frac{1}{2}$. The diffraction structure factor

$$\mathcal{S}(hkl) = f_{\mathrm{Cu}} + f_{\mathrm{Zn}}\, e^{-i\pi(h+k+l)} \ . \tag{17}$$

This cannot vanish because $f_{\mathrm{Cu}} \neq f_{\mathrm{Zn}}$; therefore all reflections of the simple cubic space lattice will occur. In the disordered structure the situation is different: the basis is equally likely to have either Zn or Cu at 000 and either Zn or Cu at $\frac{1}{2}\frac{1}{2}\frac{1}{2}$. Then the average structure factor is

$$\langle \mathcal{S}(hkl) \rangle = \langle f \rangle + \langle f \rangle e^{-i\pi(h+k+l)} \ , \tag{18}$$

where $\langle f \rangle = \frac{1}{2}(f_{\mathrm{Cu}} + f_{\mathrm{Zn}})$. Equation (18) is exactly the form of the result for the bcc lattice; the reflections vanish when $h + k + l$ is odd. We see that the ordered lattice has reflections (the superstructure lines) not present in the disordered lattice (Fig. 20).

(a)

(b)

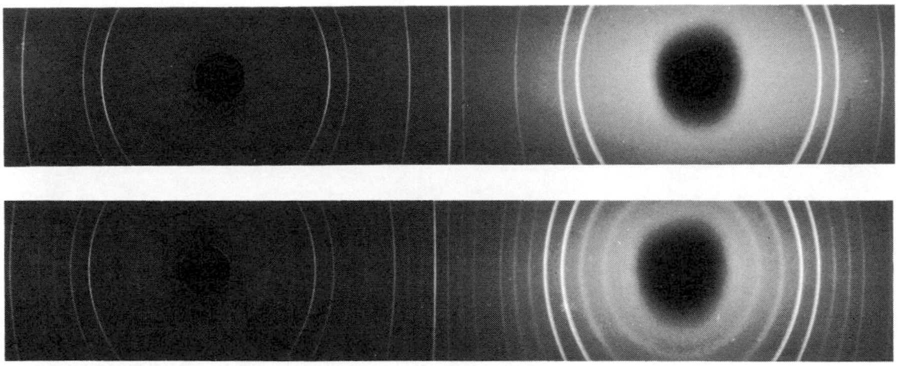

Figure 20 X-ray powder photographs in $AuCu_3$ alloy. (a) Disordered by quenching from $T > T_c$; (b) ordered by annealing at $T < T_c$. (Courtesy of G. M. Gordon.)

Elementary Theory of Order

We give a simple statistical treatment of the dependence of order on temperature for an AB alloy with a bcc structure. The case A_3B differs[5] from AB, the former having a first-order transition marked by a latent heat and the latter having a second-order transition marked by a discontinuity in the heat capacity (Fig. 21). We introduce a measure of the long-range order. We call one simple cubic lattice a and the other b: the bcc structure is composed of the two interpenetrating sc lattices, and the nearest neighbors of an atom on one lattice lie on the other lattice. If there are N atoms A and N atoms B in the alloy, the **long-range order parameter** P is defined so that the number of A's on the lattice a is equal to $\frac{1}{2}(1+P)N$. The number of A's on lattice b is equal to $\frac{1}{2}(1-P)N$. When $P=\pm 1$, the order is perfect and each lattice contains only one type of atom. When $P=0$, each lattice contains equal numbers of A and B atoms and there is no long-range order.

We consider that part of the internal energy associated with the bond energies of AA, AB, and BB nearest-neighbor pairs. The total bond energy is

$$E = N_{AA}U_{AA} + N_{BB}U_{BB} + N_{AB}U_{AB} \ , \qquad (19)$$

where N_{ij} is the number of nearest-neighbor ij bonds, and U_{ij} is the energy of an ij bond. The probability that an atom A on lattice a will have

[5]The connection between composition and the order of the phase transition is discussed by S. Strässler and C. Kittel, Phys. Rev. **139,** A758 (1965).

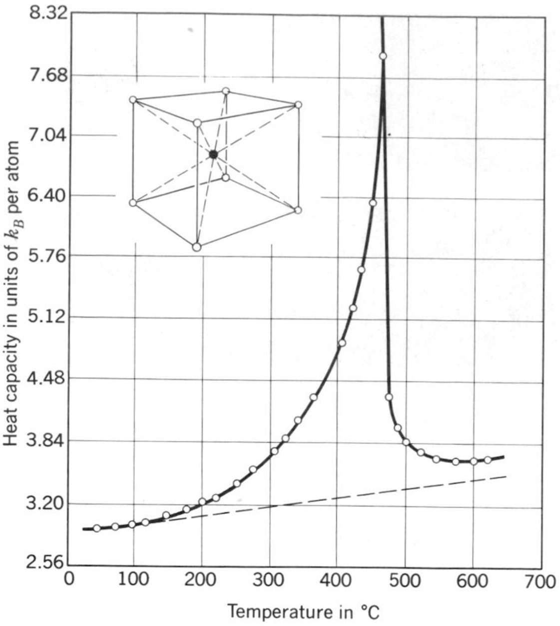

Figure 21 Heat capacity versus temperature of CuZn (β-brass) alloy. [After F. C. Nix and W. Shockley, Revs. Mod. Physics **10**, 1 (1938).]

an *AA* bond is equal to the probability that an *A* occupies a particular nearest-neighbor site on *b*, times the number of nearest-neighbor sites, which is 8 for the bcc structure. We assume that the probabilities are independent. Thus, by the preceding expressions for the number of *A*'s on *a* and *b*,

$$\begin{aligned} N_{AA} &= 8[\tfrac{1}{2}(1+P)N][\tfrac{1}{2}(1-P)] = 2(1-P^2)N \ ; \\ N_{BB} &= 8[\tfrac{1}{2}(1+P)N][\tfrac{1}{2}(1-P)] = 2(1-P^2)N \ ; \\ N_{AB} &= 8N[\tfrac{1}{2}(1+P)]^2 + 8N[\tfrac{1}{2}(1-P)]^2 = 4(1+P^2)N \ . \end{aligned} \tag{20}$$

The energy (21) becomes

$$E = E_0 + 2NP^2U \ , \tag{21}$$

where

$$E_0 = 2N(U_{AA} + U_{BB} + 2U_{AB}) \ ; \qquad U = 2U_{AB} - U_{AA} - U_{BB} \ . \tag{22}$$

We now calculate the entropy of this distribution of atoms. There are $\frac{1}{2}(1+P)N$ atoms *A* and $\frac{1}{2}(1-P)N$ atoms *B* on lattice *a*; there are $\frac{1}{2}(1-P)N$ atoms *A* and $\frac{1}{2}(1+P)N$ atoms *B* on lattice *b*. The number of arrangements *G* of these atoms is

$$G = \left[\frac{N!}{[\frac{1}{2}(1+P)N]!\,[\frac{1}{2}(1-P)N]!}\right]^2 \ . \tag{23}$$

From the Boltzmann definition of the entropy as $S = k_B \ln G$, we have, using Stirling's approximation,

$$S = 2Nk_B \ln 2 - Nk_B[(1+P)\ln(1+P) + (1-P)\ln(1-P)] \ .$$

For $P = \pm 1$, $S = 0$; for $P = 0$, $S = 2Nk_B \ln 2$.

The equilibrium order is determined by the requirement that the free energy $F = E - TS$ be a minimum with respect to the order parameter P. On differentiating F with respect to P, we have as the condition for the minimum

$$4NPU + Nk_BT \ln \frac{1+P}{1-P} = 0 \ . \tag{24}$$

The transcendental equation for P may be solved graphically; we find the smoothly decreasing curve shown in Fig. 19a. Near the transition we may expand (24), to find $4NPU + 2Nk_BTP = 0$. At the transition temperature $P = 0$, so that

$$T_c = -2U/k_B \ . \tag{25}$$

For a transition to occur, the effective interaction U must be negative.

The **short-range order parameter** r is a measure of the fraction of the average number q of nearest-neighbor bonds that are AB bonds. When completely disordered, an AB alloy has an average of four AB bonds about each atom A. The total possible is eight. We may define

$$r = \tfrac{1}{4}(q - 4) \ , \tag{26}$$

so that $r = 1$ in complete order and $r = 0$ in complete disorder. Observe that r is a measure only of the local order about an atom, whereas the long-range order parameter P refers to the purity of the entire population on a given sublattice. Above the transition temperature T_c the long-range order is rigorously zero, but the short-range order is not.

PROBLEMS

1. ***Frenkel defects.*** Show that the number n of interstitial atoms in equilibrium with n lattice vacancies in a crystal having N lattice points and N' possible interstitial positions is given by the equation

$$E_I = k_BT \ln[(N-n)(N'-n)/n^2] \ ,$$

whence, for $n \ll N, N'$, we have $n \cong (NN')^{1/2}\exp(-E_I/2k_BT)$. Here E_I is the energy necessary to remove an atom from a lattice site to an interstitial position.

2. ***Schottky vacancies.*** Suppose that the energy required to remove a sodium atom from the inside of a sodium crystal to the boundary is 1 eV. Calculate the concentration of Schottky vacancies at 300 K.

3. ***F Center.*** (a) Treat an *F* center as a free electron of mass *m* moving in the field of a point charge *e* in a medium of dielectric constant $\epsilon = n^2$; what is the 1*s*-2*p* energy difference of *F* centers in NaCl? (b) Compare from Table 2 the *F* center excitation energy in NaCl with the 3*s*-3*p* energy difference of the free sodium atom.

4. ***Superlattice lines in Cu_3Au.*** Cu_3Au alloy (75% Cu, 25% Au) has an ordered state below 400°C, in which the gold atoms occupy the 000 positions and the copper atoms the $\frac{1}{2}\frac{1}{2}0$, $\frac{1}{2}0\frac{1}{2}$, and $0\frac{1}{2}\frac{1}{2}$ positions in a face-centered cubic lattice. Give the indices of the new x-ray reflections that appear when the alloy goes from the disordered to the ordered state. List all new reflections with indices $\leqq 2$.

5. ***Configurational heat capacity.*** Derive an expression in terms of $P(T)$ for the heat capacity associated with order/disorder effects in an *AB* alloy. [The entropy (23) is called the configurational entropy.]

References

Calculations of the properties of vacancies and interstitials, National Bureau of Standards Misc. Publ. 287.

C. P. Flynn, *Point defects and diffusion*, Oxford, 1972.

W. B. Fowler, ed., *Physics of color centers*, Academic Press, 1968.

J. Friedel, "Electronic structure of primary solid solutions in metal," Advances in Physics **3**, 446 (1954).

L. A. Girifalco, *Atomic migration in crystals*, Blaisdell, New York, 1964. Elementary treatment.

L. Guttman, "Order-disorder phenomena in metals," *Solid state physics* **3**, 145 (1956).

N. B. Hannay, *Solid state chemistry*, Prentice-Hall, 1967.

R. R. Hasiguti, ed., *Lattice defects in semiconductors*, Pennsylvania State University Press, 1967.

F. A. Huntley, ed., *Lattice defects in semiconductors*, Institute of Physics, London, 1975.

D. Lazarus, "Diffusion in metals," *Solid state physics* **10**, 71 (1960).

J. J. Markham, *F centers in alkali halides*, Academic Press, 1966.

A. S. Nowick and J. J. Burton, eds., *Diffusion in solids. Recent developments*, Academic Press, 1975.

K. K. Rebane, *Impurity spectra of solids*, Plenum, 1970.

J. H. Schulman and W. D. Compton, *Color centers in solids*, Pergamon, 1962.

P. G. Shewmon, *Diffusion in solids*, McGraw-Hill, 1963.

A. M. Stoneham, *Theory of defects in solids*, Oxford, 1975.

18 DISLOCATIONS

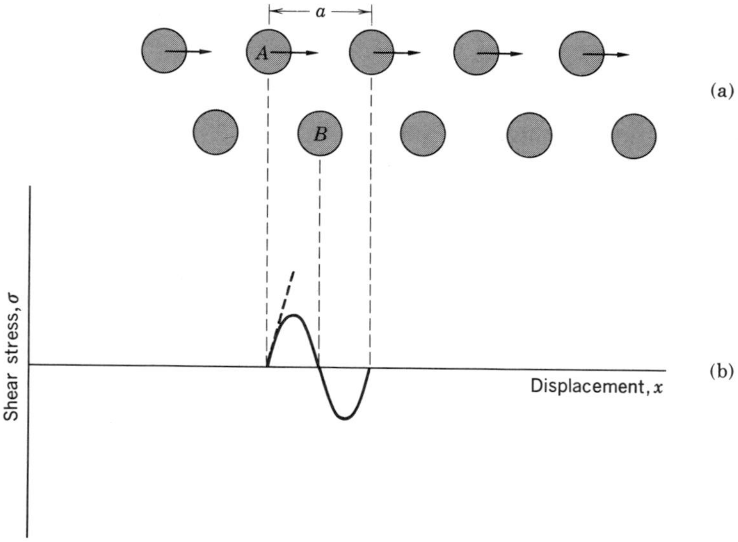

Figure 1 (a) Relative shear of two planes of atoms (shown in section) in a uniformly strained crystal; (b) shear stress as a function of the relative displacement of the planes from their equilibrium position. The heavy broken line drawn at the initial slope defines the shear modulus G.

18 DISLOCATIONS

This chapter is concerned with the interpretation of the plastic mechanical properties of crystalline solids in terms of the theory of dislocations. Plastic properties are irreversible deformations, such as yield and slip; elastic properties are reversible. The ease with which pure single crystals deform plastically is striking. This intrinsic weakness of crystals is exhibited in various ways. Pure silver chloride melts at 455°C, yet at room temperature it has a cheeselike consistency and can be rolled into sheets. Pure aluminum crystals are elastic (follow Hooke's law) only to a strain of about 10^{-5}, after which they deform plastically. Theoretical estimates of the elastic limit of perfect crystals give values 10^3 or 10^4 higher than the lowest observed values, although a factor 10^2 is more usual. There are few exceptions to the rule that pure crystals are plastic and are not strong: Crystals of high purity germanium and silicon are not plastic at room temperature and fail or yield only by fracture. Glass at room temperature fails only by fracture, but it is not crystalline. The fracture of glass is caused by stress concentration at minute cracks.

SHEAR STRENGTH OF SINGLE CRYSTALS

Frenkel gave a simple method of estimating the theoretical shear strength of a perfect crystal. We consider in Fig. 1 the force needed to make a shear displacement of two planes of atoms past each other. For small elastic strains the stress σ is related to the displacement x by

$$\sigma = Gx/d \ . \tag{1}$$

Here d is the interplanar spacing, and G denotes the appropriate shear modulus. When the displacement is large and has proceeded to the point that atom A is directly over atom B in the figure, the two planes of atoms are in a configuration of unstable equilibrium and the stress is zero. As a first approximation we represent the stress-displacement relation by

$$\sigma = \left(\frac{Ga}{2\pi d}\right) \sin \left(\frac{2\pi x}{a}\right) , \tag{2}$$

where a is the interatomic spacing in the direction of shear. This relation is constructed to reduce to (1) for small values of x/a. The critical shear stress σ_c at which the lattice becomes unstable is given by the maximum value of σ, or

$$\sigma_c = Ga/2\pi d \ . \tag{3}$$

If $a \approx d$, then $\sigma_c \approx G/2\pi$ or $G/\sigma_c \approx 2\pi$: the ideal critical shear stress is of the order of $\frac{1}{6}$ of the shear modulus.

The observations in Table 1 show the experimental values of the elastic limit are much smaller than (3) would suggest. The theoretical estimate may be improved by consideration of the actual form of the intermolecular forces and by consideration of other configurations of mechanical stability available to the lattice as it is sheared. Mackenzie has shown that these two effects may reduce the theoretical ideal shear strength to about $G/30$, corresponding to a critical shear strain angle of about 2 degrees. The observed low values of the shear strength cannot be explained without the presence of imperfections that can act as sources of mechanical weakness in real crystals. It is now known that special crystal imperfections called dislocations exist in almost all crystals, and their movement is responsible for slip at very low applied stresses.

TABLE 1 Comparison of Shear Modulus and Elastic Limit

	Shear Modulus G, in dyn/cm^2	Elastic Limit σ_c, in dyn/cm^2	G/σ_c
Sn, single crystal	1.9×10^{11}	1.3×10^{7}	15,000
Ag, single crystal	2.8×10^{11}	6×10^{6}	45,000
Al, single crystal	2.5×10^{11}	4×10^{6}	60,000
Al, pure, polycrystal	2.5×10^{11}	2.6×10^{8}	900
Al, commercial drawn	$\sim 2.5 \times 10^{11}$	9.9×10^{8}	250
Duralumin	$\sim 2.5 \times 10^{11}$	3.6×10^{9}	70
Fe, soft, polycrystal	7.7×10^{11}	1.5×10^{9}	500
Heat-treated carbon steel	$\sim 8 \times 10^{11}$	6.5×10^{9}	120
Nickel-chrome steel	$\sim 8 \times 10^{11}$	1.2×10^{10}	65

After Mott.

Slip

Plastic deformation in many crystals occurs by slip, an example of which is shown in Fig. 2. In slip one part of the crystal slides as a unit across an adjacent part. The surface on which slip takes place is often a plane, and this is known as the slip plane. The direction of motion is known as the slip direction. The great importance of lattice properties for plastic strain is indicated by the highly anisotropic nature of slip. Displacement takes place along crystallographic planes with a set of small Miller indices, such as the $\{111\}$ planes in fcc metals and the $\{110\}$, $\{112\}$, and $\{123\}$ planes in bcc metals. Under most conditions the slip direction is in the line of closest atomic packing, $\langle 110 \rangle$ in fcc metals and $\langle 111 \rangle$ in bcc metals. To maintain the crystal structure after slip, the displacement or slip vector must equal a lattice translation vector. The shortest lattice translation vector expressed in terms of the lattice constant a in a fcc structure is of the form $(a/2)(\hat{\mathbf{x}} + \hat{\mathbf{y}})$; in a bcc structure it is $(a/2)(\hat{\mathbf{x}} + \hat{\mathbf{y}} + \hat{\mathbf{z}})$. But in fcc crystals one also observes partial displacements which upset the regular sequence *ABCABC* . . . of closest-packed planes (Chapter 1), to produce a **stacking fault** such as *ABCABABC*. . . . The result is then a "mixture" of fcc and hcp stacking.

Deformation of a crystal by slip is observed to be inhomogeneous. Large shear displacements occur on a few widely separated slip planes, while parts of the crystal lying between slip planes remain essentially undeformed. An incidental property of slip is the Schmid law of the critical shear stress: slip takes place along a given slip plane and direction when the corresponding *component* of shear stress reaches the critical value.

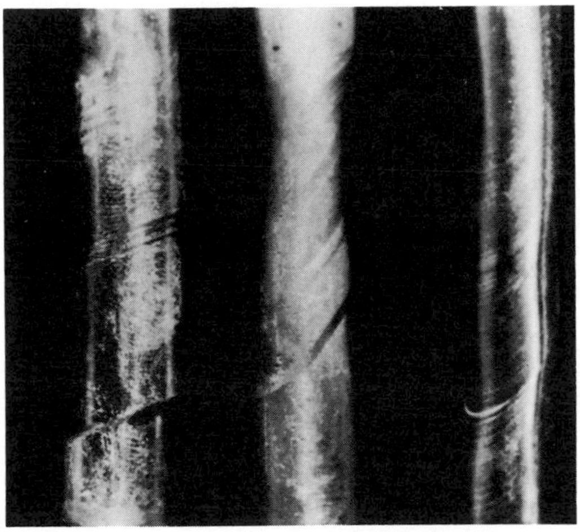

Figure 2 Translational slip in zinc single crystals. (Courtesy of E. R. Parker.)

Slip is one mode of plastic deformation. A second mode of plastic deformation, called **twinning,** is observed in some crystals, particularly in hcp and bcc structures. During the slip process a considerable amount of displacement occurs on a few widely separated slip planes. During twinning a partial displacement occurs successively on each of many neighboring crystallographic planes. After deformation by twinning, the deformed part of the crystal is a mirror image of the undeformed part. Although both slip and twinning are caused by the motion of dislocations, we shall be concerned primarily with slip.

DISLOCATIONS

The low observed values of the critical shear stress is explained in terms of the motion through the lattice of a line imperfection known as a dislocation. The idea that slip propagates by the motion of dislocations was published in 1934 independently by Taylor, Orowan, and Polanyi; the concept of dislocations was introduced somewhat earlier by Prandtl and Dehlinger. There are several basic types of dislocations. We first describe an **edge dislocation.** Figure 3 shows a simple cubic crystal in which slip of one atom distance has occurred over the left half of the slip plane but not over the right half. The boundary between the slipped and unslipped regions is called the dislocation. Its position is marked by the termination of an extra vertical half-plane of atoms crowded into the upper half of the crystal as shown in Fig. 4. Near the dislocation the crystal is highly strained. The simple edge dislocation extends indefinitely in the slip plane in a direction normal to the slip direction. In Fig. 5 we show a photograph of a dislocation in a two-dimensional soap bubble raft obtained by the method of Bragg and Nye.[1]

[1]W. L. Bragg and J. F. Nye, Proc. Roy. Soc. (London) **A190,** 474 (1947); W. L. Bragg and W. M. Lomer, Proc. Roy. Soc. (London) **A196,** 171 (1949).

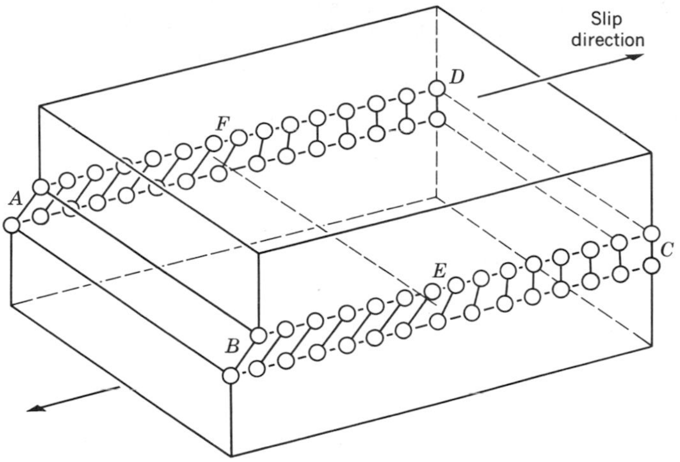

Figure 3 An edge dislocation *EF* in the glide plane *ABCD*. The figure shows the slipped region *ABEF* in which the atoms have been displaced by more than half a lattice constant and the unslipped region *FECD* with displacement less than half a lattice constant.

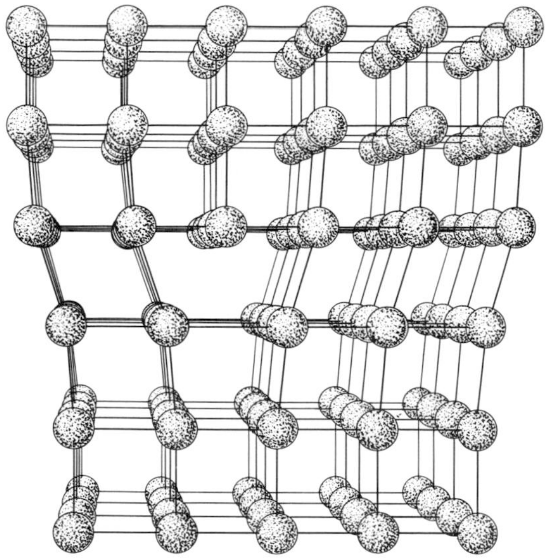

Figure 4 Structure of an edge dislocation. The deformation may be thought of as caused by inserting an extra plane of atoms on the upper half of the y axis. Atoms in the upper half-crystal are compressed by the insertion; those in the lower half are extended.

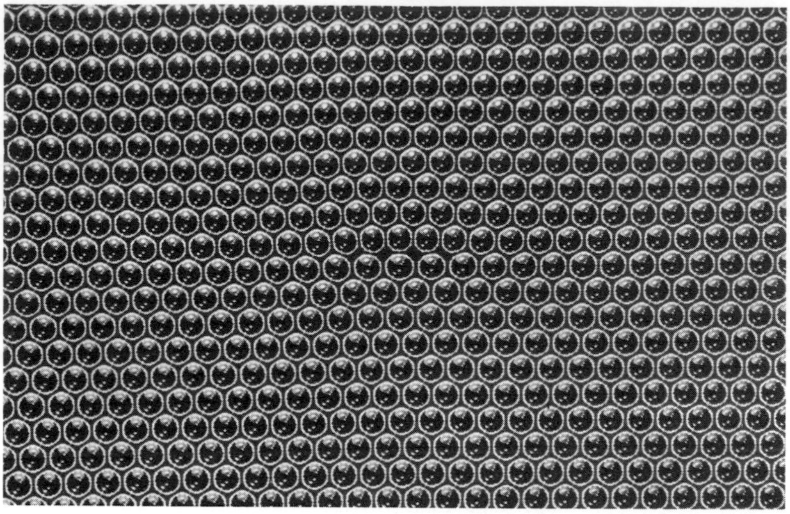

Figure 5 A dislocation in a two-dimensional bubble raft. The dislocation is most easily seen by turning the page by 30° in its plane and sighting at a low angle. (Courtesy of W. M. Lomer, after Bragg and Nye.)

The mechanism responsible for the mobility of a dislocation is shown in Fig. 6. The motion of an edge dislocation through a crystal is analogous to the passage of a ruck or wrinkle across a rug: the ruck moves more easily than the whole rug. If atoms on one side of the slip plane are moved with respect to those on the other side, atoms at the slip plane will experience repulsive forces from some neighbors and attractive forces from others across the slip plane. These forces cancel to a first approximation. The external stress required to move a dislocation has been calculated and is quite small, probably below 10^5 dyn/cm^2, provided that the bonding forces in the crystal are not highly directional. Thus dislocations may make a crystal very plastic. Passage of a dislocation through a crystal is equivalent to a slip displacement of one part of the crystal.

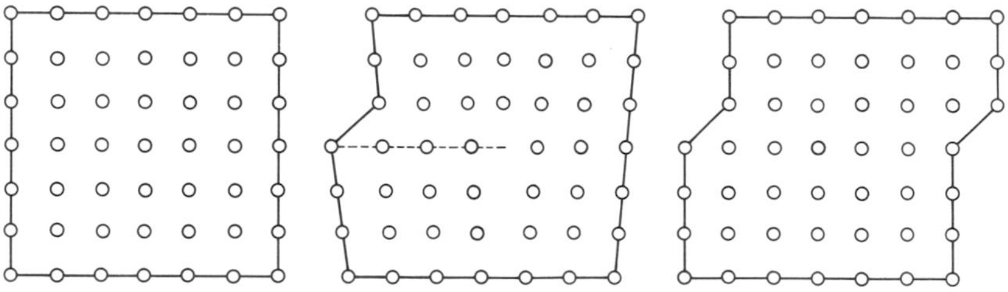

Figure 6 Motion of a dislocation under a shear tending to move the upper surface of the specimen to the right. (After Taylor.)

The second simple type of dislocation is the **screw dislocation,** sketched in Figs. 7 and 8. A screw dislocation marks the boundary between slipped and unslipped parts of the crystal. The boundary *parallels* the slip direction, instead of lying perpendicular to it as for the edge dislocation. The screw dislocation may be thought of as produced by cutting the crystal partway through with a knife and shearing it parallel to the edge of the cut by one atom spacing. A screw dislocation transforms successive atom planes into the surface of a helix; this accounts for the name of the dislocation.

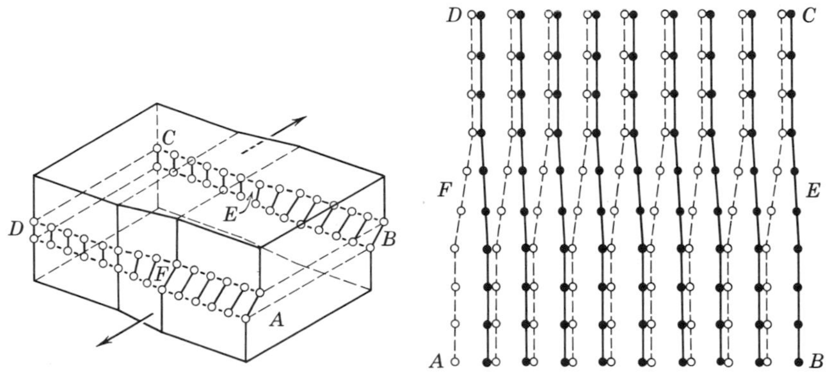

Figure 7 A screw dislocation. A part *ABEF* of the slip plane has slipped in the direction parallel to the dislocation line *EF*. A screw dislocation may be visualized as a helical arrangement of lattice planes, such that we change planes on going completely around the dislocation line. (After Cottrell.)

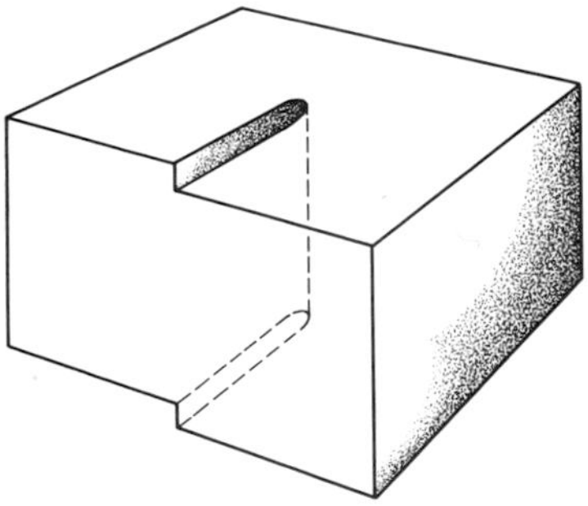

Figure 8 Another view of a screw dislocation. The broken vertical line that marks the dislocation is surrounded by strained material.

Burgers Vectors

Other dislocation forms may be constructed from segments of edge and screw dislocations. Burgers has shown that the most general form of a linear dislocation pattern in a crystal can be described as shown in Fig. 9. We consider any closed curve not necessarily planar within a crystal, or an open curve terminating on the surface at both ends: (a) Make a cut along any simple surface bounded by the line. (b) Displace the material on one side of this surface by a vector **b** relative to the other side; here **b** is called the **Burgers vector.** (c) In regions where **b** is not parallel to the cut surface, this relative displacement will either produce a gap or cause the two halves to overlap. In these cases we imagine that we either add material to fill the gap or subtract material to prevent overlap. (d) Rejoin the material on both sides. We leave the strain displacement intact at the time of rewelding, but afterwards we allow the medium to come to internal equilibrium. The resulting strain pattern is that of the dislocation characterized jointly by the boundary curve and the Burgers vector. The Burgers vector must be equal to a lattice vector in order that the rewelding process will maintain the crystallinity of the material. The Burgers vector of a screw dislocation (Figs. 7 and 8) is parallel to the dislocation line; that of an edge dislocation (Figs. 3 and 4) is perpendicular to the dislocation line and lies in the slip plane.

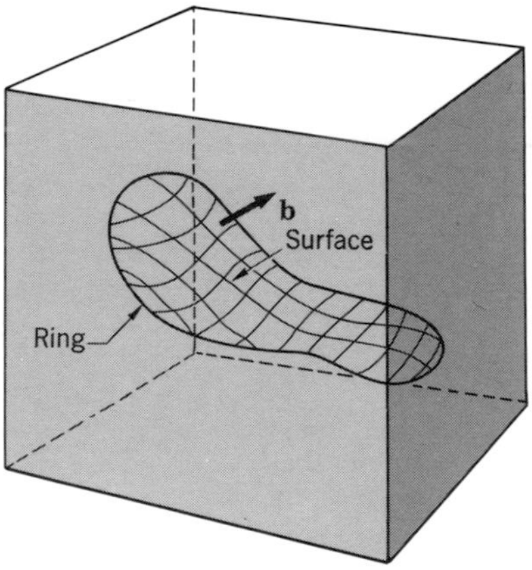

Figure 9 General method of forming a dislocation ring in a medium. The medium is represented by the rectangular block. The ring is represented by the closed curve in the interior in the block. A cut is made along the surface bounded by the curve and indicated by the contoured area. The material on one side of the cut is displaced relative to that on the other by vector distance **b**, which may be oriented arbitrarily relative to the surface. Forces will be required to effect the displacement. The medium is filled in or cut away so as to be continuous after the displacement. It is then joined in the displaced state and the applied forces are relaxed. Here **b** is the Burgers vector of the dislocation. (After Seitz.)

Stress Fields of Dislocations

The stress field of a screw dislocation is particularly simple. Figure 10 shows a shell of material surrounding an axial screw dislocation. The shell of circumference $2\pi r$ has been sheared by an amount b to give a shear strain $e = b/2\pi r$. The corresponding shear stress in an elastic continuum is

$$\sigma = Ge = Gb/2\pi r \ . \tag{4}$$

This expression cannot hold in the region immediately around the dislocation line, as the strains here are too large for continuum or linear elasticity theory to apply. The elastic energy of the shell is $dE_s = \frac{1}{2}Ge^2\, dV = (Gb^2/4\pi)\, dr/r$ per unit length. The total elastic energy per unit length of a screw dislocation is found on integration to be

$$E_s = \frac{Gb^2}{4\pi} \ln \frac{R}{r_0} \ , \tag{5}$$

where R and r_0 are appropriate upper and lower limits for the variable r. A reasonable value of r_0 is comparable to the magnitude b of the Burgers vector or to the lattice constant; the value of R cannot exceed the dimensions of the crystal. The value of the ratio R/r_0 is not very important because it enters in a logarithm term.

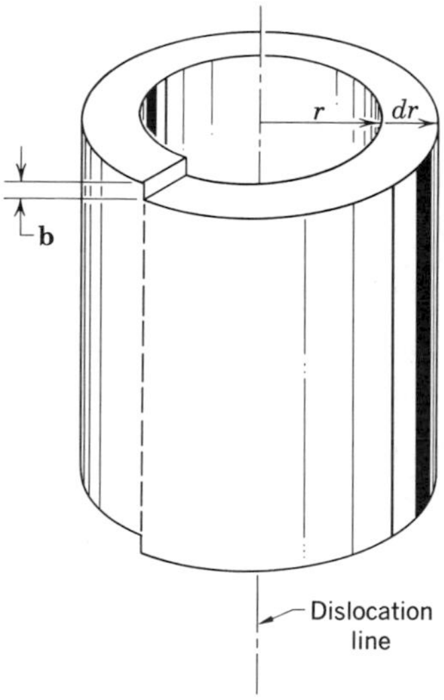

Figure 10 Shell of elastically distorted crystal surrounding screw dislocation with Burgers vector **b**.

We now calculate the energy of an edge dislocation. Let σ_{rr} and $\sigma_{\theta\theta}$ denote the tensile stresses in the radial and circumferential directions, and let $\sigma_{r\theta}$ denote the shear stress. In an isotropic elastic continuum, σ_{rr} and $\sigma_{\theta\theta}$ are proportional to $(\sin\theta)/r$: we need a function that falls off as $1/r$ and that changes sign when y is replaced by $-y$. The shear stress $\sigma_{r\theta}$ is proportional to $(\cos\theta)/r$: considering the plane $y = 0$ we see from Fig. 4 that the shear stress is an odd function of x. The constants of proportionality in the stress are proportional to the shear modulus G and to the Burgers vector b of the displacement. The final result, which is derived in books cited in the references, is

$$\sigma_{rr} = \sigma_{\theta\theta} = -\frac{Gb}{2\pi(1-\nu)}\frac{\sin\theta}{r} , \qquad \sigma_{r\theta} = \frac{Gb}{2\pi(1-\nu)}\frac{\cos\theta}{r} , \tag{6}$$

where the Poisson ratio $\nu \approx 0.3$ for most crystals. The strain energy of a unit length of edge dislocation is

$$E_e = \frac{Gb^2}{4\pi(1-\nu)} \ln\frac{R}{r_0} . \tag{7}$$

We want an expression for the shear stress component σ_{xy} on planes parallel to the slip plane in Fig. 4. From the stress components σ_{rr}, $\sigma_{\theta\theta}$, and $\sigma_{r\theta}$ evaluated on the plane a distance y above the slip plane, we find

$$\sigma_{xy} = \frac{Gb}{2\pi(1-\nu)}\frac{\sin 4\theta}{4y} . \tag{8}$$

It is shown in Problem 3 that the force caused by a uniform shear stress σ is $F = b\sigma$ per unit length of dislocation. The force that an edge dislocation at the origin exerts upon a similar one at the location (y, θ) is

$$F = b\sigma_{xy} = \frac{Gb^2}{2\pi(1-\nu)}\frac{\sin 4\theta}{4y} \tag{9}$$

per unit length. Here F is the component of force in the slip direction. There is another component of force perpendicular to the slip direction, but it is of no importance at low temperatures, where the only possible motion of a dislocation is in the slip plane.

Low-angle Grain Boundaries

Burgers suggested that low-angle boundaries between adjoining crystallites or crystal grains consist of arrays of dislocations. A simple example of the Burgers model of a grain boundary is shown in Fig. 11. The boundary occupies a (010) plane in a simple cubic lattice and divides two parts of the crystal that have a [001] axis in common. Such a boundary is called a pure tilt boundary: the misorientation can be described by a small rotation θ about the common [001] axis of one part of the crystal relative to the other. The tilt boundary is represented as an array of edge dislocations

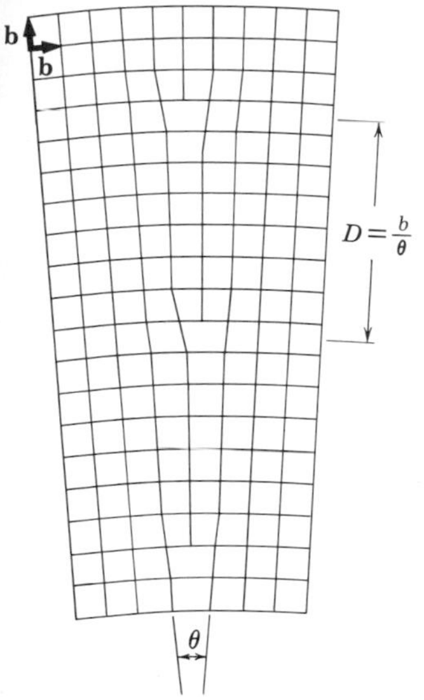

Figure 11 Low-angle grain boundary. (After Burgers.)

of spacing $D = b/\theta$, where b is the Burgers vector of the dislocations. Experiments have substantiated this model. Figure 12 shows the distribution of dislocations along small-angle grain boundaries, as observed with an electron microscope. Further, Read and Shockley derived a theory of the interfacial energy as a function of the angle of tilt, with results in excellent agreement with measurements.

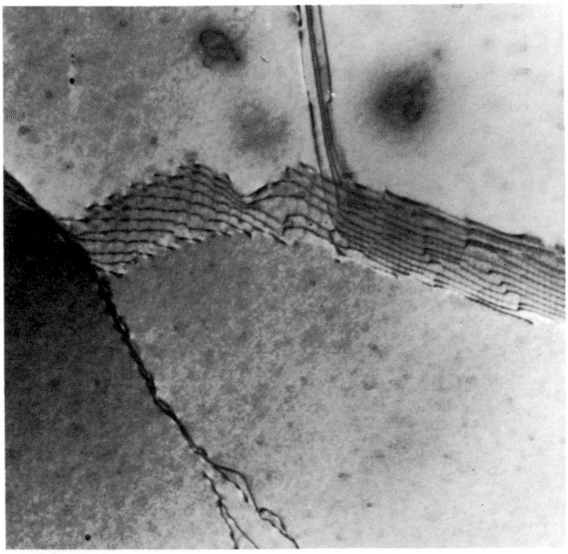

Figure 12 Electron micrograph of dislocation structures in small-angle grain boundaries in an Al–7 percent Mg solid solution. Notice the lines of small dots on the right. Mag. ×17,000. (Courtesy of R. Goodrich and G. Thomas.)

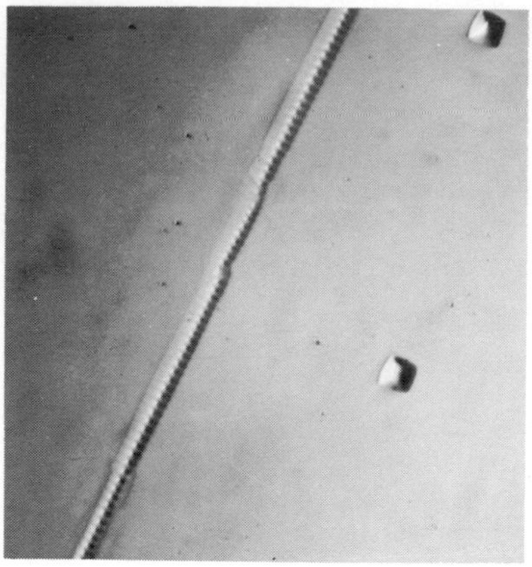

Figure 13 Dislocation etch pits in low-angle boundary on (100) face of germanium; the angle of the boundary is 27.5″. The boundary lies in a (011) plane; the line of the dislocations is [100]. The Burgers vector is the shortest lattice translation vector, or $|b| = a/\sqrt{2} = 4.0$ Å. [After F. L. Vogel, Jr., Acta Metallurgica **3**, 245 (1955).]

Direct verification of the Burgers model is provided by the quantitative x-ray and optical studies of low-angle boundaries in germanium crystals by Vogel and co-workers. By counting etch pits along the intersection of a low-angle grain boundary with an etched germanium surface (Fig. 13), they determined the dislocation spacing D. They assumed that each etch pit marked the end of a dislocation. The angle of tilt calculated from the relation $\theta = b/D$ agrees well with the angle measured directly by means of x-rays. The interpretation of low-angle boundaries as arrays of dislocations is further supported by the fact that pure tilt boundaries move normal to themselves on application of a suitable stress. The motion has been demonstrated in a beautiful experiment, Fig. 14. The specimen is a bicrystal of zinc containing a 2° tilt boundary with dislocations about 30 atomic planes apart. One side of the crystal was clamped, and a force was applied at a point on the opposite side of the boundary. Motion of the boundary took place by cooperative motion of the dislocations in the array, each dislocation moving an equal distance in its own slip plane. The motion was produced by stresses of the order of magnitude of the yield stress for zinc crystals, strong evidence that ordinary deformation results from the motion of dislocations.

Grain boundaries and dislocations offer relatively little resistance to diffusion of atoms in comparison with diffusion in perfect crystals. A dislocation is an open passage for diffusion. Diffusion is greater in plastically deformed material than in annealed crystals. Diffusion along grain boundaries controls the rates of some precipitation reactions in solids: the precipitation of tin from lead-tin solutions at room temperature proceeds about 10^8 times faster than expected from diffusion in an ideal lattice.

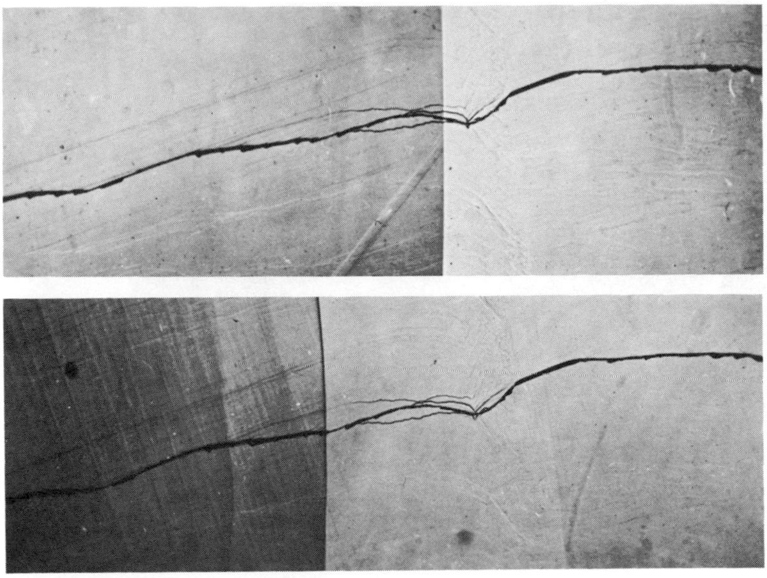

Figure 14 Motion of a small-angle grain boundary under stress. The boundary is the straight vertical line, and it is photographed under vertical illumination, thereby making evident the 2° angular change in the cleavage surface of the zinc crystal at the boundary. The irregular horizontal line is a small step in the cleavage surface which serves as a reference mark. The crystal is clamped at the left; at the right it is subject to a force normal to the plane of the page. *Top*, original position of boundary; *bottom*, moved back 0.4 mm. (Courtesy of J. Washburn and E. R. Parker.)

Dislocation Densities

The density of dislocations is the number of dislocation lines that intersect a unit area in the crystal. The density ranges from below 10^2 dislocations/cm^2 in the best germanium and silicon crystals to 10^{11} or 10^{12} dislocations/cm^2 in heavily deformed metal crystals. The methods available for estimating dislocation densities are compared in Table 2. The book by Amelinckx is an excellent reference for all present methods of observing dislocations. The actual dislocation configurations in cast or annealed (slowly cooled) crystals correspond either to a group of low-angle grain boundaries or to a three-dimensional network of dislocations arranged in cells, as shown in Fig. 15.

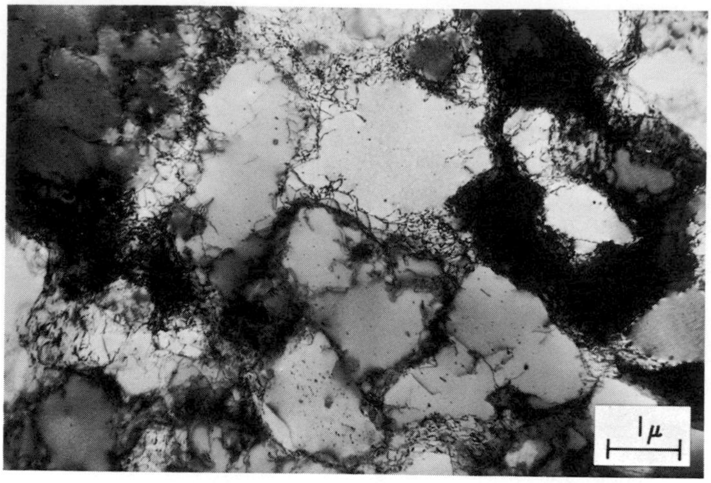

Figure 15 Cell structure of three-dimensional tangles of dislocations in deformed aluminum. (Courtesy of P. R. Swann.)

TABLE 2 Methods for Estimating Dislocation Densities[a]

Technique	Specimen Thickness	Width of Image	Maximum Practical Density, per cm^2
Electron microscopy	>1000 Å	~ 100 Å	10^{11}–10^{12}
X-ray transmission	0.1–1.0 mm	5μm	10^4–10^5
X-ray reflection	$<2\mu$m (min.) – 50μm (max.)	2μm	10^6–10^7
Decoration	$\sim 10\mu$m (depth of focus)	0.5μm	2×10^7
Etch pits	no limit	0.5μm[b]	4×10^8

[a] Based on W. G. Johnston. Prog. Ceramic Sci. **2**, 1 (1961).
[b] Limit of resolution of etch pits.

The dislocation density in heavily deformed crystals may be estimated from the increased internal energy that results from plastic deformation. From (5) and (7), the energy per unit length of a dislocation is about $(Gb^2/4\pi)\ \ln\ (R/r_0)$. We take R comparable to the dislocation spacing and $r_0 \approx b$, and so $R/r_0 \approx 10^3$. The dislocation energy then is about 5×10^{-4} erg/cm, or about 8 eV per atom plane through which the dislocation passes. The maximum energy stored in lattice distortions as a consequence of severe plastic deformation, as by twisting, filing, or compressing, has been measured thermally for several metals. If the deformation is not too great, about 10 percent of the energy expended in plastic flow is stored in the lattice. Upon continued plastic flow, however, the stored energy approaches a saturation value. The observed values of the stored energy are about 10^8 ergs/cm^3. If the energy per unit length of dislocations is 5×10^{-4} erg/cm, then there must be about 10^{11} cm of dislocation lines per cubic centimeter of crystal, or about 10^{11} dislocations/cm^2. This is about one dislocation per square 100 atoms on a side. Such a concentration is characteristic of severely deformed metals.

Lattice vacancies that precipitate along an existing edge dislocation will eat away a portion of the extra half-plane of atoms and cause the dislocation to climb, which means to move at right angles to the slip direction. If no dislocations are present, the crystal will become supersaturated with lattice vacancies; their precipitation in penny-shaped vacancy plates may be followed by collapse of the plates and formation of dislocation rings that grow with further vacancy precipitation, as in Fig. 16.

Figure 16 Electron micrograph of dislocation loops formed by aggregation and collapse of vacancies in Al-5 percent Mg quenched from 550°C. The helical dislocations are formed by the "climb" of screw dislocations as a result of the precipitation of vacancies. Mag. ×43,000. (Courtesy of A. Eikum and G. Thomas.)

Dislocation Multiplication and Slip

A problem is posed by the very great increase in dislocation density that is caused by plastic deformation. Dislocation density measurements typically show an increase from about 10^8 to about 10^{11} dislocations/cm^2 during deformation, a 1000-fold increase. Equally striking is the fact that if a dislocation were to move completely across its slip plane, an offset of only one atom spacing would be produced, whereas offsets of as much as 100 to 1000 atom spacings are actually observed.

Consider a closed circular dislocation loop of radius r surrounding a slipped area of the same radius. A dislocation loop will have to be partly edge, partly screw, and mostly intermediate in character. Because the strain energy of the dislocation loop increases in proportion to its circumference, the loop will tend to shrink. However, if a shear stress that favors slip is acting, the loop will tend to expand. A dislocation segment pinned at each end (Fig. 17) is called a **Frank-Read source,** and as in the figure it can lead to the generation of a large number of concentric dislocation loops on a single slip plane. This and related types of dislocation multiplication mechanisms give rise to slip and to the increased density of dislocations during plastic deformation. A beautiful example of a dislocation source is shown in Fig. 18.

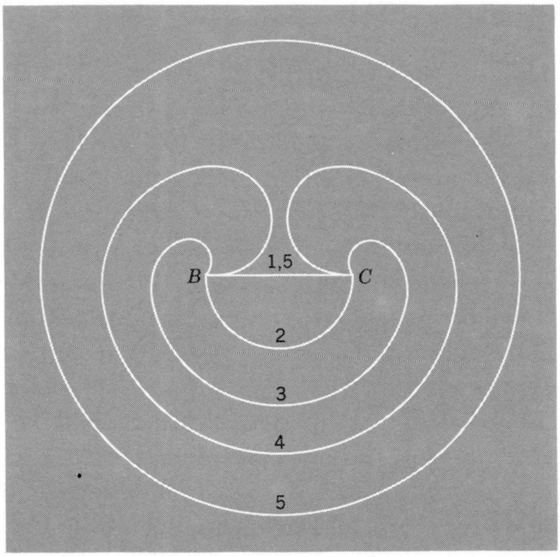

Figure 17 Frank-Read mechanism for multiplication of dislocations, showing successive stages in the generation of a dislocation loop by the segment BC of a dislocation line. The process can be repeated indefinitely.

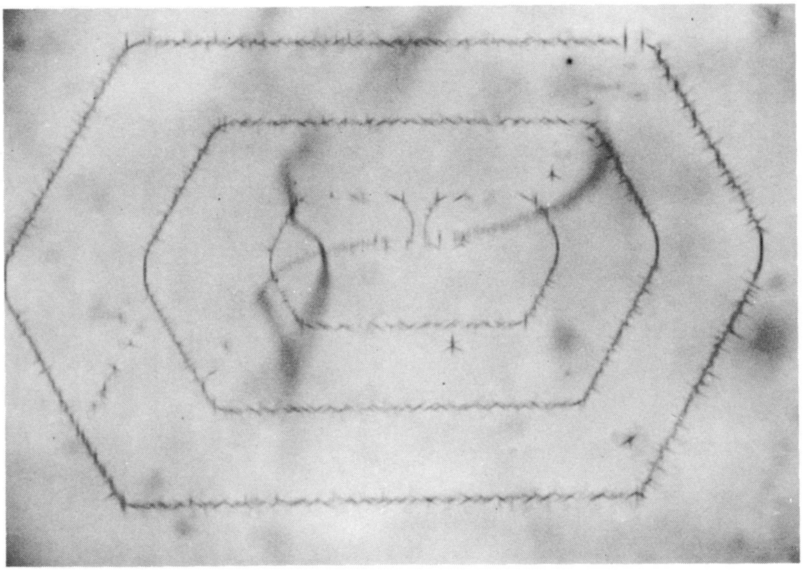

Figure 18 A Frank-Read dislocation source in silicon, decorated with copper precipitates and viewed with infrared illumination. Two complete dislocation loops are visible, and the third, innermost loop is near completion. (After W. C. Dash, *Dislocations and mechanical properties of crystals*, Fisher et al., eds., Wiley, 1957, pp. 57–68.)

STRENGTH OF ALLOYS

Pure crystals are very plastic and yield at very low stresses. There appear to be four important ways of increasing the yield strength of an alloy so that it will withstand shear stresses as high as 10^{-2} G. They are mechanical blocking of dislocation motion, pinning of dislocations by solute atoms, impeding dislocation motion by short-range order, and increasing the dislocation density so that tangling of dislocations results. All four strengthening mechanisms depend for their success upon impeding dislocation motion. A fifth mechanism, that of removing all dislocations from the crystal, may operate for certain fine hairlike crystals (whiskers) that are discussed in the section on crystal growth.

Mechanical blocking of dislocation motion can be produced most directly by introducing tiny particles of a second phase into a crystal lattice. This process is followed in the hardening of steel, where particles of iron carbide are precipitated into iron, and in hardening aluminum, where particles of Al_2Cu are precipitated. The pinning of a dislocation by impurities is shown in Fig. 19.

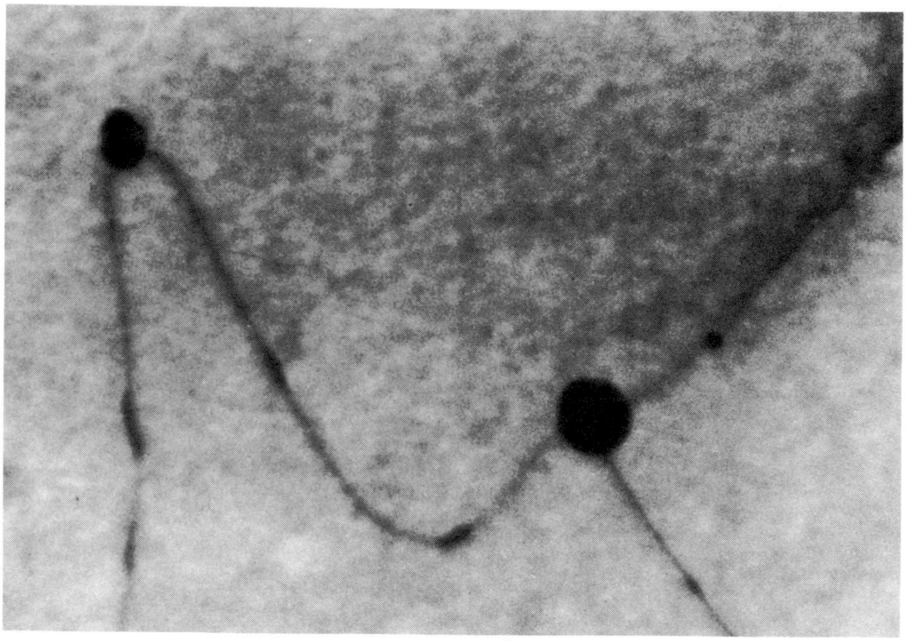

Figure 19 Dislocations pinned by impurities in magnesium oxide. (Electron micrograph courtesy of G. Thomas and J. Washburn.)

In strengthening by the addition of small particles there are two cases to be considered: either the particle can be deformed with the matrix, which requires that the particle can be traversed by the dislocation, or the particle cannot be traversed by the dislocation. If the particle cannot be cut, the stress necessary to force a dislocation between particles spaced L apart on a slip plane should be approximately

$$\frac{\sigma}{G} = \frac{b}{L} \; . \tag{10}$$

The smaller the spacing L, the higher is the yield stress σ. Before particles precipitate, L is large and the strength is low. Immediately after precipitation is complete and many small particles are present, L is a minimum and the strength is a maximum. If the alloy is then held at a high temperature, some particles grow at the expense of others, so that L increases and the strength drops. Hard intermetallic phases, such as refractory oxides, cannot be cut by dislocations.

The strength of dilute solid solutions is believed to result from the pinning of dislocations by solute atoms. Cottrell pointed out that the solubility of a foreign atom will be greater in the neighborhood of a dislocation than elsewhere in a crystal. For example, an atom that tends to expand the crystal will dissolve preferentially in the expanded region near an edge dislocation. A small atom will tend to dissolve preferentially in the contracted region near the dislocation—a dislocation offers both expanded and contracted regions. As a result of the affinity of solute atoms for dislocations, each dislocation will collect a cloud of associated solute atoms during cooling, at a time when the mobility of solute atoms is high. At still lower temperatures diffusion of solute atoms effectively ceases, and the solute atom cloud becomes fixed in the crystal. When a dislocation moves, leaving its solute cloud behind, the energy of the crystal must increase. The increase in energy can only be provided by an increased stress acting on the dislocation as it pulls away from the solute atom cloud, and so the presence of the cloud strengthens the crystal.

The passage of a dislocation across a slip plane in pure crystals does not alter the binding energy across the plane after the dislocation is gone. The internal energy of the crystal remains unaffected. The same is true for random solid solutions, because the solution is equally random across a slip plane after slip. Most solid solutions, however, have short-range order. Atoms of different species are not arranged at random on the lattice sites, but tend to have an excess or a deficiency of pairs of unlike atoms. Thus in ordered alloys dislocations tend to move in pairs: the second dislocation reorders the local disorder left by the first dislocation.

The strength of a crystalline material increases with plastic deformation. The phenomenon is called **work-hardening** or **strain-hardening.** The strength is believed to increase because of the increased density of dislocations and the greater difficulty of moving a given dislocation across a slip plane that is threaded by many dislocations. The difficulty is particularly acute when one screw dislocation tries to cross another; lattice vacancies and interstitial atoms may be generated in the process. Strain-hardening frequently is employed in the strengthening of materials, but its usefulness is limited to low enough temperatures so that annealing does not occur.

The important factor in strain-hardening is not the total density of dislocations, but their arrangement. In most metals dislocations tend to form cells (Fig. 15) of dislocation-free areas of dimensions of the order of 1 micron. But unless we can get a uniform high density of dislocations we cannot strain-harden a metal to its theoretical strength, because of slip in the dislocation-free areas. A uniform density is accomplished by explosive deformation or by special thermal-mechanical treatments.[2]

Each of the mechanisms of strengthening crystals can raise the yield strength to the range of 10^{-3} G to 10^{-2} G. All mechanisms begin to break down at temperatures where diffusion can occur at an appreciable rate. When diffusion is rapid, precipitated particles dissolve; solute clouds drift along with dislocations as they glide; short-range order repairs itself behind slowly moving dislocations; and dislocation climb and annealing tend to decrease the dislocation density. The resulting time-dependent deformation is called **creep.** This irreversible motion precedes the elastic limit. The search for alloys for use at very high temperatures is a search for reduced diffusion rates, so that the four strengthening mechanisms will survive to high temperatures. But the central problem of strong alloys is not strength, but ductility, for failure is often by fracture.

[2] V. F. Zackay, "Strength of steel," Scientific American **209,** 71 (August 1963).

DISLOCATIONS AND CRYSTAL GROWTH

It has been shown by Frank[3] and his collaborators that in some cases the presence of dislocations may be the controlling factor in crystal growth. When crystals are grown in conditions of low supersaturation, of the order of 1 percent, it has been observed that the growth rate is enormously faster than that calculated for an ideal crystal. The actual growth rate is explained in terms of the effect of dislocations on growth. The theory of growth of ideal crystals predicts that in crystal growth from vapor a supersaturation (pressure/equilibrium vapor pressure) of the order of 10 is required to nucleate new crystals, of the order of 5 to form liquid drops, and of 1.5 to form a two-dimensional monolayer of molecules on the face of a perfect crystal. Volmer and Schultze observed growth of iodine crystals at vapor supersaturations down to less than 1 percent, where the growth rate should have been down by the factor e^{-3000} from the rate defined as the minimum observable growth. The large disagreement expresses the difficulty of nucleating a new monolayer on a completed surface of an ideal crystal. But if a screw dislocation is present (Fig. 20), it is never necessary to nucleate a new layer: the crystal will grow in spiral fashion at the *edge* of the discontinuity shown. An atom can be bound to a step more strongly than to a plane. The calculated growth rates for this mechanism are in good agreement with observation. We expect that nearly all crystals in nature grown at low supersaturation will contain dislocations, as otherwise they could not have grown. Spiral growth patterns have been observed on a large number of crystals. A beautiful example of the growth pattern from a single screw dislocation is given in Fig. 21.

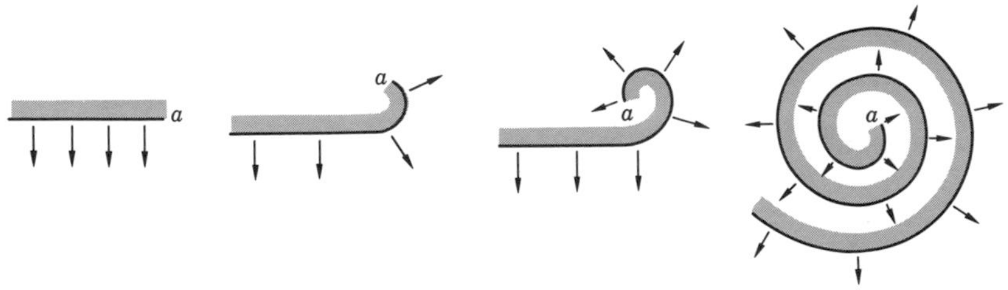

Figure 20 Development of a spiral step produced by intersection of a screw dislocation with the surface of a crystal as in Fig. 8. Each cube represents a molecule. (After F. C. Frank.)

[3]For a full review of this field see Neugebauer, ed., *Growth and perfection of crystals*, Wiley, 1959; also Chap. 7 of J. Friedel, *Dislocations*, Addison-Wesley, 1964. Long-chain polymer molecules have an unusual pattern of crystallization into flat plates; see the review by A. Keller, Physics Today **23**, 42 (May 1970).

Figure 21 Phase contrast micrograph of a hexagonal spiral growth pattern on a SiC crystal. The step height is 165 Å. (Courtesy of A. R. Verma.)

If the growth rate is independent of direction of the edge in the plane of the surface, the growth pattern is an Archimedes spiral, $r = a\theta$, where a is a constant. The limiting minimum radius of curvature near the dislocation is determined by the supersaturation. If the radius of curvature is too small, atoms on the curved edge evaporate until the equilibrium curvature is attained. Away from the origin each part of the step acquires new atoms at a constant rate, so that $dr/dt = \text{const}$.

Whiskers

Fine hairlike crystals or **whiskers** have been observed to grow under conditions of high supersaturation without the necessity for more than perhaps one dislocation. It may be that these crystals contain a single axial screw dislocation that aids their essentially one-dimensional growth. From the absence of dislocations we would expect these crystal whiskers to have high yield strengths, of the order of the calculated value $G/30$ discussed earlier in this chapter. A single axial screw dislocation, if present, could not cause yielding, because in bending the crystal the dislocation is not subjected to a shear stress parallel to its Burgers vector. That is, the stress is not in a direction that can cause slip. Herring and Galt observed

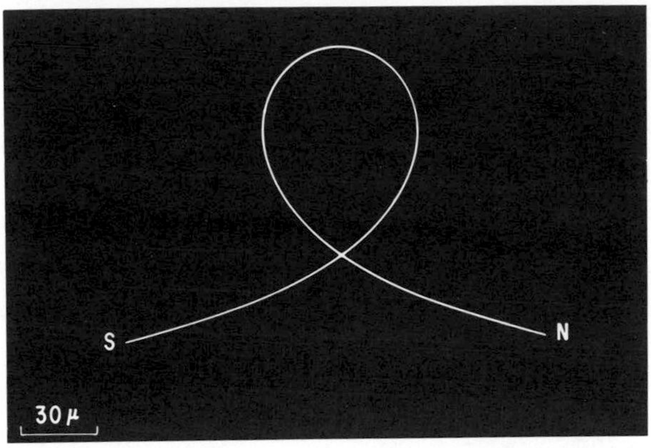

Figure 22 A nickel whisker of diameter 1000 Å bent in a loop. (Courtesy of R. W. De Blois.)

whiskers of tin of radius $\sim 10^{-4}$ cm with elastic properties near those expected from theoretically perfect crystals. They observed yield strains of the order of 10^{-2}, which correspond to shear stresses of order 10^{-2} G, about 1000 times greater than in bulk tin, confirming the early estimates of the strength of perfect crystals. Theoretical or ideal elastic properties have been observed for a number of materials. A single domain whisker of nickel is shown in Fig. 22.

PROBLEMS

1. ***Lines of closest packing.*** Show that the lines of closest atomic packing are $\langle 110 \rangle$ in fcc structures and $\langle 111 \rangle$ in bcc structures.
2. ***Dislocation pairs.*** (a) Find a pair of dislocations equivalent to a row of lattice vacancies; (b) find a pair of dislocations equivalent to a row of interstitial atoms.
3. ***Force on dislocation.*** Consider a crystal in the form of a cube of side L containing an edge dislocation of Burgers vector **b**. If the crystal is subjected to a shear stress σ on the upper and lower faces in the direction of slip, show by considering energy balance that the force acting on the dislocation is $F = b\sigma$ per unit length.

References

S. Amelinckx, *Direct observation of dislocations*, Academic Press, 1964.
C. S. Barrett and T. B. Massalski, *Structure of metals*, 3rd ed., McGraw-Hill, 1966.
J. Friedel, *Dislocations*, Addison-Wesley, 1964.
J. P. Hirth and J. Lothe, *Theory of dislocations*, McGraw-Hill, 1968.
D. Hull, *Introduction to dislocations*, Pergamon, 1965.
A. Kelly, *Strong solids*, Oxford, 1973.
F. R. N. Nabarro, *Theory of crystal dislocations*, Oxford, 1967.
G. Thomas, *Transmission electron microscopy of metals*, Wiley, 1962.
G. Thomas and J. Washburn, ed., *Electron microscopy and strength of crystals*, Wiley, 1963.
J. Weertman and J. R. Weertman, *Elementary dislocation theory*, Macmillan, 1964.

AUTHOR INDEX

SUBJECT INDEX